CAMBRIDGE MONOGRAPHS ON APPLIED AND COMPUTATIONAL MATHEMATICS

---

28 **Multiscale Methods for Fredholm Integral Equations**

The *Cambridge Monographs on Applied and Computational Mathematics* series reflects the crucial role of mathematical and computational techniques in contemporary science. The series publishes expositions on all aspects of applicable and numerical mathematics, with an emphasis on new developments in this fast-moving area of research.

State-of-the-art methods and algorithms as well as modern mathematical descriptions of physical and mechanical ideas are presented in a manner suited to graduate research students and professionals alike. Sound pedagogical presentation is a prerequisite. It is intended that books in the series will serve to inform a new generation of researchers.

A complete list of books in the series can be found at www.cambridge.org/mathematics.
Recent titles include the following:

14. Simulating Hamiltonian dynamics, *Benedict Leimkuhler & Sebastian Reich*
15. Collocation methods for Volterra integral and related functional differential equations, *Hermann Brunner*
16. Topology for computing, *Afra J. Zomorodian*
17. Scattered data approximation, *Holger Wendland*
18. Modern computer arithmetic, *Richard Brent & Paul Zimmermann*
19. Matrix preconditioning techniques and applications, *Ke Chen*
20. Greedy approximation, *Vladimir Temlyakov*
21. Spectral methods for time-dependent problems, *Jan Hesthaven, Sigal Gottlieb & David Gottlieb*
22. The mathematical foundations of mixing, *Rob Sturman, Julio M. Ottino & Stephen Wiggins*
23. Curve and surface reconstruction, *Tamal K. Dey*
24. Learning theory, *Felipe Cucker & Ding Xuan Zhou*
25. Algebraic geometry and statistical learning theory, *Sumio Watanabe*
26. A practical guide to the invariant calculus, *Elizabeth Louise Mansfield*
27. Difference equations by differential equation methods, *Peter E. Hydon*
28. Multiscale methods for Fredholm integral equations, *Zhongying Chen, Charles A. Micchelli & Yuesheng Xu*
29. Partial differential equation methods for image inpainting, *Carola-Bibiane Schönlieb*

Multiscale Methods for Fredholm Integral Equations

ZHONGYING CHEN

Sun Yat-Sen University, Guangzhou, China

CHARLES A. MICCHELLI

State University of New York, Albany

YUESHENG XU

Sun Yat-Sen University, Guangzhou, China

CAMBRIDGE
UNIVERSITY PRESS

CAMBRIDGE
UNIVERSITY PRESS

University Printing House, Cambridge CB2 8BS, United Kingdom

Cambridge University Press is part of the University of Cambridge.

It furthers the University's mission by disseminating knowledge in the pursuit of education, learning and research at the highest international levels of excellence.

www.cambridge.org
Information on this title: www.cambridge.org/9781107103474

First published 2015

A catalogue record for this publication is available from the British Library

Library of Congress Cataloguing in Publication data
Chen, Zhongying, 1946–
Multiscale methods for Fredholm integral equations / Zhongying Chen, Sun Yat-Sen University, Guangzhou, China, Charles A. Micchelli, State University of New York, Albany, Yuesheng Xu, Sun Yat-Sen University, Guangzhou, China.
pages cm. – (The Cambridge monographs on applied and computational mathematics series)
Includes bibliographical references and index.
ISBN 978-1-107-10347-4 (Hardback)
1. Fredholm equations. 2. Integral equations. I. Micchelli, Charles A.
II. Xu, Yuesheng. III. Title.
QA431.C4634 2015
515′.45–dc23 2014050239

ISBN 978-1-107-10347-4 Hardback

Contents

Preface

Fredholm equations arise in many areas of science and engineering. Consequently, they occupy a central topic in applied mathematics. Traditional numerical methods developed during the period prior to the mid-1980s include mainly quadrature, collocation and Galerkin methods. Unfortunately, all of these approaches suffer from the fact that the resulting discretization matrices are dense. That is, they have a large number of nonzero entries. This bottleneck leads to significant computational costs for the solution of the corresponding integral equations.

The recent appearance of wavelets as a new computational tool in applied mathematics has given a new direction to the area of the numerical solution of Fredholm integral equations. Shortly after their introduction it was discovered that using a wavelet basis for a singular integral equation led to numerically sparse matrix discretization. This observation, combined with a truncation strategy, then led to a fast numerical solution of this class of integral equations.

Approximately 20 years ago the authors of this book began a systematic study of the construction of wavelet bases suitable for solving Fredholm integral equations and explored their usefulness for developing fast multiscale Galerkin, Petrov–Galerkin and collocation methods. The purpose of this book is to provide a self-contained account of these ideas as well as some traditional material on Fredholm equations to make this book accessible to as large an audience as possible.

The goal of this book is twofold. It can be used as a reference text for practitioners who need to solve integral equations numerically and wish to use the new techniques presented here. At the same time, portions of this book can be used as a modern text treating the subject of the numerical solution of integral equations, which is suitable for upper-level undergraduate students as well as graduate students. Specifically, the first five chapters of this book are designed for a one-semester course, which provides students with a solid background in integral equations and fast multiscale methods for their numerical solutions.

An early version of this book was used in a summer school on applied mathematics sponsored by the Ministry of Education of the People's Republic of China. Subsequently, the authors used revised versions of this book for courses on integral equations at our respective institutions. These teaching experiences led us to make many changes in presentation, resulting from our interactions with our many students.

We are indebted to our many colleagues, who gave freely of their time and advice concerning the material in this book, and whose expertise on the subject of the numerical solution of Fredholm equations, collectively, far exceeds ours. We mention here that a preliminary version of the book was provided to Kendall Atkinson, Uday Banerjee, Hermann Brunner, Yanzhao Cao, Wolfgang Dahmen, Leslie Greengard, Weimin Han, Geroge Hsiao, Hideaki Kaneko, Rainer Kress, Wayne Lawton, Qun Lin, Paul Martin, Richard Noren, Sergei Pereverzyev, Reinhold Schneider, Johannes Tausch, Ezio Venturino and Aihui Zhou. We are grateful to them all for their constructive comments which improved our presentation.

Our special thanks go to Kendall Atkinson for his encouragement and support in writing this book. We would also like to thank our colleagues at Sun Yat-Sen University, including Bin Wu, Sirui Cheng, Xianglin Chen as well as the graduate student Jieyang Chen for their assistance in preparing this book.

Finally, we are deeply indebted to our families for their understanding, patience and continued support throughout our efforts to complete this project.

Symbols

a.e.	almost everywhere; §1.1
$\mathcal{A}^*$	adjoint operator of $\mathcal{A}$; §2.1.1
$\mathbf{A}[\mathbf{i},\mathbf{j}]$	minor of matrix $\mathbf{A}$ with lattice vectors $\mathbf{i}$ and $\mathbf{j}$; §1.2
$\mathcal{B}(\mathbb{X},\mathbb{Y})$	normed linear space of all bounded linear operators from $\mathbb{X}$ into $\mathbb{Y}$; §2.1.1
$\mathbb{C}$	set of complex numbers; §1.1
$C(D)$	linear space of all real-valued continuous functions on D; §2.1
$C^m(D)$	linear space of all real-valued m-times continuously differentiable functions on D; §2.1
$C^\infty(D)$	linear space of all real-valued infinitely differentiable functions on D; §2.1
$C_0(D)$	subspace of $C(\bar{D})$ consisting of functions with support contained inside D; §A.1
$C_0^\infty(D)$	subspace of $C^\infty(\bar{D})$ consisting of functions with support contained inside D and bounded; §A.1
$c_\sigma(D)$	positive constant defined in §2.1.2
card T	cardinality of T; §2.2.2
cond($\mathcal{A}$)	condition number of $\mathcal{A}$; §2.2.3
$D(\lambda)$	complex-valued function at λ defined by (1.18)
det($\mathbf{A}$)	determinant of matrix $\mathbf{A}$; §1.2
diag($\cdot$)	diagonal matrix; §1.2
diam(S)	diameter of set S; §1.1
$H^m(D)$	Sobolev space; §A.1
$H_0^m(D)$	Sobolev space; §A.1
$L^p(D)$	linear space of all real-valued pth power integrable functions ($1 \le p < \infty$); §2.1

$L^\infty(D)$	linear space of all real-valued essentially bounded measurable functions; §2.1
$m(D)$	positive constant defined in §2.1.2
$\mathbb{N}$	set of positive integers $\{1, 2, 3, \ldots\}$; §1.1
$\mathbb{N}_0$	set of integers $\{0, 1, 2, \ldots\}$; §1.1
$\mathbb{N}_n$	set of positive integers $\{1, 2, \ldots, n\}$ for $n \in \mathbb{N}$; §1.1
$P_{\mathbf{A}}$	characteristic polynomial of matrix $\mathbf{A}$; §1.2
$\mathbb{R}$	set of real numbers; §1.1
$\mathbb{R}^d$	d-dimensional Euclidean space; §1.1
$\operatorname{Re} f$	real part of f; §1.6
$r_q(\mathbf{A})$	minor equation of $\mathbf{A}$; (1.4)
R_λ	resolvent kernel; §1.4
$\operatorname{rank} \mathbf{A}$	rank of matrix $\mathbf{A}$; §3.3.5
$s(n)$	dimension of space $\mathbb{X}_n$; §3.3
$\operatorname{span} S$	span of set S; §3.3.1
$\mathbb{U}$	index set $\{(i,j) : i \in \mathbb{N}_0, j \in \mathbb{Z}_{w(i)}\}$; §4.1
$\mathbb{U}_n$	index set $\{(i,j) : i \in \mathbb{Z}_{n+1}, j \in \mathbb{Z}_{w(i)}\}$; §4.5.1
$\operatorname{vol}(S)$	volume of set S; §1.1
$W^{m,p}(D)$	Sobolev space; §A.1
$W_0^{m,p}(D)$	Sobolev space; §A.1
$w(n)$	dimension of space $\mathbb{W}_n$; §4.1
$\mathbb{Z}$	set of integers $\{0, \pm 1, \pm 2, \ldots\}$; §1.1
$\mathbb{Z}_n$	set of integers $\{0, 1, 2, \ldots, n-1\}$ for $n \in \mathbb{N}$; §1.1
$\Gamma(\cdot)$	gamma function; §2.1.2
∇	gradient operator; §2.1.3
Δ	Laplace operator; §2.1.3
$\rho(\mathcal{T})$	resolvent set of operator $\mathcal{T}$; §11.2
$\sigma(\mathcal{T})$	spectrum of operator $\mathcal{T}$; §11.2
ω_{d-1}	surface area of unit sphere in $\mathbb{R}^d$; §2.1.3
$\omega(K, h)$	modules of continuity of K; §1.3
$!$	factorial; for example (1.4)
$\bigcup^\perp$	union of orthogonal sets; §4.1
$\bigoplus$	direct sum of spaces; §4.1
$\bigotimes$	tensor product (direct product); §1.3
$\circ$	functional composition; §3.3.1
$\|\alpha\|$	sum of components of lattice vector α; §2.1
$\|s - t\|$	Euclidean distance between s and t; §2.1.2

$\|\mathcal{A}\|$	norm of operator $\mathcal{A}$; §2.1.1
$\|\cdot\|_{m,\infty}$	norm of $C^m(\bar{D})$; §2.1
$\|\cdot\|_{p}$,	norm of $L^p(D)$ $(1 \leq P \leq \infty)$; §2.1
$(\cdot,\cdot)$	inner product; §2.1
$\langle\cdot,\cdot\rangle$	value of a linear functional at a function; §2.1.1
$\sim$	same order; §5.1.1
$\xrightarrow{s}$	pointwise converge; §2.1.1
$\xrightarrow{u}$	uniformly converge; §2.1.1

Introduction

The equations we consider in this book are primarily Fredholm integral equations of the second kind on bounded domains in the Euclidean space. These equations are used as mathematical models for a multitude of physical problems and cover many important applications, such as radiosity equations for realistic image synthesis [18, 85, 244] and especially boundary integral equations [12, 177, 203], which themselves occur as reformulations of other problems, typically originating as partial differential equations. In practice, Fredholm integral equations are solved numerically using piecewise polynomial collocation or Galerkin methods, and when the order of the coefficient matrix (which is typically full) is large, the computational cost of generating the matrix as well as solving the corresponding linear system is large. Therefore, to enhance the range of applicability of the Fredholm equation methodology, it is critical to provide alternate algorithms which are fast, efficient and accurate. This book is concerned with this challenge: designing fast multiscale methods for the numerical solution of Fredholm integral equations.

The development and use of multiscale methods for solving integral equations is a subject of recent intense study. The history of fast multiscale solutions of integral equations began with the introduction of multiscale Galerkin (Petrov–Galerkin) methods for solving integral equations, as presented in [28, 64, 68, 88, 94, 95, 202, 260, 261] and the references cited therein. Most noteworthy is the discovery in [28] that the representation of a singular integral operator by compactly supported orthonormal wavelets produces numerically sparse matrices. In other words, most of their entries are so small in absolute value that, to some degree of precision, they can be neglected without affecting the overall accuracy of the approximation. Later, the papers [94, 95] studied Petrov–Galerkin methods using periodic multiscale bases constructed from refinement equations for periodic elliptic pseudodifferential equations, and in this restricted environment, stability, convergence and matrix compression were investigated. For a first-kind boundary integral equation, a truncation strategy for the Galerkin method using spline-based multiscale basis functions of low degree was proposed in [260]. Also, in [261], for elliptic pseudodifferential equations of order zero on a three-dimensional manifold, a Galerkin method using discontinuous piecewise linear multiscale basis functions on triangles was studied.

In another direction, a general construction of multidimensional discontinuous orthogonal and bi-orthogonal wavelets on invariant sets was presented in [200, 201]. Invariant sets include, among others, the important cases of simplices and cubes, and in the two-dimensional case L-shaped domains. A similar recursive structure was explored in [65] for multiscale function representation and approximation constructed by interpolation on invariant sets. In this regard, an essential advantage of this approach is the existence of efficient schemes for generating recursively multilevel partitions of invariant sets and their associated multiscale functions. All of these methods even extend to domains which are a finite union of invariant sets, thereby significantly expanding the range of their applicability. Therefore, the constructions given in [65, 200, 201] led to a wide variety of multiscale basis functions which, on the one hand, have desirable simple recursive structure and, on the other hand, can be used in diverse areas in which the Fredholm methodology is applied.

Subsequently, the papers [64, 68, 202] developed multiscale piecewise polynomial Galerkin, Petrov–Galerkin and discrete multiscale Petrov–Galerkin methods. An important advantage of multiscale piecewise polynomials is that their closed-form expressions are very convenient for computation. Moreover, they can easily be related to standard bases used in the conventional numerical method, thereby providing an advantage for theoretical analysis as well. Among conventional numerical methods for solving integral equations, the collocation method has received the most favorable attention in the engineering community due to its lower computational cost in generating the coefficient matrix of the corresponding discrete equations. In comparison, the implementation of the Galerkin method requires much more computational effort for the evaluation of integrals (see, for example, [19, 77] for a discussion of this point). Motivated by this issue, [69] proposed and analyzed a fast collocation algorithm for solving general multidimensional integral equations. Moreover, a matrix truncation strategy was introduced there by making a careful choice of basis functions and collocation functionals, the end result being fast, multiscale algorithms for solving the integral equations.

The development of stable, efficient and fast numerical algorithms for solving operator equations, including differential equations and integral equations, is a main focus of research in numerical analysis and scientific computation, since such algorithms are particularly important for large-scale computation. We review the three main steps in solving an operator equation. The first is at the level of approximation theory. Here, we must choose appropriate subspaces and suitable bases for them. The second step is to discretize the operator equations using these bases and to analyze convergence properties of the approximate solutions. The end result of this step of processing is a discrete linear system and its construction is considered as a main task for the numerical solution of operator equations. The third step employs methods of numerical linear algebra to design an efficient solver for the discrete linear system. The ultimate goal is, of course, to solve the discrete linear system efficiently and obtain an accurate approximate solution to the original operator equation. Theoretical considerations and practical implementations in the numerical solution of operator equations show that these three steps of processing are closely related. Therefore, designing efficient algorithms for the discrete linear system should take into consideration the choice of subspaces and their bases, the methodologies of discretization of the operator equations and the specific characteristics and advantages of the numerical solvers used to solve the resulting discrete linear system. In this book we describe how these three steps are integrated in a multiscale environment and thereby achieve our goal of providing a wide selection of fast and accurate algorithms for the second-kind integral equations. We also describe work in progress addressing related issues of eigenvalue and eigenfunction computation as well as the solution of Fredholm equations of the first kind.

This book is organized into 12 chapters plus an appendix. Chapter 1 is devoted to a review of the Fredholm approach to solving an integral equation of the second kind. In Chapter 2 we introduce essential concepts from Fredholm integral equations of the second kind and describe a general setup of projection methods for solving operator equations which will be used in later chapters. The purpose of Chapter 3 is to describe conventional numerical methods for solving Fredholm integral equations of the second kind, including the degenerate kernel method, the quadrature method, the Galerkin method, the Petrov–Galerkin method and the collocation method. In Chapter 4, a general construction of multiscale bases of piecewise polynomial spaces, including multiscale orthogonal and interpolating bases, is presented. Chapters 5, 6 and 7 use the material from Chapter 4 to construct multiscale Galerkin, Petrov–Galerkin and collocation methods. We study the discretization schemes resulting from these methods, propose truncation strategies for building fast and accurate algorithms, and give a complete analysis for the order of convergence, computational complexity, stability and condition numbers for the truncated schemes. In Chapter 8, two types of quadrature rule for the numerical integration required to generate the coefficient matrix are introduced and error control strategies are designed so that the quadrature errors will neither ruin the overall convergence order nor increase the overall computational complexity of the original multiscale methods. The goal of Chapter 9 is to investigate fast solvers for the discrete linear systems resulting from multiscale methods. We introduce multilevel augmentation methods and multilevel iteration methods based on direct sum decompositions of the range and domain of the operator equation. In Chapters 10, 11 and 12, the fast algorithms are applied to solving nonlinear integral equations of the second kind, ill-posed integral equations of the first kind and eigen-problems of compact integral operators, respectively. We summarize in the Appendix some of the standard concepts and results from functional analysis in a form which is used throughout the book. The appendix provides the reader with a convenient source of the background material needed to follow the ideas and arguments presented in other chapters of this book.

Most of the material in this book can only be found in research papers. This is the first time that it has been assembled into a book. Although this book is pronouncedly a research monograph, selected material from the initial chapters can be used in a semester course on numerical methods for integral equations which presents the multiscale point of view.

1
A review of the Fredholm approach

In this chapter we pay homage to Ivar Fredholm (April 7, 1866–August 17, 1927) and review his approach to solving an integral equation of the second kind. The methods employed in this chapter are classical and differ from the approach taken in the rest of the book. We include it here because those readers inexperienced in integral equations should be familiar with these important ideas. The basic tools of matrix theory and some complex analysis are needed, and we shall provide a reasonably self-contained discussion of the required material.

1.1 Introduction

We start by introducing the notation that will be used throughout this book. Let $\mathbb{C}$, $\mathbb{R}$, $\mathbb{Z}$ and $\mathbb{N}$ denote, respectively, the set of complex numbers, the set of real numbers, the set of integers and the set of positive integers. We also let $\mathbb{N}_0 := \{0\} \cup \mathbb{N}$. For the purpose of enumerating a nonempty finite set of objects we use the sets $\mathbb{N}_d := \{1, 2, \ldots, d\}$ and $\mathbb{Z}_d := \{0, 1, \ldots, d-1\}$, both of which consist of d distinct integers. For $d \in \mathbb{N}$, let $\mathbb{R}^d$ denote the d-dimensional Euclidean space and Ω a subset of $\mathbb{R}^d$. By $C(\Omega)$, we mean the linear space of all continuous real-valued functions defined on Ω. We usually denote matrices or vectors over $\mathbb{R}$ in boldface, for example, $\mathbf{A} := [A_{ij} : i, j \in \mathbb{N}_d] \in \mathbb{R}^{d\times d}$ and $\mathbf{u} := [u_j : j \in \mathbb{N}_d] \in \mathbb{R}^d$. When the vector has all integer coordinates, that is, $\mathbf{u} \in \mathbb{Z}^d$, we sometimes call it a *lattice vector*. Moreover, we usually denote integral operators by calligraphic letters. Especially, the *integral operator* with a *kernel* K will be denoted by $\mathcal{K}$, that is, for the kernel K defined on $\Omega \times \Omega$ and the function u defined on Ω, we define

$$(\mathcal{K}u)(s) := \int_\Omega K(s, t)u(t)\mathrm{d}t, \quad s \in \Omega.$$

The most direct approach to solving a second-kind integral equation merely replaces integrals by sums and thereby obtains a linear system of equations whose solution approximates the solution of the original equation. The study of the resulting linear system of equations leads naturally to the important notion of the Fredholm function and determinant, which remain a central tool in the theory of second-kind integral equations, see for example [183, 253]. We consider this direct approach when we are given a *continuous* kernel $K \in C(\Omega \times \Omega)$ on a *compact* subset Ω of $\mathbb{R}^d$ with positive Borel measure, a continuous function $f \in C(\Omega)$ and a nonzero complex number $\lambda \in \mathbb{C}$. The task is to find a function $u \in C(\Omega)$ such that, for $s \in \Omega$,

$$u(s) - \lambda \int_{\Omega} K(s,t)u(t)\mathrm{d}t = f(s). \tag{1.1}$$

To this end, for each positive $h > 0$, we partition Ω into nonempty compact subsets Ω_i, $i \in \mathbb{N}_n$

$$\Omega = \bigcup_{i \in \mathbb{N}_n} \Omega_i$$

such that different subsets have no overlapping interior and

$$\operatorname{diam} \Omega_i := \max\{|x - y| : x, y \in \Omega_i\} \leq h,$$

where $|x|$ is the ℓ^2-norm of the vector $x \in \mathbb{R}^d$. This partition can be constructed by first putting a large "box" around the set Ω and then decomposing this box into cubes, each of which has diameter less than or equal to h. The sets Ω_i are then formed by intersecting the set Ω with the cubes, where we discard sets of zero Borel measure. Therefore, the partition of Ω constructed in this manner is done a.e. Next, we choose *any* finite set of points $T := \{t_i : i \in \mathbb{N}_n\}$ such that, for any $i \in \mathbb{N}_n$, we have that $t_i \in \Omega_i$. With these points we now replace our integral equation (1.1) with a linear system of equations. Specifically, we choose the number $\rho := -\lambda$ and the $n \times n$ matrix $\mathbf{A}$ defined by

$$\mathbf{A} := [\operatorname{vol}(\Omega_j)K(t_i, t_j) : i, j \in \mathbb{N}_n],$$

where $\operatorname{vol}(\Omega_j)$ denotes the volume of the set Ω_j, and replace (1.1) with the system of linear equations

$$(\mathbf{I} + \rho\mathbf{A})\mathbf{u} = \mathbf{f}. \tag{1.2}$$

Here, $\mathbf{f}$ is the vector obtained by evaluating the function f on the set T.

Of course, the point of view we take here is that the vector $\mathbf{u} \in \mathbb{R}^n$ which solves equation (1.2) is an approximation to the function u on the set T. Therefore, the problem of determining the function u is replaced by the simpler one of numerically solving for the vector $\mathbf{u}$ when h is small. Certainly, an important role is played by the determinant of the coefficient matrix of the linear system (1.2). Its properties, especially as $h \to 0^+$, will be our main concern for a significant part of this chapter. We start by studying the determinant of the coefficient matrix of the linear system (1.2) and then derive a formula for the entries of the inverse of the matrix $\mathbf{I} + \rho\mathbf{A}$ in terms of the matrix $\mathbf{A}$.

1.2 Second-kind matrix Fredholm equations

We define the minor of an $n \times n$ matrix. If $\mathbf{A} = [A_{ij} : i, j \in \mathbb{N}_n]$ is an $n \times n$ matrix, q is a non-negative integer in $\mathbb{Z}_{n+1}$, $\mathbf{i} := [i_l : l \in \mathbb{N}_q]$, $\mathbf{j} := [j_l : l \in \mathbb{N}_q]$ are lattice vectors in $\mathbb{N}_n^q$ we define the corresponding *minor* by

$$A[\mathbf{i}, \mathbf{j}] := \det[A_{i_r, j_s} : r, s \in \mathbb{N}_q].$$

Sometimes, the more elaborate notation

$$A\begin{pmatrix} i_1, & i_2, & \ldots, & i_q \\ j_1, & j_2, & \ldots, & j_q \end{pmatrix} \tag{1.3}$$

is used for $A[\mathbf{i}, \mathbf{j}]$. When $\mathbf{i} = \mathbf{j}$, that is, for a *principal minor* of $\mathbf{A}$, we use the simplified notation $A[\mathbf{i}]$ in place of $A[\mathbf{i}, \mathbf{i}]$. For a positive integer $q \in \mathbb{N}_n$, we set

$$r_q(\mathbf{A}) := \frac{1}{q!} \sum_{\mathbf{i} \in \mathbb{N}_n^q} A[\mathbf{i}] \tag{1.4}$$

and also choose $r_0(\mathbf{A}) := 1$.

Lemma 1.1 *If* $\mathbf{A}$ *is an* $n \times n$ *matrix and* $\rho \in \mathbb{C}$, *then*

$$\det(\mathbf{I} + \rho\mathbf{A}) = \sum_{q \in \mathbb{Z}_{n+1}} r_q(\mathbf{A})\rho^q. \tag{1.5}$$

Before proving this lemma, we make two remarks.

Remark 1.2 *Using the extended notation for a minor as indicated in (1.3), we see that equation (1.4) is equivalent to the formula*

$$r_q(\mathbf{A}) = \frac{1}{q!} \sum_{[i_l : l \in \mathbb{N}_q] \in \mathbb{N}_n^q} A\begin{pmatrix} i_1, & \ldots, & i_q \\ i_1, & \ldots, & i_q \end{pmatrix}. \tag{1.6}$$

Certainly, if any two components of the vector $\mathbf{i} = [i_l : l \in \mathbb{N}_q]$ *are equal, then the corresponding minor has a repeated row (and column) and so is zero. These terms may be neglected. Moreover, any permutation of the components of the vector* $\mathbf{i}$ *affects both a row and a column exchange of the determinant* *appearing in (1.6), and so does not affect the value of the determinant. Since there are $q!$ such permutations, we get that*

$$r_q(\mathbf{A}) = \sum_{1 \le i_1 < i_2 < \cdots < i_q \le n} A\begin{pmatrix} i_1, & \ldots, & i_q \\ i_1, & \ldots, & i_q \end{pmatrix}. \tag{1.7}$$

Remark 1.3 *If the characteristic values of the matrix* $\mathbf{A}$ *are denoted by* $\{\lambda_j : j \in \mathbb{N}_n\}$, *then*

$$r_q(\mathbf{A}) = \sum_{1 \le i_1 < i_2 < \cdots < i_q \le n} \lambda_{i_1} \lambda_{i_2} \cdots \lambda_{i_q}. \tag{1.8}$$

The right-hand side of this equation is an elementary symmetric function of the eigenvalues of $\mathbf{A}$, *which is invariant under a similarity transformation of the matrix* $\mathbf{A}$. *This fact will be the basis of the proof of Lemma 1.1 presented below.*

We next present the proof of Lemma 1.1.

Proof By Schur's upper-triangular factorization theorem (see, for example, [142], p. 79), the matrix $\mathbf{A}$ can be factored in the form

$$\mathbf{A} = \mathbf{P}^{-1}\mathbf{T}\mathbf{P}, \tag{1.9}$$

where $\mathbf{P}$ is an orthogonal matrix and $\mathbf{T}$ is an upper-triangular matrix whose diagonal entries are the eigenvalues of $\mathbf{A}$ (chosen in any prescribed order). For an upper-triangular matrix $\mathbf{T}$ we observe from (1.8) that

$$\begin{aligned} \det(\mathbf{I} + \rho\mathbf{T}) &= \prod_{j \in \mathbb{N}_n} (1 + \rho\lambda_j) \\ &= \sum_{q \in Z_{n+1}} \rho^q \sum_{1 \le j_1 < j_2 < \cdots < j_q \le n} \lambda_{j_1} \lambda_{j_2} \cdots \lambda_{j_q} \\ &= \sum_{q \in Z_{n+1}} r_q(\mathbf{T}) \rho^q, \end{aligned}$$

thereby verifying that equation (1.5) is valid at least for an upper-triangular matrix. For a general matrix we use the reduction to the upper-triangular case by an orthogonal similarity (1.9). Since all determinants in equations (1.5) and (1.6) are unchanged under an orthogonal similarity, this comment proves the general case. □

We now add a more difficult computation, which provides an expansion of the elements of the matrix $(\mathbf{I} + \rho\mathbf{A})^{-1}$ as a rational function of ρ. To this end, we introduce some needed constants. We define for any $k, l \in \mathbb{N}_n$ and $q \in \mathbb{Z}_{n+1}$ the constants

$$u_{lkq} := \frac{1}{q!} \sum_{\mathbf{i} \in \mathbb{N}_n^q} A \left[\begin{array}{c} l, \mathbf{i} \\ k, \mathbf{i} \end{array} \right]. \tag{1.10}$$

In (1.10) there are (not necessarily principal) minors of order $q+1$. Moreover, when $q = n-1$ and $k, l \in \mathbb{N}_n$ with $k \neq l$ we have that $u_{lkq} = 0$, since $\mathbb{N}_n$ has only n distinct elements. Also, for the same reason, $u_{lkn} = 0$ for all $k, l \in \mathbb{N}_n$.

We shall relate these constants. But first we introduce for $k, l \in \mathbb{N}_n$ polynomials U_{kl} defined at ρ to be

$$U_{lk}(\rho) := \sum_{q \in \mathbb{Z}_{n+1}} u_{lkq} \rho^q.$$

Now, to relate all of these quantities, we start with the minor

$$A \begin{pmatrix} l, & i_1, & \ldots, & i_q \\ k, & i_1, & \ldots, & i_q \end{pmatrix} \tag{1.11}$$

where $l, k \in \mathbb{N}_n$ and expand it, by the *Laplace expansion* by minors, across its first row to obtain the formula

$$\begin{aligned} A \begin{pmatrix} l, & i_1, & \ldots, & i_q \\ k, & i_1, & \ldots, & i_q \end{pmatrix} &= A_{lk} A \begin{pmatrix} i_1, & \ldots, & i_q \\ i_1, & \ldots, & i_q \end{pmatrix} \\ &\quad + \sum_{m \in \mathbb{N}_q} (-1)^m A_{li_m} A \begin{pmatrix} i_1, i_2, \ldots, i_{m+1}, \ldots, i_q \\ k, i_1, \ldots, \widehat{i_m}, \cdots, i_q \end{pmatrix}. \end{aligned} \tag{1.12}$$

The symbol $\widehat{i_m}$ appearing in the minor above is to be interpreted to mean that the i_mth column of A is *deleted* in that minor. We now sum both sides of this formula over all integers $i_1, i_2, \ldots, i_q$ in $\mathbb{N}_n$, and interchange the summations of the second term on the right-hand side of the resulting equation to yield the formula

$$\begin{aligned} \sum_{\mathbf{i}_p : p \in \mathbb{N}_q \mathbf{i} \in \mathbb{N}_n^q} A \begin{pmatrix} l, & i_1, & \ldots, & i_q \\ k, & i_1, & \ldots, & i_q \end{pmatrix} &= \sum_{\mathbf{i}_p : p \in \mathbb{N}_q \mathbf{i} \in \mathbb{N}_n^q} A_{lk} A \begin{pmatrix} i_1, & \ldots, & i_q \\ i_1, & \ldots, & i_q \end{pmatrix} \\ + \sum_{m \in \mathbb{N}_q} \sum_{\mathbf{i}_p : p \in \mathbb{N}_q \mathbf{i} \in \mathbb{N}_n^q} (-1)^m A_{li_m} A &\begin{pmatrix} i_1, & i_2, & \ldots, & i_{m+1}, & \ldots, & i_q \\ k, & i_1, & \ldots, & \widehat{i_m}, & \cdots, & i_q \end{pmatrix}. \end{aligned} \tag{1.13}$$

The first term on the right-hand side of equation (1.13) is clearly $A_{lk} q! r_q(\mathbf{A})$, which follows from our definition (1.6). The value for the second term requires more explanation.

For the second term we point out that there are two sums over $i_m \in \mathbb{N}_n$. The outer sum already appears in the right-hand side of equation (1.13) and the inner one appears in the sum over all indices $i_1, i_2, \ldots, i_q$ in $\mathbb{N}_n$. In the inner sum we first fix i_m and then sum over all the other indices $i_1, \ldots, i_{m-1}, i_{m+1}, \ldots, i_q \in \mathbb{N}_n$. This leads us to the expression

$$\sum_{[i_1,\ldots,i_{m-1},i_{m+1},\ldots,i_q]\in\mathbb{N}_n^{q-1}} (-1)^m A\left(\begin{array}{ccc} i_1, & \ldots, & i_q \\ k, i_1, \ldots, i_{m-1}, i_{m+1}, \ldots, i_q \end{array}\right).$$

We now locate the i_mth row in the minor of A and move it forward to the first row. This requires $m-1$ row exchanges and gives us the expression

$$-\sum_{[i_1,\ldots,i_{m-1},i_{m+1},\ldots,i_q]\in\mathbb{N}_n^{q-1}} A\left(\begin{array}{c} i_m, i_1, \ldots, i_{m-1}, i_{m+1}, \ldots, i_q \\ k, i_1, \ldots, i_{m-1}, i_{m+1}, \ldots, i_q \end{array}\right).$$

Next, we multiply this expression by A_{li_m} and compute the (first) sum of it over $i_m \in \mathbb{N}_n$. This yields the quantity

$$-\sum_{r\in\mathbb{N}_n} A_{lr} \sum_{[i_p:p\in\mathbb{N}_{q-1}]\in\mathbb{N}_n^{q-1}} A\left(\begin{array}{cccc} r, & i_1, & \ldots, & i_{q-1} \\ k, & i_1, & \ldots, & i_{q-1} \end{array}\right).$$

But this quantity is *independent* of i_m and so appears q times in the other (second) sum over i_m. So, in summary, we get the equation

$$\sum_{[i_p:p\in\mathbb{N}_q]\in\mathbb{N}_n^q} A\left(\begin{array}{cccc} l, & i_1, & \ldots, & i_q \\ k, & i_1, & \ldots, & i_q \end{array}\right) = A_{lk} \sum_{[i_p:p\in\mathbb{N}_q]\in\mathbb{N}_n^q} A\left(\begin{array}{ccc} i_1, & \ldots, & i_q \\ i_1, & \ldots, & i_q \end{array}\right)$$
$$-q\sum_{r\in\mathbb{N}_n} A_{lr} \sum_{[i_p:p\in\mathbb{N}_{q-1}]\in\mathbb{N}_n^{q-1}} A\left(\begin{array}{cccc} r, & i_1, & \ldots, & i_{q-1} \\ k, & i_1, & \ldots, & i_{q-1} \end{array}\right),$$

which is equivalent to the formula

$$\sum_{\mathbf{i}\in\mathbb{N}_n^q} A\left[\begin{array}{c} l, \mathbf{i} \\ k, \mathbf{i} \end{array}\right] = A_{lk} \sum_{\mathbf{i}\in\mathbb{N}_n^q} A[\mathbf{i}] - q\sum_{r\in\mathbb{N}_n} A_{lr} \sum_{\mathbf{j}\in\mathbb{N}_n^{q-1}} A\left[\begin{array}{c} r, \mathbf{j} \\ k, \mathbf{j} \end{array}\right].$$

When $q = 0$ the second term on the right is zero while the first sum on the right is set to one. Likewise, the expression on the left is set to A_{lk} and so this formula is still true when $q = 0$. We now multiply both sides by $\frac{\rho^q}{q!}$ and sum over $q \in \mathbb{Z}_{n+1}$. Upon simplification we conclude, for $l, k \in \mathbb{N}_n$, that

$$U_{lk}(\rho) = A_{lk}\det(\mathbf{I} + \rho\mathbf{A}) - \rho\sum_{r\in\mathbb{N}_n} A_{lr}U_{rk}(\rho). \tag{1.14}$$

Let $\mathbf{U}$ be the $n \times n$ matrix defined by

$$\mathbf{U} := (U_{lk}(\rho))_{l,k\in\mathbb{N}_n}\,.$$

Using this matrix we re-express equation (1.14) in an equivalent matrix form as follows:

$$\mathbf{U} - \mathbf{A}\det(\mathbf{I} + \rho\mathbf{A}) + \rho\mathbf{A}\mathbf{U} = 0. \tag{1.15}$$

Likewise, by performing the Laplace expansion on (1.11) by minors, across its columns, we get the equation

$$\mathbf{U} - \mathbf{A}\det(\mathbf{I} + \rho\mathbf{A}) + \rho\mathbf{U}\mathbf{A} = 0. \tag{1.16}$$

From these formulas we get that

$$(\mathbf{I} + \rho\mathbf{A})^{-1} = \mathbf{I} - \frac{\rho\mathbf{U}}{\det(\mathbf{I} + \rho\mathbf{A})}. \tag{1.17}$$

Indeed, this can be obtained by checking the equations

$$(\mathbf{I} + \rho\mathbf{A})\left(\mathbf{I} - \tfrac{\rho\mathbf{U}}{\det(\mathbf{I}+\rho\mathbf{A})}\right) = \left(\mathbf{I} - \tfrac{\rho\mathbf{U}}{\det(\mathbf{I}+\rho\mathbf{A})}\right)(\mathbf{I} + \rho\mathbf{A}) = \mathbf{I},$$

which may be confirmed by employing equations (1.15) and (1.16).

1.3 Fredholm functions

We now apply Lemma 1.1 to Fredholm's discretization given in equation (1.2). For this purpose, we introduce the notation for the *Fredholm minors* of a continuous kernel $K \in C(\Omega \times \Omega)$.

Definition 1.4 For any two vectors $\mathbf{x} = [x_l : l \in \mathbb{N}_q]$ and $\mathbf{y} = [y_l : l \in \mathbb{N}_q]$ in Ω^q we define the corresponding qth-order minor of a continuous kernel K as

$$K\begin{bmatrix}\mathbf{x}\\ \mathbf{y}\end{bmatrix} := \det[K(x_l, y_r) : l, r \in \mathbb{N}_q].$$

If $\mathbf{x} = \mathbf{y}$ we use the simplified notation $K[\mathbf{x}]$.

Sometimes, to avoid confusion, we may use the extended notation

$$K\begin{pmatrix} x_1, & \dots, & x_q \\ y_1, & \dots, & y_q \end{pmatrix}$$

for the minor $K\left[\begin{smallmatrix}\mathbf{x}\\ \mathbf{y}\end{smallmatrix}\right]$ and similarly we might use

$$K\begin{pmatrix} x_1, & \dots, & x_q \\ x_1, & \dots, & x_q \end{pmatrix}$$

for $K[\mathbf{x}]$.

Now, returning to equation (1.2), we recall in that case the matrix $\mathbf{A}$ is given by the formula

$$\mathbf{A} = [\mathrm{vol}(\Omega_j)K(t_i, t_j) : i, j \in \mathbb{N}_n].$$

For each lattice vector $\mathbf{i} = [i_l : l \in \mathbb{N}_q]$ in $\mathbb{N}_n^q$ we form the vector $\mathbf{s_i} = [t_{i_l} : l \in \mathbb{N}_q]$ in Ω^q from the points in the set $T = \{t_i : i \in \mathbb{N}_n\}$, the subset

$$\Omega_{\mathbf{i}} := \bigotimes_{l \in \mathbb{N}_q} \Omega_{i_l}$$

of Ω^q and obtain a formula for the determinant of the coefficient matrix of the linear system (1.2), namely

$$D_h(\lambda) := \sum_{q \in \mathbb{Z}_{n+1}} \frac{(-1)^q}{q!} \lambda^q \sum_{\mathbf{i} \in \mathbb{N}_n^q} \mathrm{vol}(\Omega_{\mathbf{i}}) K[\mathbf{s_i}].$$

Note that the vector $\mathbf{s_i} = [t_{i_l} : l \in \mathbb{N}_q]$ is in $\Omega_{\mathbf{i}}$ and that

$$\Omega^q = \bigcup_{\mathbf{i} \in \mathbb{N}_n^q} \Omega_{\mathbf{i}},$$

thereby forming a partition of the set Ω^q a.e. In our expanded notation the determinant becomes

$$D_h(\lambda) = \sum_{q \in \mathbb{Z}_{n+1}} (-1)^q \lambda^q \sum_{1 \le i_1 < i_2 < \cdots < i_q \le n} \mathrm{vol}(\Omega_{\mathbf{i}}) K \begin{pmatrix} t_{i_1}, & \ldots, & t_{i_q} \\ t_{i_1}, & \ldots, & t_{i_q} \end{pmatrix}.$$

We now begin the task of identifying the limit of the polynomials D_h, as $h \to 0^+$.

Definition 1.5 If K is a continuous kernel on $\Omega \times \Omega$ and $\lambda \in \mathbb{C}$, we define the complex-valued function D at $\lambda \in \mathbb{C}$ by

$$D(\lambda) := \sum_{q \in \mathbb{N}_0} (-1)^q \frac{\lambda^q}{q!} \int_{\Omega^q} K[\mathbf{x}] \mathrm{d}\mathbf{x}. \tag{1.18}$$

Next, we point out that D is an entire function of λ. We shall refer to the function D as the *Fredholm function*. To prove that the Fredholm function D is an entire function, we recall the *Hadamard inequality* for the determinant of an $n \times n$ matrix (see, for example, [142]). When we use the Hadamard inequality it is convenient to express an $n \times n$ matrix $\mathbf{A}$ in terms of its columns. Thus, we write $\mathbf{A} = [\mathbf{a}^i : i \in \mathbb{N}_n]$ in which each $\mathbf{a}^i \in \mathbb{R}^n$ is the ith column of $\mathbf{A}$.

Lemma 1.6 *If* $\mathbf{A} = [\mathbf{a}^i : i \in \mathbb{N}_n]$ *is an* $n \times n$ *matrix then*

$$|\det \mathbf{A}| \le \prod_{i \in \mathbb{N}_n} |\mathbf{a}^i|,$$

where $|\mathbf{a}^i|$ *denotes the* ℓ^2*-norm of* $\mathbf{a}^i$.

We now set

$$\|K\|_\infty := \max\{|K(x, y)| : x, y \in \Omega\},$$

and derive the next lemma as an immediate application of the Hadamard inequality.

Lemma 1.7 *If* $K \in C(\Omega \times \Omega)$ *and* $\mathbf{x}, \mathbf{y} \in \Omega^q$ *then*

$$\left| K \begin{bmatrix} \mathbf{x} \\ \mathbf{y} \end{bmatrix} \right| \leq \|K\|_\infty^q q^{q/2}.$$

Proof Since for each $r, \ell \in \mathbb{N}_q$ and $x_r, y_\ell \in \Omega$ we have that

$$\sum_{r \in \mathbb{N}_q} K^2(x_r, y_\ell) \leq q\|K\|_\infty^2,$$

the result of this lemma follows immediately from the Hadamard inequality. □

Lemma 1.8 *The function D defined by equation (1.18) is an entire function.*

Proof For $q \in \mathbb{N}_0$ and $\lambda \in \mathbb{C}$ we have by Lemma 1.7 that

$$\left| \frac{(-1)^q \lambda^q}{q!} \int_{\Omega^q} K[\mathbf{x}] \mathrm{d}x \right| \leq \operatorname{vol}(\Omega)^q \frac{|\lambda|^q}{q!} \|K\|_\infty^q q^{q/2}.$$

By using the inequality

$$\frac{q^q}{q!} < e^q, \tag{1.19}$$

valid for all $q \in \mathbb{N}$, we obtain that

$$\operatorname{vol}(\Omega)^q \frac{|\lambda|^q}{q!} \|K\|_\infty^q q^{q/2} = \left(\frac{\operatorname{vol}(\Omega)|\lambda| \|K\|_\infty}{\sqrt{q}} \right)^q \frac{q^q}{q!} \leq \left(\frac{\operatorname{vol}(\Omega)|\lambda| \|K\|_\infty e}{\sqrt{q}} \right)^q.$$

Hence, when q is chosen to satisfy the condition

$$q > (2\operatorname{vol}(\Omega)|\lambda| \cdot \|K\|_\infty e)^2,$$

we conclude that

$$\left| \frac{(-1)^q \lambda^q}{q!} \int_{\Omega^q} K[\mathbf{x}] \mathrm{d}x \right| < 2^{-q}.$$

Thus, the series on the right-hand side of equation (1.18) converges absolutely. □

Theorem 1.9 *For $\lambda \in \mathbb{C}$,*

$$\lim_{h\to 0^+} D_h(\lambda) = D(\lambda)$$

uniformly and absolutely on any bounded subset of $\mathbb{C}$.

For the proof of Theorem 1.9 we need some ancillary lemmas. The first is about $n \times n$ matrices $\mathbf{A} := [\mathbf{a}^i : i \in \mathbb{N}_n]$ and $\mathbf{B} := [\mathbf{b}^i : i \in \mathbb{N}_n]$.

Lemma 1.10 *If there exists a constant θ such that for all $i \in \mathbb{N}_n$, $|\mathbf{a}^i| \le \theta$ and $|\mathbf{b}^i| \le \theta$ then*

$$|\det \mathbf{A} - \det \mathbf{B}| \le \theta^{n-1} \sum_{i\in\mathbb{N}_n} |\mathbf{a}^i - \mathbf{b}^i|.$$

Proof The proof follows from the Hadamard inequality applied to the formula

$$\det \mathbf{A} - \det \mathbf{B} = \sum_{i\in\mathbb{N}_n} \det[\mathbf{b}^1, \ldots, \mathbf{b}^{i-1}, \mathbf{a}^i - \mathbf{b}^i, \mathbf{a}^{i+1}, \ldots, \mathbf{a}^n].$$

□

For the next fact we need to recall the definition of the *modulus of continuity* of function K, namely, for $h > 0$

$$\omega(K, h) := \max\{|K(x, y) - K(x', y')| : |x - x'| \le h, |y - y'| \le h\}.$$

Lemma 1.11 *For any $\mathbf{x}, \mathbf{y} \in \Omega^q$ with $|\mathbf{x} - \mathbf{y}| \le h$, there holds the estimate*

$$|K[\mathbf{x}] - K[\mathbf{y}]| \le \frac{q}{\|K\|_\infty}(\|K\|_\infty \sqrt{q})^q \omega(K, h).$$

Proof We apply Lemma 1.10 to obtain the inequalities

$$\begin{aligned} |K[\mathbf{x}] - K[\mathbf{y}]| &\le (\|K\|_\infty \sqrt{q})^{q-1} \sum_{i\in\mathbb{N}_q} \left(\sum_{r\in\mathbb{N}_q} (K(x_r, x_i) - K(y_r, y_i))^2 \right)^{1/2} \\ &\le \left(\|K\|_\infty \sqrt{q}\right)^{q-1} q^{3/2} \omega(K, h), \end{aligned}$$

which gives the desired estimate. □

The next lemma estimates the difference between a multivariate integral and a finite sum which approximates it.

Lemma 1.12 *If $f \in C(\Omega^q)$ and Ω^q is partitioned as $\Omega^q = \bigcup_{\mathbf{i}\in\mathbb{N}_n^q} \Omega_{\mathbf{i}}$ a.e. and $\mathbf{s}_{\mathbf{i}} \in \Omega_{\mathbf{i}}$ for $\mathbf{i} \in \mathbb{N}_n^q$ then*

$$\left| \int_{\Omega^q} f(\mathbf{x}) \mathrm{d}\mathbf{x} - \sum_{\mathbf{i}\in\mathbb{N}_n^q} \mathrm{vol}(\Omega_{\mathbf{i}}) f(\mathbf{s}_{\mathbf{i}}) \right| \le (\mathrm{vol}(\Omega))^q \omega(f, h),$$

where h is chosen so that diam $\Omega_{\mathbf{i}} \leq h$ *for all* $\mathbf{i} \in \mathbb{N}_n^q$.

Proof We have that

$$\begin{aligned}\sum_{\mathbf{i}\in\mathbb{N}_n^q} \mathrm{vol}(\Omega_{\mathbf{i}})f(\mathbf{s}_{\mathbf{i}}) - \int_{\Omega^q} f(\mathbf{x})\mathrm{d}\mathbf{x} &= \sum_{\mathbf{i}\in\mathbb{N}_n^q} \{\mathrm{vol}(\Omega_{\mathbf{i}})f(\mathbf{s}_{\mathbf{i}}) - \int_{\Omega_{\mathbf{i}}} f(\mathbf{x})\mathrm{d}\mathbf{x}\} \\ &= \sum_{\mathbf{i}\in\mathbb{N}_n^q} \int_{\Omega_{\mathbf{i}}} (f(\mathbf{s}_{\mathbf{i}}) - f(\mathbf{x}))\mathrm{d}\mathbf{x}.\end{aligned}$$

Hence, taking absolute values of both sides of the equation gives the inequality

$$\begin{aligned}\left|\sum_{\mathbf{i}\in\mathbb{N}_n^q} \mathrm{vol}(\Omega_{\mathbf{i}})f(\mathbf{s}_{\mathbf{i}}) - \int_{\Omega^q} f(\mathbf{x})\mathrm{d}\mathbf{x}\right| &\leq \sum_{\mathbf{i}\in\mathbb{N}_n^q} \int_{\Omega_{\mathbf{i}}} |f(\mathbf{s}_{\mathbf{i}}) - f(\mathbf{x})|\mathrm{d}\mathbf{x} \\ &\leq \omega(f,h) \sum_{\mathbf{i}\in\mathbb{N}_n^q} \mathrm{vol}(\Omega_{\mathbf{i}}) \leq (\mathrm{vol}(\Omega))^q \omega(f,h).\end{aligned}$$

□

We are now ready to prove Theorem 1.9.

Proof Note that

$$|D_h(\lambda) - D(\lambda)| \leq \Delta_1 + \Delta_2,$$

where

$$\Delta_1 := \left| D_h(\lambda) - \sum_{q\in\mathbb{Z}_{n+1}} (-1)^q \frac{\lambda^q}{q!} \int_{\Omega^q} K[\mathbf{x}]\mathrm{d}\mathbf{x} \right|$$

and

$$\Delta_2 := \left| \sum_{q\in\mathbb{N}_0\setminus\mathbb{Z}_{n+1}} (-1)^q \frac{\lambda^q}{q!} \int_{\Omega^q} K[\mathbf{x}]\,\mathrm{d}\mathbf{x} \right|.$$

We estimate Δ_1 and Δ_2 separately.

We apply the above two lemmas to the function $f : \Omega^q \to \mathbb{R}$ defined for $\mathbf{x} \in \Omega^q$ as $f(\mathbf{x}) = K[\mathbf{x}]$ and obtain the inequality

$$\left| \sum_{\mathbf{i}\in\mathbb{N}_n^q} \mathrm{vol}(\Omega_{\mathbf{i}})K[\mathbf{s}_{\mathbf{i}}] - \int_{\Omega^q} K[\mathbf{x}]\mathrm{d}\mathbf{x} \right| \leq \frac{q}{\|K\|_\infty} (\mathrm{vol}(\Omega)\|K\|_\infty \sqrt{q})^q \omega(K,h).$$

Therefore, we obtain that

$$
\begin{aligned}
\Delta_1 &\le \sum_{q\in Z_{n+1}} \frac{|\lambda|^q}{q!}\left|\sum_{\mathbf{i}\in N_n^q} \mathrm{vol}(\Omega_{\mathbf{i}})K[\mathbf{s}_{\mathbf{i}}] - \int_{\Omega^q} K[\mathbf{x}]\mathrm{d}\mathbf{x}\right| \\
&\le \left\{\sum_{q\in Z_{n+1}} \frac{|\lambda|^q}{q!}\frac{q}{\|K\|_\infty}(\mathrm{vol}(\Omega)\|K\|_\infty\sqrt{q})^q\right\}\omega(K,h) \\
&\le \left\{\sum_{q\in\mathbb{N}_0} q\left(\frac{\mathrm{vol}(\Omega)|\lambda|\cdot\|K\|_\infty e}{\sqrt{q}}\right)^q\right\}\frac{\omega(K,h)}{\|K\|_\infty},
\end{aligned}
$$

where we have again used inequality (1.19) from the proof of Lemma 1.8 and we demand that q is sufficiently large so that the inequality

$$
\frac{\mathrm{vol}(\Omega)|\lambda|\cdot\|K\|_\infty e}{\sqrt{q}} < \frac{1}{2}
$$

holds. Therefore, the series above converges uniformly and absolutely on any bounded set. In fact, if we introduce the constant

$$
c := \sum_{q\in\mathbb{N}_0} q\left(\frac{\mathrm{vol}(\Omega)|\lambda|\cdot\|K\|_\infty e}{\sqrt{q}}\right)^q \Big/ \|K\|_\infty
$$

we can write the above observation as $\Delta_1 \le c\omega(K,h)$.

For Δ_2, we also have that

$$
\begin{aligned}
\Delta_2 &\le \sum_{q\in\mathbb{N}_0\setminus\mathbb{Z}_{n+1}} \frac{|\lambda|^q}{q!}\int_{\Omega^q} |K[\mathbf{x}]|\,\mathrm{d}\mathbf{x} \\
&\le \sum_{q\in\mathbb{N}_0\setminus\mathbb{Z}_{n+1}} \frac{|\lambda|^q}{q!}(\|K\|_\infty\sqrt{q})^q \mathrm{vol}(\Omega^q) \\
&\le \sum_{q\in\mathbb{N}_0\setminus\mathbb{Z}_{n+1}} \left(\frac{\mathrm{vol}(\Omega)|\lambda|\|K\|_\infty e}{\sqrt{q}}\right)^q \\
&\le \sum_{q\in\mathbb{N}_0\setminus\mathbb{Z}_{n+1}} \frac{1}{2^q}.
\end{aligned}
$$

Since $n \ge \frac{\mathrm{vol}(\Omega)}{2^d h^d}$, we see that $h \to 0^+$ implies both $n \to \infty$ and $q \to \infty$, so that the right-hand side of the above inequality goes to zero uniformly in λ in any bounded subset of the complex plane. These two inequalities prove the desired result. Indeed, when n is sufficiently large, we have

$$
|D_h(\lambda) - D(\lambda)| \le c\omega(K,h) + \sum_{q\in\mathbb{N}_0\setminus\mathbb{Z}_{n+1}} \frac{1}{2^q}.
$$

Since $\omega(K,h) \to 0$ as $h \to 0^+$, the right-hand side of the above inequality tends to zero uniformly. □

1.4 Resolvent kernels

We now explain the central role that the Fredholm determinant plays in solving a second-kind integral equation with a continuous kernel.

To this end, for each $\lambda \in \mathbb{C}$ we introduce the *resolvent kernel* R_λ defined for $s, t \in \Omega$ as

$$R_\lambda(s,t) := \sum_{q\in\mathbb{N}_0} \frac{(-1)^{q+1}}{q!}\lambda^{q+1} \int_{\Omega^q} K\begin{bmatrix} s, \mathbf{x} \\ t, \mathbf{x} \end{bmatrix} d\mathbf{x}.$$

Here, of course, the expression $K\begin{bmatrix} s,\mathbf{x} \\ t,\mathbf{x} \end{bmatrix}$ stands for the Fredholm minor

$$K\begin{pmatrix} s, x_1, \ldots, x_q \\ t, x_1, \ldots, x_q \end{pmatrix}.$$

Alternatively, we may write the resolvent kernel in the form

$$R_\lambda(s,t) = -\lambda \left\{ K(s,t) + \sum_{q\in\mathbb{N}} \frac{(-1)^q}{q!}\lambda^q \int_{\Omega^q} K\begin{bmatrix} s, \mathbf{x} \\ t, \mathbf{x} \end{bmatrix} d\mathbf{x} \right\}.$$

As in the proof of Lemma 1.8, we see for fixed $s, t \in \Omega$ that $R_\lambda(s,t)$ is an entire function of $\lambda \in \mathbb{C}$ and for a fixed $\lambda \in \mathbb{C}$ it is continuous in $s, t \in \Omega$. The next observation gives an indication of the role of the resolvent kernel in solving a second-kind integral equation.

Proposition 1.13 *For any $\lambda \in \mathbb{C}$ we have that*

$$\mathcal{R}_\lambda - \lambda\mathcal{K}\mathcal{R}_\lambda + \lambda D(\lambda)\mathcal{K} = 0 \tag{1.20}$$

and

$$\mathcal{R}_\lambda - \lambda\mathcal{R}_\lambda\mathcal{K} + \lambda D(\lambda)\mathcal{K} = 0. \tag{1.21}$$

Proof We first prove equation (1.20). For $\mathbf{x} \in \Omega^q, s, t \in \Omega$ we expand the determinant $K\begin{bmatrix} s, \mathbf{x} \\ t, \mathbf{x} \end{bmatrix}$ along its first row to obtain the formula

$$\begin{aligned} K\begin{pmatrix} s, x_1, \ldots, x_q \\ t, x_1, \ldots, x_q \end{pmatrix} &= K(s,t)K\begin{pmatrix} x_1, \ldots, x_q \\ x_1, \ldots, x_q \end{pmatrix} \\ &+ \sum_{j\in\mathbb{N}_q} (-1)^j K\begin{pmatrix} x_1, x_2, \ldots, x_j, x_{j+1}, \ldots, x_q \\ t, x_1, \ldots, x_{j-1}, x_{j+1}, \ldots, x_q \end{pmatrix} K(s, x_j). \end{aligned}$$

Now, we integrate both sides of this equation over $\mathbf{x} \in \Omega^q$, and note that the integral of each summand is independent of the summation index j. Hence, we conclude that

$$\int_{\Omega^q} K\begin{bmatrix} s, \mathbf{x} \\ t, \mathbf{x} \end{bmatrix} \mathrm{d}\mathbf{x} = K(s,t)\int_{\Omega^q} K[\mathbf{x}]\mathrm{d}\mathbf{x} - q\int_{\Omega} K(s,r)\left(\int_{\Omega^{q-1}} K\begin{bmatrix} r, \mathbf{x} \\ t, \mathbf{x} \end{bmatrix} \mathrm{d}\mathbf{x}\right)\mathrm{d}r$$

and this equation is also, by definition, valid for $q = 0$. We now substitute this equation into the power series for the resolvent kernel to obtain the equation

$$R_\lambda(s,t) = \lambda \int_\Omega K(s,r)R_\lambda(r,t)\mathrm{d}r - \lambda D(\lambda)K(s,t). \tag{1.22}$$

Next, we choose any $u \in C(\Omega)$ and multiply both sides of equation (1.22) by $u(t)$ and integrate both sides of the resulting equation over $t \in \Omega$ to obtain equation (1.20). The second equation (1.21) is proved in an analogous fashion by expanding the minor $K\begin{bmatrix} s, \mathbf{x} \\ t, \mathbf{x} \end{bmatrix}$ along its first column. □

The next result is concerned with the unique solvability of equation (1.1).

Theorem 1.14 *If $D(\lambda) \neq 0$ then $(\mathcal{I} - \lambda\mathcal{K})^{-1}$ is given by the formula*

$$(\mathcal{I} - \lambda\mathcal{K})^{-1} = \mathcal{I} - D(\lambda)^{-1}\mathcal{R}_\lambda.$$

Proof We compute

$$(\mathcal{I} - \lambda\mathcal{K})(\mathcal{I} - D(\lambda)^{-1}\mathcal{R}_\lambda) = \mathcal{I} - D(\lambda)^{-1}[\mathcal{R}_\lambda + \lambda D(\lambda)\mathcal{K} - \lambda\mathcal{K}\mathcal{R}_\lambda]$$

and now use equation (1.20) to conclude that the term inside the brackets is zero. Similarly, we verify that

$$(\mathcal{I} - D(\lambda)^{-1}\mathcal{R}_\lambda)(\mathcal{I} - \lambda\mathcal{K}) = \mathcal{I}$$

by using equation (1.21). □

It follows immediately from Theorem 1.14 that if λ is not a zero of the function D then equation (1.1) has a unique solution in $C(\Omega)$ for any $f \in C(\Omega)$. We now prove the converse of Theorem 1.14, namely, if $D(\lambda) = 0$ then $\mathcal{I} - \lambda\mathcal{K}$ is not invertible. To this end, we note the following ancillary fact.

Lemma 1.15 *For any $\lambda \in \mathbb{C}$ there holds the equation*

$$\int_\Omega R_\lambda(s,s)\mathrm{d}s = \lambda D'(\lambda). \tag{1.23}$$

The proof of this formula follows directly from a term-by-term integration and differentiation of the relevant power series.

Corollary 1.16 *If D has a zero of order m at some $\mu \neq 0$ then there is a $y \in \Omega$ for which $R_\mu(y,y)$ has a zero of order strictly less than m.*

Proof From Lemma 1.15 above and our hypothesis we have that

$$\int_\Omega \frac{\partial^{m-1}}{\partial \lambda^{m-1}} (R_\lambda(s,s)) \,|_{\lambda=\mu} \mathrm{d}s = \mu D^{(m)}(\mu) \neq 0.$$

□

Theorem 1.17 *If K is a continuous kernel such that $D(\lambda) = 0$ then $\mathcal{I}-\lambda\mathcal{K}$ is not invertible.*

Proof Choose a $y \in \Omega$ and consider the function u_y defined to be $R_\lambda(\cdot, y)$. Since $D(\lambda) = 0$ by hypothesis, equation (1.20) implies that $u_y - \lambda\mathcal{K}u_y = 0$. We would be finished with the proof if there was a $y \in \Omega$ for which $u_y \neq 0$. Otherwise, we must argue differently.

Suppose that D vanishes at μ and the multiplicity of this zero is m (note that $D(0) = 1$ and so each zero of D must be of finite multiplicity). Let l be the smallest non-negative integer such that for all $x, y \in \Omega$ the function $R_\lambda(x, y)$ has a zero of order l at μ. Corollary 1.16 says that $l < m$. We now expand $R_\lambda(x, y)$ near $\lambda = \mu$, and have that

$$R_\lambda = (\lambda - \mu)^l g + O(\lambda - \mu)^{l+1},$$

where the function g is continuous on $\Omega \times \Omega$ and never vanishes therein. Substituting this equation into equation (1.22), we obtain that

$$(\lambda - \mu)^l g(s,t) + O(\lambda - \mu)^{l+1} = \lambda \int_\Omega K(s,t')\Big[(\lambda - \mu)^l g(t',t) + O(\lambda - \mu)^{l+1}\Big]\mathrm{d}t' \\ - \lambda D(\lambda) K(s,t).$$

Dividing both sides of the above equation by $(\lambda - \mu)^l$, we get that

$$g(s,t) + O(\lambda - \mu) = \lambda \int_\Omega K(s,t')[g(t',t) + O(\lambda - \mu)]\mathrm{d}t' - \lambda \frac{D(\lambda)}{(\lambda - \mu)^l} K(s,t).$$

Letting $\lambda \to \mu$, we have that

$$g(s,t) = \mu \int_\Omega K(s,t') g(t',t)\mathrm{d}t'. \tag{1.24}$$

However, we may rewrite equation (1.24) in the equivalent form

$$(\mathcal{I} - \mu\mathcal{K})g = 0,$$

which together with the fact that g is not equal to zero establishes that $\mathcal{I} - \mu\mathcal{K}$ is not invertible, thereby proving the theorem. □

We see from the above theorems the importance of the zeros of D in the study of second-kind integral equations. Since $D(0) = 1$ the zeros of D are nonzero, countable and only have infinity as a possible accumulation point. The reciprocals of the zeros of D correspond to the nonzero eigenvalues of the operator $\mathcal{K}$. These observations prove many of the general statements we have made about compact operators in the Appendix, see Section A.2.7. We shall say more about the zeros of D later. First, we want to prove Fredholm's remarkable formula for the determinant of a product of two operators.

1.5 Fredholm determinants

In this section, we consider the Fredholm determinant and provide a result on the Fredholm determinant product.

Definition 1.18 Let $K \in C(\Omega \times \Omega)$. The linear operator $\mathcal{I} + \mathcal{K} : C(\Omega) \to C(\Omega)$ defined for $s \in \Omega$ and $u \in C(\Omega)$ by

$$((\mathcal{I} + \mathcal{K})u)(s) = u(s) + \int_{\Omega} K(s,t)u(t)\mathrm{d}t$$

has a Fredholm determinant, which is given by

$$\det(\mathcal{I} + \mathcal{K}) := \sum_{q \in \mathbb{N}_0} \frac{1}{q!} \int_{\Omega^q} K[\mathbf{x}]\mathrm{d}\mathbf{x}.$$

Alternatively, we may express the Fredholm determinant directly in terms of the Fredholm function, namely

$$\det(\mathcal{I} + \mathcal{K}) = D(-1).$$

Note that the Fredholm determinant of the operator $\mathcal{K}$ is the same as the Fredholm determinant of the operator $\mathcal{K}^*$ corresponding to the adjoint kernel, which is defined for $s, t \in \Omega$ by $K^*(s,t) = K(t,s)$. That is, for any continuous kernel K on the compact set Ω, we have the equation

$$\det(\mathcal{I} + \mathcal{K}) = \det(\mathcal{I} + \mathcal{K}^*). \tag{1.25}$$

The main result of this section is the Fredholm determinant product formula.

Theorem 1.19 *If $K, H \in C(\Omega \times \Omega)$ then*

$$\det\{(\mathcal{I} + \mathcal{H})(\mathcal{I} + \mathcal{K})\} = \det(\mathcal{I} + \mathcal{H})\det(\mathcal{I} + \mathcal{K}).$$

We start our discussion of this product formula by introducing a continuous kernel L associated with two prescribed continuous kernels K and H.

Definition 1.20 If K and H are continuous kernels, we define L to be the kernel of the operator

$$\mathcal{L} := \mathcal{K} + \mathcal{H} + \mathcal{H}\mathcal{K}.$$

Remark 1.21 *The kernel L has the characteristic property that*

$$\mathcal{I} + \mathcal{L} = (\mathcal{I} + \mathcal{H})(\mathcal{I} + \mathcal{K}). \tag{1.26}$$

There are some special cases of Theorem 1.19 that are readily proven. For example, if $\det(\mathcal{I} + \mathcal{K}) = 0$, then $\mathcal{I} + \mathcal{K}$ is not one to one. Therefore, it follows that $\mathcal{I} + \mathcal{L}$ is also not one to one. Consequently, we conclude that $\det(\mathcal{I} + \mathcal{L}) = 0$. Similarly, if $\det(\mathcal{I} + \mathcal{H}) = 0$ then $\det(\mathcal{I} + \mathcal{H}^*) = 0$, from which it follows that $\mathcal{I} + \mathcal{H}$ is not onto. Hence, the operator $\mathcal{I} + \mathcal{L}$ is also not onto and so $\det(\mathcal{I} + \mathcal{L}) = 0$. In other words, in the proof of the above theorem, we may assume that the operators $\mathcal{I} + \mathcal{L}$, $\mathcal{I} + \mathcal{H}$ and $\mathcal{I} + \mathcal{K}$ are all invertible.

To proceed further, we need the following differentiation formula. To this end, we define, for $\tau \in \mathbb{R}$ and the continuous kernel G, the one-parameter family of kernels given as $K_\tau := K + \tau G$.

Proposition 1.22 *If K and G are continuous kernels, then*

$$\frac{\mathrm{d}}{\mathrm{d}\tau}\{\det(\mathcal{I} + \mathcal{K}_\tau)\}|_{\tau=0} = \det(\mathcal{I} + \mathcal{K})\int_\Omega G(s,s)\mathrm{d}s - \int_\Omega\left(\int_\Omega R_{-1}(s,t)G(t,s)\mathrm{d}t\right)\mathrm{d}s.$$

Proof For each $\mathbf{x} \in \Omega^q$ a straightforward differentiation of the requisite determinant and a Laplace expansion yield the formula

$$\frac{\mathrm{d}}{\mathrm{d}\tau}(K_\tau[\mathbf{x}])\,|_{\tau=0} = \sum_{l,k \in \mathbb{N}_q} (-1)^{l+k} K\begin{pmatrix} x_1, \ldots, x_{k-1}, x_{k+1}, \ldots, x_q \\ x_1, \ldots, x_{l-1},\ x_{l+1}, \ldots, x_q \end{pmatrix} G(x_k, x_l). \tag{1.27}$$

We now integrate both sides of this equation over $\mathbf{x} \in \Omega^q$. The integral on the left-hand side of the equation is clear. As for the right-hand side, we distinguish the summands corresponding to $k = l$ from the remaining summands. All the summands corresponding to $k = l$ have the same integral, given by

$$\int_{\Omega^{q-1}} K[\mathbf{x}]\mathrm{d}\mathbf{x} \cdot \int_\Omega G(s,s)\mathrm{d}s. \tag{1.28}$$

There are $q(q-1)$ remaining terms corresponding to the case $k \neq l$. They also have the same integral, which is independent of k and l and is computed to be

$$-q(q-1)\int_{\Omega^2}\left(\int_{\Omega^{q-2}} K\begin{bmatrix} s, \mathbf{x} \\ t, \mathbf{x}\end{bmatrix}\mathrm{d}\mathbf{x}\right) G(t,s)\mathrm{d}s\mathrm{d}t.$$

Therefore, in summary, we obtain that

$$\int_{\Omega^q} \frac{\mathrm{d}}{\mathrm{d}\tau}\left(K_\tau[\mathbf{x}]\right)|_{\tau=0}\,\mathrm{d}\mathbf{x} = q\int_{\Omega^{q-1}} K[\mathbf{x}]\mathrm{d}\mathbf{x}\cdot\int_\Omega G(s,s)\mathrm{d}s$$
$$- q(q-1)\int_\Omega\left[\int_\Omega\left\{\left(\int_{\Omega^{q-2}} K\begin{bmatrix} s,\mathbf{x}\\ t,\mathbf{x}\end{bmatrix}\mathrm{d}\mathbf{x}\right)G(t,s)\right\}\mathrm{d}t\right]\mathrm{d}s. \tag{1.29}$$

Substituting equation (1.29) into the series expansion

$$\det(\mathcal{I}+\mathcal{K}_\tau) = \sum_{q\in\mathbb{N}_0}\frac{1}{q!}\int_{\Omega^q} K_\tau[\mathbf{x}]\mathrm{d}\mathbf{x}$$

and rearranging terms yields the desired formula

$$\frac{\mathrm{d}}{\mathrm{d}\tau}\{\det(\mathcal{I}+\mathcal{K}_\tau)\}|_{\tau=0} = D(-1)\int_\Omega G(s,s)\mathrm{d}s - \int_\Omega\left(\int_\Omega R_{-1}(s,t)G(t,s)\mathrm{d}t\right)\mathrm{d}s.$$

□

Remark 1.23 *When* $\det(\mathcal{I}+\mathcal{K})\neq 0$*, we may use equation (1.20) to express the formula in Proposition 1.22 in the equivalent form*

$$\frac{\mathrm{d}}{\mathrm{d}\tau}\{\log\det(\mathcal{I}+\mathcal{K}_\tau)\}|_{\tau=0} = \int_\Omega W(s,s)\mathrm{d}s, \tag{1.30}$$

where W is the continuous kernel corresponding to the integral operator

$$\left(\mathcal{I}-D^{-1}(-1)\mathcal{R}_{-1}\right)\mathcal{G}.$$

Indeed, by Proposition 1.22 we have that

$$\begin{aligned}
\frac{\mathrm{d}}{\mathrm{d}\tau}\{\log\det(\mathcal{I}+\mathcal{K}_\tau)\}|_{\tau=0} &= \frac{\frac{\mathrm{d}}{\mathrm{d}\tau}\{\det(\mathcal{I}+\mathcal{K}_\tau)\}|_{\tau=0}}{\det(\mathcal{I}+\mathcal{K})}\\
&= \int_\Omega G(s,s)\,\mathrm{d}s - D^{-1}(-1)\int_\Omega\left(\int_\Omega R_{-1}(s,t)G(t,s)\mathrm{d}t\right)\mathrm{d}s\\
&= \int_\Omega W(s,s)\,\mathrm{d}s,
\end{aligned}$$

where

$$W(s,s) := G(s,s) - D^{-1}(-1)\int_\Omega R_{-1}(s,t)G(t,s)\,\mathrm{d}t.$$

Moreover, for any $u \in C(\Omega), s \in \Omega$, we have that

$$\begin{aligned}\int_\Omega W(s,t)u(t)\,dt &= \int_\Omega \left\{G(s,t) - D^{-1}(-1)\int_\Omega R_{-1}(s,t')G(t',t)\,dt'\right\} u(t)\,dt \\ &= \int_\Omega G(s,t)u(t)\,dt - D^{-1}(-1)\int_\Omega\int_\Omega R_{-1}(s,t')G(t',t)u(t)\,dt'\,dt \\ &= (((\mathcal{I} - D^{-1}(-1)\mathcal{R}_{-1})\mathcal{G})u)(s),\end{aligned}$$

which ensures the desired result. By Theorem 1.14, we can rewrite this integral operator in the equivalent simpler form $(\mathcal{I}+\mathcal{K})^{-1}\mathcal{G}$. In order to make it easy to recall the association of the kernel W with this integral operator, we express equation (1.30) in the notationally suggestive form

$$\frac{d}{d\tau}\{\log\det(\mathcal{I}+\mathcal{K}_\tau)\}|_{\tau=0} = \int_\Omega \left\{(\mathcal{I}+\mathcal{K})^{-1}\mathcal{G}\right\}(s,s)ds. \tag{1.31}$$

As we already pointed out in equation (1.25), the Fredholm determinants of $\mathcal{K}$ and $\mathcal{K}^*$ are the same. Hence, since ${\mathcal{K}_\tau}^* = \mathcal{K}^* + \tau\mathcal{G}^*$, we also have that

$$\frac{d}{d\tau}\{\log\det(\mathcal{I}+\mathcal{K}_\tau)\}|_{\tau=0} = \int_\Omega \left\{(\mathcal{I}+\mathcal{K}^*)^{-1}\mathcal{G}^*\right\}(s,s)ds. \tag{1.32}$$

We are now ready to prove Theorem 1.19.

Proof We consider two one-parameter perturbations $K_\tau = K + \tau K_0$ and $H_\tau = H + \tau H_0$ and set

$$\mathcal{I}+\mathcal{L}_\tau := (\mathcal{I}+\mathcal{H}_\tau)(\mathcal{I}+\mathcal{K}_\tau).$$

Note that

$$\mathcal{L}_\tau = \mathcal{L} + \tau\mathcal{L}_0 + o(\tau^2), \quad \tau \to 0^+, \tag{1.33}$$

where $\mathcal{L}$ is defined in Definition 1.20 and $\mathcal{L}_0$ is given by the formula

$$\mathcal{L}_0 = (\mathcal{I}+\mathcal{H})\mathcal{K}_0 + \mathcal{H}_0(\mathcal{I}+\mathcal{K}). \tag{1.34}$$

Combining equations (1.31), (1.32), (1.33) and (1.34), we obtain that

$$\begin{aligned}\frac{d}{d\tau}\{\log\det(\mathcal{I}+\mathcal{L}_\tau)\}|_{\tau=0} \quad &= \quad \int_\Omega \left\{(\mathcal{I}+\mathcal{L})^{-1}(\mathcal{I}+\mathcal{H})\mathcal{K}_0\right\}(s,s)ds \\ &\quad + \int_\Omega \left\{(\mathcal{I}+\mathcal{L}^*)^{-1}(\mathcal{I}+\mathcal{K}^*)\mathcal{H}_0^*\right\}(s,s)ds.\end{aligned}$$

But we also have that

$$(\mathcal{I}+\mathcal{L})^{-1} = (\mathcal{I}+\mathcal{K})^{-1}(\mathcal{I}+\mathcal{H})^{-1}$$

and

$$(\mathcal{I}+\mathcal{L}^*)^{-1} = \left(\mathcal{I}+\mathcal{H}^*\right)^{-1}\left(\mathcal{I}+\mathcal{K}^*\right)^{-1}.$$

Therefore, we get that

$$\begin{aligned}&\frac{\mathrm{d}}{\mathrm{d}\tau}\{\log\det(\mathcal{I}+\mathcal{L}_\tau)\}|_{\tau=0}\\&=\int_\Omega\left\{(\mathcal{I}+\mathcal{H}^*)^{-1}\mathcal{H}_0^*\right\}(s,s)\mathrm{d}s+\int_\Omega\left\{(\mathcal{I}+\mathcal{K})^{-1}\mathcal{K}_0\right\}(s,s)\mathrm{d}s\\&=\frac{\mathrm{d}}{\mathrm{d}\tau}\{\log\det(\mathcal{I}+\mathcal{H}_\tau)\}|_{\tau=0}+\frac{\mathrm{d}}{\mathrm{d}\tau}\{\log\det(\mathcal{I}+\mathcal{K}_\tau)\}|_{\tau=0}.\end{aligned}$$

Let $\gamma:[0,1]\to\mathbb{R}$ be the function defined at $\tau\in[0,1]$ as

$$\gamma(\tau):=\log\det(\mathcal{I}+\mathcal{L}_\tau)-\log\{\det(\mathcal{I}+\mathcal{H}_\tau)\det(\mathcal{I}+\mathcal{K}_\tau)\}.$$

The above formula means that $\gamma'(0)=0$. Clearly, we can derive the same conclusion along any continuous path $H(\tau)$ and $K(\tau)$ as long as $H(0)=H$ and $K(0)=K$. Moreover, we can also show that the derivative of γ is zero at any point μ in $(0,1)$ provided that both the integral operators $\mathcal{I}+\mathcal{H}(\mu)$ and $\mathcal{I}+\mathcal{K}(\mu)$ have inverses.

Next, we choose *any* continuously differentiable path $\sigma:[0,1]\to[0,1]$ such that $\sigma(0)=0,\sigma(1)=1$ and then, for any $\tau\in[0,1]$ which satisfies $\det(\mathcal{I}+\tau\mathcal{H})\neq0$ and $\det(\mathcal{I}+\tau\mathcal{K})\neq0$, we introduce the operators $\mathcal{H}_\tau:=\sigma(\tau)\mathcal{H}$ and $\mathcal{K}_\tau:=\sigma(\tau)\mathcal{K}$. Consequently, we conclude by our previous remarks that the function γ defined above must be constant in the interval $[0,1]$. This, together with the fact that $\gamma(0)=0$ and $\gamma(1)=\log\det(\mathcal{I}+\mathcal{L})-\log\det(\mathcal{I}+\mathcal{H}))\det(\mathcal{I}+\mathcal{K})$, completes the proof. □

1.6 Eigenvalue estimates and a trace formula

We now turn our attention to a new topic and provide an estimate of the growth of the zero of D in the complex plane. We accomplish this only when $\Omega=[0,1]$ and the kernel K has the property that there are positive constants $\rho>0$ and $\alpha\in(0,1]$ such that, for all $u,s,t\in[0,1]$,

$$|K(u,s)-K(u,t)|\leq\rho|s-t|^\alpha.$$

When this is the case we say that K is *Hölder continuous* of order α with constant ρ (with respect to the second variable). For Hölder continuous kernels of order α we can provide a better estimate for the Fredholm minor than that given in Lemma 1.7.

Lemma 1.24 *If K is Hölder continuous of order α with constant ρ then for any $\mathbf{x},\mathbf{y}\in[0,1]^q$,*

$$\left|K\begin{bmatrix}\mathbf{x}\\\mathbf{y}\end{bmatrix}\right|\leq4^\alpha(q^q)^{1/2-\alpha}\rho^q.$$

Proof First, we reorder the columns of the minor appearing on the left-hand side of the inequality so that the components of the vector $\mathbf{y} := [y_i : i \in \mathbb{N}_n]$ are increasing. Next, for each $i \in \mathbb{N}_{q-1}$, we subtract the $(i+1)$th column from the ith column in the above minor (which does not alter its value), use the Hadamard inequality and then the hypothesis on the kernel to obtain the inequality

$$\left| K \begin{bmatrix} \mathbf{x} \\ \mathbf{y} \end{bmatrix} \right| \leq (\rho\sqrt{q})^q \prod_{i \in \mathbb{N}_{q-1}} |y_{i+1} - y_i|^\alpha.$$

We now apply the arithmetic–geometric inequality to obtain the inequalities

$$\left| K \begin{bmatrix} \mathbf{x} \\ \mathbf{y} \end{bmatrix} \right| \leq (\rho\sqrt{q})^q \left\{ \frac{1}{q-1} \sum_{i \in \mathbb{N}_{q-1}} (y_{i+1} - y_i) \right\}^{(q-1)\alpha} \leq 4^\alpha (q^q)^{1/2-\alpha} \rho^q,$$

thereby confirming the desired bound. □

Remark 1.25 *The above lemma holds if K is a Hölder continuous kernel of order α with respect to the first variable rather than the second variable.*

Note that, for the next result, we introduce the constant

$$\mu := \frac{2}{1+2\alpha}.$$

For $\alpha \in (\frac{1}{2}, 1]$ we have that $\mu \in [\frac{2}{3}, 1)$.

Theorem 1.26 *If $K(s,t)$ is a Hölder continuous kernel of order $\alpha \in (\frac{1}{2}, 1]$ with respect to either s or t in $[0,1]$, then for every $\epsilon > 0$ there is a constant $a > 0$ such that for all $\lambda \in \mathbb{C}$*

$$|D(\lambda)| \leq a e^{|\lambda|^{\mu+\epsilon}}.$$

Proof To prove this theorem, we first note that any polynomial appearing in the proof of this result can be neglected, because any polynomial is surely dominated in the complex plane by an exponential.

For every $\lambda \in \mathbb{C}$ we have that

$$|D(\lambda)| \leq 4^\alpha \sum_{m \in \mathbb{N}_0} a_m |\lambda|^m,$$

where we define the constant

$$a_m := \frac{\rho^m (m^m)^{\frac{1}{2}-\alpha}}{m!}.$$

This constant satisfies the inequality

$$a_m \leq \left(\frac{\rho e}{m^{\frac{\epsilon}{\mu+\epsilon}}} \right)^m \frac{1}{m^{\frac{m}{\mu+\epsilon}}}. \tag{1.35}$$

Now, for m large enough the expression in parentheses on the right-hand side of inequality (1.35) can be made less than one. Hence, as we can ignore polynomials of a fixed degree in $|\lambda|$, we assume without loss of generality that it suffices to bound, from above, the power series

$$\sum_{m \in \mathbb{N}} \frac{|\lambda|^m}{m^{\frac{m}{\mu+\epsilon}}}.$$

Certainly, the series above is bounded for $|\lambda| \leq 1$ and so we only have to consider it for $|\lambda| > 1$. In that case, we break the sum above into two parts. For the first part we sum over positive integers $m < (2|\lambda|)^{\mu+\epsilon}$ and for the second part over the remaining integers. For the first sum, we have that

$$\sum_{m<(2|\lambda|)^{\mu+\epsilon}} \frac{|\lambda|^m}{m^{\frac{m}{\mu+\epsilon}}} \leq c|\lambda|^{(2|\lambda|)^{\mu+\epsilon}},$$

where

$$c := \sum_{m \in \mathbb{N}} m^{-\frac{m}{\mu+\epsilon}} < \infty.$$

Since

$$\lim_{|\lambda| \to \infty} |\lambda|^{-\epsilon} \log |\lambda| = 0,$$

we conclude that there is a constant $a > 0$ such that

$$|\lambda|^{(2|\lambda|)^{\mu+\epsilon}} \leq a e^{|\lambda|^{\mu+2\epsilon}}.$$

For the other summands corresponding to $m \geq (2|\lambda|)^{\mu+\epsilon}$ we have that

$$\frac{|\lambda|^m}{m^{\frac{m}{\mu+\epsilon}}} \leq 2^{-m},$$

and so we conclude that

$$\sum_{m \geq (2|\lambda|)^{\mu+\epsilon}} \frac{|\lambda|^m}{m^{\frac{m}{\mu+\epsilon}}} \leq \sum_{m \in \mathbb{N}} 2^{-m} < \infty.$$

□

This theorem says, under the hypothesis on the kernel K, that D is an entire function of order less than or equal to μ.

Corollary 1.27 *If* $\{\lambda_n : n \in \mathbb{N}\}$ *are the zeros of* D *in* $\mathbb{C}$ *ordered so that* $0 < |\lambda_1| \leq |\lambda_2| \leq \cdots \leq |\lambda_n| \leq \cdots$ *and* $K(s,t)$ *is a Hölder continuous kernel in either* s *or* t *on* $[0,1]$ *with exponent* $\alpha \in (\frac{1}{2}, 1]$, *then*

$$\sum_{n\in\mathbb{N}} |\lambda_n|^{-1} < \infty.$$

The proof of this corollary relies on standard techniques in the study of entire functions. We briefly review some of the details.

The first step is to recall Jensen's formula (see, for example, [2], p. 208). If $\rho > 0, f$ is a function analytic in the disc $\Delta_\rho := \{z : |z| < \rho\}$ and continuous on the boundary with a finite number of zeros $\{a_j : j \in \mathbb{N}_m\}$ in the closed disc $\overline{\Delta}_\rho$, then

$$\log|f(0)| + \sum_{j\in\mathbb{N}_m} \log\frac{\rho}{|a_j|} = \frac{1}{2\pi}\int_0^{2\pi} \log|f(\rho e^{i\theta})|\mathrm{d}\theta. \tag{1.36}$$

We apply Jensen's formula to the function D in the following way. We let $\nu(\rho)$ be the number of zeros of D (counting multiplicities) in the disc Δ_ρ. Recall that $|\lambda_1| \leq |\lambda_2| \leq \cdots$ and so it follows that

$$n \leq \nu(|\lambda_n|). \tag{1.37}$$

Now, according to Jensen's formula applied to the function D on the disc $\overline{\Delta}_{2\rho}$ we have that

$$\sum_{j\in\mathbb{N}_k} \log\frac{2\rho}{|\lambda_j|} = \frac{1}{2\pi}\int_0^{2\pi} \log|D(2\rho e^{i\theta})|\mathrm{d}\theta, \tag{1.38}$$

where $\{\lambda_j : j \in \mathbb{N}_k\}$ are the zeros of D in $\overline{\Delta}_{2\rho}$. Since $|\lambda_j| \leq 2\rho$ for $j \in \mathbb{N}_k$, we have that each summand on the left-hand side of equation (1.38) is non-negative. We neglect the terms corresponding to the zeros of D in $\overline{\Delta}_{2\rho} \setminus \Delta_\rho$, thereby obtaining the inequality

$$\nu(\rho)\log 2 < \frac{1}{2\pi}\int_0^{2\pi} \log|D(2\rho e^{i\theta})|\mathrm{d}\theta. \tag{1.39}$$

Now, choose $\epsilon > 0$ so that $\mu + \epsilon < 1$ (recall that when $\alpha \in (\frac{1}{2}, 1]$, we have that $\mu < 1$). Next, we use the estimate for $D(\lambda)$ in Theorem 1.26 and get that

$$\nu(\rho)\log 2 \leq \log a + (2\rho)^{\mu+\epsilon}. \tag{1.40}$$

Consequently, for sufficiently large m, there is a positive constant c such that for $n \geq m$ we have $|\lambda_n| \geq 1$ and

$$\nu(|\lambda_n|) \leq \{c|\lambda_n|\}^{\mu+\epsilon}.$$

According to the inequality (1.37), this inequality implies for $n \geq m$ that

$$n^{\frac{1}{\mu+\epsilon}} \leq c|\lambda_n|.$$

In other words, for some other positive constant $b > 0$ we have that

$$\sum_{n\in\mathbb{N}} |\lambda_n|^{-1} \leq b \sum_{n\in\mathbb{N}} n^{-\frac{1}{\mu+\epsilon}} < \infty.$$

Theorem 1.28 *If $K(s,t)$ is a Hölder continuous kernel on $[0,1]$ in either s or t with exponent $\alpha \in (\frac{1}{2}, 1]$, then*

$$D(\lambda) = \prod_{n\in\mathbb{N}} \left(1 - \frac{\lambda}{\lambda_n}\right).$$

Proof According to the above corollary we see that the right-hand side of the above equation is an entire function of $\lambda \in \mathbb{C}$. Moreover, the function h defined at λ as

$$h(\lambda) = \frac{D(\lambda)}{\prod_{n\in\mathbb{N}} \left(1 - \frac{\lambda}{\lambda_n}\right)}$$

is free of zeros in $\mathbb{C}$. Therefore, by the Weierstrass factorization theorem there is an entire function g such that for all $\lambda \in \mathbb{C}$,

$$D(\lambda) = e^{g(\lambda)} \prod_{n\in\mathbb{N}} \left(1 - \frac{\lambda}{\lambda_n}\right). \tag{1.41}$$

We now show that g is a constant. But, since $D(0) = 1$, the result will follow. Therefore, the last remaining task is to show that g is constant. To this end, we need the following version of the Poisson–Jensen formula (see, for example, [2], p. 208). Specifically, if f is a function analytic in $\overline{\Delta}_\rho$ with only zeros $\{a_j : j \in \mathbb{N}_m\}$ in Δ_ρ and $z \in \Delta_\rho$, then

$$\log|f(z)| = -\sum_{j\in\mathbb{N}_m} \log\left|\frac{\rho^2 - \overline{a}_j z}{\rho(z - a_j)}\right| + \frac{1}{2\pi}\int_0^{2\pi} \operatorname{Re}\frac{\rho e^{i\theta} + z}{\rho e^{i\theta} - z} \log|f(\rho e^{i\theta})| d\theta. \tag{1.42}$$

We shall first differentiate both sides of this equation with respect to z appropriately and then examine the behavior of each resulting term on the right-hand side when $f = D$ and $\rho \to \infty$. For the process of differentiation we recall the following lemma.

Lemma 1.29 *If f is analytic in a neighborhood of z then*

$$\frac{\partial}{\partial \bar{z}} \log |f(z)| = \frac{f'(z)}{f(z)}$$

and

$$\frac{\partial}{\partial \bar{z}} \mathrm{Re} f(z) = f'(z).$$

To prove this lemma, we write $f(z) = u + iv$ and then by a direct application of the Cauchy–Riemann equations the result may be verified.

We now fix z, choose $\rho > 2|z|$ and apply the derivative operator $\frac{\partial}{\partial \bar{z}}$ to both sides of equation (1.42) for the choice $f = D$ to get the equation

$$\frac{D'(z)}{D(z)} = \sum_{j \in \mathbb{N}_n} \frac{\overline{\lambda}_j}{\rho^2 - \overline{\lambda}_j z} + \sum_{j \in \mathbb{N}_n} \frac{1}{z - \lambda_j} + \frac{1}{\pi} \int_0^{2\pi} \rho e^{i\theta} (\rho e^{i\theta} - z)^{-2} \log |D(\rho e^{i\theta})| \mathrm{d}\theta. \tag{1.43}$$

We first estimate the integral by using Corollary 1.27. Since $\rho > 2|z|$, we know that $|\rho e^{i\theta} - z| \geq \rho/2$, which together with Theorem 1.26 yields that

$$\begin{aligned} \left| \rho e^{i\theta} (\rho e^{i\theta} - z)^{-2} \log |D(\rho e^{i\theta})| \right| &\leq \rho |\rho e^{i\theta} - z|^{-2} (\log a + |\rho e^{i\theta}|^{\mu + \varepsilon}) \\ &\leq \frac{4}{\rho} (\log a + \rho^{\mu + \varepsilon}), \end{aligned}$$

and thus the absolute value of the third term on the right-hand side of equation (1.43) is bounded by $\frac{8}{\rho} \left(\log a + \rho^{\mu + \epsilon} \right)$, which tends to zero as $\rho \to \infty$. For the first sum on the right-hand side of equation (1.43) we note that

$$|\rho^2 - \overline{\lambda}_j z| \geq \rho^2 - \rho |z| \geq \frac{\rho^2}{2}.$$

Therefore, we see that the first sum on the right-hand side of equation (1.43) is bounded by $2\rho^{-1} \nu(\rho)$ which, according to inequality (1.40), also goes to zero as $\rho \to \infty$. Consequently, equation (1.43) yields, as $\rho \to \infty$, the equation

$$\frac{D'(z)}{D(z)} = \sum_{j \in \mathbb{N}} \frac{1}{z - \lambda_j}. \tag{1.44}$$

However, we also have that

$$D(z) = e^{g(z)} \prod_{j \in \mathbb{N}} \left(1 - \frac{z}{\lambda_j} \right). \tag{1.45}$$

Now we differentiate both sides of equation (1.45) to obtain that

$$\frac{D'(z)}{D(z)} = g'(z) + \sum_{j\in\mathbb{N}} \frac{1}{z-\lambda_j}. \tag{1.46}$$

Comparing equations (1.44) and (1.46) leads to the desired conclusion that g is constant, thereby completing the proof of the theorem. □

Our discussion in this chapter indicates that a Hölder continuous kernel $K(s,t)$ on $[0,1]$ in either s or t with exponent $\alpha \in (\frac{1}{2}, 1]$ has eigenvalues which are l^1-summable. We let $\{\mu_n : n \in \mathbb{N}\}$ be the nonzero eigenvalues of $\mathcal{K}$ so that $\lambda_n = 1/\mu_n$. We then have the following result.

Theorem 1.30 *If $K(s,t)$ is a Hölder continuous kernel on $[0,1]$ in either s or t with exponent $\alpha \in (1/2, 1]$, then*

$$\int_\Omega K(s,s)\mathrm{d}s = \sum_{n\in\mathbb{N}} \mu_n$$

and

$$\det(\mathcal{I}+\mathcal{K}) = \prod_{n\in\mathbb{N}}(1+\mu_n).$$

Proof First, we prove the second equality. Indeed, we have that

$$\det(\mathcal{I}+\mathcal{K}) = D(-1) = \prod_{n\in\mathbb{N}}(1+\lambda_n^{-1}) = \prod_{n\in\mathbb{N}}(1+\mu_n).$$

Next, we show the first equality. On the one hand, we know that

$$\frac{D'(z)}{D(z)} = \sum_{j\in\mathbb{N}} \frac{1}{z-\lambda_j},$$

which in turn yields

$$D'(0) = D(0)\sum_{j\in\mathbb{N}} -\frac{1}{\lambda_j} = -\sum_{j\in\mathbb{N}} \mu_n.$$

On the other hand, by the power series expansion of D, we have that

$$D'(\lambda) = \sum_{q\in\mathbb{N}_0} (-1)^{q+1}\frac{\lambda^q}{q!}\int_{\Omega^{q+1}} K[\mathbf{x}]\,\mathrm{d}\mathbf{x},$$

and hence it follows that

$$D'(0) = -\int_\Omega K(s,s)\mathrm{d}s.$$

Combining the above two aspects, we obtain the desired result. □

This completes our brief discussion of the classical theory of Fredholm integral equations. We next turn our attention to further essential background material on integral equations, and postpone until Chapter 4 the main subject of this book, namely, multiscale methods for the numerical solution of Fredholm integral equations.

1.7 Bibliographical remarks

Most of the material presented in this chapter is taken from the book [183]. For the important notion of the Fredholm function and the Fredholm determinant, readers are referred to [183, 253]. Readers may find a discussion of the distribution of the eigenvalues of the Fredholm integral operator in [141]. In addition, [249] is a nice reference for the topic of integral equations, where three fundamental papers on integral equations written by three outstanding mathematicians [Ivar Fredholm (1903), David Hilbert (1904) and Erhard Schmidt (1907)], published in the first decade of the twentieth century, were translated into English with commentaries.

2
Fredholm equations and projection theory

In this chapter we provide concepts and results useful for the study of the Fredholm integral equation of the second kind, especially the theory of projection methods needed for the development of the multiscale methods.

2.1 Fredholm integral equations

The main concern of this book is the Fredholm integral equation of the second kind. In this section we review concepts relevant to the study of this class of integral equations. The general form for a Fredholm integral equation of the second kind is

$$u - \mathcal{K}u = f, \tag{2.1}$$

where the linear operator $\mathcal{K}$ is defined on a normed linear space with values in another such space, the function f is given and u is a solution to be determined. Typically these spaces consist of real or complex-valued functions on a measurable subset Ω in the d-dimensional Euclidean space $\mathbb{R}^d$. The function $\mathcal{K}u$ is determined by a kernel $K(s,t), s,t \in \Omega$ and the Fredholm integral operator is defined by the formula

$$(\mathcal{K}u)(s) := \int_\Omega K(s,t)u(t)\mathrm{d}t, \quad s \in \Omega, \tag{2.2}$$

or, in short, by

$$(\mathcal{K}u)(s) := \int_\Omega K(s,\cdot)u(\cdot), \quad s \in \Omega.$$

Properties of the operator $\mathcal{K}$ are inherited from those of the kernel K. We shall restrict ourselves to kernels for which $\mathcal{K}$ is a compact operator, that is, it maps bounded subsets to relatively compact ones. Let us postpone the discussion of operators of this type until we describe our preferred notation and terminology.

For $d \in \mathbb{N}$, let $\mathbb{R}^d$ be the d-dimensional Euclidean space and Ω a domain (open set) in $\mathbb{R}^d$. By $C^m(\Omega)$, $m \in \mathbb{N}_0$, we mean the linear space of all real-valued functions defined on Ω which are m-times continuously differentiable there. That is, all derivatives up to and including all those of total order m are continuous on Ω. Therefore, the space of infinitely differentiable functions on Ω is given by $C^\infty(\Omega) := \bigcap_{m\in\mathbb{N}_0} C^m(\Omega)$. We use $\bar{\Omega}$ for the closure of Ω and denote by $C^m(\bar{\Omega})$ the subspace of all functions together with all their derivatives up to order m that are bounded and uniformly continuous on the closure of Ω. The simplified notational convention $C(\Omega)$ and $C(\bar{\Omega})$ for the spaces $C^0(\Omega)$ and $C^0(\bar{\Omega})$, respectively, will be used throughout the book. For a lattice vector $\alpha \in \mathbb{N}_0^d$ with non-negative coordinates we use the notation $|\alpha| := \sum_{i\in\mathbb{N}_d} \alpha_i$. Corresponding to any vector $t := [t_j : j \in \mathbb{N}_d] \in \mathbb{R}^d$, we denote the $|\alpha|$-derivative of a function u at t (when it exists) by

$$D^\alpha u(t) := \frac{\partial^{|\alpha|} u(t)}{\partial t_1^{\alpha_1} \cdots \partial t_d^{\alpha_d}}.$$

When the set $\bar{\Omega}$ is compact, the linear space $C^m(\bar{\Omega})$, $m \in \mathbb{N}_0$, is a Banach space with the norm

$$\|u\|_{m,\infty} := \max\{|D^\alpha u(t)| : t \in \Omega, \alpha \in \mathbb{Z}_{m+1}^d, |\alpha| \le m\},$$

and for $m = 0$ we simply denote it by $\|u\|_\infty$.

For a Lebesgue measurable subset Ω of $\mathbb{R}^d$, the linear space of all real-valued functions defined on Ω for which their pth powers, $1 \le p < \infty$, are integrable is denoted by $L^p(\Omega)$. Unless stated otherwise, all integrals are taken in the sense of the Lebesgue integration. Likewise, we use $L^\infty(\Omega)$ for the linear space of all real-valued essentially bounded (that is, bounded except on a zero measure set) measurable functions. Moreover, $L^p(\Omega)$ is a Banach space with the norm

$$\|u\|_p := \begin{cases} \left(\int_\Omega |u(t)|^p \mathrm{d}t\right)^{1/p}, & 1 \le p < \infty, \\ \inf\{\sup\{|u(t)| : t \in \Omega \setminus E\} : \mathrm{meas}(E) = 0\}, & p = \infty. \end{cases}$$

In the special case $p = 2$, $L^2(\Omega)$ is a Hilbert space equipped with the inner product

$$(u, v) := \int_\Omega u(t)v(t)\mathrm{d}t, \quad u, v \in L^2(\Omega).$$

We remark that the integral of an integrable function f over a measurable set $\Omega \in \mathbb{R}^d$ with respect to the Lebesgue measure will be denoted by $\int_\Omega f(t)\mathrm{d}t$ or, in short, by $\int_\Omega f$ when there is no independent variable indicated and no confusion can be expected. Additional details about the spaces can be found in standard texts, for example [1, 183, 236, 276].

2.1.1 Compact integral operators

Compact operators play a central role in the theory of integral equations and hence are frequently discussed in that context. We review some of their important properties here which, from time to time, we shall refer to in the text. Readers are referred to [10, 15, 47, 177, 183, 203, 236] for additional details on the material reviewed here, and to the Appendix of this book for basic elements of functional analysis.

We use the symbol $\mathcal{B}(\mathbb{X}, \mathbb{Y})$ for the normed linear space of all bounded linear operators from a normed linear space $\mathbb{X}$ into a normed linear space $\mathbb{Y}$ with operator norm $\|\mathcal{A}\| := \sup\{\|\mathcal{A}x\| : x \in \mathbb{X}, \|x\| \leq 1\}$. When $\mathbb{Y}$ is a Banach space then so is $\mathcal{B}(\mathbb{X}, \mathbb{Y})$ and in the case that $\mathbb{X} = \mathbb{Y}$ we denote it simply by $\mathcal{B}(\mathbb{X})$. Convergence of a sequence of operators $\{\mathcal{A}_n : n \in \mathbb{N}\} \subseteq \mathcal{B}(\mathbb{X}, \mathbb{Y})$ to an operator $\mathcal{A} \in \mathcal{B}(\mathbb{X}, \mathbb{Y})$ relative to the operator norm is said to be *uniform convergence* and will be denoted by $\mathcal{A}_n \xrightarrow{u} \mathcal{A}$. This notion of convergence differs from the weaker requirement that for all $x \in \mathbb{X}$ we have that $\lim_{n\to\infty} \mathcal{A}_n x = \mathcal{A}x$. This is called *pointwise convergence* and denoted by $\mathcal{A}_n \xrightarrow{s} \mathcal{A}$.

The normed linear space $\mathcal{B}(\mathbb{X}, \mathbb{R})$ is called the dual space of $\mathbb{X}$ and is denoted by $\mathbb{X}^*$. The dual space of a normed linear space is always a Banach space and consists of all continuous linear functionals on $\mathbb{X}$. For every $x \in \mathbb{X}$, $\ell \in \mathbb{X}^*$ we use the familiar bracket notation $\langle \ell, x\rangle$ (or $\langle x, \ell\rangle$) $:= \ell(x)$ and associated with any operator $\mathcal{A} \in \mathcal{B}(\mathbb{X}, \mathbb{Y})$ its *adjoint operator* $\mathcal{A}^* \in \mathcal{B}(\mathbb{Y}^*, \mathbb{X}^*)$ is defined for all $x \in \mathbb{X}$, $\ell \in \mathbb{Y}^*$ by

$$\langle \ell, \mathcal{A}x\rangle = \left\langle \mathcal{A}^*\ell, x\right\rangle.$$

An operator and its adjoint have the same norm, that is $\|\mathcal{A}\| = \|\mathcal{A}^*\|$. When $\mathbb{X}$ is a Hilbert space the Riesz representation theorem (see Theorem A.31 in the Appendix) can be used to identify $\mathbb{X}$ with its dual space. In that case, the adjoint of any $\mathcal{A} \in \mathcal{B}(\mathbb{X})$ is likewise in $\mathcal{B}(\mathbb{X})$ and $\mathcal{A}$ is said to be self-adjoint whenever $\mathcal{A} = \mathcal{A}^*$.

Definition 2.1 A linear operator $\mathcal{K}$ from a normed linear space $\mathbb{X}$ to a normed linear space $\mathbb{Y}$ is said to be compact if it maps each bounded set in $\mathbb{X}$ to a relatively compact set in $\mathbb{Y}$.

It follows from the definition of a compact operator that $\mathcal{A} \in \mathcal{B}(\mathbb{X}, \mathbb{Y})$ is compact if there is a sequence of compact operators $\{\mathcal{A}_n : n \in \mathbb{N}\} \subseteq \mathcal{B}(\mathbb{X}, \mathbb{Y})$ which uniformly converges to it, that is, $\mathcal{A}_n \xrightarrow{u} \mathcal{A}$.

We now describe some examples of compact operators which are relevant. To do this, we first recall the Arzelà–Ascoli theorem which states that a set S is relatively compact in $C(\Omega)$, when Ω is compact, if and only if S is uniformly bounded and equicontinuous (for more details, see Theorem A.43 in the Appendix). This classical result leads us to our first example.

1. The Fredholm operator defined by a continuous kernel

Proposition 2.2 *If $\Omega \subseteq \mathbb{R}^d$ is a compact set and $K \in C(\Omega \times \Omega)$, then the corresponding integral operator $\mathcal{K}$ given in (2.2) is a compact operator in $\mathcal{B}(C(\Omega))$.*

Proof First, note by the Lebesgue dominated convergence theorem (cf. [236]) that $\mathcal{K} \in \mathcal{B}(C(\Omega))$, since its operator norm satisfies the inequality

$$\|\mathcal{K}\| \leq \max\left\{\int_\Omega |K(s,t)|\mathrm{d}t : s \in \Omega\right\}.$$

Next, let $S \subseteq C(\Omega)$ be a bounded subset. Choose a constant $c > 0$ such that for any $v \in S$, we have that $\|v\|_\infty \leq c$. Therefore, for any $v \in S$, and $s, s_1, s_2 \in \Omega$, we conclude that

$$|(\mathcal{K}v)(s)| \leq c \text{ meas}(\Omega) \ \max\{|K(s,t)| : (s,t) \in \Omega\},$$

and

$$\begin{aligned}\left|(\mathcal{K}v)(s_1) - (\mathcal{K}v)(s_2)\right| &\leq c\int_\Omega \left|K(s_1,t) - K(s_2,t)\right|\mathrm{d}t \\ &\leq c \text{ meas}(\Omega) \ \max\{\left|K(s_1,t) - K(s_2,t)\right| : s_1, s_2, t \in \Omega\},\end{aligned}$$

where meas(Ω) denotes the Lebesgue measure of the domain Ω. Since the kernel K is bounded and uniformly continuous on $\Omega \times \Omega$, the right-hand side of the first inequality is finite, and that of the second inequality can be made as small as desired provided that $|s_1 - s_2|$ is small enough. Thus, not only is the image of the set S under $\mathcal{K}$, namely $\mathcal{K}(S)$, uniformly bounded and equicontinuous. Therefore, an appeal to the Arzelà–Ascoli theorem completes the proof. □

2. The Fredholm operator defined by a Schmidt kernel

Recall that a kernel K is called a Schmidt kernel provided that $K \in L^2(\Omega \times \Omega)$.

Proposition 2.3 *If $\Omega \subseteq \mathbb{R}^d$ is a measurable set and K is a Schmidt kernel then the integral operator $\mathcal{K}$ defined by (2.2) is a compact operator in $\mathcal{B}(L^2(\Omega))$.*

Proof We first show that this linear operator $\mathcal{K}$ is in $\mathcal{B}(L^2(\Omega))$. Indeed, for any $v \in L^2(\Omega)$ and any compact subset Ω_0 of Ω, by the Fubini theorem (Theorem A.14) and the Cauchy–Schwarz inequality (Section A.1.3) we have that

$$\begin{aligned}\int_{\Omega_0} |(\mathcal{K}v)(s)|\mathrm{d}s &\leq \int_{\Omega_0}\int_{\Omega} |K(s,t)v(t)|\mathrm{d}s\mathrm{d}t \\ &\leq \big(\mathrm{meas}(\Omega_0)\big)^{1/2}\left(\int_{\Omega_0}\int_{\Omega}|K(s,t)|^2\mathrm{d}s\mathrm{d}t\right)^{1/2}\|v\|_{L^2(\Omega)}.\end{aligned}$$

Therefore, again by the Fubini theorem we obtain that $\big(\mathcal{K}v\big)(s)$ exists almost everywhere for $s \in \Omega_0$, and is measurable; hence, it will be so on the entire Ω. Once again, using the Fubini theorem and the Cauchy–Schwarz inequality we obtain that

$$\begin{aligned}\|\mathcal{K}v\|_{L^2(\Omega)} &= \left(\int_{\Omega}\left|\int_{\Omega}K(s,t)v(t)\mathrm{d}t\right|^2\mathrm{d}s\right)^{1/2} \\ &\leq \left(\int_{\Omega}\int_{\Omega}|K(s,t)|^2\mathrm{d}s\mathrm{d}t\right)^{1/2}\left(\int_{\Omega}|v(t)|^2\mathrm{d}t\right)^{1/2} \qquad (2.3) \\ &= \|K\|_{L^2(\Omega\times\Omega)}\|v\|_{L^2(\Omega)},\end{aligned}$$

which proves that $\mathcal{K} \in \mathcal{B}(L^2(\Omega))$.

We next show that $\mathcal{K}$ is actually a compact operator. First, we consider the case that K is a *degenerate kernel*. That is, there exist $K_{1j}, K_{2j} \in L^2(\Omega), j \in \mathbb{N}_n$ such that the kernel K for any $s, t \in \Omega$ has the form

$$K(s,t) = \sum_{j\in\mathbb{N}_n} K_{1j}(s)K_{2j}(t).$$

In this case, we have for $v \in L^2(\Omega)$ that

$$\mathcal{K}v = \sum_{j\in\mathbb{N}_n} K_{1j}\int_{\Omega} K_{2j}(t)v(t)\mathrm{d}t,$$

which implies that the range of $\mathcal{K}$ is finite-dimensional and so $\mathcal{K}$ is indeed a compact operator. Next, we show that for any $K \in L^2(\Omega\times\Omega)$, there is always a sequence of degenerate kernels $\{K_j : j \in \mathbb{N}\} \subseteq L^2(\Omega\times\Omega)$ such that

$$\lim_{j\to\infty}\|K_j - K\|_{L^2(\Omega\times\Omega)} = 0. \qquad (2.4)$$

From this fact it will follow that $\mathcal{K}$ is a compact operator. Indeed, let $\mathcal{K}_j$ be the compact operator corresponding to the kernel K_j; thus, $\mathcal{K}_j$ is compact. From inequality (2.3) we obtain that

$$\|\mathcal{K}_j - \mathcal{K}\| \leq \|K_j - K\|_{L^2(\Omega\times\Omega)},$$

and consequently we conclude that $\mathcal{K}_j \xrightarrow{u} \mathcal{K}$. Therefore, by Proposition A.47, part (v), we see that $\mathcal{K}$ is compact.

Now, the existence of a sequence $\{K_j : j \in \mathbb{N}\}$ of degenerate kernels in $L^2(\Omega \times \Omega)$ which satisfy (2.4) follows from Theorem A.31 and Corollary A.30. Specifically, we use the Fubini theorem and conclude that the only function $h \in L^2(\Omega \times \Omega)$ with the property that

$$\int_\Omega \int_\Omega f(s)g(t)h(s,t)\mathrm{d}s\mathrm{d}t = 0$$

for all $f, g \in L^2(\Omega)$ is the zero function. Therefore, the set of degenerate kernels forms a dense subset of $L^2(\Omega \times \Omega)$. □

We remark that a constructive approximation argument can also be used to establish the existence of a sequence of degenerate kernels $\{K_j : j \in \mathbb{N}\}$ which satisfy (2.4) when Ω is compact. For example, first we approximate the Schmidt kernel K by a kernel in $C(\Omega \times \Omega)$ and then approximate the continuous kernel uniformly on Ω by bivariate polynomials. In particular, when $K \in C(\Omega \times \Omega)$, the degenerate kernel $\{K_j : j \in \mathbb{N}\}$ can also be chosen in $C(\Omega \times \Omega)$ so that

$$\lim_{j\to\infty} \|K_j - K\|_{C(\Omega\times\Omega)} = 0,$$

thereby giving an alternate proof of Proposition 2.3.

2.1.2 Weakly singular integral operators

We now turn our attention to a class of weakly singular integral operators, which is a kind of the most important compact integral operators.

Definition 2.4 Let Ω be a bounded and measurable subset of $\mathbb{R}^d$. If there exists a bounded and measurable function M defined on $\Omega \times \Omega$ such that for $s, t \in \Omega$ but $s \neq t$,

$$K(s,t) = M(s,t) \log |s - t| \tag{2.5}$$

or

$$K(s,t) = \frac{M(s,t)}{|s - t|^\sigma}, \tag{2.6}$$

where σ is a constant in the interval $(0, d)$, and $|s - t|$ is the Euclidean distance between $s, t \in \Omega$, then the function K is called a kernel with a weak singularity and the operator $\mathcal{K}$ defined by (2.2) is called a weakly singular integral operator. The case that the kernel K has a logarithmic singularity (2.5) is sometimes referred to merely by saying that $\sigma = 0$.

We introduce several constants which are convenient for our discussion of weakly singular integral operators, namely

$$c_\sigma(\Omega) := \sup\left\{\int_\Omega \frac{\mathrm{d}t}{|s-t|^\sigma} : s \in \Omega\right\},$$

$$m(\Omega) := \sup\{|M(s,t)| : s, t \in \Omega\}$$

and

$$\mathrm{diam}(\Omega) := \sup\{|s-t| : s, t \in \Omega\}.$$

Also, we let S_d be the unit sphere in $\mathbb{R}^d$ and recall that

$$\mathrm{vol}(S_d) = \frac{d\pi^{d/2}}{\Gamma(d/2+1)},$$

where Γ is the gamma function. In the next lemma, we estimate an upper bound of $c_\sigma(\bar{\Omega})$.

Lemma 2.5 *If $\Omega \subseteq \mathbb{R}^d$ is a bounded and measurable set and $\sigma \in [0, d)$, then*

$$c_\sigma(\bar{\Omega}) \le \frac{\mathrm{vol}(S_d)\mathrm{diam}(\Omega)^{d-\sigma}}{d-\sigma}. \tag{2.7}$$

Proof Fix a choice of $s \in \bar{\Omega}$. We use spherical coordinates with center at s to estimate the integral in the definition of $c_\sigma(\bar{\Omega})$. Specifically, for any $t \in \bar{\Omega}$, we have that $\mathrm{d}t = r^{d-1}\mathrm{d}r\mathrm{d}\omega_d$, where $r \in [0, \mathrm{diam}(\Omega)]$ and ω_d is the Lebesgue measure on the unit sphere S_d. Consequently, we obtain the estimate

$$\begin{aligned}\int_\Omega \frac{\mathrm{d}t}{|s-t|^\sigma} \le \int_0^{\mathrm{diam}(\Omega)} \frac{\mathrm{d}t}{r^\sigma} &= \int_{S_d}\left(\int_0^{\mathrm{diam}(\Omega)} r^{d-1-\sigma}\mathrm{d}r\right)\mathrm{d}\omega_d \\ &= \frac{\mathrm{vol}(S_d)\mathrm{diam}(\Omega)^{d-\sigma}}{d-\sigma}.\end{aligned} \tag{2.8}$$

□

Note that from inequality (2.8), it follows that every weakly singular kernel K is in $L^1(\Omega \times \Omega)$ and when $\sigma \in [0, d/2)$, K is likewise a Schmidt kernel. We consider the general weakly singular integral operator in the next result.

Proposition 2.6 *The integral operator $\mathcal{K}$ defined by (2.2) with a weak singular kernel (2.6) is in $\mathcal{B}(L^2(\Omega))$ with the norm satisfying the inequality $\|\mathcal{K}\| \le m(\Omega)c_\sigma(\Omega)$.*

Proof We first observe by the Fubini theorem for any $u \in L^2(\Omega)$ that

$$\int_\Omega \int_\Omega \frac{u^2(t)}{|s-t|^\sigma} \mathrm{d}s\mathrm{d}t = \int_\Omega u^2(t) \left(\int_\Omega \frac{\mathrm{d}s}{|s-t|^\sigma} \right) \mathrm{d}t \le c_\sigma(\Omega) \|u\|^2_{L^2(\Omega)}. \quad (2.9)$$

Therefore, the function v defined for all $s \in \Omega$ as

$$v(s) := \int_\Omega \frac{u^2(t)}{|s-t|^\sigma} \mathrm{d}t$$

exists at almost every $s \in \Omega$, and is integrable. Next, we point out for each $s, t \in \Omega$ that

$$|K(s,t)u(t)| \le m(\Omega) \frac{1}{|s-t|^{\sigma/2}} \frac{|u(t)|}{|s-t|^{\sigma/2}} \le \frac{m(\Omega)}{2} \frac{1}{|s-t|^\sigma} + \frac{m(\Omega)}{2} \frac{u^2(t)}{|s-t|^\sigma}.$$

For any $s \in \Omega$, both terms on the right-hand side of this inequality are integrable with respect to $t \in \Omega$ and so we conclude that $|K(s,t)u(t)|$ is finite for almost every $t \in \Omega$. Moreover, by the Cauchy–Schwarz inequality, we have that

$$\begin{aligned}
[(\mathcal{K}u)(s)]^2 &= \left[\int_\Omega K(s,t)u(t)\mathrm{d}t \right]^2 \\
&\le m^2(\Omega) \int_\Omega \frac{\mathrm{d}t}{|s-t|^\sigma} \int_\Omega \frac{u^2(t)}{|s-t|^\sigma} \mathrm{d}t \\
&\le m^2(\Omega) c_\sigma(\Omega) \int_\Omega \frac{u^2(t)}{|s-t|^\sigma} \mathrm{d}t,
\end{aligned}$$

which implies that $\mathcal{K}u \in L^2(\Omega)$ is square-integrable, that is, $\mathcal{K}$ is defined on $L^2(\Omega)$ and maps $L^2(\Omega)$ to $L^2(\Omega)$. Moreover, integrating both sides of the last inequality with respect to $s \in \Omega$ and employing estimate (2.9) yields the inequality

$$\|\mathcal{K}u\|^2_{L^2(\Omega)} \le m^2(\Omega) c_\sigma(\Omega)^2 \|u\|^2_{L^2(\Omega)},$$

which completes the proof. □

We next establish the compactness of a weakly singular integral operator on $L^2(\Omega)$.

Theorem 2.7 *The integral operator $\mathcal{K}$ with a weakly singular kernel (2.6) is a compact operator in $\mathcal{B}(L^2(\Omega))$.*

Proof For $\epsilon > 0$, let $\mathcal{K}_\epsilon$ and $\mathcal{K}'_\epsilon$ be the integral operators whose kernels K_ϵ, K'_ϵ are defined, respectively, for $s, t \in \Omega$ by the equations

$$K_\epsilon(s,t) = \begin{cases} K(s,t), & |s-t| \ge \epsilon \\ 0, & |s-t| < \epsilon \end{cases}$$

and

$$K'_\epsilon(s,t) = \begin{cases} 0, & |s-t| \geq \epsilon \\ K(s,t), & |s-t| < \epsilon. \end{cases}$$

These kernels were chosen to provide the decomposition

$$\mathcal{K} = \mathcal{K}_\epsilon + \mathcal{K}'_\epsilon. \tag{2.10}$$

Since for $s, t \in \Omega$, $|K_\epsilon(s,t)| \leq m(\Omega)/\epsilon^\sigma$ and Ω is bounded, it follows that $K_\epsilon \in L^2(\Omega \times \Omega)$. Consequently, we conclude from Proposition 2.3 that $\mathcal{K}_\epsilon$ is a compact operator in $\mathcal{B}(L^2(\Omega))$. Moreover, setting $\Omega_{s,\epsilon} := \{t : t \in \Omega, |s-t| < \epsilon\}$, for each $s \in \Omega$, the Cauchy–Schwarz inequality yields the inequality

$$\begin{aligned} |(\mathcal{K}'_\epsilon(u)(s)| &= \left| \int_{\Omega_{s,\epsilon}} \frac{M(s,t)}{|s-t|^\sigma} u(t) \mathrm{d}t \right| \\ &\leq m(\Omega) \left(\int_{\Omega_{s,\epsilon}} \frac{1}{|s-t|^\sigma} \mathrm{d}t \right)^{1/2} \left(\int_\Omega \frac{u^2(t)}{|s-t|^\sigma} \mathrm{d}t \right)^{1/2}. \end{aligned}$$

We bound the first integral on the right-hand side of this inequality by the method of proof used for Lemma 2.5 and then integrate both sides of the resulting inequality over $t \in \Omega$ to obtain the inequality

$$\|\mathcal{K}'_\epsilon(u)\|_{L^2(\Omega)} \leq \left[\frac{m^2(\Omega)\mathrm{vol}(S_d)(2\epsilon)^{d-\sigma}}{d-\sigma} \int_\Omega \int_\Omega \frac{u^2(t)}{|s-t|^\sigma} \mathrm{d}s\mathrm{d}t \right]^{1/2}.$$

This inequality combined with (2.9) and Lemma 2.5 leads to the estimate

$$\|\mathcal{K}'_\epsilon u\|_{L^2(\Omega)} \leq \frac{m(\Omega)\mathrm{vol}(S_d)[2\epsilon \mathrm{diam}(\Omega)]^{(d-\sigma)/2}}{d-\sigma} \|u\|_{L^2(\Omega)}.$$

In other words, we have proved that $\lim_{\epsilon \to 0} \|\mathcal{K}'_\epsilon\| = 0$ and so, being the uniform limit of the compact operators $\{\mathcal{K}_\epsilon : \epsilon > 0\}$, $\mathcal{K}$ is compact. □

The next result establishes a similar fact for $\mathcal{K}$ to be a compact operator in $\mathcal{B}(C(\Omega))$ under a modified hypothesis.

Proposition 2.8 *If $\Omega \subseteq \mathbb{R}^d$ is a compact set of positive measure and the function M in (2.6) is in $C(\Omega \times \Omega)$, then the integral operator $\mathcal{K}$ with a weakly singular kernel (2.6) is a compact operator in $\mathcal{B}(C(\Omega))$.*

Proof Let S be a bounded subset of $C(\Omega)$. By the Arzelà–Ascoli theorem it suffices to prove that the set $\mathcal{K}(S)$ is uniformly bounded and equicontinuous.

We choose a positive constant c which ensures for any $u \in S$ that $\|u\|_\infty \le c$, and so we have for any $u \in S$ that

$$|(\mathcal{K}u)(s)| = \left| \int_\Omega \frac{M(s,t)}{|s-t|^\sigma} u(t)\mathrm{d}t \right| \le c\, m(\Omega) c_\sigma(\Omega), \quad s \in \Omega.$$

Therefore, using Lemma 2.5 we conclude that the set $\mathcal{K}(S)$ is uniformly bounded.

Next, we shall show not only that $\mathcal{K}u \in C(\Omega)$ but also that the set $\mathcal{K}(S)$ is equicontinous. To this end, we choose $\epsilon > 0$, points $s+h, s \in \Omega$, and obtain the equation

$$(\mathcal{K}u)(s+h) - (\mathcal{K}u)(s) = \int_\Omega \left[\frac{M(s+h,t)}{|s+h-t|^\sigma} - \frac{M(s,t)}{|s-t|^\sigma} \right] u(t)\mathrm{d}t.$$

Let $B(s, 2\epsilon)$ be the sphere with center at s and radius 2ϵ and set $\Omega(s) := \Omega \setminus B(s, 2\epsilon)$. We have that

$$\begin{aligned} |(\mathcal{K}u)(s+h) - (\mathcal{K}u)(s)| \le c\, m(\Omega) \left(\int_{B(s,2\epsilon)} \frac{\mathrm{d}t}{|s+h-t|^\sigma} + \int_{B(s,2\epsilon)} \frac{\mathrm{d}t}{|s-t|^\sigma} \right) \\ + c \int_{\Omega(s)} \left| \frac{M(s+h,t)}{|s+h-t|^\sigma} - \frac{M(s,t)}{|s-t|^\sigma} \right| \mathrm{d}t. \end{aligned} \tag{2.11}$$

For every $s \in \Omega$ it follows from the method used to prove Lemma 2.5 that

$$\int_{B(s,2\epsilon)} \frac{\mathrm{d}t}{|s-t|^\sigma} \le \frac{\mathrm{vol}(S_d)(4\epsilon)^{d-\sigma}}{d-\sigma}.$$

When $|h| < 2\epsilon$, we are assured that $B(s, 2\epsilon) \subseteq B(s+h, 4\epsilon)$, and consequently we obtain the next inequality

$$\int_{B(s,2\epsilon)} \frac{\mathrm{d}t}{|s+h-t|^\sigma} \le \int_{B(s+h,4\epsilon)} \frac{\mathrm{d}t}{|s+h-t|^\sigma} \le \frac{\mathrm{vol}(S_d)(16\epsilon)^{d-\sigma}}{d-\sigma},$$

where in the last step we again employ the method of proof for Lemma 2.5. These two inequalities demonstrate that the first two quantities on the right-hand side of (2.11) do not exceed

$$2c\, m(\Omega) \frac{\mathrm{vol}(S_d)(16\epsilon)^{d-\sigma}}{d-\sigma}.$$

We now estimate the third integral appearing on the right-hand side of (2.11). To this end, we observe that on the compact set $W := \{(s,t) : s, t \in \Omega, |s-t| \ge \epsilon\}$ the function defined as $H(s,t) := M(s,t)/|s-t|^\sigma$ is uniformly continuous.

Hence there exists a $\delta > 0$ such that whenever $(s', t'), (s, t) \in W$ with $|s' - s| \leq \delta, |t' - t| \leq \delta$ we have that

$$\left| \frac{M(s', t')}{|s' - t'|^\sigma} - \frac{M(s, t)}{|s - t|^\sigma} \right| \leq \epsilon.$$

Now, assume that $|h| \leq \min(\epsilon, \delta)$ and $t \in \Omega(s)$. Therefore, we have that $(s + h, t), (s, t) \in W$ and so the third term on the right-hand side of (2.11) is bounded by $c\,\text{meas}(\Omega)\epsilon$ for any $s \in \Omega$ and $|h| \leq \min(\epsilon, \delta)$. Therefore, not only is the function $\mathcal{K}u$ continuous on Ω but also the set $\mathcal{K}(S)$ is equicontinuous. An application of the Arzelà–Ascoli theorem establishes that the set $\mathcal{K}(S)$ is compact and so $\mathcal{K}$ is a compact operator. □

2.1.3 Boundary integral equations

Some important boundary value problems of partial differential equations over a prescribed domain can be reformulated as equivalent integral equations over the boundary of the domain. The resulting integral equations on the boundary are called boundary integral equations (BIEs). The superiority of the BIE methodology for solving boundary value problems rests on the fact that the dimension of the domain of functions appearing in the BIE will be one lower than in the original partial differential equation. This means that the computational effort required to solve the partial differential equations can be reduced significantly by using an efficient numerical method to solve the associated BIE. In this subsection we briefly review the BIE reformulation of boundary value problems for the Laplace equation. This material will provide the reader with some concrete integral equations that supplement the general theory described throughout the book.

We begin with a discussion of the Green identities and integral representations of harmonic functions. To this end, we let $\Omega \subseteq \mathbb{R}^d$ be a bounded open domain with piecewise smooth boundary $\partial\Omega$. Throughout our discussion we use the standard notation $\nabla := \left[\frac{\partial}{\partial x_l} : l \in \mathbb{N}_d\right]$ for the gradient and the Laplace operator Δ is defined by $\Delta u := (\nabla, \nabla u)$. Let $\mathbf{A} := [a_{ij} : i, j \in \mathbb{N}_d]$ be a $d \times d$ symmetric matrix with entries in $C^2(\Omega)$, $\mathrm{b} := [b_i : i \in \mathbb{N}_d]$ a vector field with coordinates in $C^1(\bar{\Omega})$ and c a scalar-valued function in $C(\Omega)$.

The proof of the lemma below is straightforward.

Lemma 2.9 *If* $u : \mathbb{R}^d \to \mathbb{R}$ *and* $\mathrm{a} : \mathbb{R}^d \to \mathbb{R}^d$ *are in* $C^1(\Omega)$, *then*

$$\nabla \cdot (u\mathrm{a}) = u\nabla \cdot \mathrm{a} + \mathrm{a} \cdot \nabla u.$$

For the next lemma we introduce the vector field $\mathrm{P} : \mathbb{R}^d \to \mathbb{R}^d$

$$\mathrm{P} := \mathbf{A}(v\nabla u - u\nabla v) + \mathrm{b}uv,$$

the second-order elliptic partial differential operator

$$\mathcal{M}u := \nabla \cdot \mathbf{A}\nabla u + \mathrm{b} \cdot \nabla u + cu$$

and its formal adjoint operator

$$\mathcal{M}^* v := \nabla \cdot \mathbf{A}\nabla v - \nabla \cdot \mathrm{b}v + cv.$$

Lemma 2.10 *If $u, v : \mathbb{R}^d \to \mathbb{R}$ are in $C^2(\bar{\Omega})$ and $\mathbf{A}, \mathrm{b}, c$ as above, then*

$$v\mathcal{M}u - u\mathcal{M}^* v = \nabla \cdot \mathrm{P}.$$

Proof By direct computation, using Lemma 2.9, we have that

$$v\mathcal{M}u - u\mathcal{M}^* v = v\nabla \cdot (\mathbf{A}\nabla u) - u\nabla \cdot (\mathbf{A}\nabla v) + v\mathrm{b} \cdot \nabla u + u\nabla \cdot (\mathrm{b}v),$$

while the definition of P proves that the right-hand side of this equation equals $\nabla \cdot \mathrm{P}$. □

The formula appearing in the above lemma is often referred to as the adjoint identity.

Next, we write the adjoint identity in an alternate form. For this purpose, we introduce the notation

$$\mathcal{P}u := \mathbf{A}\nabla u \cdot n \ \text{ and } \mathcal{Q}v := \mathbf{A}\nabla v \cdot n - v\mathrm{b} \cdot n, \tag{2.12}$$

where n denotes the unit outer normal along $\partial\Omega$. It follows from the Gauss formula and Lemma 2.10 that

$$\int_{\Omega} (v\mathcal{M}u - u\mathcal{M}^* v) = \int_{\partial\Omega} (v\mathcal{P}u - u\mathcal{Q}v). \tag{2.13}$$

We are interested in getting a representation for the solution u of the homogeneous elliptic equation $\mathcal{M}u = 0$ in terms of its values on the boundary of the domain Ω. The standard method for doing this employs the fundamental solution to the inhomogeneous problem corresponding to the adjoint operator. We briefly describe this process. Recall that the *fundamental solution* of the linear partial differential operator $\mathcal{M}$ is a function U defined on $\Omega \times \Omega$ such that for each $x \in \Omega$ the solution u of the equation $\mathcal{M}u = f$ is given by the integral representation

$$u(x) = \int_{\mathbb{R}^d} U(x, y) f(y) \mathrm{d}y. \tag{2.14}$$

We assume that the fundamental solution of the adjoint operator $\mathcal{M}^*$ is available and is denoted by G^*. Therefore, we are ensured that the solution v of the adjoint equation $\mathcal{M}^*v = f$ is given for each $x \in \Omega$ as

$$v(x) := \int_{\mathbb{R}^d} G^*(x, y)f(y)\mathrm{d}y. \tag{2.15}$$

The function G^* leads us to the following basic result.

Proposition 2.11 *If u is the solution of the homogeneous equation*

$$\mathcal{M}u = 0, \tag{2.16}$$

then for each $x \in \Omega$

$$u(x) = \int_{\partial\Omega} \left[u(y)(\mathcal{Q}G^*(x, \cdot))(y) - G^*(x, y)\mathcal{P}u(y)\right] \mathrm{d}y, \tag{2.17}$$

where $\mathcal{P}, \mathcal{Q}$ are defined by (2.12), G^* is the fundamental solution of the operator $\mathcal{M}^*$.

Proof It follows from (2.13) that

$$\int_{\Omega} [v(y)(\mathcal{M}u)(y) - u(y)(\mathcal{M}^*v)(y)]\mathrm{d}y = \int_{\partial\Omega} [v(y)(\mathcal{P}u)(y) - u(y)(\mathcal{Q}v)(y)]\mathrm{d}y.$$

We choose $v := G^*(x, \cdot)$ in this formula and use the definition of G^* to get the desired conclusion. □

Note that (2.17) expresses u over Ω in terms of the boundary values of u and its normal derivative on $\partial\Omega$. Let us specialize this result to the Laplace operator. Thus, we choose $\mathcal{M}$ to be the Laplace operator and observe in this case that $\mathcal{M} = \mathcal{M}^* = \Delta$ and $\mathcal{P}u = \mathcal{Q}u = \nabla u \cdot n$. Therefore, we get from (2.13) the following theorem.

Theorem 2.12 *(Green theorem) If $u, v \in C^2(\bar{\Omega})$ then*

$$\int_{\Omega} (v\Delta u - u\Delta v) = \int_{\partial\Omega} \left(v\frac{\partial u}{\partial n} - u\frac{\partial v}{\partial n}\right). \tag{2.18}$$

The fundamental solution of the Laplace operator is given by the formula

$$G(x, y) = \begin{cases} -\frac{1}{2\pi} \log |x - y|, & d = 2, \\ -\frac{1}{(d-2)\omega_{d-1}} \frac{1}{|x-y|^{d-2}}, & d \geq 3, \end{cases} \tag{2.19}$$

where ω_{d-1} denotes the surface area of the unit sphere in $\mathbb{R}^d$ (cf. [7, 177]).

Definition 2.13 If $u \in C^2(\Omega)$ satisfies the Laplace equation $\Delta u = 0$ on Ω, then u is called harmonic on Ω.

It follows from direct computation that $u := G(\cdot, y)$ is harmonic on $\mathbb{R}^d \setminus \{y\}$.

Corollary 2.14 *If u is harmonic on Ω then*

$$\int_{\partial\Omega} \frac{\partial u}{\partial n} = 0.$$

Proof This result follows from Theorem 2.12 by choosing $v = 1$. □

In what follows, we review the techniques of finding solutions to the Laplace equation under various circumstances. Especially, we discuss the direct method for solving the Dirichlet and Neumann problems in both the interior and exterior of the domain Ω. Moreover, we shall also review both single and double-layer representations for the solution of these problems. All of our remarks pertain to the practically important cases of two and three dimensions. The two-dimensional case will be presented in some detail, while the corresponding three-dimensional case will be stated without the benefit of a detailed explanation, since it follows the pattern of proof of the two-dimensional case. We begin with a proposition which shows that harmonic functions have boundary integral representations.

Proposition 2.15 *If $\Omega \subseteq \mathbb{R}^2$ is a bounded open domain with smooth boundary $\partial\Omega$, $\Omega' := \mathbb{R}^2 \setminus \bar{\Omega}$ and u is a harmonic function on Ω, then*

$$u(x) = \frac{1}{2\pi} \int_{\partial\Omega} \left(u(y) \frac{\partial}{\partial n} \log|x-y| - \log|x-y| \frac{\partial u(y)}{\partial n} \right) \mathrm{d}y, \quad x \in \Omega, \tag{2.20}$$

$$u(x) = \frac{1}{\pi} \int_{\partial\Omega} \left[u(y) \frac{\partial}{\partial n} \log|x-y| - \log|x-y| \frac{\partial u(y)}{\partial n} \right] \mathrm{d}y, \quad x \in \partial\Omega, \tag{2.21}$$

$$0 = \int_{\partial\Omega} \left[u(y) \frac{\partial}{\partial n} \log|x-y| - \log|x-y| \frac{\partial u(y)}{\partial n} \right] \mathrm{d}y, \quad x \in \Omega'. \tag{2.22}$$

Proof The first equation (2.20) follows from Proposition 2.11. We now turn to the case that $x \in \partial\Omega$. To deal with the singularity on $x = y$ of the integrand on the right-hand side of equation (2.21), we choose a positive ϵ and denote by Ω_ϵ the domain obtained from Ω after removing the small disc $B(x, \epsilon) := \{y : |x-y| \le \epsilon\}$. On this punctured domain we can use the Green identity (2.18) to obtain the equation

$$\int_{\Omega_\epsilon} (v(y)\Delta u(y) - u(y)\Delta v(y))\mathrm{d}y = \int_{\partial\Omega_\epsilon} \left(v(y)\frac{\partial u(y)}{\partial n} - u(y)\frac{\partial v(y)}{\partial n} \right) \mathrm{d}y. \tag{2.23}$$

Let v be the fundamental solution (2.19) of the Laplace operator. Both of the functions u and v are harmonic functions on Ω_ϵ. Therefore, the above equation can be written as

$$\frac{1}{2\pi}\int_{\partial\Omega_\epsilon}\left(\log|x-y|\frac{\partial u(y)}{\partial n}-u(y)\frac{\partial}{\partial n}\log|x-y|\right)dy=0. \tag{2.24}$$

To evaluate the limit as $\epsilon \to 0$ of the integral on the left-hand side of the equation above, we split it into a sum of the following two integrals:

$$I_{1,\epsilon}:=\frac{1}{2\pi}\int_{\Gamma_\epsilon}\left(\log|x-y|\frac{\partial u(y)}{\partial n}-u(y)\frac{\partial}{\partial n}\log|x-y|\right)dy$$

and

$$I_{2,\epsilon}:=\frac{1}{2\pi}\int_{\partial\Omega_\epsilon\setminus\Gamma_\epsilon}\left(\log|x-y|\frac{\partial u(y)}{\partial n}-u(y)\frac{\partial}{\partial n}\log|x-y|\right)dy,$$

with $\Gamma_\epsilon := \partial B(x,\epsilon)\bigcap\Omega$. It follows that

$$I_{1,\epsilon}=\frac{1}{2\pi}\int_{\Gamma_\epsilon}\left(\log\epsilon\frac{\partial u(y)}{\partial n}+\epsilon^{-1}u(y)\right)dy, \tag{2.25}$$

from which we obtain that

$$\lim_{\epsilon\to 0} I_{1,\epsilon}=\frac{1}{2}u(x). \tag{2.26}$$

Moreover, we have that

$$\lim_{\epsilon\to 0} I_{2,\epsilon}=\frac{1}{2\pi}\int_{\partial\Omega}\left(\log|x-y|\frac{\partial u(y)}{\partial n}-u(y)\frac{\partial}{\partial n}\log|x-y|\right)dy. \tag{2.27}$$

Combining equations (2.24)–(2.27) yields (2.21). Finally, we note that when $x\in\Omega'$, both functions u and $v:=\frac{1}{2\pi}\log|x-\cdot|$ are harmonic functions on $\bar{\Omega}$. Thus, (2.22) follows from the Green identity (2.18). □

We state the corresponding result for the three-dimensional case. The proof follows the pattern of the proof for Proposition 2.15.

Proposition 2.16 *If $\Omega \subseteq \mathbb{R}^3$ is a bounded open domain with smooth boundary $\partial\Omega$, $\Omega' := \mathbb{R}^3\setminus\bar{\Omega}$ and u is a harmonic function on Ω, then*

$$u(x)=-\frac{1}{4\pi}\int_{\partial\Omega}\left[u(y)\frac{\partial}{\partial n}\frac{1}{|x-y|}-\frac{1}{|x-y|}\frac{\partial u(y)}{\partial n}\right]dy,\quad x\in\Omega, \tag{2.28}$$

$$u(x)=-\frac{1}{2\pi}\int_{\partial\Omega}\left[u(y)\frac{\partial}{\partial n}\frac{1}{|x-y|}-\frac{1}{|x-y|}\frac{\partial u(y)}{\partial n}\right]dy,\quad x\in\partial\Omega, \tag{2.29}$$

$$0=\int_{\partial\Omega}\left[u(y)\frac{\partial}{\partial n}\frac{1}{|x-y|}-\frac{1}{|x-y|}\frac{\partial u(y)}{\partial n}\right]dy,\quad x\in\Omega'. \tag{2.30}$$

Next, we make use of these boundary value formulas for harmonic functions to rewrite several boundary value problems as integral equations. We start with a description of methods for obtaining boundary integral equations of the direct type. Later we turn our attention to the indirect methods of single and double-layer potentials. First, we consider the following interior boundary value problems.

The interior Dirichlet problem: Find $u \in C(\bar{\Omega}) \bigcap C^2(\Omega)$ such that

$$\begin{cases} \Delta u(x) = 0, & x \in \Omega, \\ u(x) = u_0(x), & x \in \partial\Omega, \end{cases} \tag{2.31}$$

where $u_0 \in C(\partial\Omega)$ is a given boundary function.

The interior Neumann problem: Find $u \in C(\bar{\Omega}) \bigcap C^2(\Omega)$ such that

$$\begin{cases} \Delta u(x) = 0, & x \in \Omega, \\ \frac{\partial u(x)}{\partial n} = u_1(x), & x \in \partial\Omega, \end{cases} \tag{2.32}$$

where $u_1 \in C(\partial\Omega)$ is a given boundary function satisfying $\int_{\partial\Omega} u_1(x)\mathrm{d}x = 0$.

It is known that both of these problems have unique solutions (see, for example, Chapter 6 of [177]). We now use (2.21) and (2.29) with the boundary condition $u = u_0$ and reformulate the interior Dirichlet problem for (2.31) when $d = 2$ as the BIE of the first kind

$$\frac{1}{\pi} \int_{\partial\Omega} \log |x - y| \rho(y)\mathrm{d}y = f(x), \quad x \in \partial\Omega, \tag{2.33}$$

where

$$\rho := \frac{\partial u}{\partial n} \text{ and } f := -u_0 + \frac{1}{\pi} \int_{\partial\Omega} u_0(y) \frac{\partial}{\partial n} \log |\cdot - y| \mathrm{d}y.$$

For the case $d = 3$, we choose

$$\rho := \frac{\partial u}{\partial n} \text{ and } f := u_0 + \frac{1}{2\pi} \int_{\partial\Omega} u_0(y) \frac{\partial}{\partial n} \frac{1}{|\cdot - y|} \mathrm{d}y$$

and obtain the reformulation of the interior Dirichlet problem for (2.31) with $d = 3$ as the BIE of the first kind

$$\frac{1}{2\pi} \int_{\partial\Omega} \frac{1}{|x - y|} \rho(y)\mathrm{d}y = f(x), \quad x \in \partial\Omega. \tag{2.34}$$

In a similar manner, we treat the interior Neumann problem. First, for $d = 2$ we use (2.21) and (2.29) with the boundary condition $\frac{\partial u}{\partial n} = u_1$ and convert the interior Neumann problem (2.32) to the equivalent BIE of the second kind

$$u(x) - \frac{1}{\pi}\int_{\partial\Omega} u(y)\frac{\partial}{\partial n}\log|x-y|\mathrm{d}y = g(x), \quad x \in \partial\Omega, \tag{2.35}$$

where

$$g := -\frac{1}{\pi}\int_{\partial\Omega} u_1(y)\log|\cdot - y|\mathrm{d}y.$$

For the corresponding three-dimensional case $d = 3$, we set

$$g := \frac{1}{2\pi}\int_{\partial\Omega} \frac{u_1(y)}{|\cdot - y|}\mathrm{d}y,$$

and obtain the BIE of the second kind

$$u(x) + \frac{1}{2\pi}\int_{\partial\Omega} u(y)\frac{\partial}{\partial n}\frac{1}{|x-y|}\mathrm{d}y = g(x), \quad x \in \partial\Omega. \tag{2.36}$$

Let us now consider the Dirichlet and Neumann problems in the exterior domain $\Omega' = \mathbb{R}^d \setminus \bar{\Omega}$ for $d = 2, 3$. Specifically, we reformulate the following two problems as BIE.

The exterior Dirichlet problem: Find $u \in C(\bar{\Omega'}) \bigcap C^2(\Omega')$ such that

$$\begin{cases} \Delta u(x) = 0, & x \in \Omega', \\ u(x) = u_0(x), & x \in \partial\Omega, \end{cases} \tag{2.37}$$

where $u_0 \in C(\partial\Omega)$ is a given boundary function.

The exterior Neumann problem: Find $u \in C(\bar{\Omega'}) \bigcap C^2(\Omega')$ such that

$$\begin{cases} \Delta u(x) = 0, & x \in \Omega', \\ \frac{\partial u(x)}{\partial n'} = u_1(x), & x \in \partial\Omega, \end{cases} \tag{2.38}$$

where $u_1 \in C(\partial\Omega)$ is a given boundary function satisfying $\int_{\partial\Omega} u_1(x)dx = 0$, and n' is the outer unit normal to $\partial\Omega (= \partial\Omega')$ with respect to Ω'.

It is known (see, for example, [177]) that both problems have unique solutions under the condition that

$$\begin{cases} u(x) = \mathcal{O}\left(|x|^{-1}\right), & |x| \to \infty, \\ |\nabla u(x)| = \mathcal{O}\left(|x|^{-2}\right), & |x| \to \infty. \end{cases} \tag{2.39}$$

Below we state the analog of both Proposition 2.15 and Proposition 2.16 for the exterior domain Ω'. We begin with the case $d = 2$.

Proposition 2.17 *If $\Omega \subseteq \mathbb{R}^2$ is a bounded open domain with smooth boundary $\partial\Omega$, $\Omega' := \mathbb{R}^2 \setminus \bar{\Omega}$ and u is a harmonic function on Ω', then there holds*

$$u(x) = \frac{1}{2\pi}\int_{\partial\Omega}\left[u(y)\frac{\partial}{\partial n'}\log|x-y| - \log|x-y|\frac{\partial u(y)}{\partial n'}\right]\mathrm{d}y, \quad x \in \Omega', \tag{2.40}$$

$$u(x) = \frac{1}{\pi}\int_{\partial\Omega}\left[u(y)\frac{\partial}{\partial n'}\log|x-y| - \log|x-y|\frac{\partial u(y)}{\partial n'}\right]\mathrm{d}y, \quad x \in \partial\Omega, \tag{2.41}$$

$$0 = \int_{\partial\Omega}\left[u(y)\frac{\partial}{\partial n'}\log|x-y| - \log|x-y|\frac{\partial u(y)}{\partial n'}\right]\mathrm{d}y, \quad x \in \Omega. \tag{2.42}$$

Proof We only prove (2.40). The other two equations are similarly obtained. Let the ball $B_R := \{x : |x| < R\}$ be chosen such that $\Omega \subset B_R$ and let $\Omega'_R := \Omega' \bigcap B_R$. Consequently, it follows from (2.20) with $\Omega = \Omega'_R$ that

$$u(x) = \frac{1}{2\pi}\int_{\partial\Omega}\left[u(y)\frac{\partial}{\partial n'}\log|x-y| - \log|x-y|\frac{\partial u(y)}{\partial n'}\right]\mathrm{d}y + I_R, \quad x \in \Omega'_R, \tag{2.43}$$

where

$$I_R := \frac{1}{2\pi}\int_{\partial B_R}\left[u(y)\frac{\partial}{\partial n}\log|x-y| - \log|x-y|\frac{\partial u(y)}{\partial n}\right]\mathrm{d}y, \quad x \in \Omega'_R,$$

and n is the outer unit normal to ∂B_R. Using the condition (2.39) we have that there exists a positive constant c such that

$$\begin{aligned}|I_R| &\le \frac{1}{2\pi}\int_{\partial B_R}\left[|u(y)|\left|\frac{\partial}{\partial n}\log|x-y|\right| + |\log|x-y||\left|\frac{\partial u(y)}{\partial n}\right|\right]\mathrm{d}y \\ &\le \frac{c\log R}{2\pi R} \to 0.\end{aligned}$$

Note that the upper bound tends to zero as R tends to infinity. Therefore this estimate combined with (2.43) yields (2.40). □

The three-dimensional version of Proposition 2.16 for the exterior domain is described next. The proof is similar to that of Proposition 2.17 and so is omitted.

Proposition 2.18 *If $\Omega \subseteq \mathbb{R}^3$ is a bounded open domain with smooth boundary $\partial\Omega$, $\Omega' := \mathbb{R}^3 \setminus \bar{\Omega}$ and u is a harmonic function on Ω', then*

$$u(x) = \frac{-1}{4\pi} \int_{\partial\Omega} \left[u(y) \frac{\partial}{\partial n'} \frac{1}{|x-y|} - \frac{1}{|x-y|} \frac{\partial u(y)}{\partial n'} \right] \mathrm{d}y, \quad x \in \Omega', \tag{2.44}$$

$$u(x) = -\frac{1}{2\pi} \int_{\partial\Omega} \left[u(y) \frac{\partial}{\partial n'} \frac{1}{|x-y|} - \frac{1}{|x-y|} \frac{\partial u(y)}{\partial n'} \right] \mathrm{d}y, \quad x \in \partial\Omega, \tag{2.45}$$

$$0 = \int_{\partial\Omega} \left[u(y) \frac{\partial}{\partial n'} \frac{1}{|x-y|} - \frac{1}{|x-y|} \frac{\partial u(y)}{\partial n'} \right] \mathrm{d}y, \quad x \in \Omega. \tag{2.46}$$

We now make use of (2.41) and (2.45) to rewrite the exterior Dirichlet problem (2.37) for $d = 2$ as the BIE

$$\frac{1}{\pi} \int_{\partial\Omega} \log|x-y| \rho(y) \mathrm{d}y = f(x), \quad x \in \partial\Omega, \tag{2.47}$$

where

$$\rho := \frac{\partial u}{\partial n'} \text{ and } f := -u_0 + \frac{1}{\pi} \int_{\partial\Omega} u_0(y) \frac{\partial}{\partial n'} \log|\cdot - y| \mathrm{d}y,$$

while for $d = 3$ we have the equation

$$\frac{1}{2\pi} \int_{\partial\Omega} \frac{1}{|x-y|} \rho(y) \mathrm{d}y = f(x), \quad x \in \partial\Omega, \tag{2.48}$$

where

$$\rho := \frac{\partial u}{\partial n'} \text{ and } f := u_0 + \frac{1}{2\pi} \int_{\partial\Omega} u_0(y) \frac{\partial}{\partial n'} \frac{1}{|\cdot - y|} \mathrm{d}y.$$

For the exterior Neumann problem (2.38), the BIE is of the second kind and it is explicitly given for $d = 2$ as

$$u(x) - \frac{1}{\pi} \int_{\partial\Omega} u(y) \frac{\partial}{\partial n'} \log|x-y| \mathrm{d}y = g(x), \quad x \in \partial\Omega, \tag{2.49}$$

where

$$g(x) := -\frac{1}{\pi} \int_{\partial\Omega} u_1(y) \log|x-y| \mathrm{d}y.$$

The case $d = 3$ is covered by the following BIE:

$$u(x) + \frac{1}{2\pi} \int_{\partial\Omega} u(y) \frac{\partial}{\partial n'} \frac{1}{|x-y|} \mathrm{d}y = g(x), \quad x \in \partial\Omega, \tag{2.50}$$

where

$$g(x) := \frac{1}{2\pi} \int_{\partial\Omega} \frac{u_1(y)}{|x-y|} \mathrm{d}y.$$

In the remaining part of this section, our goal is to describe the BIE for the Laplace equation of the indirect type. First, we consider representing the unknown harmonic function u as a single-layer potential

$$u(x) = \int_{\partial\Omega} \rho(y)G(x, y)\mathrm{d}y, \quad x \in \mathbb{R}^d \setminus \partial\Omega, \tag{2.51}$$

and then later as a double-layer potential

$$u(x) = \int_{\partial\Omega} \rho(y)\frac{\partial G(x, y)}{\partial n}\mathrm{d}y, \quad x \in \mathbb{R}^d \setminus \partial\Omega, \tag{2.52}$$

where G is the fundamental solution of the Laplace operator, and $\rho \in C(\partial\Omega)$ is a function to be determined, depending on the nature of the boundary conditions. We show that the single and double-layer potentials can be used to solve interior and exterior problems for both the Dirichlet and Neumann problems.

Let us start with the single-layer method. For the interior or exterior Dirichlet problem, the boundary condition and the continuity of u on $\partial\Omega$ lead to the demand that the function ρ satisfies the first-kind Fredholm integral equation

$$\int_{\partial\Omega} \rho(y)G(x, y)\mathrm{d}y = u_0(x), \quad x \in \partial\Omega.$$

In particular, for $d = 2$, the solution u of the interior (resp., exterior) Dirichlet problem has the single-layer representation given by the equation

$$u(x) = \frac{1}{2\pi}\int_{\partial\Omega} \rho(y)\log|x - y|\mathrm{d}y, \quad x \in \Omega \ \text{(resp., } \Omega'\text{)}, \tag{2.53}$$

with ρ satisfying the requirement that

$$\frac{1}{2\pi}\int_{\partial\Omega} \rho(y)\log|x - y|\mathrm{d}y = u_0(x), \quad x \in \partial\Omega. \tag{2.54}$$

In the three-dimensional case we get the equation

$$u(x) = -\frac{1}{4\pi}\int_{\partial\Omega} \frac{\rho(y)}{|x - y|}\mathrm{d}y, \quad x \in \Omega \ \text{(resp. } \Omega'\text{)}, \tag{2.55}$$

where ρ satisfies the equation

$$-\frac{1}{4\pi}\int_{\partial\Omega} \frac{\rho(y)}{|x - y|}\mathrm{d}y = u_0(x), \quad x \in \partial\Omega. \tag{2.56}$$

For the two-dimensional interior Neumann problem, we consider the equation

$$\int_{\partial\Omega_\epsilon} \rho(y)\log|x - y|\mathrm{d}y = \int_{\Gamma_\epsilon} \rho(y)\log|x - y|\mathrm{d}y + \int_{\partial\Omega\setminus\Gamma_\epsilon} \rho(y)\log|x - y|\mathrm{d}y,$$

where $\partial\Omega_\epsilon$ is the boundary of the domain $\Omega_\epsilon := \Omega \setminus B(x,\epsilon)$, and $\Gamma_\epsilon := \partial B(x,\epsilon) \bigcap \Omega$. By taking the directional derivative in the normal direction n of both sides of this equation, letting $\epsilon \to 0$, using the boundary condition $\partial u/\partial n = u_1$ and arguments similar to that used in the proof of Proposition 2.15, we conclude that u is represented as the single-layer potential (2.53) with ρ satisfying the second-kind Fredholm integral equation

$$-\frac{\rho(x)}{2} + \frac{1}{2\pi}\int_{\partial\Omega} \rho(y)\frac{\partial}{\partial n}\log|x-y|\mathrm{d}y = u_1(x), \quad x \in \partial\Omega. \tag{2.57}$$

In a similar manner, the solution u to the three-dimensional interior Neumann problem is represented as the single-layer potential (2.55) with ρ satisfying the second-kind Fredholm integral equation

$$-\frac{\rho(x)}{2} - \frac{1}{4\pi}\int_{\partial\Omega} \rho(y)\frac{\partial}{\partial n}\frac{1}{|x-y|}\mathrm{d}y = u_1(x), \quad x \in \partial\Omega. \tag{2.58}$$

For the exterior Neumann problem, a similar argument leads to the result that the solution u in the two and three-dimensional cases is represented as (2.53) and (2.55) with ρ satisfying the second-kind Fredholm integral equations, respectively,

$$\frac{\rho(x)}{2} + \frac{1}{2\pi}\int_{\partial\Omega} \rho(y)\frac{\partial}{\partial n}\log|x-y|\mathrm{d}y = u_1(x), \quad x \in \partial\Omega \tag{2.59}$$

and

$$\frac{\rho(x)}{2} - \frac{1}{4\pi}\int_{\partial\Omega} \rho(y)\frac{\partial}{\partial n}\frac{1}{|x-y|}\mathrm{d}y = u_1(x), \quad x \in \partial\Omega. \tag{2.60}$$

We close this section with a review of the double-layer potentials for harmonic functions. We start with the two-dimensional interior Dirichlet problem. Suppose that $u = u_+ \in C(\bar{\Omega})$ is a harmonic function in Ω. Let u_- be the solution of the exterior Neumann problem with

$$\frac{\partial u_-(y)}{\partial n} = \frac{\partial u_+(y)}{\partial n}, \quad y \in \partial\Omega,$$

which satisfies (2.39). It follows from Propositions 2.15 and 2.17 that

$$u(x) = \frac{1}{2\pi}\int_{\partial\Omega} (u_+(y) - u_-(y))\frac{\partial}{\partial n}\log|x-y|\mathrm{d}y, \quad x \in \Omega.$$

This equation can be written as a double-layer potential

$$u(x) = \frac{1}{2\pi}\int_{\partial\Omega} \rho(y)\frac{\partial}{\partial n}\log|x-y|\mathrm{d}y, \quad x \in \Omega, \tag{2.61}$$

with $\rho := u_+ - u_-$. According to the proof of Proposition 2.15, we have for $x \in \partial\Omega$ that

$$\lim_{\tilde{x}\to x} \int_{\partial\Omega} \rho(y) \frac{\partial}{\partial n} \log|\tilde{x} - y| \mathrm{d}y = -\pi\rho(x) + \int_{\partial\Omega} \rho(y) \frac{\partial}{\partial n} \log|x - y| \mathrm{d}y.$$

This, with the boundary condition of the Dirichlet problem, concludes that ρ satisfies the second-kind Fredholm integral equation

$$-\frac{\rho(x)}{2} + \frac{1}{2\pi} \int_{\partial\Omega} \rho(y) \frac{\partial}{\partial n} \log|x - y| \mathrm{d}y = u_0(x), \quad x \in \partial\Omega. \tag{2.62}$$

Similarly, the solution u to the three-dimensional interior Dirichlet problem is represented as the double-layer potential

$$u(x) = -\frac{1}{4\pi} \int_{\partial\Omega} \rho(y) \frac{\partial}{\partial n} \frac{1}{|x - y|} \mathrm{d}y, \quad x \in \Omega$$

with ρ satisfying the second-kind Fredholm integral equation

$$\frac{\rho(x)}{2} - \frac{1}{4\pi} \int_{\partial\Omega} \rho(y) \frac{\partial}{\partial n} \frac{1}{|x - y|} \mathrm{d}y = u_0(x), \quad x \in \partial\Omega.$$

For the exterior Dirichlet problem, the corresponding integral equations in the two and three-dimensional cases are

$$\frac{\rho(x)}{2} + \frac{1}{2\pi} \int_{\partial\Omega} \rho(y) \frac{\partial}{\partial n} \log|x - y| \mathrm{d}y = u_0(x), \quad x \in \partial\Omega$$

and

$$-\frac{\rho(x)}{2} - \frac{1}{4\pi} \int_{\partial\Omega} \rho(y) \frac{\partial}{\partial n} \frac{1}{|x - y|} \mathrm{d}y = u_0(x), \quad x \in \partial\Omega,$$

respectively.

2.2 General theory of projection methods

The main concern in the present section is the general theory of projection methods for approximate solutions of operator equations of the form

$$\mathcal{A}u = f, \tag{2.63}$$

where $\mathcal{A} \in \mathcal{B}(\mathbb{X}, \mathbb{Y}), f \in \mathbb{Y}$ are given and $u \in \mathbb{X}$ is the solution to be determined. The case of central importance to us takes the form $\mathcal{A} = \mathcal{I} - \mathcal{K}$, where $\mathcal{I}$ is the identity operator in $\mathcal{B}(\mathbb{X})$ and $\mathcal{K}$ is a compact operator in $\mathcal{B}(\mathbb{X})$. In this case (2.63) becomes

$$(\mathcal{I} - \mathcal{K})\, u = f, \tag{2.64}$$

a Fredholm equation of the second kind.

2.2.1 Projection operators

We begin with a description of various projections, an essential tool for the development of approximation schemes for (2.63). The following notation will be used throughout the book. For a linear operator $\mathcal{A}$ not defined on all of the linear space $\mathbb{X}$, we denote by $D(\mathcal{A})$ its domain and by $N(\mathcal{A}) = \{x : x \in \mathbb{X},\ \mathcal{A}x = 0\}$ its null space. For the range of $\mathcal{A}$ we use $R(\mathcal{A}) = \{\mathcal{A}x : x \in D(\mathcal{A})\}$. Alternatively, we may sometimes write $\mathcal{A}(U)$ for the range of $\mathcal{A}$, where U is the domain of $\mathcal{A}$. We start with the following definition.

Definition 2.19 Let $\mathbb{X}$ be a normed linear space and $\mathbb{V}$ a closed linear subspace of $\mathbb{X}$. A bounded linear operator $\mathcal{P} : \mathbb{X} \to \mathbb{V}$ is called a projection from $\mathbb{X}$ onto $\mathbb{V}$ if for all $v \in \mathbb{V}$

$$\mathcal{P}v = v. \tag{2.65}$$

Note that a projection $\mathcal{P} : \mathbb{X} \to \mathbb{V}$ necessarily has the property that $\mathbb{V} = R(\mathcal{P})$. For later use, we make the following remark.

Proposition 2.20 *Let $\mathbb{X}$ be a normed linear space and $\mathcal{P} \in \mathcal{B}(\mathbb{X})$. Then $\mathcal{P}$ is a projection on $\mathbb{X}$ if and only if $\mathcal{P}^2 = \mathcal{P}$. Moreover, in this case, if $\mathcal{P} \neq 0$ then $\|\mathcal{P}\| \geq 1$.*

Proof If $\mathcal{P} : \mathbb{X} \to \mathcal{R}(\mathcal{P})$ is a projection, then for all $x \in \mathbb{X}$ we have that $\mathcal{P}^2 x = \mathcal{P}(\mathcal{P}x) = \mathcal{P}x$. Conversely, if $\mathcal{P}^2 = \mathcal{P}$ then any $v \in \mathcal{R}(\mathcal{P})$, written as $v = \mathcal{P}x$ for some $x \in \mathbb{X}$, satisfies the equation $\mathcal{P}v = \mathcal{P}^2 x = \mathcal{P}x = v$. Finally, it follows from the equation $\mathcal{P}^2 = \mathcal{P}$ that $\|\mathcal{P}\|^2 \geq \|\mathcal{P}^2\| = \|\mathcal{P}\|$, which implies that $\|\mathcal{P}\| \geq 1$ when $\mathcal{P} \neq 0$. □

Now, we describe the three kinds of projection which are most important for the practical development of approximation schemes to solve operator equations. We have in mind the well-known orthogonal and interpolation projections and perhaps the less familiar concept of a projection defined by a generalized best approximation.

1. Orthogonal projections

We recall the standard definition of the orthogonal projection on a Hilbert space.

Definition 2.21 Let $\mathbb{X}$ be a Hilbert space with inner product $(\cdot, \cdot)$. Two vectors $x, y \in \mathbb{X}$ are said to be orthogonal provided that $(x, y) = 0$. If $\mathbb{V}$ is a nontrivial closed linear subspace of $\mathbb{X}$ then a linear operator $\mathcal{P}$ from $\mathbb{X}$ onto $\mathbb{V}$ is called the orthogonal projection if for all $x \in \mathbb{X}, y \in \mathbb{V}$ it satisfies the equation

$$(\mathcal{P}x, y) = (x, y). \tag{2.66}$$

In other words, the orthogonal projection onto $\mathbb{V}$ has the property that $x-\mathcal{P}x$ is orthogonal to all $y \in \mathbb{V}$. The orthogonal projection satisfies $\|\mathcal{P}\| = 1$ and is self-adjoint, that is, $\mathcal{P}^* = \mathcal{P}$. Moreover, we have the following well-known extremal characterization of the orthogonal projection.

Proposition 2.22 *If $\mathbb{X}$ is a Hilbert space and $\mathbb{V}$ a nontrivial closed linear subspace of $\mathbb{X}$, then there exists an orthogonal projection $\mathcal{P}$ from $\mathbb{X}$ onto $\mathbb{V}$ and for all $x \in \mathbb{X}$*

$$\|x - \mathcal{P}x\| = \min\{\|x - v\| : v \in \mathbb{V}\}.$$

Moreover, the last equation uniquely characterizes $\mathcal{P}x \in \mathbb{V}$.

Proof The existence of $\mathcal{P}x$ follows from the completeness of $\mathbb{X}$ and the parallelogram law. The remaining claim follows from the definition of the orthogonal projection, which gives for $x \in \mathbb{X}$ and $v \in \mathbb{V}$ that

$$\|x - v\|^2 = \|x - \mathcal{P}x\|^2 + \|\mathcal{P}x - v\|^2.$$

□

2. Interpolating projections

We next introduce the concept of interpolating projections.

Definition 2.23 Let $\mathbb{X}$ be a Banach space and $\mathbb{V}$ a finite-dimensional subspace of $\mathbb{X}$. A subset $\{\ell_j : j \in \mathbb{N}_m\}$ of the dual space $\mathbb{X}^*$ is $\mathbb{V}$-unisolvent if, for any $\{c_j : j \in \mathbb{N}_m\} \subseteq \mathbb{R}$, there exists a unique element $v \in \mathbb{V}$ satisfying, for all $j \in \mathbb{N}_m$, the equation

$$\ell_j(v) = c_j. \tag{2.67}$$

To emphasize the pairing between $\mathbb{X}$ and $\mathbb{X}^*$, the value of a linear functional $\ell \in \mathbb{X}^*$ at $x \in \mathbb{X}$ will often be denoted by $\langle x, \ell \rangle$. That is, we define $\langle x, \ell \rangle := \ell(x)$. This convenient notation is standard and the next proposition is elementary. To state it we use the Kronecker symbol $\delta_{ij}, i,j \in \mathbb{N}_m$; that is, $\delta_{ij} = 0$ except for $i = j$ in which case $\delta_{ij} = 1$.

Proposition 2.24 *If $\mathbb{X}$ is a Banach space and $\mathbb{V}$ is an m-dimensional subspace of $\mathbb{X}$, then $\{\ell_j : j \in \mathbb{N}_m\}$ is $\mathbb{V}$-unisolvent if and only if there exists a linearly independent set $\{x_j : j \in \mathbb{N}_m\} \subseteq \mathbb{V}$ which satisfies for all $i,j \in \mathbb{N}_m$ the equation*

$$\ell_j(x_i) = \delta_{ij}. \tag{2.68}$$

In this case, the operator $\mathcal{P} : \mathbb{X} \to \mathbb{V}$ defined for each $x \in \mathbb{X}$ and $j \in \mathbb{N}_m$ by

$$\langle \mathcal{P}x, \ell_j \rangle = \langle x, \ell_j \rangle$$

is a projection from $\mathbb{X}$ *onto* $\mathbb{V}$ *and is given by the formula*

$$\mathcal{P}x = \sum_{j\in\mathbb{N}_m} \ell_j(x)x_j. \tag{2.69}$$

According to the above result any m-dimensional subspace $\mathbb{V}$ of $\mathbb{X}$ and any set of linear functionals $\{\ell_j : j \in \mathbb{N}_m\}$ in $\mathbb{X}$ which is $\mathbb{V}$-unisolvent determine the projection (2.69). An important special case occurs when the Banach space $\mathbb{X}$ consists of real-valued functions on a compact set Ω of $\mathbb{R}^d$. In this case, if there is a subset of points $\{t_j : j \in \mathbb{N}_m\}$ in Ω such that the linear functionals $\{\ell_j : j \in \mathbb{N}_m\}$ are defined for each $x \in \mathbb{X}$ by the equation

$$\ell_j(x) = x(t_j), \tag{2.70}$$

that is, ℓ_j is the point evaluation functional at t_j, then the operator $\mathcal{P} : \mathbb{X} \to \mathbb{V}$ defined by (2.69) is called the *Lagrange interpolation*. If, for some $j \in \mathbb{N}_m$, $\ell_j(x)$ is determined not only by the value of the function x at some point of Ω but also by derivatives of x, $\mathcal{P}$ is called the *Hermite interpolation*. In the case that $\mathcal{P} \in \mathcal{B}(C(\Omega))$ is a Lagrange interpolation its operator norm is given by

$$\|\mathcal{P}\| = \max\left\{\sum_{j\in\mathbb{N}_m} |x_j(t)| : t \in \Omega\right\}. \tag{2.71}$$

3. Generalized best approximation projections

As our final example, we describe the generalized best approximation projections which were introduced in [77]. Let $\mathbb{X}$ be a Banach space and $\mathbb{X}^*$ its dual space. For $n \in \mathbb{N}$ we assume that $\mathbb{X}_n$ and $\mathbb{Y}_n$ are two finite-dimensional subspaces of $\mathbb{X}$ and $\mathbb{X}^*$, respectively, with the same dimension.

Definition 2.25 For $x \in \mathbb{X}$, an element $\mathcal{P}_n x \in \mathbb{X}_n$ is called a generalized best approximation to x from $\mathbb{X}_n$ with respect to $\mathbb{Y}_n$ if for all $\ell \in \mathbb{Y}_n$ it satisfies the equation

$$\langle x - \mathcal{P}_n x, \ell\rangle = 0. \tag{2.72}$$

Similarly, given $\ell \in \mathbb{X}^*$, an element $\mathcal{P}_n'\ell \in \mathbb{Y}_n$ is called a generalized best approximation from $\mathbb{Y}_n$ to ℓ with respect to $\mathbb{X}_n$ if for all $x \in \mathbb{X}_n$ it satisfies the equation

$$\left\langle x, \ell - \mathcal{P}_n'\ell\right\rangle = 0.$$

Figure 2.1 displays schematically the generalized best approximation projection $\mathcal{P}_n x$ to $x \in \mathbb{X}$ from $\mathbb{X}_n$ with respect to $\mathbb{Y}_n$ in a Hilbert space. In this case, equation (2.72) means $(x - \mathcal{P}_n x)\perp\mathbb{Y}_n$.

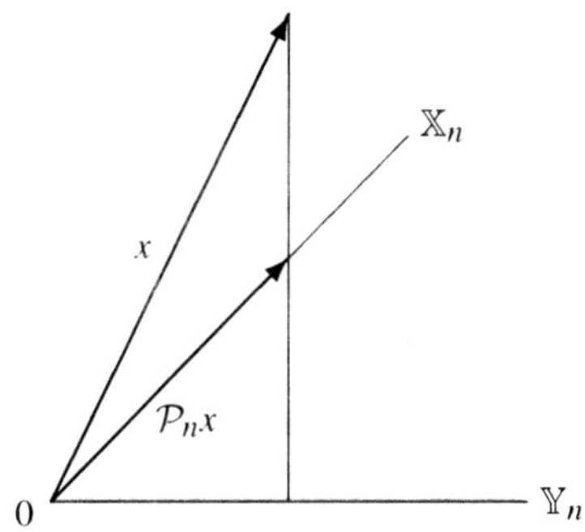

Figure 2.1 Generalized best approximation projections.

For a further explanation, we provide an example as follows. Let $I := [0, 1]$ and $\mathbb{X} := L^2(I)$. We subdivide the interval I into n subintervals by points $t_j := jh, j \in \mathbb{N}_{n-1}$, and set $t_{j-\frac{1}{2}} := (j - \frac{1}{2})h, j \in \mathbb{N}_n$, where $h := \frac{1}{n}$. We let $\mathbb{X}_n$ be the space of continuous piecewise linear polynomials with knots at $t_j, j \in \mathbb{N}_{n-1}$, and $\mathbb{Y}_n$ be the space of piecewise constant functions with knots at $t_{j-\frac{1}{2}}, j \in \mathbb{N}_n$. Clearly, $\dim \mathbb{X}_n = \dim \mathbb{Y}_n = n + 1$. For $x \in \mathbb{X}$, we define the generalized best approximation $\mathcal{P}_n x$ to x from $\mathbb{X}_n$ with respect to $\mathbb{Y}_n$ by the equation

$$\langle x - \mathcal{P}_n x, y \rangle = 0, \quad \text{for all } y \in \mathbb{Y}_n.$$

Let

$$\phi_j(t) := \begin{cases} 1 - (t_j - t)/h, & t_{j-1} \le t \le t_j, \\ 1 - (t - t_j)/h, & t_j < t \le t_{j+1}, \quad j \in \mathbb{Z}_{n+1} \\ 0, & \text{elsewhere}, \end{cases}$$

and

$$\psi_j(t) := \begin{cases} 1, & t_{j-\frac{1}{2}} \le t \le t_{j+\frac{1}{2}}, \\ 0, & \text{elsewhere}, \end{cases} \quad j \in \mathbb{Z}_{n+1}.$$

Two groups of functions $\{\phi_j : j \in \mathbb{Z}_{n+1}\}$ and $\{\psi_j : j \in \mathbb{Z}_{n+1}\}$ form the bases for $\mathbb{X}_n$ and $\mathbb{Y}_n$, respectively. Thus, $\mathcal{P}_n x$ can be written in the form $\mathcal{P}_n x = \sum_{j \in \mathbb{Z}_{n+1}} c_j \phi_j$, where the vector $\mathbf{u}_n := [c_j : j \in \mathbb{Z}_{n+1}]$ satisfies the linear equation

$$\mathbf{A}_n \mathbf{u}_n = \mathbf{f}_n,$$

in which $\mathbf{A}_n := [\langle \phi_j, \psi_i \rangle : i, j \in \mathbb{Z}_{n+1}]$ and $\mathbf{f}_n := [\langle x, \psi_i \rangle : i \in \mathbb{Z}_{n+1}]$.

We now present a necessary and sufficient condition for which each $x \in \mathbb{X}$ has a unique generalized best approximation from $\mathbb{X}_n$ with respect to $\mathbb{Y}_n$. In what follows we denote by $\mathbb{X}_n^{\perp}$ the set of all linear functionals in $\mathbb{X}^*$ which vanish on the subspace $\mathbb{X}_n$, that is, the annihilator of $\mathbb{X}_n$ in $\mathbb{X}^*$.

Proposition 2.26 *For each $x \in \mathbb{X}$, the generalized best approximation $\mathcal{P}_n x$ to x from $\mathbb{X}_n$ with respect to $\mathbb{Y}_n$ exists and is unique if and only if*

$$\mathbb{Y}_n \cap \mathbb{X}_n^{\perp} = \{0\}. \tag{2.73}$$

When this is the case and $\mathcal{P}_n x$ is the generalized best approximation of x, $\mathcal{P}_n : \mathbb{X} \to \mathbb{X}_n$ is a projection.

Proof Let $x \in \mathbb{X}$ be given and assume that spaces $\mathbb{X}_n$ and $\mathbb{Y}_n$ have bases $\{x_j : j \in \mathbb{N}_m\}$ and $\{\ell_j : j \in \mathbb{N}_m\}$, respectively. The existence and uniqueness of

$$\mathcal{P}_n x = \sum_{i \in \mathbb{N}_m} c_i x_i \in \mathbb{X}_n$$

satisfying equation (2.72) means that the linear system

$$\sum_{i \in \mathbb{N}_m} c_i \left\langle x_i, \ell_j \right\rangle = \left\langle x, \ell_j \right\rangle, \quad j \in \mathbb{N}_m \tag{2.74}$$

has a unique solution $c = [c_j : j \in \mathbb{N}_m] \in \mathbb{R}^m$ for any $x \in \mathbb{X}$. This is equivalent to the fact that the $m \times m$ matrix

$$\mathbf{A} := [\langle x_j, \ell_i \rangle : i, j \in \mathbb{N}_m]$$

is nonsingular. Moreover, a vector $b = [b_j : j \in \mathbb{N}_m] \in \mathbb{R}^m$ is in the null space of $\mathbf{A}$ if and only if the linear functional $\ell := \sum_{j \in \mathbb{N}_m} b_j \ell_j$ is in the subspace $\mathbb{Y}_n \cap \mathbb{X}_n^{\perp}$. This proves the first assertion. For the remaining claim, when $\mathbb{Y}_n \cap \mathbb{X}_n^{\perp} = \{0\}$ we solve (2.74) and define for $x \in \mathbb{X}$

$$\mathcal{P}_n x := \sum_{j \in \mathbb{N}_m} c_j x_j.$$

Clearly, $\mathcal{P}_n : \mathbb{X} \to \mathbb{X}_n$ is a linear operator that by construction satisfies (2.72).

Let us show that $\mathcal{P}_n$ is also a projection. For any $x \in \mathbb{X}$, $\mathcal{P}_n^2 x \in \mathbb{X}_n$ is a generalized best approximation to $\mathcal{P}_n x$ from $\mathbb{X}_n$ with respect to $\mathbb{Y}_n$. By Definition 2.25, we conclude for all $\ell \in \mathbb{Y}_n$ that

$$\left\langle \mathcal{P}_n x - \mathcal{P}_n^2 x, \ell \right\rangle = 0.$$

This together with (2.72) implies for all $\ell \in \mathbb{Y}_n$ that

$$\left\langle x - \mathcal{P}_n^2 x, \ell \right\rangle = 0.$$

By the uniqueness of the solution to equation (2.74), we obtain that $\mathcal{P}_n^2 x = \mathcal{P}_n x$ and so $\mathcal{P}_n$ is indeed a projection. □

We state the corresponding result for the generalized best approximation to ℓ from $\mathbb{Y}_n$ with respect to $\mathbb{X}_n$. The proof is similar to that of Proposition 2.26, where in this case the transpose of the matrix $\mathbf{A}$ is used.

Proposition 2.27 *For each $\ell \in \mathbb{Y}$, the generalized best approximation to $\ell \in \mathbb{Y}$ from $\mathbb{Y}_n$ with respect to $\mathbb{X}_n$ exists and is unique if and only if*

$$\mathbb{X}_n \cap \mathbb{Y}_n^{\perp} = \{0\}. \tag{2.75}$$

When this is the case and $\mathcal{P}'_n \ell$ is the generalized best approximation of ℓ from $\mathbb{Y}_n$ with respect to $\mathbb{X}_n$, then $\mathcal{P}'_n : \mathbb{Y} \to \mathbb{Y}_n$ is a projection.

In view of Propositions 2.26 and 2.27 we shall always assume that condition (2.73) (resp. (2.75)) holds whenever we refer to $\mathcal{P}_n$ (resp. $\mathcal{P}'_n$) as the generalized best approximation projection from $\mathbb{X}_n$ with respect to $\mathbb{Y}_n$ (resp. the generalized best approximation projection from $\mathbb{Y}_n$ with respect to $\mathbb{X}_n$). We also remark that condition (2.73) or (2.75) implies that $\dim \mathbb{Y}_n = \dim \mathbb{X}_n$.

Proposition 2.26 allows us to connect the concept of the generalized best approximation to the familiar concept of dual bases. Indeed, by the Hahn–Banach theorem every finite-dimensional subspace $\mathbb{X}_n$ of a normed linear space $\mathbb{X}$ has a dual basis (see Theorem A.32 in the Appendix). Specifically, if $\{x_j : j \in \mathbb{N}_m\}$ is a basis for $\mathbb{X}_n$, there is a subset $\{\ell_j : j \in \mathbb{N}_m\} \subseteq \mathbb{X}^*$ such that for $i, j \in \mathbb{N}_m$, $\ell_j(x_i) = \delta_{i,j}$. According to Proposition 2.26, the generalized best approximation to $x \in \mathbb{X}$ from $\mathbb{X}_n$ with respect to $\mathbb{Y}_n := \text{span}\{\ell_j : j \in \mathbb{N}_m\}$ exists and is given by

$$\mathcal{P}_n x = \sum_{j \in \mathbb{N}_m} \ell_j(x) x_j.$$

Conversely, if (2.73) holds we can find a dual basis for $\mathbb{X}_n$ in $\mathbb{Y}_n$.

Proposition 2.28 *If $\mathbb{Y}_n \cap \mathbb{X}_n^{\perp} = \{0\}$ then $\mathcal{P}'_n = \mathcal{P}^*_n$ and $\mathbb{Y}_n = \mathcal{P}^*_n \mathbb{X}^*$.*

Proof For all $x \in \mathbb{X}$ and $\ell \in \mathbb{X}^*$ we have that

$$\langle x, \mathcal{P}'_n \ell \rangle = \langle \mathcal{P}_n x, \mathcal{P}'_n \ell \rangle = \langle \mathcal{P}_n x, \ell \rangle = \langle x, \mathcal{P}^*_n \ell \rangle,$$

from which the desired result follows. □

The next proposition gives an alternative sufficient condition to ensure that every $x \in \mathbb{X}$ has a unique generalized best approximation in $\mathbb{X}_n$ with respect to $\mathbb{Y}_n$.

Proposition 2.29 *If there is a constant $c > 0$ and a linear operator $\mathcal{T}_n : \mathbb{X}_n \to \mathbb{Y}_n$ with $\mathcal{T}_n \mathbb{X}_n = \mathbb{Y}_n$ such that for all $x \in \mathbb{X}_n$,*

$$\|x\|^2 \leq c \langle x, \mathcal{T}_n x \rangle,$$

then (2.73) holds.

Proof Our hypothesis implies that for any $\ell \in \mathbb{Y}_n \cap \mathbb{X}_n^{\perp}$ there exists $x \in \mathbb{X}_n$ such that $\mathcal{T}_n x = \ell$, and so

$$\|x\| \leq \sqrt{c}\,\langle x, \mathcal{T}_n x\rangle^{1/2} = \sqrt{c}\,\langle x, \ell\rangle^{1/2} = 0.$$

Therefore, we obtain that $x = 0$ and consequently we also have that $\ell = \mathcal{T}_n x = 0$. □

The next issue that concerns us is the conditions which guarantee that a sequence of projections $\{\mathcal{P}_n : n \in \mathbb{N}\}$ converges pointwise to the identity operator in $\mathbb{X}$, that is, $\mathcal{P}_n \xrightarrow{s} \mathcal{I}$. As we shall see later, this property is crucial for the analysis of projection methods. Generally, condition (2.73) is not sufficient to ensure that this is the case. Therefore, we need to introduce the concept of a *regular pair*.

Definition 2.30 A pair of sequences of subspaces $\mathbb{X}_n \subseteq \mathbb{X}, n \in \mathbb{N}$ and $\mathbb{Y}_n \subseteq \mathbb{Y}, n \in \mathbb{N}$ is called a regular pair if there is a positive constant c such that for all $n \in \mathbb{N}$ there are linear operators $\mathcal{T}_n : \mathbb{X}_n \to \mathbb{Y}_n$ with $\mathcal{T}_n\mathbb{X}_n = \mathbb{Y}_n$ satisfying the conditions that for all $x \in \mathbb{X}_n$

(i) $\|x\| \leq c\,\langle x, \mathcal{T}_n x\rangle^{1/2}$,
(ii) $\|\mathcal{T}_n x\| \leq c\|x\|$.

In this definition it is important to realize that the constant c appearing above is independent of $n \in \mathbb{N}$.

If $\mathbb{X}$ is a Hilbert space so that $\mathbb{X}^*$ can be identified by the Riesz representation theorem (see Theorem A.31 in the Appendix) with $\mathbb{X}$ itself, and we also have for all $n \in \mathbb{N}$ that $\mathbb{X}_n = \mathbb{Y}_n$, then conditions (i) and (ii) are satisfied with $\mathcal{T}_n = \mathcal{I}, n \in \mathbb{N}$. Thus, in this case we have a regular pair. On the contrary, if $\{\mathbb{X}_n, \mathbb{Y}_n\}$ is a regular pair then from Proposition 2.29 we conclude that (2.73) holds and so $\mathcal{P}_n$ is well defined.

For the next proposition, we find it appropriate to introduce the quantity

$$\operatorname{dist}(x, \mathbb{X}_n) := \min\{\|x - u\| : u \in \mathbb{X}_n\}.$$

Proposition 2.31 *If $\{\mathbb{X}_n : n \in \mathbb{N}\}$ and $\{\mathbb{Y}_n : n \in \mathbb{N}\}$ form a regular pair and $\mathcal{P}_n : \mathbb{X} \to \mathbb{X}_n$ is the corresponding generalized best approximation projection, then for each $x \in \mathbb{X}$ and $n \in \mathbb{N}$*

(i) $\|\mathcal{P}_n\| \leq c^3$,
(ii) $\operatorname{dist}(x, \mathbb{X}_n) \leq \|x - \mathcal{P}_n x\| \leq (1 + c^3)\operatorname{dist}(x, \mathbb{X}_n)$.

Consequently, if for each $x \in \mathbb{X}$ $\lim_{n\to\infty} \operatorname{dist}(x, \mathbb{X}_n) = 0$, then $\lim_{n\to\infty} \mathcal{P}_n x = x$.

Proof For each $x \in \mathbb{X}$ and $u \in \mathbb{X}_n$ we have that

$$\begin{aligned}\|u - \mathcal{P}_n x\|^2 &\le c^2 \langle u - \mathcal{P}_n x, \mathcal{T}_n(u - \mathcal{P}_n x)\rangle \\ &= c^2 \langle u - x, \mathcal{T}_n(u - \mathcal{P}_n x)\rangle \\ &\le c^2 \|u - x\| \|\mathcal{T}_n(u - \mathcal{P}_n x)\| \\ &\le c^3 \|u - x\| \|u - \mathcal{P}_n x\|.\end{aligned}$$

Therefore, we conclude that

$$\|u - \mathcal{P}_n x\| \le c^3 \|u - x\|.$$

The choice $u = 0$ in the above inequality establishes (i). As for (ii), the lower bound is obvious while the upper bound is obtained by using the inequality that for any $u \in \mathbb{X}_n$,

$$\|x - \mathcal{P}_n x\| \le \|x - u\| + \|u - \mathcal{P}_n x\| \le (1 + c^3)\|u - x\|.$$

□

2.2.2 Projection methods for operator equations

In this subsection we describe projection methods for solving operator equations, which include the Petrov–Galerkin method, the Galerkin method, the least-squares method and the collocation method.

Let $\mathbb{X}$ and $\mathbb{Y}$ be Banach spaces, $\mathcal{A} : \mathbb{X} \to \mathbb{Y}$ be a bounded linear operator and $f \in \mathbb{Y}$. We wish to find an $u \in \mathbb{X}$ such that

$$\mathcal{A}u = f,$$

if it exists. Projection methods have the common feature of specifying a sequence $\{\mathbb{X}_n : n \in \mathbb{N}\}$ of subspaces and choosing a $u_n \in \mathbb{X}_n$ for which the residual error

$$r_n := \mathcal{A}u_n - f$$

is "small," so that u_n is a good approximation to the desired u. How this is done depends on the method used. We shall review some of the principal strategies for making r_n small.

We begin with a description of the *Petrov–Galerkin method.* The idea behind this method is to make the residual $r_n \in \mathbb{Y}$ small by choosing finite-dimensional subspaces $\{\mathbb{L}_n : n \in \mathbb{N}\}$ in $\mathbb{Y}^*$ with $\dim \mathbb{L}_n = \dim \mathbb{X}_n$ and attempting to find $u_n \in \mathbb{X}_n$ so that for all $\ell \in \mathbb{L}_n$ we have

$$\langle \mathcal{A}u_n, \ell\rangle = \langle f, \ell\rangle . \tag{2.76}$$

Specifically, we choose bases $\mathbb{X}_n = \operatorname{span}\{x_j : j \in \mathbb{N}_m\}$ and $\mathbb{L}_n = \operatorname{span}\{\ell_j : j \in \mathbb{N}_m\}$ and write u_n in the form

$$u_n = \sum_{j \in \mathbb{N}_m} c_j x_j$$

where the vector $\mathbf{u}_n := [c_j : j \in \mathbb{N}_m] \in \mathbb{R}^m$ must satisfy the linear equation

$$\mathbf{A}_n \mathbf{u}_n = \mathbf{f}_n,$$

where $\mathbf{A}_n := [\langle \mathcal{A}x_j, \ell_i \rangle : i, j \in \mathbb{N}_m]$ and $\mathbf{f}_n := [\langle f, \ell_i \rangle : i \in \mathbb{N}_m]$.

For the purpose of theoretical analysis, it is also useful to express $u_n \in \mathbb{X}_n$ as a solution of an operator equation. This can be done by specifying any sequence $\{\mathbb{Y}_n : n \in \mathbb{N}\}$ of subspaces for which there are generalized best approximation projections $\mathcal{P}_n : \mathbb{Y} \to \mathbb{Y}_n$ with respect to $\mathbb{L}_n$. This means that for all $\ell \in \mathbb{L}_n$, not only $\langle \mathcal{A}u_n - f, \ell \rangle = 0$ but also $\langle \mathcal{A}u_n - \mathcal{P}_n \mathcal{A}u_n, \ell \rangle = 0$ and $\langle f - \mathcal{P}_n f, \ell \rangle = 0$. From these three equations we conclude that $\mathcal{P}_n \mathcal{A}u_n - \mathcal{P}_n f \in \mathbb{Y}_n \cap \mathbb{L}_n^{\perp} = \{0\}$ and so we obtain that

$$\mathcal{P}_n \mathcal{A}u_n = \mathcal{P}_n f.$$

Therefore, we conclude that equation (2.76) is equivalent to the operator equation

$$\mathcal{A}_n u_n = \mathcal{P}_n f,$$

where $\mathcal{A}_n := \mathcal{P}_n \mathcal{A}|_{\mathbb{X}_n}$. Here, the symbol $\mathcal{A}|_{\mathbb{X}_n}$ stands for the operator $\mathcal{A}$ restricted to the subspace $\mathbb{X}_n$ and so $\mathcal{A}_n \in \mathcal{B}(\mathbb{X}_n, \mathbb{Y}_n)$. This means that the operator $\mathcal{A}_n$ can be realized as a square matrix because $\dim \mathbb{X}_n = \dim \mathbb{L}_n = \dim \mathbb{Y}_n$.

We remark that the commonly used *Galerkin method* and also the *least-squares method* are special cases of the Petrov–Galerkin method. Specifically, if $\mathbb{X} = \mathbb{Y}$ is a Hilbert space we identify $\mathbb{X}^*$ with $\mathbb{X}$ and choose $\mathbb{X}_n = \mathbb{Y}_n, n \in \mathbb{N}$; then the projection $\mathcal{P}_n$ above is the orthogonal projection of $\mathbb{X}$ onto $\mathbb{X}_n$, and equation (2.76) means that $\mathcal{A}u_n - f \in \mathbb{X}_n^{\perp}$ with $u_n \in \mathbb{X}_n$. Alternatively, the least-squares method chooses $\mathbb{Y}_n = \mathcal{A}(\mathbb{X}_n)$, instead of the choice $\mathbb{X}_n = \mathbb{Y}_n$ specified for the Galerkin method, which yields the requirement $\mathcal{A}u_n - f \in \mathcal{A}(\mathbb{X}_n)$ with $u_n \in \mathbb{X}_n$. This means that u_n satisfies the equation

$$\|f - \mathcal{A}u_n\| = \operatorname{dist}(f, \mathcal{A}(\mathbb{X}_n)).$$

Equivalently, u_n has the property that $\mathcal{A}^*(\mathcal{A}u_n - f) \in \mathbb{X}_n^{\perp}$. So, in particular, we see that the least-squares method is equivalent to the Galerkin method applied to the operator equation

$$\mathcal{A}^* \mathcal{A}u = \mathcal{A}^* f.$$

Our final example is the *collocation method.* This approaches the task of making the residual r_n small by making it zero on some finite set of points. Specifically, the setup requires that $\mathbb{Y} = C(\Omega)$, where Ω is a compact subset of $\mathbb{R}^d$. Choose a finite set $T \subseteq \Omega$ and demand that $r_n|_T = 0$, where $r_n|_T$ denotes the restriction of the function r_n to the finite set T. Again, to solve for $u_n \in \mathbb{X}$ we restrict our search to $u_n \in \mathbb{X}_n$, where $\dim \mathbb{X}_n = \operatorname{card} T$ and $\operatorname{card} T$ denotes the number of distinct elements in T, that is, the cardinality of T. In terms of a basis for $\mathbb{X}_n$ we have, as before, $u_n = \sum_{j \in \mathbb{N}_m} c_j x_j$, $\mathbf{u}_n = [c_j : j \in \mathbb{N}_m]$, $\mathbf{f}_n = [f(t_j) : j \in \mathbb{N}_m]$ where $T = \{t_j : j \in \mathbb{N}_m\}$ and $\mathbf{A}_n := [(\mathcal{A}x_i)(t_j) : i, j \in \mathbb{N}_m]$. These quantities are joined by the linear system of equations

$$\mathbf{A}_n \mathbf{u}_n = \mathbf{f}_n.$$

An operator version of this linear system follows by choosing a subspace $\mathbb{Y}_n \subseteq C(\Omega)$ with $\dim \mathbb{Y}_n = \operatorname{card} T$, which admits an interpolation projection $\mathcal{P}_n : \mathbb{Y} \to \mathbb{Y}_n$ corresponding to the family of linear functionals $\{\ell_t : t \in T\}$, where the linear functional ℓ_t is defined to be the "delta functional" at t, that is, for each $f \in C(\Omega)$ we have that $\ell_t(f) = f(t)$. Therefore, $(\mathcal{P}_n f)|_T = 0$ if and only if $f|_T = 0$ and so we get that

$$\mathcal{A}_n u_n = \mathcal{P}_n f,$$

where $\mathcal{A}_n := \mathcal{P}_n \mathcal{A}|_{\mathbb{X}_n}$.

2.2.3 Convergence and stability

In this subsection we discuss the convergence and stability of projection methods for operator equations. The setup is as before, namely, $\mathbb{X}$ and $\mathbb{Y}$ are normal linear spaces, $\mathcal{A} \in \mathcal{B}(\mathbb{X}, \mathbb{Y})$, $\{\mathbb{X}_n : n \in \mathbb{N}\}$ and $\{\mathbb{Y}_n : n \in \mathbb{N}\}$ are two sequences of finite-dimensional subspaces of $\mathbb{X}$ and $\mathbb{Y}$, respectively with $\dim \mathbb{X}_n = \dim \mathbb{Y}_n$, and $\mathcal{P}_n : \mathbb{Y} \to \mathbb{Y}_n$ is a projection. The projection equation for our approximate solution $u_n \in \mathbb{X}_n$ for a solution $u \in \mathbb{X}$ to the equation $\mathcal{A}u = f$ is

$$\mathcal{A}_n u_n = \mathcal{P}_n f, \tag{2.77}$$

where as before

$$\mathcal{A}_n := \mathcal{P}_n \mathcal{A}|_{\mathbb{X}_n} \tag{2.78}$$

and $\mathcal{A}_n \in \mathcal{B}(\mathbb{X}_n, \mathbb{Y}_n)$. Our goal is to clarify to what extent the sequence $\{u_n : n \in \mathbb{N}\}$, if it exists, approximates $u \in \mathbb{X}$. We start with a definition.

Definition 2.32 The projection method above is said to be convergent if there exists an integer $q \in \mathbb{N}$ such that the operator $\mathcal{A}_n \in \mathcal{B}(\mathbb{X}_n, \mathbb{Y}_n)$ is invertible for

$n \geq q$ and for each $f \in \mathcal{A}(\mathbb{X})$ the unique solution of (2.77), which we call $u_n := \mathcal{A}_n^{-1}\mathcal{P}_n f$, converges as $n \to \infty$ to a $u \in \mathbb{X}$ that satisfies the operator equation $\mathcal{A}u = f$.

We remark that convergence of the approximate solutions $\{u_n : n \in \mathbb{N}\}$ to u as defined above does not require that the operator equation $\mathcal{A}u = f$ has a unique solution, although this is often the case in applications.

We first describe a consequence of convergence.

Theorem 2.33 *If the projection method is convergent then there is an integer $q \in \mathbb{N}$ and a constant $c > 0$ such that for all $n \geq q$*

$$\|\mathcal{A}_n^{-1}\mathcal{P}_n\mathcal{A}\| \leq c. \tag{2.79}$$

If, in addition, $\mathcal{A}$ is onto, that is, $\mathcal{A}(\mathbb{X}) = \mathbb{Y}$, then the q and c above can be chosen so that for all $n \geq q$

$$\|\mathcal{A}_n^{-1}\| \leq c. \tag{2.80}$$

Proof The proof uses the uniform boundedness principle (Theorem A.25 in the Appendix). Specifically, since the projection method converges, we conclude, for each $u \in \mathbb{X}$, that the sequence $\{\mathcal{A}_n^{-1}\mathcal{P}_n\mathcal{A}u : n \in \mathbb{N}\}$ converges in $\mathbb{X}$ as $n \to \infty$ and hence is norm bounded for each $u \in \mathbb{X}$. Thus, an application of the uniform boundedness principle confirms (2.79). We note in passing that $\mathcal{A}_n^{-1}\mathcal{P}_n\mathcal{A} \in \mathcal{B}(\mathbb{X}, \mathbb{X}_n)$ and by equation (2.78) this operator is a projection of $\mathbb{X}$ onto $\mathbb{X}_n$.

As for (2.80), we argue in a similar fashion, using, in this case, the sequence $\{\mathcal{A}_n^{-1}\mathcal{P}_n f : n \geq q\}$, where f can be chosen arbitrarily in $\mathbb{Y}$ because the operator $\mathcal{A}$ is assumed to be onto, to conclude, again by the uniform boundedness principle, that $\|\mathcal{A}_n^{-1}\mathcal{P}_n\|$ is bounded uniformly in $n \in \mathbb{N}$. Here we have that $\mathcal{A}_n^{-1}\mathcal{P}_n \in \mathcal{B}(\mathbb{Y}, \mathbb{X}_n)$. However, since

$$\begin{aligned}\|\mathcal{A}_n^{-1}\| &= \sup\{\|\mathcal{A}_n^{-1}y\| : y \in \mathbb{Y}_n, \|y\| = 1\} \\ &= \sup\{\|\mathcal{A}_n^{-1}\mathcal{P}_n y\| : y \in \mathbb{Y}_n, \|y\| = 1\} \\ &\leq \sup\{\|\mathcal{A}_n^{-1}\mathcal{P}_n y\| : y \in \mathbb{Y}, \|y\| = 1\} = \|\mathcal{A}_n^{-1}\mathcal{P}_n\|,\end{aligned} \tag{2.81}$$

the claim is confirmed. □

Next, we explore to what extent (2.79) and (2.80) are also sufficient for convergence of a projection method. To this end, we introduce the notion of density of the sequences of subspaces $\{\mathbb{X}_n : n \in \mathbb{N}\}$ in $\mathbb{X}$.

Definition 2.34 We say that the sequence of subspaces $\{\mathbb{X}_n : n \in \mathbb{N}\}$ has the denseness property in $\mathbb{X}$ if, for every $x \in \mathbb{X}$,

$$\lim_{n\to\infty} \operatorname{dist}(x, \mathbb{X}_n) = 0.$$

Theorem 2.35 *If there exists a $q \in \mathbb{N}$ such that the operator $\mathcal{A}_n \in \mathcal{B}(\mathbb{X}_n, \mathbb{Y}_n)$ is invertible, a positive constant c such that for all $n \geq q$, $\|\mathcal{A}_n^{-1}\mathcal{P}_n\mathcal{A}\| \leq c$ and the sequence $\{\mathbb{X}_n : n \in \mathbb{N}\}$ has the denseness property in $\mathbb{X}$, then the operator $\mathcal{A}$ is one to one and for each $f \in R(\mathcal{A})$ the projection method converges to $u \in \mathbb{X}$, the unique solution to the equation $\mathcal{A}u = f$.*

Proof The proof uses the fact that the operator $\mathcal{A}_n^{-1}\mathcal{P}_n\mathcal{A}$ is a projection of $\mathbb{X}$ onto $\mathbb{X}_n$. Hence, we have for any $v \in \mathbb{X}$ the inequality

$$\|\mathcal{A}_n^{-1}\mathcal{P}_n\mathcal{A}v - v\| \leq (1 + c)\operatorname{dist}(v, \mathbb{X}_n) \tag{2.82}$$

and, in particular, for $f \in R(\mathcal{A})$ we have that

$$\|u_n - u\| \leq (1 + c)\operatorname{dist}(u, \mathbb{X}_n).$$

□

Although Theorem 2.35 shows that condition (2.79) nearly guarantees convergence, it is hard to apply in practice. Nevertheless, further explanation of the inequality (2.82) will lead to some improvements. Indeed, for any projection $\mathcal{P}_n : \mathbb{X} \to \mathbb{X}_n$ we have for any $x \in \mathbb{X}$ that

$$\|\mathcal{P}_n x - x\| \leq (1 + \|\mathcal{P}_n\|)\operatorname{dist}(x, \mathbb{X}_n) \tag{2.83}$$

and

$$\operatorname{dist}(x, \mathbb{X}_n) \leq \|\mathcal{P}_n x - x\|. \tag{2.84}$$

These inequalities lead to a useful criterion to ensure that a sequence of projections has the property $\mathcal{P}_n \xrightarrow{s} \mathcal{I}$.

Lemma 2.36 *A sequence of projections $\{\mathcal{P}_n : n \in \mathbb{N}\} \subseteq \mathcal{B}(\mathbb{X}, \mathbb{X}_n)$ has the property $\mathcal{P}_n \xrightarrow{s} \mathcal{I}$ if and only if the sequence $\{\|\mathcal{P}_n\| : n \in \mathbb{N}\}$ is bounded and the sequence of subspaces $\{\mathbb{X}_n : n \in \mathbb{N}\}$ has the denseness property in $\mathbb{X}$.*

Proof This result follows directly from the uniform boundedness principle and inequalities (2.83) and (2.84). □

We next comment on the denseness property. For this purpose, we introduce the subspace

$$\limsup_{n\to\infty} \mathbb{X}_n := \bigcup_{m\in\mathbb{N}} \bigcap_{n\geq m} \mathbb{X}_n. \tag{2.85}$$

Proposition 2.37 *If the subspace* $\limsup_{n\to\infty}\mathbb{X}_n$ *is dense in* $\mathbb{X}$ *then the sequence of subspaces* $\{\mathbb{X}_n : n \in \mathbb{N}\}$ *has the denseness property in* $\mathbb{X}$.

Proof If $\limsup_{n\to\infty}\mathbb{X}_n$ is dense in $\mathbb{X}$ then for any $x \in \mathbb{X}$ and any $\epsilon > 0$, there exists a positive integer $q \in \mathbb{N}$ and a $y \in \mathbb{X}$ such that $\|x - y\| < \epsilon$ and $y \in \bigcap_{n \geq q} \mathbb{X}_n$. Hence, for all $n \geq q$, we have that $\operatorname{dist}(x, \mathbb{X}_n) < \epsilon$, that is, $\{\mathbb{X}_n : n \in \mathbb{N}\}$ has the denseness property in $\mathbb{X}$. □

The next proposition presents a necessary condition for the denseness property.

Proposition 2.38 *If the sequence of subspaces* $\{\mathbb{X}_n : n \in \mathbb{N}\}$ *has the denseness property then* $\bigcup_{n\in\mathbb{N}}\mathbb{X}_n$ *is dense in* $\mathbb{X}$.

Proof This result follows from the fact that for all $x \in \mathbb{X}$,

$$\operatorname{dist}(x, \cup_{m\in\mathbb{N}}\mathbb{X}_m) \leq \operatorname{dist}(x, \mathbb{X}_n), \quad \text{for all } n \in \mathbb{N}$$

and the definition of the denseness property. □

Definition 2.39 We say that a sequence of subspaces $\{\mathbb{X}_n : n \in \mathbb{N}\}$ is nested if for all $n \in \mathbb{N}$, $\mathbb{X}_n \subseteq \mathbb{X}_{n+1}$.

Note that when the sequence of subspaces $\{\mathbb{X}_n : n \in \mathbb{N}\}$ is nested, it follows that

$$\bigcup_{n\in\mathbb{N}} \mathbb{X}_n = \limsup_{n\to\infty} \mathbb{X}_n.$$

Consequently, Lemma 2.36 gives us the following fact.

Proposition 2.40 *If the sequence of subspaces* $\{\mathbb{X}_n : n \in \mathbb{N}\}$ *is nested in* $\mathbb{X}$ *then* $\mathcal{P}_n \xrightarrow{s} \mathcal{I}$ *if and only if the sequence* $\{\|\mathcal{P}_n\| : n \in \mathbb{N}\}$ *is bounded and the subspace* $\bigcup_{n\in\mathbb{N}}\mathbb{X}_n$ *is dense in* $\mathbb{X}$.

We may express the condition that the collection of subspaces $\{\mathbb{X}_n : n \in \mathbb{N}\}$ is nested in terms of the corresponding collection of projections $\{\mathcal{P}_n : n \in \mathbb{N}\}$ such that for each $n \in \mathbb{N}$ we have that $R(\mathcal{P}_n) = \mathbb{X}_n$. Indeed, $\{\mathbb{X}_n : n \in \mathbb{N}\}$ is a nested collection of subspaces if

$$\mathcal{P}_n\mathcal{P}_m = \mathcal{P}_n \quad \text{for } m \geq n. \tag{2.86}$$

We turn our attention to the adjoint projections $\mathcal{P}_n^* : \mathbb{X}^* \to \mathbb{X}^*$. To this end, we choose a basis $\{x_i : i \in \mathbb{N}_m\}$ for $\mathbb{X}_n$ and observe that there exists a unique collection of bounded linear functionals $\{\ell_i : i \in \mathbb{N}_m\} \subseteq \mathbb{X}^*$ such that

$\ell_j(x_i) = \delta_{i,j}$, $i, j \in \mathbb{N}_m$ and for each $x \in \mathbb{X}$, we have that

$$\mathcal{P}_n x = \sum_{i \in \mathbb{N}_m} \langle x, \ell_i \rangle \, x_i. \tag{2.87}$$

In fact, for any $x \in \mathbb{X}_n$ in the form $x = \sum_{i \in \mathbb{N}_m} c_i x_i$, where $\mathbf{c} := [c_i : i \in \mathbb{N}_m] \in \mathbb{R}^m$, define bounded linear functionals $\ell_j, j \in \mathbb{N}_m$, on $\mathbb{X}_n$ by

$$\ell_j(x) := c_j,$$

which leads to $\ell_j(x_i) = \delta_{i,j}$, $i, j \in \mathbb{N}_m$ and (2.87). We then extend the functionals to the entire space $\mathbb{X}$ by the equation

$$\langle x, \ell_j \rangle = \langle \mathcal{P}_n x, \ell_j \rangle, \quad \text{for all } x \in \mathbb{X}.$$

It can easily be verified that (2.87) is valid for the collection of extended functionals, and this is unique for the requirements.

From equation (2.87) comes the formula for the adjoint projection $\mathcal{P}_n^*$, namely, for each $\ell \in \mathbb{X}^*$

$$\mathcal{P}_n^* \ell = \sum_{j \in \mathbb{N}_m} \langle x_j, \ell \rangle \, \ell_j. \tag{2.88}$$

So, we see that $\mathbb{Y}_n := R(\mathcal{P}_n^*) = \text{span}\,\{\ell_j : j \in \mathbb{N}_m\}$ and $\dim \mathbb{Y}_n = \dim \mathbb{X}_n$. By definition, for each $x \in \mathbb{X}$ and $\ell \in \mathbb{X}^*$ we have that $\langle \mathcal{P}_n x, \ell \rangle = \langle x, \mathcal{P}_n^* \ell \rangle$. It follows that $\lim_{n\to\infty} \mathcal{P}_n^* \ell = \ell$ for all $\ell \in \mathbb{X}^*$ in the weak* topology on $\mathbb{X}^*$ if and only if for every $x \in \mathbb{X}$, $\lim_{n\to\infty} \mathcal{P}_n x = x$ in the weak topology on $\mathbb{X}$ ([276], p. 111).

The next lemma prepares us for a result from [47] (p. 14) which provides a sufficient condition to ensure that $\mathcal{P}_n^* \xrightarrow{s} \mathcal{I}$ in the norm topology on $\mathbb{X}^*$. Recall that a normed linear space $\mathbb{X}$ is said to be *reflexive* if $\mathbb{X}$ may be identified with its second dual.

Lemma 2.41 *If $\mathbb{X}$ is reflexive and for every $x \in \mathbb{X}$, $\lim_{n\to\infty} \mathcal{P}_n x = x$ in the weak topology on $\mathbb{X}$ then* $\text{span}\{\cup_{n\in\mathbb{N}} R(\mathcal{P}_n^*)\}$ *is dense in $\mathbb{X}^*$ in the norm topology.*

Proof We consider the subspace of $\mathbb{X}^*$ given by

$$\mathbb{W} := \text{span}\left\{ \bigcup_{n\in\mathbb{N}} R(\mathcal{P}_n^*) \right\}$$

and choose any $F \in \mathbb{W}^\perp$. Since $\mathbb{X}$ is reflexive there is an $x \in \mathbb{X}$ such that for all $\ell \in \mathbb{X}^*$ we have that $F(\ell) = \langle x, \ell \rangle$. Choose any $m \in \mathbb{N}$ and observe that $0 = F(\mathcal{P}_m^* \ell) = \langle x, \mathcal{P}_m^* \ell \rangle = \langle \mathcal{P}_m x, \ell \rangle$. We now let $m \to \infty$ and use our hypothesis to conclude that $\langle x, \ell \rangle = 0$. Since $\ell \in \mathbb{X}^*$ is arbitrary we conclude that $x=0$, thereby establishing that $F=0$. In other words, we have confirmed that $\mathbb{X}^{\perp}=\{0\}$ and so $\mathbb{W}$ is dense in $\mathbb{X}^*$ (see Corollary A.35 in the Appendix). □

From this fact follows the next result.

Proposition 2.42 *If $\mathbb{X}$ is reflexive and for every $x \in \mathbb{X}$, $\lim_{n\to\infty} \mathcal{P}_n x = x$ in the weak topology on $\mathbb{X}$ and the projection $\mathcal{P}_n$ satisfies (2.86), then $\mathcal{P}_n^* \xrightarrow{s} \mathcal{I}$ in the norm topology on $\mathbb{X}^*$.*

Proof According to equation (2.86) we conclude that $\mathcal{P}_m^*\mathcal{P}_n^*=\mathcal{P}_n^*$, which implies that the spaces $\mathbb{Y}_n := R(\mathcal{P}_n^*)$ are nested for $n\in\mathbb{N}$. Moreover, our hypothesis ensures that for each $x \in \mathbb{X}$ the set $\{\mathcal{P}_n x : n\in\mathbb{N}\}$ is weakly bounded. Therefore, Corollary A.39 in the Appendix implies that it is norm bounded and so by the uniform boundedness principle (see Theorem A.25) the set $\{\|\mathcal{P}_n\| : n \in \mathbb{N}\}$ is bounded. But, we know that $\|\mathcal{P}_n\| = \|\mathcal{P}_n^*\|$ and therefore the claim made in this proposition follows from Proposition 2.40 and Lemma 2.41 above. □

Let us return to condition (2.80), which is usually more readily verifiable in applications. First, we comment on the relationship of (2.80) to (2.79). Our comment here is based on the following norm inequalities, the first being

$$\|\mathcal{A}_n^{-1}\mathcal{P}_n\mathcal{A}\| \le \|\mathcal{A}_n^{-1}\| \cdot \|\mathcal{P}_n\| \cdot \|\mathcal{A}\|,$$

which implies that (2.80) ensures (2.79) when $\{\|\mathcal{P}_n\| : n \in \mathbb{N}\}$ is bounded (where, of course, the constants in (2.79) and (2.80) will be different). Moreover, when $\mathcal{A} : \mathbb{X} \to \mathbb{Y}$ is one to one and onto then

$$\|\mathcal{A}_n^{-1}\mathcal{P}_n\| \le \|\mathcal{A}_n^{-1}\mathcal{P}_n\mathcal{A}\| \cdot \|\mathcal{A}^{-1}\|$$

and so, recalling inequality (2.81), we obtain the inequality

$$\|\mathcal{A}_n^{-1}\| \le \|\mathcal{A}_n^{-1}\mathcal{P}_n\mathcal{A}\|\|\mathcal{A}^{-1}\|,$$

which demonstrates that inequality (2.79) implies (2.80) at least when $\mathcal{A}$ is one to one and onto. We formalize this in the next proposition.

Proposition 2.43 *Let $\mathcal{A} : \mathbb{X} \to \mathbb{Y}$ be a bounded linear operator, $\{\mathbb{X}_n : n \in \mathbb{N}\}$ and $\{\mathbb{Y}_n : n \in \mathbb{N}\}$ finite-dimensional subspaces of $\mathbb{X}$ and $\mathbb{Y}$, respectively, and $\mathcal{P}_n : \mathbb{Y} \to \mathbb{Y}_n$ a projection. If $\{\|\mathcal{P}_n\| : n \in \mathbb{N}\}$ and $\{\|\mathcal{A}_n^{-1}\| : n \in \mathbb{N}\}$ are bounded, then so is $\{\|\mathcal{A}_n^{-1}\mathcal{P}_n\mathcal{A}\| : n \in \mathbb{N}\}$. If $\mathcal{A}$ is one to one and onto and $\{\|\mathcal{A}_n^{-1}\mathcal{P}_n\mathcal{A}\| : n \in \mathbb{N}\}$ is bounded, then $\{\|\mathcal{A}_n^{-1}\| : n \in \mathbb{N}\}$ is bounded too.*

Our final comments concerning projection methods demonstrate that under certain circumstances the existence of a unique solution of the projection equation (2.77) implies the same for the operator equation (2.63). In the next lemma, we provide conditions on the projection method which imply that $\mathcal{A}$ is one to one.

Lemma 2.44 *Let $\{\mathbb{X}_n : n \in \mathbb{N}\}$ and $\{\mathbb{Y}_n : n \in \mathbb{N}\}$ be sequences of finite-dimensional subspaces of $\mathbb{X}$ and $\mathbb{Y}$, respectively, and $\mathcal{P}_n : \mathbb{Y} \to \mathbb{Y}_n$, $\mathcal{Q}_n : \mathbb{X} \to \mathbb{X}_n$ projections. If there is a $q \in \mathbb{N}$ and a positive constant $c > 0$ such that for any $n \geq q$ both $\|\mathcal{P}_n\| \leq c$ and $\|\mathcal{A}_n^{-1}\| \leq c$ hold and also $\mathcal{Q}_n \xrightarrow{s} \mathcal{I}$, then $\mathcal{A}$ is one to one.*

Proof If $u \in \mathbb{X}$ satisfies $\mathcal{A}u = 0$ then for $n \geq q$ we have, since $\mathcal{A}_n := \mathcal{P}_n\mathcal{A}|_{\mathbb{X}_n}$, that

$$\begin{aligned} \|\mathcal{Q}_n u\|_{\mathbb{X}} &\leq c\|\mathcal{A}_n\mathcal{Q}_n u\|_{\mathbb{Y}} \\ &= c\|\mathcal{P}_n\mathcal{A}\mathcal{Q}_n u - \mathcal{P}_n\mathcal{A}u\|_{\mathbb{Y}} \\ &\leq c^2\|\mathcal{A}\|\|\mathcal{Q}_n u - u\|_{\mathbb{X}}. \end{aligned}$$

Letting $n \to \infty$ on both sides of this inequality we conclude that $u = 0$. □

We now present a similar result that implies the operator $\mathcal{A}$ is onto.

Lemma 2.45 *Let $\mathbb{X}$ and $\mathbb{Y}$ be Banach spaces with $\mathbb{X}$ reflexive and $\mathcal{A} \in \mathcal{B}(\mathbb{X}, \mathbb{Y})$. Let $\mathcal{P}_n : \mathbb{Y} \to \mathbb{Y}_n$ be a projection such that $\mathcal{P}_n \xrightarrow{s} \mathcal{I}$ on $\mathbb{Y}$ and the sequence $\{\|\mathcal{A}_n^{-1}\| : n \in \mathbb{N}\}$ is bounded. If the sequence of subsets $\{R(\mathcal{P}_n^*) : n \in \mathbb{N}\}$ is nested, then $\mathcal{A}$ is onto.*

Proof Choose any $f \in \mathbb{Y}$ and recall that $u_n := \mathcal{A}_n^{-1}\mathcal{P}_n f$. Since $\mathcal{P}_n \xrightarrow{s} \mathcal{I}$ we conclude by the uniform boundedness principle (see Theorem A.25) that the sequence $\{\|\mathcal{P}_n\| : n \in \mathbb{N}\}$ is bounded and so we obtain that $\{\|u_n\| : n \in \mathbb{N}\}$ is also bounded. Moreover, our hypothesis that $\{R(\mathcal{P}_n^*) : n \in \mathbb{N}\}$ is nested guarantees, by the proof of Proposition 2.42, that $\mathcal{P}_n^* \xrightarrow{s} \mathcal{I}$ on $\mathbb{Y}^*$. Now, since $\mathbb{X}$ is reflexive we can extract a subsequence $\{u_{n_k} : k \in \mathbb{N}\}$ which converges weakly to an element $u \in \mathbb{X}$ (see, for example, [276], p. 126). We show that $\mathcal{A}u = f$. To this end, we first observe for any $\ell \in \mathbb{Y}^*$ that

$$\lim_{k\to\infty} \langle \mathcal{A}u_{n_k}, \ell\rangle = \lim_{k\to\infty} \langle u_{n_k}, \mathcal{A}^*\ell\rangle = \langle u, \mathcal{A}^*\ell\rangle = \langle \mathcal{A}u, \ell\rangle,$$

that is, $\lim_{k\to\infty} \mathcal{A}u_{n_k} = \mathcal{A}u$ weakly in $\mathbb{Y}$. Therefore, the right-hand side of the inequality

$$|\langle \mathcal{A}u_{n_k}, \mathcal{P}_{n_k}^*\ell\rangle - \langle \mathcal{A}u, \ell\rangle| \leq \|\mathcal{A}\|\|u_{n_k}\|\|\mathcal{P}_{n_k}^*\ell - \ell\| + |\langle \mathcal{A}u_{n_k} - \mathcal{A}u, \ell\rangle|$$

goes to zero as $n \to \infty$ and so with the formula $\mathcal{A}_n u_n = \mathcal{P}_n\mathcal{A}u_n$ we obtain that $\lim_{k\to\infty} \mathcal{A}_{n_k}u_{n_k} = \mathcal{A}u$ weakly in $\mathbb{Y}$. Moreover, by definition, for all $n \in \mathbb{N}$ there holds the equation $\mathcal{A}_n u_n = \mathcal{P}_n f$ and also, by hypothesis, $\lim_{n\to\infty} \mathcal{P}_n f = f$ (in norm), from which we conclude that $\lim_{k\to\infty} \mathcal{A}_{n_k} u_{n_k} = f$ (in norm). Hence, indeed we obtain the desired conclusion that $\mathcal{A}u = f$. □

Now we turn our attention to a discussion of the numerical stability of the projection methods. The numerical *stability* of the approximate solution problem concerns how close the approximate solution of projection equation (2.77) is to that of a perturbed equation of the form

$$(\mathcal{A}_n + \mathcal{E}_n)\tilde{u}_n = \mathcal{P}_n f + g_n, \tag{2.89}$$

where $\mathcal{E}_n \in \mathcal{B}(\mathbb{X}_n, \mathbb{Y}_n)$ is a linear operator which affects a perturbation of $\mathcal{A}_n$ and $g_n \in \mathbb{Y}_n$ affects a perturbation of $\mathcal{P}_n f$, $n \in \mathbb{N}$.

We begin with a formal definition of stability.

Definition 2.46 The projection method is said to be stable if there are non-negative constants μ and ν, a positive constant δ and a positive integer q such that for any $n \geq q$, the operator $\mathcal{A}_n$ is invertible, and for any vector $g_n \in \mathbb{Y}_n$ and any linear operator $\mathcal{E}_n \in \mathcal{B}(\mathbb{X}_n, \mathbb{Y}_n)$, with $\|\mathcal{E}_n\| \leq \delta$, the perturbed equation (2.89) always has a unique solution $\tilde{u}_n \in \mathbb{X}_n$ satisfying the inequality

$$\|\tilde{u}_n - u_n\| \leq \mu \|\mathcal{E}_n\| \|u_n\| + \nu \|g_n\|. \tag{2.90}$$

We next characterize the stability of the projection method.

Theorem 2.47 *If $\mathcal{A} \in \mathcal{B}(\mathbb{X}, \mathbb{Y})$ then the projection method is stable if and only if inequality (2.80) holds.*

Proof Suppose that condition (2.80) is satisfied for $n \geq q$. Then for any $f \in \mathcal{A}(\mathbb{X})$ and $n \geq q$, the projection equation (2.77) has the unique solution $u_n \in \mathbb{X}_n$. If $n \geq q$ and the perturbation satisfies the norm inequality $\|\mathcal{E}_n\| \leq \frac{c}{2}$, then for any $x \in \mathbb{X}_n$,

$$\|(\mathcal{A}_n + \mathcal{E}_n)x\| \geq \frac{c}{2}\|x\|. \tag{2.91}$$

Hence, for any $g_n \in \mathbb{Y}_n$ and $n \geq q$, the perturbed equation (2.89) has a unique solution $\tilde{u}_n \in \mathbb{X}_n$ and it gives us the formula

$$\tilde{u}_n - u_n = (\mathcal{A}_n + \mathcal{E}_n)^{-1}(g_n - \mathcal{E}_n u_n).$$

Hence, with inequality (2.91) we get the stability estimate

$$\|\tilde{u}_n - u_n\| \leq \frac{2}{c}(\|\mathcal{E}_n\| \|u_n\| + \|g_n\|).$$

Conversely, suppose that the projection method is stable. In this case we choose the perturbation operator to be $\mathcal{E}_n = 0$. Then, for any $f \in \mathcal{A}(\mathbb{X})$ and $g_n \in \mathbb{Y}_n$, when $n \geq N$, the projection equation (2.77) and its perturbed equation (2.89) have unique solutions $u_n, \tilde{u}_n \in \mathbb{X}_n$, respectively. We now let $v_n = \tilde{u}_n - u_n$ and observe for $n \geq q$ that $\mathcal{A}_n v_n = g_n$, and so the stability inequality (2.90) gives us the desired inequality

$$\|\mathcal{A}_n^{-1} g_n\| = \|v_n\| \leq \nu \|g_n\|.$$

□

We now introduce an important concept in connection with the actual behavior of approximate methods, that is, the condition number of a linear operator, which is used to indicate how sensitive the solution of an equation may be to small relative changes in the input data.

Definition 2.48 Let $\mathbb{X}$ and $\mathbb{Y}$ be Banach spaces and $\mathcal{A} : \mathbb{X} \to \mathbb{Y}$ be a bounded linear operator with bounded inverse $\mathcal{A}^{-1} : \mathbb{Y} \to \mathbb{X}$. The condition number of $\mathcal{A}$ is defined as

$$\operatorname{cond}(\mathcal{A}) := \|\mathcal{A}\| \, \|\mathcal{A}^{-1}\|.$$

It is clear that the inequality $\operatorname{cond}(\mathcal{A}) \geq 1$ always holds. The following proposition shows that the condition number is a suitable tool for measuring the stability.

Proposition 2.49 *Suppose that $\mathbb{X}$ and $\mathbb{Y}$ are Banach spaces and $\mathcal{A} \in \mathcal{B}(\mathbb{X}, \mathbb{Y})$ has bounded inverse $\mathcal{A}^{-1}$. Let $\delta\mathcal{A} \in \mathcal{B}(\mathbb{X}, \mathbb{Y})$, $\delta f \in \mathbb{Y}$ be linear perturbations of $\mathcal{A}$ and f, and $u \in \mathbb{X}$, $u + \delta u \in \mathbb{X}$ be solutions of*

$$\mathcal{A}u = f \tag{2.92}$$

and

$$(\mathcal{A} + \delta\mathcal{A})(u + \delta u) = f + \delta f, \tag{2.93}$$

respectively. If $\|\delta\mathcal{A}\| < 1/\|\mathcal{A}^{-1}\|$ and $f \neq 0$, then

$$\frac{\|\delta u\|}{\|u\|} \leq \frac{\operatorname{cond}(\mathcal{A})}{1 - \|\mathcal{A}^{-1}\| \, \|\delta\mathcal{A}\|} \left(\frac{\|\delta\mathcal{A}\|}{\|\mathcal{A}\|} + \frac{\|\delta f\|}{\|f\|} \right).$$

Proof It follows from (2.92) and (2.93) that

$$(\mathcal{A} + \delta\mathcal{A})(u + \delta u) = \mathcal{A}u + \delta f,$$

which leads to the equation

$$\delta u = (\mathcal{I} + \mathcal{A}^{-1}\delta\mathcal{A})^{-1}\mathcal{A}^{-1}(\delta f - \delta\mathcal{A}u). \tag{2.94}$$

The inequality $\|\delta\mathcal{A}\| < 1/\|\mathcal{A}^{-1}\|$ ensures the existence of the linear operator $(\mathcal{I} + \mathcal{A}^{-1}\delta\mathcal{A})^{-1}$ and also the estimate

$$\|(\mathcal{I} + \mathcal{A}^{-1}\delta\mathcal{A})^{-1}\| \le \frac{1}{1 - \|\mathcal{A}^{-1}\|\,\|\delta\mathcal{A}\|}.$$

From this with (2.94) we conclude that

$$\begin{aligned}\frac{\|\delta u\|}{\|u\|} &\le \frac{\|\mathcal{A}^{-1}\|}{1 - \|\mathcal{A}^{-1}\|\,\|\delta\mathcal{A}\|}\left(\frac{\|\delta f\|}{\|u\|} + \|\delta\mathcal{A}\|\right)\\ &\le \frac{\|\mathcal{A}^{-1}\|\,\|\mathcal{A}\|}{1 - \|\mathcal{A}^{-1}\|\,\|\delta\mathcal{A}\|}\left(\frac{\|\delta f\|}{\|f\|} + \frac{\|\delta\mathcal{A}\|}{\|\mathcal{A}\|}\right),\end{aligned}$$

completing the proof. □

A simple fact for the condition number $\operatorname{cond}(\mathcal{A}_n) = \|\mathcal{A}_n\|\,\|\mathcal{A}_n^{-1}\|$ of the projection equation (2.77) is that, if $\mathcal{A}_n \xrightarrow{u} \mathcal{A}$ and $\mathcal{A}_n^{-1} \xrightarrow{u} \mathcal{A}_n^{-1}$, then

$$\lim_{n\to\infty} \operatorname{cond}(\mathcal{A}_n) = \operatorname{cond}(\mathcal{A}).$$

We remark that if the projection method is convergent, then for some positive constant c the inequality

$$\|\mathcal{A}_n^{-1}\| \le c\|\mathcal{A}^{-1}\|$$

holds. In fact, using Theorem 2.33 we have that

$$\begin{aligned}\|\mathcal{A}_n^{-1}\| &= \sup\{\|\mathcal{A}_n^{-1}\mathcal{P}_n y\| : y \in \mathbb{Y}_n, \|y\| = 1\}\\ &\le \sup\{\|\mathcal{A}_n^{-1}\mathcal{P}_n y\| : y \in \mathbb{Y}, \|y\| = 1\}\\ &\le \sup\{\|[\mathcal{A}_n^{-1}\mathcal{P}_n\mathcal{A}]\mathcal{A}^{-1}y\| : y \in \mathbb{Y}, \|y\| = 1\}\\ &\le c\|\mathcal{A}^{-1}\|.\end{aligned}$$

2.2.4 An abstract framework for second-kind operator equations

In this subsection, we change our perspectives somewhat to a context closer to the specific applications that we have in mind in later chapters. Specifically, in this subsection our operator $\mathcal{A}$ will have the form $\mathcal{I} - \mathcal{K}$, where $\mathcal{K}$ is compact. Projection methods for this case are well developed in the literature. Specifically, it is well known that the theory of collectively compact operators due to P. Anselone (as presented in [6, 15]) provides us with a convenient abstract setting for the analysis of many numerical schemes associated with the Fredholm operator $\mathcal{I} - \mathcal{K}$. This theory generally requires that the sequence of nested spaces $\{\mathbb{X}_n : n \in \mathbb{N}\}$ is dense in $\mathbb{X}$. For our purposes later, it is advantageous to improve upon this hypothesis.

We have in mind several concrete projection methods which will be introduced later. These include discrete Galerkin, Petrov–Galerkin, collocation and quadrature methods. We shall describe them collectively in the following context, which differs somewhat from the point of view of previous sections.

We begin with a Banach space $\mathbb{X}$ and a subspace $\mathbb{V}$ of it. Let $\mathcal{K} \in \mathcal{B}(\mathbb{X}, \mathbb{V})$ be a compact operator and consider the Fredholm equation of the second kind

$$u - \mathcal{K}u = f. \tag{2.95}$$

By the Fredholm alternative (see Theorem A.48) this equation has a unique solution for all $f \in \mathbb{V}$ if and only if the null space of $\mathcal{I} - \mathcal{K}$ is $\{0\}$, that is, as long as one is not an eigenvalue of $\mathcal{K}$. We always assume that this condition holds.

To set up our approximation methods for (2.95), unlike in the discussions earlier, we need two sequences of operators $\{\mathcal{K}_n : n \in \mathbb{N}\} \subseteq \mathcal{B}(\mathbb{X}, \mathbb{V})$ and $\{\mathcal{P}_n : n \in \mathbb{N}\} \subseteq \mathcal{B}(\mathbb{X}, \mathbb{U})$ where we require that $\mathbb{V} \subseteq \mathbb{U} \subseteq \mathbb{X}$. As before, $\mathcal{P}_n$ will approximate the identity and in the present context $\mathcal{K}_n$ will approximate the operator $\mathcal{K}$. The exact sense for which this is required shall be explained below. Postponing this issue for the moment, we associate with these two sequences of operators the approximation scheme

$$(\mathcal{I} - \mathcal{P}_n\mathcal{K}_n)u_n = \mathcal{P}_n f \tag{2.96}$$

for solving (2.95).

For the analysis of the convergence properties of (2.96) we are led to consider the existence and uniform boundedness for $n \in \mathbb{N}$ of the inverse of the operator

$$\mathcal{A}_n := \mathcal{I} - \mathcal{P}_n\mathcal{K}_n.$$

Moreover, in this section we also prepare the tools to study the phenomenon of *superconvergence*. This means that we shall approximate the solution of equation (2.96) by the function

$$\tilde{u}_n := f + \mathcal{K}_n u_n. \tag{2.97}$$

We refer to $\tilde{u}_n$ as the iterated approximation to (2.96), which is also called the Sloan iterate, and it follows directly that $\tilde{u}_n$ satisfies the equation

$$(\mathcal{I} - \mathcal{K}_n\mathcal{P}_n)\tilde{u}_n = f. \tag{2.98}$$

Therefore, we also consider in this section the existence and uniform boundedness for $n \in \mathbb{N}$ of the inverse of the operators

$$\tilde{\mathcal{A}}_n := \mathcal{I} - \mathcal{K}_n\mathcal{P}_n.$$

Our analysis of the linear operators $\mathcal{A}_n$ and $\tilde{\mathcal{A}}_n$ requires several assumptions on $\mathcal{K}_n$ and $\mathcal{P}_n$. To prepare for these conditions we introduce the following terminology.

Definition 2.50 We say that a sequence of operators $\{\mathcal{T}_n : n \in \mathbb{N}\} \subseteq \mathcal{B}(\mathbb{X}, \mathbb{Y})$ converges pointwise to an operator $\mathcal{T} \in \mathcal{B}(\mathbb{X}, \mathbb{Y})$ on the set $S \subseteq \mathbb{X}$ provided that for each $x \in S$, we have $\lim_{n\to\infty} \|\mathcal{T}_n x - \mathcal{T}x\| = 0$. Notationally, we indicate this by $\mathcal{T}_n \xrightarrow{s} \mathcal{T}$ on S. Similarly, we say that the sequence of operators $\{\mathcal{T}_n : n \in \mathbb{N}\}$ converges to the operator $\mathcal{T}$ uniformly on S provided that $\lim_{n\to\infty} \sup\{\|\mathcal{T}_n x - \mathcal{T}x\| : x \in S\} = 0$ and indicate this by $\mathcal{T}_n \xrightarrow{u} \mathcal{T}$ on S.

Clearly, when S is the unit ball in $\mathbb{X}$ and $\mathcal{T}_n \xrightarrow{u} \mathcal{T}$ on S this means that $\mathcal{T}_n \xrightarrow{u} \mathcal{T}$ on $\mathbb{X}$.

Lemma 2.51 *Let $\mathbb{X}$ be a Banach space and S a relatively compact subset of $\mathbb{X}$. If the sequence of operators $\{\mathcal{T}_n : n \in \mathbb{N}\} \subseteq \mathcal{B}(\mathbb{X})$ has uniformly bounded operator norms and $\mathcal{T}_n \xrightarrow{s} \mathcal{T}$, then $\mathcal{T}_n \xrightarrow{u} \mathcal{T}$ on S.*

Proof Since the set S is compact it is totally bounded (Theorem A.7) and so, for a given $\epsilon > 0$, there is a finite set $W \subseteq S$ such that, for each $x \in \mathbb{S}$, there is a $w \in W$ with $\|x - w\| < \epsilon$. Since W is finite, by hypothesis, there is a $q \in \mathbb{N}$ such that for any $v \in W$ and $n \geq q$ we have $\|\mathcal{T}_n v - \mathcal{T}v\| \leq \epsilon$. In particular, this inequality holds for the choice $v = w$. By hypothesis, there is a constant $c > 0$ such that for any $n \in \mathbb{N}$ we have $\|\mathcal{T}_n\| \leq c$. We now estimate the error uniformly for all $x \in S$ when $n \geq q$,

$$\begin{aligned}\|\mathcal{T}_n x - \mathcal{T}x\| &\leq \|(\mathcal{T}_n - \mathcal{T})(x - w)\| + \|\mathcal{T}_n w - \mathcal{T}w\| \\ &\leq (c + \|\mathcal{T}\| + 1)\epsilon.\end{aligned}$$

□

Let us now return to our setup for approximate schemes for solving Fredholm equations. We list below several conditions that we shall assume and investigate their consequences.

(H-1) The set of operators $\{\mathcal{K}_n : n \in \mathbb{N}\} \subseteq \mathcal{B}(\mathbb{X}, \mathbb{V})$ is *collectively compact*, that is, for any bounded set $\mathbb{B} \subseteq \mathbb{X}$ the set $\cup_{n\in\mathbb{N}}\mathcal{K}_n(\mathbb{B})$ is relatively compact in $\mathbb{V}$.

(H-2) $\mathcal{K}_n \xrightarrow{s} \mathcal{K}$ on $\mathbb{U}$.

(H-3) The set of operators $\{\mathcal{P}_n : n \in \mathbb{N}\} \subseteq \mathcal{B}(\mathbb{X}, \mathbb{U})$ is compact with norms which are uniformly bounded for $n \in \mathbb{N}$.

(H-4) $\mathcal{P}_n \xrightarrow{s} \mathcal{I}$ on $\mathbb{V}$.

As a first step, we modify Proposition 1.7 in [6] to fit our circumstances and obtain the following fact.

Lemma 2.52 *If conditions (H-1)–(H-4) hold, then*

(i) $(\mathcal{P}_n - \mathcal{I})\mathcal{K}_n \xrightarrow{u} 0$ *on* $\mathbb{X}$,
(ii) $(\mathcal{K}_n - \mathcal{K})\mathcal{P}_n\mathcal{K}_n \xrightarrow{u} 0$ *on* $\mathbb{X}$,
(iii) $(\mathcal{K}_n\mathcal{P}_n - \mathcal{K})\mathcal{K}_n\mathcal{P}_n \xrightarrow{u} 0$ *on* $\mathbb{X}$.

Proof (i) Let B denote the closed unit ball in $\mathbb{X}$, that is, $B := \{x : x \in \mathbb{X}, \|x\| \leq 1\}$ and also set $G := \{\mathcal{K}_n x : x \in B,\ n \in \mathbb{N}\}$. Condition (H-1) implies that G is a relatively compact set in $\mathbb{V}$, while hypotheses (H-3) and (H-4) coupled with Lemma 2.51 establish that $\mathcal{P}_n \xrightarrow{u} \mathcal{I}$ on G. Consequently, the inequality

$$\|(\mathcal{P}_n - \mathcal{I})\mathcal{K}_n\| = \sup\{\|(\mathcal{P}_n - \mathcal{I})\mathcal{K}_n x\| : x \in B\} \leq \sup\{\|(\mathcal{P}_n - \mathcal{I})x\| : x \in G\} \tag{2.99}$$

establishes (i).

(ii) For any $x \in \mathbb{V}$, it follows from (H-4) that $\{\mathcal{P}_n x : n \in \mathbb{N}\}$ is a relatively compact subset of $\mathbb{X}$. Therefore, by Lemma 2.51 and the hypotheses (H-1) and (H-2), we conclude that $(\mathcal{K}_n - \mathcal{K})\mathcal{P}_n \xrightarrow{s} 0$ on $\mathbb{V}$. Moreover, using the inequality

$$\begin{aligned}\|(\mathcal{K}_n - \mathcal{K})\mathcal{P}_n\mathcal{K}_n\| &= \sup\{\|(\mathcal{K}_n - \mathcal{K})\mathcal{P}_n\mathcal{K}_n y\| : y \in B\} \\ &\leq \sup\{\|(\mathcal{K}_n - \mathcal{K})\mathcal{P}_n x\| : x \in G\},\end{aligned} \tag{2.100}$$

and specializing Lemma 2.51 to the choice $\mathcal{T} = 0$, $\mathcal{T}_n = (\mathcal{K}_n - \mathcal{K})\mathcal{P}_n$ and the choice $S = G$ we conclude the validity of (ii).

(iii) Hypotheses (H-1) and (H-3) guarantee that

$$G' := \{\mathcal{K}_n\mathcal{P}_n x : x \in B, n \in \mathbb{N}\}$$

is a relatively compact subset of $\mathbb{V}$. Moreover, from the equation

$$\mathcal{K}_n\mathcal{P}_n - \mathcal{K} = (\mathcal{K}_n\mathcal{P}_n - \mathcal{K}\mathcal{P}_n) + (\mathcal{K}\mathcal{P}_n - \mathcal{K}), \tag{2.101}$$

statement (ii) and (H-4) we obtain that $\mathcal{K}_n\mathcal{P}_n - \mathcal{K} \xrightarrow{s} 0$ on $\mathbb{V}$. Thus, statement (iii) follows directly from equation (2.101) and the relative compactness of the set G'. □

We next study the existence of the inverse operator of $\mathcal{A}_n$ and $\tilde{\mathcal{A}}_n$. For this purpose, we recall a useful result about the existence and boundedness of inverse operators.

Lemma 2.53 *If* $\mathbb{X}$ *is a normed linear space with* $\mathcal{S}$ *and* $\mathcal{E}$ *in* $\mathcal{B}(\mathbb{X})$ *such that* $\mathcal{S}^{-1}$ *exists as a bounded linear operator on* $\mathcal{S}(\mathbb{X})$ *and* $\|\mathcal{E}\| < \|\mathcal{S}^{-1}\|^{-1}$, *then* *the linear operator* $\mathcal{T} := \mathcal{S} - \mathcal{E}$ *has an inverse* $\mathcal{T}^{-1}$ *as a bounded linear operator on* $\mathcal{T}(\mathbb{X})$ *and has the property that*

$$\|\mathcal{T}^{-1}\| \leq \frac{1}{\|\mathcal{S}^{-1}\|^{-1} - \|\mathcal{E}\|}.$$

Proof For any $u \in \mathbb{X}$ we have that

$$\mathcal{S}u = \mathcal{T}u + \mathcal{E}u$$

and so

$$\|\mathcal{S}u\| \leq \|\mathcal{T}u\| + \|\mathcal{E}\|\|u\|.$$

Thus

$$(\|\mathcal{S}^{-1}\|^{-1} - \|\mathcal{E}\|)\|u\| \leq \|\mathcal{S}u\| - \|\mathcal{E}\|\|u\| \leq \|\mathcal{T}u\|,$$

from which the desired result follows. □

We are now ready to prove the main result of this section.

Theorem 2.54 *If* $\mathcal{K} \in \mathcal{B}(\mathbb{X},\mathbb{X})$ *is a compact operator not having one as an eigenvalue and conditions (H-1)–(H-4) hold, then there exists a positive integer* q *such that for all* $n \geq q$*, both* $(\mathcal{I} - \mathcal{P}_n\mathcal{K}_n)^{-1}$ *and* $(\mathcal{I} - \mathcal{K}_n\mathcal{P}_n)^{-1}$ *are in* $\mathcal{B}(\mathbb{X})$ *and have norms which are uniformly bounded. Moreover, if* u, u_n *and* $\tilde{u}_n$ *are the solutions of equations (2.95), (2.96) and (2.97), respectively, and a constant* $p > 0$ *is chosen so that for all* $n \in \mathbb{N}$, $\|\mathcal{P}_n\| \leq p$*, then for all* $n \in \mathbb{N}$

$$\|u - u_n\| \leq c(\|u - \mathcal{P}_n u\| + p\|\mathcal{K}u - \mathcal{K}_n u\|) \tag{2.102}$$

and

$$\|u - \tilde{u}_n\| \leq c(\|\mathcal{K}(\mathcal{I} - \mathcal{P}_n)u\| + \|(\mathcal{K} - \mathcal{K}_n)\mathcal{P}_n u\|). \tag{2.103}$$

Proof We first note that a straightforward computation leads to the formulas

$$[\mathcal{I} + (\mathcal{I} - \mathcal{K})^{-1}\mathcal{K}_n](\mathcal{I} - \mathcal{P}_n\mathcal{K}_n) = \\ \mathcal{I} - (\mathcal{I} - \mathcal{K})^{-1}[(\mathcal{P}_n - \mathcal{I})\mathcal{K}_n + (\mathcal{K}_n - \mathcal{K})\mathcal{P}_n\mathcal{K}_n]$$

and

$$[\mathcal{I} + (\mathcal{I} - \mathcal{K})^{-1}\mathcal{K}_n\mathcal{P}_n](\mathcal{I} - \mathcal{K}_n\mathcal{P}_n) = \mathcal{I} - (\mathcal{I} - \mathcal{K})^{-1}(\mathcal{K}_n\mathcal{P}_n - \mathcal{K})\mathcal{K}_n\mathcal{P}_n.$$

By Lemma 2.52, there exists a $q > 0$ such that for all $n \geq q$, we have

$$\Delta_1 := \|(\mathcal{I} - \mathcal{K})^{-1}[(\mathcal{P}_n - \mathcal{I})\mathcal{K}_n + (\mathcal{K}_n - \mathcal{K})\mathcal{P}_n\mathcal{K}_n]\| \leq \frac{1}{2}$$

and

$$\Delta_2 := \|(\mathcal{I}-\mathcal{K})^{-1}(\mathcal{K}_n\mathcal{P}_n-\mathcal{K})\mathcal{K}_n\mathcal{P}_n\| \le \frac{1}{2}.$$

It follows from Lemma 2.53 for all $n \ge q$ that the inverse operators $(\mathcal{I}-\mathcal{P}_n\mathcal{K}_n)^{-1}$ and $(\mathcal{I}-\mathcal{K}_n\mathcal{P}_n)^{-1}$ exist and there the inequalities

$$\|(\mathcal{I}-\mathcal{P}_n\mathcal{K}_n)^{-1}\| \le \frac{1}{1-\Delta_1}(1+\|(\mathcal{I}-\mathcal{K})^{-1}\|\|\mathcal{K}_n\|)$$

and

$$\|(\mathcal{I}-\mathcal{K}_n\mathcal{P}_n)^{-1}\| \le \frac{1}{1-\Delta_2}(1+p\|(\mathcal{I}-\mathcal{K})^{-1}\|\|\mathcal{K}_n\|)$$

hold.

Since the set of operators $\{\mathcal{K}_n : n \in \mathbb{N}\}$ is collectively compact, it follows that the norms $\|\mathcal{K}_n\|$ are uniformly bounded for $n \in \mathbb{N}$, and so by the above inequalities the norms of both $(\mathcal{I}-\mathcal{P}_n\mathcal{K}_n)^{-1}$ and $(\mathcal{I}-\mathcal{K}_n\mathcal{P}_n)^{-1}$ are also uniformly bounded for $n \in \mathbb{N}$. Therefore, equations (2.96) and (2.98) have unique solutions for every $f \in \mathbb{X}$.

It remains to prove the estimates (2.102) and (2.103). To this end, we note from equations (2.95), (2.96) and (2.98) that

$$(\mathcal{I}-\mathcal{P}_n\mathcal{K}_n)u_n = \mathcal{P}_n(\mathcal{I}-\mathcal{K})u$$

and

$$(\mathcal{I}-\mathcal{K}_n\mathcal{P}_n)\tilde{u}_n = (\mathcal{I}-\mathcal{K})u.$$

Using these equations, we obtain that

$$(\mathcal{I}-\mathcal{P}_n\mathcal{K}_n)(u-u_n) = (u-\mathcal{P}_nu)+\mathcal{P}_n(\mathcal{K}u-\mathcal{K}_nu)$$

and

$$(\mathcal{I}-\mathcal{K}_n\mathcal{P}_n)(u-\tilde{u}_n) = \mathcal{K}u-\mathcal{K}_n\mathcal{P}_nu = \mathcal{K}(u-\mathcal{P}_nu)+(\mathcal{K}-\mathcal{K}_n)\mathcal{P}_nu.$$

Therefore, from what we have already proved we obtain the desired estimates. □

We remark from the estimate (2.102) that the convergence rate of u_n to u depends only on the rate of approximations of $\mathcal{P}_n$ to the identity operator and $\mathcal{K}_n$ to $\mathcal{K}$. Moreover, it is seen from (2.103) that if $\mathcal{K}_n$ approximates $\mathcal{K}$ faster than the convergence of $\mathcal{P}_n$ to the identity, superconvergence of the iterated solution will result since the first term on the right-hand side of (2.103) is more significant than the other. In fact, since for each $u \in \mathbb{X}$ we have that

$$\|\mathcal{K}(\mathcal{I}-\mathcal{P}_n)u\| = \|\mathcal{K}(\mathcal{I}-\mathcal{P}_n)(\mathcal{I}-\mathcal{P}_n)u\| \le \|\mathcal{K}(\mathcal{I}-\mathcal{P}_n)\|\|(\mathcal{I}-\mathcal{P}_n)u\|,$$

circumstances for which $\lim_{n\to\infty} \|\mathcal{K}(\mathcal{I}-\mathcal{P}_n)\| = 0$ will lead to superconvergence. Examples of this phenomenon will be described in Section 4.1.

We also remark that when $\mathbb{X} = \mathbb{V}$ and $\mathcal{K}_n = \mathcal{K}$, Theorem 2.54 leads to the following well-known theorem.

Theorem 2.55 *If $\mathbb{X}$ is a Banach space, $\{\mathbb{X}_n : n \in \mathbb{N}\}$ is a sequence of finite-dimensional subspaces of $\mathbb{X}$, $\mathcal{K} : \mathbb{X} \to \mathbb{X}$ is a compact linear operator not having one as an eigenvalue and $\mathcal{P}_n : \mathbb{X} \to \mathbb{X}_n$ is a sequence of linear projections that converges pointwise to the identity operator $\mathcal{I}$ in $\mathbb{X}$, then there exist an integer q and a positive constant c such that for all $n \geq q$, the equation*

$$u_n - \mathcal{P}_n\mathcal{K}u_n = \mathcal{P}_n f$$

has a unique solution $u_n \in \mathbb{X}_n$ and

$$\|u - u_n\| \leq c\|u - \mathcal{P}_n u\|,$$

where u is the solution of equation (2.95). Moreover, the iterated solution $\tilde{u}_n$ defined by (2.97) with $\mathcal{K}_n = \mathcal{K}$ satisfies the estimate

$$\|u - \tilde{u}_n\| \leq c\|\mathcal{K}(\mathcal{I} - \mathcal{P}_n)u\|.$$

Our final remark is that when $\mathbb{X} = \mathbb{V}$ and $\mathcal{P}_n = \mathcal{I}$, Theorem 2.54 leads to the following theorem.

Theorem 2.56 *If $\mathbb{X}$ is a Banach space, $\mathcal{K} \in \mathcal{B}(\mathbb{X}, \mathbb{X})$ is a compact operator, not having one as an eigenvalue and conditions (H-1) and (H-2) hold, then there exist an integer q and a positive constant c such that for all $n \geq q$, the equation*

$$u_n - \mathcal{K}_n u_n = f$$

has a unique solution $u_n \in \mathbb{X}$ and

$$\|u - u_n\| \leq c\|\mathcal{K}u - \mathcal{K}_n u\|,$$

where u is the solution of equation (2.95).

2.3 Bibliographical remarks

The basic concepts and results of Fredholm integral equations and the projection theory may be found in many well-written books on integral equations such as [15, 110, 121, 177, 203, 253]. In particular, for the subjects of weakly singular integral operators and boundary integral equations, we recommend the books [15, 177, 203]. The functional spaces used in this book are usually covered in standard texts (for example, [1, 183, 236, 276]). Readers are referred to [15, 47, 177, 183, 203, 236] for additional information on the notion of compact operators and weakly singular integral operators, and to the Appendix of this book for basic elements of functional analysis. Moreover, readers may find additional details on boundary integral equations in [12, 15, 22, 121, 144, 150–153, 177, 203, 267].

Regarding the projection methods, readers can see [15, 47, 175–177, 276]. Especially, the notion of generalized best approximation projections was originally introduced in [77]. For the approximate solvability of projection methods for operator equations, Theorems 2.33 and 2.35 provide respectively necessary and sufficient conditions, which can be compared with those in [47] and [177]. For the abstract framework for second-kind operator equations, the theory of collectively compact operators [6, 15] presents a convenient abstract setting for the analysis of many numerical schemes. In Section 2.2.4, we improve the framework to fit more general circumstances. For this point readers are referred to the paper [80]. More information about superconvergence of the iterated scheme may be seen in [60, 246, 247].

3

Conventional numerical methods

This chapter is designed to provide readers with a background on conventional methods for the numerical solution of the Fredholm integral equation of the second kind defined on a compact domain of a Euclidean space. Specifically, we discuss the degenerate kernel method, the quadrature method, the Galerkin method, the collocation method and the Petrov–Galerkin method.

Let Ω be a compact measurable domain in $\mathbb{R}^d$ having a piecewise smooth boundary. We present in this chapter several conventional numerical methods for solving the Fredholm integral equation of the second kind in the form

$$u - \mathcal{K}u = f, \tag{3.1}$$

where

$$(\mathcal{K}u)(s) := \int_\Omega K(s,t)u(t)\mathrm{d}t, \quad s \in \Omega.$$

We describe the principles used in the development of the numerical methods and their convergence analysis.

3.1 Degenerate kernel methods

In this section we describe the *degenerate kernel method* for solving the Fredholm integral equation of the second kind. For this purpose, we assume that $\mathbb{X}$ is either $C(\Omega)$ or $L^2(\Omega)$ with the appropriate norm $\|\cdot\|$. The integral operator $\mathcal{K}$ is assumed to be a compact operator from $\mathbb{X}$ to $\mathbb{X}$.

3.1.1 A general form of the degenerate kernel method

The degenerate kernel method approximates the original integral equation by replacing its kernel with a sequence of kernels having the form

$$K_n(s,t) = \sum_{j\in\mathbb{N}_n} K_j^1(s)K_j^2(t), \quad s,t \in \Omega, \tag{3.2}$$

where $K_j^1, K_j^2 \in \mathbb{X}$ and may depend on n. A kernel of this type is called a *degenerate kernel*. We require that integral operators $\mathcal{K}_n$ with kernels K_n uniformly converge to the integral operator $\mathcal{K}$, that is, $\mathcal{K}_n \xrightarrow{u} \mathcal{K}$. The degenerate kernel method for solving (3.1) finds $u_n \in \mathbb{X}$ such that

$$u_n - \mathcal{K}_n u_n = f. \tag{3.3}$$

For the unique existence and convergence of the approximate solution of the degenerate kernel method, we have the following theorem.

Theorem 3.1 *Let $\mathbb{X}$ be a Banach space and $\mathcal{K} \in \mathcal{B}(\mathbb{X})$ be a compact operator not having one as its eigenvalue. If the operators $\mathcal{K}_n \in \mathcal{B}(\mathbb{X})$ uniformly converge to $\mathcal{K}$, then there exists a positive integer q such that for all $n \geq q$, the inverse operators $(\mathcal{I} - \mathcal{K}_n)^{-1}$ exist from $\mathbb{X}$ to $\mathbb{X}$ and*

$$\|(\mathcal{I} - \mathcal{K}_n)^{-1}\| \leq \frac{\|(\mathcal{I} - \mathcal{K})^{-1}\|}{1 - \|(\mathcal{I} - \mathcal{K})^{-1}\|\|\mathcal{K} - \mathcal{K}_n\|}.$$

Moreover, the error estimate

$$\|u_n - u\| \leq \|(\mathcal{I} - \mathcal{K}_n)^{-1}\|\|(\mathcal{K} - \mathcal{K}_n)u\|$$

holds.

Proof Note that $(\mathcal{I} - \mathcal{K})^{-1}$ exists as a bounded linear operator on $\mathbb{X}$ (see Theorem A.48 in the Appendix). The first result of this theorem follows from Lemma 2.53 with $\mathcal{S} := \mathcal{I} - \mathcal{K}$ and $\mathcal{E} := \mathcal{K}_n - \mathcal{K}$. For the second result, we have that

$$\begin{aligned} u_n - u &= (\mathcal{I} - \mathcal{K}_n)^{-1}f - (\mathcal{I} - \mathcal{K})^{-1}f \\ &= (\mathcal{I} - \mathcal{K}_n)^{-1}(\mathcal{K}_n - \mathcal{K})(\mathcal{I} - \mathcal{K})^{-1}f \\ &= (\mathcal{I} - \mathcal{K}_n)^{-1}(\mathcal{K}_n - \mathcal{K})u, \end{aligned}$$

which yields the second estimate. □

According to the second estimate of Theorem 3.1, we obtain that

$$\|u_n - u\| \leq \|(\mathcal{I} - \mathcal{K}_n)^{-1}\|\|\mathcal{K} - \mathcal{K}_n\|\|u\|.$$

This means that the speed of convergence $\|u_n - u\|$ to zero depends on the speed of convergence $\|\mathcal{K} - \mathcal{K}_n\|$ to zero. This is determined by the choice of the kernels K_n, and is independent of the differentiability of u. It is clear that when $\mathbb{X} = C(\Omega)$, we have that

$$\|\mathcal{K} - \mathcal{K}_n\| = \max_{s\in\Omega} \int_\Omega |K(s,t) - K_n(s,t)|\mathrm{d}t, \tag{3.4}$$

and when $\mathbb{X} = L^2(\Omega)$, we have that

$$\|\mathcal{K} - \mathcal{K}_n\| \leq \left(\int_\Omega \int_\Omega |K(s,t) - K_n(s,t)|^2 \mathrm{d}s\mathrm{d}t\right)^{1/2}. \tag{3.5}$$

We now discuss the algebraic aspects of the degenerate kernel method.

Proposition 3.2 *If u_n is the solution of the degenerate kernel method (3.3) then it can be given by*

$$u_n = f + \sum_{j\in\mathbb{N}_n} v_j K_j^1, \tag{3.6}$$

in which $[v_j : j \in \mathbb{N}_n]$ is a solution of the linear system

$$v_i - \sum_{j\in\mathbb{N}_n} (K_j^1, K_i^2) v_j = (f, K_i^2), \quad i \in \mathbb{N}_n, \tag{3.7}$$

where $(\cdot,\cdot)$ denotes the $L^2(\Omega)$ inner product.

Proof If u_n is the solution of equation (3.3) then, by using (3.2), we have that

$$u_n(s) - \sum_{j\in\mathbb{N}_n} K_j^1(s) \int_\Omega K_j^2(t) u_n(t)\mathrm{d}t = f(s), \quad s \in \Omega. \tag{3.8}$$

This means that the solution u_n can be written as (3.6), with

$$v_j := \int_\Omega K_j^2(t) u_n(t)\mathrm{d}t.$$

Multiplying (3.6) by $K_i^2(s)$ and integrating over Ω, we find that $[v_j : j \in \mathbb{N}_n]$ must satisfy the linear system (3.7). □

Linear system (3.7) may be written in matrix form. To this end, we define

$$\mathbf{K}_n := [(K_j^1, K_i^2) : i,j \in \mathbb{N}_n], \quad \mathbf{v}_n := [v_j : j \in \mathbb{N}_n] \quad \text{and} \quad \mathbf{f}_n := [(f, K_i^2) : i \in \mathbb{N}_n]$$

and let $\mathbf{I}_n$ denote the identity matrix of order n. Then (3.7) can be rewritten as

$$(\mathbf{I}_n - \mathbf{K}_n)\mathbf{v}_n = \mathbf{f}_n.$$

We next consider the invertibility of matrix $\mathbf{I}_n - \mathbf{K}_n$.

Proposition 3.3 *If $\mathbb{X}$ is a Banach space, $\mathcal{K} \in \mathcal{B}(\mathbb{X})$ is a compact operator not having one as its eigenvalue and the operators $\mathcal{K}_n \in \mathcal{B}(\mathbb{X})$ uniformly converge to $\mathcal{K}$, then there exists a positive integer q such that for all $n \geq q$ the coefficient matrix $\mathbf{I}_n - \mathbf{K}_n$ of the linear system (3.7) is nonsingular.*

Proof It follows from Theorem 3.1 that there exists a positive integer q such that for all $n \geq q$, $(\mathcal{I} - \mathcal{K}_n)^{-1}$ exists. This, with Proposition 3.2, leads to the conclusion that (3.7) is solvable for any right-hand side of the form $b = [b_i : i \in \mathbb{N}_n] := [(f, K_i^2) : i \in \mathbb{N}_n]$ with $f \in \mathbb{X}$. We proceed with this proof in two cases.

Case 1: $\{K_i^2 : i \in \mathbb{N}_n\}$ is a linearly independent set of functions. To prove that the coefficient matrix of (3.7) is nonsingular, it is sufficient to prove that (3.7) with any right-hand side $b \in \mathbb{R}^n$ is solvable, or equivalently, for any $b \in \mathbb{R}^n$ there exists a function $f \in \mathbb{X}$ such that $[(f, K_i^2) : i \in \mathbb{N}_n] = b$. To do this, we let $f := \sum_{j \in \mathbb{N}_n} c_j K_j^2$ and consider the equation

$$\sum_{j \in \mathbb{N}_n} (K_j^2, K_i^2) c_j = b_i, \quad i \in \mathbb{N}_n. \tag{3.9}$$

The coefficient matrix $[(K_j^2, K_i^2) : i, j \in \mathbb{N}_n]$ is a Gram matrix, so it is positive semi-definite. Since $\{K_i^2 : i \in \mathbb{N}_n\}$ is linear independent, this matrix is positive definite. This means that (3.9) is solvable, and the function f exists indeed. Thus the coefficient matrix $\mathbf{I}_n - \mathbf{K}_n$ of (3.7) is nonsingular.

Case 2: $\{K_i^2 : i \in \mathbb{N}_n\}$ is a dependent set of functions. In this case, there is a nonsingular matrix $\mathbf{Q}_n$ such that

$$[K_1^2 \ \cdots \ K_n^2]\mathbf{Q}_n^T = [\tilde{K}_1^2 \ \cdots \ \tilde{K}_r^2 \ 0 \ \cdots \ 0],$$

where $\{\tilde{K}_i^2 : i \in \mathbb{N}_r\}$, $0 < r < n$, is a linearly independent set of functions. Let

$$[\tilde{K}_1^1 \ \cdots \ \tilde{K}_n^1] := [K_1^1 \ \cdots \ K_n^1]\mathbf{Q}_n^{-1}$$

and

$$\tilde{\mathbf{K}}_r := [(\tilde{K}_j^1, \tilde{K}_i^2) : i, j \in \mathbb{N}_r].$$

We then have that

$$\mathbf{Q}_n(\mathbf{I}_n - \mathbf{K}_n)\mathbf{Q}_n^{-1} = \begin{bmatrix} \mathbf{I}_r - \tilde{\mathbf{K}}_r & * \\ 0 & \mathbf{I}_{n-r} \end{bmatrix}.$$

Noting that $\mathbf{I}_r - \tilde{\mathbf{K}}_r$ is the coefficient matrix associated with the degenerate kernel $\tilde{K}_r(s,t) = \sum_{j \in \mathbb{N}_r} \tilde{K}_{1j}(s)\tilde{K}_{2j}(t)$, and $\{\tilde{K}_{2i} : i \in \mathbb{N}_r\}$ is a linearly independent set of functions, we conclude from case 1 that $\mathbf{I}_r - \tilde{\mathbf{K}}_r$ is nonsingular, and thus $\mathbf{I}_n - \mathbf{K}_n$ is nonsingular. □

When the hypothesis of the above proposition is satisfied, we solve (3.7) for $[v_j : j \in \mathbb{N}_n]$ and obtain from (3.6) the solution u_n of the degenerate kernel method (3.3).

3.1.2 Degenerate kernel approximations via interpolation

A natural way to construct degenerate kernel approximations is by interpolation. We may employ polynomials, piecewise polynomials, trigonometric polynomials and others as a basis to construct an interpolation for the kernel function.

We now consider the Lagrange interpolation. Let $\{t_j : j \in \mathbb{N}_n\}$ be a finite subset of the domain Ω, and $\{L_j : j \in \mathbb{N}_n\}$ be the Lagrange basis functions satisfying

$$L_j(t_i) = \delta_{ij}, \quad i, j \in \mathbb{N}_n.$$

The kernel K can be approximated by the kernel K_n interpolating K with respect to s or t. That is, we have

$$K_n(s,t) := \sum_{j \in \mathbb{N}_n} L_j(s) K(t_j, t)$$

or

$$K_n(s,t) := \sum_{j \in \mathbb{N}_n} K(s, t_j) L_j(t).$$

Using the former, the linear system (3.7) becomes

$$v_i - \sum_{j \in \mathbb{N}_n} v_j \int_\Omega L_j(t) K(t_i, t) \mathrm{d}t = \int_\Omega f(t) K(t_i, t) \mathrm{d}t, \quad i \in \mathbb{N}_n \tag{3.10}$$

and the solution is given by

$$u_n = f + \sum_{j \in \mathbb{N}_n} v_j L_j, \tag{3.11}$$

while using the latter the linear system (3.7) becomes

$$v_i - \sum_{j \in \mathbb{N}_n} v_j \int_\Omega K(s, t_j) L_i(s) \mathrm{d}s = \int_\Omega f(s) L_i(s) \mathrm{d}s, \quad i \in \mathbb{N}_n \tag{3.12}$$

and the solution is given by

$$u_n = f + \sum_{j \in \mathbb{N}_n} v_j K(\cdot, t_j). \tag{3.13}$$

As an example of the Lagrange interpolation, we consider continuous piecewise linear polynomials, that is, linear splines on $\Omega := [a,b]$. Let $t_j := a + jh$ with $h := (b-a)/n$, $j \in \mathbb{N}_n$, $n \in \mathbb{N}$. The basis functions are chosen as

$$L_j(t) := \begin{cases} 1 - |t - t_j|/h, & t \in [t_{j-1}, t_{j+1}], \\ 0, & \text{otherwise.} \end{cases}$$

We obtain the degenerate kernel approximation by interpolating K with respect to s. Specifically, for $t \in \Omega$ we have that

$$K_n(s,t) = [(t_j - s)K(t_{j-1}, t) + (s - t_{j-1})K(t_j, t)]/h, \quad s \in [t_{j-1}, t_j], j \in \mathbb{N}_n.$$

For this example, we have the following error estimate.

Proposition 3.4 *If $K(\cdot, t) \in C^2(\Omega)$ for any $t \in \Omega$, and $\frac{\partial^2 K}{\partial s^2} \in C(\Omega \times \Omega)$, then*

$$\|\mathcal{K} - \mathcal{K}_n\| \le \frac{1}{8} h^2 (b-a) \left\| \frac{\partial^2 K}{\partial s^2} \right\|_\infty .$$

Proof It can easily be derived by using the Taylor formula that

$$|K(s,t) - K_n(s,t)| \le \frac{1}{8} h^2 \left\| \frac{\partial^2 K(\cdot, t)}{\partial s^2} \right\|_\infty ,$$

for any $s \in [t_{j-1}, t_j], t \in \Omega$. This, with (3.4), leads to the desired result of this proposition. □

3.1.3 Degenerate kernel approximations via expansion

An alternative way to construct degenerate kernel approximations is by expansion, such as Taylor expansions and Fourier expansions of the kernel K. We now introduce the latter.

Let $\mathbb{X} := L^2(\Omega)$ with inner product $(\cdot, \cdot)$, which can be defined with respect to a weight function. Let $\{F_j : j \in \mathbb{N}\}$ be a complete orthonormal sequence in $\mathbb{X}$. Then for any $x \in \mathbb{X}$, we have the Fourier expansions of x with respect to $\{F_j : j \in \mathbb{N}\}$

$$x = \sum_{j \in \mathbb{N}} (x, F_j) F_j.$$

This can be used for construction of approximate degenerate kernels of K with respect to either variable. For example, we may define

$$K_n(s,t) := \sum_{j \in \mathbb{N}_n} F_j(s)(K(\cdot, t), F_j(\cdot)).$$

Let $G_j(t) := (K(\cdot, t), F_j(\cdot))$. Then the linear system (3.7) becomes

$$v_i - \sum_{j \in \mathbb{N}_n} v_j (F_j, G_i) = (f(t), G_i), \quad i \in \mathbb{N}_n \tag{3.14}$$

and the solution is given by

$$u_n = f + \sum_{j \in \mathbb{N}_n} v_j F_j. \tag{3.15}$$

Proposition 3.5 *If $K \in L^2(\Omega \times \Omega)$ and $\{F_j : j \in \mathbb{N}\}$ is a complete orthonormal set in $L^2(\Omega)$, then*

$$\|\mathcal{K} - \mathcal{K}_n\| \leq \left(\sum_{j \in \mathbb{N} \setminus \mathbb{N}_n} \left\| (K(\cdot, \bullet), F_j(\cdot)) \right\|^2 \right)^{1/2}.$$

Proof Note that

$$K(s,t) - K_n(s,t) = \sum_{j \in \mathbb{N} \setminus \mathbb{N}_n} F_j(s) (K(\cdot, t), F_j(\cdot)).$$

By employing the orthonormal property of the sequence $\{F_j : j \in \mathbb{N}\}$, it follows from (3.5) that

$$\begin{aligned} \|\mathcal{K} - \mathcal{K}_n\| &\leq \left(\int_\Omega \int_\Omega |K(s,t) - K_n(s,t)|^2 \mathrm{d}s \mathrm{d}t \right)^{1/2} \\ &= \left(\sum_{j \in \mathbb{N} \setminus \mathbb{N}_n} \left\| (K(\cdot, \bullet), F_j(\cdot)) \right\|^2 \right)^{1/2}. \end{aligned}$$

□

3.2 Quadrature methods

In this section we introduce the *quadrature* or *Nyström method* for solving the Fredholm integral equations of the second kind. This method discretizes the integral equation by directly replacing the integral appearing in the integral equation by numerical quadratures.

3.2.1 Numerical quadratures

We begin by introducing numerical integrations. Let $\Omega \in \mathbb{R}^d$ be a compact set, and assume that $g \in C(\Omega)$. To approximate the integral

$$Q(g) := \int_\Omega g(t) \mathrm{d}t,$$

we consider numerical quadrature rules of the form

$$Q_n(g) := \sum_{j \in \mathbb{N}_n} w_{n,j} g(t_{n,j}),$$

where $t_{n,j} \in \Omega, j \in \mathbb{N}_n$ are quadrature notes and $w_{n,j}, j \in \mathbb{N}_n$ are real quadrature weights.

The following are some examples of quadrature rules.

Example 3.6 Consider the trapezoidal quadrature rule

$$Q_n(g) := h\left[\frac{1}{2}g(t_0) + g(t_1) + \cdots + g(t_{n-1}) + \frac{1}{2}g(t_n)\right],$$

on $\Omega := [a, b]$, where $h := (b-a)/n, t_j := a + jh, j \in \mathbb{N}_n$. When $g \in C^2(\Omega)$, the error of the trapezoidal rule has the estimate ([172], p. 481)

$$|Q(g) - Q_n(g)| \leq \frac{b-a}{12} h^2 \|g''\|_\infty.$$

Example 3.7 Consider the Simpson quadrature rule

$$Q_n(g) := \frac{h}{3}\left[g(t_0) + 4g(t_1) + 2g(t_2) + \cdots + 2g(t_{n-2}) + 4g(t_{n-1}) + g(t_n)\right],$$

on $\Omega := [a, b]$, where $h := (b-a)/n, t_j := a + jh, j \in \mathbb{N}_n$ and n is even. When $g \in C^4(\Omega)$, the error of the Simpson rule has the estimate ([172], p. 483)

$$|Q(g) - Q_n(g)| \leq \frac{b-a}{180} h^4 \|g^{(4)}\|_\infty.$$

Example 3.8 Let $\{\psi_j : j \in \mathbb{Z}_{n+1}\}$ be a family of orthogonal polynomials of degree $\leq n$ on $\Omega := [a, b]$ with respect to a non-negative weight function ρ, and $t_j, j \in \mathbb{N}_n$ be zeros of the function ψ_n. Consider the Gaussian quadrature rule

$$Q_n(g) := \sum_{j \in \mathbb{N}_n} w_j g(t_j), \tag{3.16}$$

for the integral $Q(g)$, where

$$w_j := \int_\Omega \rho(t) L_{n,j}(t) \mathrm{d}t$$

and $L_{n,j}, j \in \mathbb{N}_n$ are the Lagrange interpolation polynomials of degree $n-1$, which have the form

$$L_{n,j}(t) := \Pi_{i \in \mathbb{N}_n, i \neq j}(t - t_i) / \Pi_{i \in \mathbb{N}_n, i \neq j}(t_j - t_i), \ \ j \in \mathbb{N}_n.$$

When $g \in C^{2n}(\Omega)$, the error of the Gaussian quadrature rule has the estimate ([172], p. 497)

$$Q(g) - Q_n(g) = \frac{g^{2n}(\eta)}{(2n)!} \int_{\Omega} \rho(t) \Pi_{i \in \mathbb{N}_n} (t - t_i)^2 \mathrm{d}t,$$

for some $\eta \in \Omega$.

When $\Omega := [-1, 1]$, $\rho = 1$ and $\{\psi_j : j \in \mathbb{Z}_{n+1}\}$ is chosen to be the set of Legendre polynomials

$$\psi_j(t) := \frac{1}{2^j j!} \frac{\mathrm{d}^j}{\mathrm{d}t^j} [(t^2 - 1)^j], \quad j \in \mathbb{Z}_{n+1},$$

formula (3.16) is called the Gauss–Legendre quadrature formula, in which

$$w_j := \frac{2}{n} \frac{1}{\psi_{n-1}(t_j) \psi_n'(t_j)}.$$

This example shows that the Nyström method using the Gaussian quadrature formula has rapid convergence. We now turn to considering convergence of a sequence of general quadrature rules.

Definition 3.9 A sequence $\{Q_n : n \in \mathbb{N}\}$ of quadrature rules is called convergent if the sequence $\{Q_n : n \in \mathbb{N}\}$ converges pointwise to the functional Q on $C(\Omega)$.

The next result characterizes convergent quadrature rules.

Proposition 3.10 *A sequence $\{Q_n : n \in \mathbb{N}\}$ of quadrature rules with the weights $w_{n,j}$, $j \in \mathbb{N}_n$ converges if and only if*

$$\sup_{n \in \mathbb{N}} \sum_{j \in \mathbb{N}_n} |w_{n,j}| < \infty$$

and

$$Q_n(g) \to Q(g), \quad n \to \infty,$$

for all g in a dense subset $U \subset C(\Omega)$.

Proof It can easily be verified that

$$\|Q_n\|_\infty = \sum_{j \in \mathbb{N}_n} |w_{n,j}|.$$

Thus, the result of this proposition follows directly from the Banach–Steinhaus theorem (Corollary A.27 in the Appendix). □

3.2.2 The Nyström method for continuous kernels

Using a sequence $\{Q_n : n \in \mathbb{N}\}$ of numerical quadrature rules, the integral operator

$$(\mathcal{K}u)(s) := \int_{\Omega} K(s,t)u(t)\mathrm{d}t, \quad s \in \Omega,$$

with a continuous kernel $K \in C(\Omega \times \Omega)$, is approximated by a sequence of summation operators

$$(\mathcal{K}_n u)(s) := \sum_{j \in \mathbb{N}_n} w_j K(s,t_j)u(t_j), \quad s \in \Omega.$$

Accordingly, the integral equation (3.1) is approximated by a sequence of discrete equations

$$u_n - \mathcal{K}_n u_n = f, \tag{3.17}$$

or

$$u_n(s) - \sum_{j \in \mathbb{N}_n} w_j K(s,t_j)u_n(t_j) = f(s), \quad s \in \Omega. \tag{3.18}$$

We specify equation (3.18) at the quadrature points t_i, $i \in \mathbb{N}_n$ and obtain the linear system

$$u_n(t_i) - \sum_{j \in \mathbb{N}_n} w_j K(t_i,t_j)u_n(t_j) = f(t_i), \quad i \in \mathbb{N}_n, \tag{3.19}$$

where the unknown is the vector $[u_n(t_j) : j \in \mathbb{N}_n]$. We summarize the above discussion in the following proposition.

Proposition 3.11 *For the solution u_n of (3.18), let $u_{n,j} := u_n(t_j)$, $j \in \mathbb{N}_n$. Then $[u_{n,j} : j \in \mathbb{N}_n]$ satisfies the linear system*

$$u_{n,i} - \sum_{j \in \mathbb{N}_n} w_j K(t_i,t_j)u_{n,j} = f(t_i), \quad i \in \mathbb{N}_n. \tag{3.20}$$

Conversely, if $[u_{n,j} : j \in \mathbb{N}_n]$ is a solution of (3.20), then the function u_n defined by

$$u_n(s) := f(s) + \sum_{j \in \mathbb{N}_n} w_j K(s,t_j)u_{n,j}, \quad s \in \Omega \tag{3.21}$$

solves equation (3.18).

Proof The first statement is trivial. Next, if $[u_{n,j} : j \in \mathbb{N}_n]$ is a solution of (3.20), then we have from (3.21) and (3.20) that

$$u_n(t_i) = f(t_i) + \sum_{j\in\mathbb{N}_n} w_j K(t_i, t_j) u_{n,j} = u_{n,i}.$$

From (3.21) and the above equation, we find that u_n satisfies (3.18). □

Formula (3.21) can be viewed as an interpolation formula, which extends the numerical solution of linear system (3.19) to all points $s \in \Omega$, and is called the *Nyström interpolation formula.*

We now consider error analysis of the Nyström method. Unlike the degenerate kernel method, we do not expect uniform convergence of the sequence $\{\mathcal{K}_n : n \in \mathbb{N}\}$ of approximate operators to the integral operator $\mathcal{K}$ in the Nyström method. In fact, $\|\mathcal{K}_n - \mathcal{K}\| \geq \|\mathcal{K}\|$. To see this, for any small positive constant ϵ we can choose a function $\phi_\epsilon \in C(\Omega)$ such that $\|\phi_\epsilon\|_\infty = 1$, $\phi_\epsilon(t_j) = 0$ for all $j \in \mathbb{N}_n$ and $\phi_\epsilon(s) = 1$ for all $s \in \Omega$ with $\min_{j\in\mathbb{N}_n} |s - t_j| \geq \epsilon$. For this choice of ϕ_ϵ, we have that

$$\begin{aligned}
\|\mathcal{K}_n - \mathcal{K}\| &= \sup\{\|(\mathcal{K}_n - \mathcal{K})v\|_\infty : v \in C(\Omega), \|v\|_\infty \leq 1\} \\
&\geq \sup\{\|(\mathcal{K}_n - \mathcal{K})(v\phi_\epsilon)\|_\infty : v \in C(\Omega), \|v\|_\infty \leq 1, \epsilon > 0\} \\
&= \sup\{\|\mathcal{K}(v\phi_\epsilon)\|_\infty : v \in C(\Omega), \|v\|_\infty \leq 1, \epsilon > 0\} \\
&= \sup\{\|\mathcal{K}v\|_\infty : v \in C(\Omega), \|v\|_\infty \leq 1\} = \|\mathcal{K}\|.
\end{aligned}$$

Although the sequence $\{\mathcal{K}_n : n \in \mathbb{N}\}$ is not uniformly convergent, it is pointwise convergent. Therefore, by using the theory of collectively compact operator approximation, we can obtain the error estimate for the Nyström method.

Theorem 3.12 *If the sequence of quadrature rules is convergent, then the sequence $\{\mathcal{K}_n : n \in \mathbb{N}\}$ of quadrature operators is collectively compact and pointwise convergent on $C(\Omega)$. Moreover, if u and u_n are the solutions of equations (3.1) and the Nyström method, respectively, then there exist a positive constant c and a positive integer q such that for all $n \geq q$,*

$$\|u_n - u\|_\infty \leq c\|(\mathcal{K}_n - \mathcal{K})u\|_\infty.$$

Proof It follows from Proposition 3.10 that

$$C := \sup_{n\in\mathbb{N}} \sum_{j\in\mathbb{N}_n} |w_{n,j}| < \infty,$$

which leads to

$$\|\mathcal{K}_n v\|_\infty \leq C \max_{s,t\in\Omega} |K(s,t)| \, \|v\|_\infty \tag{3.22}$$

and

$$|(\mathcal{K}_n v)(s_1) - (\mathcal{K}_n v)(s_2)| \leq C \max_{t\in\Omega} |K(s_1,t) - K(s_2,t)| \, \|v\|_\infty, \quad s_1, s_2 \in \Omega. \tag{3.23}$$

Noting that K is uniformly continuous on $\Omega \times \Omega$, we conclude that for any bounded set $B \subset C(\Omega)$, $\{\mathcal{K}_n v : v \in B, n \in \mathbb{N}\}$ is bounded and equicontinuous. Thus, by the Arzelà–Ascoli theorem, the sequence $\{\mathcal{K}_n : n \in \mathbb{N}\}$ is collectively compact.

Since the sequence of quadrature rules is convergent, for any $v \in C(\Omega)$,

$$(\mathcal{K}_n v)(s) \to (\mathcal{K}v)(s), \quad \text{as } n \to \infty, \quad \text{for all } s \in \Omega. \tag{3.24}$$

From (3.23) we see that $\{\mathcal{K}_n v : n \in \mathbb{N}\}$ is equicontinuous, which with (3.24) leads to the conclusion that $\{\mathcal{K}_n v : n \in \mathbb{N}\}$ is uniformly convergent, that is, $\{\mathcal{K}_n : n \in \mathbb{N}\}$ is pointwise convergent on $C(\Omega)$.

The last statement of the theorem follows from Theorem 2.56. □

It follows from equations (3.1) and (3.17) that

$$(\mathcal{I} - \mathcal{K}_n)(u_n - u) = (\mathcal{K}_n - \mathcal{K})u,$$

which yields

$$\|(\mathcal{K}_n - \mathcal{K})u\|_\infty \leq \|\mathcal{I} - \mathcal{K}_n\| \, \|u_n - u\|_\infty.$$

This, with the estimate of Theorem 3.12, means that the error $\|u_n - u\|_\infty$ converges to zero in the same order as the numerical integration error

$$\|(\mathcal{K}_n - \mathcal{K})u\|_\infty = \max_{s\in\Omega} \left| \sum_{j\in\mathbb{N}_n} w_j K(s,t_j)u(t_j) - \int_\Omega K(s,t)u(t)\mathrm{d}t \right|.$$

3.2.3 The Nyström method for weakly singular kernels

In this subsection we describe the Nyström method for the numerical solution of integral equation (3.1) with weakly singular operators defined by

$$(\mathcal{K}u)(s) := \int_\Omega K_1(s,t)K_2(s,t)u(t)\mathrm{d}t,$$

where K_1 is a weakly singular kernel and K_2 is a smooth kernel. We consider the important case

$$K_1(s,t) := \log|s-t|,$$

or

$$K_1(s,t) := \frac{1}{|s-t|^\sigma}$$

for some $\sigma \in (0, d)$. The former is often regarded as a special case of the latter, with $\sigma = 0$.

We consider a sequence $\{Q_n : n \in \mathbb{N}\}$ of numerical quadrature rules

$$(Q_n g)(s) := \sum_{j \in \mathbb{N}_n} w_{n,j}(s) g(t_{n,j}), \quad s \in \Omega \tag{3.25}$$

for the integral

$$(Qg)(s) := \int_E K_1(s,t) g(t) \mathrm{d}t, \quad s \in \Omega,$$

where the quadrature weights depend on the function K_1 and the variable s. Then, the integral operator $\mathcal{K}$ is approximated by a sequence of approximate operators defined by

$$(\mathcal{K}_n u)(s) := Q_n(K_2(s,\cdot)u(\cdot))(s), \quad s \in \Omega$$

in terms of the quadrature rules Q_n. Specifically, we have that

$$(\mathcal{K}_n u)(s) = \sum_{j \in \mathbb{N}_n} w_j(s) K_2(s, t_j) u(t_j),$$

where we use simplified notations $w_j := w_{n,j}$ and $t_j := t_{n,j}$. The integral equation (3.1) is then approximated by a sequence of linear equations

$$u_{n,i} - \sum_{j \in \mathbb{N}_n} w_j(t_i) K_2(t_i, t_j) u_{n,j} = f(t_i), \quad i \in \mathbb{N}_n \tag{3.26}$$

and the approximate solution u_n is defined by

$$u_n(s) := f(s) + \sum_{j \in \mathbb{N}_n} w_j(s) K_2(s, t_j) u_{n,j}, \quad s \in \Omega. \tag{3.27}$$

Example 3.13 Suppose that

$$(\mathcal{K}u)(s) := \int_\Omega \log|s-t| K_2(s,t) u(t) \mathrm{d}t, \quad s \in \Omega := [a,b],$$

where K_2 is a smooth function. Let $h := (b-a)/n, t_j := a + jh, j \in \mathbb{Z}_{n+1}$. For a fixed $s \in \Omega$, we choose a piecewise linear interpolation for $K_2(s,\cdot)u(\cdot)$, that is, $K_2(s,\cdot)u(\cdot)$ is approximated by

$$[(t_j - t) K_2(s, t_{j-1}) u(t_{j-1}) + (t - t_{j-1}) K_2(s, t_j) u(t_j)]/h$$

for $t \in [t_{j-1}, t_j], j \in \mathbb{N}_n$. By defining the weight functions

$$w_0(s) := \frac{1}{h} \int_{[t_0, t_1]} (t_1 - t) \log|s-t| \mathrm{d}t,$$

$$w_j(s) := \frac{1}{h} \int_{[t_{j-1}, t_j]} (t - t_{j-1}) \log|s-t| \mathrm{d}t + \frac{1}{h} \int_{[t_j, t_{j+1}]} (t_j - t)) \log|s-t| \mathrm{d}t, \ j \in \mathbb{N}_{n-1}$$

and

$$w_n(s) := \frac{1}{h}\int_{[t_{n-1},t_n]} (t - t_{n-1}) \log|s - t| \mathrm{d}t,$$

we obtain the approximate operators

$$(\mathcal{K}_n u)(s) := \sum_{j\in\mathbb{Z}_{n+1}} w_j(s) K_2(s, t_j) u(t_j), \quad s \in \Omega.$$

The error analysis for the Nyström method for weakly singular kernels can be obtained in a way similar to Theorem 3.12 for continuous kernels. Noting that in this case the quadrature weights depend on s, we need to make appropriate modifications in Theorem 3.12 to fit the current case. We first modify Proposition 3.10 to the following result.

Proposition 3.14 *The sequence $\{Q_n : n \in \mathbb{N}\}$ of quadrature rules defined as in (3.25) converges uniformly on Ω if and only if*

$$\sup_{s\in\Omega} \sup_{n\in\mathbb{N}} \sum_{j\in\mathbb{N}_n} |w_{n,j}(s)| < \infty$$

and

$$Q_n(g) \to Q(g), \quad n \to \infty,$$

uniformly on Ω for all g in some dense subset $U \subset C(\Omega)$.

With this result, we describe the main result for the Nyström method in the case of weakly singular kernels.

Theorem 3.15 *If the sequence of quadrature rules is uniformly convergent, and*

$$\lim_{t\to s} \sup_{n\in\mathbb{N}} \sum_{j\in\mathbb{N}_n} |w_{n,j}(t) - w_{n,j}(s)| = 0, \tag{3.28}$$

then the sequence $\{\mathcal{K}_n : n \in \mathbb{N}\}$ of corresponding approximate operators is collectively compact and pointwise convergent on $C(\Omega)$. Moreover, if u and u_n are the solutions of (equations (3.1)) and the Nyström method, respectively, then there exist a positive constant c and a positive integer q such that for all $n \geq q$,

$$\|u_n - u\|_\infty \leq c\|(\mathcal{K}_n - \mathcal{K})u\|_\infty.$$

Proof The proof is similar to that of Theorem 3.12. We only need to replace inequality (3.23) by the following inequality:

$$
\begin{aligned}
|(\mathcal{K}_n u)(s_1) - (\mathcal{K}_n u)(s_2)| &\le \Big| \sum_{j\in\mathbb{N}_n} w_{n,j}(s_1)[K_2(s_1,t_j) - K_2(s_2,t_j)]u(t_j) \Big| \\
&\quad + \Big| \sum_{j\in\mathbb{N}_n} [w_{n,j}(s_1) - w_{n,j}(s_2)]K_2(s_2,t_j)]u(t_j) \Big| \\
&\le C \max_{t\in\Omega} |K_2(s_1,t) - K_2(s_2,t)| \|u\|_\infty \\
&\quad + \sup_{n\in\mathbb{N}} \sum_{j\in\mathbb{N}_n} |w_{n,j}(s_1) - w_{n,j}(s_2)| \max_{s,t\in\Omega} |K_2(s,t)| \|u\|_\infty.
\end{aligned}
$$

Then, by a similar proof of Theorem 3.12, one can conclude the desired results. □

3.3 Galerkin methods

We have discussed projection methods for solving operator equations in Section 1.3. In this section and what follows, we specialize the operator equations to Fredholm integral equations of the second kind, and consider three major projection methods, namely, the Galerkin method, Petrov–Galerkin method and collocation method for solving the equations. Specifically, we present in this section the Galerkin method, the iterated Galerkin method and the discrete Galerkin method for solving Fredholm integral equations of the second kind.

As described in Section 1.3.2, for the operator equation in a Hilbert space the projection method via orthogonal projections mapped from the Hilbert space onto finite-dimensional subspaces leads to the Galerkin method.

Let $\mathbb{X} := L^2(\Omega)$, $\{\mathbb{X}_n : n \in \mathbb{N}\}$ be a sequence of subspaces of $\mathbb{X}$ satisfying $\overline{\cup_{n\in\mathbb{N}}\mathbb{X}_n} = \mathbb{X}$, and $\{\mathcal{P}_n : n \in \mathbb{N}\}$ be a sequence of orthogonal projections from $\mathbb{X}$ onto $\mathbb{X}_n$. The Galerkin method for solving (3.1) is to find $u_n \in \mathbb{X}_n$ such that

$$
(\mathcal{I} - \mathcal{P}_n\mathcal{K})u_n = \mathcal{P}_n f, \tag{3.29}
$$

or equivalently,

$$
(u_n, v) - (\mathcal{K}u_n, v) = (f, v), \quad \text{for all } v \in \mathbb{X}_n.
$$

Under the hypotheses that $s(n) := \dim \mathbb{X}_n$ and $\{\phi_j : j \in \mathbb{N}_{s(n)}\}$ is a basis for $\mathbb{X}_n$, the solution u_n of equation (3.29) can be written in the form

$$
u_n = \sum_{j\in\mathbb{N}_{s(n)}} u_j\phi_j,
$$

where the vector $\mathbf{u}_n := [u_j : j \in \mathbb{N}_{s(n)}]$ satisfies the linear system

$$\sum_{j\in\mathbb{N}_{s(n)}} u_j \left[(\phi_j, \phi_i) - (\mathcal{K}\phi_j, \phi_i) \right] = (f, \phi_i), \quad i \in \mathbb{N}_{s(n)}.$$

Setting

$$\mathbf{E}_n := \left[(\phi_j, \phi_i) : i, j \in \mathbb{N}_{s(n)} \right], \quad \mathbf{K}_n := \left[(\mathcal{K}\phi_j, \phi_i) : i, j \in \mathbb{N}_{s(n)} \right]$$

and

$$\mathbf{f}_n := \left[(f, \phi_j) : j \in \mathbb{N}_{s(n)} \right],$$

equation (3.29) can be written in the matrix form

$$(\mathbf{E}_n - \mathbf{K}_n)\mathbf{u}_n = \mathbf{f}_n. \tag{3.30}$$

We call $\mathbf{K}_n$ the *Galerkin matrix*.

Note that $\mathcal{P}_n : \mathbb{X} \to \mathbb{X}_n, n \in \mathbb{N}$ are orthogonal projections and $\overline{\bigcup_{n\in\mathbb{N}}\mathbb{X}_n} = \mathbb{X}$, $\|\mathcal{P}_n\| = 1$ and $\mathcal{P}_n$ converges pointwise to the identity operator $\mathcal{I}$ in $\mathbb{X}$. According to Theorem 2.55, if $\mathcal{K} : \mathbb{X} \to \mathbb{X}$ is a compact linear operator not having one as an eigenvalue, then there exist an integer q and a positive constant c such that for all $n \geq q$, equation (3.29) has a unique solution $u_n \in \mathbb{X}$, and

$$\|u - u_n\| \leq c\|u - \mathcal{P}_n u\|,$$

where u is the solution of equation (3.1).

3.3.1 The Galerkin method with piecewise polynomials

Piecewise polynomial bases are often used in the Galerkin method for solving equation (3.1) due to its simplicity, flexibility and excellent approximation property. In this subsection we present the standard piecewise polynomial Galerkin method for solving the Fredholm integral equation (3.1) of the second kind.

We begin with a description of piecewise polynomial subspaces of $L^2(\Omega)$. We assume that there is a partition $\{\Omega_i : i \in \mathbb{Z}_n\}$ of Ω for $n \in \mathbb{N}$, which satisfies the following conditions:

- $\Omega = \bigcup_{i\in\mathbb{Z}_n} \Omega_i$ and $\mathrm{meas}(\Omega_j \cap \Omega_{j'}) = 0, \ j \neq j'$.
- For each $i \in \mathbb{Z}_n$, there exists an invertible affine map $\phi_i : \Omega \to \Omega$ such that $\phi_i(\Omega^0) = \Omega_i$, where Ω^0 is a reference element.

For $\Omega_i \in \{\Omega_i : i \in \mathbb{Z}_n\}$, we define the parameters

$$h_{\Omega_i} := \mathrm{diam}(\Omega_i), \quad \rho_{\Omega_i} := \text{the diameter of the largest circle inscribed in } \Omega_i.$$

We also assume that the partition is *regular* in the sense that there exists a positive constant c such that for all $i \in \mathbb{Z}_n$ and for all $n \in \mathbb{N}$,

$$\frac{h_{\Omega_i}}{\rho_{\Omega_i}} \leq c.$$

Let $h_n := \max\{\text{diam}(\Omega_i) : i \in \mathbb{Z}_n\}$. For a positive integer k, we denote by $\mathbb{X}_n$ the space of the piecewise polynomials of total degree $\leq k-1$, with respect to the partition $\{\Omega_i, i \in \mathbb{Z}_n\}$. In other words, every element in $\mathbb{X}_n$ is a polynomial of total degree $\leq k-1$ on each Ω_i. Since the dimension of the space of polynomials of total degree $k-1$ is given by $m := \binom{k+d-1}{d}$, we conclude that the dimension of $\mathbb{X}_n^k$ is mn.

We next construct a basis for the space $\mathbb{X}_n$. We choose a collection $\{\tau_j : j \in \mathbb{N}_m\} \subset \Omega^0$, in a general position, that is, the Lagrange interpolation polynomial of the total degree $k-1$ at these points is uniquely defined. For $j \in \mathbb{N}_m$, we assume that a collection $\{p_j : j \in \mathbb{N}_m\}$ of polynomials of total degree $k-1$ is chosen to satisfy the equation

$$p_j(\tau_i) = \delta_{i,j}, \quad i,j \in \mathbb{N}_m.$$

For each $i \in \mathbb{Z}_n$, the functions defined by

$$\rho_{ij}(t) := \begin{cases} (p_j \circ \phi_i^{-1})(t), & t \in \Omega_i \\ 0, & t \notin \Omega_i \end{cases} \tag{3.31}$$

form a basis for the space $\mathbb{X}_n$, where "$\circ$" denotes the functional composition. We let $\mathcal{P}_n : L^2(\Omega) \to \mathbb{X}_n$ be the orthogonal projection onto $\mathbb{X}_n$. Then $\mathcal{P}_n$ is self-adjoint and $\|\mathcal{P}_n\| = 1$ for all n. Moreover, for each $x \in H^k(\Omega)$ (cf. Section A.1 in the Appendix), there exist a positive constant c and a positive integer q such that for all $n \geq q$

$$\|x - \mathcal{P}_n x\| \leq c h_n^k \|x\|_{H^k}. \tag{3.32}$$

Associated with the projection $\mathcal{P}_n$, the piecewise polynomial Galerkin method for solving equation (3.1) is described as finding $u_n \in \mathbb{X}_n$ such that

$$(\mathcal{I} - \mathcal{P}_n \mathcal{K}) u_n = \mathcal{P}_n f. \tag{3.33}$$

In terms of the basis functions described in (3.31), the Galerkin equation (3.33) is equivalent to the linear system

$$(\mathbf{E}_n - \mathbf{K}_n)\mathbf{u}_n = \mathbf{f}_n, \tag{3.34}$$

where $\mathbf{u}_n \in \mathbb{R}^{mn}$,

$$\mathbf{E}_n := [E_{ij,i'j'} : i, i' \in \mathbb{Z}_n, j, j' \in \mathbb{N}_m] \text{ with } E_{ij,i'j'} := (\rho_{i'j'}, \rho_{ij}),$$
$$\mathbf{K}_n := [K_{ij,i'j'} : i, i' \in \mathbb{Z}_n, j, j' \in \mathbb{N}_m] \text{ with } K_{ij,i'j'} := (\mathcal{K}\rho_{i'j'}, \rho_{ij})$$

and

$$\mathbf{f}_n := [f_{ij} : i \in \mathbb{Z}_n, j \in \mathbb{N}_m] \text{ with } f_{ij} := (f, \rho_{ij}).$$

The following result is concerned with the convergence order of the Galerkin method (3.33).

Theorem 3.16 *Suppose that $\mathcal{K} : L^2(\Omega) \to L^2(\Omega)$ is a compact operator not having one as its eigenvalue. If $u \in L^2(\Omega)$ is the solution of equation (3.1), then there exist a positive constant c and a positive integer q such that for each $n \geq q$, equation (3.33) has a unique solution $u_n \in \mathbb{X}_n$. Moreover, if $u \in H^k(\Omega)$ then u_n satisfies the error bound*

$$\|u - u_n\| \leq c h_n^k \|u\|_{H^k}.$$

Proof By the hypothesis that one is not an eigenvalue of the compact operator $\mathcal{K}$, the operator $\mathcal{I} - \mathcal{K}$ is one to one and onto. Since the spaces $\mathbb{X}_n$ are dense in $L^2(\Omega)$, we have that for $x \in L^2(\Omega)$

$$\lim_{h \to 0} \|\mathcal{P}_n x - \mathcal{I}x\| = 0.$$

This ensures that there exists a positive integer q such that for each $n \geq q$, equation (3.33) has a unique solution $u_n \in \mathbb{X}_n$ and the inverse operators $(\mathcal{I} - \mathcal{P}_n\mathcal{K})^{-1}$ are uniformly bounded.

It follows from equation (3.33) that

$$u - u_n = u - \mathcal{P}_n f - \mathcal{P}_n \mathcal{K} u_n. \tag{3.35}$$

Moreover, applying the projection $\mathcal{P}_n$ to both sides of equation (3.1) yields

$$\mathcal{P}_n u - \mathcal{P}_n \mathcal{K} u = \mathcal{P}_n f.$$

Substituting this equation into (3.35) leads to the equation

$$u - u_n = u - \mathcal{P}_n u + \mathcal{P}_n \mathcal{K}(u - u_n).$$

Solving for $u - u_n$ from the above equation gives

$$u - u_n = (\mathcal{I} - \mathcal{P}_n \mathcal{K})^{-1}(u - \mathcal{P}_n u).$$

Therefore, the desired estimate follows from the above equation, the uniform boundedness of $\|(\mathcal{I} - \mathcal{P}_n\mathcal{K})^{-1}\|$ and estimate (3.32). □

The last theorem shows that the Galerkin method has convergence of optimal order. That is, the order of convergence is equal to the order of approximation from the piecewise polynomial space.

We finally remark that if Ω is the boundary of a domain, the piecewise polynomial Galerkin method is called a boundary element method.

3.3.2 The Galerkin method with trigonometric polynomials

We now consider equation (3.1) with $\Omega := [0, 2\pi]$ and the functions K and f being 2π-periodic, that is, $K(s + 2\pi, t) = K(s, t + 2\pi) = K(s, t)$ and $f(s + 2\pi) = f(s)$ for $s, t \in \Omega$. In this case, trigonometric polynomials are often used as approximations for solving the equations, and projection methods of this type are referred to as spectral methods.

Let $\mathbb{X} := L^2(0, 2\pi)$ be the space of all complex-valued, 2π-periodic and square integral Lebesgue measurable functions on Ω with inner product

$$(x, y) := \int_\Omega x(t)\overline{y(t)}\mathrm{d}t.$$

For each $n \in \mathbb{N}$, let $\mathbb{X}_n$ be the subspace of $\mathbb{X}$ of all trigonometric polynomials of degree $\leq n$. That is, we set $\phi_j(s) := e^{\mathrm{i}js}$, $s \in \mathbb{R}$,

$$\mathbb{X}_n := \operatorname{span}\{\phi_j(s) : j \in \mathbb{Z}_{-n,n}\},$$

where $\mathrm{i} := \sqrt{-1}$ and $\mathbb{Z}_{-n,n} := \{-n, \ldots, 0, \ldots, n\}$. The orthogonal projection of $\mathbb{X}$ onto $\mathbb{X}_n$ is given by

$$\mathcal{P}_n x = \frac{1}{2\pi} \sum_{j \in \mathbb{Z}_{-n,n}} (x, \phi_j)\phi_j.$$

For $x \in \mathbb{X}$, it is well known that

$$\|x - \mathcal{P}_n x\| = \left(\frac{1}{2\pi} \sum_{j \in \mathbb{Z} \setminus \mathbb{Z}_{-n,n}} |(x, \phi_j)|^2 \right)^{1/2} \to 0, \quad \text{as } n \to \infty.$$

To present the error analysis, we introduce Sobolev spaces of 2π-periodic functions, which are subspaces of $L^2(0, 2\pi)$ and require for their elements a certain decay of their Fourier coefficients. That is, for $r \in [0, \infty)$,

$$H^r(0, 2\pi) := \left\{ x \in L^2(0, 2\pi) : \sum_{j \in \mathbb{Z}} (1 + j^2)^r |(x, \phi_j)|^2 < \infty \right\}.$$

Note that $H^0(0, 2\pi)$ coincides with $L^2(0, 2\pi)$, and $H^r(0, 2\pi)$ is a Hilbert space with inner product given by

$$(x, y)_r := \frac{1}{2\pi} \sum_{j \in \mathbb{Z}} (1 + j^2)^r (x, \phi_j) \overline{(y, \phi_j)}$$

and norm given by

$$\|x\|_{H^r} := \left(\frac{1}{2\pi} \sum_{j \in \mathbb{Z}} (1 + j^2)^r |(x, \phi_j)|^2 \right)^{1/2}.$$

We remark that when r is an integer, this norm is equivalent to the norm

$$\|x\|_r := \left(\sum_{j \in \mathbb{Z}_{r+1}} \|x^{(j)}\|^2 \right)^{1/2}.$$

It can easily be seen that for $x \in H^r(0, 2\pi)$,

$$\|x - \mathcal{P}_n x\| \leq \frac{1}{n^r} \left(\frac{1}{2\pi} \sum_{j \in \mathbb{Z} \setminus \mathbb{Z}_{-n,n}} (1 + j^2)^r |(x, \phi_j)|^2 \right)^{1/2} \leq \frac{1}{n^r} \|x\|_{H^r}. \quad (3.36)$$

With the help of the above estimate, we have the following error analysis.

Theorem 3.17 *Suppose that $\mathcal{K} : L^2(0, 2\pi) \to L^2(0, 2\pi)$ is a compact operator not having one as its eigenvalue. If $u \in L^2(0, 2\pi)$ is the solution of equation (3.1), then there exist an integer N_0 and a positive constant c such that for each $n \geq N_0$, the Galerkin approximate equation (3.33) has a unique solution $u_n \in \mathbb{X}_n$. Moreover, if $u \in H^r(0, 2\pi)$, then u_n satisfies the error bound*

$$\|u - u_n\| \leq c n^{-r} \|u\|_{H^r}.$$

Proof The proof of this theorem is similar to that of Theorem 3.16 with (3.32) being replaced by (3.36). □

3.3.3 The condition number for the Galerkin method

We discuss in this subsection the condition number of the linear system associated with the Galerkin equation, which depends on the choice of bases of the approximate subspace. To present the results, we need the notion of the *matrix norm* induced by a vector norm. For an $n \times n$ matrix $\mathbf{A}$, the matrix norm is defined by

$$\|\mathbf{A}\|_p := \sup\{\|\mathbf{A}x\|_p : x \in \mathbb{R}^n, \|\mathbf{x}\|_p = 1\}, \quad 1 \leq p \leq \infty.$$

It is well known that for $\mathbf{A} := [a_{ij} : i,j \in \mathbb{Z}_n]$,

$$\|\mathbf{A}\|_\infty = \max_{i\in\mathbb{Z}_n} \sum_{j\in\mathbb{Z}_n} |a_{ij}|,$$

$$\|\mathbf{A}\|_1 = \max_{j\in\mathbb{Z}_n} \sum_{i\in\mathbb{Z}_n} |a_{ij}|$$

and

$$\|\mathbf{A}\|_2 = [\rho(\mathbf{A}^T\mathbf{A})]^{1/2},$$

where $\rho(\mathbf{A})$ is called the spectral radius of $\mathbf{A}$ and is defined as the largest eigenvalue of $\mathbf{A}$.

To discuss the condition number of the coefficient matrix of the linear system (3.30) for the Galerkin method, we first provide a lemma on a change of bases for the subspace $\mathbb{X}_n$. Let $\{\phi_j : j \in \mathbb{N}_{s(n)}\}$ and $\{\psi_j : j \in \mathbb{N}_{s(n)}\}$ be two bases for the subspace $\mathbb{X}_n$, with the latter being orthonormal. These bases have the relations

$$\Psi_n = \mathbf{C}_n\Phi_n \text{ and } \Phi_n = \mathbf{D}_n\Psi_n,$$

where $\Psi_n := [\psi_j : j \in \mathbb{N}_{s(n)}]$, $\Phi_n := [\phi_j : j \in \mathbb{N}_{s(n)}]$, $\mathbf{D}_n := [(\phi_i, \psi_j) : i,j \in \mathbb{N}_{s(n)}]$ and $\mathbf{C}_n = [c_{ij} : i,j \in \mathbb{N}_{s(n)}]$ is the matrix determined by the first equation.

Lemma 3.18 *If $\{\psi_j : j \in \mathbb{N}_{s(n)}\}$ is orthonormal then*

$$\mathbf{C}_n\mathbf{D}_n = \mathbf{I}_n \text{ and } \mathbf{D}_n\mathbf{D}_n^T = \mathbf{E}_n,$$

where $\mathbf{I}_n$ is the identity matrix and $\mathbf{E}_n := [(\phi_j, \phi_i) : i,j \in \mathbb{N}_{s(n)}]$. Moreover,

$$\|\mathbf{D}_n\|_2 = \|\mathbf{D}_n^T\|_2 = \|\mathbf{E}_n\|_2^{1/2} \text{ and } \|\mathbf{D}_n^{-1}\|_2 = \|\mathbf{D}_n^{-T}\|_2 = \|\mathbf{E}_n^{-1}\|_2^{1/2}.$$

Proof The first part of this lemma follows directly from computation. For the second part, we have that

$$\|\mathbf{D}_n\|_2 = \|\mathbf{D}_n^T\|_2 = [\rho(\mathbf{D}_n\mathbf{D}_n^T)]^{1/2} = [\rho(\mathbf{E}_n)]^{1/2} = \|\mathbf{E}_n\|_2^{1/2}.$$

The second equation can be proved similarly. □

The next lemma can easily be verified.

Lemma 3.19 *For any $\mathbf{w} := [w_j : j \in \mathbb{N}_{s(n)}] \in \mathbb{R}^{s(n)}$, let $w := \mathbf{w}^T\Psi_n \in \mathbb{X}_n$. If the operator $\mathcal{Q}_n : \mathbb{R}^{s(n)} \to \mathbb{X}_n$ is defined by $\mathcal{Q}_n\mathbf{w} = w$, then $\mathcal{Q}_n$ is invertible and $\|\mathcal{Q}_n\| = \|\mathcal{Q}_n^{-1}\| = 1$.*

In the next theorem, we estimate the condition number of the coefficient matrix of the Galerkin method.

Theorem 3.20 *The condition number of the coefficient matrix of the linear system (3.30) for the Galerkin method has the bound*

$$\mathrm{cond}(\mathbf{E}_n - \mathbf{K}_n) \le \mathrm{cond}(\mathbf{E}_n)\mathrm{cond}(\mathcal{I} - \mathcal{P}_n\mathcal{K}).$$

Proof For any $\mathbf{g} := [g_j : j \in \mathbb{N}_{s(n)}]$, let

$$\mathbf{v} := (\mathbf{E}_n - \mathbf{K}_n)^{-1}\mathbf{g}.$$

It can be verified that $g := \mathbf{g}^T\Phi_n$ and $v := \mathbf{v}^T\Phi_n$ satisfy the equation

$$g = (\mathcal{I} - \mathcal{P}_n\mathcal{K})v.$$

Noting that $v = (\mathbf{D}_n^T\mathbf{v})^T\Psi_n = \mathcal{Q}_n(\mathbf{D}_n^T\mathbf{v})$ and $g = (\mathbf{D}_n^T\mathbf{g})^T\Psi_n = \mathcal{Q}_n(\mathbf{D}_n^T\mathbf{g})$, we have that

$$(\mathbf{E}_n - \mathbf{K}_n)^{-1}\mathbf{g} = \mathbf{v} = \mathbf{D}_n^{-T}\mathcal{Q}_n^{-1}v = \mathbf{D}_n^{-T}\mathcal{Q}_n^{-1}(\mathcal{I} - \mathcal{P}_n\mathcal{K})^{-1}\mathcal{Q}_n(\mathbf{D}_n^T\mathbf{g}).$$

Thus, using Lemmas 3.18 and 3.19 we conclude that

$$\begin{aligned}\|(\mathbf{E}_n - \mathbf{K}_n)^{-1}\| &\le \|\mathbf{D}_n^{-T}\|\|\mathcal{Q}_n^{-1}\|\|(\mathcal{I} - \mathcal{P}_n\mathcal{K})^{-1}\|\|\mathcal{Q}_n\|\|\mathbf{D}_n^T\| \\ &\le \mathrm{cond}(\mathbf{E}_n)^{1/2}\|(\mathcal{I} - \mathcal{P}_n\mathcal{K})^{-1}\|. \end{aligned} \tag{3.37}$$

Likewise, we conclude from

$$(\mathbf{E}_n - \mathbf{K}_n)\mathbf{v} = \mathbf{g} = \mathbf{D}_n^{-T}\mathcal{Q}_n^{-1}g = \mathbf{D}_n^{-T}\mathcal{Q}_n^{-1}(\mathcal{I} - \mathcal{P}_n\mathcal{K})\mathcal{Q}_n(\mathbf{D}_n^T\mathbf{v})$$

that

$$\|\mathbf{E}_n - \mathbf{K}_n\| \le \mathrm{cond}(\mathbf{E}_n)^{1/2}\|(\mathcal{I} - \mathcal{P}_n\mathcal{K})\|. \tag{3.38}$$

Combining estimates (3.37) and (3.38) yields the desired result of this theorem. □

We remark that if $\mathcal{K}$ is a compact linear operator not having one as its eigenvalue and $\mathcal{P}_n \stackrel{s}{\longrightarrow} \mathcal{I}$, then $\mathcal{P}_n\mathcal{K} \stackrel{u}{\longrightarrow} \mathcal{K}$ and $(\mathcal{I}-\mathcal{P}_n\mathcal{K})^{-1} \stackrel{u}{\longrightarrow} (\mathcal{I}-\mathcal{K})^{-1}$, which yields

$$\mathrm{cond}(\mathcal{I} - \mathcal{P}_n\mathcal{K}) \to \mathrm{cond}(\mathcal{I} - \mathcal{K}), \quad \text{as } n \to \infty.$$

3.3.4 The iterated Galerkin method

We presented the iterated projection scheme (2.97) in Section 2.2.4. According to Theorem 2.55, if $\mathcal{K} : \mathbb{X} \to \mathbb{X}$ is a compact linear operator not having one as an eigenvalue, then there exist a positive constant c and a positive integer q

such that for all $n \geq q$, the iterated solution defined by (2.97) with $\mathcal{K}_n = \mathcal{K}$ satisfies the estimate

$$\|u - \tilde{u}_n\| \leq c\|\mathcal{K}(\mathcal{I} - \mathcal{P}_n)u\|.$$

Since $(\mathcal{I} - \mathcal{P}_n)^2 = \mathcal{I} - \mathcal{P}_n$, we have that

$$\|\mathcal{K}(\mathcal{I} - \mathcal{P}_n)u\| \leq \|\mathcal{K}(\mathcal{I} - \mathcal{P}_n)\| \|(\mathcal{I} - \mathcal{P}_n)u\|.$$

Since $\mathcal{K}$ is compact, its adjoint $\mathcal{K}^*$ is also compact. As a result, we obtain that

$$\|\mathcal{K}(\mathcal{I} - \mathcal{P}_n)\| = \|[\mathcal{K}(\mathcal{I} - \mathcal{P}_n)]^*\| = \|(\mathcal{I} - \mathcal{P}_n)\mathcal{K}^*\| \to 0, \quad \text{as } n \to \infty.$$

Thus, we conclude that

$$\|u - \tilde{u}_n\| \leq c\|(\mathcal{I} - \mathcal{P}_n)\mathcal{K}^*\| \|(\mathcal{I} - \mathcal{P}_n)u\|,$$

and see that $\|u - \tilde{u}_n\|$ converges to zero more rapidly than $\|u - u_n\|$ does. Moreover, we have that

$$\mathcal{K}(\mathcal{I} - \mathcal{P}_n)u = (K(s, \cdot), (\mathcal{I} - \mathcal{P}_n)u(\cdot)) = ((\mathcal{I} - \mathcal{P}_n)K(s, \cdot), (\mathcal{I} - \mathcal{P}_n)u(\cdot)),$$

which leads to

$$\|u - \tilde{u}_n\| \leq c \operatorname{ess\,sup}\{\|(\mathcal{I} - \mathcal{P}_n)K(s, \cdot)\| : s \in \Omega\}\|(\mathcal{I} - \mathcal{P}_n)u\|.$$

This shows that the additional order of convergence gained from iteration is attributed to approximation of the integral kernel from the approximate subspace.

3.3.5 Discrete Galerkin methods

The implementation of the Galerkin method (3.29) requires evaluating the integrals involved in (3.30). There are two types of integral required for evaluation: the integral that defines the operator $\mathcal{K}$ and the inner product $(\cdot, \cdot)$ of the space $L^2(\Omega)$. These integrals usually cannot be evaluated exactly and thus, they require numerical integration. The Galerkin method with integrals computed using numerical quadrature is called the discrete Galerkin method.

We choose quadrature nodes τ_j, $j \in \mathbb{N}_{q_n}$ with $q_n \geq s(n)$, and discrete operators $\mathcal{K}_n$ defined by

$$(\mathcal{K}_n x)(t) := \sum_{j \in \mathbb{N}_{q_n}} w_j(t) x(\tau_j), \quad t \in \Omega$$

to approximate the operator $\mathcal{K}$. We require that the sequence $\{\mathcal{K}_n : n \in \mathbb{N}\}$ of discrete operators is collectively compact and pointwise convergent on $C(\Omega)$.

For sufficient conditions to ensure collective compactness, see Theorem 3.15. Suppose that we use a quadrature formula to approximate integrals, that is,

$$\int_{\Omega} x(t)\mathrm{d}t \approx \sum_{j\in\mathbb{N}_{q_n}} \lambda_j x(\tau_j), \quad \text{with } \lambda_j > 0, j \in \mathbb{N}_{q_n},$$

and define a discrete semi-definite inner product

$$(x, y)_n := \sum_{j\in\mathbb{N}_{q_n}} \lambda_j x(\tau_j) y(\tau_j), \quad x, y \in C(\Omega)$$

and the corresponding discrete semi-norm

$$\|x\|_n := (x, x)_n^{1/2}, \quad x \in C(\Omega).$$

We require that the rank of the matrix $\Phi_n := [\phi_i(\tau_j) : i \in \mathbb{N}_{s(n)}, j \in \mathbb{N}_{q_n}]$ is equal to $s(n)$. It follows that there is a subset of quadrature nodes, say $\{\tau_j : j \in \mathbb{N}_{s(n)}\}$, such that the matrix $[\phi_i(\tau_j) : i, j \in \mathbb{N}_{s(n)}]$ is nonsingular. Thus, for any data $\{b_j : j \in \mathbb{N}_{s(n)}\}$, there exists a unique $\phi \in \mathbb{X}_n$ satisfying

$$\phi(\tau_j) = b_j, \quad j \in \mathbb{N}_{s(n)}.$$

We see that $\|\cdot\|_n$ is a semi-norm on $C(\Omega)$ and is a norm on $\mathbb{X}_n$.

The discrete Galerkin method for solving (3.1) is to find $u_n \in \mathbb{X}_n$ such that

$$(u_n, v)_n - (\mathcal{K}_n u_n, v)_n = (f, v)_n, \quad \text{for all } v \in \mathbb{X}_n. \tag{3.39}$$

The analysis of the discrete Galerkin method requires the notation of the discrete orthogonal projection.

Definition 3.21 Let $\mathbb{X}_n$ be a subspace of $C(\Omega)$. The operator $\mathcal{P}_n : C(\Omega) \to \mathbb{X}_n$ defined for $x \in C(\Omega)$ by

$$(\mathcal{P}_n x, y)_n = (x, y)_n, \quad \text{for all } y \in \mathbb{X}_n,$$

is called the discrete orthogonal projection from $C(\Omega)$ onto $\mathbb{X}_n$.

Proposition 3.22 *If* $\lambda_j > 0, j \in \mathbb{N}_{q_n}$ *and* rank $\Phi_n = s(n)$, *then the discrete orthogonal projection* $\mathcal{P}_n : C(\Omega) \to \mathbb{X}_n$ *is well defined and is a linear projection from* $C(\Omega)$ *onto* $\mathbb{X}_n$. *If, in addition,* $q_n = s(n)$, *then* $\mathcal{P}_n$ *is an interpolating projection satisfying for* $x \in C(\Omega)$

$$(\mathcal{P}_n x)(\tau_j) = x(\tau_j), \quad j \in \mathbb{N}_{s(n)}.$$

Proof For $x \in C(\Omega)$, let $\mathcal{P}_n x = \sum_{j\in\mathbb{N}_{s(n)}} x_j \phi_j$ and consider the linear system

$$\sum_{j\in\mathbb{N}_{s(n)}} x_j (\phi_j, \phi_i)_n = (x, \phi_i)_n, \quad i \in \mathbb{N}_{s(n)}. \tag{3.40}$$

The coefficient matrix $\mathbf{G}_n := [(\phi_j, \phi_i)_n : i, j \in \mathbb{N}_{s(n)}]$ is a Gram matrix. Thus, for any $\mathbf{x} := [x_j : j \in \mathbb{N}_{s(n)}] \in \mathbb{R}^{s(n)}$ we have that

$$\mathbf{x}^T \mathbf{G}_n \mathbf{x} = \Big\| \sum_{j \in \mathbb{N}_{s(n)}} x_j \phi_j \Big\|_n^2 \geq 0.$$

Since $\lambda_j > 0$ for all $j \in \mathbb{N}_{q_n}$, $\mathbf{x}^T \mathbf{G}_n \mathbf{x} = 0$ if and only if $\sum_{j \in \mathbb{N}_{s(n)}} x_j \phi_j(\tau_i) = 0, i \in \mathbb{N}_{q_n}$. The latter is equivalent to $\mathbf{x} = 0$, since rank $\Phi_n = s(n)$. Thus we conclude that the matrix $\mathbf{G}_n$ is positive definite, and equation (3.40) is uniquely solvable, that is, $\mathcal{P}_n$ is well defined. From Definition 3.21, we can easily verify that $\mathcal{P}_n : C(\Omega) \to \mathbb{X}_n$ is a linear projection.

When $q_n = s(n)$, we define, for $x \in C(\Omega)$, $\mathcal{I}_n x := [x(\tau_j) : j \in \mathbb{N}_{s(n)}] \in \mathbb{R}^{s(n)}$. We next show that

$$\mathcal{I}_n(\mathcal{P}_n x) = \mathcal{I}_n x.$$

We first have that

$$\mathcal{I}_n(\mathcal{P}_n x) = \mathcal{I}_n\Big(\sum_{j \in \mathbb{N}_{s(n)}} x_j \phi_j \Big) = \Phi_n^T \mathbf{x}.$$

It follows from equation (3.40) that

$$\mathbf{G}_n \mathbf{x} = \Phi_n \Lambda_n \mathcal{I}_n x,$$

where Λ_n is the diagonal matrix $\mathrm{diag}(\lambda_1, \ldots, \lambda_{s(n)})$. The definition of the notation $\mathbf{G}_n$ leads to $\mathbf{G}_n = \Phi_n \Lambda_n \Phi_n^T$. Thus, combining the above equations, we obtain that

$$\mathcal{I}_n(\mathcal{P}_n x) = \Phi_n^T \mathbf{G}_n^{-1} \Phi_n \Lambda_n \mathcal{I}_n x = \mathcal{I}_n x,$$

which completes the proof. □

The discrete orthogonal projection $\mathcal{P}_n$ is self-adjoint on $C(\Omega)$ with respect to the discrete inner product, that is,

$$(\mathcal{P}_n x, y)_n = (x, \mathcal{P}_n y)_n, \quad x, y \in C(\Omega),$$

and is bounded on $C(\Omega)$ with respect to the discrete semi-norm, that is,

$$\|\mathcal{P}_n x\|_n \leq \|x\|_n, \quad x \in C(\Omega).$$

We remark that the latter does not mean the uniform boundedness of the operators $\{\mathcal{P}_n : n \in \mathbb{N}\}$. However, if $\mathbb{X}_n$ are piecewise polynomial spaces and quadrature nodes are obtained by using affine mappings from a set of quadrature nodes in a reference element, then we have the uniform boundedness: $\sup\{\|\mathcal{P}_n\|_\infty : n \in \mathbb{N}\} < \infty$ (see Section 3.5.4).

The following proposition provides convergence of the discrete orthogonal projection.

Proposition 3.23 *If the sequence of discrete orthogonal projections* $\{\mathcal{P}_n : n \in \mathbb{N}\}$ *is uniformly bounded on* $C(\Omega)$, *then there is a positive constant* c *such that for all* $x \in C(\Omega)$,

$$\|x - \mathcal{P}_n x\|_\infty \leq c \inf\{\|x - v\|_\infty : v \in \mathbb{X}_n\}.$$

Proof Since $\mathcal{P}_n$ is a linear projection from $C(\Omega)$ onto $\mathbb{X}_n$, for any $v \in \mathbb{X}_n$,

$$x - \mathcal{P}_n x = x - v - \mathcal{P}_n(x - v).$$

This yields the estimate

$$\|x - \mathcal{P}_n x\|_\infty \leq (1 + \sup\{\|\mathcal{P}_n\|_\infty : n \in \mathbb{N}\})\|x - v\|_\infty.$$

Thus, the desired result follows from the above estimate and the hypothesis that $\mathcal{P}_n$ is uniformly bounded. □

With the help of discrete orthogonal projections, the discrete Galerkin method (3.39) for solving (3.1) can be rewritten in the form of (2.96), that is,

$$(\mathcal{I} - \mathcal{P}_n \mathcal{K}_n)u_n = \mathcal{P}_n f.$$

Thus, the error analysis for this method follows from the same framework as Theorem 2.54.

3.4 Collocation methods

We consider in this section the collocation method for solving the Fredholm integral equation of the second kind. According to the description in Section 2.2.1, for the operator equation in the space $C(\Omega)$ the projection method via interpolation projections into finite-dimensional subspaces leads to the collocation method.

Let $\{\mathbb{X}_n : n \in \mathbb{N}\}$ be a sequence of subspaces of $C(\Omega)$ with $s(n) := \dim \mathbb{X}_n$, and let $\{\mathcal{P}_n : n \in \mathbb{N}\}$ be a sequence of interpolation projections from $C(\Omega)$ onto $\mathbb{X}_n$ defined for $x \in C(\Omega)$ by

$$(\mathcal{P}_n x)(t_j) = x(t_j), \quad \text{for all } j \in \mathbb{N}_{s(n)},$$

where $\{t_j : j \in \mathbb{N}_{s(n)}\}$ is a set of distinct nodes in Ω. The collocation method for solving (3.1) is to find $u_n \in \mathbb{X}_n$ such that

$$(\mathcal{I} - \mathcal{P}_n \mathcal{K})u_n = \mathcal{P}_n f, \tag{3.41}$$

or equivalently,

$$u_n(t_i) - \int_\Omega K(t_i,t)u_n(t)\mathrm{d}t = f(t_i), \quad \text{for all } i \in \mathbb{N}_{s(n)}. \tag{3.42}$$

Suppose that $\{\phi_j : j \in \mathbb{N}_{s(n)}\}$ is a basis for $\mathbb{X}_n$. Let $u_n = \sum_{j\in\mathbb{N}_{s(n)}} u_j\phi_j$. Equation (3.42) can be written in the matrix form

$$(\mathbf{E}_n - \mathbf{K}_n)\mathbf{u}_n = \mathbf{f}_n, \tag{3.43}$$

where

$$\mathbf{E}_n := \left[\phi_j(t_i) : i,j \in \mathbb{N}_{s(n)}\right], \quad \mathbf{K}_n := \left[(\mathcal{K}\phi_j)(t_i) : i,j \in \mathbb{N}_{s(n)}\right],$$
$$\mathbf{u}_n := \left[u_j : j \in \mathbb{N}_{s(n)}\right] \text{ and } \mathbf{f}_n := \left[f(t_j) : j \in \mathbb{N}_{s(n)}\right].$$

$\mathbf{K}_n$ is called the *collocation matrix*.

We remark that the set of collocation points $\{t_j : j \in \mathbb{N}_{s(n)}\}$ should be chosen such that the subspace $\mathbb{X}_n$ is unisolvent, that is, the interpolating function is uniquely determined by its values at the interpolating points. It is clear that this requirement is equivalent to the condition $\det(\mathbf{E}_n) \neq 0$. Since the collocation method is interpreted as a projection method with the interpolating operator, the general convergence results for projection methods are applicable.

When the Lagrange basis $\{L_j : j \in \mathbb{N}_{s(n)}\}$ for $\mathbb{X}_n$ is used, then

$$u_n(t) = \sum_{j\in\mathbb{N}_{s(n)}} u_jL_j(t), \quad \text{with } u_j = u_n(t_j)$$

and the linear system (3.43) becomes

$$u_i - \sum_{j\in\mathbb{N}_{s(n)}} u_j \int_\Omega K(t_i,t)L_j(t)\mathrm{d}t = f(t_i), \quad \text{for all } i \in \mathbb{N}_{s(n)}.$$

Note that the coefficient matrix is the same as that for the degenerate kernel method (3.10). In other words, the operator $\mathcal{P}_n\mathcal{K}$ is a degenerate kernel integral operator

$$\mathcal{P}_n\mathcal{K}u(t) = \int_\Omega K_n(s,t)u(t)\mathrm{d}t, \quad \text{with } K_n(s,t) = \sum_{j\in\mathbb{N}_{s(n)}} K(t_j,t)L_j(s).$$

We then have the estimate

$$\|\mathcal{K} - \mathcal{P}_n\mathcal{K}\| = \max_{s\in\Omega} \int_\Omega |K(s,t) - K_n(s,t)|\mathrm{d}t. \tag{3.44}$$

We observe that the computational cost of the collocation method is much lower than that of the Galerkin method since it reduces the calculation of integrations involved.

3.4.1 The collocation method with piecewise polynomials

In this subsection, we consider the collocation method with the subspace $\mathbb{X}_n$ being a piecewise polynomial space. Suppose that there is a regular partition $\{\Omega_i : i \in \mathbb{N}_n\}$ of Ω for $n \in \mathbb{N}$, which satisfies

$$\Omega = \bigcup_{i\in\mathbb{N}_n} \Omega_i \text{ and } \operatorname{meas}(\Omega_j \cap \Omega_{j'}) = 0,\ j \neq j',$$

and for each $i \in \mathbb{N}_n$, there exists an invertible affine map ϕ_i which maps a reference element Ω^0 onto Ω_i. For a positive integer k, let $\mathbb{X}_n$ be the space of piecewise polynomials of total degree $\leq k-1$, with respect to the partition $\{\Omega_i : i \in \mathbb{N}_n\}$. We choose m distinct points $\tau_j \in \Omega^0$, $j \in \mathbb{N}_m$, such that the Lagrange interpolation polynomial of total degree $k-1$ at these points is uniquely defined. Then we can find $p_j \in \mathbb{P}_k$, polynomials of total degree $k-1$ such that

$$p_j(\tau_i) = \delta_{ij},\quad i,j \in \mathbb{N}_m.$$

For each $n \in \mathbb{N}_n$, the functions defined by

$$\rho_{ij}(t) := \begin{cases} (p_j \circ \phi_i^{-1})(t), & t \in \Omega_i \\ 0, & t \notin \Omega_i \end{cases} \tag{3.45}$$

form a basis for the space $\mathbb{X}_n$, and the points $t_{ij} := \phi_i(\tau_j), i, j \in \mathbb{N}_m$ form a set of collocation nodes satisfying

$$\rho_{ij}(t_{i'j'}) = \delta_{ii'}\delta_{jj'}.$$

We let $\mathcal{P}_n : C(\Omega) \to \mathbb{X}_n$ be the interpolating projection onto $\mathbb{X}_n$. We then have

$$\|\mathcal{P}_n x\|_\infty \leq \max\{\|\mathcal{P}_n x\|_{\infty,\Omega_i} : i \in \mathbb{N}_n\}.$$

Noting that

$$\mathcal{P}_n x(t) = \sum_{j\in\mathbb{N}_m} x(t_{ij})\rho_{ij}(t) = \sum_{j\in\mathbb{N}_m} x(t_{ij})p_j(\tau),\quad t \in \Omega_i, \tau = \phi_i^{-1}(t) \in \Omega^0,$$

we conclude that

$$\|\mathcal{P}_n\| \leq \sum_{j\in\mathbb{N}_m} \|p_j\|_\infty,$$

which means that the sequence of projections $\{\mathcal{P}_n\}$ is uniformly bounded. Moreover, for each $x \in W^{k,\infty}(\Omega)$ (cf. Section A.1 in the Appendix), there exists a positive constant c and a positive integer q such that for all $n \geq q$

$$\|x - \mathcal{P}_n x\|_\infty \leq c \inf\{\|x - v\|_\infty : v \in \mathbb{X}_n\} \leq c h_n^k \|x\|_{W^{k,\infty}}, \tag{3.46}$$

where $h_n := \max\{\operatorname{diam}(\Omega_i) : i \in \mathbb{Z}_n\}$. We have the following theorem.

Theorem 3.24 *Suppose that $\mathcal{K}: C(\Omega) \to C(\Omega)$ is a compact operator not having one as its eigenvalue. If $u \in C(\Omega)$ is the solution of equation (3.1), then there exist a positive constant c and a positive integer q such that for each $n \geq q$, equation (3.41) has a unique solution $u_n \in \mathbb{X}_n$. Moreover, if $u \in W^{k,\infty}(\Omega)$, then there exists a positive constant c such that for all n,*

$$\|u - u_n\|_\infty \leq c h_n^k \|u\|_{W^{k,\infty}(\Omega)}.$$

3.4.2 The collocation method with trigonometric polynomials

We consider in this subsection equation (3.1), in which $\Omega := [0, 2\pi]$ and K and f are 2π-periodic, and describe the collocation method for solving the equation using trigonometric polynomials.

Let $\mathbb{X} := C_p(0, 2\pi)$ be the space of all 2π-periodic continuous functions on $\mathbb{R}$ with uniform norm $\|\cdot\|_\infty$, and choose the approximate subspace as

$$\mathbb{X}_n := \operatorname{span}\{1, \cos t, \sin t, \ldots, \cos nt, \sin nt\}, \quad n \in \mathbb{N}.$$

To define an interpolating projection from $\mathbb{X}$ onto $\mathbb{X}_n$, we recall the Dirichlet kernel

$$D_n(t) := \frac{\sin(n + \frac{1}{2})t}{2 \sin \frac{t}{2}} = \frac{1}{2} + \sum_{j \in \mathbb{N}_n} \cos jt,$$

and observe that for $t_j := \frac{j\pi}{n+\frac{1}{2}}$, $j \in \mathbb{Z}_{2n+1}$,

$$D_n(t_j) = \begin{cases} \frac{1}{2} + n, & j = 0, \\ 0, & j \in \mathbb{Z}_{2n+1} \setminus \{0\}. \end{cases}$$

This means that functions ℓ_j defined by

$$\ell_j(t) := \frac{2}{2n+1} D_n(t - t_j), \quad j \in \mathbb{Z}_{2n+1}$$

satisfy

$$\ell_j(t_i) = \delta_{ij},$$

and form a Lagrange basis for $\mathbb{X}_n$. We then define the interpolating projection $\mathcal{P}_n : \mathbb{X} \to \mathbb{X}_n$ for $x \in \mathbb{X}$ by

$$\mathcal{P}_n x := \sum_{j \in \mathbb{Z}_{2n+1}} x(t_j) \ell_j.$$

The following estimate is known (cf. [279]):

$$\|\mathcal{P}_n\| = \mathcal{O}(\log n). \tag{3.47}$$

Hence, from the principle of uniform boundedness, there exists $x \in \mathbb{X}$ for which $\mathcal{P}_n x$ does not converge to x. The bound (3.47) leads to the fact that for $x \in \mathbb{X}$,

$$\begin{aligned}\|\mathcal{P}_n x - x\|_\infty &\leq (1 + \|\mathcal{P}_n\|) \inf\{\|x - v\|_\infty : v \in \mathbb{X}_n\} \\ &\leq \mathcal{O}(\log n) \inf\{\|x - v\|_\infty : v \in \mathbb{X}_n\}.\end{aligned}$$

We next consider the estimate of $\|\mathcal{K} - \mathcal{P}_n\mathcal{K}\|$. Assume that the kernel satisfies the α-Hölder continuity condition

$$|K(s_1, t) - K(s_2, t)| \leq c|s_1 - s_2|^\alpha, \quad \text{for all } s_1, s_2, t \in \Omega,$$

for some positive constant c. Then using (3.44) we conclude that

$$\|\mathcal{K} - \mathcal{P}_n\mathcal{K}\| \leq cn^{-\alpha} \log n.$$

3.4.3 The condition number for the collocation method

We now turn our attention to consideration of the condition number cond $(\mathbf{E}_n - \mathbf{K}_n)$ of the coefficient matrix of the linear system (3.43) obtained from the collocation method. In this case, the condition number is defined in terms of the infinity norm of a matrix $\mathbf{A}$, specifically,

$$\operatorname{cond}(\mathbf{A}) := \|\mathbf{A}\|_\infty \|\mathbf{A}^{-1}\|_\infty.$$

Theorem 3.25 *If* $\det(\mathbf{E}_n) \neq 0$ *then the condition number of the linear system (3.43) of the collocation method satisfies*

$$\operatorname{cond}(\mathbf{E}_n - \mathbf{K}_n) \leq \|\mathcal{P}_n\|_\infty^2 \operatorname{cond}(\mathbf{E}_n) \operatorname{cond}(\mathcal{I} - \mathcal{P}_n\mathcal{K}). \tag{3.48}$$

Proof For $\mathbf{g} := [g_j : j \in \mathbb{N}_{s(n)}]$, let

$$\mathbf{v} := [v_j : j \in \mathbb{N}_{s(n)}] = (\mathbf{E}_n - \mathbf{K}_n)^{-1}\mathbf{g}.$$

Choose $g \in C(\Omega)$ such that

$$\|g\|_\infty = \|\mathbf{g}\|_\infty \text{ and } g(t_j) = g_j, \ j \in \mathbb{N}_{s(n)}.$$

Set

$$v := (\mathcal{I} - \mathcal{P}_n\mathcal{K})^{-1}\mathcal{P}_n g.$$

Then we have that

$$v(t_i) = \sum_{j \in \mathbb{N}_{s(n)}} v_j \phi_j(t_i), \ i \in \mathbb{N}_{s(n)}.$$

Letting $\tilde{\mathbf{v}} := [v(t_i) : i \in \mathbb{N}_{s(n)}]$, in matrix notation, the above equation can be rewritten as

$$\mathbf{E}_n\mathbf{v} = \tilde{\mathbf{v}}. \tag{3.49}$$

We then conclude from

$$(\mathbf{E}_n - \mathbf{K}_n)^{-1}\mathbf{g} = \mathbf{v} = \mathbf{E}_n^{-1}\tilde{\mathbf{v}}$$

that

$$\|(\mathbf{E}_n - \mathbf{K}_n)^{-1}\mathbf{g}\|_\infty \leq \|\mathbf{E}_n^{-1}\|_\infty\|\tilde{\mathbf{v}}\|_\infty \leq \|\mathbf{E}_n^{-1}\|_\infty\|v\|_\infty.$$

Since

$$\|v\|_\infty = \|(\mathcal{I} - \mathcal{P}_n\mathcal{K})^{-1}\mathcal{P}_n g\|_\infty \leq \|(\mathcal{I} - \mathcal{P}_n\mathcal{K})^{-1}\|\|\mathcal{P}_n\|\|g\|_\infty,$$

we conclude that

$$\|(\mathbf{E}_n - \mathbf{K}_n)^{-1}\|_\infty \leq \|\mathcal{P}_n\|\|\mathbf{E}_n^{-1}\|_\infty\|(\mathcal{I} - \mathcal{P}_n\mathcal{K})^{-1}\|. \tag{3.50}$$

Moreover, for $\mathbf{v} := [v_j : j \in \mathbb{N}_{s(n)}]$, let

$$\mathbf{g} := [g_j : j \in \mathbb{N}_{s(n)}] = (\mathbf{E}_n - \mathbf{K}_n)\mathbf{v}.$$

We choose $g \in C(\Omega)$ as before and also set $v := (\mathcal{I} - \mathcal{P}_n\mathcal{K})^{-1}\mathcal{P}_n g$. Noting that

$$\mathcal{P}_n g(t_j) = g(t_j) = g_j,\ j \in \mathbb{N}_{s(n)},$$

we have $\|\mathbf{g}\|_\infty \leq \|\mathcal{P}_n g\|_\infty$. Thus,

$$\|(\mathbf{E}_n - \mathbf{K}_n)\mathbf{v}\|_\infty = \|\mathbf{g}\|_\infty \leq \|\mathcal{P}_n g\|_\infty = \|(\mathcal{I} - \mathcal{P}_n\mathcal{K})v\|_\infty. \tag{3.51}$$

Choose $\tilde{v} \in C(\Omega)$ such that

$$\|\tilde{v}\|_\infty = \|\tilde{\mathbf{v}}\|_\infty \quad \text{and} \quad \tilde{v}(t_j) = v(t_j),\ j \in \mathbb{N}_{s(n)}.$$

Then we have that $v = \mathcal{P}_n\tilde{v}$, and thus

$$\|v\|_\infty \leq \|\mathcal{P}_n\|\|\tilde{v}\|_\infty = \|\mathcal{P}_n\|_\infty\|\tilde{\mathbf{v}}\|_\infty \leq \|\mathcal{P}_n\|\|\mathbf{E}_n\|_\infty\|\mathbf{v}\|_\infty. \tag{3.52}$$

Combining estimates (3.51) and (3.52) yields

$$\|(\mathbf{E}_n - \mathbf{K}_n)\|_\infty \leq \|\mathcal{P}_n\|\|\mathbf{E}_n\|_\infty\|(\mathcal{I} - \mathcal{P}_n\mathcal{K})\|_\infty,$$

which with (3.50) leads to the desired result of this theorem. □

3.4.4 Discrete collocation methods

Before investigating discrete collocation methods we remark on the iterated collocation method. It was known that the iterated collocation method may not lead to superconvergence. In contrast with the iterated Galerkin method,

$$\|\mathcal{K}(\mathcal{I}-\mathcal{P}_n)\| \geq \|\mathcal{K}\|$$

holds. This means that the iterated collocation method converges more rapidly only in the case that for the solution u, $\mathcal{K}(\mathcal{I}-\mathcal{P}_n)u$ has superconvergence (see Section 2.2.4). This is the case when the approximate subspaces $\mathbb{X}_n$ are chosen as piecewise polynomials of even degree, and the kernel K and solution u are sufficiently smooth (cf. [15] for details).

We now begin to discuss discrete collocation methods. This approach replaces the integrals appearing in the collocation equation (3.42) by finite sums to be chosen depending on the specific numerical methods to be used. To this end, we define

$$(\mathcal{K}_n u)(s) := \sum_{j\in\mathbb{N}_{qn}} w_j K(s,\tau_j)u(\tau_j), \quad s \in \Omega.$$

Then the discrete collocation method for solving (3.1) is to find $u_n \in \mathbb{X}_n$ such that

$$(\mathcal{I}-\mathcal{P}_n\mathcal{K}_n)u_n = \mathcal{P}_n f,$$

or equivalently,

$$u_n(t_i) - \sum_{j\in\mathbb{N}_{qn}} w_j K(t_i,\tau_j)u_n(\tau_j) = f(t_i), \quad \text{for all } i \in \mathbb{N}_{s(n)}. \tag{3.53}$$

Some assumptions should be imposed to guarantee the unique solvability of the resulting system. The iterated discrete collocation solution is defined by

$$\hat{u}_n := f + \mathcal{K}_n u_n,$$

which is the solution of the equation

$$(\mathcal{I}-\mathcal{K}_n\mathcal{P}_n)\hat{u}_n = f. \tag{3.54}$$

The analysis of the discrete collocation method can be done by using the framework given in Section 2.2.4, with $\mathbb{X} = \mathbb{V} = C(\Omega)$.

We close this subsection by giving a relationship between the iterated discrete collocation solution and the Nyström solution. That is, if $\{\tau_j : j \in \mathbb{N}_{q_n}\} \subseteq \{t_j : j \in \mathbb{N}_{s(n)}\}$, then the iterated discrete collocation solution $\hat{u}_n$ is the Nyström solution satisfying

$$(\mathcal{I}-\mathcal{K}_n)\hat{u}_n = f. \tag{3.55}$$

In fact, by the definition of the interpolating projection for $x \in C(\Omega)$,

$$\mathcal{P}_n x(\tau_j) = x(\tau_j), \quad j \in \mathbb{N}_{q_n}.$$

This leads to

$$\mathcal{K}_n \mathcal{P}_n x(s) = \sum_{j \in \mathbb{N}_{q_n}} w_j K(s, \tau_j) \mathcal{P}_n x(\tau_j) = \sum_{j \in \mathbb{N}_{q_n}} w_j K(s, \tau_j) x(\tau_j) = \mathcal{K}_n x(s),$$

which with (3.54) yields (3.55).

3.5 Petrov–Galerkin methods

In this section, we establish a theoretical framework for the analysis of convergence for the Petrov–Galerkin method and superconvergence for the iterated Petrov–Galerkin method for Fredholm integral equations of the second kind.

Unlike the standard Galerkin method, the Petrov–Galerkin method employs a sequence of finite-dimensional subspaces to approximate the solution space (the trial space) of the equation and a different sequence to approximate the image space of the integral operator (the test space). This feature provides us with great freedom in choosing a pair of space sequences in order to improve the computational efficiency of the standard Galerkin method while preserving its convergence order. However, the space of sequences cannot be chosen arbitrarily. They must be coupled properly. This motivates us to develop a theoretical framework for convergence analysis of the Petrov–Galerkin method and the iterated Petrov–Galerkin method.

It is revealed in [77] that for the Petrov–Galerkin method the roles of the trial space and test space are to approximate the solution space of the equation and the range of the integral operator (or in other words, the image space), respectively. Therefore, the convergence order of the Petrov–Galerkin method is the same as the approximation order of the trial space and it is independent of the approximation order of the test space. This leads to the following strategy of choosing the trial and test spaces. We may choose the trial space as piecewise polynomials of a *higher* degree and the test space as piecewise polynomials of a *lower* degree but keep them with the same dimension. This choice of the trial and test spaces results in a significantly less expensive numerical algorithm in comparison with the standard Galerkin method with the same convergence order, which uses the same piecewise polynomials as those for the trial space. The saving comes from computing the entries of the matrix and the right-hand-side vector of the linear system that results from the corresponding discretization. Note that an entry of the *Galerkin matrix* is the inner product of the integral operator applied to a basis function for the trial space against a basis function for the same space, which is a piecewise polynomial of a *higher* degree, while an entry of the *Petrov–Galerkin matrix* is that against a basis function for the test space, which is a piecewise polynomial of a *lower* degree. Computing the latter is less expensive than computing the former, due to the use of lower-degree polynomials for the test space. In fact, the Petrov–Galerkin method *interpolates* between the Galerkin method and the collocation method.

3.5.1 Analysis of Petrov–Galerkin and iterated Petrov–Galerkin methods

1. The Petrov–Galerkin method

Let $\mathbb{X}$ be a Banach space with the norm $\|\cdot\|$ and let $\mathbb{X}^*$ denote its dual space. Assume that $\mathcal{K} : \mathbb{X} \to \mathbb{X}$ is a compact linear operator. We consider the Fredholm equation of the second kind:

$$u - \mathcal{K}u = f, \;\; f \in \mathbb{X}, \tag{3.56}$$

where $u \in \mathbb{X}$ is the unknown to be determined.

We choose two sequences of finite-dimensional subspaces $\{\mathbb{X}_n \subset \mathbb{X} : n \in \mathbb{N}\}$ and $\{\mathbb{Y}_n \subset \mathbb{X}^* : n \in \mathbb{N}\}$ and suppose that they satisfy condition (H): For each $x \in \mathbb{X}$ and $y \in \mathbb{X}^*$, there exist $x_n \in \mathbb{X}_n$ and $y_n \in \mathbb{Y}_n$ such that $\|x_n - x\| \to 0$ and $\|y_n - y\| \to 0$ as $n \to \infty$, and

$$s(n) := \dim \mathbb{X}_n = \dim \mathbb{Y}_n, \;\; n \in \mathbb{N}. \tag{3.57}$$

The Petrov–Galerkin method for equation (3.56) is a numerical method for finding $u_n \in \mathbb{X}_n$ such that

$$(u_n - \mathcal{K}u_n, y) = (f, y), \;\; \text{for all } y \in \mathbb{Y}_n. \tag{3.58}$$

Let

$$\mathbb{X}_n = \operatorname{span}\{\phi_1, \phi_2, \ldots, \phi_{s(n)}\},$$
$$\mathbb{Y}_n = \operatorname{span}\{\psi_1, \psi_2, \ldots, \psi_{s(n)}\}$$

and

$$u_n = \sum_{j \in \mathbb{N}_{s(n)}} \alpha_j \phi_j.$$

Equation (3.58) can be written as

$$\sum_{j \in \mathbb{N}_{s(n)}} \alpha_j [(\phi_j, \psi_i) - (\mathcal{K}\phi_j, \psi_i)] = (f, \psi_i), \;\; i \in \mathbb{N}_{s(n)}.$$

If $\{\mathbb{X}_n, \mathbb{Y}_n\}$ is a regular pair (see Definition 2.30) in the sense that there is a linear operator $\Pi_n : \mathbb{X}_n \to \mathbb{Y}_n$ with $\Pi_n \mathbb{X}_n = \mathbb{Y}_n$ and satisfying the conditions

$$\|x\| \leq c_1 (x, \Pi_n x)^{1/2} \text{ and } \|\Pi_n x\| \leq c_2 \|x\|, \quad \text{for all } x \in \mathbb{X}_n,$$

where c_1 and c_2 are positive constants independent of n, then equation (3.58) may be rewritten as

$$(u_n - \mathcal{K} u_n, \Pi_n x) = (f, \Pi_n x), \quad \text{for all } x \in \mathbb{X}_n.$$

Furthermore, using the generalized best approximation projection $\mathcal{P}_n : \mathbb{X} \to \mathbb{X}_n$ (see Definition 2.25), which is defined by

$$(x - \mathcal{P}_n x, y) = 0, \quad \text{for all } y \in \mathbb{Y}_n,$$

equation (3.58) is equivalent to the operator equation

$$u_n - \mathcal{P}_n \mathcal{K} u_n = \mathcal{P}_n f. \tag{3.59}$$

This equation indicates that the Petrov–Galerkin method is a projection method. Using Theorem 2.55, we obtain the following result.

Theorem 3.26 *Let $\mathbb{X}$ be a Banach space and $\mathcal{K} : \mathbb{X} \to \mathbb{X}$ be a compact linear operator. Assume that one is not an eigenvalue of the operator $\mathcal{K}$. Suppose that $\mathbb{X}_n$ and $\mathbb{Y}_n$ satisfy condition (H) and $\{\mathbb{X}_n, \mathbb{Y}_n\}$ is a regular pair. Then there exists an $N_0 > 0$ such that for $n \geq N_0$, equation (3.59) has a unique solution $u_n \in \mathbb{X}_n$ for any given $f \in \mathbb{X}$ that satisfies*

$$\|u_n - u\| \leq c\|u - \mathcal{P}_n u\|, \quad n \geq N_0,$$

where $u \in \mathbb{X}$ is the unique solution of equation (3.56) and $c > 0$ is a constant independent of n.

2. The iterated Petrov–Galerkin method

We now turn our attention to studying superconvergence of the iterated Petrov–Galerkin method for integral equations of the second kind.

Let $\mathbb{X}$ be a Banach space and let $\{\mathbb{X}_n \subset \mathbb{X} : n \in \mathbb{N}\}$ and $\{\mathbb{Y}_n \subset \mathbb{X}^* : n \in \mathbb{N}\}$ be two sequences of finite-dimensional subspaces satisfying condition (H). Assume that $\mathcal{P}_n : \mathbb{X} \to \mathbb{X}_n$ are the linear projections of the generalized best approximation from $\mathbb{X}_n$ to $\mathbb{X}$ with respect to $\mathbb{Y}_n$. Consider the projection method (3.59). Suppose that $u_n \in \mathbb{X}_n$ is the unique solution of equation (3.59) which approximates the solution of equation (3.56). The iterated projection method is defined by

$$u_n' := f + \mathcal{K} u_n. \tag{3.60}$$

It can easily be verified that the iterated projection approximation u_n' satisfies the integral equation

$$u_n' - \mathcal{K}\mathcal{P}_n u_n' = f. \tag{3.61}$$

In order to analyze u_n' as the solution of equation (3.61), we need to understand the convergence of the approximate operator $\mathcal{K}\mathcal{P}_n$. The next lemma is helpful in this regard.

Lemma 3.27 *Suppose that $\mathbb{X}$ is a Banach space and $\mathbb{X}_n \subset \mathbb{X}$ and $\mathbb{Y}_n \subset \mathbb{X}^*$ satisfy condition (H). Let $\mathcal{P}_n : \mathbb{X} \to \mathbb{X}$ be the sequence of projections of the generalized best approximation from $\mathbb{X}$ to $\mathbb{X}_n$ with respect to $\mathbb{Y}_n$ that converges pointwise to the identity operator $\mathcal{I}$ in $\mathbb{X}$. Then the sequence of dual operators $\mathcal{P}_n^*$ converges pointwise to the identity operator $\mathcal{I}^*$ in $\mathbb{X}^*$.*

Proof It follows from condition (H) that for any $v \in \mathbb{X}^*$, there exists a sequence $v_n \in \mathbb{Y}_n$ such that $\|v_n - v\| \to 0$ as $n \to \infty$. Consequently,

$$\|\mathcal{P}_n^* v - v\| \le \|\mathcal{P}_n^* v - v_n\| + \|v_n - v\| \le (\|\mathcal{P}_n\| + 1)\|v_n - v\| \to 0,$$

where the first inequality holds because $\mathcal{P}_n^* : \mathbb{X}^* \to \mathbb{Y}_n$ are also projections. That is, $\mathcal{P}_n^* \to \mathcal{I}^*$ pointwise. The second inequality uses the general result $\|\mathcal{P}_n^*\| = \|\mathcal{P}_n\|$. □

Theorem 3.28 *Suppose that $\mathbb{X}$ is a Banach space and $\mathbb{X}_n \subset \mathbb{X}$ and $\mathbb{Y}_n \subset \mathbb{X}^*$ satisfy condition (H). Assume that $\mathcal{K}$ is a compact operator in $\mathbb{X}$. Let $\mathcal{P}_n : \mathbb{X} \to \mathbb{X}$ be the projections of the generalized best approximation from $\mathbb{X}$ to $\mathbb{X}_n$ with respect to $\mathbb{Y}_n$ that converges pointwise to the identity operator $\mathcal{I}$ in $\mathbb{X}$. Then*

$$\|\mathcal{K}\mathcal{P}_n - \mathcal{K}\| \to 0, \text{ as } n \to \infty.$$

Proof Note that

$$\|\mathcal{K}\mathcal{P}_n - \mathcal{K}\| = \|[\mathcal{K}\mathcal{P}_n - \mathcal{K}]^*\| = \|\mathcal{P}_n^*\mathcal{K}^* - \mathcal{K}^*\|.$$

Since $\mathcal{K}$ is compact, we also have that $\mathcal{K}^*$ is compact. Using Lemmas 3.27 and 2.52, we conclude the result of this theorem. □

3.5.2 Equivalent conditions in Hilbert spaces for regular pairs

From the last section, we know that the notion of regular pairs plays an essential role in the analysis of the Petrov–Galerkin method. Therefore, it is necessary to re-examine this concept from different points of view. In this subsection, we first study regular pairs from a geometric point of view and second characterize them in terms of the uniform boundedness of the sequence of projections defined by the generalized best approximation.

In what follows, we confine the space $\mathbb{X}$ to be a Hilbert space with inner product $(\cdot,\cdot)$, from which a norm $\|\cdot\|$ is induced. In this case, $\mathbb{X}^*$ is identified to be $\mathbb{X}$ via the inner product. We assume $\mathbb{X}_n, \mathbb{Y}_n \subset \mathbb{X}$ satisfy condition (H). The structure of Hilbert spaces allows us to define the angle between spaces $\mathbb{X}_n$ and $\mathbb{Y}_n$, which is done by the orthogonal projection from $\mathbb{X}$ onto $\mathbb{Y}_n$. For each $x \in \mathbb{X}$, we define the best approximation y_n^* from $\mathbb{Y}_n$ by

$$\|x - y_n^*\| = \inf\{\|x - y\| : y \in \mathbb{Y}_n\}.$$

Since $\mathbb{Y}_n$ is a finite-dimensional Hilbert subspace in $\mathbb{X}$, there exists a best approximation from $\mathbb{Y}_n$ to $x \in \mathbb{X}$. We furthermore define the best approximation operator $\mathcal{Y}_n$ by

$$\mathcal{Y}_n x = y_n^* \quad \text{for each} \quad x \in \mathbb{X}.$$

It is well known that for any $x \in \mathbb{X}$, $\mathcal{Y}_n x$ satisfies the equation

$$(x - \mathcal{Y}_n x, y) = 0, \quad \text{for all } y \in \mathbb{Y}_n. \tag{3.62}$$

In other words, the operator $\mathcal{Y}_n$ is the orthogonal projection from $\mathbb{X}$ onto $\mathbb{Y}_n$.

To define the angle between two spaces $\mathbb{X}_n$ and $\mathbb{Y}_n$, we denote

$$\gamma_n := \inf\left\{\frac{\|\mathcal{Y}_n x\|}{\|x\|} : x \in \mathbb{X}_n\right\}.$$

We call

$$\theta_n := \arccos \gamma_n$$

the *angle* between spaces $\mathbb{X}_n$ and $\mathbb{Y}_n$. The next theorem characterizes a regular pair $\{\mathbb{X}_n, \mathbb{Y}_n\}$ in a Hilbert space $\mathbb{X}$ in terms of the angles between $\mathbb{X}_n$ and $\mathbb{Y}_n$.

Theorem 3.29 *Let $\mathbb{X}$ be a Hilbert space and let $\mathbb{X}_n$ and $\mathbb{Y}_n$ be two subspaces of $\mathbb{X}$ satisfying condition (H) and* $\dim\mathbb{X}_n = \dim\mathbb{Y}_n < \infty$, *for $n \in \mathbb{N}$. Then $\{\mathbb{X}_n, \mathbb{Y}_n\}$ is a regular pair if and only if there exists a positive number $\theta_0 < \pi/2$ such that*

$$\theta_n \le \theta_0, \quad n \in \mathbb{N}.$$

Proof We first prove the sufficiency. Assume that there exists a positive number $\theta_0 < \pi/2$ such that $\theta_n \le \theta_0$ for all $n \in \mathbb{N}$. Thus

$$\gamma_n = \inf\left\{\frac{\|\mathcal{Y}_n x\|}{\|x\|} : x \in \mathbb{X}_n\right\} \ge \cos\theta_0 > 0.$$

Using the characterization of the best approximation, we have that

$$(x, \mathcal{Y}_n x) = \|\mathcal{Y}_n x\|^2 \geq \cos^2\theta_0 \|x\|^2, \quad \text{for all } x \in \mathbb{X}_n.$$

This implies that $\mathcal{Y}_n \mathbb{X}_n = \mathbb{Y}_n$ and condition (H-1) holds with $c_1 = 1/\cos\theta_0$. Moreover, since the operator $\mathcal{Y}_n$ is the orthogonal projection, we conclude that

$$\|\mathcal{Y}_n x\| \leq \|x\|, \quad \text{for all } x \in \mathbb{X}_n.$$

Hence, condition (H-2) holds with $c_2 = 1$.

We now show the necessity. It follows from the definition of a regular pair that

$$\|x\|^2 \leq c_1^2(x, \Pi_n x) \leq c_1^2\|x\|\|\Pi_n x\| \leq c_1^2 c_2 \|x\|^2, \quad \text{for all } x \in \mathbb{X}_n.$$

Thus, we obtain

$$0 < \frac{1}{c_1^2 c_2} \leq 1.$$

It can be seen that there exists an $x' \in \mathbb{X}_n$ with $x' \neq 0$ such that

$$\frac{\|\mathcal{Y}_n x'\|}{\|x'\|} = \inf\left\{\frac{\|\mathcal{Y}_n x\|}{\|x\|} : x \in \mathbb{X}_n\right\} = \cos\theta_n.$$

By the characterization of the best approximation, we obtain that

$$\|x'\|^2 \leq c_1^2(x', \Pi_n x') = c_1^2(\mathcal{Y}_n x', \Pi_n x') \leq c_1^2\|\mathcal{Y}_n x'\|\|\Pi_n x'\| \leq c_1^2 c_2 \cos\theta_n \|x'\|^2.$$

Therefore,

$$\cos\theta_n \geq \frac{1}{c_1^2 c_2} > 0$$

and

$$\theta_n \leq \arccos\frac{1}{c_1^2 c_2} < \frac{\pi}{2}.$$

The proof is complete. □

We now turn to establishing the equivalence of the regular pair and the uniform boundedness of the projections $\mathcal{P}_n$ when they are well defined. We need two preliminary results to prove this equivalence.

Lemma 3.30 *Let $\mathbb{X}$ be a Hilbert space. Assume that $\mathbb{X}_n, \mathbb{Y}_n \subset \mathbb{X}$ with* $\dim \mathbb{X}_n = \dim \mathbb{Y}_n < \infty$ *satisfy condition (2.73). Then*

$$\mathcal{Y}_n = \mathcal{Y}_n \mathcal{P}_n.$$

Proof Let $x \in \mathbb{X}$. Then $\mathcal{P}_n x$ and $\mathcal{Y}_n x$ satisfy equations (2.72) and (3.62), respectively. It follows that

$$(\mathcal{P}_n x - \mathcal{Y}_n x, y) = 0, \quad \text{for all } y \in \mathbb{Y}_n.$$

By the definition of best approximation from $\mathbb{Y}_n$ to $\mathcal{P}_n x$, we conclude that

$$\mathcal{Y}_n x = \mathcal{Y}_n \mathcal{P}_n x, \quad \text{for all } x \in \mathbb{X}.$$

The proof of the lemma is complete. □

In Hilbert spaces, we can interpret condition (2.73) from many different points of view. The next proposition lists eight equivalent statements.

Proposition 3.31 *Let $\mathbb{X}$ be a Hilbert space and $\mathbb{X}_n, \mathbb{Y}_n \subset \mathbb{X}$ with* $\dim \mathbb{X}_n = \dim \mathbb{Y}_n < \infty$. *Then the following statements are equivalent.*

(i) $\mathbb{Y}_n \cap \mathbb{X}_n^{\perp} = \{0\}$.
(ii) $\det[(\phi_i, \psi_j)] \neq 0$, *where* $\{\phi_l : l \in \mathbb{N}_m\}$ *and* $\{\psi_l : l \in \mathbb{N}_m\}$ *are bases for* $\mathbb{X}_n$ *and* $\mathbb{Y}_n$, *respectively.*
(iii) $\mathbb{X}_n \cap \mathbb{Y}_n^{\perp} = \{0\}$.
(iv) If $x \in \mathbb{X}_n$ *with* $x \neq 0$, *then* $\mathcal{Y}_n x \neq 0$.
(v) $\gamma_n > 0$.
(vi) $\mathcal{Y}_n \mathbb{X}_n = \mathbb{Y}_n$.
(vii) $\mathcal{P}_n \mathbb{Y}_n = \mathbb{X}_n$ *and* $\mathcal{P}_n|_{\mathbb{Y}_n} = (\mathcal{Y}_n|_{\mathbb{X}_n})^{-1}$.
(viii) If $y \in \mathbb{Y}_n$ *with* $y \neq 0$, *then* $\mathcal{P}_n y \neq 0$.

Proof The implications that (i) implies (ii) and (ii) implies (iii) follow from the proof of Proposition 2.26.

We prove that (iii) implies (iv). Let $x \in \mathbb{X}_n$ with $x \neq 0$. Using the definition of $\mathcal{Y}_n$ and (iii), we conclude that

$$(\mathcal{Y}_n x, y) = (x, y) \neq 0, \quad \text{for all } y \in \mathbb{Y}_n \text{ with } y \neq 0.$$

Thus, $\mathcal{Y}_n x \neq 0$.

To prove the implication that (iv) implies (v), we use (iv) to conclude that on the closed unit sphere $\{x : x \in \mathbb{X}_n, \|x\| = 1\}$, $\|\mathcal{Y}_n x\| > 0$. Thus, we have

$$\gamma_n = \inf\{\|\mathcal{Y}_n x\| : x \in \mathbb{X}_n, \|x\| = 1\} > 0,$$

and statement (v) is proved.

To establish (vi) it suffices to show that if $\{\phi_l : l \in \mathbb{N}_m\}$ is a basis for $\mathbb{X}_n$, then $\{\mathcal{Y}_n \phi_l : l \in \mathbb{Z}_m\}$ is a basis for $\mathbb{Y}_n$. Note that $\mathcal{Y}_n \phi_i \in \mathbb{Y}_n$ and $\dim \mathbb{Y}_n = N$. It remains to show that $\mathcal{Y}_n \phi_1, \ldots, \mathcal{Y}_n \phi_N$ is linearly independent. To this end, assume that there are not all zero constants $c_1, \ldots, c_N$ such that

$$\sum_{j \in \mathbb{N}_N} c_i \mathcal{Y}_n \phi_i = 0.$$

Let $x := \sum_{j \in \mathbb{N}_N} c_i \phi_i$. Then $x \in \mathbb{X}_n$ with $x \neq 0$ but $\mathcal{Y}_n x = 0$. Hence, $\gamma_n = 0$, a contradiction to (v).

We now show that (vi) implies (vii). Since $\mathcal{P}_n \mathbb{Y}_n = \mathcal{P}_n \mathcal{Y}_n \mathbb{X}_n$, it is sufficient to prove $\mathcal{P}_n \mathcal{Y}_n \mathbb{X}_n = \mathbb{X}_n$. For any $x \in \mathbb{X}_n$, applying the definition of $\mathcal{Y}_n$ gives

$$(\mathcal{Y}_n x - x, y) = 0, \quad \text{for all } y \in \mathbb{Y}_n.$$

The definition of $\mathcal{P}_n$ implies that $x = \mathcal{P}_n \mathcal{Y}_n x$ for all $x \in \mathbb{X}_n$. Hence, we conclude that $\mathcal{P}_n \mathcal{Y}_n \mathbb{X}_n = \mathbb{X}_n$ and (vii) is established.

The implication of (vii) to (viii) is obvious.

Finally, we prove that (viii) implies (i). Let $y \in \mathbb{Y}_n \cap \mathbb{X}_n^{\perp}$. By the definition of $\mathcal{P}_n$, we find

$$\|y\|^2 = (y, y) = (\mathcal{P}_n y, y) = 0.$$

This ensures that $y = 0$. □

The next theorem shows that $\{\mathbb{X}_n, \mathbb{Y}_n\}$ is a regular pair if and only if the sequence of projections $\{\mathcal{P}_n\}$ is uniformly bounded.

Theorem 3.32 *Let $\mathbb{X}$ be a Hilbert space. Assume that $\mathbb{X}_n, \mathbb{Y}_n \subset \mathbb{X}$ satisfy condition (H) and equation (2.73). Let $\{\mathcal{P}_n\}$ be a sequence of projections defined by the generalized best approximation (2.72). Then $\{\mathbb{X}_n, \mathbb{Y}_n\}$ is a regular pair if and only if there exists a positive constant c for which*

$$\|\mathcal{P}_n\| \leq c, \quad \text{for all } n \in \mathbb{N}.$$

Proof We have proved in Proposition 2.31 that if $\{\mathbb{X}_n, \mathbb{Y}_n\}$ is a regular pair, then $\mathcal{P}_n$ is uniformly bounded. It remains to prove the converse. For this purpose, we let $\mathcal{Y}_n : \mathbb{X} \to \mathbb{Y}_n$ be the orthogonal projection. By our convention, spaces $\mathbb{X}_n$ and $\mathbb{Y}_n$ satisfy condition (2.73). Thus, Proposition 3.31 ensures that $\mathcal{Y}_n \mathbb{X}_n = \mathbb{Y}_n$. The validity of condition (H-2) follows from a property of best approximation in Hilbert spaces. We now prove condition (H-1) by contradiction. Assume to the contrary that condition (H-1) is not valid. Then, to each ϵ with $0 < \epsilon < 1/c$, where c is the constant that gives the bound for $\|\mathcal{P}_n\|$, there exist $\hat{n} \in \mathbb{N}$ and $\hat{x} \in \mathbb{X}_{\hat{n}}$ such that

$$(\hat{x}, \mathcal{Y}_{\hat{n}} \hat{x}) < \epsilon^2 \|\hat{x}\|^2.$$

It follows from the characterization of best approximation in the Hilbert space $\mathbb{X}$ that

$$(\hat{x}, \mathcal{Y}_{\hat{n}}\hat{x}) = \|\mathcal{Y}_{\hat{n}}\hat{x}\|^2.$$

We then have

$$\|\mathcal{Y}_{\hat{n}}\hat{x}\| < \epsilon\|\hat{x}\|.$$

Let $x_0 := \mathcal{Y}_{\hat{n}}\hat{x}$. Clearly, $x_0 \in \mathbb{Y}_{\hat{n}}$. Since $\hat{x} \in \mathbb{X}_{\hat{n}}$ satisfies the equation

$$(x_0 - \hat{x}, y) = (\mathcal{Y}_n\hat{x} - \hat{x}, y) = 0, \quad \text{for all } y \in \mathbb{Y}_{\hat{n}},$$

we conclude that $\hat{x} = \mathcal{P}_{\hat{n}}x_0$. Consequently,

$$\|x_0\| < \epsilon\|\mathcal{P}_{\hat{n}}x_0\|.$$

In other words,

$$\|\mathcal{P}_{\hat{n}}x_0\| > c\|x_0\|,$$

which contradicts the assumption that $\|\mathcal{P}_n\| \leq c$. This contradiction shows that condition (H-1) must hold. □

In the remaining part of this subsection we discuss regular pairs from an algebraic point of view.

Definition 3.33 Let $X := \{\phi_i : i \in \mathbb{Z}_m\}, Y := \{\psi_i : i \in \mathbb{Z}_m\}$ be two finite (ordered) subsets of the Hilbert space $\mathbb{X}$. The correlation matrix between X and Y is defined to be the $m \times m$ matrix

$$\mathbf{G}(X, Y) := [(\phi_i, \psi_j) : i, j \in \mathbb{Z}_m].$$

Note that $\mathbf{G}^T(X, Y)$, the transpose of $\mathbf{G}(X, Y)$, is $\mathbf{G}(Y, X)$. For the special case $X = Y$ we use $\mathbf{G}(X)$ for $\mathbf{G}(X, X)$ and recall that $\mathbf{G}(X)$ is the *Gram matrix* for the set X. The matrix $\mathbf{G}(X)$ is positive semi-definite. Generally, the matrix $\mathbf{G}(X, Y)$ is not symmetric. We use $\mathbf{G}_+(X, Y)$ to denote the symmetric part of $\mathbf{G}(X, Y)$. Specifically, we set

$$\mathbf{G}_+(X, Y) := \frac{1}{2}[\mathbf{G}(X, Y) + \mathbf{G}(Y, X)].$$

We use the standard ordering on $m \times m$ symmetric matrices $\mathbf{A} = [a_{ij} : i, j \in \mathbb{Z}_m], \mathbf{B} = [b_{ij} : i, j \in \mathbb{Z}_m]$ and write $\mathbf{A} \leq \mathbf{B}$ provided that

$$\sum_{i \in \mathbb{Z}_m} \sum_{j \in \mathbb{Z}_m} x_i a_{ij} x_j := \mathbf{x}^T\mathbf{A}\mathbf{x} \leq \mathbf{x}^T\mathbf{B}\mathbf{x}$$

for all $\mathbf{x} := [x_i : i \in \mathbb{Z}_m] \in \mathbb{R}^m$. When the strict inequality holds above except for $\mathbf{x} = 0$, we write $\mathbf{A} < \mathbf{B}$.

Definition 3.34 Let $\mathbb{X}$ be a Hilbert space. Suppose for any $n \in \mathbb{N}$, $X_n := \{\phi_i : i \in \mathbb{Z}_{s(n)}\}$ and $Y_n := \{\psi_i : i \in \mathbb{Z}_{s(n)}\}$ are finite subsets of $\mathbb{X}$, where $s(n)$ denotes the cardinality of X_n. We say that $\{X_n, Y_n\}$ forms a regular pair provided that there are constants $\sigma > 0$ and $\sigma' > 0$ such that for all $n \in \mathbb{N}$ we have

$$0 < \mathbf{G}(X_n) \leq \sigma \mathbf{G}_+(X_n, Y_n) \tag{3.63}$$

and

$$0 < \mathbf{G}(Y_n) \leq \sigma' \mathbf{G}(X_n). \tag{3.64}$$

Thus, given any finite sets X and Y of linearly independent elements in $\mathbb{X}$ of the same cardinality, the *constant* pair $\{X, Y\}$ is regular if and only if $\mathbf{G}_+(X, Y) > 0$. Moreover, when we only have that $\det \mathbf{G}(X, Y) \neq 0$, we can form from X and Y a constant regular pair by modifying either one of the sets X and Y. To explain this we suppose that $X = \{\phi_i : i \in \mathbb{Z}_n\}$ and $Y = \{\psi_i : i \in \mathbb{Z}_n\}$. Let $W = \{\omega_i : i \in \mathbb{Z}_n\}$, where the elements of this set are defined by the formula

$$\omega_i := \sum_{j \in \mathbb{Z}_n} (\phi_j, \psi_i)\phi_j, \quad i \in \mathbb{Z}_n.$$

Then

$$\mathbf{G}(W, Y) = \mathbf{G}(X, Y)^T \mathbf{G}(X, Y)$$

and so $\{W, Y\}$ is a constant regular pair when $\det \mathbf{G}(X, Y) \neq 0$ and the elements of X and Y are linearly independent. In the special case that the elements of X are orthonormal then $\omega_i = \mathcal{X}\psi_i, i \in \mathbb{Z}_n$, where $\mathcal{X}$ is the orthogonal projection of $\mathbb{X}$ onto span$X$.

Let $\mathbb{X}_n :=$ span X_n and $\mathbb{Y}_n :=$ span Y_n. When $\{X_n, Y_n\}$ form a regular pair of finite sets and for every $x \in \mathbb{X}$, $\lim_{n\to\infty} \operatorname{dist}(x, \mathbb{X}_n) = 0$, the subspaces $\{\mathbb{X}_n, \mathbb{Y}_n\}$ form a regular pair of subspaces in the terminology of Definition 2.30. Conversely, whenever two subspaces $\{\mathbb{X}_n, \mathbb{Y}_n\}$ form a regular pair, these subspaces have bases which as sets form a regular pair. The notion of regular pairs of subspaces from Definition 2.30 is independent of the bases of the subspaces. However, Definition 3.34 is dependent upon the specific sets used and may fail to hold if these sets are transformed into others by linear transformations.

Let us observe that (3.63) and (3.64) imply

$$\mathbf{G}(Y_n) \leq \sigma\sigma' \mathbf{G}_+(X_n, Y_n). \tag{3.65}$$

Moreover, for any $\mathbf{a} = [a_i : i \in \mathbb{Z}_{s(n)}] \in \mathbb{R}^{s(n)}$, by the Cauchy–Schwarz inequality and (3.63), we have that

$$\begin{aligned}\mathbf{a}^T\mathbf{G}_+(X_n, Y_n)\mathbf{a} &= \left(\sum_{j\in\mathbb{Z}_{s(n)}} a_j\phi_j, \sum_{j\in\mathbb{Z}_{s(n)}} a_j\psi_j\right)\\ &\le [\mathbf{a}^T\mathbf{G}(X_n)\mathbf{a}]^{1/2}[\mathbf{a}^T\mathbf{G}(Y_n)\mathbf{a}]^{1/2}\\ &\le \sigma^{1/2}[\mathbf{a}^T\mathbf{G}_+(X_n, Y_n)\mathbf{a}]^{1/2}[\mathbf{a}^T\mathbf{G}(Y_n)\mathbf{a}]^{1/2}.\end{aligned}$$

This inequality implies that

$$\mathbf{G}_+(X_n, Y_n) \le \sigma\mathbf{G}(Y_n).$$

Using this inequality and (3.63), we conclude that

$$\mathbf{G}(X_n) \le \sigma\mathbf{G}_+(X_n, Y_n) \le \sigma^2\mathbf{G}(Y_n). \tag{3.66}$$

Therefore, it follows that whenever $\{X_n, Y_n\}$ is a regular pair then so is $\{Y_n, X_n\}$.

When the sets $\{X_n, Y_n\}$ form a regular pair with constants σ, σ', the generalized best approximation projection $\mathcal{P}_n : \mathbb{X} \to \mathbb{X}_n$ with respect to $\mathbb{Y}_n$ enjoys the bound

$$\|\mathcal{P}_n\| \le p := \sigma\sigma'^{1/2}. \tag{3.67}$$

To confirm this inequality for each $x \in \mathbb{X}$ we write $\mathcal{P}_n x$ in the form

$$\mathcal{P}_n x = \sum_{j\in\mathbb{Z}_{s(n)}} a_j\phi_j,$$

where the vector $\mathbf{a} = [a_j : j \in \mathbb{Z}_{s(n)}]$ is the solution of the linear equations

$$(x, \psi_i) = \Big(\sum_{j\in\mathbb{Z}_{s(n)}} a_j\phi_j, \psi_i\Big), \quad i \in \mathbb{Z}_{s(n)}.$$

Hence, multiplying both sides of these equations by a_i, summing over $i \in \mathbb{Z}_{s(n)}$ and using (3.63) and (3.64), we get that

$$\begin{aligned}\|\mathcal{P}_n x\|^2 &= \mathbf{a}^T\mathbf{G}(X_n)\mathbf{a} \le \sigma\mathbf{a}^T\mathbf{G}_+(X_n, Y_n)\mathbf{a}\\ &= \sigma\Big(\mathcal{P}_n x, \sum_{j\in\mathbb{Z}_{s(n)}} a_j\psi_j\Big) = \sigma\Big(x, \sum_{j\in\mathbb{Z}_{s(n)}} a_j\psi_j\Big)\\ &\le \sigma\|x\|\Big\|\sum_{j\in\mathbb{Z}_{s(n)}} a_j\psi_j\Big\| \le \sigma\sigma'^{1/2}\|x\|\Big\|\sum_{j\in\mathbb{Z}_{s(n)}} a_j\phi_j\Big\| = \sigma\sigma'^{1/2}\|x\|\,\|\mathcal{P}_n x\|.\end{aligned}$$

Now, we divide the first and last terms in the above inequality by $\|\mathcal{P}_n x\|$ to yield the desired inequality $\|\mathcal{P}_n\| \le p$.

Recall that

$$\gamma_n := \inf\left\{\frac{\|\mathcal{Y}_n x\|}{\|x\|} : x \in \mathbb{X}_n\right\}$$

and

$$\theta_n := \arccos \gamma_n, \quad n \in \mathbb{N}.$$

Note that $\gamma_n \in [0, 1]$ and $\theta_n \in [0, \pi/2]$. By definition, θ_n is the angle between the two subspaces $\mathbb{X}_n$ and $\mathbb{Y}_n$. Let us observe for any $x \in \mathbb{X}_n$ that the equation

$$\mathcal{P}_n \mathcal{Y}_n x = x$$

holds. This leads to

$$\cos\theta_n = \inf\left\{\frac{\|\mathcal{Y}_n x\|}{\|\mathcal{P}_n \mathcal{Y}_n x\|} : x \in \mathbb{X}_n\right\} \geq \inf\left\{\frac{\|y\|}{\|\mathcal{P}_n y\|} : y \in \mathbb{Y}_n\right\} \geq \frac{1}{\|\mathcal{P}_n\|}.$$

Therefore, we conclude that when $\{\mathbb{X}_n, \mathbb{Y}_n\}$ is a pair of subspaces with bases $\{X_n, Y_n\}$ which form a regular pair with constants σ, σ', the inequality

$$\cos\theta_n \geq p^{-1} > 0$$

holds. In other words, in this case, for all $n \in \mathbb{N}$ we have that $\theta_n \in [0, \theta^*)$, where $\theta^* < \pi/2$.

3.5.3 The discrete Petrov–Galerkin method and its iterated scheme

The Petrov–Galerkin method for Fredholm integral equations of the second kind was studied in the last sections. To use the Petrov–Galerkin method in practical computation, we have to be able to efficiently compute the integrals occurring in the method. In this subsection, we take an approach to discretizing a given integral equation by a discrete projection and a discrete inner product. The iterated solution suggested in this section is also fully discrete.

In this subsection we describe discrete Petrov–Galerkin methods for Fredholm integral equations of the second kind with weakly singular kernels. For this purpose, we consider the equation

$$(\mathcal{I} - \mathcal{K})u = f, \tag{3.68}$$

where $\mathcal{K} : L^\infty(\Omega) \to C(\Omega)$ is a compact linear integral operator defined by

$$(\mathcal{K}u)(s) = \int_\Omega K(s, t)u(t)\mathrm{d}t, \quad s \in \Omega, \tag{3.69}$$

$\Omega \subset \mathbb{R}^d$ is a bounded closed domain, and K is a function defined on $\Omega \times \Omega$ which is allowed to have weak singularities. We assume that one is not an eigenvalue of the operator $\mathcal{K}$ to guarantee the existence of a unique solution $u \in L^\infty(\Omega)$. Some additional specific assumptions will be imposed later in this subsection.

We first recall the Petrov–Galerkin method for equation (3.68). In this description, we let $\mathbb{X} := L^2(\Omega)$ with an inner product $(\cdot, \cdot)$. Let $\{\mathbb{X}_n\}$ and $\{\mathbb{Y}_n\}$ be two sequences of finite-dimensional subspaces of $\mathbb{X}$ such that

$$\begin{aligned}\dim\mathbb{X}_n &= \dim\mathbb{Y}_n = s(n),\\ \mathbb{X}_n &= \mathrm{span}\{\phi_1, \phi_2, \ldots, \phi_{s(n)}\}\end{aligned}$$

and

$$\mathbb{Y}_n = \mathrm{span}\{\psi_1, \psi_2, \ldots, \psi_{s(n)}\}.$$

We assume that $\{\mathbb{X}_n, \mathbb{Y}_n\}$ is a *regular pair*. It is known from the last section that the necessary and sufficient condition for a generalized best approximation from $\mathbb{X}_n$ to $x \in \mathbb{X}$ with respect to $\mathbb{Y}_n$ to exist uniquely is $\mathbb{Y}_n \cap \mathbb{X}_n^\perp = \{0\}$. If this condition holds, then $\mathcal{P}_n$ is a projection and $\{\mathbb{X}_n, \mathbb{Y}_n\}$ forms a regular pair if and only if $\mathcal{P}_n$ is uniformly bounded. The Petrov–Galerkin method for solving equation (3.68) is a numerical scheme to find a function

$$u_n(s) := \sum_{j \in \mathbb{N}_{s(n)}} \alpha_j \phi_j(s) \in \mathbb{X}_n$$

such that

$$((\mathcal{I} - \mathcal{K})u_n, y) = (f, y), \quad \text{for all } y \in \mathbb{Y}_n, \tag{3.70}$$

or equivalently,

$$\sum_{j \in \mathbb{N}_{s(n)}} \alpha_j \left[(\phi_j, \psi_i) - (\mathcal{K}\phi_j, \psi_i)\right] = (f, \psi_i), \quad i \in \mathbb{N}_{s(n)}. \tag{3.71}$$

Using the generalized best approximation $\mathcal{P}_n : \mathbb{X} \to \mathbb{X}_n$, we write equation (3.70) in operator form as

$$(\mathcal{I} - \mathcal{P}_n\mathcal{K})u_n = \mathcal{P}_n f. \tag{3.72}$$

It is also proved in the last section that if $\{\mathbb{X}_n, \mathbb{Y}_n\}$ is a regular pair, then for sufficiently large n, equation (3.72) has a unique solution $u_n \in \mathbb{X}_n$ which satisfies the estimate

$$\|u_n - u\| \le C \inf\{\|u - x\|, x \in \mathbb{X}_n\}.$$

Solving equation (3.72) requires solving the linear system (3.71). Of course, the entries of the coefficient matrix of (3.71) involve the integrals $(\mathcal{K}\phi_j, \psi_i)$, which are normally evaluated by a numerical quadrature formula. Roughly speaking, the discrete Petrov–Galerkin method is the scheme (3.71) with the integrals appearing in the method computed by quadrature formulas. However, we shall develop our discrete Petrov–Galerkin method independent of the Petrov–Galerkin method (3.72). In other words, we do not assume that the Petrov–Galerkin method (3.72) has been previously constructed, to avoid the "regular pair" assumption which is crucial for the solvability and convergence of the Petrov–Galerkin method. We take a one-step approach to fully discretize equation (3.68) directly. We first describe the method in "abstract" terms without specifying the bases and the concrete quadrature formulas. Later, we specialize them using the piecewise polynomial spaces. The only assumption that we have to impose later to guarantee the solvability and convergence of the resulting concrete method is a *local* condition on the reference element and thus, it is easy to verify it.

In our description, we use function values $f(t)$ at given points $t \in \Omega$ for an L^∞ function f. We follow [21] to define them precisely. Let $\tilde{C}(\Omega)$ denote the subspace of $L^\infty(\Omega)$ which consists of functions, each of which is equal to an element in $C(\Omega)$ a.e. The point evaluation functional δ_t on the space $\tilde{C}(\Omega)$ is defined by

$$\delta_t(f) := f(t), \quad t \in \Omega, \quad f \in \tilde{C}(\Omega),$$

where f on the right-hand side is chosen to be the representative function $f \in \tilde{C}(\Omega)$ which is continuous. By the Hahn–Banach theorem, the point evaluation functional δ_t can be extended from $\tilde{C}(\Omega)$ to the whole $L^\infty(\Omega)$ in such a way that the norm is preserved. We use d_t to denote such an extension and define

$$f(t) := d_t(f), \quad \text{for } f \in L^\infty(\Omega).$$

We remark that the extension is not unique, but that is usually immaterial. What is important is that it exists and preserves many of the properties naturally associated with the point evaluation functional. For example, at a point of continuity of f, the extended point evaluation is uniquely defined and has the natural value, and moreover, d_t is continuous at such points. The reader is referred to [21] for more details on this extension.

We now return to our description of the discrete Petrov–Galerkin method. As in the description of the (continuous) Petrov–Galerkin method, we choose two subspaces $\mathbb{X}_n = \text{span}\,\{\phi_j : j \in \mathbb{N}_{s(n)}\}$ and $\mathbb{Y}_n = \text{span}\,\{\psi_j : j \in \mathbb{N}_{s(n)}\}$ of the space $L^\infty(\Omega)$ such that $\dim \mathbb{X}_n = \dim \mathbb{Y}_n = s(n)$. We choose m_n points $t_i \in \Omega$ and two sets of weight functions $w_{1,i}, w_{2,i}$, $i \in \mathbb{N}_{m_n}$. We define the *discrete inner product*

$$(x, y)_n := \sum_{i \in \mathbb{N}_{m_n}} w_{1,i} x(t_i) y(t_i), \quad x, y \in L^\infty(\Omega), \tag{3.73}$$

which will be used to approximate the inner product $(x, y) := \int_D x(t)y(t)\mathrm{d}t$, and define discrete operators by

$$(\mathcal{K}_n u)(s) := \sum_{i \in \mathbb{N}_{m_n}} w_{2,i}(s) u(t_i), \quad u \in L^\infty(\Omega), \tag{3.74}$$

which will be used to approximate the operator $\mathcal{K}$. With these notations, the discrete Petrov–Galerkin method for equation (3.68) is a numerical scheme to find

$$u_n(s) := \sum_{j \in \mathbb{N}_{s(n)}} \alpha_{n,j} \phi_j(s) \tag{3.75}$$

such that

$$((\mathcal{I} - \mathcal{K}_n)u_n, y)_n = (f, y)_n, \quad \text{for all } y \in \mathbb{Y}_n. \tag{3.76}$$

In terms of basis functions, equation (3.76) is written as

$$\sum_{j \in \mathbb{N}_{s(n)}} \alpha_{n,j} \left[\sum_{\ell \in \mathbb{N}_{m_n}} w_{1,\ell} \phi_j(t_\ell) \psi_i(t_\ell) - \sum_{\ell \in \mathbb{N}_{m_n}} w_{1,\ell} \sum_{m \in \mathbb{N}_{m_n}} w_{2,m}(t_\ell) \phi_j(t_m) \psi_i(t_\ell) \right]$$
$$= \sum_{\ell \in \mathbb{N}_{m_n}} w_{1,\ell} f(t_\ell) \psi_i(t_\ell), \quad i \in \mathbb{N}_{s(n)}. \tag{3.77}$$

Upon solving the linear system (3.77), we obtain $s(n)$ values $\alpha_{n,j}$. Substituting them into (3.75) yields an approximation to the solution u of equation (3.68). Equation (3.76) can also be written in the operator form by a *discrete generalized best approximation* $\mathcal{Q}_n$, which we define next. Let $\mathcal{Q}_n : \mathbb{X} \to \mathbb{X}_n$ be defined by

$$(\mathcal{Q}_n x, y)_n = (x, y)_n, \quad \text{for all } y \in \mathbb{Y}_n. \tag{3.78}$$

If $\mathcal{Q}_n x$ is uniquely defined for every $x \in \mathbb{X}$, equation (3.76) can be written in the form

$$(\mathcal{I} - \mathcal{Q}_n \mathcal{K}_n) u_n = \mathcal{Q}_n f. \tag{3.79}$$

We postpone a discussion of the unique existence of $\mathcal{Q}_n x$ until later.

The iterated Petrov–Galerkin method has been shown to have a superconvergence property, where the additional order of convergence gained from an iteration is attributed by approximation of the kernel from the test space.

The convergence order of the iterated Petrov–Galerkin method is equal to the approximation order of space $\mathbb{X}_n$ plus the approximation order of space $\mathbb{Y}_n$. It is of interest to study the superconvergence of the *iterated discrete Petrov–Galerkin method*, which we define by

$$u_n' = f + \mathcal{K}_n u_n. \tag{3.80}$$

Equation (3.80) is a fully discrete algorithm, which can be implemented easily, involving only multiplications and additions. It can be shown that u_n' satisfies the operator equation

$$(\mathcal{I} - \mathcal{K}_n \mathcal{Q}_n) u_n' = f. \tag{3.81}$$

This form of equation allows us to treat the iterated discrete Petrov–Galerkin method as an operator equation whose analysis is covered by the theory developed in Section 2.2.4.

Up to now, the discrete Petrov–Galerkin method has been described in abstract terms without specifying the spaces $\mathbb{X}_n$ and $\mathbb{Y}_n$. In the remainder of this section, we specialize the discrete Petrov–Galerkin method by specifying the spaces $\mathbb{X}_n$ and $\mathbb{Y}_n$, and defining operators $\mathcal{Q}_n$ and $\mathcal{K}_n$ in terms of piecewise polynomials. We assume that Ω is a polyhedral region and construct a partition T_n for Ω by dividing it into N_n simplices $\Omega_{n,i}, i \in \mathbb{N}_{N_n}$, such that

$$h := \max\{\text{diam}\, \Omega_{n,i} : i \in \mathbb{N}_{N_n}\} \to 0, \text{ as } n \to \infty, \tag{3.82}$$

$$\Omega = \bigcup_{i \in \mathbb{N}_{N_n}} \Omega_{n,i}$$

and

$$\text{meas}(\Omega_{n,i} \cap \Omega_{n,j}) = 0, \quad i \neq j.$$

When the dependence of the simplex $\Omega_{n,i}$ on n is well understood, we drop the first index n in the notation and simply write it as Ω_i. For each positive integer n, the set T_n forms a partition for the domain Ω. We also require that the partition is *regular* in the sense that any vertex of a simplex in T_n is not in the interior of an edge of a face of another simplex in the set. It is well known that for each simplex there exists a unique one to one and onto affine mapping which maps the simplex onto a unit simplex Ω^0 called a *reference element*.

Let $\mathcal{F}_i, i \in \mathbb{N}_{N_n}$, denote the invertible affine mappings that map the reference element Ω^0 one to one and onto the simplices Ω_i. Then, the affine mappings $\mathcal{F}_i$ have the form

$$\mathcal{F}_i(t) = B_i t + b_i, \quad t \in \Omega^0, \tag{3.83}$$

where B_i is a $d \times d$ invertible matrix and b_i a vector in $\mathbb{R}^d$, and they satisfy

$$\Omega_i = \mathcal{F}_i(\Omega^0).$$

On the reference element Ω^0, we choose two piecewise polynomial spaces $S_{1,k_1}(\Omega^0)$ and $S_{2,k_2}(\Omega^0)$ of total degree $k_1 - 1$ and $k_2 - 1$, respectively, such that

$$\dim S_{1,k_1}(\Omega^0) = \dim S_{2,k_2}(\Omega^0) = \mu.$$

The partitions Δ_1 and Δ_2 of Ω^0 associated, respectively, with $S_{1,k_1}(\Omega^0)$ and $S_{2,k_2}(\Omega^0)$ may be different; they are arranged according to the integers k_1, k_2 and d. Assume that the numbers of sub-simplices contained in the partitions Δ_1 and Δ_2 are denoted by ν_1 and ν_2. We have to choose these pairs of integers k_1, ν_1 and k_2, ν_2 such that

$$\binom{k_1 - 1 + d}{d}\nu_1 = \binom{k_2 - 1 + d}{d}\nu_2 = \mu,$$

because the dimension of the space of polynomials of total degree k is $\binom{k+d}{d}$. We shall not provide a detailed discussion on how the partitions Δ_1 and Δ_2 are constructed. Instead, we assume that we have chosen bases for these two spaces so that

$$S_{1,k_1}(\Omega^0) := \text{span}\{\phi_j : j \in \mathbb{N}_\mu\}$$

and

$$S_{2,k_2}(\Omega^0) := \text{span}\{\psi_j : j \in \mathbb{N}_\mu\}.$$

We next map these piecewise polynomial spaces on Ω^0 to each simplex Ω_i by letting

$$\phi_{ij}(t) := \begin{cases} \phi_j \circ \mathcal{F}_i^{-1}(t), & t \in \Omega_i \\ 0, & t \notin \Omega_i \end{cases}$$

and

$$\psi_{ij}(t) := \begin{cases} \psi_j \circ \mathcal{F}_i^{-1}(t), & t \in \Omega_i \\ 0, & t \notin \Omega_i \end{cases}$$

for $i \in \mathbb{N}_{N_n}$ and $j \in \mathbb{N}_\mu$. Using these functions as bases, we define the trial space and the test space, respectively, by

$$\mathbb{X}_n = \text{span}\{\phi_{ij} : i \in \mathbb{N}_{N_n}, j \in \mathbb{N}_\mu\}$$

and

$$\mathbb{Y}_n = \text{span}\{\psi_{ij} : i \in \mathbb{N}_{N_n}, j \in \mathbb{N}_\mu\}.$$

It follows from (3.82) that

$$C(\Omega) \subseteq \overline{\bigcup \mathbb{X}_n}$$

and

$$C(\Omega) \subseteq \overline{\bigcup \mathbb{Y}_n}.$$

Moreover, we have that if $x \in W^{k_1}_\infty(\Omega)$, then there exists a constant $c > 0$ such that for all n

$$\inf\{\|x - \phi\| : \phi \in \mathbb{X}_n\} \le ch^{k_1},$$

and if $x \in W^{k_2}_\infty(\Omega)$, then likewise there exists a constant $c > 0$ such that for all n

$$\inf\{\|x - \phi\| : \phi \in \mathbb{Y}_n\} \le ch^{k_2}.$$

However, the space $\tilde{\mathbb{X}} := \overline{\bigcup \mathbb{X}_n}$ does not equal $L^\infty(\Omega)$; it is a proper subspace of $L^\infty(\Omega)$ because the space $L^\infty(\Omega)$ is not separable. Due to this fact, the existing theory of collectively compact operators (cf. [6]) does not apply directly to this setting. Some modifications of the theory are required.

We next specialize the definition of the discrete inner product (3.73) and describe a concrete construction of the approximate operators $\mathcal{K}_n$. To this end, we introduce a third piecewise polynomial space $S_{3,k_3}(\Omega^0)$ of total degree $k_3 - 1$ on Ω^0. We divide the reference element Ω^0 into ν_3 sub-simplices

$$\Delta_3 := \{e_i : i \in \mathbb{N}_{\nu_3}\}$$

and also assume that the partition Δ_3 is regular. On each of the simplices e_i, we choose $m := \binom{k_3-1+d}{d}$ points τ_{ij}, $j \in \mathbb{N}_m$, such that they admit a unique Lagrange interpolating polynomial of total degree $k_3 - 1$ on e_i. For multivariate Lagrange interpolation by polynomials of total degree, see [83] and the references cited therein. Let p_{ij} be the polynomial of total degree $k_3 - 1$ on e_i satisfying the interpolation conditions

$$p_{ij}(\tau_{i'j'}) = \delta_{ii'}\delta_{jj'}, \quad i, i' \in \mathbb{N}_{\nu_3},\ j, j' \in \mathbb{N}_m.$$

We assemble these polynomials to form a basis for the space $S_{3,k_3}(\Omega^0)$ by letting

$$\zeta_{(i-1)m+j}(t) := \begin{cases} p_{ij}(t), & t \in e_i, \\ 0, & t \notin e_i, \end{cases} \quad i \in \mathbb{N}_{\nu_3},\ j \in \mathbb{N}_m.$$

Set $\gamma := m\nu_3$, which is equal to the dimension of $S_{3,k_3}(\Omega^0)$, and

$$t_{(i-1)m+j} = \tau_{ij}, \quad i \in \mathbb{N}_{\nu_3},\ j \in \mathbb{N}_m.$$

Then $\zeta_i \in S_{3,k_3}(\Omega^0)$ and satisfy the interpolation conditions

$$\zeta_i(t_j) = \delta_{ij}, \quad i,j \in \mathbb{N}_\gamma.$$

This set of functions forms a basis for the space $S_{3,k_3}(\Omega^0)$. It can be used to introduce a piecewise polynomial space on Ω by mapping the basis $\zeta_j, j \in \mathbb{N}_\gamma$ for $S_{3,k_3}(\Omega^0)$ from Ω^0 into each Ω_i. Specifically, we define

$$\zeta_{ij}(t) := \begin{cases} \zeta_j \circ \mathcal{F}_i^{-1}(t), & t \in \Omega_i, \\ 0, & t \notin \Omega_i, \end{cases}$$

where $\mathcal{F}_i$ is the affine map defined by (3.83). Let

$$\mathbf{Z}_n := \operatorname{span}\{\zeta_{ij} : i \in \mathbb{N}_{N_n}, j \in \mathbb{N}_\gamma\}.$$

Hence, $\mathbf{Z}_n$ is a piecewise polynomial space of dimension γN_n. For each i, we define

$$t_{ij} := \mathcal{F}_i(t_j) = \mathbf{B}_i t_j + \mathbf{b}_i,$$

where $\mathbf{B}_i$ and $\mathbf{b}_i$ are respectively the matrix and vector appearing in the definition of the affine map $\mathcal{F}_i$. Furthermore, we define the linear projection $\mathcal{Z}_n : \mathbb{X} \to \mathbf{Z}_n$ by

$$\mathcal{Z}_n g = \sum_{i\in\mathbb{N}_{N_n}} \sum_{j\in\mathbb{N}_\gamma} d_{t_{ij}}(g)\zeta_{ij},$$

where d_t is the extension of the point evaluation functional δ_t satisfying $\|d_t\| = 1$, which was discussed earlier and satisfies the condition

$$d_{t_{ij}}(\zeta_{i'j'}) = \delta_{ii'}\delta_{jj'}, \quad i,i' \in \mathbb{N}_{N_n}, \quad j,j' \in \mathbb{N}_\gamma.$$

Moreover, we have that

$$\|\mathcal{Z}_n\| = \operatorname*{ess\,sup}_{t\in\Omega} \sum_{i\in\mathbb{N}_{N_n}} \sum_{j\in\mathbb{N}_\gamma} |\zeta_{ij}(t)| = \operatorname*{ess\,sup}_{t\in\Omega} \sum_{j\in\mathbb{N}_\gamma} |\zeta_j(t)|.$$

That is, $\|\mathcal{Z}_n\|$ is uniformly bounded for all n. It follows from the uniform boundedness of $\|\mathcal{Z}_n\|$ that for any $y \in W_\infty^{k_3}(\Omega)$, there holds an estimate

$$\|y - \mathcal{Z}_n y\| \leq C \inf\{\|y - \phi\|, \phi \in \mathbf{Z}_n\} \leq Ch^{k_3}. \tag{3.84}$$

Using the projection $\mathcal{Z}_n$ defined above, we have a quadrature formula

$$\int_\Omega g(t)\mathrm{d}t = \sum_{i\in\mathbb{N}_{N_n}} \sum_{j\in\mathbb{N}_\gamma} w_{ij} d_{t_{ij}}(g) + \mathcal{O}(h^{k_3}),$$

where

$$w_{ij} := \int_\Omega \zeta_{ij}(t)\mathrm{d}t.$$

If we set

$$w_i := \int_{\Omega^0} \zeta_i(t)\mathrm{d}t, \quad i \in \mathbb{N}_\gamma,$$

then we have

$$w_{ij} = \int_{\Omega_i} \zeta_j(\mathcal{F}_i^{-1}(t))\mathrm{d}t = \det(\mathbf{B}_i) \int_{\Omega^0} \zeta_j(t)\mathrm{d}t = \det(\mathbf{B}_i) w_j.$$

Without loss of generality, we assume that

$$\det(\mathbf{B}_i) > 0, \quad i \in \mathbb{N}_\gamma.$$

Employing this formula, we introduce the following discrete inner product:

$$(x, y)_n = \sum_{i \in \mathbb{N}_{N_n}} \sum_{\ell \in \mathbb{N}_\gamma} w_{i\ell} x(t_{i\ell}) y(t_{i\ell}). \tag{3.85}$$

Formula (3.85) is a concrete form for (3.73). When $x, y \in W_\infty^{k_3}(\Omega)$, we have the error estimate

$$|(x, y) - (x, y)_n| \leq Ch^{k_3}.$$

With this specific definition of the spaces $\mathbb{X}_n$, $\mathbb{Y}_n$ and the discrete inner product, we obtain a construction of the operators $\mathcal{Q}_n$ using equation (3.78).

Finally, to describe a concrete construction of the approximate operators $\mathcal{K}_n$, we impose a few additional assumptions on the kernel K of the integral operator $\mathcal{K}$. Roughly speaking, we assume that K is a product of two kernels: one of them is continuous but perhaps involves a complicated function, and the other has a simple form but has a singularity. In particular, we let

$$K(s, t) := K_1(s, t) K_2(s, t), \quad s, t \in \Omega,$$

where K_1 is continuous on $\Omega \times \Omega$ and K_2 has a singularity and satisfies the conditions

$$K_2(s, \cdot) \in L^1(\Omega), \quad s \in \Omega, \quad \sup_{s \in \Omega} \int_\Omega |K_2(s, t)|\mathrm{d}t < +\infty, \tag{3.86}$$

$$\|K_2(s, \cdot) - K_2(s', \cdot)\|_1 \to 0, \quad \text{as } s' \to s. \tag{3.87}$$

Moreover, we assume that the integration of the product of $K_2(s, t)$ and a polynomial $p(t)$ with respect to the variable t can be evaluated exactly. Many integral operators $\mathcal{K}$ that appear in practical applications are of this type.

Using the linear projection $\mathcal{Z}_n$, we define $\mathcal{K}_n : \mathbb{X} \to \mathbb{X}$ by

$$(\mathcal{K}_n x)(s) := \int_\Omega \mathcal{Z}_n(K_1(s, t) x(t)) K_2(s, t)\mathrm{d}t,$$

which approximates the operator $\mathcal{K}$. For $u_n \in \mathbb{X}_n$, we have that

$$(\mathcal{K}_n u_n)(s) = \sum_{i\in\mathbb{N}_{N_n}} \sum_{j\in\mathbb{N}_\gamma} w_{ij}(s) K_1(s, t_{ij}) u_n(t_{ij}),$$

where

$$w_{ij}(s) = \int_{\Omega_i} \zeta_{ij}(t) K_2(s, t) \mathrm{d}t.$$

This concrete construction of the trial space $\mathbb{X}_n$, the test space $\mathbb{Y}_n$ and operators $\mathcal{Q}_n$, $\mathcal{K}_n$ yields a specific discrete Petrov–Galerkin method which is described by equation (3.79). This is the method that we shall analyze in the next subsection.

3.5.4 The convergence of the discrete Petrov–Galerkin method

In this subsection, we follow the general theory developed in Section 2.2.4 to prove the convergence results of the discrete Petrov–Galerkin method when a piecewise polynomial approximation is used. Throughout the remaining part of this subsection, we let $\mathbb{X} := L^\infty(\Omega)$, $\mathbb{V} := C(\Omega)$, $\mathbb{X}_n$ and $\mathbb{Y}_n$ be the piecewise polynomial spaces defined in the last subsection, and $\tilde{\mathbb{X}} = \overline{\cup_n \mathbb{X}_n}$. Our main task is to verify that the operators $\mathcal{Q}_n$ and $\mathcal{K}_n$ with the spaces $\mathbb{X}_n$, $\mathbb{Y}_n$ defined in the last subsection by the piecewise polynomials satisfy the hypotheses (H-1)–(H-4) of Section 2.2.4 so that Theorem 2.54 can be applied. For this purpose, we define the necessary notation. Let

$$\Phi := [\phi_i(t_j) : i \in \mathbb{N}_\mu, j \in \mathbb{N}_\gamma] \text{ and } \Psi := [\psi_i(t_j) : i \in \mathbb{N}_\mu, j \in \mathbb{N}_\gamma],$$

where $\{\phi_i\}$ and $\{\psi_i\}$ are the bases we have chosen for the piecewise polynomial spaces $S_{1,k_1}(\Omega^0)$ and $S_{2,k_2}(\Omega^0)$, and $\{t_j\}$ are the interpolation points in the reference element Ω^0 chosen in the last subsection. Noting that w_i are the weights of the quadrature formula on the reference element developed in Section 4.2.1, we set

$$\mathbf{W} := \mathrm{diag}(w_1, \ldots, w_\gamma) \text{ and } \mathbf{M} := \Psi \mathbf{W} \Phi^T.$$

The next proposition presents a necessary and sufficient condition for the discrete generalized best approximation to exist uniquely.

Proposition 3.35 *For each $x \in L^\infty(\Omega)$, the discrete generalized best approximation $\mathcal{Q}_n x$ from $\mathbb{X}_n$ to x with respect to $\mathbb{Y}_n$ defined by (3.78) exists uniquely if and only if*

$$\det(\mathbf{M}) \neq 0. \tag{3.88}$$

Under this condition, $\mathcal{Q}_n$ is a projection, that is, $\mathcal{Q}_n^2 = \mathcal{Q}_n$.

Proof Let $x \in L^\infty(\Omega)$ be given. Showing that there is a unique $\mathcal{Q}_n x \in X_n$ satisfying equation (3.78) is equivalent to proving that the linear system

$$\sum_{i \in \mathbb{N}_{N_n}} \sum_{j \in \mathbb{N}_\mu} c_{ij}(\phi_{ij}, \psi_{i'j'})_n = (x, \psi_{i'j'})_n, \quad i' \in \mathbb{N}_{N_n}, \ j' \in \mathbb{N}_\mu \tag{3.89}$$

has a unique solution $[c_{11}, \ldots, c_{1\mu}, \ldots, c_{N_n,1}, \ldots, c_{N_n,\mu}]$. This in turn is equivalent to the fact that the coefficient matrix $\tilde{\mathbf{M}}$ of this system is nonsingular. It is easily seen that

$$\tilde{\mathbf{M}} = \mathrm{diag}(\det(\mathbf{B}_1)\mathbf{M}, \ldots, \det(\mathbf{B}_{N_n})\mathbf{M}).$$

Thus the first result of this proposition follows from hypothesis (3.88).

It remains to show that $\mathcal{Q}_n$ is a projection. By definition, we have for every $x \in L^\infty(\Omega)$ that

$$(\mathcal{Q}_n x, y)_n = (x, y)_n, \quad \text{for all} \ \ y \in \mathbb{Y}_n.$$

In particular, this equation holds when x is replaced by $\mathcal{Q}_n x$. That is,

$$(\mathcal{Q}_h^2 x, y)_n = (\mathcal{Q}_n x, y)_n, \quad \text{for all} \ \ y \in \mathbb{Y}_n.$$

It follows for each $x \in \mathbb{X}$ that

$$\mathcal{Q}_n^2 x = \mathcal{Q}_n x.$$

That is, $\mathcal{Q}_n$ is a projection. □

Condition (3.88) is on the choice of points $\{t_j\}$ on the reference element. They have to be selected in a careful manner so that they match with the choice of the bases $\{\phi_i\}$ and $\{\psi_i\}$. This condition has to be verified before a concrete construction of the projection $\mathcal{Q}_n$ is given. This is not a difficult task since the condition is on the reference element, it is independent of n and in practical applications, the numbers μ and γ are not too large.

The next proposition gives two useful properties of the projection $\mathcal{Q}_n$.

Proposition 3.36 *Assume that condition (3.88) is satisfied. Let $\mathcal{Q}_n$ be defined by (3.78) with the spaces $\mathbb{X}_n$, $\mathbb{Y}_n$ and the discrete inner product constructed in terms of the piecewise polynomials described in the last subsection. Then the following statements hold.*

(i) $\mathcal{Q}_n$ is uniformly bounded, that is, there exists a constant $c > 0$ such that $\|\mathcal{Q}_n\| \le c$ for all n.

(ii) There exists a constant $c > 0$ such that for all n

$$\|\mathcal{Q}_n x - x\|_\infty \le c \inf\{\|x - \phi\|_\infty, \phi \in \mathbb{X}_n\}$$

holds for all $x \in L^\infty(\Omega)$. Thus, for each $x \in C(\Omega)$, $\|\mathcal{Q}_n x - x\|_\infty \to 0$ holds as $n \to \infty$.

Proof (i) For any $x \in L^\infty(\Omega)$, we have the expression

$$\mathcal{Q}_n x = \sum_{i\in\mathbb{N}_{N_n}} \sum_{j\in\mathbb{N}_\mu} c_{ij}\phi_{ij}, \tag{3.90}$$

where the coefficients c_{ij} satisfy equation (3.89). It follows that

$$\|\mathcal{Q}_n x\|_\infty \le \|\mathbf{c}\|_\infty \operatorname{ess\,sup}_{s\in\Omega} \sum_{i\in\mathbb{N}_{N_n}} \sum_{j\in\mathbb{N}_\mu} |\phi_{ij}(s)| = \|\mathbf{c}\|_\infty \max_{s\in\Omega^0} \sum_{j\in\mathbb{N}_\mu} |\phi_j(s)|, \tag{3.91}$$

where

$$\mathbf{c} := [c_{11}, \ldots, c_{1\mu}, \ldots, c_{N_n,1}, \ldots, c_{N_n,\mu}]^T,$$

and the discrete norm of $\mathbf{c}$ is defined by $\|\mathbf{c}\|_\infty := \max\{|c_{ij}| : i \in \mathbb{N}_{N_n}, j \in \mathbb{N}_\mu\}$. By definition, the vector $\mathbf{c}$ is dependent on n although we do not specify it in the notation. However, we prove that $\|\mathbf{c}\|_\infty$ is in fact independent of n. To this end, we use system (3.89) and hypothesis (3.88) to conclude that

$$\|\mathbf{c}\|_\infty = \|\tilde{\mathbf{M}}^{-1}\mathbf{d}\|_\infty, \tag{3.92}$$

where

$$\mathbf{d} := [(x, \psi_{11})_n, \ldots, (x, \psi_{1\mu})_n, \ldots, (x, \psi_{N_n,1})_n, \ldots, (x, \psi_{N_n,\mu})_n]^T$$

and

$$\tilde{\mathbf{M}}^{-1} = \operatorname{diag}\left(\det(B_1)^{-1}\mathbf{M}^{-1}, \ldots, \det(B_{N_n})^{-1}\mathbf{M}^{-1}\right).$$

Let

$$\mathbf{d}_i := [(x, \psi_{i1})_n, \ldots, (x, \psi_{i\mu})_n]^T \in \mathbb{R}^\mu.$$

Then it follows from (3.92) that the following estimate of $\|\mathbf{c}\|_\infty$ holds in terms of blocks $\mathbf{d}_i$ and $\mathbf{M}^{-1}$:

$$\|\mathbf{c}\|_\infty \le \max_{i\in\mathbb{N}_{N_n}} \|\det(B_i)^{-1}\mathbf{M}^{-1}\mathbf{d}_i\|_\infty. \tag{3.93}$$

This inequality reduces estimating $\|\mathbf{c}\|_\infty$ to bounding each block $\mathbf{d}_i$. By the definition of the discrete inner product, we have the estimate for the norm of $\mathbf{d}_i$:

$$\|\mathbf{d}_i\|_\infty \le \|x\|_\infty \max_{j\in\mathbb{N}_\mu} \sum_{\ell\in\mathbb{N}_\gamma} w_{i\ell}|\psi_{ij}(t_{i\ell})| = \det(B_i)\|x\|_\infty \max_{j\in\mathbb{N}_\mu} \sum_{\ell\in\mathbb{N}_\gamma} w_\ell|\psi_j(t_\ell)|. \tag{3.94}$$

From (3.91)–(3.94) we conclude that

$$\|\mathcal{Q}_n x\|_\infty \le c\|x\|_\infty, \quad \text{for all } n,$$

where c is a constant independent of n with the value

$$c := \|\mathbf{M}^{-1}\|_\infty \max_{s\in E} \sum_{j\in\mathbb{N}_\mu} |\phi_j(s)| \max_{j\in\mathbb{N}_\mu} \sum_{\ell\in\mathbb{N}_\gamma} w_\ell |\psi_j(t_\ell)|.$$

(ii) Let $\phi \in \mathbb{X}_n$. Since $\mathcal{Q}_n$ is a projection, we have for each $x \in L_\infty$ that

$$\|\mathcal{Q}_n x - x\|_\infty \le \|x - \phi\|_\infty + \|\mathcal{Q}_n\phi - \mathcal{Q}_n x\|_\infty \le (1+c)\|x-\phi\|_\infty.$$

Thus, we obtain the estimate

$$\|\mathcal{Q}_n x - x\|_\infty \le c \inf\{\|x-\phi\|_\infty, \phi \in \mathbb{X}_n\}.$$

This estimate, with the relation $C(\Omega) \subseteq \overline{\cup_n \mathbb{X}_n}$, implies that $\|\mathcal{Q}_n x - x\|_\infty \to 0$ as $n \to \infty$ for each $x \in C(\Omega)$. □

In the next proposition, we verify that the operators $\mathcal{K}_n$ defined in the last subsection by the piecewise polynomial approximation satisfy hypotheses (H-1) and (H-2).

Proposition 3.37 *Suppose that $\mathcal{K}_n$ is defined as in the last subsection by the piecewise polynomial approximation. Then the following statements hold.*

(i) The set of operators $\{\mathcal{K}_n\}$ is collectively compact.
(ii) For each $x \in \tilde{\mathbb{X}}$, $\|\mathcal{K}_n x - \mathcal{K}x\|_\infty \to 0$ as $n \to \infty$.
(iii) If $x \in W_\infty^{k_3}(\Omega)$ and $K_1 \in C(\Omega) \times W_\infty^{k_3}(\Omega)$, then

$$\|\mathcal{K}x - \mathcal{K}_n x\|_\infty \le ch^{k_3}.$$

Proof (i) By the continuity of the kernel K_1 and condition (3.87), there exist constants c_1 and c_2 such that

$$\|K_1(s,\cdot)\|_\infty \le c_1 \quad \text{and} \quad \|K_2(s,\cdot)\|_1 \le c_2.$$

Thus, we have that

$$|(\mathcal{K}_n x)(s)| = \left|\int_\Omega \mathcal{Z}_n(K_1(s,t)x(t))K_2(s,t)\mathrm{d}t\right| \le c_0 c_1 c_2 \|x\|_\infty. \tag{3.95}$$

Moreover,

$$\begin{aligned}
&|(\mathcal{K}_n x)(s) - (\mathcal{K}_n x)(s')| \\
&\quad = \left|\int_\Omega \mathcal{Z}_n(K_1(s,t)x(t))K_2(s,t)\mathrm{d}t - \int_\Omega \mathcal{Z}_n(K_1(s',t)x(t))K_2(s',t)\mathrm{d}t\right| \\
&\quad \le \left|\int_\Omega \mathcal{Z}_n(K_1(s,t)x(t))[K_2(s,t) - K_2(s',t)]\mathrm{d}t\right|
\end{aligned}$$

$$+\left|\int_\Omega [\mathcal{Z}_n(K_1(s,t)x(t)) - \mathcal{Z}_n(K_1(s',t)x(t))]K_2(s',t)\mathrm{d}t\right|$$
$$\leq c_0\|x\|_\infty \left(c_1\|K_2(s,\cdot) - K_2(s',\cdot)\|_1 + c_2\|K_1(s,\cdot) - K_1(s',\cdot)\|_\infty\right).$$

Since $\|K_2(s,\cdot) - K_2(s',\cdot)\|_1$ and $\|K_1(s,\cdot) - K_1(s',\cdot)\|_\infty$ are uniformly continuous on Ω, we observe that $\{\mathcal{K}_n x\}$ is equicontinuous on Ω. By the Arzelà–Ascoli theorem we conclude that $\{\mathcal{K}_n\}$ is collectively compact.

(ii) For any $x \in \tilde{\mathbb{X}}$,

$$|(\mathcal{K}_n x)(s) - (\mathcal{K}x)(s)| = \left|\int_\Omega [\mathcal{Z}_n(K_1(s,t)x(t)) - K_1(s,t)x(t)]K_2(s,t)\mathrm{d}t\right|$$
$$\leq c_2\|\mathcal{Z}_n(K_1(s,\cdot)x) - K_1(s,\cdot)x\|_\infty.$$

Note that $K_1 x$ is piecewise continuous, as is x. By the definition of $\mathcal{Z}_n$ we have that the right-hand side of the above inequality converges to zero as $n \to \infty$. We conclude that the left-hand side converges uniformly to zero on the compact set Ω. That is, $\|\mathcal{K}_n x - \mathcal{K}x\| \to 0$ as $n \to \infty$.

(iii) If $x \in W^{k_3}_\infty(\Omega)$, by the approximate order of the interpolation projection $\mathcal{Z}_n$, we have

$$\|\mathcal{K}_n x - \mathcal{K}x\|_\infty \leq c \sup_{s\in\Omega} \|(\mathcal{Z}_n(K_1(s,\cdot)x(\cdot)))(\cdot) - K_1(s,\cdot)x(\cdot)\|_\infty \leq ch^{k_3}.$$

The estimate above follows immediately from the fact that $K_1 \in C(\Omega) \times W^{k_3}_\infty(\Omega)$ and inequality (3.84). □

Using Propositions 3.36 and 3.37 and Theorem 2.54, we obtain the following theorem.

Theorem 3.38 *The following statements are valid.*

(i) There exists $N_0 > 0$ such that for all $n > N_0$, the discrete Petrov–Galerkin method using the piecewise polynomial approximation described in Section 4.2.1 has a unique solution $u_n \in \mathbb{X}_n$.

(ii) If $u \in W^\alpha_\infty(\Omega)$ with $\alpha := \min\{k_1, k_3\}$, then

$$\|u - u_n\|_\infty \leq ch^\alpha.$$

Proof By Propositions 3.36 and 3.37, we conclude that conditions (H-1)–(H-4) are satisfied. Hence, from Theorem 2.54, statement (i) follows immediately and the estimate

$$\|u - u_n\|_\infty \leq c\,(\|u - \mathcal{Q}_n u\|_\infty + \|\mathcal{K}u - \mathcal{K}_n u\|_\infty) \tag{3.96}$$

holds. Now let $u \in W^\alpha_\infty(\Omega)$. Again, Proposition 3.36 ensures that

$$\|u - \mathcal{Q}_n u\|_\infty \leq c \inf\{\|u - \phi\|_\infty, \phi \in \mathbb{X}_n\} \leq ch^\alpha. \tag{3.97}$$

By (iii) of Proposition 3.37, we have that

$$\|\mathcal{K}u - \mathcal{K}_n u\|_\infty \le ch^\alpha. \tag{3.98}$$

Substituting estimates (3.97) and (3.98) into inequality (3.96) yields the estimate in (ii). □

3.5.5 Superconvergence of the iterated approximation

We present in this subsection a superconvergence property of the iterated discrete Petrov–Galerkin method when the kernel is smooth.

To obtain superconvergence, we require furthermore that the partitions Δ_1 and Δ_3 of Ω^0, associated with the spaces $S_{1,k}(\Omega^0)$ and $S_{3,k_3}(\Omega^0)$, respectively, are exactly the same. In the main theorem of this section, we prove that the corresponding iterated discrete Petrov–Galerkin approximation has a superconvergence property when the kernels are smooth. In particular, we assume that the kernel $K = K_1$ and $K_2 = 1$ in the notation of the last subsection. We first establish a technical lemma.

Lemma 3.39 *Let $x \in L^\infty(\Omega)$ and $k_1 \in C(\Omega) \times W^{k_3}_\infty(\Omega)$. If $\Delta_1 = \Delta_3$, then there exists a positive constant c such that for all n*

$$\|(\mathcal{K} - \mathcal{K}_n)\mathcal{Q}_n x\|_\infty \le ch^{k_3}.$$

Proof Since $\mathcal{Q}_n x$ is not a continuous function, Proposition 3.37 (iii) does not apply to this case. However, it follows from the proof of Proposition 3.37 (ii) that

$$|(\mathcal{K}_n \mathcal{Q}_n x)(s) - (\mathcal{K}\mathcal{Q}_n x)(s)| \le c\|r_s\|_\infty,$$

where

$$r_s(t) := (\mathcal{Z}_n(K_1(s,\cdot)(\mathcal{Q}_n x)(\cdot))(t) - K_1(s,t)(\mathcal{Q}_n x)(t).$$

Hence, it suffices to estimate $r_s(t)$.

Using the definition of the projection $\mathcal{Q}_n$, we write

$$(\mathcal{Q}_n x)(t) = \sum_{i\in\mathbb{N}_{N_n}} \sum_{j\in\mathbb{N}_\mu} c_{ij}\phi_{ij}(t), \quad t \in \Omega, \tag{3.99}$$

where $\phi_{i,j}$ are the basis functions for $\mathbb{X}_n$ given in Subsection 3.5.4, and the coefficients c_{ij} satisfy the linear system (3.89). Consequently, we have that

$$(\mathcal{Z}_n(K_1(s,\cdot)(\mathcal{Q}_n x)(\cdot))(t) = \sum_{i\in\mathbb{N}_{N_n}} \sum_{j\in\mathbb{N}_\mu} c_{ij}(\mathcal{Z}_n(K_1(s,\cdot)\phi_{ij}(\cdot))(t). \tag{3.100}$$

By the construction of the functions ϕ_{ij}, we have that $\phi_{ij}(t_{i'j'}) = 0$ if $i \neq i'$. Thus, it follows that

$$(\mathcal{Z}_n(K_1(s,\cdot)\phi_{ij}(\cdot))(t) = \sum_{i'\in\mathbb{N}_{N_n}}\sum_{j'\in\mathbb{N}_\gamma} K_1(s,t_{i'j'})\phi_{ij}(t_{i'j'})\zeta_{i'j'}(t)$$
$$= \sum_{j'\in\mathbb{N}_\gamma} K_1(s,t_{ij'})\phi_{ij}(t_{ij'})\zeta_{ij'}(t).$$

Substituting this equation into (3.100) yields

$$(\mathcal{Z}_n(K_1(s,\cdot)(\mathcal{Q}_n x)(\cdot))(t) = \sum_{i\in\mathbb{N}_{N_n}}\sum_{j\in\mathbb{N}_\mu} c_{ij}\sum_{j'\in\mathbb{N}_\gamma} K_1(s,t_{ij'})\phi_{ij}(t_{ij'})\zeta_{ij'}(t), \quad t\in\Omega. \tag{3.101}$$

We now assume that for some point $\hat{t}\in\Omega_{i'}$, $\|r_s\|_\infty = |r_s(\hat{t})|$. For this point $\hat{t}$, there exists a point τ in the reference element Ω^0 such that $\hat{t} = \mathcal{F}_{i'}(\tau)$. Hence,

$$\|r_s\|_\infty = \left|\sum_{j\in\mathbb{N}_\mu} c_{i'j}\left[\sum_{j'\in\mathbb{N}_\gamma} K_1(s,\mathcal{F}_{i'}(t_{j'}))\phi_j(t_{j'})\zeta_{j'}(\tau) - K_1(s,\mathcal{F}_{i'}(\tau))\phi_j(\tau)\right]\right|.$$

Because $\Omega^0 = \cup_{i\in\mathbb{N}_{\nu_3}} e_i$, the point τ must be in some $\overline{e}_i$. For each integer $j'\in\mathbb{N}_\gamma$, assume that positive integers i_0 and j_0 with $i\in\mathbb{N}_{\nu_3}$, $j_0\in\mathbb{N}_m$ are such that $(i_0-1)m + j_0 = j'$. Therefore, we have

$$\zeta_{j'}(t) = \begin{cases} p_{i_0j_0}(t), & t\in e_{i_0} \\ 0, & t\notin e_{i_0} \end{cases} \quad \text{and} \quad t_{j'} = \tau_{i_0j_0},$$

so that

$$\|r_s\|_\infty = \left|\sum_{j\in\mathbb{N}_\mu} c_{ij}\left[\sum_{i_0\in\mathbb{N}_{\nu_3}}\sum_{j_0\in\mathbb{N}_m} K_1(s,\mathcal{F}_{i'}(\tau_{i_0j_0}))\phi_j(\tau_{i_0j_0})p_{i_0j_0}(\tau) - K_1(s,\mathcal{F}_{i'}(\tau))\phi_j(\tau)\right]\right|$$
$$= \left|\sum_{j\in\mathbb{N}_\mu} c_{ij}\left[\sum_{j_0\in\mathbb{N}_m} K_1(s,\mathcal{F}_{i'}(\tau_{ij_0}))\phi_j(\tau_{ij_0})p_{ij_0}(\tau) - K_1(s,\mathcal{F}_{i'}(\tau))\phi_j(\tau)\right]\right|.$$

We identify that the function in the blanket of the last term is the error of polynomial interpolation of the function $K_1(s,\mathcal{F}_{i'}(\tau))\phi_j(\tau)$ on e_i, which we call the error term on e_i. Since $\Delta_1 = \Delta_3$, $K_1(s,\mathcal{F}_{i'}(\tau))\phi_j(\tau)$ as a function of τ is in the space $W^{k_3}_\infty(e_i)$. We conclude that the error term on e_i is bounded by a constant time $\|\Omega^{k_3}(K_1(s,\mathcal{F}_{i'}(\cdot))\phi_j(\cdot)\|_\infty$. The latter is bounded by a constant time $|\det(B_{i'})|^{k_3}\le ch^{k_3}$. Hence, we obtain

$$\|r_s\|_\infty \le c\|\mathbf{c}\|_\infty h^{k_3}.$$

By the proof of Proposition 3.36, we know that $\|\mathbf{c}\|_\infty \le c$. Therefore, we have $\|r_s\|_\infty \le ch^{k_3}$. □

We are now ready to establish the main result of this subsection concerning the superconvergence of the iterated solution.

Theorem 3.40 *If $\beta := \min\{k_1 + k_2, k_3\}$, $u \in W^\beta_\infty(\Omega)$ and $K \in C(\Omega) \times W^{k_3}_\infty(\Omega)$, then there exists a constant $c > 0$ such that for all n*

$$\|u - u_n'\|_\infty \le ch^\beta.$$

Proof It follows from Theorem 2.54 that

$$\|u - u_n'\|_\infty \le c\left(\|(\mathcal{K} - \mathcal{K}_n)\mathcal{Q}_n u\|_\infty + \|\mathcal{K}(\mathcal{I} - \mathcal{Q}_n)u\|_\infty\right). \tag{3.102}$$

Because $\Delta_1 = \Delta_3$, by applying Lemma 3.39 we have that

$$\|(\mathcal{K} - \mathcal{K}_n)\mathcal{Q}_n u\|_\infty \le ch^{k_3}. \tag{3.103}$$

Moreover, since $K(s,\cdot) \in W^{k_3}_\infty(\Omega)$ and $\Delta_1 = \Delta_3$, we conclude that

$$\|\mathcal{K}(u - \mathcal{Q}_n u)\|_\infty \le \|(K(s,t), u(t) - (\mathcal{Q}_n u)(t))_n\|_\infty + ch^{k_3}. \tag{3.104}$$

It remains to establish an upper bound for $\|(K(s,t), u(t) - (\mathcal{Q}_n u)(t))_n\|_\infty$. For this purpose, we note that for any $y \in \mathbb{Y}_n$,

$$(y, u - \mathcal{Q}_n u)_n = 0$$

holds. It follows that

$$\begin{aligned}|(K(s,t), u(t) - (\mathcal{Q}_n u)(t))_n| &= |(K(s,t) - y(t), u(t) - \mathcal{Q}_n u)(t))_n| \\ &\le \inf\{\|K(s,t) - y(t)\|_\infty, y \in \mathbb{Y}_n\}\|u - \mathcal{Q}_n u\|_\infty.\end{aligned}$$

This implies that

$$\|(K(s,t), u(t) - (\mathcal{Q}_n u)(t))_n\|_\infty \le ch^{k_2}h^{k_1} = ch^{k_1+k_2}. \tag{3.105}$$

Combining inequalities (3.102)–(3.105), we establish the estimate of this theorem. □

We remark that when $k_1 < k_3 < k_1 + k_2$, the optimal order of convergence of u_n is $\mathcal{O}(h^{k_1})$ while the iterated solution u_n' has an order of convergence $\mathcal{O}(h^{k_3})$. This phenomenon is called superconvergence.

3.5.6 Numerical examples

In this subsection, we present two numerical examples to illustrate the theoretical estimates obtained in the previous subsections. The kernel in the first example is weakly singular while the kernel in the second example is smooth. The second example is presented to show the superconvergence property of the iterated solution. We restrict ourselves to simple one-dimensional equations whose exact solutions are known.

In both examples, we use piecewise linear functions and piecewise constant functions for the spaces $\mathbb{X}_n$ and $\mathbb{Y}_n$, respectively. Specifically, we define the trial space by

$$\mathbb{X}_n := \operatorname{span}\{\phi_j : j \in \mathbb{N}_{2n}\},$$

where

$$\phi_{2j+1}(t) := \begin{cases} nt - j, & \frac{j}{n} \leq t \leq \frac{j+1}{n}, \\ 0, & \text{otherwise}, \end{cases} \quad j \in \mathbb{Z}_n$$

and

$$\phi_{2j+2}(t) := \begin{cases} j + 1 - nt, & \frac{j}{n} \leq t \leq \frac{j+1}{n}, \\ 0, & \text{otherwise}, \end{cases} \quad j \in \mathbb{Z}_n.$$

The test space is then defined by

$$\mathbb{Y}_n := \operatorname{span}\{\psi_j : j \in \mathbb{N}_{2n}\},$$

where

$$\psi_j(t) := \begin{cases} 1, & \frac{j-1}{2n} \leq t \leq \frac{j}{2n}, \\ 0, & \text{otherwise}, \end{cases} \quad j \in \mathbb{N}_{2n}.$$

Example 1. Consider the integral equation with a weakly singular kernel

$$u(s) - \int_0^\pi \log|\cos s - \cos t| u(t) \mathrm{d}t = 1, \quad 0 \leq s \leq \pi.$$

This equation is a reformulation of a third boundary value problem of the two-dimensional Laplace equation and it has the exact solution given by

$$u(s) = \frac{1}{1 + \pi \log 2}.$$

See [12] for more details on this example. By changes of variable $t = \pi t'$, $s = \pi s'$, we have an equivalent equation

$$u(\pi s) - \pi \int_0^1 \log|\cos(\pi s) - \cos(\pi t)| u(\pi t) \mathrm{d}t = 1, \quad 0 \leq s \leq 1.$$

We write the kernel

$$\log|\cos(\pi s) - \cos(\pi t)| = \sum_{i=1}^{4} K_{i,1}(s,t)K_{i,2}(s,t),$$

where

$$K_{1,1}(s,t) = \log\left(\frac{\sin\left(\frac{\pi(t-s)}{2}\right)\sin\left(\frac{\pi(t+s)}{2}\right)}{\pi^3\frac{(t-s)}{2}(t+s)(2-t-s)}\right),$$

$$K_{1,2}(s,t) = K_{2,1}(s,t) = K_{3,1}(s,t) = K_{4,1}(s,t) = 1,$$

$$K_{2,2}(s,t) = \log|\pi(s-t)|,\quad K_{3,2}(s,t) = \log(\pi(2-s-t))$$

and

$$K_{4,2}(s,t) = \log(\pi(s+t)).$$

In Table 3.1 we present the error e_n of the approximate solution and the error e_n' of the iterated approximate solution, where we use q and q' to represent the corresponding orders of approximation, respectively. In our computation, we choose $k_3 = 2$.

The order of approximation agrees with our theoretical estimate. The iteration does not improve the accuracy of the approximate solution for this example due to the nonsmoothness of the kernel.

Example 2. We consider the integral equation with a smooth kernel

$$u(s) - \int_0^1 \sin s \cos t u(t) \mathrm{d}t = \sin s(1 - e^{\sin 1}) + e^{\sin s}, \qquad 0 \le s \le 1.$$

It is not difficult to verify that $u(s) := e^{\sin s}$ is the unique solution of this equation. In the notation of Section 3.5.3, we have $K_1(s,t) := \sin s \cos t$ and $K_2(s,t) := 1$. In this case, we choose $k_3 = 3$ for the quadrature formula. The notation in Table 3.2 is the same as that in Table 3.1.

In this example, the iteration improves the accuracy of approximation by the order as estimated in Theorem 3.40.

Table 3.1

n	4	8	16	32
e_n	1.504077e-6	3.879971e-7	9.877713e-8	2.957718e-8
q		1.954761	1.973797	1.739639
e_n'	3.186220e-6	8.005914e-7	1.973337e-7	5.153006e-8
q'		1.992708	2.020429	1.93715

Table 3.2

n	4	8	16	32
e_n	1.68156e-2	4.10275e-3	1.01615e-3	3.00353e-4
q		2.035137	2.013478	1.75839
e'_n	6.78911e-5	4.16056e-6	2.58946e-7	1.61679e-8
q'		4.028373	4.006054	4.001447

3.6 Bibliographical remarks

The material presented in this chapter on conventional numerical methods for solving Fredholm integral equations of the second kind is mainly taken from the books [15, 177, 178]. Analysis of the quadrature method may be found in [6]. As related issues of the collocation method, valuation of an L^∞ function f at a given point, the reader is referred to [21] and multivariate Lagrange interpolation by polynomials may be found in [83] and the references cited therein. For the theoretical framework for analysis of the Petrov–Galerkin method, readers are referred to [64, 77]. Superconvergence of the iterated Petrov–Galerkin method was originally analyzed in [77]. The discrete Petrov–Galerkin method and its iterated scheme presented in Section 3.5.3 are taken from [68, 80]. The iterated Galerkin method, a special case of the iterated Petrov–Galerkin method for Fredholm integral equations of the second kind, was studied by many authors (see [23, 165, 246] and the references cited therein). Reference [241] gives a nice review of the iterated Galerkin method and iterated collocation method.

We would like to mention other developments on this topic not included in this book. Boundary integral equations of the second kind with periodic logarithmic kernels were solved by a Nyström scheme-based extrapolation method in [271], where asymptotic expansions for the approximate solutions obtained by the Nyström scheme were developed to analyze the extrapolation method. The generalized airfoil equation for an airfoil with a flap was solved numerically in [204]. In [49], it was shown that the dense coefficient matrix obtained from a quadrature rule for boundary integral equations with logarithm kernels can be replaced by a sparse one if appropriate graded meshes are used in the quadrature rules. A fast numerical method was developed in [266] for solving the two-dimensional Fredholm integral equation of the second kind. More information about the Galerkin method using the Fourier basis for solving boundary integral equations may be found in [25, 81]. Fast numerical algorithms for this method were developed recently in [37, 154, 155, 263]. A singularity-preserving Galerkin method was developed in [41] for the

Fredholm integral equation of the second kind with weakly singular kernels whose solutions have singularity. The method was extended in [38, 229] to solve the Volterra integral equation of the second kind with weakly singular kernels, which was also used in [111, 112] to solve fractional differential equations.

A singularity-preserving collocation method for solving the Fredholm integral equation of the second kind with weakly singular kernels was developed in [39]. In [16], a discretized Galerkin method was obtained using numerical integration for evaluation of the integrals occurring in the Galerkin method and in [23], by considering discrete inner product and discrete projections, the authors treated more appropriately kernels with discontinuous derivatives. A discrete convergence theory and its applications to the numerical solution of weakly singular integral equations were presented in [256]. Finally, we remark that it may be obtained by a similar analysis provided by [165] for the superconvergence of the iterated Galerkin method when the kernels are weakly singular.

4

Multiscale basis functions

Since a large class of physical problems is defined on bounded domains, we focus on integral equations on bounded domains. As we know, a bounded domain in $\mathbb{R}^d$ may be well approximated by a polygonal domain, which is a union of simplexes, cubes and perhaps L-shaped domains. To develop fast Galerkin, Petrov–Galerkin and collocation methods for solving the integral equations, we need multiscale bases and collocation functionals on polygonal domains. Simplexes, cubes or L-shaped domains are typical examples of invariant sets. This chapter is devoted to a description of constructions of multiscale basis functions including multiscale orthogonal bases, interpolating bases and multiscale collocation functionals. The multiscale basis functions that we construct here are discontinuous piecewise polynomials. For this reason, we describe their construction on invariant sets which can turn to bases on a polygon.

To illustrate the idea of the construction, we start with examples on $[0, 1]$, which is the simplest example of invariant sets. This will be done in Section 4.1. Constructions of multiscale basis functions and collocation functionals on invariant sets are based on self-similar partitions of the sets. Hence, we discuss such partitions in Section 4.2. Based on such self-similar partitions, we describe constructions of multiscale orthogonal bases in Section 4.3. For the construction of the multiscale interpolating basis, we require the availability of the multiscale interpolation points. Section 4.4 is devoted to the notion of refinable sets, which are a base for the construction of the multiscale interpolation points. Finally, in Section 4.5, we present the construction of multiscale interpolating bases.

4.1 Multiscale functions on the unit interval

This section serves as an illustration of the idea for the construction of orthogonal multiscale piecewise polynomial bases on an invariant set. We consider the simplest invariant set $\Omega := [0, 1]$ in this section. The essential aspect of this construction is the recursive generation of partitions of Ω and the multiscale bases based on the partitions.

Let $L^2(\Omega)$ denote the Hilbert space equipped with inner product

$$(u, v) := \int_\Omega u(t)v(t)\mathrm{d}t, \quad u, v \in L^2(\Omega),$$

and the induced norm $\|\cdot\| := \sqrt{(\cdot,\cdot)}$. We now describe a sequence of finite-dimensional subspaces of $L^2(\Omega)$. For two positive integers k and m, we let $\mathbb{S}_m^k$ denote the linear space of all functions which are polynomials of degree at most $k-1$ on

$$I_{\ell,m} := \left[\frac{\ell}{m}, \frac{\ell+1}{m}\right], \quad \ell \in \mathbb{Z}_m.$$

The functions in $\mathbb{S}_m^k$ are allowed to be discontinuous at the knots j/m, for $j \in \mathbb{N}_{m-1}$. Hence, the dimension of the space $\mathbb{S}_m^k$ is km. When m' divides m, that is, $m = \ell m'$ for some positive integer ℓ then

$$\mathbb{S}_{m'}^k \subseteq \mathbb{S}_m^k,$$

since the knot sequence $\{\ell/m' : \ell \in \mathbb{Z}_{m'}\}$ for the space $\mathbb{S}_{m'}^k$ is contained in the sequence $\{\ell/m : \ell \in \mathbb{Z}_m\}$ for the space $\mathbb{S}_m^k$. In particular, for $m := 2^k$, we have that

$$\mathbb{S}_1^k \subseteq \mathbb{S}_2^k \subseteq \cdots \subseteq \mathbb{S}_{2^n}^k. \tag{4.1}$$

In this context, we reinterpret the unit interval and its partition. Recall that the unit interval Ω is the invariant set with respect to the maps

$$\phi_\epsilon(t) := \frac{\epsilon + t}{2}, \quad t \in \Omega, \quad \epsilon \in \mathbb{Z}_2,$$

in the sense that

$$\Omega = \phi_0(\Omega) \cup \phi_1(\Omega) \text{ and } \operatorname{meas}(\phi_0(\Omega) \cap \phi_1(\Omega)) = 0,$$

where $\operatorname{meas}(A)$ denotes the Lebesgue measure of the set A. Note that the maps ϕ_0 and ϕ_1 are contractive and they map Ω onto $[0, 1/2]$ and $[1/2, 1]$, respectively. The partition $\{I_{\ell,2^k} : \ell \in \mathbb{Z}_{2^k}\}$ of Ω can be re-expressed in terms of the contractive maps ϕ_ϵ, $\epsilon \in \mathbb{Z}_2$ as

$$\left\{\phi_{\epsilon_1 \cdots \epsilon_k}(\Omega) : \epsilon_j \in \mathbb{Z}_2\right\}.$$

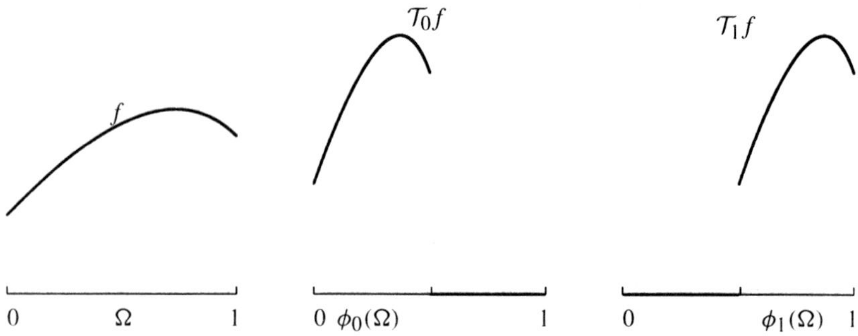

Figure 4.1 ϕ_e and $\mathcal{T}_e$.

Associated with the contractive maps ϕ_ϵ, $\epsilon \in \mathbb{Z}_2$, we introduce two mutually orthogonal isometries on $L^2(\Omega)$ that will be used to recursively generate bases for the spaces in the chain (4.1). For each $\epsilon \in \mathbb{Z}_2$ we set $\Omega_\epsilon := [\epsilon/2, (\epsilon+1)/2]$ and define the isometry $\mathcal{T}_\epsilon$ by setting, for $f \in L^2(\Omega)$,

$$\mathcal{T}_\epsilon f := \sqrt{2}\left(f \circ \phi_\epsilon^{-1}\right)\chi_{\Omega_\epsilon} = \begin{cases} \sqrt{2}f(2t-\epsilon), & t \in \Omega_\epsilon, \\ 0, & t \notin \Omega_\epsilon, \end{cases} \tag{4.2}$$

where χ_A denotes the characteristic function of the set A. Figure 4.1 illustrates the results of applications of operators $\mathcal{T}_\epsilon$ to a function.

For each $\epsilon \in \mathbb{Z}_2$, we use $\mathcal{T}_\epsilon^*$ for the adjoint operator of $\mathcal{T}_\epsilon$. We have the following result concerning the adjoint operator $\mathcal{T}_\epsilon^*$.

Proposition 4.1 *(1) If $f \in L^2(\Omega)$, then*

$$\mathcal{T}_\epsilon^* f = \frac{\sqrt{2}}{2} f \circ \phi_\epsilon.$$

(2) For any $\epsilon, \epsilon' \in \mathbb{Z}_2$,

$$\mathcal{T}_\epsilon^* \mathcal{T}_{\epsilon'} = \delta_{\epsilon\epsilon'}\mathcal{I}. \tag{4.3}$$

(3) For any $\epsilon, \epsilon' \in \mathbb{Z}_2$ and $f, g \in L^2(\Omega)$,

$$(\mathcal{T}_\epsilon f, \mathcal{T}_{\epsilon'} g) = \delta_{\epsilon\epsilon'}(f, g).$$

Proof (1) For $f, g \in L^2(\Omega)$, by the definition of $\mathcal{T}_\epsilon$, we have that

$$\int_\Omega g(x)(\mathcal{T}_\epsilon f)(x)\mathrm{d}x = \sqrt{2}\int_{\Omega_\epsilon} g(x)(f \circ \phi_\epsilon^{-1})(x)\mathrm{d}x.$$

We make a change of variables $t := \phi_\epsilon^{-1}(x)$ to conclude that

$$\int_\Omega g(x)(\mathcal{T}_\epsilon f)(x)\mathrm{d}x = \frac{\sqrt{2}}{2}\int_\Omega (g \circ \phi_\epsilon)(x)f(x)\mathrm{d}x.$$

The formula for $\mathcal{T}_\epsilon^* g$ follows.

(2) For $f \in L^2(\Omega)$, by (1) of this proposition, we observe that

$$(\mathcal{T}_{\epsilon'}^*\mathcal{T}_\epsilon f)(x) = \frac{\sqrt{2}}{2}(\mathcal{T}_\epsilon f)(\phi_{\epsilon'}(x)).$$

By the definition of the operator $\mathcal{T}_\epsilon$, if $\epsilon' \neq \epsilon$, then $\mathcal{T}_{\epsilon'}^*\mathcal{T}_\epsilon = 0$ and if $\epsilon' = \epsilon$, then $\mathcal{T}_{\epsilon'}^*\mathcal{T}_\epsilon = I$.

(3) The formula in this part follows directly from (4.3). □

It is clear from their definitions that the operators $\mathcal{T}_\epsilon$, $\epsilon \in \mathbb{Z}_2$ preserve the linear independence of a set of functions in $L^2(\Omega)$. Moreover, it follows from Proposition 4.1 (2) that functions resulting from applications of the operators $\mathcal{T}_\epsilon$ with different ϵ are orthogonal. We next show how to use the operators $\mathcal{T}_\epsilon$, $\epsilon \in \mathbb{Z}_2$ to generate, recursively, the bases for spaces

$$\mathbb{X}_n := \mathbb{S}_{2^n}^k, \quad n \in \mathbb{N}_0.$$

To this end, when S_1 and S_2 are subsets of $L^2(\Omega)$ such that $(u, v) = 0$, for all $u \in S_1, v \in S_2$, we introduce the notation $S_1 \cup^\perp S_2$, which denotes the union of S_1 and S_2.

Proposition 4.2 *If X_0 is an orthonormal basis for $\mathbb{S}_1^k$, then*

$$X_n := \bigcup_{\epsilon\in\mathbb{Z}_2}{}^{\perp} \mathcal{T}_\epsilon X_{n-1}, \quad n \in \mathbb{N} \tag{4.4}$$

is an orthonormal basis for $\mathbb{S}_{2^n}^k$.

Proof We prove by induction on n. Suppose that f_j, $j \in \mathbb{N}_{k2^{n-1}}$ form an orthonormal basis for $\mathbb{X}_{n-1}$. By Proposition 4.1, $\mathcal{T}_\epsilon f_j$, $j \in \mathbb{N}_{k2^{n-1}}$, $\epsilon \in \mathbb{Z}_2$ are also orthonormal. It can be shown that these $k2^n$ functions are elements in $\mathbb{X}_n$. Moreover, $\dim \mathbb{X}_n = k2^n$, which equals the number of these elements. Therefore, they form an orthonormal basis for the space $\mathbb{X}_n$. □

We now turn to the construction of our multiscale basis for space $\mathbb{X}_n$. Recalling $\mathbb{X}_{n-1} \subseteq \mathbb{X}_n$, we have that

$$\mathbb{X}_n = \mathbb{X}_{n-1} \oplus^\perp \mathbb{W}_n,$$

where $\mathbb{W}_n$ is the orthogonal complement of $\mathbb{X}_{n-1}$ in $\mathbb{X}_n$. Since the dimension of $\mathbb{X}_n$ is $k2^n$, the dimension of $\mathbb{W}_n$ is given by

$$\dim \mathbb{W}_n = k2^{n-1}.$$

Repeating this process leads to the decomposition

$$\mathbb{X}_n = \mathbb{X}_0 \oplus^\perp \mathbb{W}_1 \oplus^\perp \cdots \oplus^\perp \mathbb{W}_n \tag{4.5}$$

for space $\mathbb{X}_n$. In order to construct a multiscale orthonormal basis, it suffices to construct an orthonormal basis W_j for space $\mathbb{W}_j$, for each $j \in \mathbb{N}_n$. We first choose the Legendre polynomials of degree $\leq k-1$ on Ω as an orthonormal basis for $\mathbb{X}_0 = \mathbb{S}_1^k$ and denote by X_0 the basis. We then use Proposition 4.2 to construct an orthonormal basis X_1 for space $\mathbb{X}_1$, that is,

$$X_1 = \bigcup_{\epsilon \in \mathbb{Z}_2}^{\perp} \mathcal{T}_\epsilon X_0.$$

Since both $\mathbb{X}_0$ and $\mathbb{X}_1$ are finite-dimensional, we can use the Gram–Schmidt process to find an orthonormal basis W_1 for $\mathbb{W}_1$. Specifically, we form a linear combination of the basis functions in X_1 and require it to be orthogonal to all elements of X_0. This gives us k linearly independent elements which are orthogonal to $\mathbb{X}_0$. We then orthonormalize these k functions and they serve as an orthonormal basis for $\mathbb{W}_1$. For construction of basis W_j when $j \geq 2$, we appeal to the following proposition.

Proposition 4.3 *If W_1 is given as an orthonormal basis for $\mathbb{W}_1$, then*

$$W_{n+1} := \bigcup_{\epsilon \in \mathbb{Z}_2}^{\perp} \mathcal{T}_\epsilon W_n, \quad n \in \mathbb{N} \tag{4.6}$$

is an orthonormal basis for $\mathbb{W}_{n+1}$.

Proof We prove that W_n is an orthonormal basis for $\mathbb{W}_n$ by induction on n. When $n = 1$, W_1 is an orthonormal basis for $\mathbb{W}_1$ by hypothesis. Assume that W_j is an orthonormal basis for $\mathbb{W}_j$ for some $j \geq 1$; we show that W_{j+1} is an orthonormal basis for $\mathbb{W}_{j+1}$.

Let $W := \mathcal{T}_0 W_j \cup \mathcal{T}_1 W_j$. Since $W_j \subseteq \mathbb{X}_j$, by Proposition 4.2 we conclude that $W \subseteq \mathbb{X}_{j+1}$. Proposition 4.1, with the induction hypothesis that $W_j \perp X_{j-1}$, ensures $W \perp (\mathcal{T}_0 X_{j-1} \cup \mathcal{T}_1 X_{j-1}) = X_j$, which implies that $W \subseteq \mathbb{W}_{j+1}$. Because the elements in W_j are orthonormal, by Proposition 4.1, the elements in W are also orthonormal. Moreover,

$$\operatorname{card} W = \dim \mathbb{W}_{j+1}$$

holds. Therefore, W is a basis for $\mathbb{W}_{j+1}$. □

The proposition above gives a recursive generation of the multiscale basis functions for spaces $\mathbb{W}_n$, once orthonormal basis functions for $\mathbb{W}_1$ are available. It is useful for us in what follows to index the functions in the *wavelet bases* for $\mathbb{X}_n$ and to clearly have in mind the interval of their "support." To this end, we set $\mathbb{W}_0 := \mathbb{X}_0$ and define

$$w(n) := \dim \mathbb{W}_n \ \text{ and } \ s(n) := \dim \mathbb{X}_n, \quad n \in \mathbb{N}_0.$$

Thus, we have that

$$w(0) = k, \ w(n) = k2^{n-1} \ \text{ and } \ s(n) = k2^n, \quad n \in \mathbb{N}_0.$$

For $i \in \mathbb{N}_0$ we write $W_i := \{w_{ij} : j \in \mathbb{Z}_{w(i)}\}$, where we use double subscripts for the basis functions, with the first representing the level of the scale of the subspaces and the second indicating the location of its support. There are two properties of the functions in the set $\{w_{ij} : (i,j) \in \mathbb{U}\}$, where $\mathbb{U} := \{(i,j) : i \in \mathbb{N}_0, j \in \mathbb{Z}_{w(i)}\}$, which are important to us. The first is that they form a complete orthonormal system for the space $L^2(\Omega)$. In particular, we have that

$$(w_{ij}, w_{i'j'}) = \delta_{ii'}\delta_{jj'}, \quad (i,j), (i',j') \in \mathbb{U}.$$

Embodied in this fact is the useful property that the wavelet basis $\{w_{ij} : (i,j) \in \mathbb{U}\}$ has *vanishing moments* of order k, that is,

$$((\cdot)^r, w_{ij}) = 0, \quad \text{for } r \in \mathbb{Z}_k, \ j \in \mathbb{Z}_{w(i)}, \ i \in \mathbb{N}.$$

The second property is the "shrinking support" (as the level i increases) of the multiscale basis functions. To pin down this fact we take the point of view that the k functions in W_1 have "support" on Ω. Thereafter, the wavelet basis at level i will be grouped into 2^{i-1} sets of k functions each having the same "support interval." For future reference, we identify a set off which w_{ij} vanishes. For this purpose, we write $j \in \mathbb{Z}_{w(i)}, i \in \mathbb{N}$ uniquely in the form $j = vk + l$, where $l \in \mathbb{Z}_k$ and $v \in \mathbb{Z}_{2^{i-1}}, i \in \mathbb{N}$. Then,

$$w_{ij}(t) = 0, \quad t \notin I_{v,2^{i-1}}, j = vk + l \tag{4.7}$$

and therefore we have, for $j \in \mathbb{Z}_{w(i)}$, that

$$\text{meas}(\text{supp}\, w_{ij}) \leq \frac{1}{2^{i-1}}.$$

To see this fact clearly we express v in its dyadic expansion

$$v = 2^{i-2}\epsilon_1 + \cdots + 2\epsilon_{i-2} + \epsilon_{i-1},$$

where $\epsilon_1, \epsilon_2, \ldots, \epsilon_{i-1} \in \mathbb{Z}_2$. The recursion (4.6) then gives the formula

$$w_{ij} = \mathcal{T}_{\epsilon_1} \cdots \mathcal{T}_{\epsilon_{i-1}} w_{1l},$$

which confirms (4.7).

We end this section by presenting bases for spaces $\mathbb{X}_0$ and $\mathbb{W}_1$ for four concrete examples: piecewise constant, linear, quadratic and cubic polynomials.

Piecewise constant functions. This case leads to the Haar wavelet. We have a basis for $\mathbb{X}_0$ given by

$$w_{00}(t) := 1, \quad t \in [0, 1]$$

and a basis for $\mathbb{W}_1$ given by (see also Figure 4.2)

$$w_{10}(t) := \begin{cases} 1, & t \in [0, 1/2], \\ -1, & t \in (1/2, 1]. \end{cases}$$

We illustrate in Figure 4.2 the graph of the functions w_{00} and w_{10}.

Piecewise linear polynomials. In this case, $k = 2$ and $\dim \mathbb{X}_0 = \dim \mathbb{W}_1 = 2$. We have an orthonormal basis for $\mathbb{X}_0$ given by

$$w_{00}(t) := 1, \quad w_{01}(t) := \sqrt{3}(2t - 1)$$

and an orthonormal basis for $\mathbb{W}_1$ given by

$$w_{10}(t) := \begin{cases} 1 - 6t, & t \in [0, \frac{1}{2}], \\ 5 - 6t, & t \in (\frac{1}{2}, 1], \end{cases} \quad w_{11}(t) := \begin{cases} \sqrt{3}(1 - 4t), & t \in [0, \frac{1}{2}], \\ \sqrt{3}(4t - 3), & t \in (\frac{1}{2}, 1]. \end{cases}$$

We illustrate in Figure 4.3 the graph of the functions w_{00}, w_{01}, w_{10} and w_{11}.

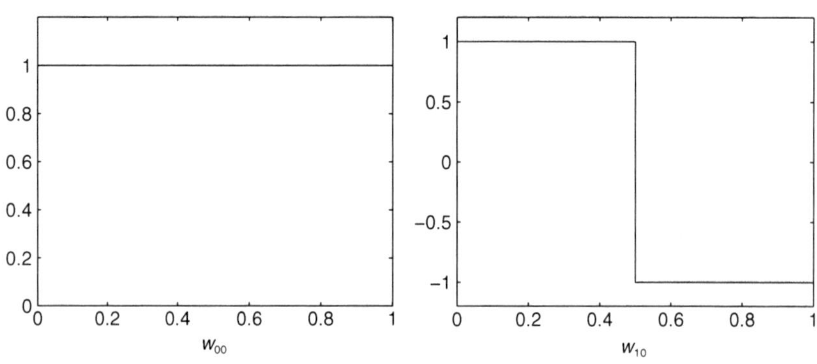

Figure 4.2 Basis functions for piecewise constant functions.

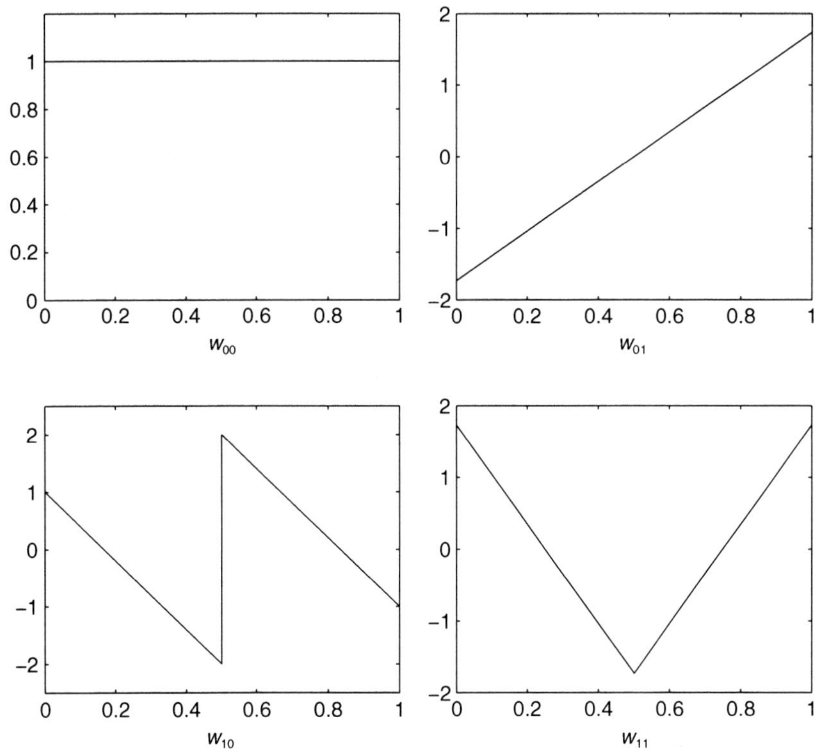

Figure 4.3 Basis functions for piecewise linear polynomials.

Piecewise quadratic polynomials. In this case, $k = 3$ and $\dim \mathbb{X}_0 = \dim \mathbb{W}_1 = 3$. An orthonormal basis for $\mathbb{X}_0$ is given by

$$w_{00}(t) := 1, \ \ w_{01}(t) := \sqrt{3}(2t - 1), \ \ w_{02}(t) := \sqrt{5}(6t^2 - 6t + 1)$$

and an orthonormal basis for $\mathbb{W}_1$ is given by

$$w_{10}(t) := \begin{cases} 1 - 6t, & t \in [0, \frac{1}{2}], \\ 5 - 6t, & t \in (\frac{1}{2}, 1], \end{cases}$$

$$w_{11}(t) := \begin{cases} \frac{\sqrt{93}}{31}(240t^2 - 116t + 9), & t \in [0, \frac{1}{2}], \\ \frac{\sqrt{93}}{31}(3 - 4t), & t \in (\frac{1}{2}, 1], \end{cases}$$

$$w_{12}(t) := \begin{cases} \frac{\sqrt{93}}{31}(4t - 1), & t \in [0, \frac{1}{2}], \\ \frac{\sqrt{93}}{31}(240t^2 - 364t + 133), & t \in (\frac{1}{2}, 1]. \end{cases}$$

In Figure 4.4, we illustrate the graph of the bases for $\mathbb{X}_0$ and $\mathbb{W}_0$ in this case.

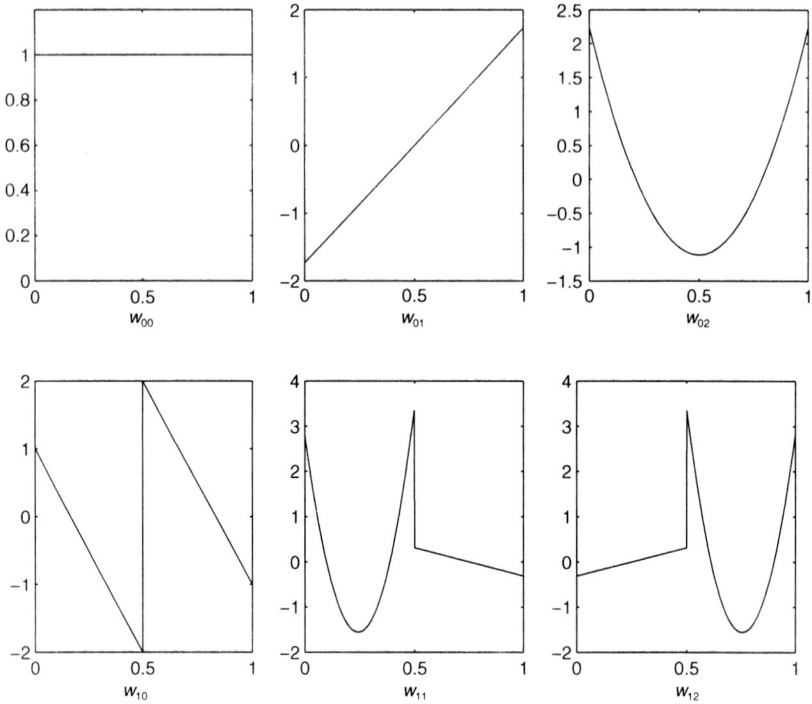

Figure 4.4 Basis functions for piecewise quadratic polynomials.

Piecewise cubic polynomials. In this case, we have that $k = 4$ and $\dim \mathbb{X}_0 = \dim \mathbb{W}_1 = 4$. An orthonormal basis for $\mathbb{X}_0$ is given by

$$w_{00}(t) := 1, \quad w_{01}(t) := \sqrt{3}(2t - 1),$$
$$w_{02}(t) := \sqrt{5}(6t^2 - 6t + 1), \quad w_{03}(t) := \sqrt{7}(20t^3 - 30t^2 + 12t - 1)$$

and a basis for $\mathbb{W}_1$ is given by

$$w_{10}(t) := \begin{cases} \frac{\sqrt{5}}{15}(240t^2 - 90t + 5), & t \in [0, \frac{1}{2}], \\ -\frac{\sqrt{5}}{15}(240t^2 - 390t + 155), & t \in (\frac{1}{2}, 1], \end{cases}$$

$$w_{11}(t) := \begin{cases} \sqrt{3}(30t^2 - 14t + 1), & t \in [0, \frac{1}{2}], \\ \sqrt{3}(30t^2 - 46t + 17), & t \in (\frac{1}{2}, 1], \end{cases}$$

$$w_{12}(t) := \begin{cases} \sqrt{7}(160t^3 - 120t^2 + 24t - 1), & t \in [0, \frac{1}{2}], \\ -\sqrt{7}(160t^3 - 360t^2 + 264t - 63), & t \in (\frac{1}{2}, 1], \end{cases}$$

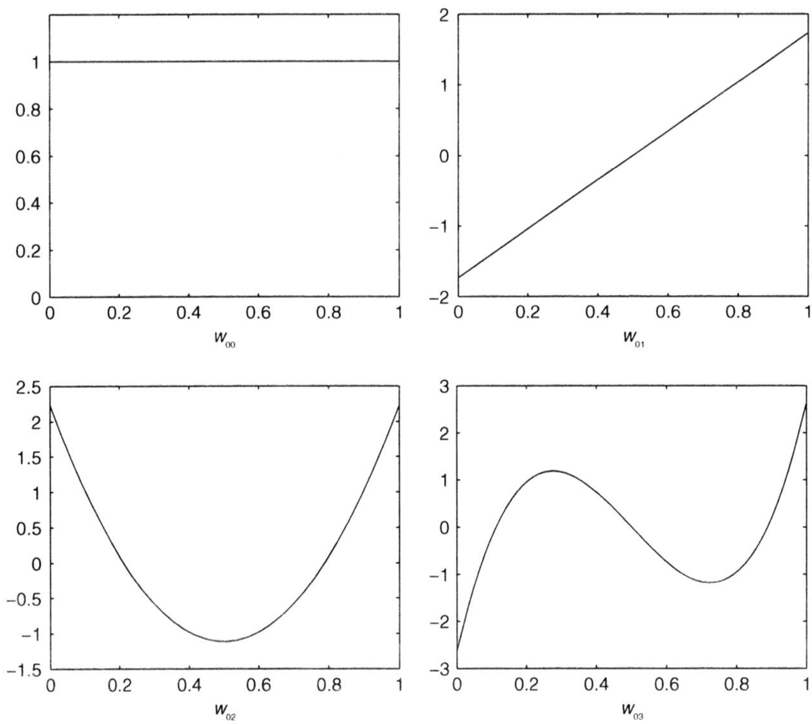

Figure 4.5 Basis functions for piecewise cubic polynomials.

$$w_{13}(t) := \begin{cases} \frac{14\sqrt{29}}{29}(160t^3 - 120t^2 + \frac{165}{7}t - \frac{13}{14}), & t \in [0, \frac{1}{2}], \\ \frac{14\sqrt{29}}{29}(160t^3 - 360t^2 + \frac{1845}{7}t - \frac{877}{14}), & t \in (\frac{1}{2}, 1]. \end{cases}$$

The bases for $\mathbb{X}_0$ and $\mathbb{W}_1$ are shown respectively in Figures 4.5 and 4.6.

4.2 Multiscale partitions

Because a polygonal domain in $\mathbb{R}^d$ is a union of a finite number of invariant sets, in this section we focus on multiscale partitioning of an invariant set in $\mathbb{R}^d$.

4.2.1 Invariant sets

We introduce the notion of invariant sets following [148]. Let $\mathbb{M}$ be a complete metric space. For any subset A of $\mathbb{M}$ and $x \in \mathbb{M}$, we define the distance of x to

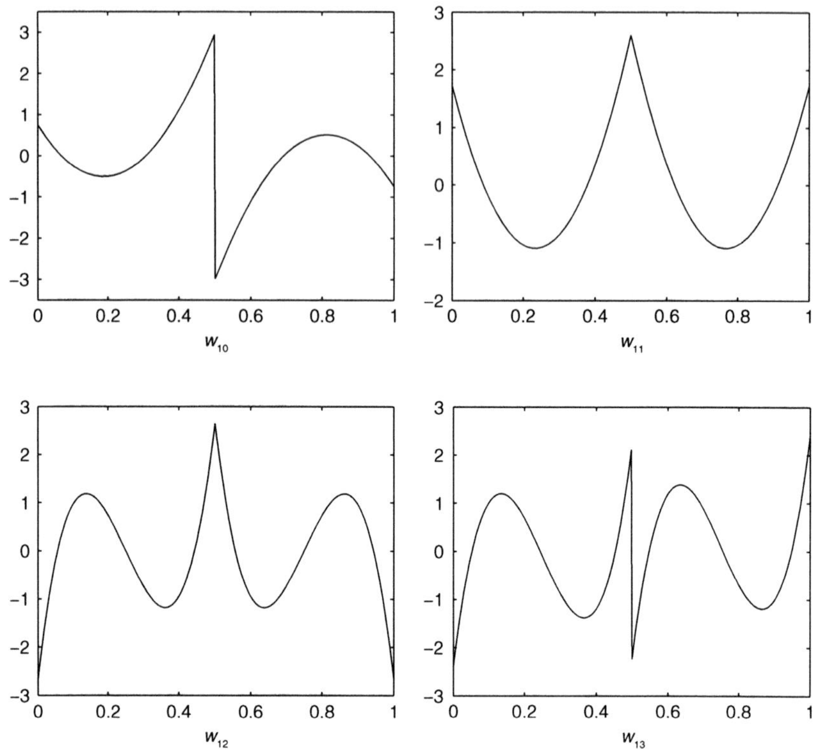

Figure 4.6 Basis functions for piecewise cubic polynomials.

A and the diameter of A, respectively, by

$$\operatorname{dist}(x, A) := \inf\{d(x, y) : y \in A\}$$

and

$$\operatorname{diam}(A) := \sup\{d(x, y) : x, y \in A\}.$$

A mapping from $\mathbb{M}$ to $\mathbb{M}$ is called contractive if there exists a $\gamma \in (0, 1)$ such that, for all subsets A of $\mathbb{M}$,

$$\operatorname{diam}(\phi_\epsilon(A)) \leq \gamma \operatorname{diam}(A), \quad \epsilon \in \mathbb{Z}_\mu. \tag{4.8}$$

For a positive integer $\mu > 1$ we suppose that $\Phi := \{\phi_\epsilon : \epsilon \in \mathbb{Z}_\mu\}$ is a family of contractive mappings on $\mathbb{M}$. We define the subset of $\mathbb{M}$ by

$$\Phi(A) := \bigcup_{\epsilon \in \mathbb{Z}_\mu} \phi_\epsilon(A).$$

According to [148], there exists a unique compact subset Ω of $\mathbb{M}$ such that

$$\Phi(\Omega) = \Omega. \tag{4.9}$$

We call the set Ω the *invariant set* relative to the family Φ of contractive mappings.

Generally, an invariant set has a complex *fractal* structure. For example, there are choices of Φ for which Ω is the Cantor subset of the interval $[0, 1]$, the Sierpinski gasket contained in an equilateral triangle or the twin dragons from wavelet analysis. In Figures 4.7 and 4.8, we illustrate the generation of the Cantor set of $[0, 1]$ and the Sierpinski gasket, respectively.

In the context of numerical solutions of integral equations, we are interested in the cases when Ω has a simple structure including, for example, the cube and simplex in $\mathbb{R}^d$. With these cases in mind, we make the following additional restrictions on the family Φ of mappings.

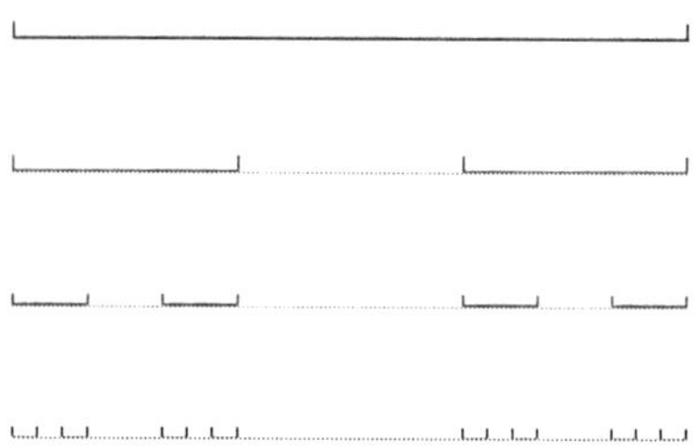

Figure 4.7 Generation of the Cantor set.

Figure 4.8 Generation of the Sierpinski gasket.

(a) For every $\epsilon \in \mathbb{Z}_\mu$, the mapping ϕ_ϵ has a continuous inverse on Ω.
(b) The set Ω has nonempty interior and

$$\text{meas}\,(\phi_\epsilon(\Omega) \cap \phi_{\epsilon'}(\Omega)) = 0, \quad \epsilon, \epsilon' \in \mathbb{Z}_\mu, \ \epsilon \neq \epsilon'.$$

We present several simple examples of invariant sets.

Example 4.4 For the metric space $\mathbb{R}$ and an integer $\mu > 1$, consider the family of contractive mappings $\Phi := \{\phi_\epsilon : \epsilon \in \mathbb{Z}_\mu\}$, where

$$\phi_\epsilon(t) := \frac{t+\epsilon}{\mu}, \quad t \in \mathbb{R}, \ \epsilon \in \mathbb{Z}_\mu.$$

The unit interval $\Omega := [0, 1]$ is the invariant set relative to Φ, which satisfies

$$\Omega = \bigcup_{\epsilon \in \mathbb{Z}_\mu} \phi_\epsilon(\Omega).$$

When $\mu = 2$, this example is discussed in Section 4.1. Figure 4.9 illustrates the case when $\mu = 3$. Note that in this case

$$\phi_0(\Omega) = \left[1, \frac{1}{3}\right], \quad \phi_1(\Omega) = \left[\frac{1}{3}, \frac{2}{3}\right], \quad \phi_2(\Omega) = \left[\frac{2}{3}, 1\right]$$

and clearly,

$$[0, 1] = \phi_0(\Omega) \cup \phi_1(\Omega) \cup \phi_2(\Omega).$$

Example 4.5 In the metric space $\mathbb{R}^2$, we consider four contractive affine mappings

$$\phi_0(s, t) := \frac{1}{2}(s, t), \quad \phi_1(s, t) := \frac{1}{2}(s + 1, t),$$

$$\phi_2(s, t) := \frac{1}{2}(s, t + 1), \quad \phi_3(s, t) := \frac{1}{2}(1 - s, 1 - t), \quad (s, t) \in \mathbb{R}^2.$$

The invariant set relative to these mappings is the unit triangle Ω with vertices at $(0, 0)$, $(1, 0)$ and $(0, 1)$, since

$$\Omega = \phi_0(\Omega) \cup \phi_1(\Omega) \cup \phi_2(\Omega) \cup \phi_3(\Omega).$$

This is illustrated in Figure 4.10.

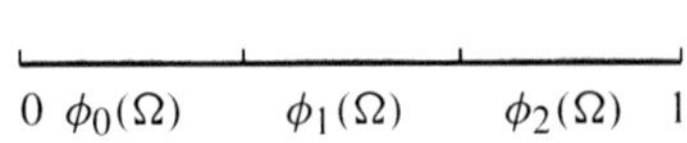

Figure 4.9 The invariant set in Example 1 with $\mu = 3$.

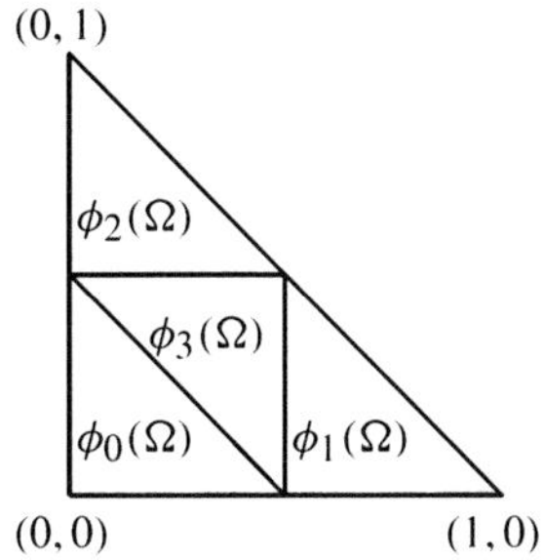

Figure 4.10 The unit triangle as an invariant set.

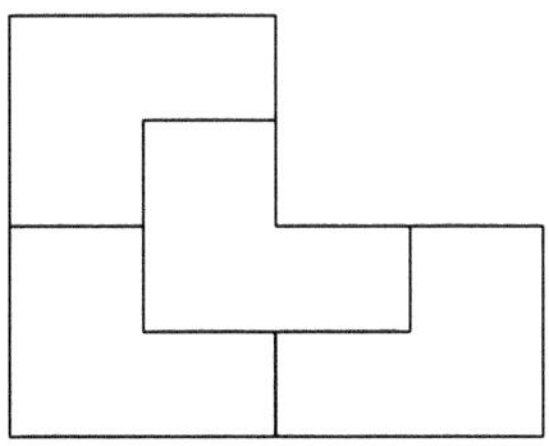

Figure 4.11 The unit L-shaped domain as an invariant set.

Example 4.6 In the metric space $\mathbb{R}^2$, we consider four contractive affine mappings

$$\phi_0(s,t) := \frac{1}{2}(s,t), \quad \phi_1(s,t) := \frac{1}{2}(2-s,t),$$
$$\phi_2(s,t) := \frac{1}{2}(s,2-t), \quad \phi_3(s,t) := \frac{1}{2}(s+1/2,t+1/2), \quad (s,t) \in \mathbb{R}^2.$$

The invariant set relative to these mappings is the L-shaped domain illustrated in Figure 4.11.

Example 4.7 As the last example, in the metric space $\mathbb{R}^3$, we consider eight contractive affine mappings

$$\phi_0(x,y,z) := \frac{1}{2}(x,y,z), \quad \phi_1(x,y,z) := \frac{1}{2}(y,z,x+1),$$
$$\phi_2(x,y,z) := \frac{1}{2}(x,z,y+1), \quad \phi_3(x,y,z) := \frac{1}{2}(x,y,z+1),$$
$$\phi_4(x,y,z) := \frac{1}{2}(x,y+1,z+1), \quad \phi_5(x,y,z) := \frac{1}{2}(y,x+1,z+1),$$
$$\phi_6(x,y,z) := \frac{1}{2}(z,x+1,y+1), \quad \phi_7(x,y,z) := \frac{1}{2}(x+1,y+1,z+1).$$

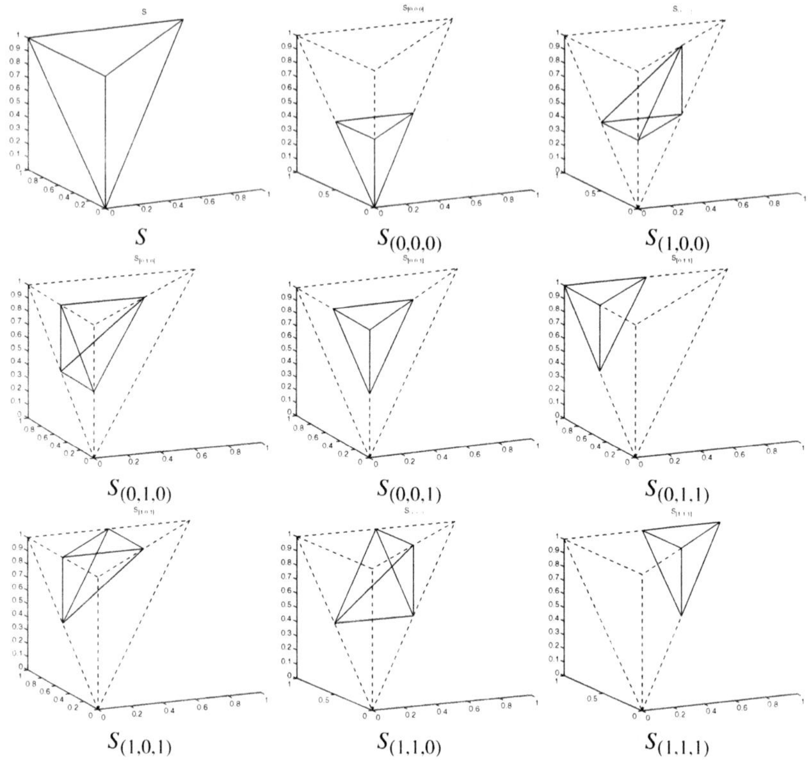

Figure 4.12 A three-dimensional unit simplex as an invariant set.

The invariant set relative to these eight mappings is the simplex in $\mathbb{R}^3$ defined by

$$S := \{(x, y, z) : 0 \leq x \leq y \leq z \leq 1\}.$$

This is illustrated in Figure 4.12.

4.2.2 Multiscale partitions by contractive mappings

The contractive mappings Φ that define the invariant set Ω naturally form a partition for the invariant set. Repeatedly applying the mappings to the invariant set generates a sequence of multiscale partitions for the invariant set.

We next show how the contractive mappings Φ are used to generate a sequence of multiscale partitions $\{\Omega_n : n \in \mathbb{N}_0\}$ of the invariant set Ω, which is defined by Φ. For notational convenience, we introduce the notation

$$\mathbb{Z}_\mu^n := \mathbb{Z}_\mu \times \cdots \times \mathbb{Z}_\mu, \quad n \text{ times.}$$

For each $\mathbf{e} := [e_j : j \in \mathbb{Z}_n] \in \mathbb{Z}_\mu^n$, we define the composition mapping

$$\phi_{\mathbf{e}} := \phi_{e_0} \circ \phi_{e_1} \circ \cdots \circ \phi_{e_{n-1}}$$

and the number

$$\mu(\mathbf{e}) := \mu^{n-1} e_0 + \cdots + \mu e_{n-2} + e_{n-1}.$$

Note that every $i \in \mathbb{Z}_{\mu^n}$ can be written uniquely as $i = \mu(\mathbf{e})$ for some $\mathbf{e} \in \mathbb{Z}_\mu^n$. From equation (4.9) and conditions (a) and (b), it follows that the collection of sets

$$\Omega_n := \{\Omega_{n,\mathbf{e}} : \Omega_{n,\mathbf{e}} = \phi_{\mathbf{e}}(\Omega),\ \mathbf{e} \in \mathbb{Z}_\mu^n\} \tag{4.10}$$

forms a partition of Ω. We require that this partition has the following property.

(c) There exist positive constants c_-, c_+ such that for all $n \in \mathbb{N}_0$

$$c_- \mu^{-n/d} \leq \max\{d(\Omega_{n,\mathbf{e}}) : \mathbf{e} \in \mathbb{Z}_\mu^n\} \leq c_+ \mu^{-n/d}, \tag{4.11}$$

where $d(A)$ represents the diameter of set A, that is, $d(A) := \sup\{|x-y| : x, y \in A\}$, with $|\cdot|$ being the Euclidean norm in the space $\mathbb{R}^d$.

If a sequence of partitions Ω_n, $n \in \mathbb{N}_0$ has property (c), we say that it forms a sequence of multiscale partitions for Ω.

Proposition 4.8 *If the Jacobian of the contractive affine mappings ϕ_e, $e \in \mathbb{Z}_\mu$, satisfies*

$$|J_{\phi_e}| = \mathcal{O}(\mu^{-1}),$$

then the sequence of partitions Ω_n, $n \in \mathbb{N}_0$, is of multiscale.

Proof For any $s, t \in \phi_e(\Omega)$, there exist $\hat{s}, \hat{t} \in \Omega$ such that $s = \phi_e(\hat{s})$ and $t = \phi_e(\hat{t})$, and thus, we have that

$$|s - t| = |J_{\phi_e}|^{1/d} |\hat{s} - \hat{t}|.$$

This, with the hypothesis on the Jacobian of the mappings, ensures that for any $e \in \mathbb{Z}_\mu$

$$d(\Omega_{1,e}) = \mathcal{O}(\mu^{-1/d}).$$

By induction, we may find that for any $\mathbf{e} \in \mathbb{Z}_\mu^n$,

$$d(\Omega_{n,\mathbf{e}}) = \mathcal{O}(\mu^{-n/d}), \tag{4.12}$$

proving the result. □

4.2.3 Multiscale partitions of a multidimensional simplex

For the purpose of solving integral equations on a polygonal domain in $\mathbb{R}^d$, we describe in this subsection multiscale partitions of a simplex in $\mathbb{R}^d$ for $d \geq 1$. For a vector $\mathbf{x} \in \mathbb{R}^d$, we write $\mathbf{x} = [x_j : x_j \in \mathbb{R}, j \in \mathbb{Z}_d]$. The unit simplex S in $\mathbb{R}^d$ is the subset

$$S := \{\mathbf{x} : \mathbf{x} \in \mathbb{R}^d, 0 \leq x_0 \leq x_1 \leq \cdots \leq x_{d-1} \leq 1\}.$$

This set is the invariant set relative to a family of μ^d contractive mappings. In order to describe these contractive mappings, for a positive integer μ, we define a family of *counting functions* $\chi_j : \mathbb{Z}_\mu^d \to \mathbb{Z}_{d+1}$, $j \in \mathbb{Z}_\mu$ for $\mathbf{e} := [e_j : j \in \mathbb{Z}_d] \in \mathbb{Z}_\mu^d$ by

$$\chi_j(\mathbf{e}) := \sum_{i \in \mathbb{Z}_d} \delta_j(e_i), \tag{4.13}$$

where $\delta_j(k) = 1$ when $j = k$ and otherwise $\delta_j(k) = 0$. Note that the value of $\chi_j(\mathbf{e})$ is exactly the number of components of $\mathbf{e}$ that equals j. Given $\mathbf{e} \in \mathbb{Z}_\mu^d$, we identify a vector $\mathbf{c}(\mathbf{e}) := [c_j : j \in \mathbb{Z}_{\mu+1}] \in \mathbb{Z}_{d+1}^{\mu+1}$ by

$$c_0 = 0, \quad c_j = \sum_{i \in \mathbb{Z}_j} \chi_i(\mathbf{e}), \quad j \in \mathbb{N}_\mu. \tag{4.14}$$

We remark that $\mathbf{c}(\mathbf{e})$ is always nondecreasing since each χ_j takes a non-negative value, and c_μ is always equal to d. For $\mathbf{e} \in \mathbb{Z}_\mu^d$ and $j < k$, we define the index set $\Psi_j^k := \{e_l : j \leq l < k, e_l = e_k\}$. Then we define the *permutation vector* $\mathbf{I_e} = [i_k : k \in \mathbb{Z}_d] \in \mathbb{Z}_d^d$ of $\mathbf{e}$ by

$$i_k = c_{e_k} + \operatorname{card}(\Psi_0^k), \tag{4.15}$$

where we assume $\operatorname{card}(\emptyset) = 0$. We have the following lemma about $\mathbf{I_e}$.

Lemma 4.9 *For any $\mathbf{e} \in \mathbb{Z}_\mu^d$, the permutation vector $\mathbf{I_e}$ has the following properties.*

(1) For $k \in \mathbb{Z}_d$, $c_m \leq i_k < c_{m+1}$ if and only if $m = e_k$.
(2) For any $j, k \in \mathbb{Z}_d$, $i_j < i_k$ if and only if $e_j < e_k$ or $e_j = e_k$ with $j < k$.
(3) The equality $i_j = i_k$ holds if and only if $j = k$.
(4) The vector $\mathbf{I_e}$ is a permutation of $\mathbf{v}_d := [j : j \in \mathbb{Z}_d]$.

Proof According to the definition of $\mathbf{I_e}$, we have for any $k \in \mathbb{Z}_k$ that

$$c_{e_k} \leq i_k < c_{e_k} + \operatorname{card}(\{e_j : j \in \mathbb{Z}_d, e_j = e_k\}) = c_{e_k+1}. \tag{4.16}$$

This implies that if $m = e_k$, $c_m \leq i_k < c_{m+1}$. On the contrary, if there is an m such that $c_m \leq i_k < c_{m+1}$, it is unique because the components of $\mathbf{c}(\mathbf{e})$ are nondecreasing. It follows from the uniqueness of m and (4.16) that $m = e_k$. Thus, property (1) is proved.

We now turn to proving property (2). If $e_j < e_k$, then $e_j + 1 \le e_k$ and hence $c_{e_j+1} \le c_{e_k}$, since the components of $\mathbf{c}(\mathbf{e})$ form a nondecreasing sequence. By (4.16) we conclude that $i_j < c_{e_j+1} \le c_{e_k} \le i_k$. If $e_j = e_k$ with $j < k$, then $i_k - i_j = \text{card}(\Psi_j^k) \ge 1$, hence $i_j < i_k$. It remains to prove that if $i_j < i_k$ then $e_j < e_k$ or $e_j = e_k$, $j < k$. Since in general for $j, k \in \mathbb{Z}_d$ one of the following cases holds: $e_j < e_k$, $e_j = e_k$ with $j < k$, $e_j = e_k$ with $j \ge k$ or $e_j > e_k$, it suffices to show that if $e_j > e_k$ or $e_j = e_k$ with $j \ge k$, then $i_j \ge i_k$. If $e_j > e_k$, by the proof we showed earlier in this paragraph, we conclude that $i_j > i_k$. If $e_j = e_k$ with $j \ge k$, we have that $i_j - i_k = \text{card}(\Psi_k^j) \ge 0$, that is, $i_j \ge i_k$. Thus, we complete a proof for property (2).

The above analysis also implies that the only possibility to have $i_j = i_k$ is $j = k$. This proves property (3).

Noticing that $e_k \in \mathbb{Z}_\mu$ for $k \in \mathbb{Z}_d$ and $0 \le c_{e_k} \le i_k < c_{e_k+1} \le d$, we conclude that $\mathbf{I_e}$ is a permutation of $\mathbf{v}_d$. □

We also need *conjugate permutations* in order to define the contractive mappings. A permutation matrix has exactly one entry in each row and column equal to one and all other entries zero. Hence, a permutation matrix is an orthogonal matrix. For any permutation $\mathbf{I_e}$ of $\mathbf{v}_d$, there is a unique permutation matrix $P_\mathbf{e}$ such that $\mathbf{I_e} = P_\mathbf{e}\mathbf{v}_d$. We call the vector

$$\mathbf{I}_\mathbf{e}^* = [i_j^* : j \in \mathbb{Z}_d] := P_\mathbf{e}^T \mathbf{v}_d$$

the *conjugate permutation* of $\mathbf{I_e}$. Thus, $\mathbf{I}_\mathbf{e}^*$ itself is also a permutation of $\mathbf{v}_d$. It follows from the definition above that for $l \in \mathbb{Z}_d$, $i_l^* = k$ if and only if $i_k = l$. We define the *conjugate vector* $\mathbf{e}^* := [e_j^* : j \in \mathbb{Z}_d]$ of $\mathbf{e}$ by setting $e_l^* = e_{i_l^*}$, $l \in \mathbb{Z}_d$. Utilizing the above notations, we define the mapping $\mathcal{G}_\mathbf{e}$ by

$$\mathcal{G}_\mathbf{e}(\mathbf{x}) := \tilde{\mathbf{x}} = \left[\tilde{x}_l = \frac{x_{i_l^*} + e_l^*}{\mu} : l \in \mathbb{Z}_d \right], \quad \mathbf{x} \in S. \tag{4.17}$$

It is clear that mappings $\mathcal{G}_\mathbf{e}$, $\mathbf{e} \in \mathbb{Z}_\mu^d$ are affine and contractive.

We next identify the set $\mathcal{G}_\mathbf{e}(S)$. To this end, associated with each $\mathbf{e} \in \mathbb{Z}_\mu^d$, we define a set in $\mathbb{R}^d$ by

$$S_\mathbf{e} := \left\{ \mathbf{x} \in \mathbb{R}^d : 0 \le x_{i_0} - \frac{e_0}{\mu} \le x_{i_1} - \frac{e_1}{\mu} \le \cdots \le x_{i_{d-1}} - \frac{e_{d-1}}{\mu} \le \frac{1}{\mu} \right\}, \tag{4.18}$$

where $i_k, k \in \mathbb{Z}_d$ are the components of the permutation vector $\mathbf{I_e}$ of $\mathbf{e}$. Since $\mathbf{I_e}$ is a permutation of $\mathbf{v}_d$, $S_\mathbf{e}$ is a simplex in $\mathbb{R}^d$. In the next lemma, we identify $\mathcal{G}_\mathbf{e}(S)$ with the simplex $S_\mathbf{e}$.

Lemma 4.10 *For all $\boldsymbol{e} \in \mathbb{Z}_\mu^d$, there holds $\mathcal{G}_{\boldsymbol{e}}(S) = S_{\boldsymbol{e}}$.*

Proof For $k \in \mathbb{Z}_d$, we let $l = i_k$ and observe by definition that $i_l^* = k, e_l^* = e_k$. Thus $\tilde{x}_l = \frac{x_k + e_k}{\mu}$, or

$$x_k = \mu \tilde{x}_l - e_k = \mu \tilde{x}_{i_k} - e_k. \tag{4.19}$$

If $\mathbf{x} \in S$, then $0 \leq x_0 \leq x_1 \leq \cdots \leq x_{d-1} \leq 1$, which implies that

$$0 \leq \mu \tilde{x}_{i_0} - e_0 \leq \mu \tilde{x}_{i_1} - e_1 \leq \cdots \leq \mu \tilde{x}_{i_{d-1}} - e_{d-1} \leq 1$$

or

$$0 \leq \tilde{x}_{i_0} - \frac{e_0}{\mu} \leq \tilde{x}_{i_1} - \frac{e_1}{\mu} \leq \cdots \leq \tilde{x}_{i_{d-1}} - \frac{e_{d-1}}{\mu} \leq \frac{1}{\mu},$$

so that $\tilde{\mathbf{x}} \in S_\mathbf{e}$. Moreover, given $\tilde{\mathbf{x}} \in S_\mathbf{e}$, we define $\mathbf{x} := [x_k : k \in \mathbb{Z}_d]$ by equation (4.19). Thus, $\mathbf{x} \in S$ and $\tilde{\mathbf{x}} = \mathcal{G}_\mathbf{e}(\mathbf{x})$. Therefore, $\mathcal{G}_\mathbf{e}(S) = S_\mathbf{e}$. □

In the following lemma we present properties of the simplices $S_\mathbf{e}$, $\mathbf{e} \in \mathbb{Z}_\mu^d$.

Lemma 4.11 *The simplices $S_{\boldsymbol{e}}$, $\boldsymbol{e} \in \mathbb{Z}_\mu^d$ have the following properties.*

(1) For any $\boldsymbol{x} \in S_{\boldsymbol{e}}$, there holds

$$\frac{k}{\mu} \leq x_{c_k} \leq x_{c_k+1} \leq \cdots \leq x_{c_{k+1}-1} \leq \frac{k+1}{\mu}, \quad k \in \mathbb{Z}_\mu. \tag{4.20}$$

(2) For any $\boldsymbol{e} \in \mathbb{Z}_\mu^d$, $S_{\boldsymbol{e}} \subset S$.
(3) If $\boldsymbol{e}_1, \boldsymbol{e}_2 \in \mathbb{Z}_\mu^d$ with $\boldsymbol{e}_1 \neq \boldsymbol{e}_2$, then $\mathrm{int}(S_{\boldsymbol{e}_1}) \cap \mathrm{int}(S_{\boldsymbol{e}_2}) = \emptyset$.
(4) For any $\boldsymbol{e} \in \mathbb{Z}_\mu^d$, $\mathrm{meas}(S_{\boldsymbol{e}}) = 1/(\mu^d d!)$, where $\mathrm{meas}(\Omega)$ denotes the Lebesgue measure of set Ω.

Proof In order to prove (4.20), it suffices to show

$$0 \leq x_{c_k} - \frac{k}{\mu} \leq x_{c_k+1} - \frac{k}{\mu} \leq \cdots \leq x_{c_{k+1}-1} - \frac{k}{\mu} \leq \frac{1}{\mu}, \quad k \in \mathbb{Z}_\mu, \tag{4.21}$$

or equivalently,

$$0 \leq x_p - \frac{k}{\mu} \leq x_q - \frac{k}{\mu} \leq \frac{1}{\mu}$$

for any $c_k \leq p < q < c_{k+1}$. In fact, since $\mathbf{I_e}$ is a permutation of $\mathbf{v}_d$, for any integers $c_k \leq p < q < c_{k+1}$, there exists a unique pair $p', q' \in \mathbb{Z}_d$ such that

$i_{p'} = p$, $i_{q'} = q$. It follows from Lemma 4.9 that $e_{p'} = e_{q'} = k$ and $p' < q'$. Thus (4.18) states that

$$0 \leq x_p - \frac{k}{\mu} = x_{i_{p'}} - \frac{e_{p'}}{\mu} \leq x_{i_{q'}} - \frac{e_{q'}}{\mu} = x_q - \frac{k}{\mu} \leq \frac{1}{\mu},$$

which concludes property (1).

Property (2) is a direct consequence of (1) and the definition of S.

For the proof of (3), we first notice that

$$\operatorname{int}(S_{\mathbf{e}}) = \left\{\mathbf{x} \in \mathbb{R}^d : 0 < x_{i_0} - \frac{e_0}{\mu} < x_{i_1} - \frac{e_1}{\mu} < \cdots < x_{i_{d-1}} - \frac{e_{d-1}}{\mu} < \frac{1}{\mu}\right\}. \tag{4.22}$$

Moreover, by a proof similar to that for (4.20), we utilize (4.22) to conclude for any $\mathbf{x} \in \operatorname{int}(S_{\mathbf{e}})$ that

$$\frac{k}{\mu} < x_{c_k} < x_{c_k+1} < \cdots < x_{c_{k+1}-1} < \frac{k+1}{\mu}, \quad k \in \mathbb{Z}_\mu. \tag{4.23}$$

For $j = 1, 2$ we let $\mathbf{e}_j := [e_k^j : k \in \mathbb{Z}_d]$, $\mathbf{I}_{\mathbf{e}_j} := [i_k^j : k \in \mathbb{Z}_d]$, $\mathbf{c}(\mathbf{e}_j) := [c_k^j : k \in \mathbb{Z}_{\mu+1}]$.

Assume to the contrary that $\operatorname{int}(S_{\mathbf{e}_1}) \cap \operatorname{int}(S_{\mathbf{e}_2})$ is not empty. We consider two cases. In case 1 that $\mathbf{c}(\mathbf{e}_1) \neq \mathbf{c}(\mathbf{e}_2)$, we let k be the smallest integer such that $c_k^1 \neq c_k^2$, and assume $c_k^1 < c_k^2$ without loss of generality. For any $\mathbf{x} \in \operatorname{int}(S_{\mathbf{e}_1}) \cap \operatorname{int}(S_{\mathbf{e}_2})$, by (4.23), we have $x_{c_k^1} > \frac{k}{\mu}$ and $x_{c_k^2-1} < \frac{k}{\mu}$. Moreover, because $\mathbf{x} \in S$, we have that $x_{c_k^1} \leq x_{c_k^2-1}$, a contradiction. In case 2 that $\mathbf{c}(\mathbf{e}_1) = \mathbf{c}(\mathbf{e}_2)$, since $\mathbf{e}_1 \neq \mathbf{e}_2$, we let k be the smallest integer such that $e_k^1 \neq e_k^2$. Hence $e_j^1 = e_j^2$ for $j < k$, and we assume that $e_k^1 < e_k^2$ without loss of generality. Thus, we have that $i_k^1 < c_{e_k^1+1}^1 \leq c_{e_k^2}^2 \leq i_k^2$. There exists a unique $p \in \mathbb{Z}_d$ such that $i_p^1 = i_k^2$ since $\mathbf{I}_{\mathbf{e}_1}$ is a permutation, and $p \geq k$ because $i_j^1 = i_j^2 \neq i_k^2$ for all $j < k$. Furthermore, it follows from Lemma 4.9, $\mathbf{c}(\mathbf{e}_1) = \mathbf{c}(\mathbf{e}_2)$ and $i_p^1 = i_k^2$ that $e_p^1 = e_k^2 \neq e_k^1$, which implies $p \neq k$. Therefore, for any $\mathbf{x} \in \operatorname{int}(S_{\mathbf{e}_1})$, there holds

$$x_{i_k^1} - \frac{e_k^1}{\mu} < x_{i_p^1} - \frac{e_p^1}{\mu} = x_{i_k^2} - \frac{e_k^2}{\mu}.$$

However, there is a unique $q \in \mathbb{Z}_d$ such that $q > k$, $i_q^2 = i_k^1$, and for any $\mathbf{x} \in \operatorname{int}(S_{\mathbf{e}_2})$,

$$x_{i_k^2} - \frac{e_k^2}{\mu} < x_{i_q^2} - \frac{e_q^2}{\mu} = x_{i_k^1} - \frac{e_k^1}{\mu},$$

again a contradiction. This completes the proof of property (3).

For property (4), we find by direct computation that $\text{meas}(S'_{\mathbf{e}}) = 1/(\mu^d d!)$, where

$$S'_{\mathbf{e}} := \left\{ \mathbf{x} \in \mathbb{R}^d : 0 \le x_{i_0} \le x_{i_1} \le \cdots \le x_{i_{d-1}} \le \frac{1}{\mu} \right\}.$$

Notice that $S_{\mathbf{e}}$ is the translation of simplex $S'_{\mathbf{e}}$ via the vector $\frac{\mathbf{e}}{\mu}$. Since the Lebesgue measure of a set is invariant under translation, we conclude property (4). □

Theorem 4.12 *The family* $S(\mathbb{Z}_\mu^d) := \{S_{\mathbf{e}} : \mathbf{e} \in \mathbb{Z}_\mu^d\}$ *is an equivolume partition of the unit simplex S.*

Proof By Lemma 4.11 we see that for any $\mathbf{e} \in \mathbb{Z}_\mu^d$, $S_{\mathbf{e}} \subset S$ and for $\mathbf{e}_1, \mathbf{e}_2 \in \mathbb{Z}_\mu^d$ with $\mathbf{e}_1 \neq \mathbf{e}_2$, $\text{int}(S_{\mathbf{e}_1}) \cap \text{int}(S_{\mathbf{e}_2}) = \emptyset$ and $\text{meas}(S_{\mathbf{e}_1}) = \text{meas}(S_{\mathbf{e}_2})$. It remains to prove that $S \subseteq \bigcup_{\mathbf{e} \in \mathbb{Z}_\mu^d} S_{\mathbf{e}}$.

To this end, for each $\mathbf{x} \in S$ we find $\mathbf{e} \in \mathbb{Z}_\mu^d$ such that $\mathbf{x} \in S_{\mathbf{e}}$. Note that for each $\mathbf{x} \in S$ we have that $0 \le x_0 \le x_1 \le \cdots \le x_{d-1} \le 1$. For each $k \in \mathbb{Z}_\mu$ we denote by c_k the subscript of the smallest component x_j greater than or equal to $\frac{k}{\mu}$. We order the elements in set $\{x_j : j \in \mathbb{Z}_d\} \cup \{\frac{k}{\mu} : k \in \mathbb{Z}_{\mu+1}\}$ in increasing order. We then obtain that

$$0 \le x_0 \le \cdots \le x_{c_1-1} < \frac{1}{\mu} \le x_{c_1} \le \cdots \le x_{c_{\mu-1}-1}$$
$$< \frac{\mu-1}{\mu} \le x_{c_{\mu-1}} \le \cdots \le x_{c_\mu -1} = x_{d-1} \le 1.$$

In other words, we have that

$$0 \le x_{c_k} - \frac{k}{\mu} \le x_{c_k+1} - \frac{k}{\mu} \le \cdots \le x_{c_{k+1}-1} - \frac{k}{\mu} \le \frac{1}{\mu}, \quad k \in \mathbb{Z}_\mu. \tag{4.24}$$

Let $p_j := \max\{k : c_k \le j\}$. It follows from (4.24) that the set $\{x_j - \frac{p_j}{\mu} : j \in \mathbb{Z}_d\} \subset [0, \frac{1}{\mu}]$. We sort the elements of this set into

$$0 \le x_{i_0} - \frac{p_{i_0}}{\mu} \le x_{i_1} - \frac{p_{i_1}}{\mu} \le \cdots \le x_{i_{d-1}} - \frac{p_{i_{d-1}}}{\mu} \le \frac{1}{\mu}. \tag{4.25}$$

Notice that the vector $\mathbf{I} := [i_k : k \in \mathbb{Z}_d]$ is a permutation of $\mathbf{v}_d$. Let $\mathbf{e} := [e_k : k \in \mathbb{Z}_d]$ be a vector such that $e_j = p_{i_j}$. It is easy to verify $i_j = c_{e_j} + |\Psi_0^j|$. Hence, $\mathbf{I} = \mathbf{I}_{\mathbf{e}}$, which together with (4.25) shows $\mathbf{x} \in S_{\mathbf{e}}$. □

The expression of the inverse mapping $\mathcal{G}_{\mathbf{e}}^{-1}$ has been given by equation (4.19), which is written formally as

$$\mathbf{x} := \mathcal{G}_{\mathbf{e}}^{-1}(\tilde{\mathbf{x}}) = [x_k = \mu \tilde{x}_{i_k} - e_k : k \in \mathbb{Z}_d], \quad \tilde{\mathbf{x}} \in S_{\mathbf{e}}. \tag{4.26}$$

For any $\mathbf{e} \in \mathbb{Z}_\mu^d$ and $\mathbf{x}', \mathbf{x}'' \in \mathbb{R}^d$,

$$\|\mathcal{G}_\mathbf{e}(\mathbf{x}') - \mathcal{G}_\mathbf{e}(\mathbf{x}'')\|_p = \frac{1}{\mu}\|\mathbf{x}' - \mathbf{x}''\|_p \tag{4.27}$$

and

$$\|\mathcal{G}_\mathbf{e}^{-1}(\mathbf{x}') - \mathcal{G}_\mathbf{e}^{-1}(\mathbf{x}'')\|_p = \mu\|\mathbf{x}' - \mathbf{x}''\|_p \tag{4.28}$$

hold, where $\|\cdot\|_p$ is the standard ℓ^p-norm on $\mathbb{R}^d$ for $1 \leq p \leq \infty$.

Proposition 4.13 *The family $S(\mathbb{Z}_\mu^d)$ is a uniform partition of the unit simplex S in the sense that all elements of $S(\mathbb{Z}_\mu^d)$ have an identical diameter.*

Proof We let $\Delta := \max_{\mathbf{x}',\mathbf{x}'' \in S} \|\mathbf{x}' - \mathbf{x}''\|_p$. It suffices to prove for any $\mathbf{e} \in \mathbb{Z}_\mu^d$ that

$$\max_{\mathbf{x}',\mathbf{x}'' \in S_\mathbf{e}} \|\mathbf{x}'_\mathbf{e} - \mathbf{x}''_\mathbf{e}\|_p = \frac{\Delta}{\mu}.$$

It follows from formula (4.28) that for any $\mathbf{x}'_\mathbf{e}, \mathbf{x}''_\mathbf{e} \in S_\mathbf{e}$,

$$\mu\|\mathbf{x}'_\mathbf{e} - \mathbf{x}''_\mathbf{e}\|_p = \|\mathcal{G}_\mathbf{e}^{-1}(\mathbf{x}'_\mathbf{e}) - \mathcal{G}_\mathbf{e}^{-1}(\mathbf{x}''_\mathbf{e})\|_p \leq \Delta.$$

Moreover, suppose that $\bar{\mathbf{x}}', \bar{\mathbf{x}}'' \in S$ such that $\|\bar{\mathbf{x}}' - \bar{\mathbf{x}}''\|_p = \Delta$ and let $\bar{\mathbf{x}}'_\mathbf{e} := \mathcal{G}_\mathbf{e}(\bar{\mathbf{x}}')$, and $\bar{\mathbf{x}}''_\mathbf{e} := \mathcal{G}_\mathbf{e}(\bar{\mathbf{x}}'')$. By (4.27), we have that

$$\|\bar{\mathbf{x}}'_\mathbf{e} - \bar{\mathbf{x}}''_\mathbf{e}\|_p = \frac{1}{\mu}\|\bar{\mathbf{x}}' - \bar{\mathbf{x}}''\|_p = \frac{\Delta}{\mu},$$

which completes the proof. □

When a partition of the unit simplex has been established, it is not difficult to obtain a corresponding partition of a general simplex in $\mathbb{R}^d$. For a nondegenerate simplex S' in $\mathbb{R}^d$, in the sense $\mathrm{Vol}(S') \neq 0$, there exists an affine mapping $\mathcal{F} : \mathbb{R}^d \to \mathbb{R}^d$ such that $\mathcal{F}(S') = S$. It can be shown that for $1 \leq p \leq \infty$ there are two positive constants c_1 and c_2 such that

$$c_1\|\mathbf{x}' - \mathbf{x}''\|_p \leq \|\mathcal{F}(\mathbf{x}') - \mathcal{F}(\mathbf{x}'')\|_p \leq c_2\|\mathbf{x}' - \mathbf{x}''\|_p, \tag{4.29}$$

for any $\mathbf{x}', \mathbf{x}'' \in S'$. For any $\mathbf{e} \in \mathbb{Z}_\mu^d$, we define $\mathcal{G}'_\mathbf{e} := \mathcal{F}^{-1} \circ \mathcal{G}_\mathbf{e} \circ \mathcal{F}$. Thus, the family of simplices $\{\mathcal{G}'_\mathbf{e}(S') : \mathbf{e} \in \mathbb{Z}_\mu^d\}$ is a partition of S'. Furthermore, for any $\mathbf{x}', \mathbf{x}'' \in \mathbb{R}^d$ and $\mathbf{e} \in \mathbb{Z}_\mu^d$,

$$\frac{c_1}{c_2\mu}\|\mathbf{x}' - \mathbf{x}''\|_p \leq \|\mathcal{G}'_\mathbf{e}(\mathbf{x}') - \mathcal{G}'_\mathbf{e}(\mathbf{x}'')\|_p \leq \frac{c_2}{c_1\mu}\|\mathbf{x}' - \mathbf{x}''\|_p$$

holds. For $\mathbf{E} := [\mathbf{e}_j : j \in \mathbb{Z}_m] \in (\mathbb{Z}_\mu^d)^m$, we define the composite mappings

$$\mathcal{G}_\mathbf{E} := \mathcal{G}_{\mathbf{e}_0} \circ \cdots \circ \mathcal{G}_{\mathbf{e}_{m-1}} \quad \text{and} \quad \mathcal{G}'_\mathbf{E} := \mathcal{G}'_{\mathbf{e}_0} \circ \cdots \circ \mathcal{G}'_{\mathbf{e}_{m-1}}$$

and observe that $\mathcal{G}'_{\mathbf{E}} = \mathcal{F}^{-1} \circ \mathcal{G}_{\mathbf{E}} \circ \mathcal{F}$. In the next theorem we show that the partition $\{\mathcal{G}'_{\mathbf{e}}(S') : \mathbf{e} \in \mathbb{Z}_\mu^d\}$ of S' is uniform. To this end, we let $S_{\mathbf{E}} := \mathcal{G}_{\mathbf{E}}(S)$ and $S'_{\mathbf{E}} := \mathcal{G}'_{\mathbf{E}}(S')$. Also, we use diam_p to denote the diameter of a domain in $\mathbb{R}^d$ with respect to the ℓ^p-norm.

Theorem 4.14 *For any $\boldsymbol{x}', \boldsymbol{x}'' \in \mathbb{R}^d$ and $\boldsymbol{E} \in (\mathbb{Z}_\mu^d)^m$, there hold*

$$\frac{c_1}{c_2}\left(\frac{1}{\mu}\right)^m \|\boldsymbol{x}' - \boldsymbol{x}''\|_p \leq \|\mathcal{G}'_{\boldsymbol{E}}(\boldsymbol{x}') - \mathcal{G}'_{\boldsymbol{E}}(\boldsymbol{x}'')\|_p \leq \frac{c_2}{c_1}\left(\frac{1}{\mu}\right)^m \|\boldsymbol{x}' - \boldsymbol{x}''\|_p,$$

$$\mathrm{diam}_p(S_{\boldsymbol{E}}) = \left(\frac{1}{\mu}\right)^m \mathrm{diam}_p(S)$$

and

$$\frac{c_1}{c_2}\left(\frac{1}{\mu}\right)^m \mathrm{diam}_p(S') \leq \mathrm{diam}_p(S'_{\boldsymbol{E}}) \leq \frac{c_2}{c_1}\left(\frac{1}{\mu}\right)^m \mathrm{diam}_p(S').$$

4.3 Multiscale orthogonal bases

In this section, we describe the recursive construction of multiscale orthogonal bases for spaces $L^2(\Omega)$ on the invariant set Ω.

4.3.1 Piecewise polynomial spaces

On the partition Ω_n, we consider piecewise polynomials in a Banach space $\mathbb{X}$ which has the norm $\|\cdot\|$. Choose a positive integer k and let $\mathbb{X}_n$ be the spaces of all functions such that their restriction to any cell $\Omega_{n,\mathbf{e}}$, $\mathbf{e} \in \mathbb{Z}_\mu^n$, is a polynomial of total degree $\leq k-1$. Here we use the convention that for $n = 0$ the set Ω is the only cell in the partition, and so

$$m := \dim \mathbb{X}_0 = \binom{k+d-1}{d}.$$

It is easily seen that

$$x(n) := \dim \mathbb{X}_n = m\mu^n, \quad n \in \mathbb{N}_0.$$

To generate the spaces $\mathbb{X}_n$ by induction from $\mathbb{X}_0$, we introduce linear operators $\mathcal{T}_e : \mathbb{X} \to \mathbb{X}, e \in \mathbb{Z}_\mu$ defined by

$$(\mathcal{T}_e v)(t) := c_0 v(\phi_e^{-1}(t))\chi_{\phi_e(\Omega)}(t), \tag{4.30}$$

where χ_A denotes the characteristic function of set A and c_0 the positive constant such that $\|\mathcal{T}_e\| = 1$. Thus we have that

$$\mathbb{X}_n = \bigoplus_{e \in \mathbb{Z}_\mu} \mathcal{T}_e \mathbb{X}_{n-1}, \quad n \in \mathbb{N}, \tag{4.31}$$

where $\mathbb{A} \oplus \mathbb{B}$ denotes the direct sum of the spaces $\mathbb{A}$ and $\mathbb{B}$. It is easily seen that the sequence of spaces has the property of nestedness, that is,

$$\mathbb{X}_{n-1} \subset \mathbb{X}_n, \quad n \in \mathbb{N}. \tag{4.32}$$

Assume that there exists a basis of elements in $\mathbb{X}_0$, denoted by $\psi_0, \psi_1, \ldots, \psi_{m-1}$ such that

$$\mathbb{X}_0 = \operatorname{span}\{\psi_j : j \in \mathbb{Z}_m\}.$$

It is clear that

$$\mathbb{X}_n = \operatorname{span}\{\mathcal{T}_{\mathbf{e}} \psi_j : j \in \mathbb{Z}_m, \mathbf{e} \in \mathbb{Z}_\mu^n\}. \tag{4.33}$$

4.3.2 A recursive construction

Noting that the subspace sequence $\{\mathbb{X}_n : n \in \mathbb{N}_0\}$ is nested, we can define for each $n \in \mathbb{N}_0$ subspaces $\mathbb{W}_{n+1} \subset \mathbb{X}_{n+1}$ such that

$$\mathbb{X}_{n+1} = \mathbb{X}_n \oplus^\perp \mathbb{W}_{n+1}, \quad n \in \mathbb{N}_0. \tag{4.34}$$

Thus, by setting $\mathbb{W}_0 := \mathbb{X}_0$, we have the multiscale space decomposition that for any $n \in \mathbb{N}$,

$$\mathbb{X}_n = \bigoplus_{i \in \mathbb{Z}_{n+1}}^\perp \mathbb{W}_i \tag{4.35}$$

and

$$L^2(\Omega) = \overline{\bigoplus_{i \in \mathbb{N}_0}^\perp \mathbb{W}_i}. \tag{4.36}$$

It can be computed that the dimension of $\mathbb{W}_n$ is given by

$$w(n) = \dim \mathbb{W}_n = x(n) - x(n-1) = m(\mu - 1)\mu^{n-1}, \quad n \in \mathbb{N}. \tag{4.37}$$

Now the family of operators $\mathcal{T}_e : L^2(\Omega) \to L^2(\Omega), e \in \mathbb{Z}_\mu$, is defined by

$$\mathcal{T}_e v := |J_{\phi_e}|^{-\frac{1}{2}} v \circ \phi_e^{-1} \chi_{\phi_e(\Omega)}.$$

In the next proposition, we see that the operators $\mathcal{T}_e, e \in \mathbb{Z}_\mu$, are isometrics from $L^2(\Omega)$ to $L^2(\Omega)$.

Proposition 4.15 *If $e, e' \in \mathbb{Z}_\mu$, then for all $u, v \in L^2(\Omega)$,*

$$(\mathcal{T}_e u, \mathcal{T}_{e'} v) = \delta_{ee'}(u, v).$$

Proof When $e \neq e'$, the intersection of the support of $\mathcal{T}_e u$ and that of $\mathcal{T}_{e'} v$ has measure zero. Hence, in this case we have that

$$(\mathcal{T}_e u, \mathcal{T}_{e'} v) = 0.$$

Now we consider the case $e = e'$. By the definition of the operator $\mathcal{T}_e$, we have that

$$(\mathcal{T}_e u, \mathcal{T}_{e'} v) = |J_{\phi_e}|^{-1} \int_{\phi_e(\Omega)} (u \circ \phi_e^{-1})(t)(v \circ \phi_e^{-1})(t)\mathrm{d}t.$$

Using a change of variable, we conclude that

$$(\mathcal{T}_e u, \mathcal{T}_{e'} v) = \int_{\Omega} u(t)v(t)\mathrm{d}t = (u, v),$$

which completes the proof. □

The following proposition shows that the spaces $\mathbb{W}_n$ can be generated recursively by $\mathbb{W}_1$.

Proposition 4.16 *It holds that*

$$\mathbb{W}_{n+1} = \bigoplus_{e \in \mathbb{Z}_\mu}^{\perp} \mathcal{T}_e \mathbb{W}_n, \quad n \in \mathbb{N}.$$

Proof We remark first that by Proposition 4.15, the direct sum above satisfies the orthogonal property. It follows from (4.34) that

$$\mathbb{W}_n \subset \mathbb{X}_n \text{ and } \mathbb{W}_n \perp \mathbb{X}_{n-1}.$$

Thus, by (4.31) and Proposition 4.15 we conclude that

$$\mathcal{T}_e \mathbb{W}_n \subset \mathbb{X}_{n+1} \text{ and } \mathcal{T}_e \mathbb{W}_n \perp \mathbb{X}_n.$$

These relations with (4.34) ensure that for any $e \in \mathbb{Z}_\mu$,

$$\mathcal{T}_e \mathbb{W}_n \subset \mathbb{W}_{n+1}.$$

Since from (4.37) we have that

$$\dim \bigoplus_{e \in \mathbb{Z}_\mu}^{\perp} \mathcal{T}_e \mathbb{W}_n = \mu \dim \mathbb{W}_n = m(\mu - 1)\mu^n = \dim \mathbb{W}_{n+1},$$

the result of this proposition holds. □

It can easily be seen from the above proposition and its proof that the following proposition holds.

Proposition 4.17 *If W_1 is a basis of space $\mathbb{W}_1$, then for $n \in \mathbb{N}$, the recursively generated set*

$$W_{n+1} = \bigcup_{e \in \mathbb{Z}_\mu}^{\perp} \mathcal{T}_e W_n = \bigcup_{\mathbf{e} \in \mathbb{Z}_\mu^n}^{\perp} \mathcal{T}_{\mathbf{e}} W_1$$

is the basis of space $\mathbb{W}_{n+1}$.

Now it is clear that as soon as we choose an orthogonal basis of $\mathbb{W}_0$, denoted by $\{w_{0j} : j \in \mathbb{Z}_m\}$, and get an orthogonal basis of $\mathbb{W}_1$, denoted by $\{w_{1j} : j \in \mathbb{Z}_r\}$ where $r = w(1)$, then we can generate recursively an orthogonal basis $\{w_{ij} : j \in \mathbb{Z}_{w(i)}\}$ of the space $\mathbb{W}_i$ by

$$w_{ij} := \mathcal{T}_{\mathbf{e}} w_{1l}, \ \ j = \mu(\mathbf{e})r + l, \ \ \mathbf{e} \in \mathbb{Z}_\mu^{i-1}, \ \ l \in \mathbb{Z}_r, \ \ i - 1 \in \mathbb{N}. \tag{4.38}$$

Functions $w_{i,j}, (i,j) \in \mathbb{U}$ are wavelet-like functions, which are also called orthogonal wavelets.

4.4 Refinable sets and set wavelets

In Section 4.3 we showed how to construct multiscale orthonormal bases on invariant sets. These wavelet-like functions are discontinuous, nonetheless they have important applications to the numerical solution of integral equations. In the next section we explore similar recursive structures for multiscale function representation and approximation by focusing on the analogous situation for interpolation on an invariant set. Thus, in this section we seek a mechanism to generate sequences of points which have a multiscale structure that can then be used to efficiently generate interpolating functions and multiscale functionals.

We first develop a notion of refinable sets, give a complete characterization of refinable sets in a general metric space setting and illustrate the general characterization with several examples of practical importance. Next we show how refinable sets lead to a multiresolution structure relative to set inclusion which is analogous to the multiresolution analysis associated with refinable functions. This set-theoretic multiresolution structure leads us to what we call set wavelets, which are generated by a successive application of the contractive mappings to an initial set wavelet. These will lead us, in particular, in Section 4.5 to the construction of interpolation that has the desired multiscale structure, and in Chapter 7 to the construction of multiscale functionals for developing fast multiscale collocation methods for solving integral equations.

4.4.1 Refinable sets

This subsection is devoted to a study of the refinable set relative to a family of contractive mappings. A complete characterization of refinable sets will be presented, which will be illustrated by several examples of practical importance.

Assume that $\mathbb{X}$ is a complete metric space, $\Phi := \{\phi_\epsilon : \epsilon \in \mathbb{Z}_\mu\}$ is a family of contractive mappings on $\mathbb{X}$ and Ω is the invariant set relative to the family Φ of mappings.

For a subset $V \subset \mathbb{X}$, we let

$$\Phi(V) := \bigcup_{\epsilon \in \mathbb{Z}_\mu} \phi_\epsilon(V).$$

Definition 4.18 A subset V of $\mathbb{X}$ is said to be refinable relative to the mappings Φ if $V \subseteq \Phi(V)$.

For every $k \in \mathbb{N}$ and $\mathbf{e}_k := [\epsilon_j : j \in \mathbb{Z}_k] \in \mathbb{Z}_\mu^k$, we define the contractive composition mapping $\phi_{\mathbf{e}_k} = \phi_{\epsilon_0} \circ \phi_{\epsilon_1} \circ \cdots \circ \phi_{\epsilon_{k-1}}$ and let $\Phi_k := \{\phi_{\mathbf{e}_k} : \mathbf{e}_k \in \mathbb{Z}_\mu^k\}$; in particular, $\Phi_1 = \Phi$. Observe that the union of any collection of refinable sets is likewise refinable. Moreover, if V is a refinable subset of $\mathbb{X}$, then $\Phi_k(V) := \bigcup_{\mathbf{e}_k \in \mathbb{Z}_\mu^k} \phi_{\mathbf{e}_k}(V)$ is also a refinable subset of $\mathbb{X}$ for all $k \in \mathbb{N}$. One of our main objectives is to identify refinable sets of finite cardinality.

Before we present a characterization of these sets, we look at some examples on the real line which will be helpful to illuminate the general result.

For the metric space $\mathbb{R}$ and integer $\mu > 1$, we consider the mappings

$$\psi_\epsilon(t) := \frac{t + \epsilon}{\mu}, \quad t \in \mathbb{R}, \ \epsilon \in \mathbb{Z}_\mu. \tag{4.39}$$

The invariant set for this family of mappings is the unit interval $[0, 1]$. Our first example of a refinable set relative to the family of mappings $\Psi := \{\psi_\epsilon : \epsilon \in \mathbb{Z}_\mu\}$ given in (4.39) comes next.

Proposition 4.19 *The set* $U_0 := \left\{\frac{j}{k} : j \in \mathbb{Z}_{k+1}\right\}$ *is refinable relative to the mappings* Ψ.

Proof It is sufficient to consider $j > 0$ since $0 = \psi_0(0)$. For every $j \in \mathbb{Z}_{k+1}$ we write the integer μj uniquely in the form $\mu j = k\epsilon + \ell$, where $\ell - 1 \in \mathbb{Z}_k$ and $\epsilon \in \mathbb{N}_0$. Since $\mu j \leq \mu k$, we conclude that $\epsilon \in \mathbb{Z}_\mu$. Moreover, we have that $\frac{j}{k} = \psi_\epsilon\left(\frac{\ell}{k}\right)$ and so U_0 is refinable. □

In some applications, the exclusion of the endpoints 0 and 1 from a refinable set is desirable. As an example of this case we present the following fact.

Proposition 4.20 *The set* $U_0 = \left\{\frac{j}{k+1} : j-1 \in \mathbb{Z}_k\right\}$ *is refinable relative to the mappings* Ψ *if and only if* μ *and* $k+1$ *are relatively prime.*

Proof Suppose that μ and $k+1$ have a common multiple $m > 1$, that is, $\mu = m\ell_1$ and $k+1 = m\ell_2$ for some integers ℓ_1 and ℓ_2 and U_0 is refinable relative to the mappings Ψ. Then we have that $\ell_2 - 1 \in \mathbb{Z}_k$ and $\ell_1 \in \mathbb{Z}_\mu$. Moreover, we have that $\psi_{\ell_1}(0) = \frac{\ell_1}{\mu} = \frac{\ell_2}{k+1}$. This equation implies that $\psi_{\ell_1}(0) \in U_0$. Since U_0 is refinable, there exist $\epsilon_0 \in \mathbb{Z}_\mu$ and $u \in U_0$ such that $\psi_{\ell_1}(0) = \psi_{\epsilon_0}(u)$. It follows from the equation above that $\ell_1 = u + \epsilon_0$. Thus, either $\ell_1 = \epsilon_0$ and $u = 0$ or $\epsilon_0 + 1 = \ell_1$ and $u = 1$. In either case we conclude that either 0 or 1 is in U_0. But this is a contradiction, since U_0 contains neither 0 nor 1. Hence, the integers μ and $k+1$ must be relatively prime.

Conversely, suppose μ and $k+1$ are relatively prime. For every $j-1 \in \mathbb{Z}_k$ there exist integers ϵ and ℓ such that $j\mu = (k+1)\epsilon + \ell$, where $\ell - 1 \in \mathbb{Z}_{k+1}$. Since $j\mu \le (k+1)\mu$, it follows that $\epsilon \in \mathbb{Z}_\mu$. Moreover, because μ and $k+1$ are relatively prime, it must also be the case that $\ell - 1 \in \mathbb{Z}_k$. Furthermore, since $\frac{j}{k+1} = \psi_\epsilon\left(\frac{\ell}{k+1}\right)$ we conclude that U_0 is refinable. □

Our third special construction of refinable sets U_0 in $[0,1]$ relative to the mappings Ψ is formed from *cyclic μ-adic expansions*. To describe this construction we introduce two additional mappings. The first mapping $\pi : \mathbb{Z}_\mu^\infty \to [0,1]$ is defined by

$$\pi(\mathbf{e}) := \sum_{j\in\mathbb{N}} \frac{\epsilon_{j-1}}{\mu^j}, \quad \mathbf{e} = [\epsilon_j : j \in \mathbb{N}_0] \in \mathbb{Z}_\mu^\infty,$$

and we also write it as $\pi(\mathbf{e}) = .\epsilon_0\epsilon_1\epsilon_2\cdots$. This mapping takes an infinite vector $\mathbf{e} \in \mathbb{Z}_\mu^\infty$ and associates it with a number in $[0,1]$ whose μ-adic expansion is read off from the components of $\mathbf{e}$. The mapping π is not invertible. Referring back to the definition (4.39), we conclude for any $\epsilon \in \mathbb{Z}_\mu$ and $\mathbf{e} \in \mathbb{Z}_\mu^\infty$ that $\psi_\epsilon(\pi(\mathbf{e})) = .\epsilon\epsilon_0\epsilon_1\cdots$. We also make use of the "shift" map $\sigma : \mathbb{Z}_\mu^\infty \to \mathbb{Z}_\mu^\infty$. Specifically, for $\mathbf{e} = [\epsilon_j : j \in \mathbb{N}_0] \in \mathbb{Z}_\mu^\infty$ we set $\sigma(\mathbf{e}) := [\epsilon_j : j \in \mathbb{N}] \in \mathbb{Z}_\mu^\infty$. Thus, the mapping σ discards the first component of $\mathbf{e}$ while the mapping ψ_ϵ restores the corresponding digit, that is,

$$\psi_{\epsilon_0}(\pi \circ \sigma(\mathbf{e})) = \pi(\mathbf{e}). \tag{4.40}$$

For any $k \in \mathbb{N}$ and $\mathbf{e}_k = [\epsilon_j : j \in \mathbb{Z}_k] \in \mathbb{Z}_\mu^k$, we let $.\overline{\epsilon_0\epsilon_1\cdots\epsilon_{k-1}}$ denote the number $\pi(\mathbf{e})$ where $\mathbf{e} = [\epsilon_j : j \in \mathbb{N}_0] \in \mathbb{Z}_\mu^\infty$ and $\epsilon_{i+k} = \epsilon_i$, $i \in \mathbb{N}_0$. Note that for such an infinite vector $\mathbf{e}$ we have that $\sigma^k(\mathbf{e}) = \mathbf{e}$, where $\sigma^k = \sigma \circ \cdots \circ \sigma$ is the k-fold composition of σ, and also the number $.\overline{\epsilon_0\epsilon_1\cdots\epsilon_{k-1}}$ is the unique fixed point of the mapping $\psi_{\mathbf{e}_k}$.

Proposition 4.21 *Choose $k \in \mathbb{N}$ and $\mathbf{e}_k = [\epsilon_j : j \in \mathbb{Z}_k] \in \mathbb{Z}_\mu^k$ such that at least two components of $\mathbf{e}_k$ are different. Let $\mathbf{e} := [\epsilon_j : j \in \mathbb{N}_0] \in \mathbb{Z}_\mu^\infty$ with $\epsilon_{i+k} = \epsilon_i$, $i \in \mathbb{N}_0$. Then the set $U_0(\pi(\mathbf{e})) := \{\pi \circ \sigma^\ell(\mathbf{e}) : \ell \in \mathbb{Z}_k\}$ is refinable relative to the mappings Ψ and has cardinality $\leq k$. Moreover, if k is the smallest positive integer such that $\epsilon_{i+k} = \epsilon_i$, $i \in \mathbb{N}_0$ then $U_0(\pi(\mathbf{e}))$ has cardinality k.*

Proof If $.\overline{\epsilon_0\epsilon_1\cdots\epsilon_{k-1}} = .\overline{\epsilon'_0\epsilon'_1\cdots\epsilon'_{k-1}}$ for $\epsilon_i, \epsilon'_i \in \mathbb{Z}_\mu$, $i \in \mathbb{Z}_k$, then $\epsilon_i = \epsilon'_i$, $i \in \mathbb{Z}_k$. Hence, it follows that all the elements of $U_0(\pi(\mathbf{e}))$ are distinct. Also, by using (4.40), for any $\ell \in \mathbb{Z}_k$ we have that $\pi \circ \sigma^\ell(\mathbf{e}) = \psi_{\epsilon_\ell}(\pi \circ \sigma^{\ell+1}(\mathbf{e}))$. Note that trivially $\pi \circ \sigma^{\ell+1}(\mathbf{e}) \in U_0(\pi(\mathbf{e}))$ for $\ell \in \mathbb{Z}_{k-1}$ and $\pi \circ \sigma^k(\mathbf{e}) = \pi(\mathbf{e}) \in U_0(\pi(\mathbf{e}))$. Thus, $U_0(\pi(\mathbf{e}))$ is indeed refinable. □

Various useful examples can be generated from this proposition. We mention the following possibilities for $\mu = 2$:

$$U_0(.\overline{01}) = \left\{\frac{1}{3}, \frac{2}{3}\right\}, \quad U_0(.\overline{001}) = \left\{\frac{1}{7}, \frac{2}{7}, \frac{4}{7}\right\}, \quad U_0(.\overline{0011}) = \left\{\frac{1}{5}, \frac{2}{5}, \frac{3}{5}, \frac{4}{5}\right\}.$$

We now present a characterization of refinable sets relative to a given family Φ of contractive mappings on *any* metric space $\mathbb{M}$. To state this result, for every $k \in \mathbb{N} := \{1, 2, \ldots\}$ and $\mathbf{e}_k := [\epsilon_j : j \in \mathbb{Z}_k] \in \mathbb{Z}_\mu^k$, we define the contractive mapping $\phi_{\mathbf{e}_k} = \phi_{\epsilon_0} \circ \phi_{\epsilon_1} \circ \cdots \circ \phi_{\epsilon_{k-1}}$ and let $\Phi_k := \{\phi_{\mathbf{e}_k} : \mathbf{e}_k \in \mathbb{Z}_\mu^k\}$; in particular, $\Phi_1 = \Phi$. We let $x_{\mathbf{e}_k}$ be the unique fixed point of the mapping $\phi_{\mathbf{e}_k}$, that is,

$$\phi_{\mathbf{e}_k}(x_{\mathbf{e}_k}) = x_{\mathbf{e}_k}$$

and set

$$\mathcal{F}_k := \{x_{\mathbf{e}_k} : \mathbf{e}_k \in \mathbb{Z}_\mu^k\}.$$

We also define $\mathbb{Z}_\mu^\infty$ to be the set of infinite vectors $\mathbf{e} = [\epsilon_j : j \in \mathbb{N}_0]$, $\epsilon_i \in \mathbb{Z}_\mu$, $i \in \mathbb{N}_0$. With every such vector $\mathbf{e} \in \mathbb{Z}_\mu^\infty$ and $k \in \mathbb{N}$ we let $\mathbf{e}_k = [\epsilon_j : j \in \mathbb{Z}_k] \in \mathbb{Z}_\mu^k$. It was shown in [148] that the limit of $x_{\mathbf{e}_k}$ as $k \to \infty$ exists and we denote this element of the metric space $\mathbb{X}$ by $x_{\mathbf{e}}$. In other words, we have that

$$\lim_{k\to\infty} x_{\mathbf{e}_k} = x_{\mathbf{e}}.$$

Moreover, we let $\mathbf{e}_{r,\ell} := [\epsilon_j : j \in \mathbb{Z}_\ell \setminus \mathbb{Z}_r]$ for $r \in \mathbb{Z}_\ell$ and use $x_{\mathbf{e}_{r,\ell}}$ to denote the fixed point of the composition mapping $\phi_{\mathbf{e}_{r,\ell}}$, where $\phi_{\mathbf{e}_{r,\ell}} := \phi_{\epsilon_r} \circ \phi_{\epsilon_{r+1}} \circ \cdots \circ \phi_{\epsilon_{\ell-1}}$ when $r \in \mathbb{Z}_\ell$ and $\phi_{\mathbf{e}_{r,\ell}}$ is the identity mapping when $r = \ell$.

Theorem 4.22 *Let $\Phi := \{\phi_\epsilon : \epsilon \in \mathbb{Z}_\mu\}$ be a family of contractive mappings on a complete metric space $\mathbb{X}$ and let $V_0 \subseteq \mathbb{X}$ be a nonempty set of cardinality $k \in \mathbb{N}$. Then V_0 is refinable relative to Φ if and only if V_0 has the following* *property. For every* $v \in V_0$ *there exist integers* $\ell, m \in \mathbb{Z}_{k+1}$ *with* $\ell < m$ *and* $\epsilon_i \in \mathbb{Z}_\mu$, $i \in \mathbb{Z}_m$, *such that* $v = \phi_{\mathbf{e}_{0,\ell}}(x_{\mathbf{e}_{\ell,m}})$ *and the points*

$$v_r := \phi_{\mathbf{e}_{r,\ell}}(x_{\mathbf{e}_{\ell,m}}) \in V_0, \quad r \in \mathbb{Z}_\ell,$$

$$v_{\ell+r} := \phi_{\mathbf{e}_{\ell+r,m}}(x_{\mathbf{e}_{\ell,m}}) \in V_0, \quad r \in \mathbb{Z}_{m-\ell}. \tag{4.41}$$

Moreover, in this case, we have that $V_i := \Phi_i(V_0) \subseteq \Omega$, $i \in \mathbb{N}$ *and also*

$$V_0 \subseteq \Phi_\ell(\mathcal{F}_{m-\ell}). \tag{4.42}$$

Proof Assume that V_0 is refinable and $v \in V_0$. Let $v_0 = v$. By the refinability of V_0, there exist points $v_{j+1} \in V_0$ and $\epsilon_j \in \mathbb{Z}_\mu$, for $j \in \mathbb{Z}_k$, such that $v_j = \phi_{\epsilon_j}(v_{j+1})$, $j \in \mathbb{Z}_k$. Therefore, we have that $v_r = \phi_{\mathbf{e}_{r,s}}(v_s)$, $r \in \mathbb{Z}_s$, $s \in \mathbb{Z}_{k+1}$. Since the cardinality of V_0 is k, there exist two integers $\ell, m \in \mathbb{Z}_{k+1}$ with $\ell < m$ for which $v_\ell = v_m$. Hence, in particular, we conclude that $v_\ell = v_m = x_{\mathbf{e}_{\ell,m}}$. It follows that

$$v_r = \phi_{\mathbf{e}_{r,\ell}}(v_\ell) = \phi_{\mathbf{e}_{r,\ell}}(x_{\mathbf{e}_{\ell,m}}), \quad r \in \mathbb{Z}_\ell$$

and

$$v_{\ell+r} = \phi_{\mathbf{e}_{\ell+r,m}}(v_m) = \phi_{\mathbf{e}_{\ell+r,m}}(x_{\mathbf{e}_{\ell,m}}), \quad r \in \mathbb{Z}_{m-\ell}.$$

These remarks establish the necessity and also the fact that $v_0 \in \Phi_\ell(\mathcal{F}_{m-\ell}) \subseteq \Omega$.

Conversely, let V_0 be a set of points with the property and let v be a typical element of V_0. Then we have that either $v = \phi_{\mathbf{e}_{0,\ell}}(x_{\mathbf{e}_{\ell,m}})$ with $\ell > 0$ or $v = x_{\mathbf{e}_{0,m}}$ with $\ell = 0$. In the first case, since $v = \phi_{\epsilon_0}(\phi_{\mathbf{e}_{1,\ell}}(x_{\mathbf{e}_{\ell,m}}))$ and $\phi_{\mathbf{e}_{1,\ell}}(x_{\mathbf{e}_{\ell,m}}) \in V_0$, we have that $v \in \Phi(V_0)$. In the second case, since $x_{\mathbf{e}_{0,m}}$ is the unique fixed point of the mapping $\phi_{\mathbf{e}_{0,m}}$, we write $v = \phi_{\epsilon_0}(\phi_{\mathbf{e}_{1,m}}(x_{\mathbf{e}_{0,m}}))$. By our hypothesis, $\phi_{\mathbf{e}_{1,m}}(x_{\mathbf{e}_{0,m}}) \in V_0$ and thus, in this case, we also have that $v \in \Phi(V_0)$. Therefore, in either case, $v \in \Phi(V_0)$ and so V_0 is refinable. These comments complete the proof of the theorem. □

We next derive two consequences of this observation. To present the first we go back to the definition of the point $x_{\mathbf{e}}$, in the metric space $\mathbb{X}$, where $\mathbf{e} = [\epsilon_j : j \in \mathbb{N}_0] \in \mathbb{Z}_\mu^\infty$, and observe that when the vector $\mathbf{e}$ is s-periodic, that is, its coordinates have the property that s is the smallest positive integer such that $\epsilon_i = \epsilon_{i+s}$, $i \in \mathbb{N}_0$, we have $x_{\mathbf{e}} = x_{\mathbf{e}_s}$, where $\mathbf{e}_s = [\epsilon_j : j \in \mathbb{Z}_s]$. Conversely, given any $\mathbf{e}_s \in \mathbb{Z}_\mu^s$ we can extend it as an s-periodic vector $\mathbf{e} \in \mathbb{Z}_\mu^\infty$ and conclude that $x_{\mathbf{e}} = x_{\mathbf{e}_s}$.

Let us observe that the powers of the shift operator σ act on s-periodic vectors in $\mathbb{Z}_\mu^\infty$ as a cyclic permutation of vectors in $\mathbb{Z}_\mu^s$. Also, the s-periodic orbits of σ, that is, vectors $\mathbf{e} \in \mathbb{Z}_\mu^\infty$ such that $\sigma^s(\mathbf{e}) = \mathbf{e}$ are exactly the s-periodic vectors in $\mathbb{Z}_\mu^\infty$. With this viewpoint in mind we can draw the following conclusion from Theorem 4.22.

Theorem 4.23 *A finite set V_0 in a metric space $\mathbb{X}$ is refinable relative to the mappings Φ if and only if, for every $v \in V_0$, there exists an $\mathbf{e} \in \mathbb{Z}_\mu^\infty$ such that $v = x_{\mathbf{e}}$ and $x_{\sigma^k(\mathbf{e})} \in V_0$ for all $k \in \mathbb{N}$.*

Proof For convenience, we define the notation $\pi^*(.\epsilon_0\epsilon_1\epsilon_2\cdots) := [\epsilon_j : j \in \mathbb{N}_0] \in \mathbb{Z}_\mu^\infty$ for $\epsilon_i \in \mathbb{Z}_\mu$. The proof requires a formula from [148] (p. 727) which in our notation takes the form

$$\phi_\epsilon(x_{\mathbf{e}}) = x_{\pi^*(\psi_\epsilon(\pi(\mathbf{e})))}, \quad \epsilon \in \mathbb{Z}_\mu, \quad \mathbf{e} \in \mathbb{Z}_\mu^\infty,$$

where ψ_ϵ are the concrete mappings defined by (4.39). Using this formula, the number $\pi(\mathbf{e})$ associated with the vector $\mathbf{e}$ in Theorem 4.22 is identified as

$$\pi(\mathbf{e}) = .\epsilon_0\epsilon_1\cdots\epsilon_{\ell-1}\overline{\epsilon_\ell\cdots\epsilon_{m-1}}.$$

□

An immediate corollary of this result characterizes refinable sets on $\mathbb{R}$ relative to the mappings defined by (4.39).

Theorem 4.24 *Let U_0 be a subset of $\mathbb{R}$ having cardinality k. Then U_0 is refinable relative to the mappings (4.39) if and only if, for every point $u \in U_0$, there exist integers $\ell, m \in \mathbb{Z}_{k+1}$ with $\ell < m$ and $\epsilon_i \in \mathbb{Z}_\mu$, $i \in \mathbb{Z}_m$ such that $u = .\epsilon_0\cdots\epsilon_{\ell-1}\overline{\epsilon_\ell\cdots\epsilon_{m-1}}$ and for any cyclic permutation $\eta_\ell,\ldots,\eta_{m-1}$ of $\epsilon_\ell,\ldots,\epsilon_{m-1}$ and $r \in \mathbb{Z}_\ell$ the point $.\epsilon_r\cdots\epsilon_{\ell-1}\overline{\eta_\ell\cdots\eta_{m-1}}$ is in U_0.*

It is the vectors $\mathbf{e} \in \mathbb{Z}_\mu^\infty$ which are *pre-orbits* of σ, that is, for some $k \in \mathbb{N}_0$ the vector $\sigma^k(\mathbf{e})$ is periodic, which characterize refinable sets. Thus, there is an obvious way to build from refinable sets U_0 relative to the mappings (4.39) on $\mathbb{R}$ refinable sets relative to *any finite contractive mappings on a metric space*. For example, let U_0 be a finite subset of cardinality k in the interval $[0, 1]$. We require for each number u in this set that there is an $\mathbf{e} \in \mathbb{Z}_\mu^\infty$ such that $u = \pi(\mathbf{e})$ and for every $j \in \mathbb{N}_0$, $\pi(\sigma^j(\mathbf{e})) \in U_0$. In other words, U_0 is refinable relative to the mappings (4.39). We define a set V_0 in $\mathbb{X}$, associated with U_0, by the formula

$$V_0 := \{x_{\mathbf{e}} : \pi(\mathbf{e}) \in U_0\},$$

where $x_{\mathbf{e}} \in \mathbb{X}$ is the limit of $x_{\mathbf{e}_k}$. This set is a refinable subset of $\mathbb{X}$ relative to the contractive mappings Φ. We may use this association to construct examples of practical importance in the finite element method and boundary integral equation method.

Example Let $\Delta \subset \mathbb{R}^2$ be the triangle with vertices at $y_0 := (0,0)$, $y_1 := (1,0)$ and $y_2 := (0,1)$. Set $y_3 := (1,1)$ and consider four contractive affine mappings

$$\phi_\epsilon(x) := \frac{1}{2}(y_\epsilon + (-1)^{\tau(\epsilon)}x), \quad \epsilon \in \mathbb{Z}_4, \quad x \in \mathbb{R}^2, \tag{4.43}$$

where $\tau(\epsilon) = 0, \epsilon \in \mathbb{Z}_3$ and $\tau(3) = 1$. The invariant subset of $\mathbb{R}^2$ relative to these mappings is the triangle Δ and the following sets are refinable:

$$\left\{\left(\frac{1}{3}, \frac{1}{3}\right)\right\}, \quad \left\{\left(\frac{1}{7}, \frac{4}{7}\right), \left(\frac{2}{7}, \frac{1}{7}\right), \left(\frac{4}{7}, \frac{2}{7}\right)\right\},$$
$$\left\{\left(\frac{1}{15}, \frac{2}{15}\right), \left(\frac{2}{15}, \frac{4}{15}\right), \left(\frac{4}{15}, \frac{8}{15}\right), \left(\frac{8}{15}, \frac{1}{15}\right)\right\}$$

with respect to these mappings. Also, we record for any $\mathbf{e}_k = [\epsilon_j : j \in \mathbb{Z}_k] \in \mathbb{Z}_\mu^k$ and $x \in \mathbb{R}^2$ that

$$\phi_{\mathbf{e}_k}(x) = \frac{1}{2^k}\left[(-1)^{\tau_k}x + \sum_{j \in \mathbb{Z}_k}(-1)^{\tau_j}2^{k-j}y_{\epsilon_j}\right],$$

where $\tau_j := \sum_{\ell=0}^{j-1}\tau(\epsilon_\ell)$, $j \in \mathbb{Z}_{k+1}$. From this equation it follows that

$$x_{\mathbf{e}_k} = \frac{1}{2^k + (-1)^{\tau_k+1}}\sum_{j \in \mathbb{Z}_k}(-1)^{\tau_j}2^{k-j}y_{\epsilon_j}.$$

These formulas can be used to generate the above sets.

4.4.2 Set wavelets

In this subsection we generate a sequence $\mathcal{W} := \{W_n : n \in \mathbb{N}_0\}$ of finite sets of a metric space $\mathbb{X}$ which have a wavelet-like multiresolution structure. We call an element of $\mathcal{W}$ a *set wavelet* and demonstrate in subsequent sections that set wavelets are crucial for the construction of interpolating wavelets on certain compact subsets of $\mathbb{R}^d$.

The generation of set wavelets begins with an initial finite subset of distinct points $V_0 := \{v_j : j \in \mathbb{Z}_m\}$ in $\mathbb{X}$. We use this subset and the finite set of contractive mappings Φ to define a sequence of subsets of $\mathbb{X}$ given by

$$V_i := \Phi(V_{i-1}), \quad i \in \mathbb{N}. \tag{4.44}$$

Assume that a compact set Ω in $\mathbb{X}$ is the unique invariant set relative to the mappings Φ. When $V_0 \subseteq \Omega$, it follows for each $i \in \mathbb{N}$ that $V_i \subseteq \Omega$. Furthermore, using the set $\Phi_k := \{\phi_{\mathbf{e}_k} : \mathbf{e}_k \in \mathbb{Z}_\mu^k\}$ of contractive mappings introduced in the last subsection, and for every subset A of $\mathbb{X}$, we define the set

$$\Phi_k(A) := \bigcup_{\mathbf{e}_k \in \mathbb{Z}_\mu^k} \phi_{\mathbf{e}_k}(A),$$

and so, in particular, $\Phi_1(A) = \Phi(A)$. Therefore, equation (4.44) implies that $V_i = \Phi_i(V_0)$, $i \in \mathbb{N}$.

The next lemma is useful to us.

Lemma 4.25 *Let Φ be a finite family of contractive mappings on $\mathbb{X}$. Assume that $\Omega \subseteq \mathbb{X}$ is the invariant set relative to the mappings Φ. If V_0 is a nonempty finite subset of $\mathbb{X}$ then*

$$\Omega \subseteq \overline{\bigcup_{i \in \mathbb{N}_0} V_i},$$

where V_i is generated by the mappings Φ by (4.44).

Proof Let $x \in \Omega$ and $\delta > 0$. Since Ω is a compact set in $\mathbb{X}$, we choose an integer $n > 0$ such that $\gamma^n \mathrm{diam}\,(\Omega \cup V_0) < \delta$, where γ is the contraction parameter appearing in equation (4.8). According to the defining property (4.9) of the set Ω, there exists an $\mathbf{e}_n \in \mathbb{Z}_\mu^n$ such that $x \in \phi_{\mathbf{e}_n}(\Omega) \subseteq \phi_{\mathbf{e}_n}(\Omega \cup V_0)$. Since V_0 is a nonempty set of $\mathbb{X}$, there exists a $y \in \phi_{\mathbf{e}_n}(V_0) \subseteq \phi_{\mathbf{e}_n}(\Omega \cup V_0)$. Moreover, by the contractivity (4.8) of the family Φ, we have that

$$d(x, y) \leq \mathrm{diam}\,\phi_{\mathbf{e}_n}(\Omega \cup V_0) \leq \gamma^n \mathrm{diam}\,(\Omega \cup V_0) < \delta.$$

This inequality proves the result. □

Proposition 4.26 *Let V_0 be a nonempty refinable set of $\mathbb{X}$ relative to a finite family Φ of contractive mappings and $\{V_i : i \in \mathbb{N}_0\}$ be the collection of sets generated by definition (4.44). Then*

$$\Omega = \overline{\bigcup_{i \in \mathbb{N}_0} V_i}.$$

Proof This result follows directly from Lemma 4.25 and Theorem 4.43. □

Let us recall the construction of the invariant set Ω given a family of contractive mappings Φ (see [148]). The invariant set Ω is given by either one of the formulas

$$\Omega = \{x_{\mathbf{e}} : \mathbf{e} \in \mathbb{Z}_\mu^\infty\}$$

or

$$\Omega = \overline{\bigcup_{k \in \mathbb{N}} \mathcal{F}_k}.$$

The above proposition provides another way to construct the unique invariant set Ω relative to a finite family Φ of contractive mappings. In other words, we start with a refinable set V_0 and then form V_i, $i \in \mathbb{N}$, recursively by (4.44).

We say that a sequence of sets $\{A_i : i \in \mathbb{N}_0\}$ is nested (resp., strictly nested) provided that $A_{i-1} \subseteq A_i$, $i \in \mathbb{N}$ (resp., $A_{i-1} \subset A_i$, $i \in \mathbb{N}$). The next lemma shows the importance of the notion of the refinable set.

Lemma 4.27 *Let Ω be the invariant set in $\mathbb{X}$ relative to a finite family Φ of contractive mappings. Suppose that Ω is not a finite set and V_0 is a nonempty finite subset of $\mathbb{X}$. Then the collection of sets $\{V_i : i \in \mathbb{N}_0\}$ defined by (4.44) is strictly nested if and only if the set V_0 is refinable relative to Φ.*

Proof Suppose that V_0 is refinable relative to Φ. Then, it follows by induction on $i \in \mathbb{N}$ that $V_{i-1} \subseteq V_i$.

It remains to prove that this inclusion is strict for all $i \in \mathbb{N}$. Assume to the contrary that for some $i \in \mathbb{N}$ we have that $V_{i-1} = V_i$. By the definition of V_i, we conclude that $V_{i-1} = V_j$ for all $j \geq i$ and thus, we have that

$$\overline{\bigcup_{j \in \mathbb{N}_0} V_j} = V_{i-1}.$$

This conclusion contradicts Proposition 4.26 and the fact that Ω does not have finite cardinality. □

When the sequence of sets $\{V_i : i \in \mathbb{N}_0\}$ is strictly nested, we let $W_i := V_i \setminus V_{i-1}$, $i \in \mathbb{N}$, that is, $V_i = V_{i-1} \cup^{\perp} W_i$, $i \in \mathbb{N}$, where we use the notation $A \cup^{\perp} B$ to denote $A \cup B$ when $A \cap B = \phi$. By Lemma 4.27, if the set V_0 is refinable relative to Φ, we have that $W_i \neq \emptyset$, $i \in \mathbb{N}$. Similarly, we use the notation

$$\Phi^{\perp}(A) := \bigcup_{\epsilon \in \mathbb{Z}_\mu} {}^{\perp} \phi_\epsilon(A),$$

when $\phi_\epsilon(A) \cap \phi_{\epsilon'}(A) = \emptyset$, $\epsilon, \epsilon' \in \mathbb{Z}_\mu$, $\epsilon \neq \epsilon'$. The sets W_i, $i \in \mathbb{N}_0$, give us the decomposition

$$V_n = \bigcup_{i \in \mathbb{Z}_{n+1}} {}^{\perp} W_i, \tag{4.45}$$

where $W_0 := V_0$. The next theorem shows that when the set W_1 is specified, the sets W_i, $i \in \mathbb{N}$, can be constructed recursively and the set Ω has a decomposition in terms of these sets. This result provides a multiresolution decomposition for the invariant set Ω. For this reason, we call the sets W_i, $i \in \mathbb{N}$, *set wavelets*, the set W_1 the *initial set wavelet* and the decomposition of Ω in terms of W_i, $i \in \mathbb{N}$, the *set wavelet decomposition* of Ω.

Theorem 4.28 *Let Ω be the invariant set in $\mathbb{X}$ relative to a finite family Φ of contractive mappings. Suppose that each of the contractive mappings ϕ_ϵ, $\epsilon \in \mathbb{Z}_\mu$, in Φ has a continuous inverse on $\mathbb{X}$ and they have the property that*

$$\phi_\epsilon(\text{int}\,\Omega) \cap \phi_{\epsilon'}(\text{int}\,\Omega) = \emptyset, \quad \epsilon, \epsilon' \in \mathbb{Z}_\mu, \ \epsilon \neq \epsilon'. \tag{4.46}$$

Let V_0 be refinable with respect to Φ and $W_1 \subset \text{int}\,\Omega$. Then

$$W_{i+1} = \Phi^\perp(W_i), \quad i \in \mathbb{N} \tag{4.47}$$

and the compact set Ω has the set wavelet decomposition

$$\Omega = \overline{\bigcup_{n \in \mathbb{N}_0} {}^\perp W_n}, \tag{4.48}$$

where $W_0 := V_0$.

Proof Our hypotheses on the contractive mappings ϕ_ϵ, $\epsilon \in \mathbb{Z}_\mu$ guarantee that they are topological mappings. Hence, for any subsets A and B of $\mathbb{X}$, we have for any $\epsilon \in \mathbb{Z}_\mu$ that

$$\text{int}\,\phi_\epsilon(A) = \phi_\epsilon(\text{int}\,A) \tag{4.49}$$

and

$$\phi_\epsilon(A) \cap \phi_\epsilon(B) = \phi_\epsilon(A \cap B). \tag{4.50}$$

Let us first establish that when $W_1 \subset \text{int}\,\Omega$, the sets W_i, $i \in \mathbb{N}$, defined by the recursion (4.47), are all in $\text{int}\,\Omega$. We prove this fact by induction on $i \in \mathbb{N}$. To this end, we suppose that $W_i \subseteq \text{int}\,\Omega$, $i \in \mathbb{N}$. Then the invariance property (4.9) of Ω and (4.49) imply that

$$\Phi(W_i) \subseteq \Phi(\text{int}\,\Omega) = \text{int}\,\Phi(\Omega) = \text{int}\,\Omega.$$

Therefore, we have advanced the induction hypothesis and proved that $W_{i+1} \subseteq \text{int}\,\Omega$ for all $i \in \mathbb{N}$. Using the fact that $W_{i+1} \subseteq \text{int}\,\Omega$, we conclude from our hypothesis (4.46) for any $i \in \mathbb{N}$, $\epsilon, \epsilon' \in \mathbb{Z}_\mu$, $\epsilon \neq \epsilon'$ that $\phi_\epsilon(W_i) \cap \phi_{\epsilon'}(W_i) = \emptyset$, which justifies the "$\perp$" in formula (4.47). It follows from (4.44), (4.47) and $W_1 = V_1 \setminus V_0$ that $W_{i+1} \subseteq V_{i+1}$, $i \in \mathbb{N}_0$.

Next, we wish to confirm that

$$V_i \setminus V_{i-1} = W_i, \quad i \in \mathbb{N}. \tag{4.51}$$

Again, we rely upon induction on i and assume that

$$V_i \setminus V_{i-1} = W_i. \tag{4.52}$$

Therefore, we obtain that

$$\begin{aligned} V_i \cup W_{i+1} = \Phi(V_{i-1}) \cup \Phi(W_i) &= \bigcup_{\epsilon \in \mathbb{Z}_\mu} (\phi_\epsilon(V_{i-1}) \cup \phi_\epsilon(W_i)) \\ &= \bigcup_{\epsilon \in \mathbb{Z}_\mu} \phi_\epsilon(V_i) = V_{i+1}, \end{aligned}$$

which implies that $V_{i+1} \setminus V_i \subseteq W_i$. To confirm that equality holds, we observe that

$$V_i \cap W_{i+1} = \Phi(V_{i-1}) \cap \Phi(W_i) = \bigcup_{\epsilon \in \mathbb{Z}_\mu} \bigcup_{\epsilon' \in \mathbb{Z}_\mu} (\phi_\epsilon(V_{i-1})) \cap (\phi_{\epsilon'}(W_i)). \tag{4.53}$$

For $\epsilon \neq \epsilon'$ we can use (4.49) and hypothesis (4.46) to conclude that $\phi_\epsilon(\Omega) \cap \phi_{\epsilon'}(\operatorname{int}\Omega) = \emptyset$. To see this, we assume to the contrary that there exists $x \in \phi_\epsilon(\Omega) \cap \phi_{\epsilon'}(\operatorname{int}\Omega)$. Then there exist $y \in \Omega$ and $y' \in \operatorname{int}\Omega$ such that $x = \phi_\epsilon(y) = \phi_{\epsilon'}(y')$. Condition (4.46) insures that $y \in \Omega \setminus \operatorname{int}\Omega$. Hence, by equation (4.49), it follows from the first equality that $x \in \Omega \setminus \operatorname{int}\Omega$ and from the second equality that $x \in \operatorname{int}\Omega$, a contradiction. Consequently, we have that

$$(\phi_\epsilon(V_{i-1})) \cap (\phi_{\epsilon'}(W_i)) = \emptyset. \tag{4.54}$$

When $\epsilon = \epsilon'$ we use (4.50) to obtain that (4.54) still holds. Hence equation (4.53) implies that $V_i \cap W_{i+1} = \emptyset$. This establishes (4.51), advances the induction hypothesis and proves the result. □

We end this section by considering the following converse question to the one we have considered so far. Given a finite set in a metric space, is it refinable relative to some finite set of contractive mappings? The motivation for this question comes from practical considerations. As is often the case in certain numerical problems associated with interpolation and approximation, we begin on an interval of the real line with prescribed points, for example, Gaussian points or the zeros of Chebyshev polynomials. We then want to find mappings to make these prescribed points refinable relative to them. We shall only address this question in the generality of the space $\mathbb{R}^d$ relative to the ℓ^∞-norm. It is easy to see that given any subset $V_0 := \{v_i : i \in \mathbb{Z}_k\}$ of $\mathbb{R}^d$, there is a family of contractive mappings on $\mathbb{R}^d$ such that V_0 is refinable relative to them. For example, the mappings $\phi_i(x) = \frac{1}{2}(x + v_i)$, $i \in \mathbb{Z}_k$, $x \in \mathbb{R}^d$ will do since clearly the fixed point of the mapping ϕ_i is v_i for $i \in \mathbb{Z}_k$. However, almost surely the associated invariant set will have an empty interior and therefore Theorem 4.28 will not apply. For instance, in the example of a triangle mentioned above, the general prescription described applied to the vertices of the triangle will yield the Serpinski gasket. This invariant set is a Cantor set and is formed by successively applying the maps (4.43) to the triangles, throwing away the middle triangle which is the image of the fourth map used in the example. To overcome this we must add to the above family of mappings another set of contractive mappings "which fill the holes." To describe this process we review some facts about parallelepipeds.

A finite set $\mathcal{I} = \{t_i : i \in \mathbb{Z}_{n+1}\}$ with $t_0 < t_1 < \cdots < t_{n-1} < t_n$ is called a partition of the interval $I := [t_0, t_n]$ and divides it into subintervals $I_i := [t_i, t_{i+1}]$, $i \in \mathbb{Z}_n$, where the points in $\mathcal{I} \cap (t_0, t_n)$ appear as endpoints of two adjacent subintervals. For every finite set U_0 of $(0, 1)$ there exists a partition $\mathcal{I}$ such that the points of U_0 lie in the interior of the corresponding subintervals. The lengths of these subintervals can be chosen as small as desired.

Likewise, for any two vectors $x := [x_i : i \in \mathbb{Z}_d]$, $y := [y_i : i \in \mathbb{Z}_d]$ in $\mathbb{R}^d$, where $x_i < y_i$, $i \in \mathbb{Z}_d$ which we denote by $x \prec y$ (also $x \preceq y$ when $x_i \leq y_i$, $i \in \mathbb{Z}_d$), we can partition the set $\bigotimes_{i \in \mathbb{Z}_d} [x_i, y_i]$ – called a *parallelepiped* and denoted by $\langle x, y \rangle$ – into (sub)parallelepipeds formed from the partition

$$\mathcal{I}^d := \bigotimes_{i \in \mathbb{Z}_d} \mathcal{I}_i := \{[t_j : j \in \mathbb{Z}_d] : t_i \in \mathcal{I}_i, i \in \mathbb{Z}_d\}, \tag{4.55}$$

where each $\mathcal{I}_i$ is a partition of the interval $[x_i, y_i]$, $i \in \mathbb{Z}_d$. If $\{I_{i,j} : j \in \mathbb{Z}_{n_i}\}$ is the set of subintervals associated with the partition $\mathcal{I}_i$, then a typical parallelepiped associated with the partition $\mathcal{I}^d$ corresponds to a lattice point $i = [i_j : j \in \mathbb{Z}_d]$ where $i_j \in \mathbb{Z}_{n_j}$, $j \in \mathbb{Z}_d$ defined by

$$\mathcal{I}_i = \bigotimes_{j \in \mathbb{Z}_d} I_{i_j, j}. \tag{4.56}$$

Given any finite set $V_0 \subset \mathbb{R}^d$ contained in the interior of a parallelepiped P, we can partition it as above so that the interior of the subparallelepipeds contains the vectors of V_0. We can, if required, choose the volume of these subparallelepipeds to be as small as desired.

The set of all parallelepipeds is closed under translation as the simple rule

$$\langle x, y \rangle + z := \{w + z : w \in \langle x, y \rangle\} = \langle x + z, y + z \rangle,$$

valid for any $x, y, z \in \mathbb{R}^d$ with $x \prec y$. For any $x, y \in \mathbb{R}^d$ we associate an affine mapping $\mathcal{A}$ on $\mathbb{R}^d$ defined by the formula

$$\mathcal{A}t := \mathbf{X}t + y, \quad t \in \mathbb{R}^d, \tag{4.57}$$

where $\mathbf{X} := \operatorname{diag}(x_0, x_1, \ldots, x_{d-1})$. Such an affine map takes a parallelepiped one to one and onto a parallelepiped (as long as the vector x has no zero components). Conversely, given any two parallelepipeds P and P' there exists an affine mapping of the form (4.57) which takes P one to one and onto P'. Moreover, if there exists a $z \in \mathbb{R}^d$ such that $P' + z \subset \operatorname{int} P$, then $\mathcal{A}$ is a contraction relative to the ℓ^∞-norm on $\mathbb{R}^d$ given by

$$\|[x_i : i \in \mathbb{Z}_d]\|_\infty := \max\{|x_i| : i \in \mathbb{Z}_d\}.$$

For any two parallelepipeds $P = \langle x, y \rangle$ and $P' = \langle x', y' \rangle$ with $P' \subseteq P$, we can partition their set-theoretic difference into parallelepipeds in the following way. For each $i \in \mathbb{Z}_d$ we partition the interval $[x_i, y_i]$ into three subintervals by using the partition $\mathcal{I}_i := \{x_i, x_i', y_i', y_i\}$. The associated partition $\mathcal{I}^d$ decomposes P into subparallelepipeds such that one and only one of them corresponds to P' itself. In other words, if $P_i, i \in \mathbb{Z}_N$ with $N = 3^d$ are the subparallelepipeds which partition P and $P_{N-1} = P'$, then we have

$$P \setminus P' = \bigcup_{i \in \mathbb{Z}_{N-1}} P_i.$$

We can now state the theorem.

Theorem 4.29 *Let m be a positive integer and V_0 a finite subset of cardinality m in the metric space $(\mathbb{R}^d, \|\cdot\|_\infty)$. There exists a finite set Φ of contractive mappings of the form (4.57) such that V_0 is refinable relative to Φ and the invariant set for Φ is a parallelepiped.*

Proof First we put the set V_0 into the interior of a parallelepiped P, which we partition as described above into subparallelepipeds so that the vectors in V_0 are in the union of the interior of these subparallelepipeds. Specifically, we suppose that

$$V_0 = \{v_i : i \in \mathbb{Z}_m\}, \quad P = \bigcup_{i \in \mathbb{Z}_M} P_i,$$

with $m < M$, $v_i \in \operatorname{int} P_i$, $i \in \mathbb{Z}_m$ and $V_0 \cap \operatorname{int} P_i = \emptyset$, $i \in \mathbb{Z}_M \setminus \mathbb{Z}_m$.

For each $i \in \mathbb{Z}_m$ we choose a vector $z_i := [z_{i,j} : j \in \mathbb{Z}_d] \in (0,1)^d$ with sufficiently small components $z_{i,j}$ so that the affine mapping

$$\mathcal{A}_i t = \mathbb{Q}_i(t - v_i) + v_i, \quad t \in \mathbb{R}^d, \tag{4.58}$$

where $\mathbb{Q}_i := \operatorname{diag}(z_{i,0}, z_{i,1}, \ldots, z_{i,d-1})$, has the property that the parallelepiped $Q_i := \mathcal{A}_i P$ is contained in P_i. Since $\mathcal{A}_i v_i = v_i$, $i \in \mathbb{Z}_m$, the set V_0 is refinable relative to any set of mappings, including those in (4.58). We wish to append to these m mappings another collection of one to one and onto contractive affine mappings of the type (4.57), so that the extended family has P as the invariant set.

To this end, for each $i \in \mathbb{Z}_m$ we partition the difference set $P_i \setminus Q_i$ into parallelepipeds in the manner described above:

$$P_i \setminus Q_i = \bigcup_{j \in \mathbb{Z}_{N-1}} P_{i,j},$$

where $N = 3^d$. Thus, we have succeeded in decomposing P into subparallelepipeds so that exactly m of them are the subparallelepipeds Q_i, $i \in \mathbb{Z}_m$. In other words, we have

$$P = \bigcup_{i \in \mathbb{Z}_k} W_i,$$

where $m < k$ and $W_i = Q_i$, $i \in \mathbb{Z}_m$. Finally, for every $i \in \mathbb{Z}_k \setminus \mathbb{Z}_m$ we choose a one to one and onto contractive affine mapping $\mathcal{A}_i$ such that $\mathcal{A}_i P = W_i$. This implies that

$$P = \bigcup_{i \in \mathbb{Z}_k} \mathcal{A}_i P$$

and therefore P is the invariant set relative to the one to one and onto contractive mappings $\mathcal{A}_i$, $i \in \mathbb{Z}_k$. □

In the remainder of this section we look at the above result for the real line and try to economize on the number of affine mappings needed to make a given set V_0 refinable.

Theorem 4.30 *Let k be a positive integer and $V_0 := \{v_l : l \in \mathbb{Z}_k\}$ a subset of distinct points in $[0, 1]$ of cardinality k. Then there exists a family of one-to-one and onto contractive affine mappings $\{\phi_\epsilon : \epsilon \in \mathbb{Z}_\mu\}$ of the type (4.57) for some $2 \leq \mu \leq 4$ when $k = 1, 2$ and $3 \leq \mu \leq 2k - 1$ when $k \geq 3$ such that V_0 is refinable relative to these mappings.*

Proof Since the mappings $\phi_0(t) := \frac{t}{2}$ and $\phi_1(t) = \frac{t+1}{2}$ have the fixed points $t = 0$ and $t = 1$, respectively, we conclude that when $k = 1$ and V_0 consists of either 0 or 1, and when $k = 2$ and V_0 consists of 0 and 1, these two mappings are the desired mappings. When V_0 consists of one interior point v_0, we need at least three mappings. For example, we choose

$$\phi_0(t) = \frac{v_0}{2} t, \quad \phi_1(t) = \frac{1}{2}(t - v_0) + v_0$$

and

$$\phi_2(t) = \frac{1 - v_0}{2} t + \frac{v_0 + 1}{2}, \quad \text{for } t \in [0, 1].$$

When V_0 consists of two interior points of $(0, 1)$, we need four mappings constructed by following the spirit of the construction for the case $k \geq 3$, which is given below.

When $k \geq 3$, regardless of the location of the points, there exist $2k - 1$ mappings that do the job. We next construct these mappings specifically. Without loss of generality, we assume that $v_0 < v_1 < \cdots < v_{k-1}$. We first choose a parameter γ_1 such that

$$0 < \gamma_1 < \min\left\{\frac{v_1 - v_0}{v_1}, \frac{v_2 - v_1}{1 - v_1}\right\}$$

and consider the mapping $\phi_1(t) := \gamma_1(t - v_1) + v_1$, $t \in [0, 1]$. Therefore, if we let $\alpha_1 := \phi_1(0)$ and $\beta_1 := \phi_1(1)$, then $v_0 < \alpha_1 < \beta_1 < v_2$. Next, we let $\gamma_0 := (\alpha_1 - v_0)/(1 - v_0)$ and introduce the mapping $\phi_0(t) := \gamma_0(t - v_0) + v_0$, $t \in [0, 1]$. Clearly, by letting $\alpha_0 := \phi_0(0)$ and $\beta_0 := \phi_0(1)$, we have $0 \leq \alpha_0 < \beta_0 = \alpha_1$.

The remaining steps in the construction proceed inductively on k. For this purpose, we assume that the affine mapping ϕ_{j-2} has been constructed. We let $\beta_{j-2} := \phi_{j-2}(1)$ and define

$$\phi_{j-1}(t) := \gamma_{j-1}(t - v_{j-1}) + v_{j-1}, \quad t \in [0, 1], \quad j = 3, 4, \ldots, k - 1,$$

where the parameters γ_{j-1} are chosen to satisfy the conditions

$$0 < \gamma_{j-1} < \min\left\{\frac{v_{j-1} - \beta_{j-2}}{v_{j-1}}, \frac{v_j - v_{j-1}}{1 - v_{j-1}}\right\}, \quad j = 3, 4, \ldots, k - 1.$$

It is not difficult to verify that $\phi_{j-1}([0, 1]) \subset (\beta_{j-2}, v_j)$, or equivalently $\beta_{j-2} < \alpha_{j-1} < \beta_{j-1} < v_j$, by letting $\alpha_{j-1} := \phi_{j-1}(0)$ and $\beta_{j-1} := \phi_{j-1}(1)$. Next, we let $\phi_{k-1}(t) := \gamma_{k-1}(t - v_{k-1}) + v_{k-1}$, $t \in [0, 1]$, where $0 < \gamma_{k-1} = \frac{v_{k-1} - \beta_{k-2}}{v_{k-1}}$ and let $\alpha_{k-1} := \phi_{k-2}(0)$ and $\beta_{k-1} := \phi_{k-1}(1)$. Then, we have that $\beta_{k-2} = \alpha_{k-1} < \beta_{k-1} \leq 1$. By the construction above, we find two sets $\{\alpha_i : i \in \mathbb{Z}_k\}$ and $\{\beta_i : i \in \mathbb{Z}_k\}$ of numbers that satisfy the condition

$$0 \leq \alpha_0 < \beta_0 = \alpha_1 < \beta_1 < \cdots < \beta_{k-2} = \alpha_{k-1} < \beta_{k-1} \leq 1$$

and the union of the images of the interval $[0, 1]$ under mappings $\{\phi_j : j \in \mathbb{Z}_k\}$ is

$$U := \bigcup_{j \in \mathbb{Z}_k} [\alpha_j, \beta_j].$$

Notice that the set U is not the whole interval $[0, 1]$. There are at most $k - 1$ gaps which need to be covered. It is straightforward to construct these $k - 1$ additional mappings.

The family of mappings of cardinality at most $2k - 1$ that we have constructed above has $[0, 1]$ as the invariant set and V_0 is a refinable set relative to them. □

When the points in a given set V_0 have special structure, the number of mappings may be reduced.

4.5 Multiscale interpolating bases

In this section we present the construction of multiscale bases for both Lagrange interpolation and Hermite interpolation based on refinable sets and set wavelets developed in the previous section.

4.5.1 Multiscale Lagrange interpolation

In this subsection we describe a construction of the Lagrange interpolating wavelet-like basis using the set wavelets constructed previously. For this purpose, we let $\mathbb{X} := \mathbb{R}^d$ and assume that $\Phi := \{\phi_\epsilon : \epsilon \in \mathbb{Z}_\mu\}$ is a family of contractive mappings that satisfies the hypotheses of Theorem 4.49. We also assume that $\Omega \subset \mathbb{X}$ is the invariant set relative to Φ with $\operatorname{meas}(\Omega \setminus \operatorname{int}\Omega) = 0$. Let k be a positive integer and assume that

$$V_0 := \{v_l : l \in \mathbb{Z}_k\} \subset \operatorname{int}\Omega$$

is refinable relative to Φ. Note that in this construction of discontinuous wavelets, we restrict the choice of the points in the set V_0 to interior points of Ω.

As in [200, 201], we choose a refinable curve $\mathbf{f} := [f_l : l \in \mathbb{Z}_k] : \Omega \to \mathbb{R}^k$ which satisfies a refinement equation

$$\mathbf{f} \circ \phi_i = \mathbf{A}_i \mathbf{f}, \quad i \in \mathbb{Z}_\mu \tag{4.59}$$

for some prescribed $k \times k$ matrices $\mathbf{A}_i$, $i \in \mathbb{Z}_\mu$. We remark that if there is $\mathbf{g} : \Omega \to \mathbb{R}^k$ and a $k \times k$ nonsingular matrix $\mathbf{B}$ such that $\mathbf{f} = \mathbf{B}\mathbf{g}$, then $\mathbf{g}$ is also a refinable curve. We let

$$\mathbb{F}_0 := \operatorname{span}\{f_l : l \in \mathbb{Z}_k\}$$

and suppose that $\dim \mathbb{F}_0 = k$. Furthermore, we require that for any $\mathbf{b} := [b_l : l \in \mathbb{Z}_k] \in \mathbb{R}^k$, there exists a unique element $f \in \mathbb{F}_0$ such that $f(v_i) = b_i$, $i \in \mathbb{Z}_k$. In other words, there exist k elements in $\mathbb{F}_0$ which we also denote by $f_0, f_1, \ldots, f_{k-1}$ such that $f_i(v_j) = \delta_{ij}$, $i, j \in \mathbb{Z}_k$. Refinable sets that admit a unique Lagrange interpolating polynomial were constructed in [198]. When this condition holds we say $\{f_i : i \in \mathbb{Z}_k\}$ *interpolates on the set* V_0 and that f_j interpolates at v_j, $j \in \mathbb{Z}_k$. Under this condition, any element $f \in \mathbb{F}_0$ has a representation of the form

$$f = \sum_{i \in \mathbb{Z}_k} f(v_i) f_i.$$

A set $V_0 \subseteq \mathbb{X}$ is called *(Lagrange) admissible* relative to $(\Phi, \mathbb{F}_0)$ if it is refinable relative to Φ and there is a basis of functions f_i, $i \in \mathbb{Z}_k$ for $\mathbb{F}_0$ which interpolate on the set V_0. In this subsection we shall always assume that V_0 is (Lagrange) admissible. We record in the next proposition the simple fact of the Lagrange admissibility of any set of cardinality k for the special case when $\Phi = \Psi$ defined by (4.39), $\Omega = [0, 1]$ and $\mathbb{F}_0 = \mathbb{P}_{k-1}$, the space of polynomials of degree $\leq k - 1$.

Proposition 4.31 *If* $V_0 \subset [0, 1]$ *is refinable relative to* Ψ *and has cardinality* k*, then* V_0 *is Lagrange admissible relative to* $(\Psi, \mathbb{P}_{k-1})$.

Proof It is a well-known fact that the polynomial basis functions satisfy the refinement equation (4.59) with $\phi_i = \psi_i$ for some matrices $\mathbf{A}_i$. Hence, this result follows immediately from the unique solvability of the univariate Lagrange interpolation. □

In a manner similar to the construction of orthogonal wavelets in Section 4.3, we define linear operators $\mathcal{T}_\epsilon : L^\infty(\Omega) \to L^\infty(\Omega)$, $\epsilon \in \mathbb{Z}_\mu$ by

$$(\mathcal{T}_\epsilon x)(t) := \begin{cases} x(\phi_\epsilon^{-1}(t)), & t \in \phi_\epsilon(\Omega) \\ 0, & t \notin \phi_\epsilon(\Omega) \end{cases}$$

and set

$$\mathbb{F}_{i+1} := \bigoplus_{\epsilon \in \mathbb{Z}_\mu} \mathcal{T}_\epsilon \mathbb{F}_i, \quad i \in \mathbb{N}_0.$$

This sequence of spaces is nested, that is, $\mathbb{F}_i \subseteq \mathbb{F}_{i+1}$, $i \in \mathbb{N}_0$ and $\dim \mathbb{F}_i = k\mu^i$, $i \in \mathbb{N}_0$.

We next construct a convenient basis for each of the spaces $\mathbb{F}_i$. For this purpose, we let $F_0 := \{f_j : j \in \mathbb{Z}_k\}$, where f_j, $j \in \mathbb{Z}_k$, interpolate the set V_0 and

$$\begin{aligned} F_i &:= \bigcup\nolimits_{\epsilon \in \mathbb{Z}_\mu}^{\perp} \mathcal{T}_\epsilon F_{i-1} \\ &= \{\mathcal{T}_{\epsilon_0} \circ \cdots \circ \mathcal{T}_{\epsilon_{i-1}} f_j : j \in \mathbb{Z}_k, \ \ \epsilon_\ell \in \mathbb{Z}_\mu, \ \ \ell \in \mathbb{Z}_i\}, \ \ i \in \mathbb{N}. \end{aligned} \tag{4.60}$$

Since the functions f_j, $j \in \mathbb{Z}_k$ interpolate the set V_0, we conclude that the elements in F_i interpolate the set V_i. In other words, the functions in the set F_i

satisfy the condition

$$(\mathcal{T}_{\epsilon_0} \circ \cdots \circ \mathcal{T}_{\epsilon_{i-1}} f_j)(\phi_{\epsilon'_0} \circ \cdots \circ \phi_{\epsilon'_{i-1}}(v_{j'})) = \delta_{(\epsilon_0,\ldots,\epsilon_{i-1},j),(\epsilon'_0,\ldots,\epsilon'_{i-1},j')}, \tag{4.61}$$

where we use the notation

$$\delta_{\mathbf{a},\mathbf{a}'} = \begin{cases} 1, & \mathbf{a} = \mathbf{a}', \\ 0, & \mathbf{a} \neq \mathbf{a}', \end{cases} \quad \mathbf{a}, \mathbf{a}' \in \mathbb{N}_0^i, \;\; i \in \mathbb{N}.$$

For ease of notation, we let $\mathbf{e}_i := [\epsilon_j : j \in \mathbb{Z}_i]$ and $\mathcal{T}_{\mathbf{e}_i} f_j := \mathcal{T}_{\epsilon_0} \circ \cdots \circ \mathcal{T}_{\epsilon_{i-1}} f_j$. By equation (4.61), this function interpolates at $\phi_{\mathbf{e}_i}(v_j)$. Moreover, the relation

$$\mathbb{F}_i = \operatorname{span} F_i, \;\; i \in \mathbb{N}_0 \tag{4.62}$$

holds. Now, for each $n \in \mathbb{N}_0$, we decompose the space $\mathbb{F}_{n+1}$ as the direct sum of the space $\mathbb{F}_n$ and its complement space $\mathbb{G}_{n+1}$, which consists of the elements in $\mathbb{F}_{n+1}$ vanishing at all points in V_n, that is,

$$\mathbb{F}_{n+1} = \mathbb{F}_n \oplus \mathbb{G}_{n+1}, \;\; n \in \mathbb{N}_0. \tag{4.63}$$

This decomposition is analogous to the orthogonal decomposition in the construction of the orthogonal wavelet-like basis in Section 4.3 and can be viewed as an interpolatory decomposition in the sense that we describe below.

We first label the points in the set V_n according to the set wavelet decomposition for V_n given in Section 4.4. We assume that the initial set wavelet is given by $W_1 = \{w_j : j \in \mathbb{Z}_r\}$ with $r := k(\mu - 1)$ and we let

$$t_{0j} := v_j, \;\; j \in \mathbb{Z}_k; \;\; t_{1j} := w_j, \;\; j \in \mathbb{Z}_r;$$

$$t_{ij} := \phi_{\mathbf{e}} w_\ell, \;\; j = \mu(\mathbf{e})r + \ell, \;\; \mathbf{e} \in \mathbb{Z}_\mu^{i-1}, \;\; \ell \in \mathbb{Z}_r, \;\; i = 2, 3, \ldots, n.$$

Then, we conclude that $V_n = \{t_{ij} : (i,j) \in \mathbb{U}_n\}$, where $\mathbb{U}_n := \{(i,j) : i \in \mathbb{Z}_{n+1} j \in \mathbb{Z}_{w(i)},\}$, with

$$w(i) := \begin{cases} k, & i = 0, \\ k(\mu - 1)\mu^{i-1}, & i \geq 1. \end{cases}$$

The Lagrange interpolation problem for $\mathbb{F}_n$ relative to V_n is to find, for a vector $\mathbf{b} := [b_{ij} : (i,j) \in \mathbb{U}_n]$, an element $f \in \mathbb{F}_n$ such that

$$f(t_{ij}) = b_{ij}, \;\; (i,j) \in \mathbb{U}_n. \tag{4.64}$$

The following fact is useful in this regard.

Lemma 4.32 *If V_0 is Lagrange admissible relative to $(\Phi, \mathbb{F}_0)$, then for each $n \in \mathbb{N}_0$ the set V_n is also Lagrange admissible relative to $(\Phi, \mathbb{F}_n)$.*

Proof This result follows immediately from equations (4.61). □

Lemma 4.32 insures that each $f \in \mathbb{F}_{n+1}$ has the representation $f = \mathcal{P}_n f + g_n$, where $\mathcal{P}_n f$ is the Lagrange interpolant to f from $\mathbb{F}_n$ relative to V_n and $g_n = f - \mathcal{P}_n f$ is the error of the interpolation. Therefore, we have that

$$\mathbb{G}_{n+1} = \{g_n : g_n = f - \mathcal{P}_n f,\ f \in \mathbb{F}_{n+1}\}. \tag{4.65}$$

The fact that the subspace decomposition (4.63) is a direct sum also follows from equation (4.65) and the unique solvability of the Lagrange interpolation problem (4.64). For this reason, the spaces $\mathbb{G}_n$ are called the *interpolating wavelet spaces*, and in particular the space $\mathbb{G}_1$ is called the initial interpolating wavelet space. Direct computation yields the dimension of the wavelet space $\mathbb{G}_n$ $\dim \mathbb{G}_n = k\mu^{n-1}(\mu - 1)$. Also, we have an interpolating wavelet decomposition for $\mathbb{F}_{n+1}$:

$$\mathbb{F}_{n+1} = \mathbb{F}_0 \oplus \mathbb{G}_1 \oplus \cdots \oplus \mathbb{G}_{n+1}. \tag{4.66}$$

In the next theorem, we describe a recursive construction for the wavelet spaces $\mathbb{G}_n$. To establish the theorem, we need the following lemma regarding the distributivity of the linear operators $\mathcal{T}_\epsilon$, $\epsilon \in \mathbb{Z}_\mu$, relative to a direct sum of two subspaces of $L^\infty(\Omega)$.

Lemma 4.33 *Let $\mathbb{B}, \mathbb{C} \subset L^\infty(\Omega)$ be two subspaces. If $\mathbb{B} \oplus \mathbb{C}$ is a direct sum, then for each $\epsilon \in \mathbb{Z}_\mu$, $\mathcal{T}_\epsilon(\mathbb{B} \oplus \mathbb{C}) = (\mathcal{T}_\epsilon \mathbb{B}) \oplus (\mathcal{T}_\epsilon \mathbb{C})$.*

Proof It is clear that $\mathcal{T}_\epsilon(\mathbb{B} \oplus \mathbb{C}) = (\mathcal{T}_\epsilon \mathbb{B}) + (\mathcal{T}_\epsilon \mathbb{C})$. Therefore, it remains to verify that the sum on the right-hand side is a direct sum. To this end, we let $x \in (\mathcal{T}_\epsilon \mathbb{B}) \cap (\mathcal{T}_\epsilon \mathbb{C})$ and observe that there exist $f \in \mathbb{B}$ and $g \in \mathbb{C}$ such that

$$x = \mathcal{T}_\epsilon f = \mathcal{T}_\epsilon g. \tag{4.67}$$

By the definition of the operators $\mathcal{T}_\epsilon$, we have that $x(t) = 0$, for $t \in \Omega \setminus \phi_\epsilon(\Omega)$. Now, for each $t \in \phi_\epsilon(\Omega)$, there exists $\tau \in \Omega$ such that $t = \phi_\epsilon(\tau)$ and thus using equation (4.67), we observe that

$$x(t) = f(\phi_\epsilon^{-1}(t)) = f(\tau) \in \mathbb{B},\ \ x(t) = g(\phi_\epsilon^{-1}(t)) = g(\tau) \in \mathbb{C}.$$

Since $\mathbb{B} \oplus \mathbb{C}$ is a direct sum, we conclude that $x(t) = 0$, for $t \in \phi_\epsilon(\Omega)$. It follows that $x = 0$. □

We also need the following fact for the proof of our main theorem.

Lemma 4.34 *If $Y \subseteq L^\infty(\Omega)$, then there holds*

$$\mathcal{T}_\epsilon Y \cap \mathcal{T}_{\epsilon'} Y = \{0\},\ \ \epsilon, \epsilon' \in \mathbb{Z}_\mu,\ \ \epsilon \neq \epsilon'.$$

Proof Let $x \in \mathcal{T}_\epsilon Y \cap \mathcal{T}_{\epsilon'} Y$. There exist $y_1, y_2 \in Y$ such that $x = \mathcal{T}_\epsilon y_1 = \mathcal{T}_{\epsilon'} y_2$. By the definition of the operators $\mathcal{T}_\epsilon$, we conclude from the first equality that

$x(t) = 0$ for $t \in \Omega \backslash \phi_\epsilon(\Omega)$ and from the second that $x(t) = 0$ for $t \in \Omega \backslash \phi_{\epsilon'}(\Omega)$. Since $\epsilon \neq \epsilon'$, we have that meas $(\phi_\epsilon(\Omega) \cap \phi_{\epsilon'}(\Omega)) = 0$. This implies that $x = 0$ a.e. in Ω and therefore establishes the result in this lemma. □

We are now ready to prove the main result of this section.

Theorem 4.35 *Let V_0 be Lagrange admissible relative to $(\Phi, \mathbb{F}_0)$ and W_n, $n \in \mathbb{N}$, be the set wavelets generated from V_0. Then*

$$\mathbb{G}_1 = \operatorname{span}\{\mathcal{T}_\epsilon f_j : \epsilon \in \mathbb{Z}_\mu, j \in \mathbb{Z}_k, \mathcal{T}_\epsilon f_j \text{ interpolates at } \phi_\epsilon(v_j) \in W_1\},$$

$$\mathbb{G}_{n+1} = \bigoplus_{\epsilon \in \mathbb{Z}_\mu} \mathcal{T}_\epsilon \mathbb{G}_n, \quad n \in \mathbb{N} \tag{4.68}$$

and $\mathbb{G}_n = \operatorname{span} G_n$, where

$$G_n := \{\mathcal{T}_{\boldsymbol{e}_n} f_j : \boldsymbol{e}_n \in \mathbb{Z}_\mu^n, j \in \mathbb{Z}_k, \mathcal{T}_{\boldsymbol{e}_n} f_j \text{ interpolates at } \phi_{\boldsymbol{e}_n}(v_j) \in W_n\}.$$

Proof Let $\mathcal{T}_\epsilon f_j$ interpolate at $\phi_\epsilon(v_j) \in W_1$. Then, $\mathcal{T}_\epsilon f_j$ has the property

$$(\mathcal{T}_\epsilon f_j)(\phi_{\epsilon'}(v_{j'})) = 0, \quad \epsilon, \epsilon' \in \mathbb{Z}_\mu, \epsilon' \neq \epsilon, \quad \text{or } j, j' \in \mathbb{Z}_k, j' \neq j.$$

By the definition of the set wavelet, $W_1 = V_1 \setminus V_0$, we conclude for all $v_{j'} \in V_0$ that we have $(\mathcal{T}_\epsilon f_j)(v_{j'}) = 0$. Thus, by the definition of $\mathbb{G}_1$, we have that corresponding to each point $\phi_\epsilon(v_j) \in W_1$, the basis function $\mathcal{T}_\epsilon f_j$ is in $\mathbb{G}_1$. Note that the cardinality of W_1 is given by the formula card $W_1 =$ card $V_1 -$ card $V_0 = k(\mu - 1)$. It follows that the number of basis functions for which $\mathcal{T}_\epsilon f_j$ interpolate at $\phi_\epsilon(v_j)$ in W_1 is r, the dimension of $\mathbb{G}_1$. Because these r functions are linearly independent, they constitute a basis for $\mathbb{G}_1$.

We next prove equation (4.68) by induction on n. For this purpose, we assume that equation (4.68) holds for $n \leq m$ and consider the case when $n = m + 1$. By the definition of $\mathbb{F}_{m+1}$ and $\mathbb{G}_m$, we have that

$$\mathbb{F}_{m+1} = \bigoplus_{\epsilon \in \mathbb{Z}_\mu} \mathcal{T}_\epsilon \mathbb{F}_m = \bigoplus_{\epsilon \in \mathbb{Z}_\mu} \mathcal{T}_\epsilon (\mathbb{F}_{m-1} \oplus \mathbb{G}_m).$$

Using Lemma 4.33, we obtain that

$$\mathbb{F}_{m+1} = \bigoplus_{\epsilon \in \mathbb{Z}_\mu} [(\mathcal{T}_\epsilon \mathbb{F}_{m-1}) \oplus (\mathcal{T}_\epsilon \mathbb{G}_m)] = \left(\bigoplus_{\epsilon \in \mathbb{Z}_\mu} \mathcal{T}_\epsilon \mathbb{F}_{m-1} \right) \oplus \left(\bigoplus_{\epsilon \in \mathbb{Z}_\mu} \mathcal{T}_\epsilon \mathbb{G}_m \right).$$

It then follows from the definition of $\mathbb{F}_m$ that

$$\mathbb{F}_{m+1} = \mathbb{F}_m \oplus \left(\bigoplus_{\epsilon \in \mathbb{Z}_\mu} \mathcal{T}_\epsilon \mathbb{G}_m \right).$$

Let

$$\mathbb{G} := \bigoplus_{\epsilon\in\mathbb{Z}_\mu} \mathcal{T}_\epsilon \mathbb{G}_m$$

and assume that $f \in \mathbb{G}$. Then, there exist $g_0, \ldots, g_{\mu-1} \in \mathbb{G}_m$ such that

$$f = \sum_{\epsilon\in\mathbb{Z}_\mu} \mathcal{T}_\epsilon g_\epsilon.$$

For each $v \in V_m$, there exist $v' \in V_{m-1}$ and $\epsilon' \in \mathbb{Z}_\mu$ such that $v = \phi_{\epsilon'}(v')$. By the definition of the linear operators $\mathcal{T}_\epsilon$, $\epsilon \in \mathbb{Z}_\mu$ and the fact that $g_\epsilon \in \mathbb{G}_m$, $\epsilon \in \mathbb{Z}_\mu$, direct computation leads to the condition for each $v \in V_m$ that

$$f(v) = \sum_{\epsilon\in\mathbb{Z}_\mu} \mathcal{T}_\epsilon g_\epsilon(\phi_{\epsilon'}(v')) = g_{\epsilon'}(\phi_{\epsilon'}^{-1} \circ \phi_{\epsilon'}(v')) = g_{\epsilon'}(v') = 0.$$

Hence, $\mathbb{G} \subseteq \mathbb{G}_{m+1}$. Moreover, it is easy to see that $\dim \mathbb{G} = \dim \mathbb{G}_{m+1}$, which implies that $\mathbb{G} = \mathbb{G}_{m+1}$.

To prove the second part of the theorem, it suffices to establish the recursion

$$G_{n+1} = \bigcup_{\epsilon\in\mathbb{Z}_\mu} {}^{\perp}\mathcal{T}_\epsilon G_n.$$

The "$\perp$" on the right-hand side of this equation is justified by Lemma 4.34. To establish its validity, we let

$$G := \bigcup_{\epsilon\in\mathbb{Z}_\mu} {}^{\perp}\mathcal{T}_\epsilon G_n.$$

Hence, the set G consists of the elements $\mathcal{T}_{\epsilon_n}\mathcal{T}_{\mathbf{e}_n} f_j$, where $\mathcal{T}_{\mathbf{e}_n} f_j$ interpolates at $\phi_{\mathbf{e}_n}(v_j) \in W_n$, $\epsilon_n \in \mathbb{Z}_\mu$. By Theorem 4.28, we have that

$$\{\phi_{\mathbf{e}_{n+1}}(v_j) := \phi_{\epsilon_n} \circ \phi_{\mathbf{e}_n}(v_j) : \epsilon_n \in \mathbb{Z}_\mu,\ \phi_{\mathbf{e}_n}(v_j) \in W_n\} \subseteq \{\phi_{\mathbf{e}_{n+1}}(v_j) \in W_{n+1}\}.$$

Hence, $G \subseteq G_{n+1}$. Since $\operatorname{card} G = \operatorname{card} G_{n+1} = \operatorname{card} W_{n+1}$, we conclude that $G = G_{n+1}$. □

Theorem 4.36 *It holds that*

$$L^2(\Omega) = \overline{\bigcup_{n\in\mathbb{N}_0} \mathbb{F}_n} = \overline{\mathbb{F}_0 \oplus \left(\bigoplus_{n\in\mathbb{N}} \mathbb{G}_n\right)}.$$

Proof Since the mappings ϕ_ϵ, $\epsilon \in \mathbb{Z}_\mu$, are contractive, the condition of Theorem 4.7 of [201] is satisfied. The finite-dimensional spaces $\mathbb{F}_n$ appearing here are the same as those generated by the family of mutually orthogonal isometries in [201] if we begin with the same initial space $\mathbb{F}_1$. Therefore, the first equality holds. An examination of the proof for Theorem 4.7 of [201] shows that the same proof proves the second equality. □

As a result of the decomposition obtained in Theorems 4.35 and 4.36, we present a multiscale algorithm for the Lagrange interpolation. To this end, we let $g_j, j \in \mathbb{Z}_r$, be a basis for $\mathbb{G}_1$ so that

$$g_j(t_{0i}) = 0, \;\; i \in \mathbb{Z}_k, \;\; j \in \mathbb{Z}_r, \;\; g_j(t_{1j'}) = \delta_{j,j'}, \;\; j, j' \in \mathbb{Z}_r.$$

We label those functions according to points in V_n in the following way. Let

$$g_{0j} := f_j, \;\; j \in \mathbb{Z}_k; \;\; g_{1j} := g_j, \;\; j \in \mathbb{Z}_r,$$

$$g_{ij} := \mathcal{T}_{\mathbf{e}} g_\ell, \;\; j = \mu(\mathbf{e})r + \ell, \;\; \mathbf{e} \in \mathbb{Z}_\mu^{i-1}, \;\; \ell \in \mathbb{Z}_r, \;\; i = 2, 3, \ldots, n.$$

With this labeling, we see that $g_{ij}(t_{i'j'}) = \delta_{ii'}\delta_{jj'}$, $(i, j), (i', j') \in \mathbb{U}_n$, $i \geq i'$ and

$$\mathbb{F}_n = \operatorname{span}\{g_{ij} : (i, j) \in \mathbb{U}_n\}.$$

Functions g_{ij}, $(i, j) \in \mathbb{U}$ are also called interpolating wavelets. Now, we express the interpolation projection in terms of this basis. For each $x \in C(\Omega)$, the interpolation projection $\mathcal{P}_n x$ of x is given by

$$\mathcal{P}_n x = \sum_{(i,j) \in \mathbb{U}_n} x_{ij} g_{ij}. \tag{4.69}$$

The coefficients x_{ij} in (4.69) can be obtained from the recursive formula

$$x_{0j} = x(t_{0j}), \; j \in \mathbb{Z}_k;$$

$$x_{ij} = x(t_{ij}) - \sum_{(i',j') \in \mathbb{U}_{i-1}} x_{i'j'} g_{i'j'}(t_{ij}), \;\; (i, j) \in \mathbb{U}_n.$$

This recursive formula allows us to interpolate a given function efficiently by functions in $\mathbb{F}_n$. When we increase the level from n to $n+1$, we *do not need to* recompute the coefficients x_{ij} for $0 \leq i \leq n$. We describe this important point with the formula $\mathcal{P}_{n+1}x = \mathcal{P}_n x + \mathcal{Q}_{n+1}x$, where $\mathcal{Q}_{n+1}x \in \mathbb{G}_{n+1}$ and

$$\mathcal{Q}_{n+1}x := \sum_{j \in \mathbb{Z}_{J(n+1)}} x_{n+1,j} g_{n+1,j}.$$

The coefficients $x_{n+1,j}$ are computed by the previous recursive formula using the coefficients obtained for the previous levels, that is,

$$x_{n+1,j} = x(t_{n+1,j}) - \sum_{(i',j') \in \mathbb{U}_n} x_{i'j'} g_{i'j'}(t_{n+1,j}).$$

4.5.2 Multiscale Hermite interpolation

In the last section we showed how refinable sets lead to a multiresolution structure and result in what we call set wavelets. In the last subsection we then used this *recursive structure* of the points to construct the Lagrange interpolation that has a much desired *multiscale structure*. In this subsection we describe a similar construction for Hermite piecewise polynomial interpolation on invariant sets.

Let $\mathbb{X}$ be Euclidean space $\mathbb{R}^d$ and $\Phi := \{\phi_\epsilon : \epsilon \in \mathbb{Z}_\mu\}$ be a family of contractive mappings on $\mathbb{X}$, Ω is the invariant set relative to the family of contractive mappings Φ. Let V_0 be a nonempty finite subset of distinct points in $\mathbb{X}$ and recursively define

$$V_i := \Phi(V_{i-1}), \quad i \in \mathbb{N}.$$

It was shown in the last section that the collection of sets $\{V_i : i \in \mathbb{N}_0\}$ is strictly nested if and only if the set V_0 is refinable relative to Φ. Denote

$$W_i := V_i \setminus V_{i-1}, \quad i \in \mathbb{N}.$$

When the contractive mappings have a continuous inverse on $\mathbb{X}$,

$$\phi_\epsilon(\operatorname{int}\Omega) \cap \phi_{\epsilon'}(\operatorname{int}\Omega) = \emptyset, \quad \epsilon, \epsilon' \in \mathbb{Z}_\mu, \ \epsilon \neq \epsilon'$$

and W_1 is chosen to be a subset of int Ω, then the sets W_{i+1}, $i \in \mathbb{N}$ can be generated recursively from W_1 by the formula

$$W_{i+1} = \Phi^\perp(W_i) := \bigcup_{\epsilon \in \mathbb{Z}_\mu}{}^{\perp} \phi_\epsilon(W_i), \quad i \in \mathbb{N}$$

and the invariant set Ω has the decomposition

$$\Omega = \overline{V_0 \cup^\perp \left(\bigcup_{n \in \mathbb{N}}{}^{\perp} W_n \right)}.$$

In the following, we first describe a construction of multiscale discontinuous Hermite interpolation, and then a construction of multiscale *smooth* Hermite interpolation on the interval $[0, 1]$.

1. Multiscale discontinuous Hermite interpolation

We start with nonempty finite sets $V \subset \operatorname{int}\Omega$ and $U \subset \mathbb{N}_0^d$. For $u = [u_i : i \in \mathbb{Z}_d] \in U$, we set

$$D^u := \frac{\partial^{|u|}}{\partial t_1^{u_0} \cdots \partial t_d^{u_{d-1}}},$$

where $|u| := \sum_{i \in \mathbb{Z}_d} u_i$.

Let $\mathbb{P}$ be a linear space of functions on Ω. We say that $(\mathbb{P}, U, V)$ is *Hermite admissible* provided that V is refinable and for any given real numbers c_r, $r = (u, v) \in U \times V$, there exists a unique element $p \in \mathbb{P}$ such that

$$\mathcal{D}_r(p) := (D^u p)(v) = c_r. \tag{4.70}$$

When this is the case the dimension of $\mathbb{P}$ is (card V)(card U) and there exists a basis $\{p_r : r \in U \times V\}$ for $\mathbb{P}$ such that for every $r = (u, v) \in U \times V$ and $r' = (u', v') \in U \times V$,

$$(D^{u'} p_r)(v') = \delta_{r,r'}. \tag{4.71}$$

Moreover, for any function $p \in \mathbb{P}$ the representation

$$p = \sum_{r \in U \times V} \mathcal{D}_r(p) p_r$$

holds. We call $\{p_r : r \in U \times V\}$ a Hermite basis for $\mathbb{P}$ relative to $U \times V$. To proceed further we must restrict the family Φ to have the form

$$\phi_\epsilon(t) = a_\epsilon t + b_\epsilon, \quad t \in \Omega, \tag{4.72}$$

where $a_\epsilon t$ is the vector formed by the *componentwise* product of the vectors a_ϵ and t. We also require linear operators $\mathcal{T}_\epsilon : L^\infty(\Omega) \to L^\infty(\Omega)$, $\epsilon \in \mathbb{Z}_\mu$ defined by

$$\mathcal{T}_\epsilon f := f \circ \phi_\epsilon^{-1} \chi_{\phi_\epsilon(\Omega)},$$

and for every $\mathbf{e}_n = [\epsilon_j : j \in \mathbb{Z}_n] \in \mathbb{Z}_\mu^n$, we define constants

$$a_{\mathbf{e}_n}^{-u} := a_{\epsilon_0}^{-u} a_{\epsilon_1}^{-u} \cdots a_{\epsilon_{n-1}}^{-u}.$$

Lemma 4.37 *If Φ is a family of contractive mappings of the form (4.72), then for all $\mathbf{e}_n \in \mathbb{Z}_\mu^n$ and $u \in \mathbb{N}_0^d$ the following formula holds:*

$$D^u \mathcal{T}_{e_n} = a_{\mathbf{e}_n}^{-u} \mathcal{T}_{e_n} D^u, \quad n \in \mathbb{N}.$$

Proof We prove this lemma by induction on n and the proof begins by first verifying the case when $n = 1$ by the chain rule. The induction hypothesis is then advanced by again using the chain rule and the case $n = 1$. □

We suppose that $(\mathbb{P}, U, V_0)$ is admissible, $\mathbb{P}$ is a subspace of *polynomials* with Hermite basis

$$F_0 := \{p_r : r \in U \times V_0\}$$

relative to $U \times V_0$ and use the operators $\mathcal{T}_\epsilon$, $\epsilon \in \mathbb{Z}_\mu$ to recursively define the sets

$$F_n := \{a_{\mathbf{e}_n}^u \mathcal{T}_{\mathbf{e}_n} p_r : r = (u, v) \in U \times V_0, \mathbf{e}_n \in \mathbb{Z}_\mu^n\}.$$

Since the polynomials p_r, $r \in U \times V_0$, were chosen to be a Hermite basis for $\mathbb{P}$ relative to $U \times V_0$, the functions in F_n form a Hermite basis for $\mathbb{F}_n := \operatorname{span} F_n$ relative to $U \times V_n$. That is, the function $a^u_{\mathbf{e}_n} \mathcal{T}_{\mathbf{e}_n} p_r$ satisfies the condition

$$D^{u'}(a^u_{\mathbf{e}_n} \mathcal{T}_{\mathbf{e}_n} p_r)(\phi_{\mathbf{e}'_n}(v')) = \delta_{(r,\mathbf{e}_n),(r',\mathbf{e}'_n)}, \tag{4.73}$$

where $r' = (u', v') \in U \times V_0$.

Since $\mathbb{F}_n \subseteq \mathbb{F}_{n+1}$, we can decompose $\mathbb{F}_{n+1}$ as the direct sum of the space $\mathbb{F}_n$ and $\mathbb{G}_{n+1}$ defined to be the elements in $\mathbb{F}_{n+1}$ whose uth derivatives, for $u \in U$, vanish at all points in V_n. We let $\mathcal{P}_n f \in \mathbb{F}_n$ be uniquely defined by the conditions

$$(D^u \mathcal{P}_n f)(v) = D^u f(v), \quad v \in V_n, \quad u \in U. \tag{4.74}$$

Hence, each $f \in \mathbb{F}_{n+1}$ has the representation $f = \mathcal{P}_n f + g_n$, where $g_n \in \mathbb{G}_{n+1}$ with $\mathbb{G}_{n+1} = \{f - \mathcal{P}_n f : \ f \in \mathbb{F}_{n+1}\}$ and we have the decomposition

$$\mathbb{F}_n = \mathbb{F}_0 \oplus \mathbb{G}_1 \oplus \cdots \oplus \mathbb{G}_n.$$

Most importantly, the spaces $\mathbb{G}_n$ can be generated recursively, the proof of which follows the pattern of those given for Theorems 4.35 and 4.36. To state the next result, we make use of the following notation. For each $n \in \mathbb{N}_0$ and a subset $A \subseteq V_n$, we let $\mathbb{Z}^n_\mu(A)$ denote the subset of $\mathbb{Z}^n_\mu$ consisting of the indices $\mathbf{e}_n \in \mathbb{Z}^n_\mu$ such that there exists $r := (u, v) \in U \times V_0$ for which equation (4.73) holds and $\phi_{\mathbf{e}_n}(v) \in A$.

Theorem 4.38 *If $\mathbb{P}$ is a subspace of polynomials, $(\mathbb{P}, U, V_0)$ is admissible and W_n, $n \in \mathbb{N}$ is the set wavelets generated by V_0, then*

$$\mathbb{G}_1 = \operatorname{span}\{a^u_\epsilon \mathcal{T}_\epsilon p_r : r = (u, v) \in U \times V_0, \ \epsilon \in \mathbb{Z}_\mu(W_1)\}$$

and

$$\mathbb{G}_{n+1} = \bigoplus_{\epsilon \in \mathbb{Z}_\mu} \mathcal{T}_\epsilon \mathbb{G}_n, \quad n \in \mathbb{N}.$$

Moreover, we have that $\mathbb{G}_n = \operatorname{span} G_n$, where for $n \in \mathbb{N}$,

$$G_n := \{a^u_{e_n} \mathcal{T}_{e_n} p_r : \ r = (u, v) \in U \times V_0, \ \boldsymbol{e}_n \in \mathbb{Z}^n_\mu(W_n)\}$$

and the formula

$$L^2(\Omega) = \overline{\bigcup_{n \in \mathbb{N}_0} \mathbb{F}_n} = \overline{\mathbb{F}_0 \oplus \left(\bigoplus_{n \in \mathbb{N}} \mathbb{G}_n \right)}$$

holds.

2. Multiscale smooth Hermite interpolation on $[0, 1]$

In the following we focus on a construction of *smooth* multiscale Hermite interpolating polynomials on the interval $[0, 1]$, which generate finite-dimensional spaces dense in the Sobolev space $W^{m,p}[0, 1]$, where m is a positive integer and $1 \leq p < \infty$. To this end, we choose affine mappings $\Phi := \{\phi_\epsilon : \epsilon \in \mathbb{Z}_\mu\}$,

$$\phi_\epsilon(t) = (t_{\epsilon+1} - t_\epsilon)t + t_\epsilon, \ \ \epsilon \in \mathbb{Z}_\mu,$$

where $0 = t_0 < t_1 < \cdots < t_{\mu-1} < t_\mu = 1$ and $\mu > 1$. The invariant set of Φ is $[0, 1]$ and this family of mappings has all the properties delineated earlier. We let V_0 be a refinable set containing the *endpoints* of $[0, 1]$, that is, $V_0 := \{v_0, v_1, \ldots, v_{k-1}\}$, where $0 = v_0 < v_1 < \cdots < v_{k-2} < v_{k-1} = 1$. Since the endpoints are the fixed points of the first and last mappings, respectively, $W_1 := V_1 \setminus V_0 \subset (0, 1)$.

We let $\mathbb{F}_n$ be the space of piecewise polynomials of degree $\leq km - 1$ in $W^{m,p}[0, 1]$ with knots at $\{\phi_{\mathbf{e}_n}(\{0, 1\}) : \mathbf{e}_n \in \mathbb{Z}_\mu^n\}$. In particular, $\mathbb{F}_0$ is the space of polynomials of degree $\leq km - 1$ on $[0, 1]$ and

$$\dim \mathbb{F}_n = \mu^n(k-1)m + m, \ \ \mathbb{F}_n \subseteq \mathbb{F}_{n+1}, \ \ n \in \mathbb{N}_0.$$

This sequence of spaces is dense in $W^{m,p}[0, 1]$, for $1 \leq p < \infty$.

We construct multiscale bases for these spaces $\mathbb{F}_n$ using the solution of the Hermite interpolation problem

$$p^{(i)}(\phi_{\mathbf{e}_n}(v)) = f^{(i)}(\phi_{\mathbf{e}_n}(v)), \ \ \mathbf{e}_n \in \mathbb{Z}_\mu^n, \ \ v \in V_0, \ \ i \in \mathbb{Z}_m, \tag{4.75}$$

which has a unique solution $p \in \mathbb{F}_n$ for any $f \in W^{m,p}[0, 1]$. Hence, in this special case, the refinability of V_0 insures that $(\mathbb{F}_n, \mathbb{Z}_m, V_n)$ is admissible.

Let $\mathbb{G}_{n+1}$ be the space of all functions in $\mathbb{F}_{n+1}$ such that $g^{(i)}$, $i \in \mathbb{Z}_m$, vanish at all points in V_n. A basis for the space $\mathbb{G}_n$ can be constructed recursively, starting with interpolating bases for $\mathbb{F}_0$ and $\mathbb{F}_1$. To this end, for $r := (i, j) \in \mathbb{Z}_m \times \mathbb{Z}_k$ we let $p_r \in \mathbb{F}_0$ satisfy the conditions

$$p_r^{(i')}(v_{j'}) = \delta_{rr'}, \ \ r' := (i', j') \in \mathbb{Z}_m \times \mathbb{Z}_k.$$

Then the set of functions

$$F_0 = \{p_r : r \in \mathbb{Z}_m \times \mathbb{Z}_k\}$$

constitutes a Hermite basis for the space $\mathbb{F}_0$ relative to $\mathbb{Z}_m \times \mathbb{Z}_k$. To construct a basis for $\mathbb{F}_1$, we recall the linear operators $\mathcal{T}_\epsilon$, which in this case have the special forms

$$(\mathcal{T}_\epsilon f)(t) = f\left(\frac{t - t_\epsilon}{t_{\epsilon+1} - t_\epsilon}\right) \chi_{[t_\epsilon, t_{\epsilon+1}]}(t), \ \ t \in [0, 1].$$

We remark that in general the range of operators $\mathcal{T}_\epsilon$ is not contained in the Sobolev space $W^{m,p}[0,1]$. However, we do have the following fact whose statement uses for $1 \le p < \infty$ the spaces

$$W^{m,p}_{0,-}[0,1] := \{f \in W^{m,p}[0,1] : f^{(i)}(0) = 0,\ i \in \mathbb{Z}_{m+1}\},$$

$$W^{m,p}_{0,+}[0,1] := \{f \in W^{m,p}[0,1] : f^{(i)}(1) = 0,\ i \in \mathbb{Z}_{m+1}\}$$

and

$$W^{m,p}_{0}[0,1] := W^{m,p}_{0,-}[0,1] \cap W^{m,p}_{0,+}[0,1].$$

Lemma 4.39 *The following inclusions hold:*

$$\mathcal{T}_{\mu-1}(W^{m,p}_{0,-}[0,1]) \subseteq W^{m,p}_{0,-}[0,1],$$

$$\mathcal{T}_{0}(W^{m,p}_{0,+}[0,1]) \subseteq W^{m,p}_{0,+}[0,1]$$

and

$$\mathcal{T}_\epsilon(W^{m,p}_{0}[0,1]) \subseteq W^{m,p}_{0}[0,1],\quad \epsilon \in \mathbb{Z}_\mu.$$

Next, we show how to use the functions p_r in F_0 and the operators $\mathcal{T}_\epsilon$ to construct a basis of F_1. The operators $\mathcal{T}_\epsilon$ may introduce discontinuities when applied to a function in F_0, thereby leading to an unacceptable basis for F_1. Lemma 4.39 reveals exactly what happens when we apply $\mathcal{T}_\epsilon$ to p_r. Using the operators $\mathcal{T}_\epsilon$, $\epsilon \in \mathbb{Z}_\mu$, we define for $r := (i,\ell) \in \mathbb{Z}_m \times \mathbb{Z}_k$ and $j = \epsilon(k-1)+\ell$,

$$q_{ij} = (t_{\epsilon+1} - t_\epsilon)^i \mathcal{T}_\epsilon p_r,$$

when $\epsilon = 0, \ell = 0$ or $\epsilon \in \mathbb{Z}_\mu$, $\ell - 1 \in \mathbb{Z}_{k-2}$ or $\epsilon = \mu - 1, \ell = k-1$ and

$$q_{ij} = (t_\epsilon - t_{\epsilon-1})^i \mathcal{T}_{\epsilon-1} p_{(i,k-1)} + (t_{\epsilon+1} - t_\epsilon)^i \mathcal{T}_\epsilon p_{(i,0)},$$

when $\epsilon - 1 \in \mathbb{Z}_{\mu-1}$, $\ell = 0$. Set

$$F_1 := \{q_{ij} : i \in \mathbb{Z}_m,\ j \in \mathbb{Z}_{\mu(k-1)+1}\}.$$

In the next lemma, we state some properties of this set of functions. To this end, we number the points in V_1 according to the scheme

$$z_j := \phi_\epsilon(v_\ell),\ \ j = \epsilon(k-1)+\ell,\ \ \epsilon \in \mathbb{Z}_\mu,\ \ \ell - 1 \in \mathbb{Z}_{k-1}.$$

Lemma 4.40 *The set F_1 forms a basis for $\mathbb{F}_1$ such that*

$$q^{(i')}_{ij}(z_{j'}) = \delta_{rr'},\quad r := (i,j),\quad r' := (i',j') \in \mathbb{Z}_m \times \mathbb{Z}_{\mu(k-1)+1}. \tag{4.76}$$

Proof Lemma 4.39 insures that these functions are in $W^{m,p}[0,1]$. Hence, it is clear that they are elements in $\mathbb{F}_1$. Moreover, a direct verification leads to the conclusion that these functions satisfy the conditions (4.76). □

For a general n, we follow the same process as described earlier to construct a basis for space $\mathbb{F}_n$ from the basis for $\mathbb{F}_{n-1}$. At each level, it requires a procedure of eliminating discontinuities introduced by the operators $\mathcal{T}_\epsilon$. However, we do not construct the basis for $\mathbb{F}_n$ directly for $n > 1$. Instead, we turn our attention to the construction of bases for the complement spaces $\mathbb{G}_1, \mathbb{G}_2, \ldots, \mathbb{G}_n$. Surprisingly, the next theorem shows that we can choose $(\mu - 1)(k - 1)m$ functions from the set F_1 to form a basis for $\mathbb{G}_1$ and recursively generate bases for the spaces $\mathbb{G}_n$ from this basis of $\mathbb{G}_1$ by applying the operators $\mathcal{T}_\epsilon$. We see that the construction of bases for $\mathbb{G}_n$ for $n \geq 2$ does not require the process of eliminating discontinuities, which is required for the direct construction of bases for $\mathbb{F}_n$ because $\mathbb{G}_1 \subseteq W_0^{m,p}[0,1]$.

Theorem 4.41 *If V_0 is refinable relative to the affine mappings Φ and*

$$G_1 := \{q_{ij} : j \in \mathbb{Z}_{\mu(k-1)+1}(W_1), i \in \mathbb{Z}_m\},$$

then

$$\mathbb{G}_1 = \text{span}\, G_1.$$

Moreover, if

$$G_{n+1} := \{a_{e_n}^i \mathcal{T}_{e_n} q_{ij} : q_{ij} \in G_1,\ i \in \mathbb{Z}_m,\ e_n \in \mathbb{Z}_\mu^n\}$$

where

$$a_{e_n} = \prod_{i \in \mathbb{Z}_n} (t_{\epsilon+1} - t_\epsilon),$$

then

$$\mathbb{G}_{n+1} = \text{span}\, G_{n+1},\quad n \in \mathbb{N}.$$

Proof Since the cardinality of W_1 equals the dimension of the space $\mathbb{G}_1$, the set G_1 consists of $(\mu - 1)(k - 1)$ linearly independent functions. It remains to show that $G_1 \subset \mathbb{G}_1$. To this end, we prove that the functions in G_1 vanish at all points in V_0. Let $q_{ij} \in G_1$. By (4.76), we obtain that $q_{ij}^{(i)}(z_j) = 1$ for $z_j \in W_1$ and $q_{ij}^{(i')}(z_{j'}) = 0$ for $(i', j') \neq (i, j)$. Since $V_0 \subset V_1$ and $z_j \notin V_0$, we conclude that $q_{ij}^{(i)}$, $i \in \mathbb{Z}_m$ vanish at all points in V_0, and thus $q_{ij} \in \mathbb{G}_1$.

We now prove the second statement of the theorem. It follows from Lemma 4.39 that the functions $a_{\mathbf{e}_n}^i \mathcal{T}_{\mathbf{e}_n} q_{ij} \in W_0^{m,p}[0,1]$, since $q_{ij} \in W_0^{m,p}[0,1]$. In addition, we have that

$$\begin{aligned}(a_{\mathbf{e}_n}^i \mathcal{T}_{\mathbf{e}_n} q_{ij})^{(i')}(\phi_{\mathbf{e}'_n}(z_{j'})) &= a_{\mathbf{e}_n}^{i-i'} \mathcal{T}_{\mathbf{e}_n} q_{ij}^{(i')}(\phi_{\mathbf{e}'_n}(z_{j'}))\\ &= a_{\mathbf{e}_n}^{i-i'} \delta_{\mathbf{e}_n,\mathbf{e}'_n} q_{ij}^{(i')}(z_{j'})\\ &= \delta_{\mathbf{e}_n,\mathbf{e}'_n} \delta_{ii'} \delta_{jj'}.\end{aligned}$$

This equation implies that these functions are linearly independent in $\mathbb{G}_{n+1}$.

Next, we observe that the cardinality of the set G_{n+1} is equal to the dimension of $\mathbb{G}_{n+1}$. Consequently, we conclude that $\mathbb{G}_{n+1} = \operatorname{span} G_{n+1}$. □

Since

$$\mathbb{F}_n = \mathbb{F}_0 \oplus \mathbb{G}_1 \oplus \cdots \oplus \mathbb{G}_n$$

and the sequence of spaces $\mathbb{F}_n$, $n \in \mathbb{N}_0$, is dense in the space $W^{m,p}[0,1]$ for $1 \leq p < \infty$, we obtain the following result.

Theorem 4.42 *The equation*

$$\overline{\mathbb{F}_0 \oplus \left(\bigoplus_{n\in\mathbb{N}} \mathbb{G}_n \right)} = W^{m,p}[0,1]$$

holds for $1 \leq p < \infty$.

In the finite element method, the space $W_0^{m,p}[0,1]$ has a special importance. For this reason, we define

$$\mathbb{F}_0^0 := \{f \in \mathbb{F}_0 : f^{(i)}(0) = f^{(i)}(1) = 0,\ i \in \mathbb{Z}_m\}$$

and observe that $\dim \mathbb{F}_0^0 = (k-2)m$, where $\mathbb{F}_0^0 = \operatorname{span} F_0^0$ and

$$F_0^0 := \{p_j^i : i \in \mathbb{Z}_m,\ j-1 \in \mathbb{Z}_{k-1}\}.$$

Corollary 4.43 *The equation*

$$\overline{\mathbb{F}_0^0 \oplus \left(\bigoplus_{n\in\mathbb{N}} \mathbb{G}_n \right)} = W_0^{m,p}[0,1]$$

holds for $1 \leq p < \infty$.

4.6 Bibliographical remarks

The material presented in this chapter regarding the multiscale bases was mainly taken from [65, 196, 200, 201]. The construction of orthogonal wavelets on invariant sets was originally introduced in [200]. Then, the construction was extended in [201] to a general bounded domain and bi-orthogonal wavelets. In particular, the construction of the initial wavelet space was formulated in [201] in terms of a general solution of a matrix completion problem. Later, [65] gave a construction of interpolating wavelets on invariant sets. The concept of a refinable set relative to a family of contractive mappings on a metric space which define the invariant set is introduced in [65], and a recursive structure was explored in the paper for multiscale function representation and approximation constructed by interpolation on invariant sets. For the notion of invariant sets, the reader is referred to [148]. The material about multiscale partitions of a multidimensional simplex was originally developed in [74]. Paper [198] constructed refinable sets that admit a unique Lagrange interpolating polynomial (see also [199]). The description for multiscale Hermite interpolation in Section 4.5.2 follows [66]. Moreover, [69] presented a construction of multiscale basis functions and the corresponding multiscale collocation functionals, both having vanishing moments (see also Section 7.1).

For wavelets on an unbounded domain, the reader is referred to [43, 48, 82, 84, 92, 93, 97, 98, 100, 101, 232] and the references cited therein.

5
Multiscale Galerkin methods

The main purpose of this chapter is to present fast multiscale Galerkin methods for solving the second-kind Fredholm integral equations

$$u - \mathcal{K}u = f, \tag{5.1}$$

defined on a compact domain Ω in $\mathbb{R}^d$. The classical Galerkin method using the piecewise polynomial applied to equation (5.1) leads to a linear system of equations with a dense coefficient matrix. Hence, the numerical solution of this equation is computationally costly. The multiscale Galerkin method, to be described in this chapter, makes use of the multiscale feature and the vanishing moment property of the multiscale piecewise polynomial basis and results in a linear system with a numerically sparse coefficient matrix. As a result, fast algorithms may be designed based on a truncation of the coefficient matrix. Specifically, the multiscale Galerkin method uses the L^2-orthogonal projection for a discretization principle with the multiscale basis functions whose construction is described in Chapter 4.

The fast multiscale Galerkin method is based on a matrix compression scheme. We show that the matrix compression scheme preserves almost the optimal convergence order of the standard Galerkin method, while it reduces the number of nonzero entries of its coefficient matrix from $\mathcal{O}(N^2)$ to $\mathcal{O}(N \log^{\sigma} N)$, where N is the size of the matrix and σ may be 1 or 2. We also prove that the condition number of the compressed matrix is uniformly bounded, independent of the size of the matrix.

The kernels of the integral operators in which we are interested in this chapter are weakly singular or smooth. We present theoretical results for the weakly singular case in detail and only give comments for the smooth case.

5.1 The multiscale Galerkin method

In this section we present the multiscale Galerkin method for solving equation (5.1). For this purpose, we first describe the properties of multiscale bases, required necessarily for developing the multiscale Galerkin method. These properties are satisfied for the multiscale bases constructed in Chapter 3. However, the multiscale bases constructed in Chapter 3 have other properties that are not essential for developing the multiscale Galerkin method.

5.1.1 Multiscale bases

The multiscale basis requires a multiscale partition of the domain Ω. We assume that there is a family of partitions $\{\Omega_i : i \in \mathbb{N}_0\}$ such that for each scale $i \in \mathbb{N}_0$, $\Omega_i := \{\Omega_{ij}, j \in \mathbb{Z}_{e(i)}\}$, where $e(i)$ denotes the cardinality of Ω_i, has the properties

(1) $\bigcup_{j\in\mathbb{Z}_{e(i)}} \Omega_{ij} = \Omega$,

(2) $\text{meas}(\Omega_{ij} \bigcap \Omega_{ij'}) = 0, \;\; j,j' \in \mathbb{Z}_{e(i)}, j \neq j'$,

(3) $\text{meas}(\Omega_{ij}) \sim d_i^d$ for all $j \in \mathbb{Z}_{e(i)}$,

where $d_i := \max\{d(\Omega_{ij}) : j \in \mathbb{Z}_{e(i)}\}$. Here the notation $a_i \sim b_i$ for $i \in \mathbb{N}_0$ means that there are positive constants c_1 and c_2 such that $c_1 a_i \leq b_i \leq c_2 a_i$ for all $i \in \mathbb{N}_0$.

In addition, we assume that

(4) the sets $\Omega_{ij}, j \in \mathbb{Z}_{e(i)}$ are star-shaped.

We remark that a set $A \subset \mathbb{R}^d$ is called a *star-shaped set* if it contains a point for which the line segment connecting this point and any other point in the set is contained in the set. Such a point is called a *center* of the set.

We further suppose that there is a nested sequence of finite-dimensional subspaces $\{\mathbb{X}_n : n \in \mathbb{N}_0\}$ of $\mathbb{X}$, that is,

$$\mathbb{X}_{n-1} \subset \mathbb{X}_n, \; n \in \mathbb{N}.$$

Thus, for each $n \in \mathbb{N}_0$, a subspace $\mathbb{W}_n \subset \mathbb{X}_n$ can be defined such that $\mathbb{X}_n$ is an orthogonal direct sum of $\mathbb{X}_{n-1}$ and $\mathbb{W}_n$. Moreover, we assume that $\mathbb{X}_n$, $n \in \mathbb{N}_0$ is ultimately dense in $L^2(\Omega)$ in the sense that

$$\overline{\bigcup_{n\in\mathbb{N}_0} \mathbb{X}_n} = L^2(\Omega).$$

We then have an orthogonal decomposition of space $L^2(\Omega)$:

$$L^2(\Omega) = \overline{\bigoplus_{n\in\mathbb{N}_0}^{\perp} \mathbb{W}_n}, \tag{5.2}$$

where we have used the notation $\mathbb{W}_0 := \mathbb{X}_0$. Set $w(n) := \dim \mathbb{W}_n$ and $s(n) := \dim \mathbb{X}_n$, for $n \in \mathbb{N}_0$. It follows that

$$s(n) = \sum_{i \in \mathbb{Z}_{n+1}} w(i).$$

For each $n \in \mathbb{N}_0$, we introduce an index set $\mathbb{U}_n := \{(i,j) : i \in \mathbb{Z}_{n+1}, j \in \mathbb{Z}_{w(i)}\}$. We also use the notation $\mathbb{U} := \{(i,j) : i \in \mathbb{N}_0, j \in \mathbb{Z}_{w(i)}\}$. We assume that there is a family of basis functions $\{w_{ij} : (i,j) \in \mathbb{U}\} \subset \mathbb{X}$ such that

$$\mathbb{W}_n = \operatorname{span}\{w_{nj} : j \in \mathbb{Z}_{w(n)}\}, \quad n \in \mathbb{N}_0.$$

Thus,

$$\mathbb{X}_n = \operatorname{span}\{w_{ij} : (i,j) \in \mathbb{U}_n\}.$$

We require that the following multiscale properties hold for the partitions and the basis functions.

(I) There is a positive integer $\mu > 1$ such that for $i \in \mathbb{N}_0$,

$$d_i \sim \mu^{-i/d}, \quad w(i) \sim \mu^i \text{ and } s(i) \sim \mu^i.$$

(II) There exist positive integers ρ and γ such that for every $i > \gamma$ and $j \in \mathbb{Z}_{w(i)}$ written in the form $j = \nu\rho + s$, where $s \in \mathbb{Z}_\rho$ and $\nu \in \mathbb{N}_0$,

$$w_{ij}(t) = 0, \quad t \notin \Omega_{i-\gamma,\nu}.$$

Setting $S_{ij} := \Omega_{i-\gamma,\nu}$, we see that the support of w_{ij} is contained in S_{ij}. It can easily be verified that

$$d_i \sim \max\{d(S_{ij}) : j \in \mathbb{Z}_{e(i)}\}.$$

Because of this property, we shall not distinguish d_i from the right-hand side of the above equation.

(III) For any $(i,j) \in \mathbb{U}$ with $i \geq 1$, and polynomial p of total degree less than a positive integer k,

$$(p, w_{ij}) = 0,$$

where $(\cdot,\cdot)$ denotes the L^2-inner product.

(IV) There is a constant θ_0 such that for any $(i,j) \in \mathbb{U}$,

$$\|w_{ij}\| = 1 \quad \text{and} \quad \|w_{ij}\|_\infty \leq \theta_0 \mu^{i/2},$$

where $\|\cdot\|$ and $\|\cdot\|_\infty$ denote the L^2-norm and the L^∞-norm, respectively.

(V) There is a positive constant θ_1 such that for all $n \in \mathbb{N}_0, v = \sum_{(i,j)\in\mathbb{U}_n} v_{ij}w_{ij}$,

$$\|\mathbf{E}_n\mathbf{v}\|_2 \sim \|\mathbf{v}\|_2 \text{ and } \|\mathbf{v}\|_2 \leq \theta_1\|v\|,$$

where $\mathbf{v} := [v_{ij} : (i,j) \in \mathbb{U}_n]$, $\mathbf{E}_n = \left[(w_{i'j'}, w_{ij}) : (i',j'), (i,j) \in \mathbb{U}_n\right]$ and the notation $\|\mathbf{x}\|_p$, $1 \leq p \leq \infty$, for a vector $\mathbf{x} := [x_j : j \in \mathbb{Z}_n]$ denotes the ℓ_p-norm defined by

$$\|\mathbf{x}\|_p := \begin{cases} \left(\sum_{j\in\mathbb{Z}_n} |x_j|^p\right)^{1/p}, & 1 \leq p < \infty, \\ \max\{|x_j| : j \in \mathbb{Z}_n\}, & p = \infty. \end{cases}$$

(VI) If $\mathcal{P}_n$ is the orthogonal projection from $\mathbb{X}$ onto $\mathbb{X}_n$, then there exists a positive constant c such that for any $u \in H^k(\Omega)$,

$$\|u - \mathcal{P}_n u\| \leq c d_n^k \|u\|_{H^k}.$$

All of these properties are fulfilled by the multiscale basis functions constructed in Chapter 3. In general, the matrix $\mathbf{E}_n$ is a block diagonal matrix. Moreover, if w_{ij}, $(i,j) \in \mathbb{U}$, is a sequence of orthonormal basis functions, then $\mathbf{E}_n$ is the identity matrix and property (V) holds with $\|\mathbf{v}\|_2 = \|v\|$. Furthermore, if $\mathbb{X}_n$, $n \in \mathbb{N}_0$, are spaces of piecewise polynomials of total degree less than k, then the vanishing moment property (III) and the approximation property (VI) hold naturally.

5.1.2 Formulation of the multiscale Galerkin method

As we have discussed in Section 3.1, the Galerkin method for equation (5.1) is to find $u_n \in \mathbb{X}_n$ that satisfies the operator equation

$$(\mathcal{I} - \mathcal{K}_n)u_n = \mathcal{P}_n f, \tag{5.3}$$

where $\mathcal{K}_n := \mathcal{P}_n\mathcal{K}|_{\mathbb{X}_n}$. It is clear that the following theoretical results hold for the Galerkin method (see Section 3.3).

Theorem 5.1 *Let $\mathcal{K}$ be a linear compact operator not having one as its eigenvalue. Then, there exists $N > 0$ such that for all $n \geq N$, the Galerkin scheme (5.3) has a unique solution $u_n \in \mathbb{X}_n$, and there is a constant $c > 0$ such that for all $n \geq N$,*

$$\|(\mathcal{I} - \mathcal{K}_n)^{-1}\| \leq c.$$

Moreover, if the solution u of equation (5.1) satisfies $u \in H^k(\Omega)$, then there exists a positive constant c such that for all $n \geq N$,

$$\|u - u_n\| \leq c\mu^{-kn/d}\|u\|_{H^k}.$$

Using the multiscale bases for spaces $\mathbb{X}_n$ described in the last section, the above Galerkin method (5.3) is to seek

$$u_n := \sum_{(i,j)\in\mathbb{U}_n} u_{ij} w_{ij} \in \mathbb{X}_n,$$

such that

$$\sum_{(i,j)\in\mathbb{U}_n} u_{ij}(w_{i'j'}, w_{ij} - \mathcal{K}w_{ij}) = (w_{i'j'}, f), \quad (i',j') \in \mathbb{U}_n. \tag{5.4}$$

Because the multiscale basis is used, in order to distinguish it from the traditional Galerkin method, we call (5.4) the *multiscale* Galerkin method.

To write (5.4) in a matrix form, we use the lexicographic ordering on $\mathbb{Z}_{n+1} \times \mathbb{Z}_{n+1}$ and define the matrix

$$\mathbf{K}_n := [(w_{i'j'}, \mathcal{K}w_{ij}) : (i',j'), (i,j) \in \mathbb{U}_n]$$

and vectors

$$\mathbf{f}_n := [(w_{i'j'}, f) : (i',j') \in \mathbb{U}_n], \quad \mathbf{u}_n := [u_{ij} : (i,j) \in \mathbb{U}_n].$$

Note that these vectors have length $s(n)$. With these notations, equation (5.4) takes the equivalent matrix form

$$(\mathbf{E}_n - \mathbf{K}_n)\mathbf{u}_n = \mathbf{f}_n. \tag{5.5}$$

Even though the coefficient matrix $\mathbf{K}_n$ is a full matrix, it differs significantly from the matrix $\mathbf{K}_n$ in Section 3.3.1. The use of multiscale basis functions makes the matrix $\mathbf{K}_n$ *numerically sparse*. By the numerically sparse matrix, we mean a matrix with significantly large number of entries being very small in magnitude. This forms a base for developing the fast multiscale Galerkin method. We illustrate this observation by the following example.

Example 5.2 Consider $\Omega := [0, 1]$ and the compact integral operator with kernel

$$K(s,t) := \log|s - t|, \quad s, t \in [0, 1].$$

We choose $\mathbb{X}_n$ as the piecewise linear functions with knots $j/2^n$, $j \in \mathbb{N}_{2^n-1}$. In this case, $k = 2$. The Galerkin matrix of this operator with respect to the Lagrange interpolating basis is illustrated in Figure 5.1 with $n = 6$.

We can see that generating the full matrix and then solving the corresponding linear system requires large computational cost when its order is large. The idea to overcome this computational deficiency is to change the basis for the piecewise polynomial space so that the projection of the integral operator $\mathcal{K}$ to the space has a numerically sparse Galerkin matrix under the new basis.

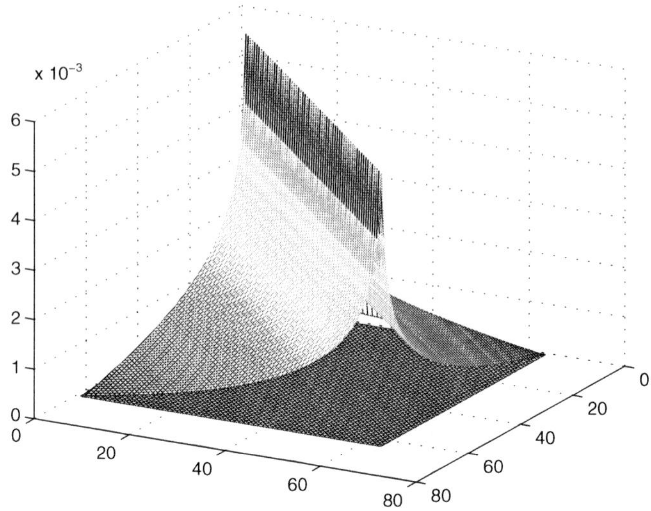

Figure 5.1 The Galerkin matrix with respect to the piecewise linear polynomial basis.

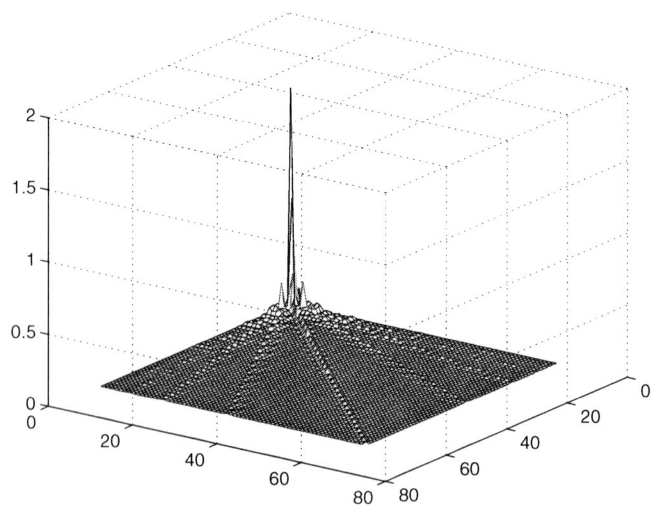

Figure 5.2 The Galerkin matrix with respect to the piecewise linear polynomial multiscale basis.

The Galerkin matrix of this operator with respect to the piecewise linear polynomial multiscale basis described in the last section is illustrated in Figure 5.2 with $n = 6$. It can be seen that the absolute value of the entries off the diagonals of the blocks corresponding to different scales of spaces is very small. We can set the entries small in magnitude to zero and obtain a sparse matrix, which leads to a fast Galerkin method. We present this fast method and its analysis in the next several sections.

5.2 The fast multiscale Galerkin method

In this section, we develop the fast multiscale Galerkin method based on a matrix *truncation strategy*. We consider two classes of kernels. Class one consists of kernels having weak singularity along the diagonal. Specifically, for $\sigma \in [0, d)$ and integer $k \geq 1$, we define $\mathbb{S}_{\sigma,k}$. We say $K \in \mathbb{S}_{\sigma,k}$ if for $s, t \in \Omega$, $s \neq t$, K has continuous partial derivatives $D_s^\alpha D_t^\beta K(s,t)$ for $|\alpha| \leq k$, $|\beta| \leq k$, and there exists a constant $c > 0$ such that for $|\alpha| = |\beta| = k$,

$$|D_s^\alpha D_t^\beta K(s,t)| \leq \frac{c}{|s-t|^{\sigma+2k}}, \quad s, t \in \Omega. \tag{5.6}$$

Related to the kernel on the right-hand side of (5.6) we remark that when $\sigma = 0$, the function $1/x^\sigma$ is understood as $\log x$. Class two consists of kernels $K \in C^k(\Omega \times \Omega)$. Kernels in this class are smooth.

Set

$$K_{i'j',ij} := (w_{i'j'}, \mathcal{K} w_{ij}), \quad (i,j), (i',j') \in \mathbb{U}_n$$

and observe that $K_{i'j',ij}$ are entries of matrix $\mathbf{K}_n$. In the next lemma, we estimate the bound of $K_{i'j',ij}$.

Lemma 5.3 *Suppose that conditions (I)–(IV) hold.*

(1) If $K \in \mathbb{S}_{\sigma,k}$ for some $\sigma \in [0, d)$ and a positive integer k, and there is a constant $r > 1$ such that

$$\operatorname{dist}(S_{ij}, S_{i'j'}) \geq \max\{r d_i, r d_{i'}\},$$

then there exists a positive constant c such that for all $(i,j), (i',j') \in \mathbb{U}$, $i, i' \in \mathbb{N}$,

$$|K_{i'j',ij}| \leq c(d_i d_{i'})^{k-\frac{d}{2}}$$
$$\min\left\{ d_{i'}^d \max_{s \in S_{i'j'}} \int_{S_{ij}} \frac{\mathrm{d}t}{|s-t|^{2k+\sigma}}, \; d_i^d \max_{t \in S_{ij}} \int_{S_{i'j'}} \frac{\mathrm{d}s}{|s-t|^{2k+\sigma}} \right\}.$$

(2) If $K \in C^k(\Omega \times \Omega)$, then there exists a positive constant c such that for all $i, i' \in \mathbb{N}$,

$$|K_{i'j',ij}| \leq c d_{i'}^{k+d/2} d_i^{k+d/2}.$$

Proof We present a proof for part (1) only since the proof for part (2) is similar. This is done by using the Taylor theorem. By hypothesis, for each $(i,j) \in \mathbb{U}$, the set S_{ij} is star-shaped. Let s_0 and t_0 be centers of the sets $S_{i'j'}$ and S_{ij}, respectively. It follows from the Taylor theorem that

$$K(s,t) = p(s,t) + q(s,t) + \sum_{|\alpha|=k} \sum_{|\beta|=k} \frac{(s-s_0)^\alpha (t-t_0)^\beta}{\alpha!\beta!} r_{\alpha\beta}(s,t),$$

where $p(s,\cdot)$ and $q(\cdot,t)$ are polynomials of total degree less than k in t and in s, respectively, and

$$r_{\alpha,\beta}(s,t) := \int_0^1 \int_0^1 D_s^\alpha D_t^\beta K(s_0 + \theta_1(s-s_0), t_0 + \theta_2(t-t_0))(1-\theta_1)^{k-1}(1-\theta_2)^{k-1} \mathrm{d}\theta_1 \mathrm{d}\theta_2.$$

By conditions (II) and (III) we have that

$$K_{i'j',ij} = \sum_{|\alpha|=k} \sum_{|\beta|=k} \int_{S_{i'j'}} \int_{S_{ij}} \frac{(s-s_0)^\alpha (t-t_0)^\beta}{\alpha!\beta!} r_{\alpha,\beta}(s,t) w_{i'j'}(s) w_{ij}(t) \mathrm{d}s\mathrm{d}t.$$

This, with conditions (I) and (IV), yields the bound

$$|K_{i'j',ij}| \le c d_i^{k-\frac{d}{2}} d_{i'}^{k-\frac{d}{2}} \sum_{|\alpha|=k} \sum_{|\beta|=k} \frac{1}{\alpha!\beta!} \int_{S_{i'j'}} \int_{S_{ij}} |r_{\alpha,\beta}(s,t)| \mathrm{d}s\mathrm{d}t. \tag{5.7}$$

We conclude from the mean-value theorem and the hypothesis $K \in \mathbb{S}_{\sigma,k}$ that there exist $s' \in S_{i'j'}$ and $t' \in S_{ij}$ such that

$$|r_{\alpha,\beta}(s,t)| = k^{-2} |D_s^\alpha D_t^\beta K(s',t')| \le \frac{c}{|s'-t'|^{2k+\sigma}}.$$

The assumption of this lemma yields

$$|s'-t'| \ge |s'-t| - d_i \ge (1-r^{-1})|s'-t|.$$

Thus, for a new constant c,

$$|r_{\alpha,\beta}(s,t)| \le \frac{c}{|s'-t|^{2k+\sigma}}.$$

This inequality, with (5.7) and the relationship $\text{meas}(S_{i'j'}) \sim d_{i'}^d$, leads to the desired estimate. □

The above lemma shows that most of the entries are so small that they can be neglected without affecting the overall accuracy of the approximation scheme. This observation leads to a matrix truncation strategy. To present it, we partition matrix $\mathbf{K}_n$ into a block matrix

$$\mathbf{K}_n = [\mathbf{K}_{i'i} : i', i \in \mathbb{Z}_{n+1}] \text{ with } \mathbf{K}_{i'i} = [K_{i'j',ij} : j' \in \mathbb{Z}_{w(i')}, j \in \mathbb{Z}_{w(i)}].$$

For each $i', i \in \mathbb{Z}_{n+1}$, we choose a truncation parameter $\delta^n_{i'i}$, which will be specified later. We define for the weakly singular case

$$\tilde{K}_{i'j',ij} = \begin{cases} K_{i'j',ij}, & \operatorname{dist}(S_{i'j'}, S_{ij}) \le \delta^n_{i'i}, \\ 0, & \text{otherwise} \end{cases} \tag{5.8}$$

and obtain a truncation matrix

$$\tilde{\mathbf{K}}_n = [\tilde{\mathbf{K}}_{i'i} : i', i \in \mathbb{Z}_{n+1}],$$

where

$$\tilde{\mathbf{K}}_{i'i} := \mathbf{K}(\delta^n_{i'i})_{i'i} = [\tilde{K}_{i'j',ij} : j' \in \mathbb{Z}_{w(i')}, j \in \mathbb{Z}_{w(i)}].$$

Likewise, for the smooth case we define for each $i', i \in \mathbb{N}$

$$\tilde{\mathbf{K}}_{i'i} = \begin{cases} \mathbf{K}_{i'i}, & i + i' \le n, \\ 0, & \text{otherwise.} \end{cases} \tag{5.9}$$

This truncation strategy leads to the fast multiscale Galerkin method, which is to find $\tilde{\mathbf{u}}_n = [\tilde{u}_{ij} : (i,j) \in \mathbb{U}_n] \in \mathbb{R}^{s(n)}$ such that

$$(\mathbf{E}_n - \tilde{\mathbf{K}}_n)\tilde{\mathbf{u}}_n = \mathbf{f}_n. \tag{5.10}$$

Example 5.4 We again consider the compact integral operator with the kernel

$$K(s,t) := \log |s - t|, \quad s, t \in [0, 1]$$

and choose $\mathbb{X}_n$ as the piecewise linear functions ($k = 2$) with knots $j/2^n$, $j \in \mathbb{N}_{2^n - 1}$. The truncated Galerkin matrix of this operator with respect to the piecewise linear polynomial multiscale basis is illustrated in Figure 5.3 with $n = 6$.

The analysis of the fast multiscale Galerkin method requires the availability of an operator form of equation (5.10). To this end, we first introduce the concept of the matrix representation of an operator.

Definition 5.5 The matrix $\mathbf{B}$ is said to be the matrix representation of the linear operator $\mathcal{A}$ relative to the basis $\Phi := \{\phi_j : j \in \mathbb{N}_n\}$ if

$$\Phi^T \mathbf{B} = \mathcal{A}(\Phi^T).$$

Proposition 5.6 *The matrix representation of the operator $\mathcal{K}$ relative to the basis $W_n := \{w_{ij} : (i,j) \in \mathbb{U}_n\}$ is $\mathbf{B}_n := \boldsymbol{E}_n^{-1}\mathbf{K}_n$.*

Proof Let $\mathbf{B}_n := [b_{i'j',ij} : (i,j) \in \mathbb{U}_n]$ be the matrix representation of the operator $\mathcal{K}$ relative to the basis W_n. According to Definition 5.5, we have that

$$\mathcal{K}w_{ij} = \sum_{(k,l) \in \mathbb{U}_n} b_{kl,ij} w_{kl}, \quad \text{for all } (i,j) \in \mathbb{U}_n.$$

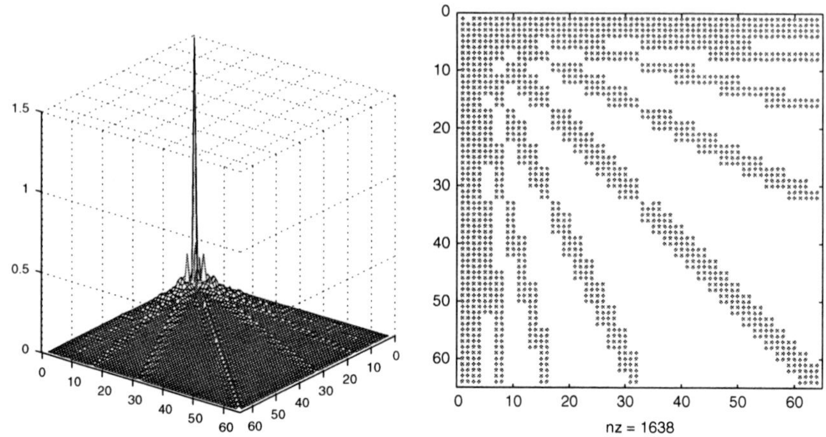

Figure 5.3 (a) The truncated Galerkin matrix with respect to the piecewise linear polynomial multiscale basis. (b) The figure of nonzero entries of the truncated matrix.

This leads to

$$(w_{i'j'}, \mathcal{K}w_{ij}) = \sum_{(k,l)\in\mathbb{U}_n} b_{kl,ij}(w_{i'j'}, w_{kl}), \quad \text{for all } (i,j),(i',j') \in \mathbb{U}_n,$$

which means $\mathbf{K}_n = \mathbf{E}_n\mathbf{B}_n$ and completes the proof. □

We next convert the linear system (5.10) to an abstract operator equation form. Let $\beta_{i'j',ij}, (i,j),(i',j') \in \mathbb{U}_n$ denote the entries of matrix $\mathbf{E}_n^{-1}\tilde{\mathbf{K}}_n\mathbf{E}_n^{-1}$ and let

$$\tilde{K}_n(s,t) = \sum_{(i,j),(i',j')\in\mathbb{U}_n} \beta_{i'j',ij} w_{i'j'}(s) w_{ij}(t).$$

We denote by $\tilde{\mathcal{K}}_n$ the integral operator defined by the kernel $\tilde{K}_n(s,t)$.

Proposition 5.7 *Solving the linear system (5.10) is equivalent to finding*

$$\tilde{u}_n = \sum_{(i,j)\in\mathbb{U}_n} \tilde{u}_{ij} w_{ij} \in \mathbb{X}_n$$

such that

$$(\mathcal{I} - \tilde{\mathcal{K}}_n)\tilde{u}_n = \mathcal{P}_n f.$$

Proof It follows for $(i,j),(i',j') \in \mathbb{U}_n$ that

$$
\begin{aligned}
(w_{i'j'}, \tilde{\mathcal{K}}_n w_{ij}) &= \left(w_{i'j'}, \int_\Omega \tilde{K}_n(\cdot, t) w_{ij}(t) \mathrm{d}t \right) \\
&= \sum_{(k,l),(k',l') \in \mathbb{U}_n} \beta_{k'l',kl} (w_{kl}, w_{ij})(w_{i'j'}, w_{k'l'}) \\
&= \sum_{(k,l),(k',l') \in \mathbb{U}_n} (\mathbf{E}_n)_{i'j',k'l'} \beta_{k'l',kl} (\mathbf{E}_n)_{kl,ij} \\
&= (\tilde{\mathbf{K}}_n)_{i'j',ij},
\end{aligned} \tag{5.11}
$$

which means

$$
\tilde{\mathbf{K}}_n = [(w_{i'j'}, \tilde{\mathcal{K}}_n w_{ij}) : (i,j),(i',j') \in \mathbb{U}_n],
$$

and leads to the desired result of this proposition. □

The analysis of the fast multiscale Galerkin method with an appropriate choice of the truncation parameters $\delta^n_{i',i}$ will be discussed in the next section.

5.3 Theoretical analysis

In this section, we analyze the fast multiscale Galerkin method. Specifically, we show that the number of nonzero entries of the truncated matrix is of linear order up to a logarithm factor, prove that the method is stable and that it gives almost optimal order of convergence. We also prove that the condition number of the truncated matrix is uniformly bounded. We consider the weakly singular case in Sections 5.3.1–5.3.3. Special results for the smooth case will be presented in the last subsection without proof.

5.3.1 Computational complexity

The computational complexity of the fast multiscale Galerkin method is measured in terms of the number of nonzero entries of the truncated matrix. In this subsection, we estimate the number of nonzero entries of matrix $\tilde{\mathbf{K}}_n$. For a matrix $\mathbf{A}$, we denote by $\mathcal{N}(\mathbf{A})$ the number of its nonzero entries.

Lemma 5.8 *If conditions (I) and (II) hold, then there exists a constant $c > 0$ such that for all $i', i \in \mathbb{N}_0$ and for all $n \in \mathbb{N}$,*

$$
\mathcal{N}(\tilde{\boldsymbol{K}}_{i'i}) \leq c\mu^{i+i'} \left(d_i^d + d_{i'}^d + (\delta^n_{i'i})^d \right).
$$

Proof For fixed i, i', j' and an arbitrarily fixed point s_0 in $S_{i',j'}$, we let

$$S(i, i', j') := \{s \in \mathbb{R}^d : |s - s_0| \leq d_i + d_{i'} + \delta^n_{i'i}\}.$$

If $\tilde{K}_{i'j',ij} \neq 0$, then $\operatorname{dist}(S_{i'j'}, S_{ij}) \leq \delta^n_{i'i}$. Thus, $S_{ij} \subset S(i, i', j')$. Let $\mathcal{N}_{i,i',j'}$ denote the number of indices (i, j) such that S_{ij} is contained in $S(i, i', j')$. Property (3) of the partition Ω_i and condition (I) imply that there exists a constant $c > 0$ such that

$$\mathcal{N}_{i,i',j'} \leq \frac{\operatorname{meas}(S(i, i', j'))}{\min\{\operatorname{meas}(S_{ij}) : S_{ij} \subset S(i, i', j')\}} \leq c\mu^i (d_i + d_{i'} + \delta^n_{i',i})^d.$$

It follows from condition (II) that the number of functions w_{ij} having supports contained in S_{ij} is bounded by ρ. Since $w(i') \sim \mu^{i'}$,

$$\mathcal{N}(\tilde{\mathbf{K}}_{i'i}) \leq \rho \sum_{j' \in \mathbb{Z}_{w(i')}} \mathcal{N}_{i,i',j'} \leq c\mu^{i+i'} (d_i + d_{i'} + \delta^n_{i'i})^d,$$

proving the desired result. □

To continue estimating the number of nonzero entries of matrix $\tilde{\mathbf{K}}_n$, we now specify choices of the truncation parameters $\delta^n_{i'i}$. Specifically, for each $i, i' \in \mathbb{Z}_{n+1}$ and for arbitrarily chosen constants $a > 0$ and $r > 1$, we choose the truncation parameter $\delta^n_{i'i}$ such that

$$\delta^n_{i'i} \leq \max\left\{a\mu^{[-n+\alpha(n-i)+\alpha'(n-i')]/d}, rd_i, rd_{i'}\right\}, \tag{5.12}$$

where α and α' are any numbers in $(-\infty, 1]$. The lemma above and the choice of truncation parameters lead to the following estimate of the number of nonzero entries of matrix $\tilde{\mathbf{K}}_n$.

Theorem 5.9 *If the truncation parameters $\delta^n_{i'i}$ are chosen according to (5.12) and if conditions (I) and (II) hold, then*

$$\mathcal{N}(\tilde{\boldsymbol{K}}_n) = \begin{cases} \mathcal{O}(s(n) \log^2 s(n)), & \alpha = \alpha' = 1, \\ \mathcal{O}(s(n) \log s(n)), & \text{otherwise}. \end{cases}$$

Proof Because

$$\mathcal{N}(\tilde{\mathbf{K}}_n) = \sum_{i' \in \mathbb{Z}_{n+1}} \sum_{i \in \mathbb{Z}_{n+1}} \mathcal{N}(\tilde{\mathbf{K}}_{i'i}), \tag{5.13}$$

we use Lemma 5.8 to estimate $\mathcal{N}(\tilde{\mathbf{K}}_n)$. The choice (5.12) of truncation parameters ensures that

$$\delta^n_{i'i} \leq a\mu^{[-n+\alpha(n-i)+\alpha'(n-i')]/d} + rd_i + rd_{i'}.$$

Using (5.13) and substituting the above estimate into the inequality in Lemma 5.8, we have that

$$\mathcal{N}(\tilde{\mathbf{K}}_n) \leq c \sum_{i\in\mathbb{Z}_{n+1}} \sum_{i'\in\mathbb{Z}_{n+1}} \mu^{i+i'} \left(2\mu^{-i} + 2\mu^{-i'} + a^d \mu^{-n+\alpha(n-i)+\alpha'(n-i')}\right)$$

$$= c\left[4(n+1)\sum_{i\in\mathbb{Z}_{n+1}} \mu^i + a^d\mu^n \left(\sum_{i\in\mathbb{Z}_{n+1}} \mu^{(\alpha-1)(n-i)}\right)\right.$$

$$\left.\times\left(\sum_{i'\in\mathbb{Z}_{n+1}} \mu^{(\alpha'-1)(n-i')}\right)\right]$$

$$= \begin{cases} \mathcal{O}(\mu^n(n+1)^2), & \alpha = \alpha' = 1, \\ \mathcal{O}(\mu^n(n+1)), & \text{otherwise}, \end{cases}$$

as $n \to \infty$. This leads to the desired result of this theorem. □

5.3.2 Stability and convergence

In this subsection we show that the fast multiscale Galerkin method is stable and it has an almost optimal convergence order.

The first lemma that we present here gives an estimate for the discrepancy between the block $\mathbf{K}_{i'i}$ and $\tilde{\mathbf{K}}_{i'i} := \mathbf{K}(\delta)_{i'i}$, where the latter is obtained by using the truncation strategy with parameter $\delta = \delta^n_{i'i}$.

Lemma 5.10 *Suppose that $\tilde{\mathbf{K}}_{i'i}$ is obtained from the truncation strategy (5.8) with truncation parameter δ. If conditions (I)–(IV) hold and $K \in \mathbb{S}_{\sigma,k}$ for some $\sigma \in [0, d)$ and a positive integer k, then for any $r > 1$ and $\delta > 0$, there exists a constant c such that when $\delta \geq \max\{rd_i, rd_{i'}\}$,*

$$\|\mathbf{K}_{i'i} - \tilde{\mathbf{K}}_{i'i}\|_2 \leq c(d_i d_{i'})^k \delta^{-\eta},$$

where $\eta := 2k - d + \sigma > 0$.

Proof By the definition of $\tilde{\mathbf{K}}_{i',i}$ we have that

$$\|\mathbf{K}_{i'i} - \tilde{\mathbf{K}}_{i'i}\|_\infty = \max_{j'\in\mathbb{Z}_{w(i')}} \sum_{j\in\mathbb{Z}_\delta} |K_{i'j',ij}|,$$

where

$$\mathbb{Z}_\delta := \{j : j \in \mathbb{Z}_{w(i)}, \operatorname{dist}(S_{ij}, S_{i'j'}) > \delta\}.$$

It follows from Lemma 5.3 that

$$\left\|\mathbf{K}_{i'i}-\tilde{\mathbf{K}}_{i'i}\right\|_{\infty}\le c(d_id_{i'})^{k-\frac{d}{2}}d_{i'}^{d}\max_{j'\in\mathbb{Z}_{w(i')}}\max_{s\in S_{i'j'}}\sum_{j\in\mathbb{Z}_\delta}\int_{S_{ij}}\frac{\mathrm{d}t}{|s-t|^{2k+\sigma}}$$
$$\le c(d_id_{i'})^{k-\frac{d}{2}}d_{i'}^{d}\int_{|t|>\delta}\frac{\mathrm{d}t}{|t|^{2k+\sigma}}$$
$$\le c(d_id_{i'})^{k-\frac{d}{2}}d_{i'}^{d}\delta^{-\eta}.$$

Likewise, we have that

$$\left\|\mathbf{K}_{i'i}-\tilde{\mathbf{K}}_{i'i}\right\|_{1}\le c(d_id_{i'})^{k-\frac{d}{2}}d_{i}^{d}\delta^{-\eta}.$$

Since the spectral radius of a matrix $\mathbf{A}$ is less than or equal to any of its matrix norms,

$$\|\mathbf{A}\|_2^2=\rho(\mathbf{A}^T\mathbf{A})\le\|\mathbf{A}^T\mathbf{A}\|_\infty\le\|\mathbf{A}^T\|_\infty\|\mathbf{A}\|_\infty=\|\mathbf{A}\|_1\|\mathbf{A}\|_\infty.$$

Using the above inequality, we have that

$$\left\|\mathbf{K}_{i'i}-\tilde{\mathbf{K}}_{i'i}\right\|_2^2\le\left\|\mathbf{K}_{i'i}-\tilde{\mathbf{K}}_{i'i}\right\|_1\left\|\mathbf{K}_{i'i}-\tilde{\mathbf{K}}_{i'i}\right\|_\infty.$$

Substituting both estimates obtained earlier into the right-hand side of the above inequality proves the desired result. □

We now describe a second criterion for the choice of truncation parameters $\delta_{i'i}^n$. For each $i,i'\in\mathbb{Z}_{n+1}$ and for arbitrarily chosen constants $a>0$ and $r>1$, we choose the truncation parameter $\delta_{i'i}^n$ such that

$$\delta_{i'i}^n\ge\max\left\{a\mu^{[-n+\alpha(n-i)+\alpha'(n-i')]/d},rd_i,rd_{i'}\right\},\tag{5.14}$$

where α and α' are any numbers in $(-\infty,1]$. For real numbers a and b, we set

$$\mu[a,b;n]:=\sum_{i\in\mathbb{Z}_{n+1}}\mu^{ai/d}\sum_{i'\in\mathbb{Z}_{n+1}}\mu^{bi'/d}.$$

We next estimate the error $\mathcal{R}_n:=\mathcal{K}_n-\tilde{\mathcal{K}}_n$ of the truncation operator in terms of the function $\mu[\cdot,\cdot;n]$.

Lemma 5.11 *Let $u\in H^m(\Omega)$ with $0\le m\le k$ and $K\in\mathbb{S}_{\sigma,k}$ for some $\sigma\in[0,d)$ and a positive integer k. If the truncation parameters $\delta_{i'i}^n$ are chosen according to (5.14) and conditions (I)–(VI) hold, then there exists a positive constant c such that for all $n\in\mathbb{N}_0$,*

$$\|(\mathcal{K}_n-\tilde{\mathcal{K}}_n)\mathcal{P}_nu\|\le c\mu[k+m-\alpha\eta,k-\alpha'\eta;n]\mu^{-(m+d-\sigma)n/d}\|u\|_{H^m}.$$

Proof For any $u, v \in \mathbb{X}$, we project them into the subspace $\mathbb{X}_n$. Hence,

$$\mathcal{P}_n u = \sum_{i\in\mathbb{Z}_{n+1}} (\mathcal{P}_i - \mathcal{P}_{i-1})u = \sum_{(i,j)\in\mathbb{U}_n} u_{ij}w_{ij},$$

for some constants u_{ij} and

$$\mathcal{P}_n v = \sum_{i\in\mathbb{Z}_{n+1}} (\mathcal{P}_i - \mathcal{P}_{i-1})v = \sum_{(i,j)\in\mathbb{U}_n} v_{ij}w_{ij},$$

for some constants v_{ij}, where $\mathcal{P}_{-1} = 0$. By the definitions of operators $\mathcal{K}_n$ and $\tilde{\mathcal{K}}_n$, we have that

$$\begin{aligned}\left((\mathcal{K}_n - \tilde{\mathcal{K}}_n)\mathcal{P}_n u, \mathcal{P}_n v\right) &= \sum_{i,i'\in\mathbb{Z}_{n+1}} \left((\mathcal{K}_n - \tilde{\mathcal{K}}_n)(\mathcal{P}_i - \mathcal{P}_{i-1})u, (\mathcal{P}_{i'} - \mathcal{P}_{i'-1})v\right) \\ &= \sum_{i,i'\in\mathbb{Z}_{n+1}} \sum_{j\in\mathbb{Z}_{w(i)}} \sum_{j'\in\mathbb{Z}_{w(i')}} (K_{i'j',ij} - \tilde{K}_{i'j',ij})u_{ij}v_{i'j'}.\end{aligned}$$

Set

$$e_n := \left|\left((\mathcal{K}_n - \tilde{\mathcal{K}}_n)\mathcal{P}_n u, \mathcal{P}_n v\right)\right|.$$

Using the Cauchy–Schwarz inequality and condition (V), we conclude that

$$e_n \le c \sum_{i,i'\in\mathbb{Z}_{n+1}} \|\mathbf{K}_{i'i} - \tilde{\mathbf{K}}_{i'i}\|_2 \|(\mathcal{P}_i - \mathcal{P}_{i-1})u\| \|(\mathcal{P}_{i'} - \mathcal{P}_{i'-1})v\|.$$

It follows from condition (VI) that for $u \in H^m(\Omega)$ with $0 \le m \le k$,

$$\|(\mathcal{P}_i - \mathcal{P}_{i-1})u\| \le c d_{i-1}^m \|u\|_{H^m}.$$

Combining the above estimates and using Lemma 5.10, we have for $u \in H^m(\Omega)$ and $v \in H^{m'}(\Omega)$ with $0 \le m, m' \le k$ that

$$e_n \le c \sum_{i,i'\in\mathbb{Z}_{n+1}} (d_i d_{i'})^k (\delta_{i',i}^n)^{-\eta} d_{i-1}^m d_{i'-1}^{m'} \|u\|_{H^m} \|v\|_{H^{m'}}.$$

Using $d_i \sim \mu^{-i/d}$ and the choice of $\delta_{i',i}^n$, we conclude that

$$\begin{aligned}e_n \le ca^{-\eta} &\sum_{i,i'\in\mathbb{Z}_{n+1}} \mu^{(k+m-\alpha\eta)(n-i)/d+(k+m'-\alpha'\eta)(n-i')/d} \\ &\mu^{-(m+m'+d-\sigma)n/d} \|u\|_{H^m} \|v\|_{H^{m'}} \\ = ca^{-\eta} &\mu[k+m-\alpha\eta, k+m'-\alpha'\eta; n] \cdot \mu^{-(m+m'+d-\sigma)\frac{n}{d}} \|u\|_{H^m} \|v\|_{H^{m'}}.\end{aligned}$$

Since $(\mathcal{K}_n - \tilde{\mathcal{K}}_n) = \mathcal{P}_n(\mathcal{K}_n - \tilde{\mathcal{K}}_n)$, we have for $u \in \mathbb{X}$ that

$$\left\|(\mathcal{K}_n - \tilde{\mathcal{K}}_n)\mathcal{P}_n u\right\| = \sup_{v\in\mathbb{X}, \|v\|\ne 0} \left|\left((\mathcal{K}_n - \tilde{\mathcal{K}}_n)\mathcal{P}_n u, \mathcal{P}_n v\right)\right| / \|v\|_2.$$

Combining this equation with the inequality above with $m' = 0$ yields the desired result of this lemma. □

The next theorem provides a stability estimate for operator $\mathcal{I} - \tilde{\mathcal{K}}_n$. Recall that for the standard Galerkin method, there exist positive constants c_0 and N_0 such that for all $n > N_0$,

$$\|(\mathcal{I} - \mathcal{K}_n)v\| \geq c_0\|v\|, \quad \text{for all } v \in \mathbb{X}_n. \tag{5.15}$$

Theorem 5.12 *Let $K \in \mathbb{S}_{\sigma,k}$ for some $\sigma \in [0, d)$ and a positive integer k. Suppose that the truncation parameters $\delta^n_{i'i}$ are chosen according to (5.14) with*

$$\alpha > \frac{1}{2} - \frac{d-\sigma}{2\eta}, \quad \alpha' > \frac{1}{2} - \frac{d-\sigma}{2\eta}, \quad \alpha + \alpha' > 1.$$

If conditions (I)–(VI) hold, then there exist a positive constant c and a positive integer N such that for all $n \geq N$ and $v \in \mathbb{X}_n$,

$$\|(\mathcal{I} - \tilde{\mathcal{K}}_n)v\| \geq c\|v\|.$$

Proof Note that for any real numbers a, b and e,

$$\lim_{n\to\infty} \mu[a, b; n]\mu^{-en/d} = 0$$

when $e > \max\{0, a, b, a + b\}$. Thus, the choice of $\delta^n_{i',i}$ ensures that there exists a positive integer N such that for all $n \geq N$,

$$c\mu[k - \alpha\eta, k - \alpha'\eta; n]\mu^{-(d-\sigma)n/d} \leq c_0/2.$$

This, with the estimate in Lemma 5.11, leads to

$$\|(\mathcal{K}_n - \tilde{\mathcal{K}}_n)v\| \leq \frac{c_0}{2}\|v\|, \quad \text{for all } v \in \mathbb{X}_n. \tag{5.16}$$

Combining (5.16) and the stability estimate (5.15) of the standard Galerkin method yields

$$\|(\mathcal{I} - \tilde{\mathcal{K}}_n)v\| \geq \|(\mathcal{I} - \mathcal{K}_n)v\| - \|(\mathcal{K}_n - \tilde{\mathcal{K}}_n)v\| \geq \frac{c_0}{2}\|v\|,$$

for any $v \in \mathbb{X}_n$. This completes the proof. □

The above stability estimate ensures that $(\mathcal{I} - \tilde{\mathcal{K}}_n)^{-1}$ exists and is uniformly bounded. As a result, the fast multiscale Galerkin method (5.10) has a unique solution for a sufficiently large n.

Theorem 5.13 *Let $u \in H^k(\Omega)$ and $K \in \mathbb{S}_{\sigma,k}$ for some $\sigma \in [0, d)$ and a positive integer k. Suppose that the truncation parameters $\delta^n_{i'i}$ are chosen according to (5.14), with α and α' satisfying one of the following conditions:*

(i) $\alpha \geq 1, \alpha' > \frac{1}{2} - \frac{d-\sigma}{2\eta}, \alpha + \alpha' > 1 + \frac{k}{\eta}$ *or* $\alpha > 1, \alpha' \geq \frac{1}{2} - \frac{d-\sigma}{2\eta}, \alpha + \alpha' > 1 + \frac{k}{\eta}$ *or* $\alpha > 1, \alpha' > \frac{1}{2} - \frac{d-\sigma}{2\eta}, \alpha + \alpha' \geq 1 + \frac{k}{\eta}$.

(ii) $\alpha = 1, \alpha' = \frac{k}{\eta}$ *or* $\alpha = \frac{2k}{\eta}, \alpha' = \frac{1}{2} - \frac{d-\sigma}{2\eta}$.

If conditions (I)–(VI) hold, then there exist a positive constant c and a positive integer N such that for all $n \geq N$,

$$\|u - \tilde{u}_n\| \leq c s(n)^{-k/d} (\log s(n))^{\tau} \|u\|_{H^k(\Omega)},$$

where $\tau = 0$ in case (i) and $\tau = 1$ in case (ii).

Proof It follows from Theorem 5.12 that there exist a positive constant c and a positive integer N such that for all $n \geq N$,

$$\|\mathcal{P}_n u - \tilde{u}_n\| \leq c \|(\mathcal{I} - \tilde{\mathcal{K}}_n)(\mathcal{P}_n u - \tilde{u}_n)\|. \tag{5.17}$$

Since

$$\mathcal{P}_n(\mathcal{I} - \mathcal{K})u = (\mathcal{I} - \tilde{\mathcal{K}}_n)\tilde{u}_n = \mathcal{P}_n f,$$

we have that

$$(\mathcal{I} - \tilde{\mathcal{K}}_n)(\mathcal{P}_n u - \tilde{u}_n) = \mathcal{P}_n(\mathcal{I} - \mathcal{K})(\mathcal{P}_n u - u) + (\mathcal{K}_n - \tilde{\mathcal{K}}_n)\mathcal{P}_n u. \tag{5.18}$$

Now, by the triangle inequality, we have that

$$\|u - \tilde{u}_n\| \leq \|u - \mathcal{P}_n u\| + \|\mathcal{P}_n u - \tilde{u}_n\|. \tag{5.19}$$

Using inequality (5.17) and equation (5.18), we obtain that

$$\|\mathcal{P}_n u - \tilde{u}_n\| \leq c\|\mathcal{I} - \mathcal{K}\| \|\mathcal{P}_n u - u\| + c\|(\mathcal{K}_n - \tilde{\mathcal{K}}_n)\mathcal{P}_n u\|.$$

Substituting this estimate into the right-hand side of (5.19) yields

$$\|u - \tilde{u}_n\| \leq (1 + c\|\mathcal{I} - \mathcal{K}\|)\|\mathcal{P}_n u - u\| + c\|(\mathcal{K}_n - \tilde{\mathcal{K}}_n)\mathcal{P}_n u\|.$$

It follows from Lemma 5.11 that

$$\|(\mathcal{K}_n - \tilde{\mathcal{K}}_n)\mathcal{P}_n u\| \leq c\mu[2k - \alpha\eta, k - \alpha'\eta; n]\mu^{-(d-\sigma)n/d}\mu^{-kn/d}\|u\|_{H^k}.$$

Observing that

$$\mu[a, b; n]\mu^{-en/d} = \begin{cases} \mathcal{O}(1), & \text{if } e \geq a, e > b, e > a + b \\ & \text{or } e > a, e \geq b, e > a + b \\ & \text{or } e > a, e > b, e \geq a + b, \\ \mathcal{O}(n), & \text{if } e = a, b = 0 \text{ or } e = b, a = 0, \end{cases}$$

as $n \to \infty$, we obtain that

$$\mu[2k - \alpha\eta, k - \alpha'\eta; n]\mu^{-(d-\sigma)n/d} = \begin{cases} \mathcal{O}(1), & \text{in case (i)}, \\ \mathcal{O}(n), & \text{in case (ii)}, \end{cases}$$

with $a := 2k - \alpha\eta, b := k - \alpha'\eta$ and $e := d - \sigma$. This yields the desired result. □

5.3.3 The condition number of the truncated matrix

We show in this subsection that the condition number of the truncated matrix is uniformly bounded. To this end, we need a norm equivalence result, which is presented below.

Lemma 5.14 *If conditions (II), (IV) and (V) hold, then for any $n \in \mathbb{N}$ and $v = \sum_{(i,j)\in\mathbb{U}_n} v_{ij}w_{ij}$,*

$$\|v\| \sim \|\mathbf{v}\|_2,$$

where $\mathbf{v} := [v_{ij} : (i,j) \in \mathbb{U}_n]$.

Proof Since condition (V) holds, it suffices to prove that there is a positive constant θ_2 such that for all v,

$$\|v\| \le \theta_2 \|\mathbf{v}\|_2.$$

It follows from the orthogonal decomposition (5.2) that

$$\|v\|^2 = \sum_{i\in\mathbb{Z}_{n+1}} \Big\| \sum_{j\in\mathbb{Z}_{w(i)}} v_{ij}w_{ij} \Big\|^2.$$

According to the construction of the partition of Ω and condition (II), for all $i > \gamma$,

$$\Big\| \sum_{j\in\mathbb{Z}_{w(i)}} v_{ij}w_{ij} \Big\|^2 = \sum_{\nu\in\mathbb{Z}_{e(i-\gamma)}} \Big\| \sum_{j\in\mathbb{Z}_{(\nu)}} v_{ij}w_{ij} \Big\|^2,$$

where $\mathbb{Z}_{(\nu)} := \{j : \text{supp } w_{ij} \subseteq S_{ij} = \Omega_{i-\gamma,\nu}\}$. Using the Cauchy–Schwarz inequality and condition (II), we have that

$$\Big\| \sum_{j\in\mathbb{Z}_{(\nu)}} v_{ij}w_{ij} \Big\|^2 \le \int_\Omega \sum_{j\in\mathbb{Z}_{(\nu)}} v_{ij}^2 \sum_{j\in\mathbb{Z}_{(\nu)}} w_{ij}^2(t)\mathrm{d}t \le \rho \sum_{j\in\mathbb{Z}_{(\nu)}} v_{ij}^2.$$

The last inequality holds because the cardinality of $\mathbb{Z}_{(v)}$ is less than or equal to ρ and the L^2-norm of w_{ij} is equal to 1. Hence, we conclude that there is a positive constant θ_2 such that

$$\|v\|^2 \leq \theta_2^2 \sum_{(i,j)\in\mathbb{U}_n} v_{ij}^2 = \theta_2^2 \|\mathbf{v}\|_2^2.$$

This completes the proof. □

With the help of the above lemma, we are ready to show that the condition number of the coefficient matrix

$$\tilde{\mathbf{A}}_n := \mathbf{E}_n - \tilde{\mathbf{K}}_n$$

is uniformly bounded.

Theorem 5.15 *Suppose that $K \in \mathbb{S}_{\sigma,k}$ for some $\sigma \in [0,d)$ and a positive integer k and the truncation parameters $\delta_{i',i}^n$ are chosen according to (5.14) with α and α' satisfying the following conditions:*

$$\alpha > \frac{1}{2} - \frac{d-\sigma}{2\eta}, \quad \alpha' > \frac{1}{2} - \frac{d-\sigma}{2\eta}, \quad \alpha + \alpha' > 1.$$

If conditions (I)–(VI) hold, then the condition number of the coefficient matrix of the truncated approximate equation (5.10) is bounded, that is, there exists a positive constant c such that for all $n \in \mathbb{N}$,

$$\operatorname{cond}_2(\tilde{\mathbf{A}}_n) \leq c.$$

Proof For any $\mathbf{v} := [v_{ij} : (i,j) \in \mathbb{U}_n] \in \mathbb{R}^{s(n)}$, let

$$v := \sum_{(i,j)\in\mathbb{U}_n} v_{ij} w_{ij}$$

and

$$g := (\mathcal{I} - \tilde{\mathcal{K}}_n)v.$$

Thus $g \in \mathbb{X}_n$ and it can be written as

$$g = \sum_{(i,j)\in\mathbb{U}_n} g_{ij} w_{ij}.$$

Set

$$\mathbf{g} := [g_{ij} : (i,j) \in \mathbb{U}_n].$$

It can be verified that

$$\mathbf{g} = (\mathbf{E}_n - \tilde{\mathbf{K}}_n)\mathbf{v}.$$

It follows from Theorem 5.12, Lemma 5.14 and the above equations that there exist a positive constant c and positive integer N such that for all $n \geq N$,

$$\|\mathbf{v}\|_2 \leq c\|v\| \leq c\|(\mathcal{I} - \tilde{\mathcal{K}}_n)v\| = c\|g\| \leq c\|\mathbf{g}\|_2 = c\|(\mathbf{E}_n - \tilde{\mathbf{K}}_n)\mathbf{v}\|_2.$$

This means that

$$\|(\mathbf{E}_n - \tilde{\mathbf{K}}_n)^{-1}\|_2 \leq c. \tag{5.20}$$

Conversely, we have that

$$\|(\mathbf{E}_n - \tilde{\mathbf{K}}_n)\mathbf{v}\|_2 = \|\mathbf{g}\|_2 \leq c\|g\| = c\|(\mathcal{I} - \tilde{\mathcal{K}}_n)v\|.$$

Note that

$$\|(\mathcal{I} - \tilde{\mathcal{K}}_n)v\| \leq \|(\mathcal{I} - \mathcal{K}_n)v\| + \|(\mathcal{K}_n - \tilde{\mathcal{K}}_n)v\|.$$

This, with (5.16), implies that

$$\|(\mathcal{I} - \tilde{\mathcal{K}}_n)v\| \leq (1 + \|\mathcal{K}\|)\|v\| + \frac{c_0}{2}\|v\| \leq c\|\mathbf{v}\|_2.$$

Thus

$$\|\mathbf{E}_n - \tilde{\mathbf{K}}_n\|_2 \leq c. \tag{5.21}$$

The result of this theorem follows from (5.20) and (5.21). □

To close this section, we would like to know if we can choose appropriate truncation parameters such that the optimal results about the order of convergence and computational complexity can be reached. Combining Theorems 5.9, 5.12, 5.13 and 5.15 leads to the following.

Theorem 5.16 *Let $u \in H^k(\Omega)$ and $K \in \mathbb{S}_{\sigma,k}$ for some $\sigma \in [0, d)$ and a positive integer k. If conditions (I)–(VI) hold, and $\delta^n_{i',i}$ are chosen as*

$$\delta^n_{i'i} := \max\left\{a\mu^{[-n+\alpha(n-i)+\alpha'(n-i')]/d}, rd_i, rd_{i'}\right\},$$

with $\alpha = 1$ and $1 - \frac{k}{\eta} < \alpha' \leq 1$, then the following hold: the stability estimate

$$\|(\mathcal{I} - \tilde{\mathcal{K}}_n)v\| \geq c\|v\|, \quad \text{for all } v \in \mathbb{X}_n,$$

the boundedness of the condition number

$$\text{cond}_2(\tilde{\mathbf{A}}_n) \leq c,$$

the optimal convergence order

$$\|u - \tilde{u}_n\| \leq cs(n)^{-k/d}\|u\|_{H^k(\Omega)}$$

and the optimal (up to a logarithmic factor) order of the complexity

$$\mathcal{N}(\tilde{\boldsymbol{K}}_n) = \begin{cases} \mathcal{O}(s(n)\log^2 s(n)), & \alpha = \alpha' = 1, \\ \mathcal{O}(s(n)\log s(n)), & \text{otherwise.} \end{cases}$$

5.3.4 Remarks on the smooth kernel case

In this subsection, we present special results for the smooth kernel case. Since the proofs are similar to those for the weakly singular case, we omit the details of the proof except for Lemma 5.19, whose results have something different from Lemma 5.11.

Lemma 5.17 *If conditions (I)–(IV) hold and $K \in C^k(\Omega \times \Omega)$, then there exists a positive constant c such that for $i, i' \in \mathbb{N}$ and for all $n \in \mathbb{N}$,*

$$\|\boldsymbol{K}_{i'i}\|_2 \leq c d_i^k d_{i'}^k.$$

To avoid computing the entries whose values are nearly zero, we make a special block truncation strategy, that is, setting

$$\tilde{\mathbf{K}}_{i'i} = \begin{cases} \mathbf{K}_{i'i}, & i + i' \leq n, \\ 0, & \text{otherwise,} \end{cases} \quad i', i \in \mathbb{N} \tag{5.22}$$

to obtain a sparse truncation matrix

$$\tilde{\mathbf{K}}_n := [\tilde{\mathbf{K}}_{i'i} : i', i \in \mathbb{Z}_{n+1}].$$

The following theorems provide the computational complexity, the convergence estimate and the stability of the truncation scheme for integral equations with smooth kernels.

Theorem 5.18 *Suppose that condition (I) holds and $K \in C^k(\Omega \times \Omega)$. If the truncated matrix $\tilde{\boldsymbol{K}}_n$ is chosen as (5.22), then*

$$\mathcal{N}(\tilde{\boldsymbol{K}}_n) = \mathcal{O}(s(n)\log s(n)).$$

Lemma 5.19 *Suppose that conditions (I)–(VI) hold, and $K \in C^k(\Omega \times \Omega)$. If the truncated matrix $\tilde{\boldsymbol{K}}_n$ is chosen as (5.22), then there exists a constant c such that for all $u \in H^k(\Omega)$ and for all $n \in \mathbb{N}$,*

$$\|(\mathcal{K}_n - \tilde{\mathcal{K}}_n)\mathcal{P}_n u\| \leq c\mu^{-kn/d}\|u\|_{H^k},$$

and for $u \in L^2(\Omega)$,

$$\|(\mathcal{K}_n - \tilde{\mathcal{K}}_n)\mathcal{P}_n u\| \leq c(n+1)\mu^{-kn/d}\|u\|.$$

Proof Similar to the proof of Theorem 5.11, for any $u, v \in \mathbb{X}$,

$$\mathcal{P}_n u = \sum_{(i,j)\in\mathbb{U}_n} u_{ij} w_{ij}, \quad \mathcal{P}_n v = \sum_{(i,j)\in\mathbb{U}_n} v_{ij} w_{ij},$$

we have

$$\left((\mathcal{K}_n - \tilde{\mathcal{K}}_n)\mathcal{P}_n u, v\right) = \sum_{i,i'\in\mathbb{Z}_{n+1}} \sum_{j\in\mathbb{Z}_{w(i)}} \sum_{j'\in\mathbb{Z}_{w(i')}} (K_{i'j',ij} - \tilde{K}_{i'j',ij}) u_{ij} v_{i'j'}.$$

Using the Cauchy–Schwarz inequality and condition (V), we conclude that its absolute value is bounded by

$$c \sum_{i,i'\in\mathbb{Z}_{n+1}} \|\mathbf{K}_{i'i} - \tilde{\mathbf{K}}_{i'i}\|_2 \|(\mathcal{P}_i - \mathcal{P}_{i-1})u\| \, \|(\mathcal{P}_{i'} - \mathcal{P}_{i'-1})v\|.$$

It follows from condition (VI) that for $u \in H^m(\Omega)$ with $0 \le m \le k$,

$$\|(\mathcal{P}_i - \mathcal{P}_{i-1})u\| \le c d_{i-1}^m \|u\|_{H^m}.$$

Denote $\mathbb{Z}'_{(i)} := \{i' \in \mathbb{Z}_{n+1} : i' > n - i\}$. Combining the above estimates, using Lemma 5.17 and the truncation strategy (5.22), we have that for $u \in H^m(\Omega)$ and $v \in L^2(\Omega)$,

$$\left|\left((\mathcal{K}_n - \tilde{\mathcal{K}}_n)\mathcal{P}_n u, v\right)\right| \le c \sum_{i\in\mathbb{Z}_{n+1}} \sum_{i'\in\mathbb{Z}'_{(i)}} (d_i d_{i'})^k d_{i-1}^m \|u\|_{H^m} \|v\|. \tag{5.23}$$

Since $d_i \sim \mu^{-i/d}$, a simple computation yields that

$$\begin{aligned}\sum_{i\in\mathbb{Z}_{n+1}} \sum_{i'\in\mathbb{Z}'_{(i)}} (d_i d_{i'})^k d_{i-1}^m &\le c \sum_{i\in\mathbb{Z}_{n+1}} \sum_{i'\in\mathbb{Z}'_{(i)}} \mu^{-k(i+i')/d - m(i-1)/d} \\ &= c\mu^{-kn/d} \sum_{i\in\mathbb{Z}_{n+1}} \mu^{-m(i-1)/d} \sum_{i'\in\mathbb{Z}'_{(i)}} \mu^{-k(i+i'-n)/d}.\end{aligned}$$

For any $i \in \mathbb{Z}_{n+1}$,

$$\sum_{i'\in\mathbb{Z}'_{(i)}} \mu^{-k(i+i'-n)/d} \le \sum_{l\in\mathbb{N}} \mu^{-kl/d} \le \frac{\mu^{-k/d}}{1 - \mu^{-k/d}},$$

which leads to the fact that there exists a constant c such that

$$\sum_{i\in\mathbb{Z}_{n+1}} \sum_{i'\in\mathbb{Z}'_{(i)}} (d_i d_{i'})^k d_{i-1}^m \le \begin{cases} c\mu^{-kn/d}, & \text{if } 0 < m \le k, \\ c(n+1)\mu^{-kn/d}, & \text{if } m = 0. \end{cases} \tag{5.24}$$

Combining the above inequalities (5.23), (5.24) and

$$\|(\mathcal{K}_n - \tilde{\mathcal{K}}_n)\mathcal{P}_n u\| = \sup_{v\in\mathbb{X}, \|v\|\neq 0} |((\mathcal{K}_n - \tilde{\mathcal{K}}_n)\mathcal{P}_n u, v)| / \|v\|,$$

we obtain the estimates of this lemma. □

The compactness of $\mathcal{K}$ and the property of the orthogonal projection $\mathcal{P}_n$ lead to the stability estimate of the operator equation. This, with the second estimate of Lemma 5.19, yields the following theorem about the stability of the truncation equation.

Theorem 5.20 *Suppose that conditions (I)–(VI) hold, and $K \in C^k(\Omega \times \Omega)$. If the truncated matrix $\tilde{\boldsymbol{K}}_n$ is chosen as (5.22), then there exist a positive constant c_0 and an integer N such that for all $n \geq N$ and $x \in \mathbb{X}_n$,*

$$\|(\mathcal{I} - \tilde{\mathcal{K}}_n)x\| \geq c_0\|x\|.$$

We have the following convergence estimate similar to Theorem 5.13.

Theorem 5.21 *Suppose that conditions (I)–(VI) hold, and $K \in C^k(\Omega \times \Omega)$. If the truncated matrix $\tilde{\boldsymbol{K}}_n$ is chosen as (5.22), then there exist a positive constant c and an integer N such that for all $n \geq N$,*

$$\|u - \tilde{u}_n\| \leq cs(n)^{-k/d}\|u\|_{H^k}.$$

We also have that the condition number of the coefficient matrix $\tilde{\mathbf{A}}_n := \mathbf{E}_n - \tilde{\mathbf{K}}_n$ of the truncated scheme is bounded by a constant independent of n.

Theorem 5.22 *Suppose that conditions (I)–(VI) hold, and $K \in C^k(\Omega \times \Omega)$. If the truncated matrix $\tilde{\boldsymbol{K}}_n$ is chosen as (5.22), then the condition number of the coefficient matrix of the truncated approximate equation is bounded, that is, there exists a positive constant c such that for all $n \in \mathbb{N}$,*

$$\mathrm{cond}_2(\tilde{\mathbf{A}}_n) \leq c.$$

5.4 Bibliographical remarks

Since the 1990s, wavelet and multiscale methods have been developed for solving the Fredholm integral equation of the second kind. The history of fast multiscale solutions of the equation began with the remarkable discovery in [28] that the matrix representation of a singular Fredholm integral operator under a wavelet basis is numerically sparse. This fact was then used in developing the multiscale Galerkin (Petrov–Galerkin) method for solving the Fredholm integral equation, see [5, 64, 68, 88–91, 94, 95, 135, 136, 139, 140, 202, 251, 260, 261] and references cited therein. Readers are referred to the Introduction of this book for more information. The multiscale piecewise polynomial Petrov–Galerkin discrete multiscale Petrov–Galerkin, and multiscale collocation methods were developed in [64, 68, 69]. We give an in-depth discussion of these methods in the next two chapters. A numerical implementation issue of the multiscale Galerkin method was considered in [109].

The convergence results presented in this chapter are for the smooth solution. However, solutions of the Fredholm integral equation of the second kind with weakly singular kernels may not be smooth. When the solution is not smooth, a fast singularity-preserving multiscale Galerkin method was developed in [46] for solving weakly singular Fredholm integral equations of the second kind. This method was designed based on the singularity-preserving Galerkin method introduced originally in [41] and a matrix truncation strategy similar to what we have discussed in Section 5.2.

There are several fast methods in the literature for solving the Fredholm integral equation of the second kind which are closely related to the fast multiscale method. They include the fast multipole method, the panel clustering method and the method of sparse grids. The fast multipole method [114, 115, 235, 250] was originally introduced by V. Rokhlin and L. Greengard based on the multipole expansion. It effectively reduces the computational complexity involving a certain type of the dense matrix which can arise out of many physical systems. The panel clustering method, proposed by W. Hackbusch and Z. Nowak, also significantly lessens the computational complexity (see, for example, [124, 125]). For the method of sparse grids, readers are referred to [36] and the references cited therein. Fast Fourier–Galerkin methods developed in [37, 53, 154, 155, 263] for solving boundary integral equations are special cases of the method of sparse grids. Fast methods for solving Fredholm integral equations of the second kind in high dimensions were developed in [272] and [102], respectively, based on a combination technique and the lattice integration.

6
Multiscale Petrov–Galerkin methods

This chapter is devoted to presenting multiscale Petrov–Galerkin methods for solving Fredholm integral equations of the second kind. In a manner similar to the Galerkin method, the Petrov–Galerkin method also suffers from the density of the coefficient matrix of its resulting linear system. We show that with the multiscale basis, the Petrov–Galerkin method leads to a linear system having a numerically sparse coefficient matrix. We propose a matrix compression scheme for solving the linear system and prove that it almost preserves the optimal convergence order of the numerical solution that the original Petrov–Galerkin method enjoys and it reduces the computational complexity from the square order to the quasi-linear order. We also present the discrete version of the multiscale Petrov–Galerkin method, which further treats the nonzero entries of the compressed coefficient matrix that results from the multiscale Petrov–Galerkin method by using the product integration method. We call this method the discrete multiscale Petrov–Galerkin method.

In Section 6.1 we first present the development of the multiscale Petrov–Galerkin method and its analysis. We then discuss in Section 6.2 the discrete multiscale Petrov–Galerkin method.

6.1 Fast multiscale Petrov–Galerkin methods

In this section we describe the construction of two sequences of multiscale bases for trial and test spaces, and use them to develop multiscale Petrov–Galerkin methods for solving the second-kind integral equations.

6.1.1 Multiscale bases for Petrov–Galerkin methods

We review first a special case of the recursive construction given in Chapter 4 for piecewise polynomial spaces on $\Omega = [0, 1]$, which can be used to develop a multiscale Petrov–Galerkin scheme.

We start with positive integers k, k', ν and μ which satisfy $k\nu = k'\mu$ and $k' \leq k$. We choose our initial trial space and test space to be $\mathbb{X}_0 := \mathbb{S}_\nu^k$ and $\mathbb{Y}_0 := \mathbb{S}_\mu^{k'}$ and thereafter, we recursively divide the corresponding subintervals into μ pieces to obtain two sequences of subspaces

$$\mathbb{X}_n = \mathbb{S}^k_{\nu\mu^n}, \quad \mathbb{Y}_n = \mathbb{S}^{k'}_{\mu^{n+1}}, \quad n \in \mathbb{N}_0.$$

These spaces are referred to as the (k, k') *element* spaces.

We have that

$$\dim \mathbb{X}_n = \dim \mathbb{Y}_n, \quad n \in \mathbb{N}_0,$$
$$\mathbb{X}_n \subset \mathbb{X}_{n+1}, \quad \mathbb{Y}_n \subset \mathbb{Y}_{n+1}, \quad n \in \mathbb{N}_0$$

and

$$\overline{\bigcup_{n\in\mathbb{N}_0} \mathbb{X}_n} = \overline{\bigcup_{n\in\mathbb{N}_0} \mathbb{Y}_n} = L^2(\Omega).$$

Moreover, $\{\mathbb{X}_n, \mathbb{Y}_n\}$ forms a regular pair (see Definition 2.30).

We use $X_n := \{f_{ij} : (i,j) \in \mathbb{U}_n\}$ and $Y_n := \{h_{ij} : (i,j) \in \mathbb{U}_n\}$ for the associated multiscale bases for $\mathbb{X}_n$ and $\mathbb{Y}_n$, respectively, where $\mathbb{U}_n = \{(i,j) : i \in \mathbb{Z}_{n+1}, j \in \mathbb{Z}_{w(i)}\}$, with $w(0) := k\nu = k'\mu$, $w(i) := k\nu(\mu - 1)\mu^{i-1} = k'(\mu - 1)\mu^i, i \in \mathbb{N}$, for given $k, \nu, k', \mu \in \mathbb{N}$. These bases can be constructed recursively by the method described in Section 4.1 such that both $\{f_{ij} : (i,j) \in \mathbb{U}\}$ and $\{h_{ij} : (i,j) \in \mathbb{U}\}$ are orthonormal bases in $\mathbb{X} = L^2[0, 1]$, having some important properties such as vanishing moment conditions

$$\int_0^1 t^\ell f_{ij}(t)\mathrm{d}t = 0, \quad \ell \in \mathbb{Z}_k, j \in \mathbb{Z}_{w(i)}, i \in \mathbb{N},$$
$$\int_0^1 t^\ell h_{ij}(t)\mathrm{d}t = 0, \quad \ell \in \mathbb{Z}_{k'}, j \in \mathbb{Z}_{w(i)}, i \in \mathbb{N}$$

and compact support properties

$$\mathrm{meas}(\mathrm{supp} f_{ij}) \leq 1/\mu^{i-1}, \quad \mathrm{meas}(\mathrm{supp} h_{ij}) \leq 1/\mu^{i-1}, \quad j \in \mathbb{Z}_{w(i)}, i \in \mathbb{N}.$$

The vanishing moment conditions play an important role in developing truncated schemes (see Chapter 5). Therefore, it is expected to raise the order of the vanishing moments of h_{ij} to k when $k' < k$. This can be done as follows.

We first choose basis $\{g_{0j} : j \in \mathbb{Z}_{w(0)}\}$ for $\mathbb{Y}_0$, which is bi-orthogonal to $\{f_{0j} : j \in \mathbb{Z}_{w(0)}\}$, that is,

$$(f_{0j'}, g_{0j}) = \delta_{jj'}, \ \ j, j' \in \mathbb{Z}_{w(0)}.$$

Then for $j \in \mathbb{Z}_{w(1)}$, we find a vector $[c_{js} : s \in \mathbb{Z}_{s(1)}] \in \mathbb{R}^{s(1)}$ where $s(i) := \dim \mathbb{Y}_i, i \in \mathbb{N}_0$, such that

$$g_{1j} := \sum_{s \in \mathbb{Z}_{w(0)}} c_{js} h_{0s} + \sum_{s \in \mathbb{Z}_{w(1)}} c_{j,w(0)+s} h_{1s}, \ \ j \in \mathbb{Z}_{w(1)}$$

satisfies the equations

$$(f_{0j'}, g_{1j}) = 0, \ \ j' \in \mathbb{Z}_{w(0)}$$

and

$$(f_{1j'}, g_{1j}) = \delta_{jj'}, \ \ j' \in \mathbb{Z}_{w(1)}.$$

Noting that the matrix of order $s(1)$ for this linear system of equations is

$$\mathbf{H} = [(f_{i'j'}, h_{ij}) : (i', j'), (i, j) \in \mathbb{U}_1],$$

and $\{\mathbb{X}_n, \mathbb{Y}_n\}$ forms a regular pair, we conclude that $\mathbf{H}$ is nonsingular. Thus there exists a unique solution which satisfies the above equations. It can easily be verified that these functions $g_{1j}, j \in \mathbb{Z}_{w(1)}$ are linearly independent, and $\mathbb{Y}_1 = \text{span}\{g_{ij} : (i, j) \in \mathbb{U}_1\}$. Using the isometry operator $\mathcal{T}_\epsilon$ (see (4.2)), we define recursively for $i \in \mathbb{N}$ that

$$g_{i+1,j} := \mathcal{T}_\epsilon g_{il},$$

where $j := \epsilon w(i) + l, \epsilon \in \mathbb{Z}_\mu, l \in \mathbb{Z}_{w(i)}$. Then we have that $\mathbb{Y}_n = \text{span}\{g_{ij} : (i, j) \in \mathbb{U}_n\}$ for $n \in \mathbb{N}_0$. Defining

$$\mathbb{W}_i := \text{span}\{f_{ij} : j \in \mathbb{Z}_{w(i)}\} \ \text{and} \ \mathbb{V}_i := \text{span}\{g_{ij} : j \in \mathbb{Z}_{w(i)}\}, \ \ i \in \mathbb{N}_0,$$

we have that

$$\mathbb{X}_n = \bigoplus_{i \in \mathbb{Z}_{n+1}}{}^{\perp} \mathbb{W}_i \ \text{and} \ \mathbb{Y}_n = \bigoplus_{i \in \mathbb{Z}_{n+1}} \mathbb{V}_i, \ \ n \in \mathbb{N}_0.$$

Proposition 6.1 *The multiscale bases $\{f_{ij} : (i, j) \in \mathbb{U}\}$ and $\{g_{ij} : (i, j) \in \mathbb{U}\}$ have the following properties.*

(I) There exist positive integers ρ and r such that for every $i > r$ and $j \in \mathbb{Z}_{w(i)}$ written in the form $j = \nu\rho + s$ where $s \in \mathbb{Z}_\rho$ and $\nu \in \mathbb{N}_0$,

$$f_{ij}(x) = 0, \ \ g_{ij}(x) = 0, \ \ x \notin \Omega_{i-r,\nu}.$$

Setting $S_{ij} := \Omega_{i-r,\nu}$, the supports of f_{ij} and g_{ij} are then contained in S_{ij}.

(II) *For any* $(i,j),(i',j') \in \mathbb{U}$,

$$(f_{ij}, f_{i'j'}) = \delta_{i'i}\delta_{j'j}.$$

(III) *For any* $(i,j),(i',j') \in \mathbb{U}$ *and* $i' \geq i$,

$$(f_{ij}, g_{i'j'}) = \delta_{i'i}\delta_{j'j}.$$

(IV) *For any* $(i,j) \in \mathbb{U}$ *with* $i \geq 1$, *and polynomial* p *of total degree less than* k,

$$(f_{ij}, p) = 0, \quad (g_{ij}, p) = 0.$$

(V) *There is a positive constant* c *such that for any* $(i,j) \in \mathbb{U}$,

$$\|f_{ij}\|_\infty \leq c\mu^{i/2} \text{ and } \|g_{ij}\|_\infty \leq c\mu^{i/2}.$$

Set

$$\mathbf{E}_n := [(g_{i'j'}, f_{ij}) : (i',j'),(i,j) \in \mathbb{U}_n].$$

It is useful to make the construction of the matrix $\mathbf{E}_n$ clear.

Lemma 6.2 *For any* $n \in \mathbb{N}$ *the following statements hold.*

(i) *The matrix* $\mathbf{E}_n$ *has the form*

$$\boldsymbol{E}_n = \begin{bmatrix} \boldsymbol{I}_0 & \boldsymbol{G}_0 & & & \\ & \boldsymbol{I}_1 & \boldsymbol{G}_1 & & \\ & & \ddots & \ddots & \\ & & & \ddots & \boldsymbol{G}_{n-1} \\ & & & & \boldsymbol{I}_n \end{bmatrix},$$

where $\boldsymbol{I}_i, i \in \mathbb{N}_0$ *is the* $w(i) \times w(i)$ *identity matrix,*

$$\boldsymbol{G}_0 := [(g_{0j'}, f_{1j}) : j' \in \mathbb{Z}_{w(0)}, j \in \mathbb{Z}_{w(1)}],$$
$$\boldsymbol{G}_1 := [(g_{1j'}, f_{2j}) : j' \in \mathbb{Z}_{w(1)}, j \in \mathbb{Z}_{w(2)}]$$

and $\boldsymbol{G}_i, i \in \mathbb{N}$ *is the block diagonal matrix* $\operatorname{diag}(\boldsymbol{G}_1, \boldsymbol{G}_1, \ldots, \boldsymbol{G}_1)$ *with* μ^{i-1} *diagonal blocks.*

(ii) *There exists a positive constant* c *such that*

$$\|\boldsymbol{E}_n\|_2 \leq c.$$

Proof (i) We first partition the matrix $\mathbf{E}_n$ into a block matrix

$$\mathbf{E}_n = [\mathbf{E}_{i'i} : i', i \in \mathbb{Z}_{n+1}],$$

where

$$\mathbf{E}_{i'i} = [(g_{i'j'}, f_{ij}) : j' \in \mathbb{Z}_{w(i')}, j \in \mathbb{Z}_{w(i)}].$$

It follows from property (III) that

$$\mathbf{E}_{i'i} = \begin{cases} \mathbf{I}_i, & i' = i, \\ 0, & i' > i. \end{cases}$$

When $i \geq i' + 2$, it follows from $g_{i'j'} \in \mathbb{Y}_{i'}, f_{ij} \in \mathbb{W}_i$ and the fact that $\mathbb{Y}_{i'} \subseteq \mathbb{X}_{i'+1} \perp \mathbb{W}_i$ that

$$(g_{i'j'}, f_{ij}) = 0,$$

which means that

$$\mathbf{E}_{i'i} = 0, \quad \text{for all } i \geq i' + 2.$$

We finally consider the case $i = i' + 1$. When $i' = 0$ and $i = 1$, it is clear that $\mathbf{E}_{01} = \mathbf{G}_0$. When $i' \geq 1$ and $i = i' + 1$, assume that $g_{i'j'} = \mathcal{T}_{\mathbf{e}'} g_{1l'}$ and $f_{ij} = \mathcal{T}_{\mathbf{e}} f_{2l}$, where $j' = \mu(\mathbf{e}')w(1) + l'$ and $j = \mu(\mathbf{e})w(2) + l$ with $\mathbf{e}', \mathbf{e} \in \mathbb{Z}_\mu^{i'-1}$, $l' \in \mathbb{Z}_{w(1)}$ and $l \in \mathbb{Z}_{w(2)}$. Using Proposition 4.15, we conclude that

$$(g_{i'j'}, f_{ij}) = \delta_{\mathbf{e}'\mathbf{e}}(g_{1l'}, f_{2l}).$$

This means that for $i' \geq 1$, $\mathbf{E}_{i',i'+1} = \mathbf{G}_i$ is the block diagonal matrix $\operatorname{diag}(\mathbf{G}_1, \mathbf{G}_1, \ldots, \mathbf{G}_1)$.

(ii) It is clear from (i) that

$$\|\mathbf{E}_n\|_\infty = \max\{\|\mathbf{G}_i\|_\infty + 1 : i \in \{0, 1\}\}$$

and

$$\|\mathbf{E}_n\|_1 = \max\{\|\mathbf{G}_i\|_1 + 1 : i \in \{0, 1\}\}.$$

Thus, we obtain that

$$\begin{aligned} \|\mathbf{E}_n\|_2 \leq c :&= \max\{\|\mathbf{E}_n\|_\infty, \|\mathbf{E}_n\|_1\} \\ &= \max\{\|\mathbf{G}_i\|_l + 1 : i \in \{0, 1\}, l \in \{1, \infty\}\}. \end{aligned}$$

This completes the proof. □

To estimate the norm of an element $u \in \mathbb{X}_n$ or $v \in \mathbb{Y}_n$, we introduce a sequence of functions $\{\xi_{ij} : (i,j) \in \mathbb{U}\}$ which is bi-orthogonal to $\{g_{ij} : (i,j) \in \mathbb{U}\}$. To obtain the sequence we can find $\{\xi_{ij} \in \mathbb{X}_i : (i,j) \in \mathbb{U}_1\}$ such that

$$(g_{i'j'}, \xi_{ij}) = \delta_{i'i}\delta_{j'j}, \quad (i',j'), (i,j) \in \mathbb{U}_1,$$

and then set

$$\xi_{ij} := \mathcal{T}_{\mathbf{e}}\xi_{1l}, \quad j = \mu(\mathbf{e})w(1) + l, \mathbf{e} \in \mathbb{Z}_\mu^{i-1}, l \in \mathbb{Z}_{w(1)}.$$

Using this sequence we have, for $v \in \mathbb{Y}_n$, that

$$v = \sum_{(i,j)\in\mathbb{U}_n} v_{ij} g_{ij},$$

with $v_{ij} = \langle v, \xi_{ij}\rangle$. Let

$$\Xi_n := [(\xi_{i'j'}, \xi_{ij}) : (i', j'), (i, j) \in \mathbb{U}_n].$$

Lemma 6.3 *There exists a positive constant c such that*

$$\|\Xi_n\|_2 \le c.$$

Proof We first estimate the entries of matrix $\Xi_{\mathbf{n}}$. The fact that $\{\xi_{ij} : (i, j) \in \mathbb{U}\}$ is bi-orthogonal to $\{g_{ij} : (i, j) \in \mathbb{U}\}$ implies that $\xi_{ij}, i \in \mathbb{N}$ has vanishing moments of order k'. For $i' \ge i$ and $i \in \mathbb{Z}_2$, let t_0 be the center of the set $S_{i'j'}$, and write $\xi_{ij} = \sum_{m\in\mathbb{Z}_k} c_m(t - t_0)^m$ on $S_{i'j'}$. There exists a positive constant c such that

$$|(\xi_{i'j'}, \xi_{ij})| \le cd(S_{i'j'})^{k'} \int_{S_{i'j'}} |\xi_{i'j'}(t)|\mathrm{d}t \le cd(S_{i'j'})^{k'+1/2}\|\xi_{i'j'}\| \le c\mu^{-i'(k'+1/2)}.$$

When $i' \ge i > 1$, there exist $\mathbf{e}', \mathbf{e} \in \mathbb{Z}_\mu^{i-1}$, $l' \in \mathbb{Z}_{w(i'-i+1)}$ and $l \in \mathbb{Z}_{w(1)}$ such that

$$j' = \mu(\mathbf{e}')w(i' - i + 1) + l', \;\; j = \mu(\mathbf{e})w(1) + l$$

and

$$\xi_{i'j'} = \mathcal{T}_{\mathbf{e}'}\xi_{i'-i+1,l'}, \;\; \xi_{ij} = \mathcal{T}_{\mathbf{e}}\xi_{1l}.$$

Thus

$$|(\xi_{i'j'}, \xi_{ij})| = \delta_{\mathbf{e}'\mathbf{e}}|(\xi_{i'-i+1,l'}, \xi_{1l})| \le c\delta_{\mathbf{e}'\mathbf{e}}\mu^{-(i'-i+1)(k'+1/2)}.$$

Combining the above estimates, we obtain for $(i, j), (i', j') \in \mathbb{U}_n$ that

$$|(\xi_{i'j'}, \xi_{ij})| \le c\delta_{\mathbf{e}'\mathbf{e}}\mu^{-|i'-i|(k'+1/2)},$$

where $\mathbf{e}', \mathbf{e} \in \mathbb{Z}_\mu^{|i'-i|}$.

We next partition Ξ_n into a block matrix

$$\Xi_n = [\Xi_{i'i} : i', i \in \mathbb{Z}_{n+1}],$$

with

$$\Xi_{i'i} := [(\xi_{i'j'}, \xi_{ij}) : j' \in \mathbb{Z}_{w(i')}, j \in \mathbb{Z}_{w(i)}],$$

and estimate the norm of these blocks. It can be seen that

$$\|\Xi_{i'i}\|_\infty = \max_{j'\in\mathbb{Z}_{w(i')}} \sum_{j\in\mathbb{Z}_{w(i)}} |(\xi_{i'j'}, \xi_{ij})| \le cw(|i'-i|)\mu^{-|i'-i|(k'+1/2)}$$
$$\le c\mu^{-|i'-i|(k'-1/2)}.$$

We next estimate matrix Ξ_n. Using the above inequality, we have that

$$\|\Xi_n\|_1 = \|\Xi_n\|_\infty \le \max_{i'\in\mathbb{Z}_{n+1}} \sum_{i\in\mathbb{Z}_{n+1}} \|\Xi_{i'i}\|_\infty \le \frac{2c}{1-\mu^{-(k'-1/2)}},$$

which leads to the desired result of this lemma. □

Using the above lemmas we can verify the following proposition.

Proposition 6.4 *There exist two positive constants c_- and c_+ such that for all $n \in \mathbb{N}_0$, $u \in \mathbb{X}_n$ having form $u = \sum_{(i,j)\in\mathbb{U}_n} u_{ij}f_{ij} = \sum_{(i,j)\in\mathbb{U}_n} \hat{u}_{ij}\xi_{ij}$, and $v \in \mathbb{Y}_n$ having form $v = \sum_{(i,j)\in\mathbb{U}_n} v_{ij}g_{ij}$,*

$$\|u\| = \|\boldsymbol{u}\|_2, \tag{6.1}$$
$$c_-\|u\| \le \|\hat{\boldsymbol{u}}\|_2 \le c_+\|u\| \tag{6.2}$$

and

$$c_-\|v\| \le \|\boldsymbol{v}\|_2 \le c_+\|v\|, \tag{6.3}$$

where $\boldsymbol{u} := [u_{ij} : (i,j) \in \mathbb{U}_n]$, $\hat{\mathbf{u}} := [\hat{u}_{ij} : (i,j) \in \mathbb{U}_n]$ and $\boldsymbol{v} := [v_{ij} : (i,j) \in \mathbb{U}_n]$.

Proof Recall that $\{f_{ij} : (i,j) \in \mathbb{U}\}$ is an orthonormal basis in $\mathbb{X}$, and $\{\xi_{ij} : (i,j) \in \mathbb{U}\}$ is bi-orthogonal to $\{g_{ij} : (i,j) \in \mathbb{U}\}$. Therefore, for $(i,j) \in \mathbb{U}_n$, $u_{ij} := (u, f_{ij})$, $\hat{u}_{ij} := (u, g_{ij})$ and $v_{ij} := (v, \xi_{ij})$. Moreover, equation (6.1) holds. It can easily be verified that $\|u\|^2 = \hat{\mathbf{u}}^T\Xi_n\hat{\mathbf{u}}$ and $\hat{\mathbf{u}} = \mathbf{E}_n\mathbf{u}$. Using Lemmas 6.2, 6.3 and (6.1), we have that

$$\|u\| \le (\|\Xi_n\|_2\|\hat{\mathbf{u}}\|_2^2)^{1/2} \le c\|\hat{\mathbf{u}}\|_2 \text{ and } \|\hat{\mathbf{u}}\|_2 \le \|\mathbf{E}_n\|\,\|\mathbf{u}\|_2 \le c\|u\|,$$

which yield (6.2).

Noting that $\mathbb{Y}_n \subseteq \mathbb{X}_{n+1}$, any $v \in \mathbb{Y}_n$ can be expressed as

$$v = \sum_{(i,j)\in\mathbb{U}_{n+1}} (v, g_{ij})\xi_{ij}.$$

Thus, we have that

$$\mathbf{v} = \Xi_{n+1}\hat{\mathbf{v}},$$

where $\mathbf{v} := [(v, \xi_{ij}) : (i,j) \in \mathbb{U}_{n+1}]$ and $\hat{\mathbf{v}} := [(v, g_{ij}) : (i,j) \in \mathbb{U}_{n+1}]$. By Lemma 3.37 and (6.2) we conclude that

$$\|\mathbf{v}\|_2 \leq \|\Xi_{n+1}\|_2 \|\hat{\mathbf{v}}\|_2 \leq c\|v\|. \tag{6.4}$$

On the contrary,

$$\begin{aligned} \|v\|^2 &= \left(\sum_{(i,j)\in\mathbb{U}_n} (v, \xi_{ij}) g_{ij}, \sum_{(i,j)\in\mathbb{U}_{n+1}} (v, g_{ij}) \xi_{ij} \right) \\ &= \sum_{(i,j)\in\mathbb{U}_n} (v, \xi_{ij})(v, g_{ij}) \\ &\leq \|\mathbf{v}\|_2 \|\hat{\mathbf{v}}\|_2 \leq c\|\mathbf{v}\|_2 \|v\|. \end{aligned}$$

This, with (6.4), yields (6.3). □

6.1.2 Multiscale Petrov–Galerkin methods

We now formulate the Petrov–Galerkin method using multiscale bases for Fredholm integral equations of the second kind given in the form

$$u - \mathcal{K}u = f, \tag{6.5}$$

where

$$(\mathcal{K}u)(s) := \int_\Omega K(s,t)u(t)\mathrm{d}t,$$

the function $f \in \mathbb{X} = L^2(\Omega)$, the kernel $K \in L^2(\Omega \times \Omega)$ are given and $u \in \mathbb{X}$ is the unknown function to be determined.

We assume that there are two sequences of multiscale functions $\{f_{ij} : (i,j) \in \mathbb{U}\}$ and $\{g_{ij} : (i,j) \in \mathbb{U}\}$, where $\mathbb{U} = \{(i,j) : j \in \mathbb{Z}_{w(i)}, i \in \mathbb{N}_0\}$, such that the subspaces

$$\mathbb{X}_n = \operatorname{span}\{f_{ij} : (i,j) \in \mathbb{U}_n\} \text{ and } \mathbb{Y}_n = \operatorname{span}\{g_{ij} : (i,j) \in \mathbb{U}_n\}$$

satisfy condition (H), and $\{\mathbb{X}_n, \mathbb{Y}_n\}$ forms a regular pair. These bases may not be those constructed in the above subsection, but they are demanded to satisfy the properties listed in Propositions 6.1 and 6.4.

The Petrov–Galerkin method for solving equation (6.5) seeks a vector $\mathbf{u}_n := [u_{ij} : (i,j) \in \mathbb{U}_n]$, such that the function

$$u_n := \sum_{(i,j)\in\mathbb{U}_n} u_{ij} f_{ij} \in \mathbb{X}_n$$

satisfies

$$(g_{i'j'}, u_n - \mathcal{K}u_n) = (g_{i'j'}, f), \quad (i', j') \in \mathbb{U}_n. \tag{6.6}$$

Equivalently, we obtain the linear system of equations

$$(\mathbf{E}_n - \mathbf{K}_n)\mathbf{u}_n = \mathbf{f}_n,$$

where

$$\mathbf{K}_n := [(g_{i'j'}, \mathcal{K}f_{ij}) : (i', j'), (i, j) \in \mathbb{U}_n],$$
$$\mathbf{E}_n := \left[(g_{i'j'}, f_{ij}) : (i', j'), (i, j) \in \mathbb{U}_n\right]$$

and

$$\mathbf{f}_n := [(g_{ij}, f) : (i, j) \in \mathbb{U}_n].$$

The truncated scheme and its analysis of convergence and computational complexity are nearly the same as for the multiscale Galerkin method, we leave them to the reader. Readers are also referred to the discrete version of the multiscale Petrov–Galerkin methods in the next section.

6.2 Discrete multiscale Petrov–Galerkin methods

One can find that the compression strategy for the design of the fast multiscale Petrov–Galerkin method is similar to that of the fast multiscale Galerkin method, and the practical use of the fast multiscale method requires the numerical computation of integrals appearing in the method. Therefore, in this section we turn our attention to discrete multiscale schemes. We develop a *discrete multiscale Petrov–Galerkin* (DMPG) method for integral equations of the second kind with weakly singular kernels. A compression strategy for designing fast algorithms is suggested. Estimates for the order of convergence and computational complexity of the method are provided.

We consider in this section the following Fredholm integral equations:

$$u - \mathcal{K}u = f, \tag{6.7}$$

where $\mathcal{K}$ is an integral operator with a weakly singular kernel.

The idea that we use to develop our DMPG method is to combine the discrete Petrov–Galerkin (DPG) method with multiscale bases to exploit the vanishing moment property of the multiscale bases and the computational algorithms for computing singular integrals of the DPG method. Note that the analysis of the DPG method was done in [80] in the L^∞-norm, since this norm is natural for discrete methods which use interpolatory projections. However, for our DMPG method, in order to make use of the vanishing moment property of the multiscale bases, we have to switch back and forth between the L^∞-norm and the L^2-norm, to obtain the necessary estimates. We give special attention to this issue.

6.2.1 DPG methods and L^p-stability

We review the abstract framework outlined in [80] for analysis of discrete numerical methods of Fredholm integral equations of the second kind with weakly singular kernels. To this end, we let $\mathbb{X}$ be a Banach space with norm $\|\cdot\|$ and $\mathbb{V}$ be a subspace of $\mathbb{X}$. We require that $\mathcal{K}:\mathbb{X}\to\mathbb{V}$ be a compact linear operator and that the integral equation (1.4) be uniquely solvable in $\mathbb{X}$ for all $f\in\mathbb{X}$. Note that whenever $f\in\mathbb{V}$, the unique solution of (1.4) is in $\mathbb{V}$. Let $\{\mathbb{X}_n : n\in\mathbb{N}\}$ be a sequence of finite-dimensional subspaces of $\mathbb{X}$ satisfying

$$\mathbb{V}\subseteq\tilde{\mathbb{X}}:=\overline{\bigcup_{n\in\mathbb{N}}\mathbb{X}_n}\subseteq\mathbb{X}.$$

Suppose that the operators $\mathcal{K}$ and $\mathcal{I}$ (the identity from $\mathbb{X}$ to $\mathbb{X}$) are approximated by operators $\mathcal{K}_n:\mathbb{X}\to\mathbb{V}$ and $\mathcal{Q}_n:\mathbb{X}\to\mathbb{X}_n$, respectively. Specifically, we assume that $\mathcal{K}_n$ and $\mathcal{Q}_n$ converge pointwise to $\mathcal{K}$ and $\mathcal{I}$, respectively. An approximation scheme for solving equation (1.4) is defined by the equation

$$(\mathcal{I}-\mathcal{Q}_n\mathcal{K}_n)u_n=\mathcal{Q}_nf,\quad n\in\mathbb{N}. \tag{6.8}$$

This approximate scheme includes the discrete and nondiscrete versions of the Petrov–Galerkin method, collocation method and quadrature method as special cases. Under various conditions elucidated in [80], for n large enough equation (6.8) has a unique solution. We discuss this issue later. Instead, we turn to specifying the operators and other related quantities needed for the definition of the DPG method for our current context. In this section we fix $\mathbb{X}:=L^\infty(\Omega)$ and $\mathbb{V}:=C(\Omega)$ with $\Omega:=[0,1]$, and use the following terminology about the singularity.

Definition 6.5 We say a kernel $K(s,t)$, $s,t\in\Omega:=[0,1]$, is quasi-weakly singular provided that

$$\sup_{s\in\Omega}\|K(s,\cdot)\|_1<\infty$$

and

$$\lim_{s'\to s}\|K(s,\cdot)-K(s',\cdot)\|_1=0,$$

where $\|\cdot\|_1$ is the $L^1(\Omega)$-norm on Ω.

It can easily be verified that the weakly singular kernel in the sense of Definition 2.4 with a continuous function M is quasi-weakly singular. Every quasi-weakly singular kernel determines, by the formula

$$(\mathcal{K}u)(s) := \int_{\Omega} K(s,t)u(t)\mathrm{d}t, \quad s \in \Omega, \quad u \in L^{\infty}(\Omega), \tag{6.9}$$

a compact operator from $L^{\infty}(\Omega)$ into $C(\Omega)$.

For $n \in \mathbb{N}$, we partition Ω into N (depending on n) subintervals $\Delta_0 = \{\Omega_i : i \in \mathbb{Z}_N\}$. That is, we have

$$\Omega = \bigcup_{r \in \mathbb{Z}_N} \Omega_r, \quad \operatorname{meas}(\Omega_i \cap \Omega_j) = 0, \quad i \neq j, \quad i, j \in \mathbb{Z}_N.$$

Moreover, we assume that as $n \to \infty$ the sequence of partition lengths

$$h := \max\{|\Omega_i| : i \in \mathbb{Z}_N\}$$

goes to zero. For each $i \in \mathbb{Z}_N$, let $\mathcal{F}_i$ denote the linear function that maps the interval Ω one to one and onto Ω_i. Thus, $\mathcal{F}_i$ has the form

$$\mathcal{F}_i(t) = |\Omega_i|t + b_i, \quad t \in \Omega, \quad i \in \mathbb{Z}_N \tag{6.10}$$

for some constant b_i. For every partition Δ_0 of Ω described above and any positive integer k, we let $\mathbb{S}_k(\Delta_0)$ be the space of all functions defined on Ω which are continuous from the right and on each subinterval Ω_i it is a polynomial of degree at most $k-1$ (at the right-most endpoint of Ω we require that the functions in $\mathbb{S}_k(\Delta_0)$ are left continuous).

We use the following mechanism to refine a given fixed partition $\Delta = \{J_j : j \in \mathbb{Z}_\ell\}$ chosen independently of n. For any $i \in \mathbb{Z}_{N\ell}$ written in the form $i = k\ell + j, j \in \mathbb{Z}_\ell, k \in \mathbb{Z}_N$ we define the intervals

$$H_i := \mathcal{F}_k(J_j)$$

which collectively determine the partition

$$\Delta_0 \circ \Delta := \{H_i : i \in \mathbb{Z}_{N\ell}\}.$$

This partition consists of N “copies” of Δ, each of which is put on the subintervals Ω_i, $i \in \mathbb{Z}_N$. Given two partitions Δ_1 and Δ_2 of Ω (*independent* of n) and positive integers k_1 and k_2, we introduce the following trial and test spaces:

$$\mathbb{X}_n := \{f : f \circ \mathcal{F}_i \in \mathbb{S}_{k_1}(\Delta_1), \ i \in \mathbb{Z}_N\} = \mathbb{S}_{k_1}(\Delta_0 \circ \Delta_1)$$

and

$$\mathbb{Y}_n := \mathbb{S}_{k_2}(\Delta_0 \circ \Delta_2),$$

respectively. These are also spaces of piecewise polynomials of degree $k_1 - 1$, $k_2 - 1$ on finer partitions induced by Δ_1, Δ_2 and Δ_0, respectively. To insure that the spaces $\mathbb{X}_n$ and $\mathbb{Y}_n$ have the same dimension, we require that

$$\dim \mathbb{S}_{k_1}(\Delta_1) = \dim \mathbb{S}_{k_2}(\Delta_2) = \lambda.$$

We choose bases in $\mathbb{X}_n$ and $\mathbb{Y}_n$ in the following manner. Starting with spaces

$$\mathbb{S}_{k_1}(\Delta_1) := \operatorname{span}\{\hat{\xi}_i : i \in \mathbb{Z}_\lambda\}$$

and

$$\mathbb{S}_{k_2}(\Delta_2) := \operatorname{span}\{\hat{\eta}_i : i \in \mathbb{Z}_\lambda\},$$

for $j = \lambda i + \ell$ where $i \in \mathbb{Z}_N$ and $\ell \in \mathbb{Z}_\lambda$, we define functions

$$\xi_j := (\hat{\xi}_\ell \circ \mathcal{F}_i^{-1})\chi_{\Omega_i} \text{ and } \eta_j := (\hat{\eta}_\ell \circ \mathcal{F}_i^{-1})\chi_{\Omega_i}.$$

These functions form a basis for spaces $\mathbb{X}_n$ and $\mathbb{Y}_n$, respectively, that is, $\mathbb{X}_n = \operatorname{span}\{\xi_j : j \in \mathbb{Z}_{\lambda N}\}$ and $\mathbb{Y}_n = \operatorname{span}\{\eta_j : j \in \mathbb{Z}_{\lambda N}\}$.

To construct a quadrature formula, we introduce a third piecewise polynomial space $\mathbb{S}_{k_3}(\Delta_3)$, of dimension γ, where Δ_3 is yet another partition of Ω (independent of n) and choose distinct points $\hat{t}_j$, $j \in \mathbb{Z}_\gamma$, in Ω such that there exist unique functions $\hat{\zeta}_i \in \mathbb{S}_{k_3}(\Delta_3)$, $i \in \mathbb{Z}_\gamma$, satisfying the interpolation conditions $\hat{\zeta}_i(\hat{t}_j) = \delta_{ij}$, $i, j \in \mathbb{Z}_\gamma$. The functions $\hat{\zeta}_i$, $i \in \mathbb{Z}_\gamma$ form a basis for the space $\mathbb{S}_{k_3}(\Delta_3)$. As above, for $j = \gamma i + \ell$ where $i \in \mathbb{Z}_N$ and $\ell \in \mathbb{Z}_\gamma$, we define functions

$$\zeta_j := (\hat{\zeta}_\ell \circ \mathcal{F}_i^{-1})\chi_{\Omega_i}$$

and points

$$t_j := \mathcal{F}_i(\hat{t}_\ell).$$

We also introduce the subspace

$$\mathbb{Q}_n := \mathbb{S}_{k_3}(\Delta_0 \circ \Delta_3)$$

and observe that

$$\mathbb{Q}_n = \operatorname{span}\{\zeta_j : j \in \mathbb{Z}_{\gamma N}\}.$$

We require the linear projection $\mathcal{Z}_n : \mathbb{X} \to \mathbb{Q}_n$ by

$$\mathcal{Q} = \mathcal{Z}_n g = \sum_{j \in \mathbb{Z}_{\gamma N}} g(t_j)\zeta_j, \tag{6.11}$$

where for a function $g \in L^\infty(\Omega)$, $g(t)$ is defined in the sense described in Section 3.5.3, that is, as any norm-preserving bounded linear functional which extends point evaluation at t from $C(\Omega)$ to $L^\infty(\Omega)$ (cf. [21]). For any $x, y \in \mathbb{X}$, we introduce the following discrete inner product:

$$(x, y)_n = \sum_{j \in \mathbb{Z}_{\gamma N}} w_j x(t_j) y(t_j), \tag{6.12}$$

where

$$w_j = \int_\Omega \zeta_j(t)\mathrm{d}t.$$

Note that

$$w_j = |\Omega_i| \int_\Omega \hat{\zeta}_\ell(t)\mathrm{d}t = |\Omega_i|\hat{w}_\ell,$$

where $j = \gamma i + \ell$, with $i \in \mathbb{Z}_N$ and $\ell \in \mathbb{Z}_\gamma$ and for every $\ell \in \mathbb{Z}_\gamma$, $\hat{w}_\ell := \int_\Omega \hat{\zeta}_\ell(t)\mathrm{d}t$. Henceforth, we assume that $\hat{w}_\ell > 0$, $\ell \in \mathbb{Z}_\gamma$. This way, $\|x\|_n := (x, x)_n^{1/2}$ is a semi-norm on $\mathbb{X}$.

We now define a pair of operators using the discrete inner product. Specifically, we define the operator $\mathcal{Q}_n : L^\infty(\Omega) \to \mathbb{X}_n$ by requiring

$$(\mathcal{Q}_n x, y)_n = (x, y)_n, \quad y \in \mathbb{Y}_n. \tag{6.13}$$

An element $\mathcal{Q}_n x \in \mathbb{X}_n$ satisfying (6.13) is called the *discrete generalized best approximation* (DGBA) to x from $\mathbb{X}_n$ with respect to $\mathbb{Y}_n$. Similarly, we let $\mathcal{Q}'_n : L^\infty(\Omega) \to \mathbb{Y}_n$ be the *discrete generalized best approximation projection* from $L^\infty(\Omega)$ onto $\mathbb{Y}_n$ with respect to $\mathbb{X}_n$, defined by the equation

$$(v, \mathcal{Q}'_n x)_n = (v, x)_n, \quad v \in \mathbb{X}_n. \tag{6.14}$$

The following lemma, proved in [80], presents a necessary and sufficient condition for $\mathcal{Q}_n$ and $\mathcal{Q}'_n$ to be well defined. To state this lemma, we introduce a matrix notation. Let

$$\Phi := [\hat{\xi}_i(\hat{t}_j) : i \in \mathbb{Z}_\lambda, j \in \mathbb{Z}_\gamma], \quad \Psi := [\hat{\eta}_i(\hat{t}_j) : i \in \mathbb{Z}_\lambda, j \in \mathbb{Z}_\gamma],$$
$$\mathbf{W} := \mathrm{diag}(\hat{w}_j : j \in \mathbb{N}_\gamma)$$

and define the square matrix of order λ

$$\mathbf{M} := \Phi \mathbf{W} \Psi^T.$$

Lemma 6.6 *Let $x \in L^\infty(\Omega)$. Then the following statements are equivalent.*

(i) The discrete generalized best approximation to x from $\mathbb{X}_n$ with respect to $\mathbb{Y}_n$ is well defined.

(ii) The discrete generalized best approximation x from $\mathbb{Y}_n$ with respect to $\mathbb{X}_n$ is well defined.

(iii) *The functions* $\{\hat{\xi}_i : i \in \mathbb{Z}_\lambda\}$, $\{\hat{\eta}_i : i \in \mathbb{Z}_\lambda\}$ *and the points* $\{\hat{t}_i : i \in \mathbb{Z}_\gamma\}$ *have the property that* $\mathbf{M}$ *is nonsingular.*

Moreover, under any one of these conditions, the operators $\mathcal{Q}_n$ *and* $\mathcal{Q}'_n$ *are uniformly bounded projections with* $\mathcal{Q}_n(\mathbb{Y}_n) = \mathbb{X}_n$ *and* $\mathcal{Q}'_n(\mathbb{X}_n) = \mathbb{Y}_n$.

It remains to define the operator $\mathcal{K}_n$. For this purpose, we express the quasi-weakly singular kernel K as K_1K_2, where $K_1 \in C(\Omega \times \Omega)$ and K_2 is quasi-weakly singular. Using this factorization, we develop a product integration formula that discretizes the kernel K_1 but not the kernel K_2. Specifically, we define the operator $\mathcal{K}_n : L^\infty(\Omega) \to C(\Omega)$ by the formula

$$(\mathcal{K}_n x)(s) := \int_\Omega \mathcal{Z}_n(K_1(s,\cdot)x(\cdot))(t)K_2(s,t)\mathrm{d}t, \quad s \in \Omega, \quad x \in L^\infty(\Omega) \tag{6.15}$$

and note that

$$(\mathcal{K}_n x)(s) = \sum_{j\in\mathbb{Z}_{\gamma N}} W_j(s)K_1(s,t_j)x(t_j), \quad s \in \Omega$$

where, for all $j \in \mathbb{Z}_{\gamma N}$, we define the function

$$W_j(s) := \int_\Omega K_2(s,t)\zeta_j(t)\mathrm{d}t, \quad s \in \Omega.$$

With the approximate operators $\mathcal{K}_n$ and $\mathcal{Q}_n$ as defined above, equation (6.8) specifies a DPG scheme for solving (6.7).

We describe two special constructions of the triple of spaces $\mathbb{S}_{k_i}(\Delta_i)$, $i = 1,2,3$, such that the matrix $\mathbf{M}$ is nonsingular. In the first construction, we assume that $k_1 \le k_3$ and $k_1 = rk_2$, where r is a positive integer, and choose the partitions $\Delta_1 = \Delta_3 = \Omega$, that is, $\mathbb{S}_{k_1}(\Delta_1)$ and $\mathbb{S}_{k_3}(\Delta_3)$ are spaces of polynomials of degree $k_1 - 1$ and $k_3 - 1$, respectively. We then choose k_3 points $\hat{t}_0 < \hat{t}_1 < \cdots < \hat{t}_{k_3-1}$, so that the weights $\hat{w}_i > 0$ for $i \in \mathbb{Z}_{k_3}$. Thus, in this case, $\lambda = k_1$ and $\gamma = k_3$. Now we define the partition $\Delta_2 := \{[x_i, x_{i+1}] : i \in \mathbb{Z}_r\}$ of the interval Ω by letting $x_0 = 0$, $x_r = 1$ and choosing $x_i \in (\hat{t}_{ik_2-1}, \hat{t}_{ik_2})$, $i - 1 \in \mathbb{Z}_{r-1}$. Thus, for $i \in \mathbb{Z}_{r-1}$, each of the first $r - 1$ subintervals $[x_i, x_{i+1}]$ contains exactly k_2 points $\hat{t}_{ik_2+j}$, $j \in \mathbb{Z}_{k_2}$ and the last subinterval $[x_{r-1}, x_r]$ contains exactly $k_3 - k_1 + k_2$ points $\hat{t}_{(r-1)k_2+j}$, $j \in \mathbb{Z}_{k_3-k_1+k_2}$. We call this a type I construction for the triple of spaces $\mathbb{S}_{k_i}(\Delta_i)$, $i = 1,2,3$.

Proposition 6.7 *For type I spaces* $\mathbb{S}_{k_i}(\Delta_i)$, $i = 1,2,3$, $\det(\mathbf{M}) \neq 0$ *holds.*

Proof Let

$$\Phi\begin{pmatrix} 0,\ldots,k_1-1 \\ j_1,\ldots,j_{k_1}\end{pmatrix} \quad \text{and} \quad \Psi\begin{pmatrix} 0,\ldots,k_1-1 \\ j_1,\ldots,j_{k_1}\end{pmatrix}$$

denote the minors of matrices Φ and Ψ, corresponding to the columns $j_1, \ldots, j_{k_1}$, respectively. We have by the Cauchy–Binet formula that

$$\det(\mathbf{M}) = \sum_{0 \leq j_1 < \cdots < j_{k_1} \leq k_3 - 1} w_{j_1} \cdots w_{j_{k_1}} \Phi \begin{pmatrix} 0, \ldots, k_1 - 1 \\ j_1, \ldots, j_{k_1} \end{pmatrix} \Psi \begin{pmatrix} 0, \ldots, k_1 - 1 \\ j_1, \ldots, j_{k_1} \end{pmatrix}.$$

Let Φ_1 and Ψ_1 be the submatrices of Φ and Ψ consisting of the first k_1 columns, respectively. Since the triple spaces $\mathbb{S}_{k_i}(\Delta_i)$, $i = 1, 2, 3$, are type I constructions, both these matrices are invertible and hence it follows that

$$\det(\mathbf{M}) \geq \det(\Phi_1) \det(\Psi_1) w_1 \cdots w_{k_1} > 0. \qquad \square$$

Note that for a type I construction the partition Δ_2 may not be *equally spaced*. As our multiscale construction requires this property for both Δ_1 and Δ_2, we present a second construction which guarantees that this is the case. For this purpose, we again assume that $k_1 = rk_2$, where r is a positive integer and $\mathbb{S}_{k_1}(\Delta_1)$ is the space of polynomials of degree $k_1 - 1$ on Ω. For the remaining spline spaces we divide the interval Ω into r equally space subintervals, that is,

$$\Delta_2 = \Delta_3 := \left\{ \left[\frac{i}{r}, \frac{i+1}{r} \right] : i \in \mathbb{Z}_r \right\}.$$

Hence, in this case, $\lambda = k_1$ and $\gamma = rk_3$. We choose the points $\hat{t}_j$, $j \in \mathbb{Z}_r$ in the following way. On each interval $[\frac{i}{r}, \frac{i+1}{r}]$, $i \in \mathbb{Z}_r$ we choose the zeros of the Legendre polynomial of degree k_3 ($\geq k_1/2$) shifted from $[-1, 1]$ to the interval $[\frac{i}{r}, \frac{i+1}{r}]$ and we number these points increasingly by $\hat{t}_i$, $i \in \mathbb{Z}_{rk'}$. In this choice, we have $\hat{w}_i > 0$, $i \in \mathbb{Z}_{rk'}$. We call this a type II construction.

Proposition 6.8 *For type II spaces* $\mathbb{S}_{k_i}(\Delta_i)$, $i = 1, 2, 3$, $\det(\mathbf{M}) \neq 0$ *holds.*

Proof By construction, we conclude from a similar argument used in the proof of Proposition 6.7 that the determinant of the matrix $\mathbf{M}$ is positive. $\square$

We say that the triple of spaces $\mathbb{S}_{k_i}(\Delta_i)$, $i = 1, 2, 3$, forms an acceptable triple provided that the matrix $\mathbf{M}$ is nonsingular.

According to Lemma 6.6, the discrete approximate operators $\mathcal{Q}_n$ and $\mathcal{Q}'_n$ are bounded uniformly in $n \in \mathbb{N}$ on $L^\infty(\Omega)$. For the analysis of DMPG, we need similar properties on $L^2(\Omega)$. Our first comment is of a general nature. To this end, let $\mathbb{P} := \mathbb{S}_k(\Delta_0 \circ \Delta)$, where Δ is a fixed partition of Ω chosen independent of n. For any linear operator $\mathcal{A}$ such that $\mathcal{A} : L^p(\Omega) \to L^\infty(\Omega)$, $1 \leq p \leq \infty$, we set

$$\|\mathcal{A}\|_{p,\mathbb{P}} := \sup\{\|\mathcal{A}x\|_p : x \in \mathbb{P}, \|x\|_p = 1\}.$$

The next lemma gives a condition on an operator $\mathcal{A}$ such that $\|\mathcal{A}\|_{p,\mathbb{P}}$ can be bounded by a constant independent of n times $\|\mathcal{A}\|_{\infty,\mathbb{P}}$.

Lemma 6.9 *Let $\mathcal{A} : L^\infty(\Omega) \to L^\infty(\Omega)$ be a linear operator such that for $x \in \mathbb{P}$ there holds*

$$(\mathcal{A}x)\chi_{I_i} = \mathcal{A}(x\chi_{I_i}), \quad i \in \mathbb{Z}_N \tag{6.16}$$

where χ_{Ω_i} is the characteristic function of the set Ω_i. Then for any $1 \le p \le \infty$, there exists a positive constant α depending only on k and Δ such that

$$\|\mathcal{A}\|_{p,\mathbb{P}} \le \alpha\|\mathcal{A}\|_{\infty,\mathbb{P}}.$$

Proof For $v \in L^2(\Omega_i)$ we have that

$$\|v\|_{L^p(\Omega_i)} = |\Omega_i|^{1/p}\|v \circ \mathcal{F}_i\|_{L^p(\Omega)}. \tag{6.17}$$

We conclude from (6.17) that for $x \in \mathbb{P}$,

$$\|\mathcal{A}x\|_{L^p(\Omega_i)} = |\Omega_i|^{1/p}\|(\mathcal{A}x) \circ \mathcal{F}_i\|_{L^p(\Omega)} \le |\Omega_i|^{1/p}\|(\mathcal{A}x) \circ \mathcal{F}_i\|_{L^\infty(\Omega)}.$$

Moreover, using assumption (6.16), we obtain that

$$\|(\mathcal{A}x) \circ \mathcal{F}_i\|_{L^\infty(\Omega)} = \|(\mathcal{A}x)\chi_{\Omega_i}\|_{L^\infty(\Omega)} = \|\mathcal{A}(x\chi_{\Omega_i})\|_{L^\infty(\Omega)}.$$

Combining this fact with (6.17) gives us the inequality

$$\|\mathcal{A}x\|_{L^p(\Omega_i)} \le \|\mathcal{A}\|_{\infty,\mathbb{P}}|\Omega_i|^{1/p}\|x \circ \mathcal{F}_i\|_{L^\infty(\Omega)}.$$

For any $i \in \mathbb{Z}_N$, $\{x \circ \mathcal{F}_i : x \in \mathbb{P}\} = \mathbb{S}_k(\Delta)$ and so there is a positive constant α, depending only on k and Δ, such that

$$\|x \circ \mathcal{F}_i\|_{L^\infty(\Omega)} \le \alpha\|x \circ \mathcal{F}_i\|_{L^p(\Omega)}.$$

Therefore, we have confirmed for $x \in \mathbb{P}$ and $i \in \mathbb{Z}_N$ that

$$\|\mathcal{A}x\|_{L^p(\Omega_i)} \le \alpha\|\mathcal{A}\|_{\infty,\mathbb{P}}\|x\|_{L^p(\Omega_i)}.$$

Summing both sides of the inequality over all $i \in \mathbb{Z}_N$ completes the proof. □

Lemma 6.10 *For any $1 \le p \le \infty$, there exists a constant $c > 0$ such that for $n \in \mathbb{N}$*

$$\|\mathcal{Q}_n\|_{p,\mathbb{X}_n} + \|\mathcal{Q}_n\|_{p,\mathbb{Y}_n} \le c, \quad \|\mathcal{Q}'_n\|_{p,\mathbb{X}_n} + \|\mathcal{Q}'_n\|_{p,\mathbb{Y}_n} \le c.$$

Proof We employ Lemma 6.9 to prove this result. In view of Lemma 6.6, the operators $\mathcal{Q}_n$ are uniformly bounded in L^∞. Therefore, it suffices to verify that the following conditions are satisfied:

$$(\mathcal{Q}_n x) \cdot \chi_{\Omega_i} = \mathcal{Q}_n(x \cdot \chi_{\Omega_i}), \quad x \in \mathbb{X}_n$$

and

$$(\mathcal{Q}_n y) \cdot \chi_{\Omega_i} = \mathcal{Q}_n(y \cdot \chi_{\Omega_i}), \quad y \in \mathbb{Y}_n.$$

To prove this fact, it is useful to introduce another discrete inner product by the formula

$$[x, y] := \sum_{\ell \in \mathbb{Z}_\gamma} \hat{w}_\ell x(\hat{t}_\ell) y(\hat{t}_\ell)$$

and corresponding to this inner product the DGBA $\hat{\mathcal{Q}}x$ from $\mathbb{S}_{k_1}(\Delta_1)$ to $x \in L^\infty(\Omega)$ with respect to $\mathbb{S}_{k_2}(\Delta_2)$. Therefore, by definition we conclude for any $i \in \mathbb{Z}_N$ and $x \in L^\infty(\Omega)$ that

$$(\mathcal{Q}_n x) \circ \mathcal{F}_i = \hat{\mathcal{Q}}(x \circ \mathcal{F}_i).$$

This proves the first estimate and the second estimate can be obtained similarly. □

Let $\mathcal{X}_n : L^2(\Omega) \to \mathbb{X}_n$ and $\mathcal{Y}_n : L^2(\Omega) \to \mathbb{Y}_n$ be the orthogonal projections from $L^2(\Omega)$ onto $\mathbb{X}_n$ and $\mathbb{Y}_n$, respectively.

Lemma 6.11 *Let K be a quasi-weakly singular kernel in the factored form $K = K_1 K_2$ where K_2 is quasi-weakly singular,*

$$\sup_{t \in \Omega} \|K_2(\cdot, t)\|_1 < \infty$$

and K_1 is continuous on $\Omega \times \Omega$. Then the set of operators $\{\mathcal{K}_n \mathcal{X}_n : n \in \mathbb{N}\}$ is uniformly bounded in the space $L^2(\Omega)$.

Proof To prove the uniform boundedness of the sequence of operators $\mathcal{K}_n \mathcal{X}_n$, for every $f \in L^2(\Omega)$ we introduce the function

$$e(t) := \sum_{j \in \mathbb{Z}_{\gamma N}} |(\mathcal{X}_n f)(t_j)||\zeta_j(t)|, \quad t \in \Omega,$$

which may be rewritten as

$$e(t) = \sum_{i \in \mathbb{Z}_N} \sum_{\ell \in \mathbb{Z}_\gamma} |(\mathcal{X}_n f)(\mathcal{F}_i(\hat{t}_\ell))||(\hat{\zeta}_\ell \circ \mathcal{F}_i^{-1}) \chi_{\Omega_i}(t)|.$$

We observe for any $s \in \Omega$ that

$$|(\mathcal{K}_n \mathcal{X}_n f)(s)| \leq \|K_1\|_\infty |(\mathcal{G} e)(s)|, \tag{6.18}$$

where $\|K_1\|_\infty$ is the L^∞-norm of K_1 on $\Omega \times \Omega$ and $\mathcal{G}$ is the integral operator with kernel $G = |K_2|$. By the Cauchy–Schwarz inequality, we conclude that

$$\|\mathcal{G} e\|_2 \leq \beta \|e\|_2, \tag{6.19}$$

where

$$\beta := \frac{1}{2}\left[\sup_{s\in\Omega}\|K_2(s,\cdot)\|_1 + \sup_{t\in\Omega}\|K_2(\cdot,t)\|_1\right].$$

It remains to bound $\|e\|_2$. To this end, we note that

$$(\hat{\zeta}_\ell \circ \mathcal{F}_i^{-1})\chi_{\Omega_i}(\hat{\zeta}_{\ell'} \circ \mathcal{F}_{i'}^{-1})\chi_{\Omega_{i'}} = 0, \ \text{ if } i \neq i'.$$

It follows that

$$(e(t))^2 = \sum_{i\in\mathbb{Z}_N}\left(\sum_{\ell\in\mathbb{Z}_\gamma}\left|(\mathcal{X}_n f)(\mathcal{F}_i(\hat{t}_\ell))\right|\left|(\hat{\zeta}_\ell \circ \mathcal{F}_i^{-1})\chi_{\Omega_i}(t)\right|\right)^2,$$

and thus by the Cauchy–Schwarz inequality we have that

$$\|e\|_2^2 \leq \sum_{i\in\mathbb{Z}_N}\sum_{\ell\in\mathbb{Z}_\gamma}\left|(\mathcal{X}_n f)(\mathcal{F}_i(\hat{t}_\ell))\right|^2 \sum_{\ell\in\mathbb{Z}_\gamma}\int_\Omega \left|(\hat{\zeta}_\ell \circ \mathcal{F}_i^{-1})\chi_{\Omega_i}(t)\right|^2 \mathrm{d}t.$$

Note that for each $i \in \mathbb{Z}_N$ and $f \in L^2(\Omega)$,

$$(\mathcal{X}_n f) \circ \mathcal{F}_i = \hat{\mathcal{X}}(f \circ \mathcal{F}_i),$$

where $\hat{\mathcal{X}}$ is the orthogonal projection of $L^2(\Omega)$ onto $\mathbb{S}_{k_1}(\Delta_1)$. We conclude that

$$\|e\|_2^2 \leq \mu \sum_{i\in\mathbb{Z}_N}\sum_{j\in\mathbb{Z}_\gamma}|\hat{\mathcal{X}}(f \circ \mathcal{F}_i)(\hat{t}_j)|^2|\Omega_i|,$$

where

$$\mu := \sum_{j\in\mathbb{Z}_\gamma}\int_\Omega |\hat{\zeta}_j(t)|^2 \mathrm{d}t.$$

We let $\|\hat{\mathcal{X}}\|_{2,\infty}$ denote its norm as an operator from $L^2(\Omega)$ into $L^\infty(\Omega)$. Therefore,

$$\|e\|_2^2 \leq \gamma\mu\|\hat{\mathcal{X}}\|_{2,\infty}^2 \sum_{i\in\mathbb{Z}_N}\|f \circ \mathcal{F}_i\|_2^2|\Omega_i| = \gamma\mu\|\hat{\mathcal{X}}\|_{2,\infty}^2\|f\|_2^2, \tag{6.20}$$

which establishes the uniform boundedness of the operators $\{\mathcal{K}_n\mathcal{X}_n : n \in \mathbb{N}\}$. □

We need to prove that the set of operators $\{\mathcal{K}_n\mathcal{X}_n : n \in \mathbb{N}\}$ with a special requirement on the kernel K_2 is collectively compact on $L^2(\Omega)$. We recall that a set $\tau := \{\mathcal{A}\}$ of linear operators mapping a normed space $\mathbb{X}$ into a normed space $\mathbb{Y}$ is called *collectively compact* if for each bounded set $S \subseteq \mathbb{X}$, the image set $\{\mathcal{A}x : x \in S, \mathcal{A} \in \tau\}$ is relatively compact (see [6]).

We further require that the kernel K_2 has the *α-property*, namely

$$K_2(s,t) = \frac{A(s,t)}{|s-t|^\alpha}, \quad s,t \in \Omega,$$

where A is a continuous function on $\Omega \times \Omega$ and α is a constant satisfying $0 \leq \alpha < 1$.

We now prove the collective compactness of the set $\{\mathcal{K}_n\mathcal{X}_n : n \in \mathbb{N}\}$.

Theorem 6.12 *Let K be a quasi-weakly singular kernel in the factored form $K = K_1K_2$, where K_2 has the α-property and K_1 is continuous on $\Omega \times \Omega$. Let $\mathcal{A}_n := \mathcal{K}_n\mathcal{X}_n$. Then the set of operators $\{\mathcal{A}_n : n \in \mathbb{N}\}$ is collectively compact on the space $L^2(\Omega)$.*

Proof By the definition of collective compactness, we are required to confirm that for any bounded set $S := \{f \in L^2(\Omega) : \|f\|_2 \leq c_0\}$ where c_0 is a positive constant, the set $\{\mathcal{A}_nf : f \in S, n \in \mathbb{N}\}$ is relatively compact. This will be done using the Kolmogorov theorem (Theorem A.44). The first condition in the theorem, that is, the uniform boundedness of the set $\{\mathcal{A}_n : n \in \mathbb{N}\}$, has been verified in Lemma 6.11. It remains to verify conditions (ii) and (iii) in Theorem A.44. We first show that the sequence of operators $\mathcal{A}_n$ has the property that

$$\lim_{h\to 0}\left(\int_0^{1-h} |(\mathcal{A}_nf)(s+h) - (\mathcal{A}_nf)(s)|^2 \mathrm{d}s\right)^{\frac{1}{2}} = 0, \tag{6.21}$$

uniformly in $n \in \mathbb{N}$ and $f \in S$. To this end, we note that

$$\left(\int_0^{1-h} |(\mathcal{A}_nf)(s+h) - (\mathcal{A}_nf)(s)|^2 \mathrm{d}s\right)^{\frac{1}{2}} \leq r_1 + r_2,$$

where

$$r_1 := \left(\int_0^{1-h}\left|\int_0^1 \mathcal{Z}_n((K_1(s+h,\cdot) - K_1(s,\cdot))(\mathcal{X}_nf)(\cdot))(t)K_2(s+h,t)\mathrm{d}t\right|^2 \mathrm{d}s\right)^{\frac{1}{2}}$$

and

$$r_2 := \left(\int_0^{1-h}\left|\int_0^1 \mathcal{Z}_n(K_1(s,\cdot)(\mathcal{X}_nf)(\cdot))(t)[K_2(s+h,t) - K_2(s,t)]\mathrm{d}t\right|^2 \mathrm{d}s\right)^{\frac{1}{2}}.$$

We first bound the term r_1. By the definition of $\mathcal{Z}_n$, we have that

$$r_1 \leq \sup_{t\in\Omega}\sup_{s\in\Omega_h} |K_1(s+h,t) - K_1(s,t)|\|\mathcal{G}e\|_2,$$

where $\Omega_h := [0, 1-h]$. Using (6.19) and (6.20), we conclude that

$$r_1 \le \beta\gamma^{\frac{1}{2}}\mu^{\frac{1}{2}}\|\hat{\mathcal{X}}\|_{2,\infty}\|f\|_2 \sup_{t\in\Omega}\sup_{s\in\Omega_h}|K_1(s+h,t)-K_1(s,t)|.$$

To estimate the term r_2, we make use of the property of the kernel K_2 that K_2 is continuous off the diagonal. Specifically, for any $s \in (0,1)$ and $\delta > 0$, we let $U(s,\delta) := (s-\delta, s+\delta)$ and $U'(s,\delta) := \Omega \setminus U(s,\delta)$ and write

$$r_2 \le r' + r'' + r''',$$

where

$$r' := \left[\int_0^{1-h}\left(\int_{U(s,2\delta)}|\mathcal{Z}_n(K_1(s,\cdot)(\mathcal{X}_n f)(\cdot))(t)||K_2(s,t)|\mathrm{d}t\right)^2\mathrm{d}s\right]^{\frac{1}{2}},$$

$$r'' := \left[\int_0^{1-h}\left(\int_{U(s,2\delta)}|\mathcal{Z}_n(K_1(s,\cdot)(\mathcal{X}_n f)(\cdot))(t)||K_2(s+h,t)|\mathrm{d}t\right)^2\mathrm{d}s\right]^{\frac{1}{2}}$$

and

$$r''' := \left[\int_0^{1-h}\left|\int_{U'(s,2\delta)}|\mathcal{Z}_n(K_1(s,\cdot)(\mathcal{X}_n f)(\cdot))(t)[K_2(s+h,t)-K_2(s,t)]\mathrm{d}t\right|^2\mathrm{d}s\right]^{\frac{1}{2}}.$$

We observe by using the α-property of the kernel K_2 and by a straightforward computation that

$$\begin{aligned} r' &\le \|K_1\|_\infty\left[\int_0^{1-h}\left(\int_{U(s,2\delta)}|K_2(s,t)|e(t)\mathrm{d}t\right)^2\mathrm{d}s\right]^{\frac{1}{2}} \\ &\le A\frac{2^{1-\frac{\alpha}{2}}}{1-\alpha}\|K_1\|_\infty\mu^{\frac{1}{2}}\beta^{\frac{1}{2}}\|\hat{\mathcal{X}}\|_{2,\infty}\delta^{\frac{1-\alpha}{2}}\|f\|_2 \\ &\le C\delta^{\frac{1-\alpha}{2}}\|f\|_2, \end{aligned}$$

where

$$A := \sup_{s,t\in\Omega}|A(s,t)|.$$

To bound r'' we note that if $h < \delta$, then $U(s-h, 2\delta) \subset U(s, 3\delta)$ and thus by change of variables we conclude that

$$r'' \le \left[\int_0^1\left(\int_{U(s,3\delta)}|\mathcal{Z}_n(K_1(s,\cdot)(\mathcal{X}_n f)(\cdot))(t)||K_2(s,t)|\mathrm{d}t\right)^2\mathrm{d}s\right]^{\frac{1}{2}}.$$

Likewise, we have that

$$r'' \leq c\delta^{\frac{1-\alpha}{2}} \|f\|_2.$$

For any $\epsilon > 0$, we find a $\delta > 0$ such that $c\delta^{\frac{1-\alpha}{2}} < \frac{\epsilon}{4}$. Hence, we observe that

$$r' + r'' < \frac{1}{2}\epsilon \|f\|_2.$$

Finally, we deal with the term r'''. By the α-property of the kernel K_2, K_2 is continuous off the diagonal and hence

$$\begin{aligned} r''' &\leq \|K_1\|_\infty \|e\|_2 \left(\int_0^1 \int_{U'(s,2\delta)} |K_2(s+h,t) - K_2(s,t)|^2 \mathrm{d}t\mathrm{d}s \right)^{\frac{1}{2}} \\ &\leq c\|f\|_2 \left(\int_0^1 \int_{U'(s,2\delta)} |K_2(s+h,t) - K_2(s,t)|^2 \mathrm{d}t\mathrm{d}s \right)^{\frac{1}{2}}. \end{aligned}$$

When $s \in (0,1), t \in U'(s,2\delta)$ and $h < \delta$, we have $|s+h-t| \geq \delta$ so that both points (s,t) and $(s+h,t)$ are contained in the set

$$D_\delta := \{(s,t) \in \Omega \times \Omega : |s-t| \geq \delta\}.$$

For fixed number $\delta > 0$, the function $K_2(s,t)$ is bounded and uniformly continuous with respect to the totality of s and t on D_δ. Therefore, there exists a constant σ_1 with $0 < \sigma_1 \leq \delta$ such that, when $h < \sigma_1$, $(s,t), (s+h,t) \in D_\delta$,

$$|K_2(s,t) - K_2(s+h,t)| < \frac{\epsilon}{4c}.$$

Thus,

$$\left(\int_\Omega \|K_2(s,\cdot) - K_2(s+h,\cdot)\|^2_{L^2(U'(s,2\delta))} \mathrm{d}s \right)^{1/2} \leq \frac{\epsilon}{4c}.$$

In summary, we have established the estimate

$$\left(\int_0^{1-h} |(\mathcal{A}_n f)(s+h) - (\mathcal{A}_n f)(s)|^2 \mathrm{d}s \right)^{\frac{1}{2}} < \epsilon \|f\|_2$$

and thus proved equation (6.21).

Finally, we verify that

$$\lim_{h\to 0} \left(\int_{1-h}^1 |(\mathcal{A}_n f)(s)|^2 \mathrm{d}s \right)^{\frac{1}{2}} = 0 \tag{6.22}$$

uniformly in $n \in \mathbb{N}$ and $f \in S$. To do this, note that (6.18) and (6.19) give

$$\left(\int_{1-h}^{1} |(\mathcal{A}_n f)(s)|^2 \mathrm{d}s\right)^{\frac{1}{2}} \leq \|K_1\|_\infty \|\mathcal{G}e\|_{L^2[1-h,1]} \leq \beta \|K_1\|_\infty \|e\|_{L^2[1-h,1]},$$

where $\mathcal{G}$, e and β are all defined in the proof of Lemma 6.11. Similar arguments used to prove estimate (6.20) lead to

$$\|e\|_{L^2[1-h,1]}^2 \leq \gamma \|\hat{\mathcal{X}}\|_{2,\infty}^2 \|f\|_2^2 \int_{1-h}^{1} \sum_{j\in\mathbb{Z}_\gamma} |\hat{\eta}_j(t)|^2 \mathrm{d}t.$$

Note that

$$\lim_{h\to 0} \int_{1-h}^{1} \sum_{j\in\mathbb{Z}_\gamma} |\hat{\eta}_j(t)|^2 \mathrm{d}t = 0$$

uniformly in $n \in \mathbb{N}$ and $f \in S$. We then conclude from the estimate

$$\left(\int_{1-h}^{1} |(\mathcal{A}_n f)(s)|^2 \mathrm{d}s\right)^{\frac{1}{2}} \leq \gamma\beta \|K_1\|_\infty \|\hat{\mathcal{X}}\|_{2,\infty} \|f\|_2 \left(\int_{1-h}^{1} \sum_{j\in\mathbb{Z}_\gamma} |\hat{\eta}_j(t)|^2 \mathrm{d}t\right)^{\frac{1}{2}}$$

that (6.22) holds uniformly in $n \in \mathbb{N}$ and $f \in S$. □

The following is a result from [80], which generalizes Proposition 1.7 of [6].

Lemma 6.13 *Let $\mathbb{X}$ be a Banach space and $S \subset \mathbb{X}$ a relatively compact set. Assume that $\mathcal{T}, \mathcal{T}_n$ are bounded linear operators from $\mathbb{X}$ to $\mathbb{X}$ satisfying*

$$\|\mathcal{T}_n\| \leq C, \quad \text{for all } n$$

and for each $x \in S$,

$$\|\mathcal{T}_n x - \mathcal{T}x\| \to 0, \quad \text{as } n \to \infty,$$

where C is a constant independent of n. Then, $\|\mathcal{T}_n x - \mathcal{T}x\| \to 0$ uniformly for all $x \in S$.

Lemma 6.14 *Let K be a quasi-weakly singular kernel in the factored form $K = K_1K_2$, where K_2 has the α-property and K_1 is continuous on $\Omega \times \Omega$. Then*

(i) $\|(\mathcal{Q}_n - \mathcal{I})\mathcal{K}_n\mathcal{X}_n\|_2 \to 0$, as $n \to \infty$,
(ii) $\|(\mathcal{K}_n - \mathcal{K})\mathcal{Q}_n\mathcal{K}_n\mathcal{X}_n\|_2 \to 0$, as $n \to \infty$.

Proof (i) Let B denote the unit ball in $L^2(\Omega)$, that is,

$$B := \{x \in L^2(\Omega) : \|x\|_2 \leq 1\}$$

and let

$$A := \{\mathcal{K}_n\mathcal{X}_n x : x \in B, n \in \mathbb{N}\}.$$

Since $\{\mathcal{K}_n\mathcal{X}_n\}$ is collectively compact on $L^2(\Omega)$, we conclude that A is a relatively compact set in $L^2(\Omega)$. Note that $A \subset C(\Omega)$ and for each $x \in C(\Omega)$,

$$\|\mathcal{Q}_n x - x\|_2 \leq c\|\mathcal{Q}_n x - x\|_\infty \to 0, \quad \text{as } n \to \infty.$$

Using Lemma 6.13 with $\mathcal{T}_n = \mathcal{Q}_n$ and $\mathcal{T} = \mathcal{I}$, we have that

$$\begin{aligned}\|(\mathcal{Q}_n - \mathcal{I})\mathcal{K}_n\mathcal{X}_n\|_2 &= \sup_{x\in B} \|(\mathcal{Q}_n - \mathcal{I})\mathcal{K}_n\mathcal{X}_n x\|_2 \\ &= \sup_{x\in A} \|(\mathcal{Q}_n - \mathcal{I})x\|_2 \to 0.\end{aligned}$$

(ii) For a fixed $x \in C(\Omega)$, $\{\mathcal{Q}_n x : n \in \mathbb{N}\}$ is a relatively compact set in $L^\infty(\Omega)$. Since $\{\mathcal{K}_n : n \in \mathbb{N}\}$ is collectively compact on $L^\infty(\Omega)$ and $\mathcal{K}_n$ converges pointwise to $\mathcal{K}$ on the set $\tilde{\mathbb{X}}$, by Lemma 6.13 we conclude that

$$\|(\mathcal{K}_n - \mathcal{K})\mathcal{Q}_n x\|_2 \leq C\|(\mathcal{K}_n - \mathcal{K})\mathcal{Q}_n x\|_\infty \to 0, \quad \text{as } n \to \infty.$$

Using Lemma 6.13 with the bounded linear operators $\mathcal{T}_n := (\mathcal{K}_n - \mathcal{K})\mathcal{Q}_n$ and $\mathcal{T} := 0$, we have that

$$\begin{aligned}\|(\mathcal{K}_n - \mathcal{K})\mathcal{Q}_n\mathcal{K}_n\mathcal{X}_n\|_2 &= \sup_{x\in B} \|(\mathcal{K}_n - \mathcal{K})\mathcal{Q}_n\mathcal{K}_n\mathcal{X}_n x\|_2 \\ &\leq \sup_{x\in A} \|(\mathcal{K}_n - \mathcal{K})\mathcal{Q}_n x\|_2 \to 0, \quad \text{as } n \to \infty.\end{aligned}$$

This completes the proof. □

Lemma 6.15 *Let K be a quasi-weakly singular kernel in the factored form $K = K_1K_2$, where K_2 has the α-property and K_1 is continuous on $\Omega \times \Omega$. Then there exist positive constants c and N such that for all $n > N$, the inverses $(\mathcal{I} - \mathcal{Q}_n\mathcal{K}_n\mathcal{X}_n)^{-1}$ exist as linear operators defined on $L^2(\Omega)$ and*

$$\|(\mathcal{I} - \mathcal{Q}_n\mathcal{K}_n\mathcal{X}_n)^{-1}\|_2 \leq c.$$

Proof A straightforward computation yields

$$\begin{aligned}&[\mathcal{I} + (\mathcal{I} - \mathcal{K})^{-1}\mathcal{K}_n\mathcal{X}_n](\mathcal{I} - \mathcal{Q}_n\mathcal{K}_n\mathcal{X}_n) \\ &\quad = \mathcal{I} - (\mathcal{I} - \mathcal{K})^{-1}[(\mathcal{Q}_n - \mathcal{I})\mathcal{K}_n\mathcal{X}_n + (\mathcal{K}_n - \mathcal{K})\mathcal{Q}_n\mathcal{K}_n\mathcal{X}_n].\end{aligned}$$

By Lemma 6.14, there exists $N > 0$ such that for all $n > N$,

$$\Delta := \|(\mathcal{I} - \mathcal{K})^{-1}[(\mathcal{Q}_n - \mathcal{I})\mathcal{K}_n\mathcal{X}_n + (\mathcal{K}_n - \mathcal{K})\mathcal{Q}_n\mathcal{K}_n\mathcal{X}_n]\| \leq \frac{1}{2}.$$

Therefore, the inverse operators $(\mathcal{I} - \mathcal{Q}_n\mathcal{K}_n\mathcal{X}_n)^{-1}$ exist and it holds that

$$\|(\mathcal{I} - \mathcal{Q}_n\mathcal{K}_n\mathcal{X}_n)^{-1}\|_2 \le \frac{1}{1-\Delta}(1 + \|(\mathcal{I} - \mathcal{K})^{-1}\|_2\|\mathcal{K}_n\mathcal{X}_n\|_2).$$

Thus the results of this lemma follow from Lemma 6.11. Moreover, the Fredholm theory allows us to conclude that the inverse operators are defined on $L^2(\Omega)$. □

Proposition 6.16 *Let K be a quasi-weakly singular kernel in the factored form $K = K_1K_2$, where K_2 has the α-property and K_1 is continuous on $\Omega \times \Omega$. Then there exist positive constants c_0 and N such that for $n > N$ and $x \in \mathbb{X}_n$,*

$$\|(\mathcal{I} - \mathcal{Q}_n\mathcal{K}_n)x\|_2 \ge c_0\|x\|_2.$$

Proof For all $x \in \mathbb{X}_n$, it follows from Lemma 6.15 that

$$\begin{aligned}\|x\|_2 &= \|(\mathcal{I} - \mathcal{Q}_n\mathcal{K}_n\mathcal{X}_n)^{-1}(\mathcal{I} - \mathcal{Q}_n\mathcal{K}_n\mathcal{X}_n)x\|_2\\ &\le C\|(\mathcal{I} - \mathcal{Q}_n\mathcal{K}_n\mathcal{X}_n)x\|_2\\ &= C\|(\mathcal{I} - \mathcal{Q}_n\mathcal{K}_n)x\|_2\end{aligned}$$

and thus the lemma follows. □

6.2.2 A truncated DMPG scheme

In this subsection, we describe the DMPG method and a truncated scheme for equation (6.7).

We first use type II spaces $\mathbb{S}_{k_i}(\Delta_i), i = 1, 2, 3$, to generate $\mathbb{X}_n$, $\mathbb{Y}_n$ and $\mathbb{Q}_n$. Specifically, we choose an integer $r > 1$, define the partitions Δ_0 of Ω by

$$\Delta_0 := \left\{\left[\frac{j}{r^n}, \frac{j+1}{r^n}\right] : j \in \mathbb{Z}_{r^n}\right\}$$

and define $\mathbb{X}_n := \mathbb{S}_{k_1}(\Delta_0 \circ \Delta_1)$, $\mathbb{Y}_n := \mathbb{S}_{k_2}(\Delta_0 \circ \Delta_2)$ and $\mathbb{Q}_n := \mathbb{S}_{k_3}(\Delta_0 \circ \Delta_3)$, where $k_1 = rk_2, k_2 \le k_1 \le k_3$. We next describe our multiscale bases. For this purpose, we denote by $\mathbb{S}_m^k$ the space of piecewise polynomials of degree $\le k - 1$ defined on Ω, corresponding to the equally spaced partition $\Omega_j := [j/m, (j+1)/m], j \in \mathbb{Z}_m$. We assume that k, k', ν, μ are positive integers with $\mu > 1$ such that $\lambda := k\nu = k'\mu$, $k' \le k$ and μ/ν is an integer. We define the trial space $\mathbb{X}_n$ as $\mathbb{S}^k_{\nu\mu^n}$ and the test space $\mathbb{Y}_n$ as $\mathbb{S}^{k'}_{\mu^{n+1}}$. For $i \in \mathbb{Z}_{n+1}$, we set

$$w(i) := \begin{cases}\lambda, & i = 0,\\ \lambda(\mu - 1)\mu^{i-1}, & i \ge 1.\end{cases}$$

The orthogonal multiscale bases ϕ_{ij}, $(i,j) \in \mathbb{U}_n$ for $\mathbb{X}_n$ and ψ_{ij}, $(i,j) \in \mathbb{U}_n$ for $\mathbb{Y}_n$ can be constructed using the general construction described in Section 4.3 (see also [200, 201] and [64]). The orthonormal multiscale basis ϕ_{ij} enjoys the following properties:

$$\mathbb{X}_n = \operatorname{span}\{\phi_{ij} : (i,j) \in \mathbb{U}_n\},$$

$$\int_\Omega t^\ell \phi_{ij}(t)\mathrm{d}t = 0, \ \ \ell \in \mathbb{Z}_k,\ j \in \mathbb{Z}_{w(i)},\ i-1 \in \mathbb{Z}_n,$$

and the length of the support $\operatorname{supp}(\phi_{ij})$ is given by

$$\operatorname{meas}(\operatorname{supp}\phi_{ij}) \le \frac{1}{\mu^{i-1}}, \ \ i \ge 1.$$

Similarly, the multiscale basis ψ_{ij} also has the properties

$$\mathbb{Y}_n = \operatorname{span}\{\psi_{ij} : (i,j) \in \mathbb{U}_n\},$$

$$\int_\Omega t^\ell \psi_{ij}(t)\mathrm{d}t = 0, \ \ \ell \in \mathbb{Z}_{\hat{k}},\ j \in \mathbb{Z}_{w(i)},\ i-1 \in \mathbb{Z}_n,$$

where $\hat{k}$ is an integer between k' and k, and there is a constant c' depending only on $\hat{k}$ for which

$$\operatorname{meas}(\operatorname{supp}\psi_{ij}) \le \frac{c'}{\mu^{i-1}}, \ \ i \ge 1.$$

Consequently, any function $x_n \in \mathbb{X}_n$ has the representation

$$x_n = \sum_{(i,j)\in\mathbb{U}_n} x_{ij}\phi_{ij},$$

where

$$x_{ij} = (x_n, \phi_{ij}), \ \ (i,j) \in \mathbb{U}_n.$$

We have described multiscale bases for the trial and test spaces $\mathbb{X}_n$ and $\mathbb{Y}_n$. It remains to describe the third space $\mathbb{Q}_n$, which is used for integration. We do this next by choosing $k'' \ge k$ and letting the space $\mathbb{Q}_n$ be $\mathbb{S}^{k''}_{\mu^{n+1}}$. Specifically, we choose k'' points in Ω: $0 < \tau_0 < \tau_1 < \cdots < \tau_{k''-1} < 1$, and let $\hat{q}_0, \hat{q}_1, \ldots, \hat{q}_{k''-1}$ be the Lagrange interpolating polynomials satisfying $\deg \hat{q}_j \le k''-1$ and $\hat{q}_i(\tau_j) = \delta_{ij}$, $i,j \in \mathbb{Z}_{k''}$. We let $\varphi_{\epsilon,\mu}$ be the linear function mapping from Ω bijectively onto $[\frac{\epsilon}{\mu}, \frac{\epsilon+1}{\mu}]$ for $\epsilon \in \mathbb{Z}_\mu$, and set

$$\hat{t}_j = \varphi_{\epsilon,\mu}(\tau_l), \ \ \hat{\zeta}_j = \chi_{I_{\epsilon,\mu}} \hat{q}_l \circ \varphi^{-1}_{\epsilon,\mu}, \ \ j = \epsilon k'' + l, \ \ \epsilon \in \mathbb{Z}_\mu, \ \ \ell \in \mathbb{Z}_{k''}.$$

It can easily be seen that $\hat{\zeta}_i(\hat{t}_j) = \delta_{ij}$, $i,j \in \mathbb{Z}_\gamma$, where $\gamma := k''\mu \ge \lambda$. Following the last section, we use affine mappings $\mathcal{F}_i : \Omega_i = [\frac{i}{\mu^n}, \frac{i+1}{\mu^n}] \to \Omega$ defined by

$\mathcal{F}_i(t) = \mu^n t - i$, $i \in \mathbb{Z}_N$ where $N = \mu^n$, to define basis functions ζ_{ij} for the space $\mathbb{Q}_n$. The three spaces $\mathbb{X}_n$, $\mathbb{Y}_n$ and $\mathbb{Q}_n$ were chosen so that

$$\mathbb{X}_n \subseteq \mathbb{Q}_n \text{ and } \mathbb{Y}_n \subseteq \mathbb{Q}_n.$$

These inclusions are crucial for developing a compressed scheme, which will be discussed in the next section. We also define the discrete inner product $(\cdot, \cdot)_n$ and operators $\mathcal{Q}_n$ and $\mathcal{K}_n$ according to (6.13) and (6.15).

With the multiscale bases for spaces $\mathbb{X}_n$ and $\mathbb{Y}_n$, the DMPG scheme for equation (6.7) becomes

$$u_n = \sum_{(i,j)\in\mathbb{U}_n} u_{ij}\phi_{ij} \in \mathbb{X}_n,$$

where the coefficients of the function u_n satisfy

$$\sum_{(i,j)\in\mathbb{U}_n} u_{ij}(\psi_{i'j'}, \phi_{ij} - \mathcal{K}_n\phi_{ij})_n = (\psi_{i'j'}, f)_n,\ (i', j') \in \mathbb{U}_n. \tag{6.23}$$

To write (6.23) in a matrix form, we use lexicographic ordering on $\mathbb{Z}_{n+1} \times \mathbb{Z}_{n+1}$ and define matrices

$$\mathbf{E}_n := [(\psi_{ij}, \phi_{i'j'})_n : (i', j'), (i, j) \in \mathbb{U}_n],$$
$$\mathbf{K}_n := [(\psi_{ij}, \mathcal{K}_n\phi_{i'j'})_n : (i', j'), (i, j) \in \mathbb{U}_n]$$

and vectors

$$\mathbf{f}_n := [(\psi_{ij}, f)_n : (i, j) \in \mathbb{U}_n], \quad \mathbf{u}_n := [u_{ij} : (i, j) \in \mathbb{U}_n].$$

Note that the vectors have length $s(n) = \lambda\mu^{n-1}$. With these notations, equation (6.23) takes the equivalent form

$$(\mathbf{E}_n - \mathbf{K}_n)\mathbf{u}_n = \mathbf{f}_n. \tag{6.24}$$

To present a convergence result about the DMPG scheme, we let

$$\Phi := [\phi_{0i}(t_j) : i \in \mathbb{Z}_\lambda, j \in \mathbb{Z}_\gamma],$$
$$\Psi := [\psi_{0i}(t_j) : i \in \mathbb{Z}_\lambda, j \in \mathbb{Z}_\gamma] \text{ and } \mathbf{M} := \Phi\mathbf{W}\Psi^T,$$

where $\mathbf{W} := \text{diag}(w_j : j \in \mathbb{N}_\gamma)$ and $w_i := \int_\Omega \hat{\zeta}_i(t)\mathrm{d}t$, $i \in \mathbb{Z}_\gamma$.

Theorem 6.17 *Let K be a quasi-weakly singular kernel in the factored form $K = K_1K_2$, where K_2 has the α-property and K_1 is continuous on $\Omega \times \Omega$. Then there exists $N > 0$ such that for all $n \geq N$, the DMPG scheme (6.23) has a unique solution $u_n \in \mathbb{X}_n$. Moreover, if the solution u of equation (6.7) satisfies $u \in W^k_\infty(\Omega)$, then there exists a positive constant c such that for all $n \in \mathbb{N}$,*

$$\|u - u_n\|_2 \leq c\mu^{-kn}.$$

Proof By the construction of spaces $\mathbb{X}_n$ and $\mathbb{Y}_n$ and the choice of the quadrature nodes $\{t_j\}$, we conclude from Proposition 6.8 that $\det(\mathbf{M}) \neq 0$ holds. Therefore, the conclusion of this theorem follows directly from Theorem 3.38. □

We now develop a matrix compression strategy for the DMPG method. This compression strategy will lead us to a fast algorithm for the approximate solution to equation (6.7).

Throughout this section, we suppose that the following additional conditions on the kernel K hold: $K = K_1K_2$, and K_1 and K_2 have derivatives

$$K_r^{(l,m)}(s,t) := \frac{\partial^{l+m}K_r(s,t)}{\partial s^l \partial t^m}$$

for $l \in \mathbb{Z}_{\tilde{k}+1}, m \in \mathbb{Z}_{\tilde{k}'+1}$ when $r = 1, s,t \in I$ and $r = 2, s,t \in I$ with $s \neq t$, where $\tilde{k} := \min\{k'' - k + 1, k\}$ and $\tilde{k}' := \min\{k'' - \hat{k} + 1, \hat{k}\}$, and there exists a positive constant c_0 such that for $s,t \in I, s \neq t$,

$$|K_2^{(l,m)}(s,t)| \leq \frac{c_0}{|s-t|^{\alpha+l+m}}. \tag{6.25}$$

We denote the entries of matrix $\mathbf{K}_n$ by $K_{i'j',ij} := (\psi_{i'j'}, \mathcal{K}_n\phi_{ij})_n$, $(i,j), (i',j') \in \mathbb{U}_n$. The entries of the matrix $\mathbf{K}_n$ are discrete inner products, which we shall show in the next lemma have a similar estimate of continuous inner products presented in Lemma 7.1 of [64] with k, k' replaced by $\tilde{k}, \tilde{k}'$, respectively, which show the influence of the full discretization. To this end, we let $S_{(ij)}$ and $S'_{(i'j')}$ denote the support of ϕ_{ij} and $\psi_{i'j'}$, respectively. Then

$$|S_{(ij)}| := \text{meas}(S_{(ij)}) \leq \frac{1}{\mu^{i-1}} \qquad \text{for } i > 1$$

and

$$|S'_{(i'j')}| := \text{meas}(S'_{(i'j')}) \leq \frac{c'}{\mu^{i'-1}} \qquad \text{for } i' > 1.$$

We first estimate the entries of matrix $\mathbf{K}_n$ in the following lemma.

Lemma 6.18 *The estimate*

$$|K_{i'j',ij}| \leq c\mu^{-(\tilde{k}+\frac{1}{2})i-(\tilde{k}'+\frac{1}{2})i'} \text{dist}(S_{(ij)}, S'_{(i'j')})^{-\alpha-\tilde{k}-\tilde{k}'}$$

holds, where c is a positive constant independent of n, i, i', j and j'.

Proof The entries of matrix $\mathbf{K}_n$ can be written in the following way:

$$K_{i'j',ij} = \int_\Omega (\mathcal{Z}_n g)(s)\text{d}s,$$

where the function g is defined by

$$g(s) := \psi_{i'j'}(s)\int_\Omega K_2(s,t)(\mathcal{Z}_n(K_1(s,\cdot)\phi_{ij}(\cdot)))(t)\mathrm{d}t.$$

Let t_0 and s_0 be the midpoints of the intervals $S_{(ij)}$ and $S'_{(i'j')}$, respectively. By the Taylor theorem, for $(s,t)\in S_{(i'j')}\times S_{(ij)}$ and $r=1,2$, we have that

$$\begin{aligned}K_r(s,t) = &\sum_{l\in\mathbb{Z}_{\tilde{k}}}\frac{1}{l!}K_r^{(0,l)}(s,t_0)(t-t_0)^l\\ &+\frac{1}{(\tilde{k}-1)!}\int_\Omega K_r^{(0,\tilde{k})}(s,t_0+\theta_r(t-t_0))(t-t_0)^{\tilde{k}}(1-\theta_r)^{\tilde{k}-1}\mathrm{d}\theta_r.\end{aligned}$$

Likewise, we have that

$$\begin{aligned}K_r^{(0,\tilde{k})}(s,t') = &\sum_{l'\in\mathbb{Z}_{\tilde{k}'}}\frac{1}{l'!}K_r^{(l',\tilde{k})}(s_0,t')(s-s_0)^{l'}\\ &+\frac{1}{(\tilde{k}'-1)!}\int_\Omega K_r^{(\tilde{k}',\tilde{k})}(s',t')(s-s_0)^{\tilde{k}'}(1-\theta_{\tilde{k},r})^{\tilde{k}'-1}\mathrm{d}\theta_{\tilde{k},r},\end{aligned}$$

where $s' := s_0+\theta_{\tilde{k},r}(s-s_0)$ and $t' = t_0+\theta_r(t-t_0)$. Thus, we obtain that

$$K_r(s,t) = \sum_{m\in\mathbb{Z}_4} T_{r,m},$$

where

$$\begin{aligned}T_{r,0} :=& \sum_{l\in\mathbb{Z}_{\tilde{k}}}\sum_{l'\in\mathbb{Z}_{\tilde{k}'}}\frac{1}{l!l'!}K_r^{(l',l)}(s_0,t_0)(s-s_0)^{l'}(t-t_0)^l,\\ T_{r,1} :=& \sum_{l\in\mathbb{Z}_{\tilde{k}}}\frac{1}{l!(\tilde{k}'-1)!}(s-s_0)^{\tilde{k}'}(t-t_0)^l\\ &\times\int_\Omega K_r^{(\tilde{k}',\tilde{k})}(s_0+\theta_{l,r}(s-s_0),t_0)(1-\theta_{l,r})^{\tilde{k}'-1}\mathrm{d}\theta_{l,r},\\ T_{r,2} :=& \frac{1}{(\tilde{k}-1)!}\sum_{l'\in\mathbb{Z}_{\tilde{k}'}}\frac{1}{l'}(s-s_0)^{l'}(t-t_0)^{\tilde{k}}\\ &\times\int_\Omega K_r^{(l',\tilde{k})}(s_0,t_0+\theta_r(t-t_0))(1-\theta_r)^{\tilde{k}-1}\mathrm{d}\theta_r\end{aligned}$$

and

$$\begin{aligned}T_{r,3} :=& \frac{1}{(\tilde{k}-1)!(\tilde{k}'-1)!}(s-s_0)^{\tilde{k}'}(t-t_0)^{\tilde{k}}\\ &\times\int_\Omega\int_\Omega K_r^{(\tilde{k}',\tilde{k})}(s',t')(1-\theta_{\tilde{k},r})^{\tilde{k}'-1}(1-\theta_r)^{\tilde{k}-1}\mathrm{d}\theta_{\tilde{k},r}\mathrm{d}\theta_r,\end{aligned}$$

with $(s', t') = (s_0 + \theta_{\tilde{k},r}(s - s_0), t_0 + \theta_r(t - t_0))$. Note that the polynomials of degree $k'' - 1$ are invariant under the projectors $\mathcal{Z}_n$, and $\mathcal{Z}_n : L^\infty(\Omega) \to L^\infty(\Omega)$ is uniformly bounded. Using the vanishing moment conditions of bases ϕ_{ij} and $\psi_{i'j'}$, that is,

$$\int_{S_{(ij)}} (t - t_0)^l \phi_{ij}(t) dt = 0, \ \ l \in \mathbb{Z}_{\tilde{k}}, \ \ i - 1 \in \mathbb{Z}_n, \ \ j \in \mathbb{Z}_{w(i)}$$

and

$$\int_{S'_{(i'j')}} (s - s_0)^l \psi_{i'j'}(s) ds = 0, \ \ l \in \mathbb{Z}_{\tilde{k}'}, \ \ i' - 1 \in \mathbb{Z}_n, \ \ j' \in \mathbb{Z}_{w(i')},$$

we conclude that

$$|K_{i'j',ij}| \le C \sup(|K_1^{(l',l)}(s,t)||K_2^{(l',l)}(s,t)|) \\ \times \int_{S_{(ij)}} |(t - t_0)^{\tilde{k}} \phi_{ij}(t)| dt \int_{S'_{(i'j')}} |(s - s_0)^{\tilde{k}'} \psi_{i'j'}(s)| ds,$$

where the supremum is taken over $(s, t) \in S'_{(i'j')} \times S_{(ij)}$ and $(l', l) \in \mathbb{Z}_{\tilde{k}'+1} \times \mathbb{Z}_{\tilde{k}+1}$. In fact, to see the estimate, it is sufficient to consider the cases $K_r(s, t) = T_{r,m}, r = 1, 2, m \in \mathbb{Z}_4$. We assume, without loss of generality, that in the sum of $T_{r,m}$ there is only one term, which means that $K_r(s, t)$ is a product of an integral and two factors $(t - t_0)^{l_r}$ and $(s - s_0)^{l'_r}$. Obviously, if $l_1 + l_2 < \tilde{k}$ (or $l'_1 + l'_2 < \tilde{k}'$), the operator $\mathcal{Z}_n$ can be discarded, and by using the vanishing moment conditions of bases ϕ_{ij} and $\psi_{i'j'}$, we can obtain $K_{i'j',ij} = 0$. Otherwise, $l_1 + l_2 \ge \tilde{k}$ and $l'_1 + l'_2 \ge \tilde{k}'$, then the desired estimate follows by using the uniform boundedness of $\mathcal{Z}_n$. Now, using $\|\phi_{ij}\|_2 = \|\psi_{ij}\|_2 = 1$, we conclude from the Cauchy–Schwarz inequality that

$$|K_{i'j',ij}| \le C \operatorname{dist}(S_{(ij)}, S'_{(i'j')})^{-\alpha - \tilde{k} - \tilde{k}'} |S_{(ij)}|^{\tilde{k} + \frac{1}{2}} |S'_{(i'j')}|^{\tilde{k}' + \frac{1}{2}} \\ \le C\mu^{-(\tilde{k} + \frac{1}{2})i - (\tilde{k}' + \frac{1}{2})i'} \operatorname{dist}(S_{(ij)}, S'_{(i'j')})^{-\alpha - \tilde{k} - \tilde{k}'}.$$

□

Lemma 6.18 leads us to a truncation strategy, which we describe below. To truncate the matrix $\mathbf{K}_n$, we first partition it into a block matrix $\mathbf{K}_n = [\mathbf{K}_{i'i} : (i', i) \in \mathbb{Z}_{n+1} \times \mathbb{Z}_{n+1}]$ according to the decomposition of spaces $\mathbb{X}_n$ and $\mathbb{Y}_n$, where

$$\mathbf{K}_{i'i} := [K_{i'j',ij} : j' \in \mathbb{Z}_{w(i')}, j \in \mathbb{Z}_{w(i)}].$$

For a given positive number $\delta^n_{i'i}$, we truncate the block $\mathbf{K}_{i',i}$ to obtain

$$\tilde{\mathbf{K}}_{i'i} := [\tilde{K}_{i'j',ij} : j' \in \mathbb{Z}_{w(i')}, j \in \mathbb{Z}_{w(i)}],$$

where

$$\tilde{K}_{i'j',ij} := \begin{cases} K_{i'j',ij}, & \operatorname{dist}(S_{(ij)}, S'_{(i'j')}) \le \delta^n_{i'i}, \\ 0, & \text{otherwise}, \end{cases}$$

and let $\tilde{\mathbf{K}}_n := [\tilde{\mathbf{K}}_{i'i} : \mathbb{Z}_{n+1} \times \mathbb{Z}_{n+1}]$. Using the truncation matrix $\tilde{\mathbf{K}}_n$ to replace the matrix $\mathbf{K}_n$ in equation (6.24), we obtain a new linear system

$$(\mathbf{E}_n - \tilde{\mathbf{K}}_n)\tilde{\mathbf{u}}_n = \mathbf{f}_n, \tag{6.26}$$

where $\tilde{\mathbf{u}}_n := [\tilde{u}_{ij}] \in \mathbb{R}^{s(n)}$. Equation (6.26) is a compressed scheme which provides us with a fast algorithm.

It is convenient to work with a functional analytic approach in our development. For this purpose, we convert the linear system (6.26) to an abstract operator equation form. Let $b_{i'j',ij}$ denote the entries of matrix $\mathbf{B}_n := \mathbf{E}_n^{-1}\tilde{\mathbf{K}}_n\mathbf{E}_n^{-1}$ and let

$$\tilde{K}(s,t) := \sum_{(i,j),(i',j')\in\mathbb{U}_n} b_{i'j',ij}\psi_{ij}(s)\phi_{i'j'}(t).$$

We denote by $\tilde{\mathcal{K}}_n$ the discrete integral operator defined by the kernel $\tilde{K}$. Then, solving the linear system (6.26) is equivalent to finding

$$\tilde{u}_n = \sum_{(i,j)\in\mathbb{U}_n} \tilde{u}_{ij}\phi_{ij} \in \mathbb{X}_n$$

such that

$$(\mathcal{I} - \tilde{\mathcal{K}}_n)\tilde{u}_n = \mathcal{Q}_n f. \tag{6.27}$$

The analysis of this truncated scheme and the choice of the truncation parameter $\delta^n_{i',i}$ will be discussed in the next subsection.

6.2.3 Analysis of convergence and complexity

In this subsection we present the analysis for the order of convergence and the order of computational complexity of the compressed scheme developed in the last subsection.

We first estimate the truncated matrix $\tilde{\mathbf{K}}_n$. To do this, we denote $\eta := \alpha + \tilde{k} + \tilde{k}' - 1$.

Lemma 6.19 *Let J be any integral contained in Ω. Suppose that n and s are positive integers and $s > 1$. Then*

$$\sum_{j\in J_n} \operatorname{dist}(J, \Omega_{j,n})^{-s} \le \frac{4n}{s-1}\delta^{-s} \max\left\{\frac{s-1}{n}, \delta\right\},$$

where $\Omega_{j,n} := [j/n, (j+1)/n], j \in \mathbb{Z}_n$ and $J_n := \{j : j \in \mathbb{Z}_n, \operatorname{dist}(J, \Omega_{j,n}) > \delta\}$.

Proof Suppose that $J = [a, b]$. Choose the greatest integer p with $p + 1 \in \mathbb{Z}_n$ such that $p/n < a - \delta \le (p+1)/n$ and the least integer q with $q + 1 \in \mathbb{Z}_{n+1}$ such that $q/n \le b + \delta < (q+1)/n$. Therefore

$$\sum_{j \in J_n} \operatorname{dist}(J, \Omega_{j,n})^{-s} = \sum_{j=0}^{p-1} \left(a - \frac{j+1}{n}\right)^{-s} + \sum_{j=q+2}^{n-1} \left(\frac{j}{n} - b\right)^{-s}.$$

When $a - \delta \le 1/n$, the first sum is zero and likewise when $b + \delta \ge 1 - 1/n$, the second sum is zero. In any case, we have that

$$\begin{aligned}\sum_{j \in J_n} \operatorname{dist}(J, \Omega_{j,n})^{-s} &\le 2\delta^{-s} + \sum_{j=0}^{p-2} \left(a - \frac{j+1}{n}\right)^{-s} + \sum_{j=q+2}^{n-1} \left(\frac{j}{n} - b\right)^{-s} \\ &\le 2\delta^{-s} + n \int_{a-p/n}^{\infty} t^{-s} \mathrm{d}t + n \int_{(q+1)/n-b}^{\infty} t^{-s} \mathrm{d}t.\end{aligned}$$

Evaluating the integrals and using our choice of p and q gives the bound for the right-hand side of the inequality above:

$$2\delta^{-s} + \frac{2n}{s-1}\delta^{-s+1} = \frac{2n\delta^{-s}}{s-1}\left(\frac{s-1}{n} + \delta\right) \le \frac{4n\delta^{-s}}{s-1} \max\left\{\frac{s-1}{n}, \delta\right\}.$$

□

Lemma 6.20 *For any $i, i' \in \mathbb{Z}_{n+1}$ and positive constant $\delta^n_{i'i}$, the estimates*

$$\|\mathbf{K}_{i'i} - \tilde{\mathbf{K}}_{i'i}\|_\infty \le c\mu^{-(\tilde{k}-\frac{1}{2})i-(\tilde{k}'+\frac{1}{2})i'} (\delta^n_{i'i})^{-\alpha-\tilde{k}-\tilde{k}'} \max\left\{\frac{\eta}{\mu^{i-1}}, \delta^n_{i'i}\right\}$$

and

$$\|\mathbf{K}_{i'i} - \tilde{\mathbf{K}}_{i'i}\|_1 \le c\mu^{-(\tilde{k}+\frac{1}{2})i-(\tilde{k}'-\frac{1}{2})i'} (\delta^n_{i'i})^{-\alpha-\tilde{k}-\tilde{k}'} \max\left\{\frac{c'\eta}{\mu^{i'-1}}, \delta^n_{i'i}\right\}$$

hold, where c is a positive constant independent of n, i and i'.

Proof Using the estimate given in Lemma 6.18 on the entries $|K_{i',j';i,j}|$, noting the definition of the truncated matrix $\tilde{K}_{i',j';i,j}$, we have that

$$\|\mathbf{K}_{i'i} - \tilde{\mathbf{K}}_{i'i}\|_\infty \le c\mu^{-(\tilde{k}+\frac{1}{2})i-(\tilde{k}'+\frac{1}{2})i'} \max_{j' \in \mathbb{Z}_{w(i')}} \sum_{j \in \mathbb{Z}_{\delta^n_{i'i}}} \operatorname{dist}(S_{(ij)}, S'_{(i'j')})^{-\alpha-\tilde{k}-\tilde{k}'},$$

where $\mathbb{Z}_{\delta^n_{i'i}} := \{j : j \in \mathbb{Z}_{w(i)}, \operatorname{dist}(S_{(ij)}, S'_{(i'j')}) > \delta^n_{i'i}\}$. Using Lemma 6.19 (with n replaced by μ^{i-2}, s replaced by $\alpha+\tilde{k}+\tilde{k}'$, $\Omega_{j,n}$ replaced by $S_{(ij)}$ and J replaced by $S'_{(i'j')}$), we get the first estimate of this lemma. The second estimate follows likewise from Lemma 6.19. □

We recall a useful result of Schur.

Lemma 6.21 (Schur's lemma) *Let* $\mathbf{A} := [a_{ij} : i,j \in \mathbb{Z}_n], n \in \mathbb{N}$ *be a matrix such that there exist positive constants* $\gamma_i, i \in \mathbb{Z}_n$ *and a positive constant* c *independent of* n *satisfying, for all* $i,j \in \mathbb{Z}_n$,

$$\sum_{\ell \in \mathbb{Z}_n} |a_{\ell j}| \gamma_\ell \leq c\gamma_j$$

and

$$\sum_{\ell \in \mathbb{Z}_n} |a_{i\ell}| \gamma_\ell \leq c\gamma_i.$$

Then, $\|\mathbf{A}\|_2 \leq c$ *for all* $n \in \mathbb{N}$.

We are now ready to estimate the truncation operator $\mathcal{R}_n := \mathcal{Q}_n\mathcal{K}_n - \tilde{\mathcal{K}}_n$ in terms of the function $\mu[\cdot,\cdot;n]$.

Lemma 6.22 *Let* b, b' *be real numbers and* m *a non-negative integer. Choose the truncation parameter* $\delta^n_{i'i}, i, i' \in \mathbb{Z}_{n+1}$ *such that*

$$\delta^n_{i'i} \geq \max\left\{\frac{c'\eta}{\mu^{i-1}}, \frac{\eta}{\mu^{i'-1}}, \mu^{-n+b(n-i)+b'(n-i')}\right\}. \tag{6.28}$$

Then, for any $u \in H^m(I)$ *with* $0 < m \leq k$,

$$\|\mathcal{R}_n\mathcal{X}_n u\|_{L^2} \leq c\mu[\tilde{k} + m - b\eta, \tilde{k}' - b'\eta; n](n+1)^{1/2}\mu^{-(1-\alpha+m)n}\|u\|_{H^m},$$

and for $u \in L^2[0,1]$,

$$\|\mathcal{R}_n\mathcal{X}_n u\|_{L^2} \leq c\mu[\tilde{k} - b\eta, \tilde{k}' - b'\eta; n]\mu^{-(1-\alpha)n}\|u\|_{L^2},$$

where c *is a positive constant independent of* n.

Proof It is clear that for $\phi \in \mathbb{X}_n$,

$$\|\phi\|_n = \sup_{v \in \mathbb{X}_n} |(\phi, v)_n| / \|v\|_n.$$

It follows that for $u \in H^m[0,1]$ with $0 \leq m \leq k$,

$$\|\mathcal{R}_n\mathcal{X}_n u\|_n = \sup_{v \in \mathbb{X}_n} |(\mathcal{R}_n\mathcal{X}_n u, v)_n| / \|v\|_n. \tag{6.29}$$

We need to pass the estimate (6.29) in terms of the discrete norm to an estimate in terms of the L^2-norm. To this end, we need a norm equivalence result. That is, there exist positive constants A and B such that, for all functions $x \in \mathbb{X}_n$ and $n \in \mathbb{N}$, the inequality

$$A\|x\|_n \leq \|x\|_{L^2} \leq B\|x\|_n \tag{6.30}$$

holds. A proof of (6.30) is in order. Because $\mathbb{X}_n \subseteq \mathbb{Q}_n$, any function $x \in \mathbb{X}_n$ has the representation

$$x(t) = \sum_{j\in\mathbb{Z}_{\gamma N}} x(t_j)\zeta_j(t).$$

It follows that

$$\|x\|_{L^2}^2 = \sum_{i\in\mathbb{Z}_N} |\Omega_i| \mathbf{x}_i^T \tilde{\mathbf{W}} \mathbf{x}_i,$$

where $\mathbf{x}_i = [x(\mathcal{F}_i(\hat{t}_j)) : j \in \mathbb{Z}_\gamma]$ and

$$\tilde{\mathbf{W}} = \left[\int_\Omega \hat{\zeta}_\ell(\tau)\hat{\zeta}_m(\tau)\mathrm{d}\tau : \ell, m \in \mathbb{Z}_\gamma \right].$$

Moreover, we have that

$$\|x\|_n^2 = \sum_{j\in\mathbb{Z}_{\gamma N}} w_j x^2(t_j) = \sum_{i\in\mathbb{Z}_N} |\Omega_i| \mathbf{x}_i^T \mathbf{W} \mathbf{x}_i.$$

Because both $\tilde{\mathbf{W}}$ and $\mathbf{W}$ are positive definite matrices, (6.30) follows. Using (6.30) in (6.29), we have that

$$\begin{aligned}\|\mathcal{R}_n\mathcal{X}_n u\|_2 &\le c \sup_{v\in\mathbb{X}_n} |(\mathcal{R}_n\mathcal{X}_n u, v)_n| / \|v\|_2 \\ &= c \sup_{v\in\mathbb{X}_n} |(\mathcal{R}_n\mathcal{X}_n u, \mathcal{Q}_n' v)_n| / \|v\|_2.\end{aligned}$$

Using the second estimate in Lemma 6.10, we conclude that

$$\|\mathcal{R}_n\mathcal{X}_n u\|_2 \le c \sup_{v\in\mathbb{X}_n} |(\mathcal{R}_n\mathcal{X}_n u, \mathcal{Q}_n' v)_n| / \|\mathcal{Q}_n' v\|_2.$$

Since $\mathcal{Q}_n'(\mathbb{X}_n) = \mathcal{Y}_n(\mathbb{X}_n) = \mathbb{Y}_n$, we obtain that

$$\|\mathcal{R}_n\mathcal{X}_n u\|_2 \le c \sup_{v\in\mathbb{X}_n} |(\mathcal{R}_n\mathcal{X}_n u, \mathcal{Y}_n v)_n| / \|\mathcal{Y}_n v\|_2. \tag{6.31}$$

Now we estimate $|(\mathcal{R}_n\mathcal{X}_n u, \mathcal{Y}_n v)_n|$. Note that

$$\mathcal{X}_n u = \sum_{(i,j)\in\mathbb{U}_n} (u, \phi_{ij})\phi_{ij} \quad \text{and} \quad \mathcal{Y}_n v = \sum_{(i,j)\in\mathbb{U}_n} (v, \psi_{ij})\psi_{ij}.$$

Let $R_{i'j',ij} := K_{i'j',ij} - \tilde{K}_{i'j',ij}$. We have that

$$(\mathcal{R}_n\mathcal{X}_n u, \mathcal{Y}_n v)_n = \sum_{(i,j),(i',j')\in\mathbb{U}_n} R_{i'j',ij}(u, \phi_{ij})(v, \psi_{i'j'}).$$

Denoting $\Gamma := \mu[\tilde{k} + m - b\eta, \tilde{k}' - b'\eta; n]$, $\Delta_{i'j',ij} := \Gamma^{-1}\mu^{(1-\alpha+m)n-mi}R_{i'j',ij}$ and defining the matrix $\Delta_n := [\Delta_{i'j',ij}]$, we see that

$$|(\mathcal{R}_n\mathcal{X}_n u, \mathcal{Y}_n v)_n| \le \Gamma\mu^{-(1-\alpha+m)n}\left(\sum_{i\in\mathbb{Z}_{n+1}}\mu^{2mi}\sum_{j\in\mathbb{Z}_{w(i)}}|(u,\phi_{ij})|^2\right)^{1/2}\|\Delta_{\mathbf{n}}\|_2\|\mathcal{Y}_n v\|_2.$$

Now, for $u \in H^m(\Omega)$, we have that

$$\mu^{2mi}\sum_{j\in\mathbb{Z}_{w(i)}}|(u,\phi_{ij})|^2 = \mu^{2mi}\|\mathcal{X}_i u - \mathcal{X}_{i-1}u\|_{L^2}^2 \le C\|u\|_{H^m}^2$$

so

$$\sum_{i\in\mathbb{Z}_{n+1}}\mu^{2mi}\sum_{j\in\mathbb{Z}_{w(i)}}|(u,\phi_{ij})|^2 \le C(n+1)\|u\|_{H^m}^2.$$

Thus, we conclude that for $u \in H^m(\Omega)$,

$$|(\mathcal{R}_n\mathcal{X}_n u, \mathcal{Y}_n v)_n| \le C(n+1)^{\sigma(m)}\Gamma\mu^{-(1-\alpha+m)n}\|\Delta_{\mathbf{n}}\|_2\|u\|_{H^m}\|\mathcal{Y}_n v\|_2 \quad (6.32)$$

holds, where

$$\sigma(m) := \begin{cases} 1/2, & 0 < m \le k, \\ 0, & m = 0. \end{cases}$$

We now estimate $\|\Delta_{\mathbf{n}}\|_2$ using Lemma 6.21 with the choice $\gamma_{ij} = \mu^{-i/2}$. We have from Lemma 6.20 that

$$\begin{aligned}\sum_{(i,j)\in\mathbb{U}_n}|\Delta_{i'j',ij}|\gamma_{ij} &\le \sum_{i\in\mathbb{Z}_{n+1}}\Gamma^{-1}\mu^{(1-\alpha+m)n-mi}\|\mathbf{K}_{i'i} - \tilde{\mathbf{K}}_{i'i}\|_\infty\mu^{-i/2}\\ &\le c\Gamma^{-1}\mu^{(1-\alpha)n}\sum_{i\in\mathbb{Z}_{n+1}}\mu^{m(n-i)}\mu^{-(\tilde{k}-1/2)i-(\tilde{k}'+1/2)i'}(\delta^n_{i'i})^{-\eta}\mu^{-i/2}.\end{aligned}$$

The right-hand side of this inequality can be bounded by

$$c\Gamma^{-1}\gamma_{i'j'}\sum_{i\in\mathbb{Z}_{n+1}}\mu^{(\tilde{k}+m-b\eta)(n-i)}\mu^{(\tilde{k}'-b'\eta)(n-i')}.$$

This inequality and our hypothesis imply that

$$\sum_{(i,j)\in\mathbb{U}_n}|\Delta_{i'j',ij}|\gamma_{ij} \le c\gamma_{i'j'}.$$

Similarly, using the second estimate in Lemma 6.20, we have that

$$\begin{aligned}\sum_{(i',j')\in\mathbb{U}_n} |\Delta_{i'j',ij}|\gamma_{i'j'} &\le \sum_{i'\in\mathbb{Z}_{n+1}} \Gamma^{-1}\mu^{(1-\alpha+m)n-mi}\|\mathbf{K}_{i'i}-\tilde{\mathbf{K}}_{i'i}\|_1\mu^{-i'/2}\\ &\le c\Gamma^{-1}\mu^{(1-\alpha)n}\sum_{i'\in\mathbb{Z}_{n+1}}\mu^{m(n-i)}\mu^{-(\tilde{k}+1/2)i-(\tilde{k}'-1/2)i'}\\ &\quad\times(\delta^n_{i'i})^{-\eta}\mu^{-i'/2}\\ &\le c\Gamma^{-1}\gamma_{ij}\sum_{i'\in\mathbb{Z}_{n+1}}\mu^{(\tilde{k}+m-b\eta)(n-i)}\mu^{(\tilde{k}'-b'\eta)(n-i')},\end{aligned}$$

which implies that

$$\sum_{(i',j')\in\mathbb{U}_n} |\Delta_{i'j',ij}|\gamma_{i'j'} \le c\gamma_{ij}.$$

Hence, Lemma 6.21 yields

$$\|\Delta_n\|_2 \le c. \tag{6.33}$$

Combining the inequalities (6.31)–(6.33) yields the estimates of this lemma. □

The next result shows that the truncated scheme is uniquely solvable and stable.

Theorem 6.23 *Let $\delta^n_{i'i}$ be chosen according to (6.28) with $b > \frac{\tilde{k}+\alpha-1}{\eta}$, $b' > \frac{\tilde{k}'+\alpha-1}{\eta}$ and $b+b' > 1$, then there exist a positive constant c and a positive integer N such that, for $n \ge N$ and $x \in \mathbb{X}_n$,*

$$\|(\mathcal{I}-\tilde{\mathcal{K}}_n)x\|_2 \ge c\|x\|_2.$$

In particular, equation (6.27) has a unique solution $\tilde{u}_n \in \mathbb{X}_n$ for $n \ge N$ and the truncated scheme (6.27) is stable.

Proof By Proposition 6.16,

$$\|(\mathcal{I}-\mathcal{Q}_n\mathcal{K}_n)x\|_2 \ge c_0\|x\|_2 \text{ for all } x\in\mathbb{X}_n.$$

Thus, using the second estimate in Lemma 6.22 we have that

$$\|(\mathcal{I}-\tilde{\mathcal{K}}_n)x\|_2 \ge \|(\mathcal{I}-\mathcal{Q}_n\mathcal{K}_n)x\|_2 - \|\mathcal{R}_nx\|_2 \ge (c_0-\epsilon_n)\|x\|_2,$$

where $\epsilon_n := \mu[\tilde{k}-b\eta,\tilde{k}'-b'\eta;n]\mu^{-(1-\alpha)n} \to 0$ as $n\to\infty$. The result of this theorem follows. □

In the next theorem, we present a result on the order of convergence for the approximate solution $\tilde{u}_n$ obtained from the truncated scheme. It is convenient to introduce an interpolation projection $\mathcal{I}_n$ from $C(\Omega)$ to $\mathbb{X}_n$. For $x \in C(\Omega)$, we let $\mathcal{I}_n x$ be the interpolation to x from $\mathbb{X}_n$ such that $(\mathcal{I}_n x)(t_{ij}) = x(t_{ij})$, for $j \in \mathbb{Z}_{w(i)}$, $i \in \mathbb{Z}_{n+1}$. We assume that the interpolatory points t_{ij} are chosen so that the interpolation problem has a unique solution. We also need the following error estimates. For $x \in W^{m,\infty}(\Omega)$, the following estimates hold:

$$\|x - \mathcal{X}_n x\|_2 \le c\mu^{-mn}\|x\|_{H^m},$$
$$\|x - \mathcal{I}_n x\|_\infty \le c\mu^{-mn}\|x\|_{W^{m,\infty}}$$

and

$$\|(\mathcal{K} - \mathcal{K}_n)x\|_\infty \le c\mu^{-mn}\|x\|_{W^{m,\infty}}.$$

Theorem 6.24 *Let $\delta^n_{i'i}$ be chosen according to (6.28) with $b \ge \frac{m+\tilde{k}+\alpha-1}{\eta}$, $b' > \frac{\tilde{k}'+\alpha-1}{\eta}$, $b + b' \ge 1 + \frac{m}{\eta}$ and $0 < m \le k$, then there exist a positive constant c and a positive integer N such that, for $n \ge N$,*

$$\|u - \tilde{u}_n\|_{L^2} \le cs(n)^{-m}(\log(s(n)))^{\tau+1/2}\|u\|_{W^{m,\infty}},$$

where $\tau = 0$ except for $(b, b') = (\frac{m+\tilde{k}+\alpha-1}{\eta}, \frac{\tilde{k}'}{\eta})$, in which case $\tau = 1$.

Proof To estimate the error of the solution $\tilde{u}_n$ of the truncated equation (6.27), we observe that Theorem 6.23 yields

$$\begin{aligned}\|\mathcal{X}_n u - \tilde{u}_n\|_2 &\le c^{-1}\|(\mathcal{I} - \tilde{\mathcal{K}}_n)(\mathcal{X}_n u - \tilde{u}_n)\|_2 \\ &\le c^{-1}(\|\mathcal{R}_n \mathcal{X}_n u\|_2 + \|\mathcal{Q}_n(\mathcal{K} - \mathcal{K}_n)u\|_2 \\ &\quad + \|\mathcal{Q}_n(\mathcal{I} - \mathcal{K}_n)(\mathcal{X}_n u - u)\|_2).\end{aligned}$$

The first term has been estimated in Lemma 6.22. We deal with the last two terms. Recalling that $\|\mathcal{Q}_n\|_\infty$ is uniformly bounded, we conclude that

$$\begin{aligned}\|\mathcal{Q}_n(\mathcal{K} - \mathcal{K}_n)u\|_2 &\le \|\mathcal{Q}_n(\mathcal{K} - \mathcal{K}_n)u\|_\infty \\ &\le \|\mathcal{Q}_n\|_\infty\|(\mathcal{K} - \mathcal{K}_n)u\|_\infty \\ &\le c\mu^{-mn}\|u\|_{W^{m,\infty}}.\end{aligned}$$

We next estimate the last term. Note that

$$\begin{aligned}\|\mathcal{Q}_n(\mathcal{I} - \mathcal{K}_n)(\mathcal{X}_n u - u)\|_2 &\le \|\mathcal{Q}_n\|_\infty(1 + \|\mathcal{K}_n\|_\infty)\|\mathcal{X}_n u - u\|_\infty \\ &\le c(\|\mathcal{X}_n(u - \mathcal{I}_n u)\|_\infty + \|\mathcal{I}_n u - u\|_\infty).\end{aligned}$$

There is a point $\hat{t} \in \Omega_j$ such that $\hat{t} = \mathcal{F}_j(\tau)$, where $\tau \in \Omega$ and

$$\|\mathcal{X}_n(u - \mathcal{I}_n u)\|_\infty = |\mathcal{X}_n(u - \mathcal{I}_n u)(\hat{t})|.$$

It follows that

$$\|\mathcal{X}_n(u-\mathcal{I}_n u)\|_\infty = |((\mathcal{X}_n(u-\mathcal{I}_n u))\circ\mathcal{F}_j)(\tau)| = |(\hat{\mathcal{X}}((u-\mathcal{I}_n u)\circ\mathcal{F}_j))(\tau)|,$$

where the notation $\hat{\mathcal{X}}$ was introduced in the proof of Lemma 6.11. Therefore,

$$\begin{aligned}\|\mathcal{X}_n(u-\mathcal{I}_n u)\|_\infty &\leq \|\hat{\mathcal{X}}((u-\mathcal{I}_n u)\circ F_j)\|_\infty\\ &\leq \|\hat{\mathcal{X}}\|_{2,\infty}\|(u-\mathcal{I}_n u)\circ F_j\|_2\\ &\leq \|\hat{\mathcal{X}}\|_{2,\infty}\|(u-\mathcal{I}_n u)\circ F_j\|_\infty\\ &\leq \|\hat{\mathcal{X}}\|_{2,\infty}\|u-\mathcal{I}_n u\|_\infty.\end{aligned}$$

Consequently,

$$\|\mathcal{X}_n u - u\|_\infty \leq (\|\hat{\mathcal{X}}\|_{2,\infty}+1)\|u-\mathcal{I}_n u\|_\infty \leq c\mu^{-mn}\|u\|_{W^{m,\infty}}.$$

Using these estimates and Lemma 6.22, we complete the proof. □

The next result shows that the condition number of the coefficient matrix of the truncated scheme (6.27) is bounded by a constant independent of n.

Theorem 6.25 *Let $\delta^n_{i'i}$ be chosen according to (6.28) with $b > \frac{\tilde{k}+\alpha-1}{\eta}$, $b' > \frac{\tilde{k}'+\alpha-1}{\eta}$ and $b+b'>1$, then the condition number of the coefficient matrix of the truncated approximate equation (6.27) is bounded, that is, there exists a positive constant c such that, for $n\in\mathbb{N}$,*

$$\operatorname{cond}(\boldsymbol{E}_n - \tilde{\mathbf{K}}_n) \leq c.$$

Proof For $\mathbf{e} := [e_{ij} : (i,j)\in\mathbb{U}_n]\in\mathbb{R}^{s(n)}$ and $\mathbf{e}' := [e'_{ij} : (i,j)\in\mathbb{U}_n]\in\mathbb{R}^{s(n)}$, set

$$x := \sum_{(i,j)\in\mathbb{U}_n} e_{ij}\phi_{ij} \text{ and } y := \sum_{(i,j)\in\mathbb{U}_n} e'_{ij}\psi_{ij}.$$

It follows from (6.30) and $\|y\|_2 = \|\mathbf{e}'\|_2$ that

$$\begin{aligned}\|(\mathbf{E}_n - \tilde{\mathbf{K}}_n)\mathbf{e}\|_2 &= \sup_{\mathbf{e}'\in\mathbb{R}^{s(n)},\|\mathbf{e}'\|_2=1} |\mathbf{e}'^T(\mathbf{E}_n-\tilde{\mathbf{K}}_n)\mathbf{e}|\\ &= \sup_{y\in\mathbb{Y}_n,\|y\|_2=1} |(y,(\mathcal{I}-\tilde{\mathcal{K}}_n)x)_n|\\ &\leq c(\|(\mathcal{I}-\mathcal{Q}_n\mathcal{K}_n\mathcal{X}_n)x\|_2 + \|\mathcal{R}_n x\|_2).\end{aligned}$$

From Lemma 6.11 we know that $\mathcal{K}_n\mathcal{X}_n : L^2(\Omega)\to L^2(\Omega)$ are uniformly bounded. By Lemma 6.14 (i), we conclude that $\|\mathcal{I}-\mathcal{Q}_n\mathcal{K}_n\mathcal{X}_n\|_2$ are uniformly bounded. This fact, with the second estimate in Lemma 6.22, yields

$$\|(\mathbf{E}_n - \tilde{\mathbf{K}}_n)\mathbf{e}\|_2 \leq c\|x\|_2 = c\|\mathbf{e}\|_2,$$

that is, $\|\mathbf{E}_n - \tilde{\mathbf{K}}_n\| \leq c$.

Moreover, for any $\mathbf{e}' \in \mathbb{R}^{s(n)}$ we find $\mathbf{e} \in \mathbb{R}^{s(n)}$ such that $\mathbf{e}' = (\mathbf{E}_n - \tilde{\mathbf{K}}_n)\mathbf{e}$ and choose the unique $x \in \mathbb{X}_n$ and $y \in \mathbb{Y}_n$ such that, for all $(r, \ell) \in \mathbb{U}_n$, $\langle x, \phi_{r\ell} \rangle = e_{r\ell}$ and $\langle y, \psi_{r\ell} \rangle = e'_{r\ell}$. Therefore, we have that

$$\mathcal{Q}_n y = (\mathcal{I} - \tilde{\mathcal{K}}_n)x.$$

Since $\|x\|_2 = \|\mathbf{e}\|_2$, $\|y\|_2 = \|\mathbf{e}'\|_2$, we have by Theorem 6.23 that

$$\begin{aligned} c\|(\mathbf{E}_n - \tilde{\mathbf{K}}_n)^{-1}\mathbf{e}'\|_2 = c\|\mathbf{e}\|_2 = c\|x\|_2 &\le \|(\mathcal{I} - \tilde{\mathcal{K}}_n)x\|_2 = \|\mathcal{Q}_n y\|_2 \\ &\le \|\mathcal{Q}_n\|_{2,\mathbb{Y}_n}\|y\|_2, \end{aligned}$$

which proves that

$$\|(\mathbf{E}_n - \tilde{\mathbf{K}}_n)^{-1}\|_2 \le c^{-1}\|\mathcal{Q}_n\|_{2,\mathbb{Y}_n}.$$

It follows from Lemma 6.10 that $\|(\mathbf{E}_n - \tilde{\mathbf{K}}_n)^{-1}\|_2$ is bounded by a constant. □

For any matrix $\mathbf{A}$ we use $\mathcal{N}(\mathbf{A})$ to denote the number of nonzero entries in $\mathbf{A}$. Employing a standard argument (see, for example, Section 5.3.1 and [64]), we conclude that when b, b' are real numbers, and $\delta^n_{r,r'}$ is chosen according to (6.28) with $\eta = \alpha + k + \hat{k} - 1$, there exists a positive constant c such that for all $n \in \mathbb{N}$

$$\mathcal{N}(\mathbf{E}_n - \tilde{\mathbf{K}}_n) = c\mu^n(n + \mu[b - 1, b' - 1; n]).$$

The choice $b = 1$ and $b' \in (\frac{\hat{k}+\alpha-1}{\eta}, 1)$ results in

$$\mu[b - 1, b' - 1; n] = \mathcal{O}(n), \quad n \to \infty.$$

Theorem 6.26 *Suppose that $k'' \ge 2k - 1$, $m = k$ and $\hat{k} = k$. Choose $b = 1$ and $b' \in (\frac{\hat{k}+\alpha-1}{\eta}, 1)$. Then there hold the stability estimate in Theorem 6.23, the error estimate in Theorem 6.24 with $\tau = 0$ the boundedness of the condition number in Theorem 6.25 and the following asymptotic result of the complexity:*

$$\mathcal{N}(\mathbf{E}_n - \tilde{\mathbf{K}}_n) = \mathcal{O}(s(n)\log(s(n))),$$

as $n \to \infty$.

6.2.4 A numerical example

In this subsection, we present a numerical example of the DMPG algorithm applied to a boundary integral equation to illustrate the convergence order, matrix compression and computational complexity of the method.

We consider the reformulation of the following boundary value problem for the Laplace equation as an integral equation. The boundary value problem is given by

$$\Delta u(P) = 0, \quad P \in D,$$
$$\frac{\partial u(P)}{\partial n_p} = -cu(P) + g(P), \quad P \in \Gamma := \partial D,$$

where D is a bounded, simply connected open region in $\mathbb{R}^2$ with a smooth boundary Γ, n_P the exterior unit normal to Γ at P, $g(P)$ a given continuous function on the boundary Γ and c a positive constant. We seek a solution $u \in C^2(D) \cap C^1(\overline{D})$ for the boundary value problem. Following Section 2.1.3 (see also [12, 270]), we employ the Green representation formula for harmonic functions and rewrite the above problem as a boundary integral equation

$$u(P) - (Au)(P) - (Bu)(P) = -(Bg)(P), \quad P \in \Gamma,$$

where

$$(Au)(P) = \frac{1}{\pi}\int_\Gamma u(Q)\frac{\partial}{\partial n_Q}\log|P-Q|\mathrm{d}\sigma(Q), \quad P \in \Gamma$$

and

$$(Bu)(P) = \frac{1}{\pi}\int_\Gamma u(Q)\log|P-Q|\mathrm{d}\sigma(Q), \quad P \in \Gamma.$$

To convert the above boundary integral equation to an integral equation on an interval, we introduce a parametrization $\mathbf{r}(t) = (\xi(t), \eta(t))$, $0 \le t \le 2\pi$, for the boundary Γ. We assume that each component of $\mathbf{r}$ is a 2π-periodic function in C^2 and $|\mathbf{r}'(t)| = \sqrt{\xi'(t)^2 + \eta'(t)^2} \neq 0$ for $0 \le t \le 2\pi$. Using this parametrization, we convert the above equation to the following equivalent one:

$$(u \circ \mathbf{r})(t) - (K(u \circ \mathbf{r}))(t) = -\frac{1}{c}(B(g \circ \mathbf{r}))(t), \quad t \in [0, 2\pi],$$

where

$$(Kv)(t) := \int_0^{2\pi} k(t,s)v(s)\mathrm{d}s,$$
$$k(t,s) = \frac{c}{\pi}|\mathbf{r}'(s)|\log|\mathbf{r}(t)-\mathbf{r}(s)| + \frac{1}{\pi}\frac{\eta'(s)[\xi(s)-\xi(t)] - \xi'(s)[\eta(s)-\eta(t)]}{[\xi(s)-\xi(t)]^2 + [\eta(s)-\eta(t)]^2}$$

and "$\circ$" stands for the function composition. This kernel has a weak singularity along its diagonal and a detailed discussion of its regularity is given in [270].

In the numerical example presented below, we consider the boundary value problem described above with

$$D := \left\{(x,y) : x^2 + \left(\frac{y}{2}\right)^2 < 1\right\}.$$

For comparison purposes, we consider the boundary value problem with the known exact solution $u(P) = e^x \cos y$, $P \in D$ and compute the function g accordingly. This leads to

$$g(P) = \frac{\partial}{\partial n_P}(e^x \cos y) + e^x \cos y, \;\; P = (x,y) \in \Gamma.$$

The equivalent boundary integral equation is given by (for details, see [270])

$$u(s) - \int_\Omega K(s,t)u(t)\mathrm{d}t = f(s), \;\; s \in \Omega := [0,1],$$

where $u(s)$ represents the value $u(P)$ at the point $P = (\cos 2\pi s, 2\sin 2\pi s)$, $s \in \Omega$, the right-hand-side function f is also determined by the exact solution u and the kernel is given by

$$K(s,t) = \sqrt{1+3\cos^2 2\pi t}\log\left[(\cos 2\pi t - \cos 2\pi s)^2 + 4(\sin 2\pi t - \sin 2\pi s)^2\right] \\ + \frac{8\sin^2 \pi(t-s)}{4\sin^2 \pi(t-s) + 3(\sin 2\pi t - \sin 2\pi s)^2}.$$

To apply the DMPG method, we choose $k = \mu = 2$, $k' = \nu = 1$, $k'' = 3$ (so that $\lambda = 2$) and choose the initial trial and test spaces as

$$\mathbb{X}_0 := \operatorname{span}\{\phi_{00}, \phi_{01}\} \text{ and } \mathbb{Y}_0 := \operatorname{span}\{\psi_{00}, \psi_{01}\},$$

respectively, where

$$\phi_{00}(t) := 1, \;\; \phi_{01}(t) := \sqrt{3}(2t-1)$$

and

$$\psi_{00}(t) := \begin{cases} \sqrt{2}, & 0 \le t \le \frac{1}{2}, \\ 0, & \frac{1}{2} < t \le 1, \end{cases} \quad \psi_{01}(t) := \begin{cases} 0, & 0 \le t \le \frac{1}{2}, \\ \sqrt{2}, & \frac{1}{2} < t \le 1. \end{cases}$$

The initial spaces

$$\mathbb{W}_0 := \operatorname{span}\{\phi_{10}, \phi_{11}\}, \;\; \mathbb{W}'_0 := \operatorname{span}\{\psi_{10}, \psi_{11}\}$$

are given for $t \in \Omega$ by the equations

$$\phi_{10}(t) := \begin{cases} 6t-1, & 0 \le t \le \frac{1}{2}, \\ 6t-5, & \frac{1}{2} < t \le 1, \end{cases} \quad \phi_{11}(t) := \begin{cases} \sqrt{3}(4t-1), & 0 \le t \le \frac{1}{2}, \\ \sqrt{3}(3-4t), & \frac{1}{2} < t \le 1, \end{cases}$$

$$\psi_{10}(t) := \begin{cases} \sqrt{2}, & 0 \le t \le \frac{1}{4}, \\ -\sqrt{2}, & \frac{1}{4} < t \le \frac{1}{2}, \\ 0, & \frac{1}{2} < t \le 1, \end{cases} \quad \psi_{11}(t) := \begin{cases} 0, & 0 \le t \le \frac{1}{2}, \\ \sqrt{2}, & \frac{1}{2} < t \le \frac{3}{4}, \\ -\sqrt{2}, & \frac{3}{4} < t \le 1. \end{cases}$$

The bases for $\mathbb{W}_n$, $\mathbb{W}'_n$, $n \in \mathbb{N}$ are generated recursively as described in Chapter 4.

We report numerical results of the truncated DMPG scheme applied to the integral equation given above in the following table. We define the compression rate by the number of nonzero entries in the matrix $\tilde{\mathcal{K}}_n$ divided by the total number of entries in the matrix before truncation. In the table, we use $\mathcal{O}_n$, C_n and R_n to denote, respectively, the order of convergence in the L^2-norm, the condition number of the truncated coefficient matrix and the compression rate.

n	7	8	9
$\mathcal{O}_n$		1.877	1.829
C_n	14.02	15.07	16.04
R_n	0.43	0.35	0.19

6.3 Bibliographical remarks

The Petrov–Galerkin and iterated Petrov–Galerkin methods for Fredholm integral equations of the second kind were originally studied in [77] (cf. Section 3.5), where the useful notions of the *generalized best approximation* and *regular pair* of subspaces were introduced to analyze convergence of the methods. One advantage of the Petrov–Galerkin method is that it allows us to achieve the same order of convergence as the Galerkin method for the same trial space with much less computational cost by choosing the test spaces as spaces of piecewise polynomials of degree lower than that for the trial space. However, the practical use of the Petrov–Galerkin method requires the numerical computation of integrals appearing in the resulting linear system. In general, the entries of the coefficient matrix are singular integrals and they may not be evaluated exactly. To overcome this difficulty, a theoretical framework for the analysis of a large class of numerical schemes, including the discrete

Galerkin, Petrov–Galerkin, collocation methods and quadrature methods, was developed in [80] (cf. Section 3.5.3).

The multiscale Petrov–Galerkin schemes presented in Section 6.1 were originally developed in [64] to solve weakly singular Fredholm integral equations of the second kind. They are based on *discontinuous orthogonal multiscale basis functions* constructed in [200] and [201], and lead to a fast algorithm for the numerical solution of equations. Additional information on multiscale Petrov–Galerkin algorithms can be found in [146, 147].

The discrete multiscale Petrov–Galerkin method for the integral equations presented in Section 6.2 was introduced in [68] by employing the product integration method. Computational complexity, stability and convergence of the method based on the compression strategy are analyzed using the framework developed in [80]. For the classical product integration method solving Fredholm integral equations of the second kind, the reader is referred to [15]. Some early work on the discrete Galerkin method can be found in [16] and [24].

7
Multiscale collocation methods

The purpose of this chapter is to present a multiscale collocation method for solving Fredholm integral equations of the second kind with weakly singular kernels. Among conventional numerical methods for solving integral equations, the collocation method receives more favorable attention from engineering applications due to its lower computational cost in generating the coefficient matrix of the corresponding discrete equations. In comparison, the implementation of the Galerkin method requires much more computational effort for the evaluation of integrals (see, for example, [12, 19, 77]). Nonetheless, it seems that the most attention in multiscale and wavelet methods for boundary integral equations has been paid to Galerkin methods or Petrov–Galerkin methods (see [28, 64, 95] and the references cited therein). These methods are amenable to the L^2 analysis and therefore the vanishing moments of the multiscale basis functions naturally lead to matrix truncation techniques. For collocation methods, the appropriate context to work in is the L^∞ space, and this provides challenging technical obstacles for the identification of good matrix truncation strategies. Following [69], we present a construction of multiscale basis functions and the corresponding multiscale collocation functionals, both having vanishing moments. These basis functions and collocation functionals lead to a numerically sparse matrix presentation of the Fredholm integral operator. A proper truncation of such a numerically sparse matrix will result in a fast numerical algorithm for solving the equation, which preserves the optimal convergence order up to a logarithmic factor.

In Section 7.1, we describe multiscale basis functions and the corresponding functionals needed for developing the fast algorithm. Section 7.2 is devoted to a presentation of the multiscale collocation method. We analyze the proposed method in Section 7.3, giving estimates for the convergence order, computational costs and a bound of the condition number of the related coefficient matrix.

7.1 Multiscale basis functions and collocation functionals

Multiscale collocation methods require availability of multiscale basis functions and collocation functionals which have vanishing moments of certain degrees. In this section, we present a construction of multiscale basis functions and the corresponding collocation functionals for the multiscale collocation methods. We also study properties of these functions and functionals.

7.1.1 A construction of multiscale basis functions and functionals

The solution spaces of the integral equation will be piecewise polynomials on multiscale partitions. Let us first recall the method we used in previous chapters to generate multiscale partitions of an invariant set Ω in d-dimensional Euclidean space $\mathbb{R}^d$. We start with a positive integer μ and a family $\Phi := \{\phi_e : e \in \mathbb{Z}_\mu\}$ of contractive affine mappings on $\mathbb{R}^d$. There exists a unique invariant set Ω of $\mathbb{R}^d$ associated with the family Φ of mappings such that

$$\Phi(\Omega) = \Omega, \tag{7.1}$$

where

$$\Phi(\Omega) := \bigcup_{e \in \mathbb{Z}_\mu} \phi_e(\Omega).$$

We are interested in the cases where Ω has a simple structure including, for example, the cube and simplex in $\mathbb{R}^d$. With these cases in mind, we make the following additional restriction on the family Φ of mappings.

(a) For every $e \in \mathbb{Z}_\mu$, the mapping ϕ_e has a continuous inverse on Ω.

(b) The set Ω has nonempty interior and

$$\text{meas}\,(\phi_e(\Omega) \cap \phi_{e'}(\Omega)) = 0, \quad e, e' \in \mathbb{Z}_\mu, e \neq e'.$$

We use Φ to obtain a sequence of multiscale partitions $\{\Omega_n : n \in \mathbb{N}_0\}$ of the set Ω in the following way. Given any $\mathbf{e} := (e_j : j \in \mathbb{Z}_n) \in \mathbb{Z}_\mu^n$, we define the mapping

$$\phi_{\mathbf{e}} := \phi_{e_0} \circ \phi_{e_1} \circ \cdots \circ \phi_{e_{n-1}}$$

and the number

$$\mu(\mathbf{e}) := \mu^{n-1} e_0 + \cdots + \mu e_{n-2} + e_{n-1}.$$

Note that every $i \in \mathbb{Z}_{\mu^n}$ can be written uniquely as $i = \mu(\mathbf{e})$ for some $\mathbf{e} \in \mathbb{Z}_\mu^n$. From equation (7.1) and conditions (a) and (b), it follows that the collection of sets

$$\Omega_n := \{\Omega_{n,\mathbf{e}} : \Omega_{n,\mathbf{e}} = \phi_{\mathbf{e}}(\Omega),\ \mathbf{e} \in \mathbb{Z}_\mu^n\}$$

forms a partition of Ω. We require that this partition has the following property.

(c) There exist positive constants c_-, c_+ such that for all $n \in \mathbb{N}_0$,

$$c_-\mu^{-n/d} \leq \max\{d(\Omega_{n,\mathbf{e}}) : \mathbf{e} \in \mathbb{Z}_\mu^n\} \leq c_+\mu^{-n/d}. \tag{7.2}$$

On this partition, we consider piecewise polynomials. Choose a positive integer k and for $n \in \mathbb{N}$, let $\mathbb{X}_n$ be the space of all functions such that their restrictions to any cell $\Omega_{n,\mathbf{e}}$, $\mathbf{e} \in \mathbb{Z}_\mu^n$, are polynomials of total degree $\leq k-1$. Here we use the convention that for $n = 0$ the set Ω is the only cell in the partition and so

$$m := \dim \mathbb{X}_0 = \binom{k+d-1}{d}.$$

We must generate a suitable multiscale decomposition of $\mathbb{X}_n$. To this end, let $G_0 := \{t_j : j \in \mathbb{Z}_m\}$ be a finite set of distinct points in Ω, which is refinable relative to the mappings Φ, that is, G_0 satisfies

$$G_0 \subseteq \Phi(G_0).$$

Set

$$G_1 := \Phi(G_0),\quad V_1 := G_1 \setminus G_0 = \{t_{m+j} : j \in \mathbb{Z}_r\}$$

with $r := (\mu - 1)m$. Now we require that there exists a basis of elements in $\mathbb{X}_0$, denoted by $\psi_j, j \in \mathbb{Z}_m$, such that

$$\mathbb{X}_0 := \operatorname{span}\{\psi_j : j \in \mathbb{Z}_m\}$$

and they satisfy the Lagrange interpolation conditions

$$\psi_i(t_j) = \delta_{ij},\quad i, j \in \mathbb{Z}_m. \tag{7.3}$$

A construction of refinable points $\{t_j : j \in \mathbb{Z}_m\} \in \Omega$ that admit a unique d-dimensional Lagrange interpolation is presented in [198].

With this basis of $\mathbb{X}_0$ at hand, we generate a multiscale basis for $\mathbb{X}_n$ in the following way. For this purpose, we introduce linear operators $\mathcal{T}_e : \mathbb{X} \to \mathbb{X}, e \in \mathbb{Z}_\mu$ defined by

$$(\mathcal{T}_e x)(t) := x(\phi_e^{-1}(t))\chi_{\phi_e(\Omega)}(t),\quad t \in \Omega,$$

where χ_S denotes the characteristic function of set S. Therefore, it follows that

$$\mathbb{X}_n = \bigoplus_{e \in \mathbb{Z}_\mu} \mathcal{T}_e \mathbb{X}_{n-1},\quad n \in \mathbb{N},$$

where $\mathbb{A} \oplus \mathbb{B}$ denotes the direct sum of spaces $\mathbb{A}$ and $\mathbb{B}$. We assume that the functions $\psi_{m+j} \in \mathbb{X}_1, j \in \mathbb{Z}_r$ satisfy

$$\psi_{m+j}(t_i) = 0, \quad i \in \mathbb{Z}_m, \ j \in \mathbb{Z}_r, \quad \psi_{m+j}(t_{m+j'}) = \delta_{jj'}, \ \ j, j' \in \mathbb{Z}_r. \tag{7.4}$$

Then, the functions $\psi_j, j \in \mathbb{Z}_q$ form a basis for $\mathbb{X}_1$, where $q := m + r$.

We require another basis for $\mathbb{X}_1$ consisting of functions with vanishing moments. To this end, we set

$$w_{0j} := \psi_j, \ \ j \in \mathbb{Z}_m.$$

For $j \in \mathbb{Z}_r$ we find a vector $[c_{js} : s \in \mathbb{Z}_q] \in \mathbb{R}^q$ such that

$$w_{1j} := \sum_{s \in \mathbb{Z}_q} c_{js}\psi_s, \ \ j \in \mathbb{Z}_r \tag{7.5}$$

satisfies the equation

$$\left(w_{1j}, \psi_{j'}\right) = 0, \ \ j' \in \mathbb{Z}_m, \ j \in \mathbb{Z}_r, \tag{7.6}$$

where $(\cdot, \cdot)$ denotes the inner product in $L^2(\Omega)$. Since (7.6) is a linear system of rank m with m equations and q unknowns, there exist r linearly independent solutions of this system, which we denote by $w_{1j}, j \in \mathbb{Z}_r$. These functions form a basis for the orthogonal complement of $\mathbb{X}_0$ in $\mathbb{X}_1$, denoted by $\mathbb{W}_1$. Let

$$\mathbb{W}_{i+1} := \bigoplus_{e \in \mathbb{Z}_\mu} \mathcal{T}_e \mathbb{W}_i, \ \ i \in \mathbb{N}.$$

Then $\mathbb{X}_n$ has the decomposition (see Chapter 4)

$$\mathbb{X}_n = \mathbb{X}_0 \oplus^\perp \mathbb{W}_1 \oplus^\perp \cdots \oplus^\perp \mathbb{W}_n.$$

We denote $\mathbb{W}_0 := \mathbb{X}_0$, $w(i) := \dim \mathbb{W}_i$, $i \in \mathbb{N}_0$ and $s(n) := \dim \mathbb{X}_n$, $n \in \mathbb{N}_0$ as before.

Let us now turn our attention to the construction of multiscale collocation functionals. We start by recalling our notational conventions. For a compact subset Ω of the d-dimensional Euclidean space $\mathbb{R}^d$, we let $\mathbb{X} := L^\infty(\Omega)$, $\mathbb{V} := C(\Omega)$ and denote the dual space of $\mathbb{V}$ by $\mathbb{V}^*$. For any $s \in \Omega$, we use δ_s to denote the linear functional in $\mathbb{V}^*$ defined for $v \in \mathbb{V}$ by the equation $\langle \delta_s, v \rangle = v(s)$. We need to evaluate δ_s on functions in $\mathbb{X}$. Therefore, as in [21], we take any norm-preserving extension of δ_s to $\mathbb{X}$ and use the same notation for the extension. In particular, this convention allows us to evaluate piecewise polynomials *anywhere* on Ω. We begin by defining

$$\ell_{0j} := \delta_{t_j}, \ \ j \in \mathbb{Z}_m$$

and for $j' \in \mathbb{Z}_r$, we find the vector $[c'_{j's} : s \in \mathbb{Z}_q]$ such that

$$\ell_{1j'} := \sum_{s\in\mathbb{Z}_q} c'_{j's}\delta_{t_s}, \quad j' \in \mathbb{Z}_r \tag{7.7}$$

satisfies the equations

$$\langle \ell_{1j'}, w_{0j} \rangle = 0, \quad j \in \mathbb{Z}_m, \quad j' \in \mathbb{Z}_r, \tag{7.8}$$

$$\langle \ell_{1j'}, w_{1j} \rangle = \delta_{jj'}, \quad j \in \mathbb{Z}_r, \quad j' \in \mathbb{Z}_r. \tag{7.9}$$

For $j \in \mathbb{Z}_r$, the $q \times q$ coefficient matrix of this linear system is given by

$$\mathbf{A} := \left[\left\langle \delta_{t_{i'j'}}, w_{ij} \right\rangle : (i,j), (i',j') \in \mathbb{U}_1 \right],$$

where $t_{1j} = t_{m+j}, j \in \mathbb{Z}_r$. Let us prove that the matrix $\mathbf{A}$ is nonsingular. To this end, we assume that there are constants a_{ij}, $(i,j) \in \mathbb{U}_1$ such that

$$\sum_{(i,j)\in\mathbb{U}_1} a_{ij} \left\langle \delta_{t_{i'j'}}, w_{ij} \right\rangle = 0, \quad (i',j') \in \mathbb{U}_1,$$

that is,

$$\left\langle \delta_{t_{i'j'}}, \sum_{(i,j)\in\mathbb{U}_1} a_{ij} w_{ij} \right\rangle = 0, \quad (i',j') \in \mathbb{U}_1.$$

Since the set G_1 is Lagrange admissible relative to $(\Phi, \mathbb{X}_1)$ (see Section 4.5.1), we conclude that

$$\sum_{(i,j)\in\mathbb{U}_1} a_{ij} w_{ij} = 0,$$

and therefore $a_{ij} = 0$, $(i,j) \in \mathbb{U}_1$. This proves that $\mathbf{A}$ is nonsingular.

We find it convenient to write equations (7.8) and (7.9) in a matrix form. For this purpose, we introduce the matrices

$$\tilde{\mathbf{B}} := \left[\langle \delta_{t_i}, \psi_j \rangle : i,j \in \mathbb{Z}_q \right], \quad \mathbf{B} := \left[\langle \delta_{t_{m+i}}, \psi_j \rangle : i \in \mathbb{Z}_r, j \in \mathbb{Z}_m \right],$$

$$\mathbf{C}_1 := \left[c_{js} : j \in \mathbb{Z}_r, s \in \mathbb{Z}_m \right], \quad \mathbf{C}_2 := \left[c_{j,m+s} : j \in \mathbb{Z}_r, s \in \mathbb{Z}_r \right],$$

$$\mathbf{C}'_1 = \left[c'_{js} : j \in \mathbb{Z}_r, s \in \mathbb{Z}_m \right], \quad \mathbf{C}'_2 := \left[c'_{j,m+s} : j \in \mathbb{Z}_r, s \in \mathbb{Z}_r \right]$$

and

$$\mathbf{C} := [\mathbf{C}_1, \mathbf{C}_2], \quad \mathbf{C}' := [\mathbf{C}'_1, \mathbf{C}'_2].$$

The next lemma gives a relationship between the matrices $\mathbf{C}$ and $\mathbf{C}'$.

Lemma 7.1 *The following relation holds:*

$$\mathbf{C}'_1 = -\mathbf{C}'_2\mathbf{B}, \ \ \mathbf{C}'_2 = (\mathbf{C}_2^T)^{-1}.$$

Proof It follows from equations (7.8) and (7.9) that

$$\mathbf{C}'\tilde{\mathbf{B}}[\mathbf{I}_m, \mathbf{O}_{m\times r}]^T = \mathbf{O}_{r\times m}$$

and

$$\mathbf{C}'\tilde{\mathbf{B}}\mathbf{C}^T = \mathbf{I}_r,$$

where $\mathbf{O}_{m\times r}$ denotes the $m \times r$ zero matrix and $\mathbf{I}_m$ denotes the $m \times m$ identity matrix. The properties of the basis $\{\psi_j : j \in \mathbb{Z}_q\}$ and the functionals $\{\delta_{t_j} : j \in \mathbb{Z}_q\}$ described above in equations (7.3) and (7.4) imply that

$$\tilde{\mathbf{B}} = \begin{bmatrix} \mathbf{I}_m & \mathbf{O}_{m\times r} \\ \mathbf{B} & \mathbf{I}_r \end{bmatrix}$$

and

$$\mathbf{C}'_1 + \mathbf{C}'_2\mathbf{B} = \mathbf{O}_{r\times m}, \ \ (\mathbf{C}'_1 + \mathbf{C}'_2\mathbf{B})\mathbf{C}_1^T + \mathbf{C}'_2\mathbf{C}_2^T = \mathbf{I}_r,$$

from which the result follows. □

We next describe the construction of a basis for $\mathbb{W}_i$, $i \in \mathbb{N}$. To this end, for $\mathbf{e} := [e_j : j \in \mathbb{Z}_n] \in \mathbb{Z}_\mu^n$, we introduce a composition operator $\mathcal{T}_\mathbf{e}$ by

$$\mathcal{T}_\mathbf{e} := \mathcal{T}_{e_0} \circ \cdots \circ \mathcal{T}_{e_{n-1}}.$$

For $i = 2, 3, \ldots, n$ we let

$$w_{ij} := \mathcal{T}_\mathbf{e} w_{1l}, \ \ j = \mu(\mathbf{e})r + l, \mathbf{e} \in \mathbb{Z}_\mu^{i-1}, \ \ l \in \mathbb{Z}_r \tag{7.10}$$

and

$$\mathbb{W}_i := \operatorname{span}\{w_{ij} : j \in \mathbb{Z}_{w(i)}\}.$$

Observe that the support of w_{ij} is contained in $S_{ij} := \phi_\mathbf{e}(\Omega), j \in \mathbb{Z}_{w(i)}$.

To generate a multiscale collocation functional, we introduce for any $e \in \mathbb{Z}_\mu$ a linear operator $\mathcal{L}_e : \mathbb{X}^* \to \mathbb{X}^*$ defined by the equation

$$\langle \mathcal{L}_e \ell, v \rangle = \langle \ell, v \circ \phi_e \rangle, \quad v \in \mathbb{X}, \ \ell \in \mathbb{X}^*.$$

Moreover, for $\mathbf{e} := [e_j : j \in \mathbb{Z}_n] \in \mathbb{Z}_\mu^n$, we define the composition operator

$$\mathcal{L}_\mathbf{e} := \mathcal{L}_{e_0} \circ \cdots \circ \mathcal{L}_{e_{n-1}}.$$

Consequently, for any $\mathbf{e}, \mathbf{e}' \in \mathbb{Z}_\mu^i$, $w \in \mathbb{X}$ and $\ell \in \mathbb{X}^*$, we have that

$$\langle \mathcal{L}_\mathbf{e} \ell, \mathcal{T}_{\mathbf{e}'} w \rangle = \langle \ell, w \rangle \, \delta_{\mathbf{e}\mathbf{e}'}. \tag{7.11}$$

In addition, for $i > 1, j = \mu(\mathbf{e})r + l$, $\mathbf{e} \in \mathbb{Z}_\mu^{i-1}$, $l \in \mathbb{Z}_r$ we define

$$\ell_{ij} := \mathcal{L}_{\mathbf{e}}\ell_{1l} \tag{7.12}$$

and observe that

$$\left\langle \ell_{ij}, v \right\rangle = \langle \ell_{1l}, v \circ \phi_{\mathbf{e}} \rangle = \sum_{s\in\mathbb{Z}_q} c'_{ls} v(\phi_{\mathbf{e}}(t_s)).$$

Note that the "support" of ℓ_{ij} is also contained in S_{ij}.

7.1.2 Properties of basis functions and collocation functionals

We discuss the properties of the multiscale functions and their corresponding collocation functionals constructed in the last subsection. To this end, we partition the matrix $\mathbf{E}_n$ into a block matrix

$$\mathbf{E}_n := [\mathbf{E}_{i'i} : i', i \in \mathbb{Z}_{n+1}],$$

where

$$\mathbf{E}_{i'i} := [\Omega_{i'j',ij} : j' \in \mathbb{Z}_{w(i')}, j \in \mathbb{Z}_{w(i)}], \quad \Omega_{i'j',ij} := \left\langle \ell_{i'j'}, w_{ij} \right\rangle$$

and in the next lemma we relate the norm of the matrix $\mathbf{E}_{i'i}$ to that of $\mathbf{E}_{1,i-i'+1}$.

Lemma 7.2 *If $i', i \in \mathbb{N}$ with $i > i'$, then*

$$\|\boldsymbol{E}_{i'i}\|_\infty = \|\boldsymbol{E}_{1,i-i'+1}\|_\infty. \tag{7.13}$$

Proof From the definition of $\ell_{i'j'}$ and w_{ij}, for $(i', j'), (i, j) \in \mathbb{U}_n$ with $i > i'$, we obtain that there exist $\mathbf{e} \in \mathbb{Z}_\mu^{i-1}$, $\mathbf{e}' \in \mathbb{Z}_\mu^{i'-1}$, $l, l' \in \mathbb{Z}_r$ such that

$$\left\langle \ell_{i'j'}, w_{ij} \right\rangle = \langle \mathcal{L}_{\mathbf{e}'}\ell_{1l'}, \mathcal{T}_{\mathbf{e}} w_{1l} \rangle .$$

We introduce the vectors

$$\mathbf{e}_1 := [e_j : j \in \mathbb{Z}_{i'-1}], \quad \mathbf{e}_2 := [e_j : j \in \mathbb{Z}_{i-1} \setminus \mathbb{Z}_{i'-1}]$$

and conclude from (7.11) that

$$\left\langle \ell_{i'j'}, w_{ij} \right\rangle = \left\langle \ell_{1l'}, \mathcal{T}_{\mathbf{e}_2} w_{1l} \right\rangle \delta_{\mathbf{e}'\mathbf{e}_1}.$$

Let $j_0 := \mu(\mathbf{e}_2)r + l$ and obtain that

$$\left\langle \ell_{i'j'}, w_{ij} \right\rangle = \left\langle \ell_{1l'}, w_{i-i'+1,j_0} \right\rangle \delta_{\mathbf{e}'\mathbf{e}_1}.$$

Consequently, we have that

$$\sum_{j\in\mathbb{Z}_{w(i)}} \left| \left\langle \ell_{i'j'}, w_{ij} \right\rangle \right| = \sum_{j\in\mathbb{Z}_{w(i-i'+1)}} \left| \left\langle \ell_{1l'}, w_{i-i'+1,j} \right\rangle \right|,$$

which proves the lemma. □

Lemma 7.3 *The condition*

$$\sum_{m \in \mathbb{Z}_{w(n)}} |\langle \ell_{n'm'}, w_{nm} \rangle| \le \gamma, \quad (n', m') \in \mathbb{U}, \quad n > n'$$

is satisfied with

$$\gamma := \max\{\|\mathbf{C}_1\|_1, \|\mathbf{C}'\|_\infty \|\mathbf{C}_1\|_1\}.$$

Proof By Lemma 7.2, it suffices to prove for $i \in \mathbb{N}$ that

$$\|\mathbf{E}_{0i}\|_\infty \le \|\mathbf{C}_1\|_1 \tag{7.14}$$

and

$$\|\mathbf{E}_{1,i+1}\|_\infty \le \|\mathbf{C}'\|_\infty \|\mathbf{C}_1\|_1. \tag{7.15}$$

Recall the definition

$$\mathbf{E}_{1,i+1} := \left[\langle \ell_{1l'}, w_{i+1,j} \rangle : l' \in \mathbb{Z}_r, j \in \mathbb{Z}_{w(i+1)}\right].$$

We need to decompose this matrix. This is done by using equation (7.7) to write

$$\ell_{1l'} = \sum_{s' \in \mathbb{Z}_q} c'_{l's'} \delta_{t_{s'}}, \quad l' \in \mathbb{Z}_r.$$

Thus, it follows from (7.5) and (7.10) for any $j \in \mathbb{Z}_{w(i+1)}$ that there exists a unique pair $\mathbf{e}_j \in \mathbb{Z}_\mu^i$ and $l \in \mathbb{Z}_r$ such that

$$w_{i+1,j} = \sum_{s \in \mathbb{Z}_q} c_{ls} \mathcal{T}_{\mathbf{e}_j} \psi_s.$$

Since for any $s' \in \mathbb{Z}_q, \mathbf{e}_j \in \mathbb{Z}_\mu^i, i \in \mathbb{N}$ and $s = m, \ldots, q-1$,

$$\langle \delta_{t_{s'}}, \mathcal{T}_{\mathbf{e}_j} \psi_s \rangle = 0,$$

we conclude for $l' \in \mathbb{Z}_r, j = \mu(\mathbf{e}_j)r + l, \mathbf{e}_j \in \mathbb{Z}_\mu^i, l \in \mathbb{Z}_r$ that

$$\langle \ell_{1l'}, w_{i+1,j} \rangle = \sum_{s' \in \mathbb{Z}_q} \sum_{s \in \mathbb{Z}_m} c'_{l's'} c_{ls} \langle \delta_{t_{s'}}, \mathcal{T}_{\mathbf{e}_j} \psi_s \rangle.$$

We write this equation in a matrix form by introducing for each $\mathbf{e} \in \mathbb{Z}_\mu^i$ the matrix

$$\mathbf{D}_\mathbf{e} := \left[\langle \delta_{t_{s'}}, \mathcal{T}_\mathbf{e} \psi_s \rangle : s' \in \mathbb{Z}_q, s \in \mathbb{Z}_m\right],$$

and from these matrices we build the matrix

$$\mathbf{D} := \left[\mathbf{D}_{\mathbf{e}_0}, \mathbf{D}_{\mathbf{e}_1}, \ldots, \mathbf{D}_{\mathbf{e}_{\mu^i - 1}}\right].$$

This notation allows us to write

$$\mathbf{E}_{1,i+1} = \mathbf{C}'\mathbf{D}\mathrm{diag}(\mathbf{C}_1^T, \ldots, \mathbf{C}_1^T),$$

where the right-most matrix is a block diagonal matrix with μ^i identical blocks of $\mathbf{C}_1^T$. This formula will allow us to estimate the norm of the matrix $\mathbf{E}_{1,i+1}$.

Because the set G_0 is refinable relative to the contractive affine mappings Φ, we assume that for any $s' \in \mathbb{Z}_m$ there exist unique $\mathbf{e}'' \in \mathbb{Z}_\mu^i$ and $s'' \in \mathbb{Z}_m$ such that $t_{s'} = \phi_{\mathbf{e}''}(t_{s''})$. Thus,

$$\left\langle \delta_{t_{s'}}, \mathcal{T}_{\mathbf{e}}\psi_s \right\rangle = \left\langle \mathcal{L}_{\mathbf{e}''}\delta_{t_{s''}}, \mathcal{T}_{\mathbf{e}}\psi_s \right\rangle = \left\langle \delta_{t_{s''}}, \psi_s \right\rangle \delta_{\mathbf{e}''\mathbf{e}} = \delta_{ss''}\delta_{\mathbf{e}''\mathbf{e}},$$

which implies that $\|\mathbf{D}\|_\infty = 1$. Consequently, we conclude inequality (7.15). Similarly, inequality (7.14) follows from

$$\mathbf{E}_{0i} = \left[\left\langle \ell_{0l'}, w_{ij} \right\rangle : l' \in \mathbb{Z}_m, j \in \mathbb{Z}_{w(i)} \right].$$

This completes the proof of this lemma. □

We next show that the pair (W, L) of basis functions and collocation functionals constructed in the last subsection has some important properties. To this end, we let $\mathcal{P}_n$ be the projection from $\mathbb{X}$ onto $\mathbb{X}_n$ defined by the requirement that

$$\left\langle \ell_{ij}, \mathcal{P}_n x \right\rangle = \left\langle \ell_{ij}, x \right\rangle, \quad (i,j) \in \mathbb{U}_n.$$

Proposition 7.4 *The following properties hold.*

(I) There exist positive integers ρ and h such that for every $n > h$ and $m \in \mathbb{Z}_{w(n)}$ written in the form $m = j\rho + s$, where $s \in \mathbb{Z}_\rho$ and $j \in \mathbb{N}_0$,

$$w_{nm}(x) = 0, \quad x \notin \Omega_{n-h,j}.$$

(II) For any $n, n' \in \mathbb{N}_0$

$$\langle \ell_{n'm'}, w_{nm} \rangle = \delta_{nn'}\delta_{mm'}, \quad (n,m), (n',m') \in \mathbb{U}, \quad n \le n'$$

and there exists a positive constant γ for which

$$\sum_{m \in \mathbb{Z}_{w(n)}} |\langle \ell_{n'm'}, w_{nm} \rangle| \le \gamma, \quad (n',m') \in \mathbb{U}, \quad n > n'.$$

(III) There exists a positive integer k such that for all $p \in \pi_k$, where π_k denotes the space of polynomials of total degree less than k,

$$\langle \ell_{nm}, p \rangle = 0, \quad (w_{nm}, p) = 0, \quad (n,m) \in \mathbb{U}.$$

(IV) There exists a positive constant θ_0 such that for all $(n,m) \in \mathbb{U}$,

$$\|\ell_{nm}\| + \|w_{nm}\|_\infty \le \theta_0.$$

(V) *There exists a positive integer* $\mu > 1$ *such that*

$$\dim \mathbb{X}_n = \mathcal{O}(\mu^n), \ \dim \mathbb{W}_n = \mathcal{O}(\mu^n), \ d_n = \mathcal{O}(\mu^{-n/d}), \ n \to \infty.$$

(VI) *The operators* $\mathcal{P}_n$ *are well defined and converge pointwise to the identity operator* $\mathcal{I}$ *in* $L^\infty(\Omega)$ *as* $n \to \infty$. *In other words, for any* $x \in L^\infty(\Omega)$,

$$\lim_{n\to\infty} \|\mathcal{P}_n x - x\|_\infty = 0.$$

(VII) *There exists a positive constant c such that for* $u \in W^{k,\infty}(\Omega)$,

$$\operatorname{dist}(u, \mathbb{X}_n) \le c\mu^{-kn/d}\|u\|_{k,\infty}.$$

Proof Property (I) is satisfied because for $(i,j) \in \mathbb{U}$, with $i > 1$, the support of w_{ij} is contained in $S_{ij} = \phi_{\mathbf{e}}(\Omega) = \Omega_{i-1,\mathbf{e}}$, where $j = \mu(\mathbf{e})r + l$, $l \in \mathbb{Z}_r$, $\mathbf{e} \in \mathbb{Z}_\mu^{i-1}$.

We now prove that the pair (W, L) satisfies property (II). For $(i,j) \in \mathbb{U}$, there exists a unique pair of $\mathbf{e} \in \mathbb{Z}_\mu^{i-1}$ and $l \in \mathbb{Z}_r$ such that $j = \mu(\mathbf{e})r + l$ and $w_{ij} = \mathcal{T}_{\mathbf{e}} w_{1l}$. Likewise, for $(i',j') \in \mathbb{U}$, there exists a unique pair of $\mathbf{e}' \in \mathbb{Z}_\mu^{i'-1}$ and $l' \in \mathbb{Z}_r$ such that $j' = \mu(\mathbf{e}')r + l'$ and $\ell_{i'j'} = \mathcal{L}_{\mathbf{e}'}\ell_{1l'}$. When $i = i'$, it follows from (7.11) and (7.9) that

$$\langle \ell_{i'j'}, w_{ij}\rangle = \langle \mathcal{L}_{\mathbf{e}'}\ell_{1l'}, \mathcal{T}_{\mathbf{e}} w_{1l}\rangle = \langle \ell_{1l'}, w_{1l}\rangle\, \delta_{\mathbf{e}'\mathbf{e}} = \delta_{l'l}\delta_{\mathbf{e}'\mathbf{e}} = \delta_{j'j}.$$

When $i < i'$, let $\mathbf{e}'_1 = [e'_j : j \in \mathbb{Z}_{i-1}]$, $\mathbf{e}'_2 = [e'_j : j \in \mathbb{Z}_{i'-1} \setminus \mathbb{Z}_{i-1}]$. Then

$$\langle \ell_{i'j'}, w_{ij}\rangle = \left\langle \mathcal{L}_{\mathbf{e}'_2}\ell_{1l'}, w_{1l}\right\rangle \delta_{\mathbf{e}'_1\mathbf{e}} = \left\langle \ell_{1l'}, w_{1l} \circ \phi_{\mathbf{e}'_2}\right\rangle \delta_{\mathbf{e}'_1\mathbf{e}}.$$

Since $\phi_{\mathbf{e}'_2} : \Omega \to \phi_{\mathbf{e}'_2}(\Omega)$ is an affine mapping, we conclude that $w_{1l} \circ \phi_{\mathbf{e}'_2}$ is a polynomial of total degree $\le k - 1$ in $\mathbb{X}_0$. By using (7.8), we have that

$$\langle \ell_{i'j'}, w_{ij}\rangle = 0, \ \ (i,j), (i',j') \in \mathbb{U}, i < i'.$$

When $i > i'$, Lemma 7.3 ensures that the second equation of (II) is satisfied. This proves property (II).

Next, we verify that property (III) is satisfied. Again, it follows from (7.8) that

$$\langle \ell_{i'j'}, \psi_j\rangle = \langle \ell_{1l'}, \psi_j \circ \phi_{\mathbf{e}'}\rangle = 0, \ \ j \in \mathbb{Z}_m.$$

This proves the first equation of property (III). To prove the second equation, we consider $\mathcal{T}_{\mathbf{e}}$ as an operator from $L^2(\Omega)$ to $L^2(\Omega)$ and denote by $\mathcal{T}_{\mathbf{e}}^*$ the conjugate operator of $\mathcal{T}_{\mathbf{e}}$, which is defined by

$$(\mathcal{T}_{\mathbf{e}} x, y) = \left(x, \mathcal{T}_{\mathbf{e}}^* y\right).$$

It can be shown that for $y \in L^2(\Omega)$,

$$\mathcal{T}_{\mathbf{e}}^* y = J_{\phi_{\mathbf{e}}} y \circ \phi_{\mathbf{e}},$$

where $J_{\Phi_{\mathbf{e}}}$ is the Jacobi of mapping $\phi_{\mathbf{e}}$. Therefore, we have that

$$\left(w_{ij}, \psi_{j'}\right) = \left(\mathcal{T}_{\mathbf{e}} w_{1l}, \psi_{j'}\right) = \left(w_{1l}, \mathcal{T}_{\mathbf{e}}^* \psi_{j'}\right) = 0.$$

The last equality holds because $\mathcal{T}_{\mathbf{e}}^* \psi_{j'}$ is a polynomial of total degree $\leq k-1$ and w_{1l} satisfies condition (7.6).

From (7.12), (7.7), (7.10) and (7.5), we have for $(i,j) \in \mathbb{U}, j = \mu(\mathbf{e})r + l$ that

$$|\langle \ell_{ij}, v\rangle| = |\langle \ell_{1l}, v \circ \phi_{\mathbf{e}}\rangle| \leq \|\mathbf{C}'\|_\infty \|v\|_\infty$$

and

$$\|w_{ij}\|_\infty \leq \|w_{1l}(\phi_{\mathbf{e}}^{-1}(t))\chi(\phi_{\mathbf{e}}(\Omega))\|_\infty \leq \|\mathbf{C}\|_\infty \max_{j \in \mathbb{Z}_q} \|\psi_j\|_\infty,$$

which confirm property (IV).

By our construction, it is the case that

$$\dim \mathbb{X}_n = m\mu^n, \quad \dim \mathbb{W}_n = m(\mu - 1)\mu^{n-1}.$$

These equations with (7.2) imply that property (V) is satisfied.

It follows from the first equation of (II) that $\mathcal{P}_n$ is well defined. The pointwise convergence condition (VI) of the interpolating projections $\mathcal{P}_n$ follows from a result of [21]. Finally, property (VII) holds, since the $\mathbb{X}_n$ are spaces of piecewise polynomials of total degree $\leq k-1$. □

Proposition 7.5 *For any $k, d \in \mathbb{N}$, there exists an integer $\mu > 1$ such that the following property holds.*

(VIII) The constant γ in property (II) satisfies the condition

$$(1+\gamma)\mu^{-\frac{k}{d}} < 1.$$

Proof We must show that there exists an integer $\mu > 1$ such that

$$1 + \gamma \leq \mu^{\frac{k}{d}},$$

where γ is defined in Lemma 7.3. This will be done by proving that γ is bounded from above independently by μ. For this purpose, we consider the matrices

$$\mathbf{H}_1 := [\left(\psi_i, \psi_j\right) : i \in \mathbb{Z}_m, j \in \mathbb{Z}_m],$$
$$\mathbf{H}_2 := [\left(\psi_i, \psi_{m+j}\right) : i \in \mathbb{Z}_m, j \in \mathbb{Z}_r], \quad \mathbf{H} := [\mathbf{H}_1, \mathbf{H}_2].$$

Therefore, from equation (7.6) it follows that

$$\mathbf{C}\mathbf{H}^T = \mathbf{C}_1\mathbf{H}_1^T + \mathbf{C}_2\mathbf{H}_2^T = \mathbf{0},$$

where $\mathbf{C}_2$ is an arbitrary $r \times r$ nonsingular matrix. We choose $\mathbf{C}_2 := \mathbf{I}_r$, from which we have that

$$\mathbf{C}_1 = -\mathbf{H}_2^T(\mathbf{H}_1^T)^{-1}. \tag{7.16}$$

Moreover, from Lemma 7.1 we have that

$$\mathbf{C}' = [-\mathbf{B}, \mathbf{I}]$$

and thus

$$\|\mathbf{C}'\|_\infty = \|\mathbf{B}\|_\infty + 1. \tag{7.17}$$

For $j \in \mathbb{Z}_m$, the functions ψ_j are polynomials and therefore continuous. Thus, there exists a positive constant ρ such that

$$\max_{j\in\mathbb{Z}_m} \|\psi_j\|_\infty \le \rho.$$

Hence, recalling the definition of matrix $\mathbf{B}$ and equation (7.17), we have that

$$\|\mathbf{C}'\|_\infty = 1 + \max_{i\in\mathbb{Z}_r} \sum_{j\in\mathbb{Z}_m} |\psi_j(t_{m+i})| \le 1 + m \max_{j\in\mathbb{Z}_m} \|\psi_j\|_\infty \le 1 + m\rho.$$

Moreover, we have by equation (7.17) that

$$\|\mathbf{C}_1\|_1 = \|\mathbf{H}_1^{-1}\mathbf{H}_2\|_\infty \le \|\mathbf{H}_1^{-1}\|_\infty \|\mathbf{H}_2\|_\infty.$$

Since $\|\mathbf{H}_1^{-1}\|_\infty$ is independent of μ, it remains to estimate $\|\mathbf{H}_2\|_\infty$ from the above independent of μ. Therefore, we recall for $j \in \mathbb{Z}_r$ that

$$\psi_{m+j}(t) = \psi_l(\phi_e^{-1}(t))\chi_{\phi_e(\Omega)}(t), \quad t \in \Omega,$$

for some $l \in \mathbb{Z}_m$ and $e \in \mathbb{Z}_\mu$. Consequently, from (7.2), we conclude that

$$|(\psi_i, \psi_{m+j})| \le \int_{\phi_e(\Omega)} \left|\psi_i(t)\psi_l(\phi_e^{-1}(t))\right| dt \le \rho^2 \text{meas}(\phi_e(\Omega)) \le \frac{\rho^2}{\mu}.$$

Noting that $r = (\mu - 1)m$, we obtain the desired estimate

$$\|\mathbf{H}_2\|_\infty = \max_{i\in\mathbb{Z}_m} \sum_{j\in\mathbb{Z}_r} |(\psi_i, \psi_{m+j})| \le \frac{\rho^2}{\mu}(\mu - 1)m \le \rho^2 m,$$

thereby proving the result. □

We next examine property (VIII) in several cases of practical importance. We consider the cases when $d = 1$ and $\Omega = [0, 1]$, as well as $d = 2$ and $\Omega = \Delta$, where Δ is the triangle with vertices $(0,0),(1,0),(1,0)$. When $d = 1$ and $\Omega = [0, 1]$, property (VIII) is satisfied for the following choices:

(1) $k = 2$, $\mu = 2$,

$$\phi_e(t) = (t + e)/2, \quad i \in \Omega, e = 0, 1$$

and $t_i = (i + 1)/3$, for $i = 0, 1$;

(2) $k = 3$, $\mu = 2$,

$$\phi_e(t) = (t + e)/2, \quad t \in \Omega, e = 0, 1$$

and $t_i = 2^i/7$, for $i = 0, 1, 2$;

(3) $k = 3$, $\mu = 3$,

$$\phi_e(t) = (t + e)/3, \quad t \in \Omega, e = 0, 1, 2$$

and $t_i = (i + 1)/4$, for $i = 0, 1, 2$;

(4) $k = 4$, $\mu = 2$,

$$\phi_e(t) = (t + e)/2, \quad t \in \Omega, e = 0, 1$$

and $t_i = (i + 1)/5$, for $i = 0, 1, 2, 3$.

In the other case (VIII) is also satisfied when $k = 2$, $\mu = 4$ for $(x, y) \in \Delta$:

$$\phi_0(x, y) = (x/2, y/2), \quad \phi_1(x, y) = ((x + 1)/2, y/2),$$
$$\phi_2(x, y) = (x/2, (y + 1)/2), \quad \phi_3(x, y) = ((1 - x)/2, (1 - y)/2)$$

and $t_0 = (1/7, 4/7)$, $t_1 = (2/7, 1/7)$, $t_2 = (4/7, 2/7)$.

Finally, we turn our attention to another property.

(IX) There exist positive constants θ_1 and θ_2 such that for all $n \in \mathbb{N}_0$ and $v \in \mathbb{X}_n$ having the form $v = \sum_{(i,j)\in\mathbb{U}_n} v_{ij} w_{ij}$,

$$\theta_1 \|\mathbf{v}\|_\infty \le \|v\|_\infty \le \theta_2(n + 1)\|\mathbf{E}_n\mathbf{v}\|_\infty,$$

where $\mathbf{v} = [v_{ij} : (i,j) \in \mathbb{U}_n]$.

To show this, we consider the sequence of functions $\{\zeta_{ij} : (i,j) \in \mathbb{U}\}$, bi-orthogonal to the linear functionals $\{\ell_{ij} : (i,j) \in \mathbb{U}\}$ and having the property that for all $i \in \mathbb{N}_0$,

$$\sup_{t\in\Omega} \sum_{j\in\mathbb{Z}_{w(i)}} |\zeta_{ij}(t)| \le \theta_2. \tag{7.18}$$

Let

$$\zeta_{0j} := w_{0j}, \ \ j \in \mathbb{Z}_m$$

and observe that

$$\langle \ell_{0j}, \zeta_{0j'} \rangle = \delta_{jj'}, \ \ j, j' \in \mathbb{Z}_m.$$

For each $j \in \mathbb{Z}_r$, we find vectors $\mathbf{c}'' := [c''_{js} : s \in \mathbb{Z}_q]$ such that the function

$$\zeta_{1j} := \sum_{s \in \mathbb{Z}_q} c''_{js} \psi_s$$

satisfies the system of linear equations

$$\langle \ell_{0j'}, \zeta_{1j} \rangle = 0, \ \ j' \in \mathbb{Z}_m \tag{7.19}$$

and

$$\langle \ell_{1j'}, \zeta_{1j} \rangle = \delta_{jj'}, \ \ j' \in \mathbb{Z}_r. \tag{7.20}$$

Let us confirm that $\mathbf{c}''$ exists and is unique. The coefficient matrix for equations (7.19) and (7.20) is

$$\tilde{\mathbf{A}} := [\langle \ell_{i'j'}, \psi_j \rangle : j \in \mathbb{Z}_q, (i', j') \in \mathbb{U}_1]. \tag{7.21}$$

Because $\{\psi_j : j \in \mathbb{Z}_q\}$ is a basis for the space $\mathbb{X}_1$, we conclude that matrix $\tilde{\mathbf{A}}$ is nonsingular since $\mathbf{A}$ is nonsingular. Thus, there exists a unique solution $\mathbf{c}''$ for equations (7.19) and (7.20).

For $i > 1, j = \mu(\mathbf{e})r + l, \mathbf{e} \in \mathbb{Z}_\mu^{i-1}, l \in \mathbb{Z}_r$, we define functions

$$\zeta_{ij} := \mathcal{T}_\mathbf{e} \zeta_{1l}.$$

These functions will be used in the proof of the next result.

Proposition 7.6 *The pair (W, L) has property (IX).*

Proof It suffices to verify that the sequences of functions $\{\ell_{ij} : (i, j) \in \mathbb{U}\}$ and functionals $\{\zeta_{ij} : (i, j) \in \mathbb{U}\}$ are bi-orthogonal, that is, they satisfy the condition

$$\langle \ell_{i'j'}, \zeta_{ij} \rangle = \delta_{ii'} \delta_{jj'}, \ \ (i, j), (i', j') \in \mathbb{U} \tag{7.22}$$

and, in addition, there exists a positive constant θ_2 such that for any $i \in \mathbb{N}_0$, condition (7.18) is satisfied.

The proof of (7.22) for the case $i \leq i'$ is similar to that for (II) in Proposition 7.4. Hence, we only present the proof for the case $i' < i$. In this case, we have that

$$\langle \ell_{i'j'}, \zeta_{ij} \rangle = \langle \mathcal{L}_{\mathbf{e}'} \ell_{1l'}, \mathcal{T}_\mathbf{e} \zeta_{1l} \rangle = \langle \ell_{1l'}, \mathcal{T}_{\mathbf{e}_2} \zeta_{1l} \rangle \delta_{\mathbf{e}'\mathbf{e}_1},$$

where $j' = \mu(\mathbf{e}')r + l', j = \mu(\mathbf{e})r + l, \mathbf{e}_1 = [e_j : j \in \mathbb{Z}_{i'-1}]$ and $\mathbf{e}_2 = [e_j : j \in \mathbb{Z}_{i-1} \setminus \mathbb{Z}_{i'-1}]$. From this, it follows that $\langle \ell_{i'j'}, \zeta_{ij} \rangle = 0$, except for $\mathbf{e}' = \mathbf{e}_1$, in which case

$$\langle \ell_{i'j'}, \zeta_{ij} \rangle = \left\langle \ell_{1l'}, \zeta_{1l} \circ \phi_{\mathbf{e}_2}^{-1} \chi_{\phi_{\mathbf{e}_2}(\Omega)} \right\rangle. \tag{7.23}$$

Since G_0 is a refinable set, we have that $\phi_e^{-1}(t) \in G_0$, when $t \in G_1 \cap \phi_e(\Omega)$, $e \in \mathbb{Z}_\mu$ and thus $\phi_{\mathbf{e}_2}^{-1}(t_s) \in G_0$, when $t_s \in \phi_{\mathbf{e}_2}(\Omega), s \in \mathbb{Z}_q$. This observation with (7.20) yields the equation

$$\zeta_{1l}(\phi_{\mathbf{e}_2}^{-1}(t_s)) = \left\langle \delta_{\phi_{\mathbf{e}_2}^{-1}(t_s)}, \zeta_{1l} \right\rangle = 0, \tag{7.24}$$

whenever $t_s \in \phi_{\mathbf{e}_2}(\Omega)$, $s \in \mathbb{Z}_q$. We appeal to (7.23) and (7.24) to conclude that $\langle \ell_{i'j'}, \zeta_{ij} \rangle = 0$.

Next, we show that condition (7.18) is satisfied. Without loss of generality, we consider only the case when $i \geq 1$. In this case, the definition of ζ_{ij}, for $i \geq 1$, guarantees

$$\begin{aligned}
\sup_{t\in\Omega} \sum_{j\in\mathbb{Z}_{w(i)}} |\zeta_{ij}(t)| &= \sup_{t\in\Omega} \sum_{\mathbf{e}\in\mathbb{Z}_\mu^{i-1}} \sum_{l\in\mathbb{Z}_r} |\mathcal{T}_{\mathbf{e}}\zeta_{1l}(t)| \\
&= \sup_{t\in\Omega} \sum_{\mathbf{e}\in\mathbb{Z}_\mu^{i-1}} \sum_{l\in\mathbb{Z}_r} |\zeta_{1l}(\phi_{\mathbf{e}}^{-1}(t))\chi_{\phi_{\mathbf{e}}(\Omega)}(t)| \\
&\leq \sum_{l\in\mathbb{Z}_r} \|\zeta_{1l}\|_\infty,
\end{aligned}$$

and therefore (7.18) holds with $\theta_2 := \sum_{l\in\mathbb{Z}_r} \|\zeta_{1l}\|_\infty$.

Finally, we verify the first inequality in property (IX). To this end, we note that for $v := \sum_{(i,j)\in\mathbb{U}_n} v_{ij}w_{ij}$ and $\mathbf{v} := [v_{ij} : (i,j) \in \mathbb{U}_n]$, there exists $(i_0, j_0) \in \mathbb{U}_n$ with $j_0 = \mu(\mathbf{e}_0)r + l_0, \mathbf{e}_0 \in \mathbb{Z}_\mu^{i_0-1}, l_0 \in \mathbb{Z}_r$, such that

$$\|\mathbf{v}\|_\infty = |v_{i_0j_0}|. \tag{7.25}$$

For $l \in \mathbb{Z}_r$ we denote $\tilde{v}_l := v_{i_0j}$ and $\tilde{w}_l := w_{i_0j}$, where $j = \mu(\mathbf{e}_0)r + l$ and $\mathbf{e}_0 \in \mathbb{Z}_\mu^{i_0-1}$, and observe that

$$|v_{i_0j_0}| \leq \left(\sum_{l\in\mathbb{Z}_r} |\tilde{v}_l|^2 \right)^{1/2}. \tag{7.26}$$

Recalling that $w_{i_0j} = \mathcal{T}_{\mathbf{e}_0} w_{1l}, l \in \mathbb{Z}_r$ and $\phi_e, e \in \mathbb{Z}_\mu$ are affine, we conclude that

$$(\tilde{w}_{l'}, \tilde{w}_l) = J_{\phi_{\mathbf{e}_0}}(w_{1l'}, w_{1l}), \tag{7.27}$$

where $J_{\phi_{\mathbf{e}_0}}$ denotes the Jacobi of the mapping $\phi_{\mathbf{e}_0}$. We introduce an $r \times r$ matrix

$$\mathbf{W} := [(w_{1l'}, w_{1l}) : l', l \in \mathbb{Z}_r]$$

and note that it is the Gram matrix of the basis $w_{1l}, l \in \mathbb{Z}_r$, and thus it is positive definite. It follows that there exists a positive constant c_0 such that for $\tilde{v} := \sum_{l \in \mathbb{Z}_r} \tilde{v}_l \tilde{w}_l$ and $\tilde{\mathbf{v}} := [\tilde{v}_l : l \in \mathbb{Z}_r]$,

$$c_0 \sum_{l \in \mathbb{Z}_r} |\tilde{v}_l|^2 \leq \tilde{\mathbf{v}}^T \mathbf{W} \tilde{\mathbf{v}}. \tag{7.28}$$

By formula (7.27), we have that

$$\|\tilde{v}\|_2^2 = (\tilde{v}, \tilde{v}) = J_{\phi_{\mathbf{e}_0}} \tilde{\mathbf{v}}^T \mathbf{W} \tilde{\mathbf{v}}.$$

Combining this equation with (7.28) yields

$$\sum_{l \in \mathbb{Z}_r} |\tilde{v}_l|^2 \leq \frac{1}{c_0 J_{\phi_{\mathbf{e}_0}}} \|\tilde{v}\|_2^2. \tag{7.29}$$

Since the basis $\{w_{ij} : (i,j) \in \mathbb{U}_n\}$ constructed in this section has property (III), we obtain that

$$\|\tilde{v}\|_2^2 = \int_{\phi_{\mathbf{e}_0}(\Omega)} \tilde{v}(t) v(t) \mathrm{d}t \leq J_{\phi_{\mathbf{e}_0}} \|\tilde{v}\|_\infty \|v\|_\infty \leq J_{\phi_{\mathbf{e}_0}} |v_{i_0 j_0}| \sum_{l \in \mathbb{Z}_r} \|\tilde{w}_l\|_\infty \|v\|_\infty.$$

Using property (IV), we conclude that there exists a positive constant c_0 such that

$$\sum_{l \in \mathbb{Z}_r} \|\tilde{w}_l\|_\infty \leq c_0,$$

which implies with the last inequality that

$$\|\tilde{v}\|_2^2 \leq c_0 J_{\phi_{\mathbf{e}_0}} \|\mathbf{v}\|_\infty \|v\|_\infty. \tag{7.30}$$

Combining (7.25), (7.26), (7.29) and (7.30) yields that there exists a positive constant c such that

$$\|\mathbf{v}\|_\infty \leq c \|\mathbf{v}\|_\infty^{1/2} \|v\|_\infty^{1/2},$$

and thus

$$\|\mathbf{v}\|_\infty \leq c \|v\|_\infty.$$

We have proved the first inequality of (IX) with $\theta_1 := 1/c$. □

7.2 Multiscale collocation methods

In the last section we presented a concrete construction of multiscale bases on an invariant set in $\mathbb{R}^d$ and multiscale collocation functionals needed for multiscale collocation methods. In this section we develop a general collocation scheme for solving Fredholm integral equations of the second kind using multiscale basis functions and multiscale collocation functionals having the properties described in the last section.

7.2.1 The collocation scheme

For a set $A \subset \mathbb{R}^d$, $d(A)$ represents the diameter of A, that is,

$$d(A) := \sup\{|x - y| : x, y \in A\}, \tag{7.31}$$

where $|\cdot|$ denotes the Euclidean norm on the space $\mathbb{R}^d$. We use $\alpha := [\alpha_i \in \mathbb{N}_0 : i \in \mathbb{Z}_d]$ to denote a lattice point in $\mathbb{N}_0^d$. As is usually the case, we set $|\alpha| := \sum_{i \in \mathbb{Z}_d} \alpha_i$.

As usual, for a positive integer k, $W^{k,\infty}(\Omega)$ will denote the set of all functions v on Ω such that $D^\alpha v \in \mathbb{X}$ for $|\alpha| \leq k$, where we use the standard multi-index notation for derivatives

$$D^\alpha v(x) = \frac{\partial^{|\alpha|} v(x)}{\partial x_0^{\alpha_0} \cdots \partial x_{d-1}^{\alpha_{d-1}}}, \quad x \in \mathbb{R}^d$$

and the norm

$$\|v\|_{k,\infty} := \max\{\|D^\alpha v\|_\infty : |\alpha| \leq k\}$$

on $W^{k,\infty}(\Omega)$. For a star-shaped set Ω it is easy to estimate the distance of a function $v \in W^{k,\infty}(\Omega)$ from the space π_k. Specifically, there is a positive constant c such that

$$\operatorname{dist}(v, \pi_k) \leq c(d(\Omega))^k \|v\|_{k,\infty}. \tag{7.32}$$

Throughout the following sections, c will always stand for a generic constant whose value will change with the context. Its meaning will be clear from the order of the qualifiers used to describe its role in our estimates.

There are several ingredients required in the development of the fast collocation algorithms for solving the integral equation. First, we require a *multiscale* of finite-dimensional subspaces of $\mathbb{X}$ denoted by $\mathbb{X}_n$, $n \in \mathbb{N}_0$, in which we make our approximation. These spaces are required to have the property that

$$\mathbb{X}_n \subseteq \mathbb{X}_{n+1}, \quad n \in \mathbb{N}_0 \tag{7.33}$$

and

$$\mathbb{V} \subseteq \overline{\bigcup_{n \in \mathbb{N}_0} \mathbb{X}_n}. \tag{7.34}$$

For efficient computation relative to a scale of spaces, we express them as a direct sum of subspaces

$$\mathbb{X}_n = \mathbb{W}_0 \oplus \mathbb{W}_1 \oplus \cdots \oplus \mathbb{W}_n. \tag{7.35}$$

These spaces serve as *multiscale* subspaces of $\mathbb{X}$ and will be constructed as piecewise polynomial functions on Ω.

We need a multiscale partition of the set Ω. It consists of a family of partitions $\{\Omega_n : n \in \mathbb{N}_0\}$ of Ω such that for each scale $n \in \mathbb{N}_0$, the partition Ω_n consists of a family of subsets $\{\Omega_{ni} : i \in \mathbb{Z}_{e(n)}\}$ of Ω with the properties that

$$\operatorname{meas}(\Omega_{ni} \cap \Omega_{ni'}) = 0, \quad i, i' \in \mathbb{Z}_{e(n)}, i \neq i' \tag{7.36}$$

and

$$\bigcup_{i \in \mathbb{Z}_{e(n)}} \Omega_{ni} = \Omega. \tag{7.37}$$

At the appropriate time later, we shall adjust the number $e(n)$ of elements and the maximum diameter of the cells in the nth partition to be commensurate with $\dim \mathbb{W}_n$.

A construction of multiscale partitions for an invariant set has been described in Section 7.1.1. For unions Ω of invariant sets, their multiscale partitions can be constructed from multiscale partitions of the invariant sets which form the unions. For example, a polygonal domain in $\mathbb{R}^2$ is a union of triangles which are invariant sets. Hence, multiscale partitions of each of the triangles consisting of the polygon form a multiscale partition of the polygon. The partition $\{\Omega_n : n \in \mathbb{N}_0\}$ is used in two ways. First, we demand that there is a basis $W_n := \{w_{nm} : m \in \mathbb{Z}_{w(n)}\}$ for the spaces

$$\mathbb{W}_n := \operatorname{span} W_n, \quad n \in \mathbb{N}_0 \tag{7.38}$$

having the property (I) (stated in Proposition 7.4).

For $n > h$ we use the notation $S_{nm} := \Omega_{n-h,j}$, so that the support of the function w_{nm} is contained in the set S_{nm}. Note that the supports of the basis functions at the nth level are *not* disjoint. However, for every $n > h$ and every function w_{nm}, there are at most ρ other functions at level n whose support overlaps the support of w_{nm}.

To define the collocation method we need a set of linear functionals in $\mathbb{V}^*$ given by

$$L_n := \{\ell_{nm} : m \in \mathbb{Z}_{w(n)}\}, \quad n \in \mathbb{N}_0.$$

The multiscale partitions $\{\Omega_n : n \in \mathbb{N}_0\}$ are also used to specify the supports of the linear functionals by the requirement that the linear functional ℓ_{nm} is a finite sum of point evaluations

$$\ell_{nm} = \sum_{s \in \hat{\Omega}_{n-h,j}} c_s \delta_s, \tag{7.39}$$

where c_s are constants and $\hat{\Omega}_{ni}$ is a finite subset of distinct points in Ω_{ni} with the cardinality bounded *independent* of $n \in \mathbb{N}$ and $i \in \mathbb{Z}_{w(n)}$. We set $\hat{S}_{nm} := \hat{\Omega}_{n-h,j}$ and consider it as the "support" of the functionals ℓ_{nm}.

The linear functionals and multiscale basis are tied together by the requirement that property (II) holds. We do not require the linear functionals and the multiscale basis functions to be bi-orthogonal. Instead, we require them to have a "semi-bi-orthogonality" property imposed by the first equation of (II) with a controllable perturbation from the bi-orthogonality, which is ensured by the second equation of (II). Specifically, the first one means that the basis functions vanish when they are applied by collocation functionals of higher levels. We denote by $\mathbf{E}$ the semi-infinite matrix with entries

$$\mathbf{E}_{n'm',nm} := \langle \ell_{n'm'}, w_{nm} \rangle, \quad (n', m'), (n, m) \in \mathbb{U}.$$

We note by the first equation of property (II) that the matrix $\mathbf{E}$ can be viewed as a block upper triangular matrix with the diagonal blocks equal to identity matrices. Consequently, the infinite matrix $\mathbf{E}$ has an inverse $\mathbf{E}^{-1}$ of the same type, that is,

$$(\mathbf{E}^{-1})_{n'm',nm} = \delta_{nn'} \delta_{mm'}, \quad n \leq n', \quad m \in \mathbb{Z}_{w(n)}, \quad m' \in \mathbb{Z}_{w(n')}.$$

To introduce the collocation method for solving integral equations of the second kind, we suppose that K is a weakly singular kernel such that the operator $\mathcal{K} : \mathbb{X} \to \mathbb{V}$ defined by

$$(\mathcal{K}u)(s) := \int_{\Omega} K(s,t) u(t) \mathrm{d}t, \quad s \in \Omega$$

is compact in $\mathbb{X}$. We consider Fredholm integral equations of the second kind in the form

$$u - \mathcal{K}u = f, \tag{7.40}$$

where $f \in \mathbb{X}$ is a given function and $u \in \mathbb{X}$ is the unknown to be determined. When one is not an eigenvalue of $\mathcal{K}$, equation (7.40) has a unique solution in $\mathbb{X}$. The collocation scheme for solving equation (7.40) seeks a vector $\mathbf{u}_n := [u_{ij} : (i,j) \in \mathbb{U}_n]$, where $\mathbb{U}_n$ is the set of lattice points in $\mathbb{R}^2$ defined as $\{(i,j) : j \in \mathbb{Z}_{w(i)}, i \in \mathbb{Z}_{n+1}\}$, such that the function

$$u_n := \sum_{(i,j)\in\mathbb{U}_n} u_{ij} w_{ij}$$

in $\mathbb{X}_n$ has the property that

$$\langle \ell_{i'j'}, u_n - \mathcal{K}u_n \rangle = \langle \ell_{i'j'}, f \rangle, \quad (i',j') \in \mathbb{U}_n. \tag{7.41}$$

Equivalently, we obtain the linear system of equations

$$(\mathbf{E}_n - \mathbf{K}_n)\mathbf{u}_n = \mathbf{f}_n,$$

where

$$\mathbf{K}_n := [\langle \ell_{i'j'}, \mathcal{K}w_{ij} \rangle : (i',j'), (i,j) \in \mathbb{U}_n],$$
$$\mathbf{E}_n := [\langle \ell_{i'j'}, w_{ij} \rangle : (i',j'), (i,j) \in \mathbb{U}_n]$$

and

$$\mathbf{f}_n := [\langle \ell_{ij}, f \rangle : (i,j) \in \mathbb{U}_n].$$

By definition, we have that $(\mathbf{E}_n)_{i'j',ij} = \mathbf{E}_{i'j',ij}$ for $(i',j'), (i,j) \in \mathbb{U}_n$ and by (II) we see that

$$(\mathbf{E}_n^{-1})_{i'j',ij} = (\mathbf{E}^{-1})_{i'j',ij}, \quad (i',j'), (i,j) \in \mathbb{U}_n. \tag{7.42}$$

Let us use condition (II) to estimate the *inverse* of the matrix $\mathbf{E}_n$. To this end, we introduce a weighted norm on the vector $\mathbf{x} := [x_{ij} : (i,j) \in \mathbb{U}_n]$. For any $i \in \mathbb{Z}_{n+1}$ we set $\mathbf{x}_i := [x_{ij} : j \in \mathbb{Z}_{w(i)}]$,

$$\|\mathbf{x}_i\|_\infty := \max\{|x_{ij}| : j \in \mathbb{Z}_{w(i)}\}$$

and whenever $\nu \in (0,1)$ we define

$$\|\mathbf{x}\|_\nu := \max\{\|\mathbf{x}_i\|_\infty \nu^{-i} : i \in \mathbb{Z}_{n+1}\}.$$

We also use the notation $\|\mathbf{x}\|_\infty := \max\{\|\mathbf{x}_i\|_\infty : i \in \mathbb{Z}_{n+1}\}$ for the maximum norm of the vector $\mathbf{x}$.

Lemma 7.7 *If condition (II) holds,* $0 < \nu < 1$ *and* $(1+\gamma)\nu < 1$, *then for any integer* $n \in \mathbb{N}_0$ *and vector* $\boldsymbol{x} \in \mathbb{R}^{s(n)}$,

$$\|\boldsymbol{x}\|_\nu \le \frac{1-\nu}{1-(1+\gamma)\nu}\|\boldsymbol{E}_n\boldsymbol{x}\|_\nu.$$

Proof Let $\mathbf{y} := \mathbf{E}_n\mathbf{x}$ so that

$$y_{ij} = \sum_{(i',j')\in\mathbb{U}_n} \langle \ell_{ij}, w_{i'j'} \rangle x_{i'j'}.$$

In particular, for $i = n$, we have that $y_{ij} = x_{ij}$. For $0 \le l \le n-1$ we have, from the first equation of (II), that

$$x_{n-l-1,j} = y_{n-l-1,j} - \sum_{n-l\le i'\le n, j'\in\mathbb{Z}_{w(i')}} \langle \ell_{n-l-1,j}, w_{i'j'} \rangle x_{i'j'}, \quad j \in \mathbb{Z}_{w(n-l-1)}.$$

Using the second equation of (II), we conclude that

$$\|\mathbf{x}_{n-l-1}\|_\infty \le \|\mathbf{y}_{n-l-1}\|_\infty + \gamma \sum_{i=0}^{l} \|\mathbf{x}_{n-i}\|_\infty.$$

By induction on j, it readily follows that

$$\|\mathbf{x}_{n-j}\|_\infty \le \sum_{l=0}^{j-1} \gamma(1+\gamma)^l \|\mathbf{y}_{n-j+l+1}\|_\infty + \|\mathbf{y}_{n-j}\|_\infty.$$

Thus, we have that

$$\begin{aligned}
\|\mathbf{x}_{n-j}\|_\infty \nu^{-(n-j)} &\le \gamma\nu \sum_{l=0}^{j-1} [(1+\gamma)\nu]^l \|\mathbf{y}_{n-j+l+1}\|_\infty \nu^{-(n-j+l+1)} \\
&\quad + \|\mathbf{y}_{n-j}\|_\infty \nu^{-(n-j)} \\
&\le \left[1 + \gamma\nu \sum_{l=0}^{j-1} [(1+\gamma)\nu]^\ell \right] \|\mathbf{E}_n\mathbf{x}\|_\nu \\
&\le \frac{1-\nu}{1-(1+\gamma)\nu} \|\mathbf{E}_n\mathbf{x}\|_\nu,
\end{aligned}$$

from which the result is proved. □

7.2.2 Estimates for matrices and a truncated scheme

In this subsection, our goal is to obtain estimates for the entries of the matrix $\mathbf{K}_n$. This requires conditions on the support of the basis functions for $\mathbb{W}_n$ and vanishing moments for both the basis functions and the collocation functionals, which have been described in (III) and (IV). We also need the regularity of kernel K.

(X) For $s, t \in \Omega, s \neq t$, the kernel K has continuous partial derivatives $D_s^\alpha D_t^\beta K(s,t)$ for $|\alpha| \leq k$, $|\beta| \leq k$. Moreover, there exist positive constants σ and θ_3 with $\sigma < d$ such that for $|\alpha| = |\beta| = k$,

$$\left|D_s^\alpha D_t^\beta K(s,t)\right| \leq \frac{\theta_3}{|s-t|^{\sigma+|\alpha|+|\beta|}}. \tag{7.43}$$

In the next lemma, we present an estimate of the entries of the matrix $\mathbf{K}_n$. Such an estimate forms the basis for a truncation strategy. In the statement of the next lemma we use the quantities

$$d_i := \max\{d(S_{ij}) : j \in \mathbb{Z}_{w(i)}\}, \quad i \in \mathbb{N}_0.$$

Lemma 7.8 *If (I), (III), (IV) and (X) hold and there is a constant* $r > 1$ *such that*

$$\operatorname{dist}(S_{ij}, S_{i'j'}) \geq r(d_i + d_{i'}), \tag{7.44}$$

then there exists a positive constant c *such that*

$$|K_{i'j',ij}| \leq c(d_i d_{i'})^k \sum_{s \in \hat{S}_{i'j'}} \int_{S_{ij}} \frac{1}{|s-t|^{2k+\sigma}} \mathrm{d}t.$$

Proof Let s_0, t_0 be centers of the sets $S_{i'j'}$ and S_{ij}, respectively. Using the Taylor theorem with remainder we write $K = K_1 + K_2 + K_3$, where $K_1(s,\cdot)$ and $K_2(\cdot, t)$ are polynomials of total degree $\leq k-1$ in t and s, respectively,

$$|K_3(s,t)| \leq d_i^k d_{i'}^k v(s,t), \quad s \in S_{i'j'}, \ t \in S_{ij},$$

where

$$v(s,t) := \sum_{|\alpha|=k} \sum_{|\beta|=k} \frac{|r_{\alpha\beta}(s,t)|}{\alpha!\beta!} \tag{7.45}$$

and

$$\begin{aligned} r_{\alpha\beta}(s,t) = \int_0^1 \int_0^1 & D_s^\alpha D_t^\beta K(s_0 + t_1(s-s_0), \\ & t_0 + t_2(t-t_0))(1-t_1)^{k-1}(1-t_2)^{k-1} \mathrm{d}t_1 \mathrm{d}t_2. \end{aligned} \tag{7.46}$$

Applying the vanishing moment conditions yields the bound

$$|K_{i'j',ij}| \leq \|\ell_{i'j'}\| \|w_{ij}\| d_i^k d_{i'}^k \sum_{s \in S_{i'j'}} \int_{S_{ij}} |v(s,t)| \mathrm{d}t. \tag{7.47}$$

It follows from the mean-value theorem and condition (X) that

$$|r_{\alpha\beta}(s,t)| = k^{-2} |D_s^\alpha D_t^\beta K(s',t')| \leq \frac{\theta_3}{k^2 |s'-t'|^{2k+\sigma}}$$

holds for some $s' \in S_{i'j'}, t' \in S_{ij}$. For $s \in S_{i'j'}$, $t \in S_{ij}$ the assumption (7.44) yields

$$|s' - t'| \geq |s - t| - d_i - d_{i'} \geq (1 - r^{-1})|s - t|,$$

from which it follows that

$$|r_{\alpha\beta}(s,t)| \leq \frac{c_1}{|s - t|^{2k+\sigma}},$$

where

$$c_1 := \frac{\theta_3}{k^2(1 - r^{-1})^{2k+\sigma}}.$$

Substituting the above inequality into (7.47) completes the proof, with

$$c := \frac{\theta_3 \theta_0^2 e^{\frac{2d}{1-r^{-1}}}}{k^2(1 - r^{-1})^{\sigma}}.$$

□

To present the truncation strategy we partition matrix $\mathbf{K}_n$ into a block matrix

$$\mathbf{K}_n = [\mathbf{K}_{i'i} : i', i \in \mathbb{Z}_{n+1}],$$

with

$$\mathbf{K}_{i'i} := [K_{i'j',ij} : j' \in \mathbb{Z}_{w(i')}, j \in \mathbb{Z}_{w(i)}].$$

We truncate the block $\mathbf{K}_{i'i}$ by using a given positive number ϵ to form a matrix

$$\mathbf{K}(\epsilon)_{i'i} = [K(\epsilon)_{i'j',ij} : j' \in \mathbb{Z}_{w(i')}, j \in \mathbb{Z}_{w(i)}],$$

with

$$K(\epsilon)_{i'j',ij} := \begin{cases} K_{i'j',ij}, & \text{dist}(S_{i'j'}, S_{ij}) \leq \epsilon, \\ 0, & \text{otherwise}, \end{cases}$$

where ϵ may depend on i', i and n. In the next lemma, we use the estimate for the entries of $\mathbf{K}_n$ presented in Lemma 7.8 to obtain an estimate for the discrepancy between the blocks of $\mathbf{K}(\epsilon)$ and $\mathbf{K}_n$.

Lemma 7.9 *If (I), (III), (IV) and (X) hold then given any constant $r > 1$ and $0 \leq \sigma' < \max\{2k, d - \sigma\}$, there exists a positive constant c such that whenever $\epsilon \geq r(d_i + d_{i'})$,*

$$\|\mathbf{K}_{i'i} - \mathbf{K}(\epsilon)_{i'i}\|_\infty \leq c\epsilon^{-\eta}(d_i d_{i'})^k,$$

where $\eta := 2k - \sigma'$.

Proof We first note that

$$\|\mathbf{K}_{i'i}-\mathbf{K}(\epsilon)_{i'i}\|_\infty = \max_{j'\in\mathbb{Z}_{w(i')}} \sum_{j\in\mathbb{Z}_{i'j'}(\epsilon)} |K_{i'j',ij}|,$$

where

$$\mathbb{Z}_{i'j'}(\epsilon) := \{j : j\in\mathbb{Z}_{w(i)}, \operatorname{dist}(S_{ij}, S_{i'j'}) > \epsilon\}.$$

Therefore, by using Lemma 7.8, we have that

$$\|\mathbf{K}_{i'i}-\mathbf{K}(\epsilon)_{i'i}\|_\infty \le c(d_i d_{i'})^k \max_{j'\in\mathbb{Z}_{w(i')}} \sum_{s\in\hat{S}_{i'j'}} \sum_{j\in\mathbb{Z}_{i'j'}(\epsilon)} \int_{S_{ij}} \frac{1}{|s-t|^{2k+\sigma}}\mathrm{d}t.$$

Although the sets S_{ij} are not disjoint, we can use property (I) to conclude that

$$\sum_{j\in\mathbb{Z}_{i'j'}(\epsilon)} \int_{S_{ij}} \frac{1}{|s-t|^{2k+\sigma}}\mathrm{d}t \le \rho\epsilon^{-\eta} \int_{\Omega} \frac{1}{|s-t|^{\sigma+\sigma'}}\mathrm{d}t.$$

Since $\sigma+\sigma' < d$ and Ω is a compact set,

$$\max_{s\in\Omega} \int_{\Omega} \frac{1}{|s-t|^{\sigma+\sigma'}}\mathrm{d}t < \infty.$$

We employ the above inequalities to obtain the desired estimate. □

7.3 Analysis of the truncation scheme

In this section, we discuss the truncation strategy for the collocation method proposed in the previous section. We analyze the order of convergence, stability and computational complexity of the truncation algorithm.

7.3.1 Stability and convergence

We introduce the operator from $\mathbb{X}_n$ into itself defined by the equation $\mathcal{K}_n := \mathcal{P}_n\mathcal{K}|_{\mathbb{X}_n}$ and note that its matrix representation relative to the basis W_n is given by $\mathbf{E}_n^{-1}\mathbf{K}_n$. For each block $\mathbf{K}_{i'i}$, $i, i' \in \mathbb{Z}_{n+1}$ of $\mathbf{K}_n$ we shall specify later truncation parameters $\epsilon^n_{i'i}$ and reassemble the block to form a truncation matrix

$$\tilde{\mathbf{K}}_n = [\mathbf{K}(\epsilon^n_{i'i})_{i'i} : i', i \in \mathbb{Z}_{n+1}].$$

Using this truncation matrix, we let $\tilde{\mathcal{K}}_n : \mathbb{X}_n \to \mathbb{X}_n$ be the linear operator from $\mathbb{X}_n$ into itself relative to the basis W_n having the matrix representation $\mathbf{E}_n^{-1}\tilde{\mathbf{K}}_n$.

Our goal is to provide an essential estimate for the difference of two operators $\mathcal{K}_n$ and $\tilde{\mathcal{K}}_n$. To this end, for $v \in L^\infty(\Omega)$ we set

$$\mathcal{P}_n v = \sum_{(i,j)\in\mathbb{U}_n} v_{ij} w_{ij}.$$

The quantities v_{ij} are linear functionals of v. We estimate them in the next lemma.

Lemma 7.10 *Suppose that conditions (I)–(V) and (VIII) hold. If $v \in W^{k,\infty}(\Omega)$ then there exists a positive constant c such that*

$$|v_{ij}| \leq c\mu^{-ki/d}\|v\|_{k,\infty}, \quad (i,j) \in \mathbb{U}_n. \tag{7.48}$$

Proof For $v \in W^{k,\infty}(\Omega)$, we write

$$\mathcal{P}_n v = \sum_{(i,j)\in\mathbb{U}_n} v_{ij} w_{ij}$$

and let $\mathbf{v} := [v_{ij} : (i,j) \in \mathbb{U}_n]$. By the definition of the projection $\mathcal{P}_n$, we have that

$$\mathbf{E}_n\mathbf{v} = \left[\left\langle \ell_{ij}, \sum_{(i',j')\in\mathbb{U}_n} v_{i'j'} w_{i'j'})\right\rangle : (i,j) \in \mathbb{U}_n\right] = [\langle \ell_{ij}, v\rangle : (i,j) \in \mathbb{U}_n].$$

Meanwhile, using Lemma 7.7 with $\nu := \mu^{-k/d}$ and condition (VIII), we conclude that

$$\|\mathbf{v}\|_{\mu^{-k/d}} \leq c\|\mathbf{E}_n\mathbf{v}\|_{\mu^{-k/d}},$$

where

$$c := \frac{1-\mu^{-k/d}}{1-(1+\gamma)\mu^{-k/d}} > 0$$

is a constant. Hence,

$$\|\mathbf{v}\|_{\mu^{-k/d}} \leq c \max_{(i,j)\in\mathbb{U}_n} |\mu^{ik/d}\langle \ell_{ij}, v\rangle|. \tag{7.49}$$

Moreover, recalling that the "support" of the functional ℓ_{ij} is the set $\hat{S}_{ij} \subseteq S_{ij}$, we use the Taylor theorem with remainder on the set S_{ij} for $v \in W^{k,\infty}(\Omega)$ and conditions (III)–(V) to conclude that there exists a positive constant c such that

$$|\langle \ell_{ij}, v\rangle| \leq c d_i^k \|v\|_{k,\infty} \leq c\mu^{-ki/d}\|v\|_{k,\infty}.$$

Combining this inequality with (7.49), we obtain the estimate

$$\|\mathbf{v}\|_{\mu^{-k/d}} \leq c\|v\|_{k,\infty}.$$

Again, using the definition of the weighted norms, we have that

$$\|\mathbf{v}_i\|_\infty \le c\mu^{-ki/d}\|v\|_{k,\infty},$$

which proves the estimate of this lemma. □

Lemma 7.10 ensures that for a function $v \in W^{k,\infty}(\Omega)$, the coefficients of its expansion in basis W_n and functionals L_n decay in order $\mathcal{O}(\mu^{-ik/d})$. This is an extension of a well-known result for orthogonal multiscale bases to the multiscale interpolating piecewise polynomials constructed in this chapter.

For positive numbers α and β, we make use of the notation

$$\mu[\alpha, \beta; n] := \sum_{i\in\mathbb{Z}_{n+1}} \mu^{\alpha i/d} \sum_{i'\in\mathbb{Z}_{n+1}} \mu^{\beta i'/d}$$

to state the next lemma, which will play an important role in the analysis for the order of convergence and stability of the multiscale collocation method. To prove the next lemma, we need to estimate the L^∞-norm of a typical element in $\mathbb{X}_n$ given by

$$v := \sum_{(i,j)\in\mathbb{U}_n} v_{ij} w_{ij} \tag{7.50}$$

in terms of the norm $\|\mathbf{v}\|_\infty$ of its coefficients $\mathbf{v} := [v_{ij} : (i,j) \in \mathbb{U}_n]$. Specifically, we require the condition (IX). One way to satisfy the condition is to consider the sequence of functions $\{\zeta_{ij} : (i,j) \in \mathbb{U}\}$ defined by the equation

$$\zeta_{ij} := \sum_{(i',j')\in\mathbb{U}} (\mathbf{E}^{-1})_{i'j',ij} w_{i'j'}, \quad (i,j) \in \mathbb{U}. \tag{7.51}$$

These functions are bi-orthogonal relative to the set of linear functionals $\{\ell_{ij} : j \in \mathbb{Z}_{w(i)}, i \in \mathbb{N}_0\}$, that is,

$$\langle \ell_{i'j'}, \zeta_{ij} \rangle = \delta_{ii'}\delta_{jj'}, \quad (i,j), (i',j') \in \mathbb{U}.$$

If in addition, for all $i \in \mathbb{N}_0$,

$$\sup_{t\in\Omega} \sum_{j\in\mathbb{Z}_{w(i)}} |\zeta_{ij}(t)| \le \theta_2, \tag{7.52}$$

then the second inequality of (IX) follows.

In the next lemma, we estimate the difference of operators $\mathcal{K}_n$ and $\tilde{\mathcal{K}}_n$ applying to $\mathcal{P}_n v$. It is an important step for both stability analysis and the convergence estimate of the multiscale collocation method.

Lemma 7.11 *Suppose that conditions (I)–(V) and (VIII)–(X) hold,* $0 < \sigma' < \min\{2k, d - \sigma\}$ *and* $\eta := 2k - \sigma'$. *Let* b *and* b' *be real numbers, and let the* *truncation parameters* $\epsilon^n_{i'i}$, $i', i \in \mathbb{Z}_{n+1}$ *be chosen such that*

$$\epsilon^n_{i'i} \geq \max\left\{a\mu^{[-n+b(n-i)+b'(n-i')]/d}, r(d_i + d_{i'})\right\}, \quad i, i' \in \mathbb{Z}_{n+1},$$

for some constants $a > 0$ *and* $r > 1$. *Then there exists a positive constant* c *such that for all* $n \in \mathbb{N}$ *and* $v \in W^{k,\infty}(\Omega)$,

$$\|(\mathcal{K}_n - \tilde{\mathcal{K}}_n)\mathcal{P}_n v\|_\infty \leq c\mu[2k - b\eta, k - b'\eta; n](n+1)\mu^{-(k+\sigma')n/d}\|v\|_{k,\infty} \tag{7.53}$$

and for $v \in L^\infty(\Omega)$,

$$\|(\mathcal{K}_n - \tilde{\mathcal{K}}_n)\mathcal{P}_n v\|_\infty \leq c\mu[k - b\eta, k - b'\eta; n](n+1)\mu^{-\sigma' n/d}\|v\|_\infty. \tag{7.54}$$

Proof Since

$$\mathcal{P}_n v = \sum_{(i,j)\in\mathbb{U}_n} v_{ij} w_{ij},$$

we conclude that

$$(\mathcal{K}_n - \tilde{\mathcal{K}}_n)\mathcal{P}_n v = \sum_{(i,j)\in\mathbb{U}_n} h_{ij} w_{ij},$$

where

$$\mathbf{h} := \mathbf{E}_n^{-1}(\mathbf{K}_n - \tilde{\mathbf{K}}_n)\mathbf{v}.$$

Thus, by hypothesis (IX), we conclude that

$$\|(\mathcal{K}_n - \tilde{\mathcal{K}}_n)\mathcal{P}_n v\|_\infty \leq \theta_2(n+1)\|(\mathbf{K}_n - \tilde{\mathbf{K}}_n)\mathbf{v}\|_\infty. \tag{7.55}$$

We next estimate $\|(\mathbf{K}_n - \tilde{\mathbf{K}}_n)\mathbf{v}\|_\infty$. To this end, we introduce the matrix

$$\mathbf{\Delta}_n := [\Delta_{i'j',ij} : (i,j), (i',j') \in \mathbb{U}_n],$$

whose elements are given by

$$\Delta_{i'j',ij} := \nu\mu^{k(n-i)/d+\sigma' n/d}(K_{i'j',ij} - \tilde{K}_{i'j',ij}), \quad (i,j), (i',j') \in \mathbb{U}_n,$$

where $\nu = 1/\mu[2k - b\eta, k - b'\eta; n]$ and the vector

$$\mathbf{v}' := [v'_{ij} : (i,j) \in \mathbb{U}_n],$$

whose components are

$$v'_{ij} := \mu^{ki/d} v_{ij}, \quad (i,j) \in \mathbb{U}_n.$$

In this notation, we observe that

$$\|(\mathbf{K}_n - \tilde{\mathbf{K}}_n)\mathbf{v}\|_\infty \leq \nu^{-1}\mu^{-(k+\sigma')n/d}\|\mathbf{\Delta}_n\|_\infty\|\mathbf{v}'\|_\infty. \tag{7.56}$$

By Lemma 7.10 there exists a positive constant c such that for all $n \in \mathbb{N}$ and all $v \in W^{k,\infty}(\Omega)$,

$$\|\mathbf{v}'\|_\infty \leq c\|v\|_{k,\infty}. \tag{7.57}$$

Moreover, from Lemma 7.9, there exists a positive constant c such that

$$\begin{aligned}\sum_{(i,j)\in\mathbb{U}_n} \left|\Delta_{i'j',ij}\right| &\leq \nu \sum_{i\in\mathbb{Z}_{n+1}} \mu^{k(n-i)/d+\sigma' n/d}\|\mathbf{K}_{i'i} - \tilde{\mathbf{K}}_{i'i}\|_\infty \\ &\leq c\nu \sum_{i\in\mathbb{Z}_{n+1}} \mu^{k(n-i)/d+\sigma' n/d-k(i+i')/d}(\epsilon^n_{i'i})^{-\eta}.\end{aligned}$$

Consequently, by the choice of $\epsilon_{i'i}$, we conclude that

$$\|\boldsymbol{\Delta}_n\|_\infty := \max_{(i',j')\in\mathbb{U}_n} \sum_{(i,j)\in\mathbb{U}_n} \left|\Delta_{i'j',ij}\right| \leq c. \tag{7.58}$$

Combining this inequality with (7.56)–(7.58) yields the first estimate.

To prove the second estimate, we proceed similarly and introduce the matrix

$$\boldsymbol{\Delta}'_n := [\Delta'_{i'j',ij}]_{s(n)\times s(n)},$$

whose entries are given by

$$\Delta'_{i'j',ij} := \nu'\mu^{\sigma' n/d}(K_{i'j',ij} - \tilde{K}_{i'j',ij}), \quad (i,j),(i',j') \in \mathbb{U}_n,$$

where $\nu' = 1/\mu[k-b\eta, k-b'\eta; n]$. With these quantities, we have the estimate

$$\|(\mathbf{K}_n - \tilde{\mathbf{K}}_n)\mathbf{v}\|_\infty \leq (\nu')^{-1}\mu^{-\sigma' n/d}\|\boldsymbol{\Delta}'_n\|_\infty\|\mathbf{v}\|_\infty. \tag{7.59}$$

Condition (IX) provides a positive constant c such that for $v \in L^\infty(\Omega)$,

$$\|\mathbf{v}\|_\infty \leq c\|v\|_\infty. \tag{7.60}$$

As before, Lemma 7.9 and the choice of $\epsilon_{i'i}$, $i, i' \in \mathbb{Z}_{n+1}$ ensure that there exists a positive constant c such that

$$\|\boldsymbol{\Delta}'_n\|_\infty \leq c. \tag{7.61}$$

Combining this inequality with (7.59) and (7.60) yields the second estimate. □

We now turn our attention to the stability of the multiscale collocation method. For this purpose, we require property (VI). This property follows trivially if $\mathbb{X}_n$ is a space of piecewise polynomials. Because of this property and the fact that $\mathcal{K}$ is compact, we conclude for sufficiently large n that the operators $(\mathcal{I} - \mathcal{K}_n)^{-1}$ exist and are uniformly bounded in $L^\infty(\Omega)$ (see, for example, [6, 7]). From this fact follows the stability estimate, that is, there exists a positive constant ρ and a positive integer m such that for $n \geq m$ and $x \in \mathbb{X}_n$,

$$\|(\mathcal{I} - \mathcal{K}_n)x\|_\infty \geq \rho\|x\|_\infty.$$

We establish a similar estimate for $\mathcal{I} - \tilde{\mathcal{K}}_n$.

Theorem 7.12 *Suppose that* $0 < \sigma' < \min\{2k, d - \sigma\}$ *and* $\eta = 2k - \sigma'$. *If conditions (I)–(VI) and (VIII)–(X) hold and* $\epsilon^n_{i'i}$, $i, i' \in \mathbb{Z}_{n+1}$ *are chosen as in Lemma 7.11 with*

$$b > \frac{k - \sigma'}{\eta}, \quad b' > \frac{k - \sigma'}{\eta}, \quad b + b' > 1,$$

then there exist a positive constant c *and a positive integer* m *such that for all* $n \geq m$ *and* $x \in \mathbb{X}_n$,

$$\|(\mathcal{I} - \tilde{\mathcal{K}}_n)x\|_\infty \geq c\|x\|_\infty.$$

Proof Note that for any real numbers α, β and e,

$$\lim_{n\to\infty} \mu[\alpha, \beta; n](n + 1)\mu^{-en/d} = 0$$

when $e > \max\{0, \alpha, \beta, \alpha + \beta\}$. Thus, our hypothesis ensures that there exists a positive integer m such that when $n \geq m$,

$$c\mu[k - b\eta, k - b'\eta; n](n + 1)\mu^{-\sigma' n/d} < \rho/2, \tag{7.62}$$

where the constant c is that appearing in (7.54). The stability of the collocation scheme and the second estimate in Lemma 7.11, together with (7.62), yield for $x \in \mathbb{X}_n$ that

$$\|(\mathcal{I} - \tilde{\mathcal{K}}_n)x\|_\infty \geq \|(\mathcal{I} - \mathcal{K}_n)x\|_\infty - \|(\mathcal{K}_n - \tilde{\mathcal{K}}_n)\mathcal{P}_n x\|_\infty \geq \frac{\rho}{2}\|x\|_\infty.$$

This completes the proof. □

In particular, this theorem ensures for $n \geq m$ that the equation

$$(\mathcal{I} - \tilde{\mathcal{K}}_n)\tilde{u}_n = \mathcal{P}_n f \tag{7.63}$$

has a unique solution given by

$$\tilde{u}_n := \sum_{(i,j)\in\mathbb{U}_n} \tilde{u}_{ij} w_{ij}.$$

This equation is equivalent to the matrix equation

$$(\mathbf{E}_n - \tilde{\mathbf{K}}_n)\tilde{\mathbf{u}}_n = \mathbf{f}_n,$$

where $\tilde{\mathbf{u}}_n = [\tilde{u}_{ij} : (i,j) \in \mathbb{U}_n]$. The next theorem provides an error bound for $\|u - \tilde{u}_n\|_\infty$.

Theorem 7.13 *Suppose that conditions (I)–(X) hold and that* $0 < \sigma' < \min\{2k, d - \sigma\}$ *and* $\eta = 2k - \sigma'$. *Let* $\epsilon^n_{i'i}$, $i, i' \in \mathbb{Z}_{n+1}$ *be chosen as in Lemma 7.11 with* b *and* b' *satisfying one of the following three conditions:*

(i) $b > 1$, $b' > \frac{k-\sigma'}{\eta}$, $b + b' > 1 + \frac{k}{\eta}$.

(ii) $b = 1$, $b' > \frac{k-\sigma'}{\eta}$, $b + b' > 1 + \frac{k}{\eta}$; $b > 1, b' = \frac{k-\sigma'}{\eta}, b + b' > 1 + \frac{k}{\eta}$; *or* $b > 1, b' > \frac{k-\sigma'}{\eta}, b + b' = 1 + \frac{k}{\eta}$.

(iii) $b = 1, b' = \frac{k}{\eta}$ *or* $b = \frac{2k}{\eta}, b' = \frac{k-\sigma'}{\eta}$.

Then there exist a positive constant c *and positive integer* m *such that for all* $n \geq m$,

$$\|u - \tilde{u}_n\|_\infty \leq cs(n)^{-k/d}(\log s(n))^\tau \|u\|_{k,\infty},$$

where $\tau = 0$ *in case (i),* $\tau = 1$ *in case (ii) and* $\tau = 2$ *in case (iii).*

Proof It follows from Theorem 7.12 that there exists a positive constant c such that

$$\|u - \tilde{u}_n\|_\infty \leq \|u - \mathcal{P}_n u\|_\infty + c\|(\mathcal{I} - \tilde{\mathcal{K}}_n)(\mathcal{P}_n u - \tilde{u}_n)\|_\infty. \tag{7.64}$$

Using the equation

$$\mathcal{P}_n(\mathcal{I} - \mathcal{K})u = (\mathcal{I} - \tilde{\mathcal{K}}_n)\tilde{u}_n,$$

we find that

$$(\mathcal{I} - \tilde{\mathcal{K}}_n)(\mathcal{P}_n u - \tilde{u}_n) = \mathcal{P}_n(\mathcal{I} - \mathcal{K})(\mathcal{P}_n u - u) + (\mathcal{K}_n - \tilde{\mathcal{K}}_n)\mathcal{P}_n u. \tag{7.65}$$

From (7.64), (7.65), hypothesis (VI) and Lemma 7.11, there exist positive constants c, p such that

$$\|u - \tilde{u}_n\|_\infty \leq (1 + p\|\mathcal{I} - \mathcal{K}\|)\|\mathcal{P}_n u - u\|_\infty + c\mu'\mu^{-kn/d}\|u\|_{k,\infty},$$

where

$$\mu' := \mu[2k - b\eta, k - b'\eta; n](n + 1)\mu^{-\sigma' n/d}.$$

We estimate each term in the inequality above separately. For the first term, we note that conditions (VI) and (VII) provide a positive constant c such that

$$\|\mathcal{P}_n u - u\|_\infty \leq c\mu^{-kn/d}\|u\|_{k,\infty}.$$

Now we turn our attention to estimating the quantity μ'. To this end, we observe for any real numbers α, β and e, with $e > 0$, the asymptotic order

$$\mu[\alpha,\beta;n](n+1)\mu^{-en/d} = \begin{cases} o(1) & \text{if } e > \max\{\alpha,\beta,\alpha+\beta\}, \\ \mathcal{O}(n) & \text{if } \alpha = e, \beta < e, \alpha+\beta < e \\ & \text{or if } \alpha < e, \beta = e, \alpha+\beta < e \\ & \text{or if } \alpha < e, \beta < e, \alpha+\beta = e, \\ \mathcal{O}(n^2) & \text{if } \alpha = 0, \beta = e \text{ or } \alpha = e, \beta = 0, \end{cases}$$

as $n \to \infty$. Using this fact with $\alpha := 2k - b\eta, \beta := k - b'\eta$ and $e := \sigma'$, we conclude that

$$\mu' = \begin{cases} o(1) & \text{in case } (i), \\ \mathcal{O}(n) & \text{in case } (ii), \\ \mathcal{O}(n^2) & \text{in case } (iii), \end{cases}$$

which establishes the result of this theorem by noting that $n \sim \log s(n)$. □

We see from this theorem that the convergence order of the approximate solution $\tilde{u}_n$ obtained from the truncated collocation method is optimal up to a logarithmic factor.

7.3.2 The condition number of the truncated matrix and complexity

We next estimate the condition number of the matrix $\tilde{\mathbf{A}}_n := \mathbf{E}_n - \tilde{\mathbf{K}}_n$.

Theorem 7.14 *If the conditions of Theorem 7.12 hold, then there exists a positive constant c such that the condition number of the matrix $\tilde{\mathbf{A}}_n$ satisfies the estimate*

$$\operatorname{cond}_\infty(\tilde{\mathbf{A}}_n) \le c \log^2(s(n)),$$

where $\operatorname{cond}_\infty(\mathbf{A})$ *denotes the condition number of a matrix* $\mathbf{A}$ *in the* ℓ^∞ *matrix norm.*

Proof For any $\mathbf{v} := [v_{ij} : (i,j) \in \mathbb{U}_n] \in \mathbb{R}^{s(n)}$, we define the vector $\mathbf{g} := [g_{ij} : (i,j) \in \mathbb{U}_n] \in \mathbb{R}^{s(n)}$ by the equation

$$\tilde{\mathbf{A}}_n \mathbf{v} = \mathbf{g} \tag{7.66}$$

and the function

$$g := \sum_{(i,j)\in\mathbb{U}_n} g_{ij}\zeta_{ij}.$$

Therefore, we have that

$$g_{ij} = \langle \ell_{ij}, g \rangle = \langle \ell_{ij}, \mathcal{P}_n g \rangle, \quad (i,j) \in \mathbb{U}_n.$$

It follows from (IV) that

$$\|\tilde{\mathbf{A}}_n \mathbf{v}\|_\infty \leq \theta_0 \|\mathcal{P}_n g\|_\infty. \tag{7.67}$$

Let

$$v := \sum_{(i,j)\in\mathbb{U}_n} v_{ij} w_{ij},$$

and observe the equation

$$(\mathcal{I} - \tilde{\mathcal{K}}_n)v = \mathcal{P}_n g. \tag{7.68}$$

We conclude from (7.67) and (7.68) that there exists a positive constant c such that

$$\begin{aligned} \|\tilde{\mathbf{A}}_n \mathbf{v}\|_\infty &\leq \theta_0 \|(\mathcal{I} - \tilde{\mathcal{K}}_n)v\|_\infty \\ &\leq \theta_0 \left(\|(\mathcal{I} - \mathcal{K}_n)v\|_\infty + \|(\mathcal{K}_n - \tilde{\mathcal{K}}_n)v\|_\infty \right) \\ &\leq c\|v\|_\infty, \end{aligned}$$

where the last inequality holds because of (7.54) and (7.62). Next, appealing to hypotheses (I) and (IV), we observe for any $t \in \Omega$ and $i \in \mathbb{Z}_{n+1}$ that

$$\left| \sum_{j\in\mathbb{Z}_{w(i)}} v_{ij} w_{ij}(t) \right| \leq \rho\theta_0 \|v\|_\infty,$$

because there are at most ρ values of $j \in \mathbb{Z}_{w(i)}$ such that functions $w_{ij}(t) \neq 0$. Therefore, we conclude that

$$\|v\|_\infty \leq \rho\theta_0(n+1)\|\mathbf{v}\|_\infty. \tag{7.69}$$

Consequently, there exists a positive constant c such that

$$\|\tilde{\mathbf{A}}_n\|_\infty \leq c(n+1). \tag{7.70}$$

Conversely, for any $\mathbf{g} \in \mathbb{R}^{s(n)}$, there exists a vector $\mathbf{v} \in \mathbb{R}^{s(n)}$ such that equation (7.66) holds. We argue that there exists a positive constant c such that

$$\|g\|_\infty \leq c(n+1)\|\mathbf{g}\|_\infty.$$

Hence, we obtain from condition (IX) the inequality

$$\|\mathbf{v}\|_\infty \leq c\|v\|_\infty \leq c\|(\mathcal{I} - \tilde{\mathcal{K}}_n)v\|_\infty = c\|g\|_\infty \leq c(n+1)\|\mathbf{g}\|_\infty,$$

from which it follows that there exists a positive constant c such that for all $n \in \mathbb{N}$,

$$\|\tilde{\mathbf{A}}_n^{-1}\|_\infty \le c(n+1). \tag{7.71}$$

Recalling hypothesis (V) we combine the estimates (7.70) and (7.71) to obtain the desired result, namely

$$\operatorname{cond}_\infty(\tilde{\mathbf{A}}_n) = \mathcal{O}\left((n+1)^2\right) = \mathcal{O}\left(\log^2(s(n))\right), \quad n \to \infty.$$

□

In the remainder of this section, we estimate the number of nonzero entries of matrix $\tilde{\mathbf{A}}_n := \mathbf{E}_n - \tilde{\mathbf{K}}_n$, which shows that the truncation strategy embodied in Lemma 7.11 can lead to a fast numerical algorithm for solving equation (7.40) while preserving nearly optimal order of convergence. For any matrix $\mathbf{A}$, we denote by $\mathcal{N}(\mathbf{A})$ the number of nonzero entries in $\mathbf{A}$.

Theorem 7.15 *Suppose that hypotheses (I) and (V) hold. Let b and b' be real numbers not larger than one, and the truncation parameters $\epsilon^n_{i'i}, i', i \in \mathbb{Z}_{n+1}$ be chosen such that*

$$\epsilon^n_{i'i} \le \max\{a\mu^{[-n+b(n-i)+b'(n-i')]/d}, r(d_i + d_{i'})\}, \quad i, i' \in \mathbb{Z}_{n+1}$$

for some constants $a > 0$ and $r > 1$. Then

$$\mathcal{N}(\tilde{\mathbf{A}}_n) = \mathcal{O}(s(n)\log^\tau s(n)),$$

where $\tau = 1$ except for $b = b' = 1$, in which case $\tau = 2$.

Proof We first estimate the number $\mathcal{N}(\tilde{\mathbf{A}}_{i'i})$. For fixed i, i' and j', if $\tilde{A}_{i'j',ij} \ne 0$, then $\operatorname{dist}(S_{i'j'}, S_{ij}) \le \epsilon^n_{i'i}$, so that

$$S_{ij} \subseteq S(i,i') := \{v : v \in \mathbb{R}^d, |v - v_0| \le d_i + d_{i'} + \epsilon^n_{i'i}\},$$

where v_0 is an arbitrary point in the set $S_{i'j'}$. Let $\mathcal{N}_{i,i'j'}$ be the number of such sets which are contained in $S(i,i')$. Using condition (V), we conclude that there exists a positive constant c such that

$$\mathcal{N}_{i,i'j'} \le \frac{\operatorname{meas}(S(i,i'))}{\min\{\operatorname{meas}(S_{ij}) : S_{ij} \subseteq S(i,i')\}} \le c\mu^i(d_i + d_{i'} + \epsilon^n_{i'i})^d.$$

Next, we invoke condition (I) to conclude that the number of functions w_{ij} having support contained in S_{ij} is bounded by ρ, and appealing to condition (V) we have that $w(i) = \mathcal{O}(\mu^i)$, $i \to \infty$. Consequently, there exists a positive constant c such that

$$\mathcal{N}(\tilde{\mathbf{A}}_{i'i}) \le \rho \sum_{j' \in \mathbb{Z}_{w(i')}} \mathcal{N}_{i,i'j'} \le c\mu^{i+i'}(d_i + d_{i'} + \epsilon^n_{i'i})^d, \quad i, i' \in \mathbb{Z}_{n+1},$$

from which it follows that

$$\mathcal{N}(\tilde{\mathbf{A}}_n) \leq c \sum_{i,i' \in \mathbb{Z}_{n+1}} \mu^{i+i'} \left[(d_i)^d + (d_{i'})^d + (\epsilon^n_{i'i})^d \right].$$

This inequality and condition (I) imply that if the truncation parameters have the bound

$$\epsilon^n_{i'i} \leq a\mu^{[-n+b(n-i)+b'(n-i')]/d},$$

then

$$\begin{aligned}
\mathcal{N}(\tilde{\mathbf{A}}_n) &\leq c \sum_{i' \in \mathbb{Z}_{n+1}} \sum_{i \in \mathbb{Z}_{n+1}} \mu^{i+i'} \left(\mu^{-i} + \mu^{-i'} + a^d \mu^{-n+b(n-i)+b'(n-i')} \right) \\
&\leq c \left[2(n+1) \sum_{i \in \mathbb{Z}_{n+1}} \mu^i + a^d \mu^n \left(\sum_{i \in \mathbb{Z}_{n+1}} \mu^{(b-1)(n-i)} \right) \right. \\
&\qquad \left. \left(\sum_{i' \in \mathbb{Z}_{n+1}} \mu^{(b'-1)(n-i')} \right) \right] \\
&= \mathcal{O}\left(\mu^n (n+1)^\tau \right) = \mathcal{O}\left(s(n) \log^\tau s(n) \right),
\end{aligned}$$

as $n \to \infty$. If $\epsilon^n_{i'i} \leq r(d_i + d_{i'})$, a similar argument leads to

$$\mathcal{N}(\tilde{\mathbf{A}}_n) = \mathcal{O}(s(n) \log s(n)), \quad n \to \infty.$$

This completes the proof. □

It follows from Theorems 7.12–7.15 that for the truncation scheme to have *all* the desired properties of stability, convergence and complexity, we have to choose the truncation parameters to satisfy the equation

$$\epsilon^n_{i'i} = \max\{a\mu^{[-n+b(n-i)+b'(n-i')]/d}, r(d_i + d_{i'})\}, \quad i, i' \in \mathbb{Z}_{n+1},$$

with $b = 1, b' > \frac{k-\sigma}{\eta}, b + b' \geq 1 + \frac{k}{\eta}$ or with $b = 1, b' = \frac{k}{\eta}, \sigma' < k$.

7.4 Bibliographical remarks

This chapter lays out the foundation of the fast multiscale collocation method for solving the Fredholm integral equation of the second kind with a weakly singular kernel. Most of the material presented in this chapter is taken from the paper [69]. The construction of multiscale basis functions and the corresponding multiscale collocation functionals, both having vanishing moments, was described in [69], based on ideas developed in [65, 200, 201]. The analysis of the collocation method for solving the integral equation is based on the theory of collectively compact operators described in [6, 15]. For the definition of function values $f(t)$ at given points $t \in \Omega$ for an L^∞ function f, readers are referred to [21]. We remark that the fast multiscale collocation method was realized for integral equations of one dimension, two dimensions and higher dimensions, respectively, in [75], [264] and [74]. Another wavelet collocation method was presented in [105] for the solution of boundary integral equations of order $r = 0, 1$ over a closed and smooth boundary manifold, where the trial space is the space of all continuous and piecewise linear functions defined over a uniform triangular grid and the collocation points are the grid points. For more wavelet collocation methods for solving integral equations, readers are referred to [225, 226]. A quadrature algorithm for the piecewise linear wavelet collocation applied to boundary integral equations can be found in [227]. Wavelet collocation methods for a first-kind boundary integral equation in acoustic scattering were developed in [143].

Numerical integrations with error control strategies in fast collocation methods described in this chapter were originally presented in [72]. An iterated fast collocation method was developed for solving integral equations of the second kind in [62]. Multiscale collocation methods were applied in [40] to solve stochastic integral equations and in [52, 76, 158, 160] to solve Hammerstein equations and nonlinear boundary integral equations. Moreover, multiscale collocation methods were applied to solve ill-posed integral equations of the first kind and inverse boundary value problems in [56, 57, 79, 107], to identify Volterra kernels of high order in [33] and to solve eigen-problems of weakly singular integral operators in [70].

8
Numerical integrations and error control

In the last three chapters, multiscale Galerkin methods, multiscale Petrov–Galerkin methods and multiscale collocation methods were developed for solving the Fredholm integral equation of the second kind with a weakly singular kernel on a domain in $\mathbb{R}^d$. These methods, which use multiscale bases having vanishing moments, lead to compression strategies for the coefficient matrices of the resulting linear systems. They provide fast algorithms for solving the integral equations with an optimal order of convergence and quasi-linear (up to a logarithmic factor) order in computational complexity. However, it should be pointed out that there is still a challenging problem to solve: computation of the entries of the compressed coefficient matrix which are weakly singular integrals. The purpose of this chapter is to introduce error control strategies for numerical integrations in generating the coefficient matrix of these multiscale methods. The error control techniques are so designed that quadrature errors will not ruin the overall convergence order of the approximate solution of the integral equation and will not increase the overall computational complexity order of the original multiscale method. Specifically, we discuss the problems in the setting of multiscale collocation methods. Two types of quadrature rule are used, and the corresponding error control techniques are discussed in this chapter. The numerical integration issue of the other two types of multiscale method can be handled similarly and thus, we leave it to the interested reader.

8.1 Discrete systems of the multiscale collocation method

We begin this chapter with a brief review on discrete systems of linear equations resulting from the multiscale collocation method introduced in Chapter 7.

We consider solving the Fredholm integral equation of the second kind in the form

$$u - \mathcal{K}u = f, \tag{8.1}$$

where $f \in L^\infty(\Omega)$ is a given function, $u \in L^\infty(\Omega)$ is the unknown to be determined, $\Omega := [0,1]$ and the operator $\mathcal{K} : L^\infty(\Omega) \to L^\infty(\Omega)$ is defined by

$$(\mathcal{K}u)(s) := \int_\Omega K(s,t)u(t)\mathrm{d}t, \quad s \in \Omega. \tag{8.2}$$

The multiscale collocation scheme for solving (8.1) seeks a vector $\mathbf{u}_n := [u_{ij} : (i,j) \in \mathbb{U}_n]$ such that the function

$$u_n := \sum_{(i,j)\in\mathbb{U}_n} u_{ij} w_{ij}$$

satisfies the equation

$$\langle \ell_{i'j'}, u_n - \mathcal{K}u_n \rangle = \langle \ell_{i'j'}, f \rangle, \quad (i',j') \in \mathbb{U}_n,$$

or equivalently,

$$(\mathbf{E}_n - \mathbf{K}_n)\mathbf{u}_n = \mathbf{f}_n, \tag{8.3}$$

where

$$\mathbf{E}_n := \left[\langle \ell_{i'j'}, w_{ij} \rangle : (i',j'),(i,j) \in \mathbb{U}_n\right], \tag{8.4}$$

$$\mathbf{K}_n := [\langle \ell_{i'j'}, \mathcal{K}w_{ij} \rangle : (i',j'),(i,j) \in \mathbb{U}_n] \tag{8.5}$$

and

$$\mathbf{f}_n := \left[\langle \ell_{i'j'}, f \rangle : (i',j') \in \mathbb{U}_n\right].$$

The coefficient matrix $\mathbf{A}_n := \mathbf{E}_n - \mathbf{K}_n$ is a full matrix. However, due to the special properties of the multiscale bases and functionals, the matrix $\mathbf{E}_n$ is sparse and $\mathbf{K}_n$ is numerically sparse in the sense that most of its entries have small absolute values. To avoid computing all of these entries, a truncation strategy is proposed so that the numerical solution obtained from the truncated sparse matrix is as accurate as that from the full matrix (see Chapter 7). Let

$$\mathbf{K}_n := \left[\mathbf{K}_{i'i} : i', i \in \mathbb{Z}_{n+1}\right],$$

where

$$\mathbf{K}_{i'i} := \left[K_{i'j',ij} : j' \in \mathbb{Z}_{w(i')}, j \in \mathbb{Z}_{w(i)}\right], \quad \text{with } K_{i'j',ij} := \langle \ell_{i'j'}, \mathcal{K}w_{ij} \rangle.$$

The truncation parameters $\epsilon^n_{i'i}$ are chosen for a pair of level indices (i',i) by

$$\epsilon^n_{i'i} := \max\{\nu(d_i + d_{i'}), a\mu^{b'(n-i')-i}\}$$

for some constants $a > 0$ and $\nu > 1$. The truncated matrix $\tilde{\mathbf{K}}_n$ is then defined by

$$\tilde{\mathbf{K}}_n := \left[\tilde{\mathbf{K}}_{i'i} : i', i \in \mathbb{Z}_{n+1}\right],$$

where

$$\tilde{\mathbf{K}}_{i'i} := \left[\tilde{K}_{i'j',ij} : j' \in \mathbb{Z}_{w(i')}, j \in \mathbb{Z}_{w(i)}\right],$$

with

$$\tilde{K}_{i'j',ij} := \begin{cases} K_{i'j',ij}, & \text{dist}(S_{i'j'}, S_{ij}) \leq \epsilon^n_{i'i}, \\ 0, & \text{otherwise}. \end{cases}$$

We see from (8.5) that each entry of $\mathbf{K}_n$ is a weakly singular integral and has to be computed numerically. When an elementary quadrature rule is chosen for the evaluation of $K_{i'j',ij}$, the bound of the numerical error can be obtained. Then, we are able to utilize the bound to gauge how the accumulation of these errors influences the accuracy of the resulting numerical solutions. The problem is how to choose the quadrature rules with their parameter values so that the convergence order of the multiscale collocation scheme can be preserved with low computational complexity.

8.2 Quadrature rules with polynomial order of accuracy

We first present a class of quadrature rules with polynomial order of accuracy.

8.2.1 Quadrature rule I

The first quadrature rule that we present here was introduced in [164] for a class of weakly singular univariate functions. For a fixed positive integer k', let $h \in C^{2k'}(0, 1]$ satisfy the property that there exists a positive constant c such that

$$|h^{(2k')}(t)| \leq ct^{-\sigma-2k'}, \ t \in (0, 1],$$

where $\sigma \in [0, 1)$. Note that the function h is integrable on $[0, 1]$ but may have a singularity at $t = 0$. We wish to compute a numerical value of the integral

$$I(h) := \int_0^1 h(t)\mathrm{d}t$$

in accuracy $\mathcal{O}(m^{-2k'})$ by using $\mathcal{O}(m)$ number of functional evaluations. For this purpose, we assume that $g_{k'}$ is the Legendre polynomial of degree k' on

[0, 1], that is,

$$\int_0^1 g_{k'}(t)t^{\ell}\mathrm{d}t = 0, \quad \ell \in \mathbb{Z}_{k'}$$

and denote by τ_ℓ, $\ell \in \mathbb{Z}_{k'}$ the k' zeros of $g_{k'}$ in an order given by $0 < \tau_0 < \cdots < \tau_{k'-1} < 1$. To compute $I(h)$, we set $q := \frac{2k'+1}{1-\sigma}$ and according to this parameter q, we choose m points

$$t_j := \left(\frac{j}{m}\right)^q, \quad j \in \mathbb{Z}_{m+1} \tag{8.6}$$

so that the set of subintervals $\Omega_j := [t_j, t_{j+1}]$, $j \in \mathbb{Z}_m$ form a partition for interval [0, 1]. Let

$$\tau_\ell^j := t_j + (t_{j+1} - t_j)\tau_\ell, \quad \ell \in \mathbb{Z}_{k'}, j \in \mathbb{Z}_m \tag{8.7}$$

and note that τ_ℓ^j, $\ell \in \mathbb{Z}_{k'}$ are the k' zeros of the Legendre polynomial of degree k' on Ω_j. We now use these points to define a piecewise polynomial $S(h)$ over [0, 1] with knots t_j, $j = 1, 2, \ldots, m-1$. Set $S(h)(t) = 0, t \in [t_0, t_1)$ and let $S(h)$ be the Lagrange interpolation polynomial of degree $k'-1$ to h at nodes τ_ℓ^j, $\ell \in \mathbb{Z}_{k'}$, for $x \in [t_j, t_{j+1})$, $j = 1, 2, \ldots, m-2$ and $x \in [t_{m-1}, t_m]$. We use the value

$$I(S(h)) := \sum_{j-1 \in \mathbb{Z}_{m-1}} \sum_{\ell \in \mathbb{Z}_{k'}} \omega_\ell^j h(\tau_\ell^j),$$

where

$$\omega_\ell^j := \int_{t_j}^{t_{j+1}} \prod_{i \in \mathbb{Z}_{k'}, i \neq \ell} \frac{t - \tau_i^j}{\tau_\ell^j - \tau_i^j} \mathrm{d}t,$$

to approximate the integral $I(h)$. Let $E_{m,k'}(h) := I(h) - I(S(h))$ denote the error of the approximation and it is proved in [164] that there exists a positive constant c', which might depend on h, such that

$$|E_{m,k'}(h)| \leq c' m^{-2k'}, \quad \text{for all } m \in \mathbb{N}.$$

We next consider computing the entries of $\tilde{\mathbf{K}}_n$ using this integration method. The entries of matrix $\tilde{\mathbf{K}}_n$ are integrals of integrands in the form

$$h_{ij}(s, t) := K(s, t)w_{ij}(t), \tag{8.8}$$

for some $s \in \Omega$ and $(i, j) \in \mathbb{U}_n$. Note that these functions h_{ij} have a singularity at the point s, whose supports are given by the support of w_{ij} and they are piecewise smooth. To apply the integration method to these functions, we define a function class and extend the integration method to this class of functions h. A function h is said to be in class $\mathcal{A}$ if it has the properties that

(I) $\mathrm{supp}(h)$ is a subinterval of Ω;

(II) there exists a set of nodes $\pi(h) := \{s_j : j-1 \in \mathbb{Z}_{m'-1}\}$ such that $h \in C^{2k'}(\Omega\backslash(\{s\} \cup \pi(h)))$; and

(III) there exists a positive constant θ' such that

$$|h^{(2k')}(t)| \le \theta' |t-s|^{-(\sigma+2k')}, \quad t \in \Omega\backslash(\{s\} \cup \pi(h)).$$

For a function in class $\mathcal{A}$ we choose the canonical partition of Ω with respect to m, as described by (8.6), and associated with the singular point s we pick two collections of nodes

$$\pi_t^r := \{t_j^r := s + t_j : j \in \mathbb{Z}_{m+1}\} \text{ and } \pi_t^l := \{t_j^l := s - t_j : j \in \mathbb{Z}_{m+1}\}.$$

Let $[q', q''] := \mathrm{supp}(h)$. We rearrange the elements of

$$(\pi(h) \cup \pi_t^r \cup \pi_t^l \cup \{q', q''\}) \cap \mathrm{supp}(h)$$

in the increasing order and write them as a new sequence $q' = q_0 < q_1 < \cdots < q_{m''} = q''$ with an integer m'' that depends on m and has the bound $m'' \le 2m + m' + 1$. Define a partition $\Pi(h)$ of $\mathrm{supp}(h)$ by

$$\Pi(h) := \{Q_\alpha := [q_\alpha, q_{\alpha+1}) : \alpha \in \mathbb{Z}_{m''}\}$$

and define a piecewise polynomial $S(h)$ of order k' on $\mathrm{supp}(h)$ by the rule that $S(h) = 0$ on Q_α if $Q_\alpha \subset [t_1^l, t_1^r)$ and otherwise, on Q_α, $S(h)$ is the Lagrange interpolation polynomial of order k' which interpolates h at the k' zeros $\tau_\ell^\alpha := q_\alpha + (q_{\alpha+1} - q_\alpha)\tau_\ell$, $\ell \in \mathbb{Z}_{k'}$ of the Legendre polynomial of degree k' defined on Q_α. We compute the value $I(S(h))$ and use it to approximate $I(h)$. In the next lemma we analyze the order of convergence of this integration method. For this purpose, we set

$$\tilde{E}_{m,k'}(h) := \sum_{Q_\alpha \in \Pi(h)} \left| \int_{Q_\alpha} [h(t) - S(h)(t)]\mathrm{d}t \right|.$$

Lemma 8.1 *Let h be a function in class $\mathcal{A}$. Then, there exists a positive constant c_1 independent of h, s or $\Pi(h)$ such that*

$$\tilde{E}_{m,k'}(h) \le c_1 \theta' m^{-2k'}, \tag{8.9}$$

where θ' is the constant appearing in the definition of class $\mathcal{A}$.

Proof The proof is done by modifying the proof of Theorem 2.2 in [164]. For $j \in \mathbb{Z}_m$, we introduce two index sets

$$\Gamma_j^r := \{\alpha \in \mathbb{Z}_{m''} : Q_\alpha \in \Pi(h), Q_\alpha \subset [t_j^r, t_{j+1}^r]\}$$

and

$$\Gamma_j^l := \{\alpha \in \mathbb{Z}_{m''} : Q_\alpha \in \Pi(h), Q_\alpha \subset [t_{j+1}^l, t_j^l]\}.$$

Associated with these two index sets, we set

$$E_{k',j}^r(h) := \sum_{\alpha \in \Gamma_j^r} \left| \int_{Q_\alpha} [h(t) - S(h)(t)] \mathrm{d}t \right|$$

and

$$E_{k',j}^l(h) := \sum_{\alpha \in \Gamma_j^l} \left| \int_{Q_\alpha} [h(t) - S(h)(t)] \mathrm{d}t \right|,$$

and observe that

$$\tilde{E}_{m,k'}(h) = \sum_{\Gamma_j^l \neq \emptyset} E_{k',j}^l(h) + \sum_{\Gamma_j^r \neq \emptyset} E_{k',j}^r(h).$$

We first estimate $E_{k',j}^r$. By the definition of $S(h)$, we have that

$$E_{k',0}^r(h) \leq \int_{t_0^r}^{t_1^r} |h(t)| \mathrm{d}t \leq \theta' \int_{t_0^r}^{t_1^r} |s - t|^{-\sigma} \mathrm{d}t = \frac{\theta'}{1 - \sigma} m^{-(2k'+1)}. \tag{8.10}$$

For $j \geq 1$, it follows from the error estimate of the Gaussian quadrature that there exists $\eta_\alpha \in Q_\alpha$ such that

$$E_{k',j}^r(h) = \sum_{\alpha \in \Gamma_j^r} \frac{|h^{(2k')}(\eta_\alpha)|}{(2k')!} \left| \int_{Q_\alpha} (t - \tau_0^\alpha)^2 \cdots (t - \tau_{k'-1}^\alpha)^2 \mathrm{d}t \right|.$$

Using condition (III) and noting that

$$|t - \tau_\ell^\alpha| < t_{j+1}^r - t_j^r, \text{ for any } \alpha \in \Gamma_j^r, \ t \in Q_\alpha \text{ and } \sum_{\alpha \in \Gamma_j^r} \int_{Q_\alpha} \mathrm{d}t \leq t_{j+1}^r - t_j^r,$$

we conclude that

$$E_{k',j}^r(h) \leq \frac{\theta'}{(2k')!} |s - t_j^r|^{-(\sigma+2k')} (t_{j+1}^r - t_j^r)^{2k'+1}.$$

By the definition of t_j^r and (8.6), we observe that

$$E_{k',j}^r(h) \leq \frac{\theta' m^{-(2k'+1)}}{(2k')!} j^{-q(\sigma+2k')} q^{2k'+1} (j+1)^{(q-1)(2k'+1)}.$$

Since $q(\sigma + 2k') = (q-1)(2k'+1)$, it follows that

$$E^r_{k',j}(h) \leq \frac{\theta'}{(2k')!} q^{2k'+1} 2^{(q-1)(2k'+1)} m^{-(2k'+1)}. \tag{8.11}$$

Likewise, we obtain estimates similar to (8.10) and (8.11) for $E^l_{k',0}$ and $E^l_{k',j}$. Therefore, we conclude (8.9) with

$$c_1 := 2 \max \left\{ \frac{q^{2k'+1} 2^{(q-1)(2k'+1)}}{(2k')!}, \frac{1}{1-\sigma} \right\},$$

proving the lemma. □

We now apply the integration method described above for functions in class $\mathcal{A}$ to h_{ij} defined by (8.8), which appear in the compressed matrix $\tilde{\mathbf{K}}_n$. Recall that the compressed matrix is obtained by a truncation strategy defined with the truncation parameters $\epsilon^n_{i'i}$, $i', i \in \mathbb{Z}_{n+1}$ from the full matrix $\mathbf{K}_n$. For the given truncation parameters $\epsilon^n_{i'i}$, we introduce an index set

$$\mathbb{Z}_{i'j',i} := \{j \in \mathbb{Z}_{w(i)} : \operatorname{dist}(S_{i'j'}, S_{ij}) \leq \epsilon^n_{i'i}\},$$

for $(i', j') \in \mathbb{U}_n$ and define, for $\ell \in \mathbb{Z}_r$,

$$\mathbb{Z}^\ell_{i'j',i} := \{j \in \mathbb{Z}_{i'j',i} : j = \mu(\mathbf{e})r + \ell\}.$$

We observe that $\mathbb{Z}^\ell_{i'j',i} \subseteq \mathbb{Z}_{i'j',i}$ and for $j_1, j_2 \in \mathbb{Z}^\ell_{i'j',i}$ with $j_1 \neq j_2$,

$$\operatorname{meas}(\operatorname{supp}(w_{ij_1}) \cap \operatorname{supp}(w_{ij_2})) = 0$$

and for any $\ell \in \mathbb{Z}_r$,

$$\bigcup_{j \in \mathbb{Z}^\ell_{i'j',i}} \operatorname{supp}(w_{ij}) \subset \Omega.$$

Therefore, we define for $\ell \in \mathbb{Z}_r$ and $(i', j') \in \mathbb{U}_n$,

$$\bar{w}_{i'j',i\ell}(t) := \begin{cases} w_{ij}(t), & \text{if } t \in \operatorname{int}(\operatorname{supp}(w_{ij})) \text{ for some } j \in \mathbb{Z}^\ell_{i'j',i}, \\ 0, & \text{otherwise} \end{cases}$$

and set

$$\bar{h}_{i'j',i\ell}(s,t) := K(s,t)\bar{w}_{i'j',i\ell}(t).$$

The next lemma presents estimates for the error of the integration method applied to functions $\bar{h}_{i'j',i\ell}$.

Lemma 8.2 *Suppose that there exists a positive constant* θ *such that*

$$|D_t^{\beta} K(s,t)| \le \theta |s-t|^{-(\sigma+\beta)}$$

for any $0 \le \beta \le 2k'$ *and* $s, t \in \Omega$, $s \ne t$. *Then, there exists a positive constant* c_2 *such that for all* $i \in \mathbb{Z}_{n+1}$, $\ell \in \mathbb{Z}_r$, $(i', j') \in \mathbb{U}_n$ *and* $s \in \Omega$,

$$\tilde{E}_{m,k'}(\bar{h}_{i'j',i\ell}) \le c_2 m^{-2k'} \left[\left(\mu^{-i+1} + \mu^{-i'+1} + \epsilon_{i'i}^{n} \right) \mu^{i-1} \right]^{k_0}, \tag{8.12}$$

where $k_0 := \min\{k, 2k'+1\}$.

Proof It suffices to prove that $\bar{h}_{i'j',i\ell}$ is in class $\mathcal{A}$ and to compute the constant θ' for this function.

It is clear that condition (I) is satisfied with this function. According to the construction of w_{ij}, for any $(i, j) \in \mathbb{U}_n$, there exist $\mathbf{e} \in \mathbb{Z}_{\mu}^{i-1}$ and $l \in \mathbb{Z}_r$ such that $j = \mu(\mathbf{e})r + l$ and

$$w_{ij} = \mathcal{T}_{\mathbf{e}} w_{1l} := w_{1l} \circ \phi_{\mathbf{e}}^{-1} \chi_{\phi_{\mathbf{e}}(\Omega)}.$$

Note that w_{1l} is a piecewise polynomial with a finite set of knots. This set of knots forms a set $\pi(h_{ij})$ of knots for the function h_{ij} required by condition (II) in the definition of class $\mathcal{A}$. Observing that

$$\pi(\bar{h}_{i'j',i\ell}) = \bigcup_{j \in \mathbb{Z}_{i'j',i}^{\ell}} \pi(h_{ij}), \tag{8.13}$$

we confirm that $\bar{h}_{i'j',i\ell}$ satisfies condition (II) with the set $\pi(\bar{h}_{i'j',i\ell})$ of knots. It remains to show that it also satisfies condition (III). Again, by noting that each $\bar{w}_{i'j',i\ell}$ is a piecewise polynomial of order k with the knots $\pi(\bar{h}_{i'j',i\ell})$, it follows from the hypothesis on kernel K that for $t \in \operatorname{supp}(w_{ij}) \setminus (\{s\} \cup \pi(h_{ij}))$,

$$\begin{aligned} |D_t^{2k'} \bar{h}_{i'j',i\ell}(s,t)| &= \left| \sum_{\beta \in \mathbb{Z}_{k_0}} \binom{\beta}{2k'} D_t^{2k'-\beta} K(s,t) w_{ij}^{(\beta)}(t) \right| \\ &\le \theta \sum_{\beta \in \mathbb{Z}_{k_0}} \binom{\beta}{2k'} |s-t|^{-(\sigma+2k'-\beta)} \mu^{\beta(i-1)} |w_{1l}^{(\beta)}(\phi_{\mathbf{e}}^{-1}(t))|. \end{aligned}$$

Introducing a constant

$$\Delta := \max_{\beta \in \mathbb{Z}_{k_0}, l \in \mathbb{Z}_r} \sup\{|w_{1l}^{(\beta)}(t)| : t \in \Omega\} < \infty,$$

we obtain the estimate for $t \in \operatorname{supp}(w_{ij}) \setminus (\{s\} \cup \pi(h_{ij}))$:

$$|D_t^{2k'} \bar{h}_{i'j',i\ell}(t)| \le \theta \Delta (2k')! \sum_{\beta \in \mathbb{Z}_{k_0}} [|s-t| \mu^{i-1}]^{\beta} |s-t|^{-(\sigma+2k')}. \tag{8.14}$$

We now compute the constant θ' associated with function $\bar{h}_{i'j',i\ell}$. For any $j \in \mathbb{Z}_{i'j',i}$, we have that

$$|s-t| \leq d_i + d_{i'} + \epsilon^n_{i'i}, \quad \text{for any} \quad t \in \operatorname{supp}(w_{ij}).$$

This implies that for $t \in \operatorname{supp}(w_{ij})$,

$$\sum_{\beta \in \mathbb{Z}_{k_0}} [|s-t|\mu^{i-1}]^{\beta} \leq k_0 \left[\left(\mu^{-i+1} + \mu^{-i'+1} + \epsilon^n_{i'i}\right)\mu^{i-1}\right]^{k_0}.$$

Noticing that

$$\operatorname{supp}(\bar{w}_{i'j',i\ell}) \subset \bigcup_{j \in \mathbb{Z}_{i'j',i}} \operatorname{supp}(w_{ij}),$$

we observe that condition (III) holds for function $\bar{h}_{i'j',i\ell}$ with a constant

$$\theta' := \theta \Delta k_0 (2k')! \left[\left(\mu^{-i+1} + \mu^{-i'+1} + \epsilon^n_{i'i}\right)\mu^{i-1}\right]^{k_0}.$$

Finally, using Lemma 8.1 we conclude the estimate (8.12) with $c_2 := c_1 \theta \Delta k_0 (2k')!$. □

We now use the integration method described above to compute the integrals involved in the nonzero entries

$$K_{i'j',ij} := \sum_{s \in \hat{S}_{i'j'}} c_s \int_{S_{ij}} h_{ij}(s,t)\mathrm{d}t \tag{8.15}$$

of $\tilde{\mathbf{K}}_{i'i}$. In other words, we use

$$\tilde{K}_{i'j',ij} := \sum_{s \in \hat{S}_{i'j'}} c_s I(S(h_{ij}(s,\cdot)))$$

to approximate $K_{i'j',ij}$ given by (8.15). For a given set of truncation parameters we let

$$\tilde{\tilde{\mathbf{K}}}_{i'i} := \left[\tilde{\tilde{K}}(\epsilon)_{i'j',ij} : j' \in \mathbb{Z}_{w(i')}, j \in \mathbb{Z}_{w(i)}\right],$$

where

$$\tilde{\tilde{K}}(\epsilon)_{i'j',ij} := \begin{cases} \tilde{K}_{i'j',ij}, & \operatorname{dist}(S_{i'j'}, S_{ij}) \leq \epsilon^n_{i'i}, \\ 0, & \text{otherwise}. \end{cases} \tag{8.16}$$

In the next proposition, we estimate the L_∞-norm of error $\tilde{\mathbf{K}}_{i'i} - \tilde{\tilde{\mathbf{K}}}_{i'i}$.

Lemma 8.3 *Let m be a positive integer. Then, there exists a positive constant $c_3 > 0$ such that for all $i', i \in \mathbb{Z}_{n+1}$ and $n \in \mathbb{N}$,*

$$\|\tilde{\mathbf{K}}_{i'i} - \tilde{\tilde{\mathbf{K}}}_{i'i}\|_\infty \leq c_3 \left[\left(\mu^{-i+1} + \mu^{-i'+1} + \epsilon^n_{i'i}\right) \mu^{i-1}\right]^{k_0} m^{-2k'}. \tag{8.17}$$

Proof For $(i', j') \in \mathbb{U}_n$, we set $C_{i'j'} := \max\{|c_s| : s \in \hat{S}_{i'j'}\}$. By the construction of collocation functionals, there exist positive integers c_2', c_2'' such that for all $(i', j') \in \mathbb{U}_n$ and for all $n \in \mathbb{N}$,

$$\operatorname{card}(\hat{S}_{i'j'}) \leq c_2' \text{ and } \max\left\{C_{i'j'} : (i', j') \in \mathbb{U}_n\right\} \leq c_2''. \tag{8.18}$$

Using (8.15) and (8.18), we see that there exists a positive constant c such that for all $i', i \in \mathbb{Z}_{n+1}$,

$$\|\tilde{\mathbf{K}}_{i'i} - \tilde{\tilde{\mathbf{K}}}_{i'i}\|_\infty \leq c \max_{j' \in \mathbb{Z}_{w(i')}} \left\{ \sum_{s \in \hat{S}_{i'j'}} \sum_{j \in \mathbb{Z}_{i'j',i}} \tilde{E}_{m,k'}(h_{ij}) \right\}.$$

According to the definition of $\tilde{E}_{m,k'}(h)$, we conclude that

$$\sum_{j \in \mathbb{Z}_{i'j',i}} \tilde{E}_{m,k'}(h_{ij}) = \sum_{\ell \in \mathbb{Z}_r} \sum_{j \in \mathbb{Z}^\ell_{i'j',i}} \sum_{Q_\alpha \in \Pi(h_{ij})} \left| \int_{Q_\alpha} [h_{ij}(t) - (S(h_{ij}))(t)]\mathrm{d}t \right|.$$

Recalling (8.13), we obtain that the right-hand side of the equation above is equal to

$$\sum_{\ell \in \mathbb{Z}_r} \sum_{Q_\alpha \in \Pi(\bar{h}_{i'j',i\ell})} \left| \int_{Q_\alpha} [\bar{h}_{i'j',i\ell}(t) - (S(\bar{h}_{i'j',i\ell}))(t)]\mathrm{d}t \right| = \sum_{\ell \in \mathbb{Z}_r} \tilde{E}_{m,k'}(\bar{h}_{i'j',i\ell}).$$

It follows that

$$\|\tilde{\mathbf{K}}_{i'i} - \tilde{\tilde{\mathbf{K}}}_{i'i}\|_\infty \leq c \max \left\{ \sum_{\ell \in \mathbb{Z}_r} \tilde{E}_{m,k'}(\bar{h}_{i'j',i\ell}) : j' \in \mathbb{Z}_{w(i')} \right\}. \tag{8.19}$$

By (8.19) and Lemma 8.2, we obtain the desired estimate. □

8.2.2 Convergence order and computational complexity

To ensure that the numerical integration will not ruin the convergence order of the collocation method, we are required to choose different integers m, the number of functional evaluations used in numerical integration of the integrals involved in the entries of different blocks $\tilde{\mathbf{K}}_{i'i}$. We now denote these integers by $m_{i'i}$, $i', i \in \mathbb{Z}_{n+1}$ to indicate their dependence on different blocks. Specifically, we choose $m_{i'i}$ to satisfy the inequality

$$m_{i'i} \geq c_0 \left(\epsilon^n_{i'i}\right)^{\lambda} \mu(i', i), \quad i', i \in \mathbb{Z}_{n+1}, \tag{8.20}$$

for some positive constant c_0, where

$$\lambda := \frac{2k + k_0 - \sigma'}{2k'} \quad \text{and} \quad \mu(i', i) := \mu^{\frac{k(i'+i)+k_0(i-1)}{2k'}}.$$

Suppose that the numerical values of blocks $\tilde{\tilde{\mathbf{K}}}_{i'i}$ are computed accordingly. We solve the linear system

$$(\mathbf{E}_n - \tilde{\tilde{\mathbf{K}}}_n)\tilde{\tilde{\mathbf{u}}}_n = \mathbf{f}_n \tag{8.21}$$

for $\tilde{\tilde{\mathbf{u}}}_n := [\tilde{\tilde{u}}_{ij} : (i, j) \in \mathbb{U}_n]$ and denote

$$\tilde{\tilde{u}}_n := \sum_{(i,j)\in\mathbb{U}_n} \tilde{\tilde{u}}_{ij} w_{ij}.$$

Our next theorem shows that the integers $m_{i'i}$ so chosen allow the approximate solution $\tilde{\tilde{u}}_n$ to preserve the convergence order that $\tilde{u}_n$ has.

Theorem 8.4 *Suppose that the condition of Lemma 8.2 holds, $u \in W^{k,\infty}(\Omega)$ and that the integrals in $\tilde{\tilde{\mathbf{K}}}_n$ are computed by the integration formula described above using $m_{i'i}$ functional evaluations, where $m_{i'i}$ satisfy (8.20) and $\tilde{\tilde{u}}_n$ is solved accordingly. Then, there exist a positive constant c and a positive integer N such that for all $n > N$,*

$$\|u - \tilde{\tilde{u}}_n\|_\infty \leq c(s(n))^{-k}(\log s(n))^{\tau} \|u\|_{k,\infty}, \tag{8.22}$$

where $\tau = 1$ if $b' > \frac{k}{2k-\sigma'}$, $\tau = 2$ if $b' = \frac{k}{2k-\sigma'}$.

Proof By the proof of Theorem 7.13, the estimate (8.22) holds if there exists a positive constant c such that for all $i', i \in \mathbb{Z}_{n+1}$ and for all $n \in \mathbb{N}$,

$$\|\mathbf{K}_{i'i} - \tilde{\tilde{\mathbf{K}}}_{i'i}\|_\infty \leq c(\epsilon^n_{i'i})^{-(2k-\sigma')}\mu^{-k(i'+i)}. \tag{8.23}$$

By estimate (8.23) and the triangle inequality, it suffices to prove that there exists a positive constant c such that for all $i', i \in \mathbb{Z}_{n+1}$ and for all $n \in \mathbb{N}$, (8.23) holds with $\mathbf{K}_{i'i}$ replaced by $\tilde{\mathbf{K}}_{i'i}$. To this end, we recall that the definition of $\epsilon^n_{i'i}$ ensures that

$$\mu^{-i'+1} + \mu^{-i+1} \leq \epsilon^n_{i'i}.$$

It follows from Lemma 8.3 and the choice of $m_{i'i}$ (8.20) that

$$\|\tilde{\mathbf{K}}_{i'i} - \tilde{\tilde{\mathbf{K}}}_{i'i}\|_\infty \le 2^{k_0} c_0^{-2k'} c \left(\epsilon^n_{i'i}\right)^{-(2k-\sigma')} \mu^{-k(i'+i)},$$

proving the claim. □

Now we turn to analyzing computational complexity for the generating matrix $\tilde{\tilde{\mathbf{K}}}_n$. For $i', i \in \mathbb{Z}_{n+1}$, we denote by $\mathcal{M}_{i'i}$ the number of functional evaluations for computing the entries of $\tilde{\tilde{\mathbf{K}}}_{i'i}$. Thus,

$$\mathcal{M}^U_n := \sum_{i\in\mathbb{Z}_{n+1}} \sum_{i'\in\mathbb{Z}_{i+1}} \mathcal{M}_{i'i} \tag{8.24}$$

and

$$\mathcal{M}^L_n := \sum_{i'\in\mathbb{Z}_{n+1}} \sum_{i\in\mathbb{Z}_{i'}} \mathcal{M}_{i'i} \tag{8.25}$$

are the number of functional evaluations used for computing the upper and lower triangular entries of $\tilde{\tilde{\mathbf{K}}}_n$, respectively, and $\mathcal{M}_n := \mathcal{M}^U_n + \mathcal{M}^L_n$ represents the total number of functional evaluations used for computing all entries of $\tilde{\tilde{\mathbf{K}}}_n$. In the next theorem we estimate $\mathcal{M}^U_n$ and $\mathcal{M}^L_n$.

Theorem 8.5 *Suppose that the matrix $\tilde{\tilde{\mathbf{K}}}_n$ is generated by using the integration formula described above. Let $m_{i'i}, i', i \in \mathbb{Z}_{n+1}$ be the smallest integers satisfying (8.20). Choose*

$$k' \ge k \text{ and } \frac{2k'}{2k'+1}(1-\sigma) \le \sigma' < 1-\sigma. \tag{8.26}$$

Then, there exists a positive constant c such that for all $n \in \mathbb{N}$,

$$\mathcal{M}^U_n \le c(s(n))^{1+\lambda'} \tag{8.27}$$

and

$$\mathcal{M}^L_n \le c(s(n))^{1+\lambda''}, \tag{8.28}$$

where $\lambda' := \frac{\sigma'}{2k'} - \frac{1-\sigma}{2k'+1}$ and $\lambda'' := \frac{k}{2k'}$.

Proof For $i', i \in \mathbb{Z}_{n+1}$, we let $\mathcal{M}_{i'j',i}$ denote the number of functional evaluations used in computing the j'th row of the block $\tilde{\tilde{\mathbf{K}}}_{i'i}$. Recalling that the number of rows in this block is $w(i')$, we have that

$$\mathcal{M}_{i'i} = w(i')\mathcal{M}_{i'j',i}. \tag{8.29}$$

To estimate $\mathcal{M}_{i'j',i}$, we let $\mathcal{M}(h)$ be the number of functional evaluations used in computing $I(S(h))$. Recalling the definition of the function $\bar{h}_{i'j',i\ell}$, we have that

$$\mathcal{M}_{i'j',i} = \sum_{j\in\mathbb{Z}_{i'j',i}} \mathcal{M}(h_{ij}) = \sum_{\ell\in\mathbb{Z}_r} \sum_{j\in\mathbb{Z}^{\ell}_{i'j',i}} \mathcal{M}(h_{ij}) = \sum_{\ell\in\mathbb{Z}_r} \mathcal{M}(\bar{h}_{i'j',i\ell}). \tag{8.30}$$

Note that we actually integrate $S(\bar{h}_{i'j',i\ell})$ for an approximate value of the integral $I(\bar{h}_{i'j',i\ell})$. Since $S(\bar{h}_{i'j',i\ell})$ is a piecewise polynomial of order k', the number of functional evaluations used in integrating it between two consecutive knots is exactly k'. Setting

$$N_1 := \operatorname{card}(\pi(\bar{h}_{i'j',i\ell})) \text{ and } N_2 := \operatorname{card}(\pi^{r}_{s} \cup \pi^{l}_{s}),$$

we find that

$$\mathcal{M}(\bar{h}_{i'j',i\ell}) \le k'(N_1 + N_2). \tag{8.31}$$

For $j = 1, 2$, we let

$$\mathcal{M}^U_{n,j} := \sum_{i\in\mathbb{Z}_{n+1}} \sum_{i'\in\mathbb{Z}_{i+1}} w(i') \sum_{\ell\in\mathbb{Z}_r} k'N_j \tag{8.32}$$

and

$$\mathcal{M}^L_{n,j} := \sum_{i'\in\mathbb{Z}_{n+1}} \sum_{i\in\mathbb{Z}_{i'}} w(i') \sum_{\ell\in\mathbb{Z}_r} k'N_j. \tag{8.33}$$

From (8.24), (8.25) and (8.29)–(8.31), we conclude that $\mathcal{M}^U_n \le \mathcal{M}^U_{n,1} + \mathcal{M}^U_{n,2}$ and $\mathcal{M}^L_n \le \mathcal{M}^L_{n,1} + \mathcal{M}^L_{n,2}$.

We now estimate $\mathcal{M}^U_{n,1}$. From the construction of the basis functions w_{ij}, it is clear that the functions w_{ij} are piecewise polynomials with sets of knots whose cardinality is uniformly bounded. As a result, according to the definition of $\bar{w}_{i'j',i\ell}$, there exists a positive constant c' such that

$$N_1 \le c' \operatorname{card}(\mathbb{Z}^{\ell}_{i'j',i}). \tag{8.34}$$

By (8.34) and Theorem 7.15, we observe that there exists a positive constant c such that for all $n \in \mathbb{N}$,

$$\mathcal{M}^U_{n,1} \le c'k' \sum_{i\in\mathbb{Z}_{n+1}} \sum_{i'\in\mathbb{Z}_{i+1}} \mathcal{N}(\tilde{\mathbf{K}}_{i'i}) \le ck's(n)\log^{\tau} s(n), \tag{8.35}$$

where $\tau = 2$ if $b' = 1$ and $\tau = 1$ if $\frac{k-\sigma'}{2k-\sigma'} < b' < 1$. Likewise, we have the estimate for $\mathcal{M}^L_{n,1}$:

$$\mathcal{M}^L_{n,1} \le ck's(n)\log^{\tau} s(n).$$

Now we turn to estimating $\mathcal{M}^U_{n,2}$ and $\mathcal{M}^L_{n,2}$. By (8.6) and the truncation strategy, a sufficient condition for $t^r_\ell \notin \operatorname{supp}(\bar{w}_{i'j',i\ell})$ or $t^l_\ell \notin \operatorname{supp}(\bar{w}_{i'j',i\ell})$ is

$$\left(\frac{\ell}{m_{i'i}}\right)^{\frac{2k'+1}{1-\sigma}} \geq d_i + d_{i'} + \epsilon^n_{i'i}. \tag{8.36}$$

The smallest ℓ that satisfies (8.36) is an upper bound of the number of elements in π^r_s or π^l_s which are also located in $\operatorname{supp}(\bar{h}_{i'j',i\ell})$. Therefore, by the choice of $m_{i'i}$ there exists a positive constant c such that for all $i', i \in \mathbb{Z}_{n+1}$ and for all $n \in \mathbb{N}$,

$$N_2 \leq 2c_0 \left(d_i + d_{i'} + \epsilon^n_{i'i}\right)^{\frac{1-\sigma}{2k'+1}} \left(\epsilon^n_{i'i}\right)^{\lambda} \mu(i', i) \leq c \left(\epsilon^n_{i'i}\right)^{\lambda_0} \mu(i', i),$$

where $\lambda_0 := \frac{2k+k_0}{2k'} - \lambda'$. When $i' \leq i$, if $\epsilon^n_{i'i} = \nu(\mu^{-i'+1} + \mu^{-i+1})$, then

$$N_2 \leq c \left(\nu(\mu^{-i'+1} + \mu^{-i+1})\right)^{\frac{2k+k_0}{2k'} - \lambda'} \mu(i', i) \leq c\mu^{-\lambda_0 i' + \frac{k+k_0}{2k'} i}$$

and if $\epsilon^n_{i'i} = a\mu^{b'(n-i')-i}$, then

$$N_2 \leq c(a\mu^{b'(n-i')-i})^{\lambda_0} \mu(i', i) = c\mu^{b'\lambda_0 n} \mu^{(\frac{k}{2k'} - b'\lambda_0)i' - (\frac{k}{2k'} - \lambda')i}.$$

When $i' > i$, if $\epsilon^n_{i'i} = \nu(\mu^{-i'+1} + \mu^{-i+1})$, then

$$N_2 \leq c \left(\nu(\mu^{-i'+1} + \mu^{-i+1})\right)^{\lambda_0} \mu(i', i) \leq c\mu^{\frac{k}{2k'} i' - (\frac{k}{2k'} - \lambda')i}$$

and if $\epsilon^n_{i'i} = a\mu^{b'(n-i')-i}$, then

$$N_2 \leq c(a\mu^{b'(n-i')-i})^{\lambda_0} \mu(i', i) = c\mu^{b'\lambda_0 n} \mu^{(\frac{k}{2k'} - b'\lambda_0)i' - (\frac{k}{2k'} - \lambda')i}.$$

Note that for $i - 1 \in \mathbb{Z}_n$, $w(i) = r\mu^{i-1}$. By the definition of $\mathcal{M}^U_{n,2}$, we observe that if $\epsilon^n_{i'i} = \nu(\mu^{-i'+1} + \mu^{-i+1})$, then

$$\mathcal{M}^U_{n,2} \leq c \sum_{i \in \mathbb{Z}_{n+1}} \sum_{i' \in \mathbb{Z}_{i+1}} \mu^{(1 - \frac{k+k_0}{2k'} + \lambda')i' + \frac{k+k_0}{2k'} i}.$$

Hence,

$$\mathcal{M}^U_{n,2} \leq \begin{cases} c\mu^{\frac{k+k_0}{2k'} n}, & \lambda' < \frac{k+k_0}{2k'} - 1, \\ c\mu^{(1+\lambda')n}, & \lambda' \geq \frac{k+k_0}{2k'} - 1. \end{cases} \tag{8.37}$$

Similarly, if $\epsilon^n_{i'i} = a\mu^{b'(n-i')-i}$, we introduce a new parameter $\tilde{\lambda} := (1 + \frac{k}{2k'})/\lambda_0$ and observe that

$$\mathcal{M}^U_{n,2} \leq \begin{cases} c\mu^{b'\lambda_0 n}, & b' > \tilde{\lambda}, \\ c\mu^{(1+\lambda')n}, & b' \leq \tilde{\lambda}. \end{cases} \tag{8.38}$$

For an estimate of $\mathcal{M}_{n,2}^L$, if $\epsilon_{i'i}^n = \nu(\mu^{-i'+1} + \mu^{-i+1})$, since $\lambda' < \frac{k}{2k'}$ we have

$$\mathcal{M}_{n,2}^L \leq c\mu^{(1+\frac{k}{2k'})n} \tag{8.39}$$

and if $\epsilon_{i'i}^n = a\mu^{b'(n-i')-i}$, we see

$$\mathcal{M}_{n,2}^L \leq \begin{cases} c\mu^{b'\lambda_0 n}, & b' > \tilde{\lambda}, \\ c\mu^{(1+\frac{k}{2k'})n}, & b' \leq \tilde{\lambda}. \end{cases} \tag{8.40}$$

Now using the assumption (8.26), we have that $k_0 = k$ and $0 < \lambda' < \frac{\sigma}{2k'}$ and thus, $\tilde{\lambda} > 1$. Noting that $b' \leq 1$, we conclude the estimates (8.27) and (8.28) from (8.32), (8.33) and (8.37)–(8.40). The proof is complete. □

8.3 Quadrature rules with exponential order of accuracy

In this section we study a class of quadrature rules which have an exponential order of accuracy.

8.3.1 Quadrature rule II

In this section we present another integration method when the kernel is a C^∞ function off the diagonal. This stronger assumption on the kernel allows us to use different orders of polynomials in different subintervals to achieve an exponential order of convergence for the integration method. This idea was used in [242] in a different context. As a result, the computational complexity for numerical integration is improved considerably. Specifically, we assume that for any $s \in \Omega$, $K(s,\cdot) \in C^\infty(\Omega \setminus \{s\})$ and there exists a positive constant θ such that

$$|D_t^\beta K(s,t)| \leq \theta |s-t|^{-(\sigma+\beta)} \tag{8.41}$$

for any $\beta \in \mathbb{N}_0$ and $s, t \in \Omega, s \neq t$. Instead of using the knots described by (8.6), for any $\gamma \in (0,1)$ we set

$$t_0 := 0, \quad t_\iota := \gamma^{m-\iota}, \quad \iota = 1, 2, \ldots, m.$$

As before, we define the sets π_ι^r, π_ι^l of knots, subintervals Q_α and partition $\Pi(h)$ for function $h(s,\cdot) \in C^\infty(\Omega \setminus (\{s\} \cup \pi(h)))$. In the present case we define the piecewise polynomial $S(h)$ by the following rule. Note that $\tau_j^\alpha := q_\alpha + (q_{\alpha+1} - q_\alpha)\tau_j$, $j \in \mathbb{Z}_{k_\iota}$ are the k_ι zeros of the Legendre polynomial of degree k_ι defined on Q_α. If $Q_\alpha \subset [t_1^l, t_1^r)$, $S(h) = 0$ on Q_α and if $Q_\alpha \subset [t_\iota^r, t_{\iota+1}^r)$ or $[t_{\iota+1}^l, t_\iota^l)$, on Q_α, $S(h)$ is the Lagrange interpolating polynomial of order k_ι

to h at the k_ι zeros τ_j^α. Note that k_ι varies depending on ι and $S(h)$ depends on the vector $\mathbf{k} := [k_\iota : \iota = 1, 2, \ldots, m]$. We use $I(S(h))$ to approximate $I(h)$. For a constant $a \in \mathbb{R}$, $\lceil a \rceil$ denotes the smallest integer not less than a.

Lemma 8.6 *Let $\epsilon > 0$, $\gamma \in (0, 1)$ and choose*

$$k_\iota := \lceil \iota\epsilon \rceil, \quad \iota = 1, 2, \ldots, m. \tag{8.42}$$

Then, there exists a positive constant c such that for all integers m and for $\iota = 1, 2, \ldots, m$,

$$\left[(2k_\iota - k)!\gamma^{(1-\sigma)\iota+2k_\iota+1}\right]^{-1} \leq c.$$

Proof Note that $n! \sim (n/e)^{n+1/2}$ as $n \to +\infty$. It suffices to prove that there exists a positive integer n_0 such that when $\iota \geq n_0$,

$$\left[\left(\frac{2k_\iota - k}{e}\right)^{2k_\iota - k + \frac{1}{2}} \gamma^{(1-\sigma)\iota+2k_\iota+1}\right]^{-1} \leq 1.$$

Let $\zeta := \max\{\frac{1-\sigma}{\epsilon}, 4\}$. Since $\frac{2k_\iota - k}{e} \to +\infty$ as $\iota \to +\infty$, there exists a positive integer n_0 such that for all $\iota \geq n_0$,

$$\frac{2k_\iota - k}{e} \geq \gamma^{-\zeta} \quad \text{and} \quad \frac{1}{2}k_\iota > k.$$

Thus, for any $\gamma \in (0, 1)$,

$$\left(\frac{2k_\iota - k}{e}\right)^{2k_\iota - k + \frac{1}{2}} \geq \gamma^{-\zeta(2k_\iota - k + \frac{1}{2})} \geq \gamma^{-\frac{1-\sigma}{\epsilon}k_\iota - 4(k_\iota - k + \frac{1}{2})} \geq \gamma^{-(1-\sigma)\iota - 2k_\iota - 1}.$$

This completes the proof. □

To analyze the convergence of this integral method, we let

$$E_{m,\mathbf{k}}(h) := \sum_{Q_\alpha \in \Pi(h)} \left| \int_{Q_\alpha} [h(t) - S(h)(t)]\mathrm{d}t \right|.$$

Lemma 8.7 *Suppose that the kernel K satisfies (8.41). Then, there exists a positive constant c such that for any $i \in \mathbb{Z}_{n+1}, l \in \mathbb{Z}_r, (i', j') \in \mathbb{U}_n$ and $s \in E$,*

$$E_{m,\mathbf{k}}(\bar{h}_{i'j',il}) \leq c\left[\left(\mu^{-i+1} + \mu^{-i'+1} + \epsilon_{i'i}^{n}\right)\mu^{i-1}\right]^k \gamma^{(1-\sigma)m}.$$

Proof Define $\Gamma_\iota^r, \Gamma_\iota^l$ as in the proof of Lemma 8.1, with $E_{k',\iota}^r(h)$ and $E_{k',\iota}^l(h)$ replaced by $E_\iota^r(h)$ and $E_\iota^l(h)$, respectively. Now we have that

$$E_{m,\mathbf{k}}(h) = \sum_{\Gamma_\iota^l \neq \emptyset} E_\iota^l(h) + \sum_{\Gamma_\iota^r \neq \emptyset} E_\iota^r(h).$$

We first estimate $E_\iota^r(\bar{h}_{i'j',il})$. According to the definition of $S(h)$, we have that

$$E_0^r(\bar{h}_{i'j',il}) \le \int_{t_0^r}^{t_1^r} |\bar{h}_{i'j',il}(s,t)|\mathrm{d}t \le c\theta \int_{t_0^r}^{t_1^r} |s-t|^{-\sigma}\mathrm{d}t \le \frac{c\theta}{(1-\sigma)\gamma^{1-\sigma}}\gamma^{(1-\sigma)m}.$$

For $\iota \ge 1$, by the error estimate of the Gaussian quadrature, there exist $\xi_\alpha \in Q_\alpha$ such that

$$E_\iota^r(\bar{h}_{i'j',il}) = \sum_{\alpha\in\Gamma_\iota^r} \frac{|D_t^{2k_\iota}\bar{h}_{i'j',il}(s,\xi_\alpha)|}{(2k_\iota)!}\left|\int_{Q_\alpha}(t-\tau_0^\alpha)^2\cdots(t-\tau_{k_\iota-1}^\alpha)^2\mathrm{d}t\right|.$$

Note that w_{ij} is a piecewise polynomial of order k. By using assumption (8.41), we have for $t \in \operatorname{supp}(w_{ij}) \setminus (\{s\} \cup \pi(h_{ij}))$ that

$$\begin{aligned}
|D_t^{2k_\iota}\bar{h}_{i'j',il}(s,t)| &= \left|\sum_{\beta\in\mathbb{Z}_k}\binom{\beta}{2k_\iota} D_t^{2k_\iota-\beta}K(s,t)w_{ij}^{(\beta)}(t)\right| \\
&\le \theta\sum_{\beta\in\mathbb{Z}_k}\binom{\beta}{2k_\iota}|s-t|^{-(\sigma+2k_\iota-\beta)}\mu^{\beta(i-1)}\left|w_{1l}^{(\beta)}(\phi_{\mathbf{e}}^{-1}(t))\right|.
\end{aligned}$$

Moreover, we have that $t_\iota = \gamma^{m-\iota} \le |\xi_\alpha - s| \le d_i + d_{i'} + \epsilon_{i'i}^n$. Thus, we conclude that

$$\begin{aligned}
E_\iota^r(\bar{h}_{i'j',il}) &\le \frac{\theta\Delta(2k_\iota)(2k_\iota-1)\cdots(2k_\iota-k+1)}{(2k_\iota)!t_\iota^{(\sigma+2k_\iota)}}(t_{\iota+1}-t_\iota)^{2k_\iota+1} \\
&\quad\times\sum_{\beta\in\mathbb{Z}_k}\left[\left(d_i+d_{i'}+\epsilon_{i'i}^n\right)\mu^{i-1}\right]^\beta \\
&\le \frac{k\theta\Delta(1-\gamma)^{2k_\iota+1}}{(2k_\iota-k)!\gamma^{(1-\sigma)\iota+2k_\iota+1}}\gamma^{(1-\sigma)m}\left[\left(d_i+d_{i'}+\epsilon_{i'i}^n\right)\mu^{i-1}\right]^k.
\end{aligned}$$

It follows from Lemma 8.6 that there exists a positive constant c such that

$$E_\iota^r(\bar{h}_{i'j',il}) \le c(1-\gamma)^{2\epsilon\iota}\gamma^{(1-\sigma)m}\left[\left(d_i+d_{i'}+\epsilon_{i'i}^n\right)\mu^{i-1}\right]^k.$$

Therefore,

$$\sum_{\Gamma_\iota^r\ne\emptyset} E_\iota^r(\bar{h}_{i'j',il}) \le c\gamma^{(1-\sigma)m}\left[\left(d_i+d_{i'}+\epsilon_{i'i}^n\right)\mu^{i-1}\right]^k.$$

Likewise, we obtain the same estimate for $\sum_{\Gamma_\iota^l\ne\emptyset} E_\iota^l(\bar{h}_{i'j',il})$ and complete the proof of this lemma. □

8.3.2 Convergence order and computational complexity

Using Lemma 8.7 and similar arguments used in the proofs for Lemma 8.3 and Theorem 8.4, we have the following results.

Lemma 8.8 *Let m be a positive integer. Then, there exists a positive constant $c > 0$ such that for all $i', i \in \mathbb{Z}_{n+1}$ and $n \in \mathbb{N}$,*

$$\|\tilde{\mathbf{K}}_{i'i} - \tilde{\tilde{\mathbf{K}}}_{i'i}\|_\infty \le c\left[\left(\mu^{-i+1} + \mu^{-i'+1} + \epsilon^n_{i'i}\right)\mu^{i-1}\right]^k \gamma^{(1-\sigma)m}.$$

Theorem 8.9 *Let $u \in W^{k,\infty}(\Omega)$. For $i', i \in \mathbb{Z}_{n+1}$ choose*

$$m_{i'i} \ge \frac{-k\log\mu}{(1-\sigma)\log\gamma}(2i+i'). \tag{8.43}$$

Suppose that the kernel K satisfies (8.41), the integrals $K_{i'j',ij}$ in matrix $\tilde{\tilde{\mathbf{K}}}_n$ are computed by the integration methods described earlier with $m := m_{i'i}$ and k_ι determined by (8.42) and $\tilde{\tilde{u}}_n$ is solved accordingly. Then, there exist a positive constant c and a positive integer N such that for all $n > N$,

$$\|u - \tilde{\tilde{u}}_n\|_\infty \le c(s(n))^{-k}(\log s(n))^\tau \|u\|_{k,\infty}, \tag{8.44}$$

where $\tau = 1$ if $b' > \frac{k}{2k-\sigma'}$ and $\tau = 2$ if $b' = \frac{k}{2k-\sigma'}$.

Proof As in the proof of Theorem 8.4, it suffices to prove that there holds

$$\left[\left(\mu^{-i+1} + \mu^{-i'+1} + \epsilon^n_{i'i}\right)\mu^{i-1}\right]^k \gamma^{(1-\sigma)m} \le c\left(\epsilon^n_{i'i}\right)^{-(2k-\sigma')}\mu^{-k(i'+i)}.$$

Since $\mu^{-i'+1} + \mu^{-i+1} \le \epsilon^n_{i'i}$, we only need to show

$$\gamma^{(1-\sigma)m_{i'i}} \le c\left(\epsilon^n_{i'i}\right)^{-(3k-\sigma')}\mu^{-(2ki+ki')}.$$

This holds with the choice of $m_{i'i}$ and thus, the conclusion of this theorem follows. □

In the next theorem we present an estimate of the number of functional evaluations used for computing the entries of $\tilde{\tilde{\mathbf{K}}}_n$.

Theorem 8.10 *Suppose that $m_{i'i}, i', i \in \mathbb{Z}_{n+1}$ are chosen to be the smallest integer satisfying condition (8.43). Then, there exists a positive constant c such that for all $n \in \mathbb{N}$,*

$$\mathcal{M}_n \le cs(n)(\log s(n))^3.$$

Proof As in the proof of Theorem 8.5, we have that

$$\mathcal{M}_n = \sum_{i\in\mathbb{Z}_{n+1}} \sum_{i'\in\mathbb{Z}_{n+1}} \mathcal{M}_{i'i} = \sum_{i\in\mathbb{Z}_{n+1}} \sum_{i'\in\mathbb{Z}_{n+1}} w(i')\mathcal{M}_{i'j',i},$$

where

$$\mathcal{M}_{i'j',i} := \sum_{\ell\in\mathbb{Z}_r} \mathcal{M}(\bar{h}_{i'j',i\ell}).$$

In the present case, we obtain that

$$\operatorname{card}\{Q_\alpha : Q_\alpha \in [\gamma^{m_{i'i}-\iota}, \gamma^{m_{i'i}-\iota-1})\} \le \frac{\gamma^{m_{i'i}-\iota-1} - \gamma^{m_{i'i}-\iota}}{\mu^{-i+1}} + 2.$$

Thus, there is a positive constant c such that

$$\mathcal{M}_{i'j',i} \le c\sum_{\iota=1}^{l} k_\iota \left(\frac{\gamma^{m_{i'i}-\iota-1} - \gamma^{m_{i'i}-\iota}}{\mu^{-i+1}} + 2\right).$$

Since $k_\iota < \epsilon\iota + 1$,

$$\mathcal{M}_{i'j',i} \le c\left(\frac{1}{\gamma} - 1\right)\mu^{i-1}\left(\epsilon\sum_{\iota=1}^{l} \iota\gamma^{m_{i'i}-\iota} + \sum_{\iota=1}^{l} \gamma^{m_{i'i}-\iota}\right) + 2c\sum_{\iota=1}^{l}(\epsilon\iota + 1).$$

According to the truncation strategy we have that $t_\iota = \gamma^{m_{i'i}-\iota} \le \epsilon_{i'i}^n + d_{i'} + d_i$, where $i_0 := \min\{i', i\}$. By the choice of $\epsilon_{i'i}^n$ and $m_{i'i}$, we conclude that

$$\mathcal{M}_{i'j',i} \le c\left[n\mu^i\left(\mu^{b'(n-i')-i} + \mu^{\iota+1} + \mu^{-i'+1}\right) + n^2\right].$$

It follows from this inequality and $w(i') = k(\mu - 1)\mu^{i'-1}$ that

$$\mathcal{M}_n \le c\sum_{i\in\mathbb{Z}_{n+1}} \sum_{i'\in\mathbb{Z}_{n+1}} \left\{n\mu^{b'n}\mu^{(1-b')i'} + \mu^i + \mu^{i'} + n^2\mu^{i'}\right\}.$$

A simple computation yields the estimate of this theorem. □

8.4 Numerical experiments

In this section we use numerical experiments to verify our theoretical estimates. We consider equation (8.1) with

$$K(s,t) := \log|\cos(\pi s) - \cos(\pi t)|, \quad t, s \in \Omega,$$

and choose

$$f(s) := \sin(\pi s) + \frac{1}{\pi}\big[2 - (1 - \cos(\pi s)) \log(1 - \cos(\pi s)) \\ - (1 + \cos(\pi s)) \log(1 + \cos(\pi s))\big]$$

so that the exact solution is $u(s) = \sin(\pi s)$, for comparison purposes. In the experiment, we apply linear bases to discretize the equation.

To verify the computational complexity of the quadrature rules, we report the time for establishing the matrix $\tilde{\mathbf{K}}_n$. In the first experiment we let $k' = 2$, that is, we use piecewise linear polynomials to approximate the integrands. The values of other related parameters are identified as follows. Let $k' = 2$, $\sigma' = 0.8$, $a = 0.25$, $b' = 0.8$, $\nu = 1.01$ and

$$m_{i'i} \geq 1.4\left(\epsilon_{i'i}^{n}\right)^{1.3} 2^{\frac{i'+2i-1}{2}}.$$

We use the notations $\mathbf{T}_n^U$ and $\mathbf{T}_n^L$ for the time to evaluate the upper and lower triangles of $\tilde{\mathbf{K}}_n$. According to our theoretical estimates, there should hold

$$r_n^U := \log_2\left(\mathbf{T}_{n+1}^U / \mathbf{T}_n^U\right) = 1 + \frac{\sigma}{5} \quad \text{and} \quad r_n^L := \log_2\left(\mathbf{T}_{n+1}^L / \mathbf{T}_n^L\right) = 1.5.$$

The computed results are listed in Table 8.1.

We see that most of the values of r_n^L are around 1.5, and those of r_n^U tend to be lower than 1.2 when n increases. We also include the errors of the numerical solutions to the true solution as well as the convergence order, which is shown to maintain the optimal convergence rate.

In order to observe the influence of the values of k and k' on the order of time complexity, we now choose $k = 2$, $k' = 4$, that is, we continue to use the linear basis to discretize the integral equation, while piecewise cubic polynomials are

Table 8.1 *Numerical results for quadrature rule I: Linear quadrature case*

n	$s(n)$	$\mathbf{T}_n^U$	r_n^U	$\mathbf{T}_n^L$	r_n^L	$\|u - \tilde{u}_n\|_\infty$	Conv. rate
4	32	0.02		0.03		4.162345e-3	
5	64	0.05	1.32	0.07	1.22	1.302214e-3	1.6764
6	128	0.14	1.48	0.23	1.72	3.324927e-4	1.9696
7	256	0.33	1.24	0.69	1.59	6.300048e-5	2.3999
8	512	0.82	1.31	1.99	1.53	1.481033e-5	2.0888
9	1024	1.99	1.28	5.75	1.53	3.409346e-6	2.1190
10	2048	4.65	1.22	16.33	1.51	8.696977e-7	1.9709
11	4096	10.56	1.18	47.01	1.52	2.161024e-7	2.0088
12	8192	23.69	1.16	132.11	1.49	5.410341e-8	1.9979

Table 8.2 *Numerical results for quadrature rule I: Cubic quadrature case*

n	$s(n)$	$\mathbf{T}_n^U$	r_n^U	$\mathbf{T}_n^L$	r_n^L	$\|u - \tilde{u}_n\|_\infty$	Conv. rate
4	32	0.02		0.02		3.911817e-3	
5	64	0.05	1.32	0.06	1.58	1.112090e-3	1.8146
6	128	0.11	1.14	0.17	1.50	2.887823e-4	1.9452
7	256	0.28	1.35	0.46	1.44	6.471414e-5	2.1578
8	512	0.63	1.17	1.21	1.39	1.454796e-5	2.1533
9	1024	1.44	1.19	3.08	1.35	3.578654e-6	2.0233
10	2048	3.29	1.19	7.65	1.31	8.812394e-7	2.0218
11	4096	7.39	1.17	19.01	1.31	2.195702e-7	2.0049
12	8192	16.62	1.17	46.79	1.30	5.795965e-8	1.9216

Table 8.3 *Numerical results for quadrature rule II*

n	$s(n)$	CT	r_n	$2\left(\frac{n}{n-1}\right)^3$	$\|u - \tilde{u}_n\|_\infty$	Conv. rate
3	16	0.07			1.520796e-2	
4	32	0.19	2.71	4.74	3.816222e-3	1.994610
5	64	0.55	2.89	3.91	9.462863e-4	2.011796
6	128	1.81	3.29	3.46	2.475222e-4	1.934719
7	256	4.76	2.63	3.18	5.903598e-5	2.067892
8	512	12.31	2.59	2.99	1.479396e-5	1.996586
9	1024	31.57	2.56	2.85	3.686005e-6	2.004878
10	2048	84.45	2.68	2.74	8.944827e-7	2.042933

applied in the quadrature rule. The values of other parameters remain the same, except that

$$m_{i'i} \geq 2.4 \left(\epsilon_{i'i}^n\right)^{0.64} 2^{\frac{i'+2i-1}{4}}.$$

The corresponding numerical results are listed in Table 8.2.

In the last experiment we implement the quadrature rule described in Section 8.3.1, using the same parameters for truncation. The time for evaluation is listed in Table 8.3, in which "CT" stands for the time for the computation of matrix $\tilde{\mathbf{K}}_n$, r_n is defined as the ratio of the two successive times, that is,

$$r_n := \frac{\mathrm{CT}_n}{\mathrm{CT}_{n-1}}.$$

We also list in this table the theoretical value $2(\frac{n+1}{n})^3$ for comparison, and the L^∞-norms of the numerical errors $\|u - \tilde{u}_n\|_\infty$ as well as the convergence order.

8.5 Bibliographical remarks

The main developments along the direction of multiscale methods for solving Fredholm integral equations can be found in [28, 64, 68, 69, 71, 88–90, 95, 202, 226, 241, 243, 260, 261]. Appropriate bases for multiscale Galerkin, Petrov–Galerkin and collocation methods were constructed in [200, 201] and [65, 69]. To develop error control techniques for numerical integrations in generating the coefficient matrix for the one-dimensional case, we propose using the graded quadrature methods. When the integrand has only a polynomial order of smoothness except at the singular points, we use a quadrature method, suggested in [164], having a polynomial order of accuracy. When the integrand has an infinite order of smoothness except at the singular points, we use the idea of [242] to develop a quadrature having an exponential order of accuracy. Readers are referred to [72] for more information on error control strategies for numerical integrations in one dimension and [75] for those in higher dimensions.

9

Fast solvers for discrete systems

The goal of this chapter is to develop efficient solvers for the discrete linear systems resulting from discretization of the Fredholm integral equation of the second kind by using the multiscale methods discussed in previous chapters. We introduce the multilevel augmentation method (MAM) and the multilevel iteration method (MIM) for solving operator equations based on multilevel decompositions of the approximate subspaces. Reflecting the direct sum decompositions of the subspaces, the coefficient matrix of the linear system has a special structure. Specifically, the matrix corresponding to a *finer* level of approximate spaces is obtained by augmenting the matrix corresponding to a *coarser* level with submatrices that correspond to the *difference* spaces between the spaces of the finer level and the coarser level. The main idea is to split the matrix into a sum of two matrices, with one reflecting its lower frequency and the other reflecting its higher frequency. We are required to choose the splitting in such a way that the inverse of the lower-frequency matrix either has an explicit form or can easily be computed with a lower computational cost.

In this chapter we introduce the MAM and MIM and provide a complete analysis of their convergence and stability.

9.1 Multilevel augmentation methods

In this section, we describe a general setting of the MAM for solving operator equations. This method is based on a standard approximation method at a coarse level and updates the resulting approximate solutions by adding details corresponding to higher levels in a direct sum decomposition. We prove that this method provides the same order of convergence as the original approximation method.

9.1.1 Multilevel augmentation methods for solving operator equations

We begin with a description of the general setup for the operator equations under consideration. Let $\mathbb{X}$ and $\mathbb{Y}$ be two Banach spaces, and $\mathcal{A} : \mathbb{X} \to \mathbb{Y}$ be a bounded linear operator. For a function $f \in \mathbb{Y}$, we consider the operator equation

$$\mathcal{A}u = f, \tag{9.1}$$

where $u \in \mathbb{X}$ is the solution to be determined. We assume that equation (9.1) has a unique solution in $\mathbb{X}$. To solve the equation, we choose two sequences of finite-dimensional subspaces $\mathbb{X}_n, n \in \mathbb{N}_0 := \{0, 1, \ldots\}$ and $\mathbb{Y}_n, n \in \mathbb{N}_0$ of $\mathbb{X}$ and $\mathbb{Y}$, respectively, such that

$$\overline{\bigcup_{n\in\mathbb{N}_0} \mathbb{X}_n} = \mathbb{X}, \qquad \overline{\bigcup_{n\in\mathbb{N}_0} \mathbb{Y}_n} = \mathbb{Y}$$

and

$$\dim \mathbb{X}_n = \dim \mathbb{Y}_n, \quad n \in \mathbb{N}_0.$$

We suppose that equation (9.1) has an approximate operator equation

$$\mathcal{A}_n u_n = f_n, \tag{9.2}$$

where $\mathcal{A}_n : \mathbb{X}_n \to \mathbb{Y}_n$ is an approximate operator of $\mathcal{A}$, $u_n \in \mathbb{X}_n$ and $f_n \in \mathbb{Y}_n$ is an approximation of f. Examples of such equations include projection methods such as Galerkin methods and collocation methods. In particular, for solving integral equations they also include approximate operator equations obtained from quadrature methods and degenerate kernel methods. Wavelet compression schemes using both orthogonal projection (Galerkin methods) and interpolation projection (collocation methods) are also examples of this type.

Our method is based on an additional hypothesis that the subspaces are nested, that is,

$$\mathbb{X}_n \subset \mathbb{X}_{n+1}, \quad \mathbb{Y}_n \subset \mathbb{Y}_{n+1}, \quad n \in \mathbb{N}_0 \tag{9.3}$$

so that we can define two subspaces $\mathbb{W}_{n+1} \subset \mathbb{X}_{n+1}$ and $\mathbb{Q}_{n+1} \subset \mathbb{Y}_{n+1}$ such that $\mathbb{X}_{n+1}$ becomes a direct sum of $\mathbb{X}_n$ and $\mathbb{W}_{n+1}$ and likewise, $\mathbb{Y}_{n+1}$ is a direct sum of $\mathbb{Y}_n$ and $\mathbb{Q}_{n+1}$. Specifically, we assume that two direct sums $\oplus_1$ and $\oplus_2$ are defined so that we have the decompositions

$$\mathbb{X}_{n+1} = \mathbb{X}_n \oplus_1 \mathbb{W}_{n+1} \text{ and } \mathbb{Y}_{n+1} = \mathbb{Y}_n \oplus_2 \mathbb{Q}_{n+1}, \quad n \in \mathbb{N}_0. \tag{9.4}$$

In practice, the finer-level subspaces $\mathbb{X}_{n+1}$ and $\mathbb{Y}_{n+1}$ are obtained respectively from the coarse-level subspaces $\mathbb{X}_n$ and $\mathbb{Y}_n$ by local or global subdivisions. It follows from (9.4) for a fixed $k \in \mathbb{N}_0$ and any $m \in \mathbb{N}_0$ that

$$\mathbb{X}_{k+m} = \mathbb{X}_k \oplus_1 \mathbb{W}_{k+1} \oplus_1 \cdots \oplus_1 \mathbb{W}_{k+m} \tag{9.5}$$

and

$$\mathbb{Y}_{k+m} = \mathbb{Y}_k \oplus_2 \mathbb{Q}_{k+1} \oplus_2 \cdots \oplus_2 \mathbb{Q}_{k+m}. \tag{9.6}$$

As in [67], for $g_0 \in \mathbb{X}_k$ and $g_i \in \mathbb{W}_{k+i}$, $i = 1, 2, \ldots, m$, we identify the vector $[g_0, g_1, \ldots, g_m]^T$ in $\mathbb{X}_k \times \mathbb{W}_{k+1} \times \cdots \times \mathbb{W}_{k+m}$ with the sum $g_0 + g_1 + \cdots + g_m$ in $\mathbb{X}_k \oplus_1 \mathbb{W}_{k+1} \oplus_1 \cdots \oplus_1 \mathbb{W}_{k+m}$. Similarly, for $g_0 \in \mathbb{Y}_k$ and $g_i \in \mathbb{Q}_{k+i}$, for $i = 1, 2, \ldots, m$, we also identify the vector $[g_0, g_1, \ldots, g_m]^T$ in $\mathbb{Y}_k \times \mathbb{Q}_{k+1} \times \cdots \times \mathbb{Q}_{k+m}$ with the sum $g_0 + g_1 + \cdots + g_m$ in $\mathbb{Y}_k \oplus_2 \mathbb{Q}_{k+1} \oplus_2 \cdots \oplus_2 \mathbb{Q}_{k+m}$. In this notation, we describe the multilevel method for solving equation (9.2) with $n := k + m$, which has the form

$$\mathcal{A}_{k+m} u_{k+m} = f_{k+m}. \tag{9.7}$$

According to decomposition (9.5), we write the solution $u_{k+m} \in \mathbb{X}_{k+m}$ as

$$u_{k+m} = u_{k,0} + \sum_{i=1}^{m} v_{k,i}, \tag{9.8}$$

where $u_{k,0} \in \mathbb{X}_k$ and $v_{k,i} \in \mathbb{W}_{k+i}$ for $i = 1, 2, \ldots, m$. Hence, u_{k+m} is identified as $u_k(m) := [u_{k,0}, v_{k,1}, \ldots, v_{k,m}]^T$. We use both of these notations exchangeably.

We let $\mathcal{F}_{k,k+j} : \mathbb{W}_{k+j} \to \mathbb{Y}_k$, $\mathcal{G}_{k+i,k} : \mathbb{X}_k \to \mathbb{Q}_{k+i}$ and $\mathcal{H}_{k+i,k+j} : \mathbb{W}_{k+j} \to \mathbb{Q}_{k+i}$, $i, j = 1, 2, \ldots, m$ be given and assume that the operator $\mathcal{A}_{k+m}$ is identified as the matrix of operators

$$\mathcal{A}_{k,m} := \begin{bmatrix} \mathcal{A}_k & \mathcal{F}_{k,k+1} & \cdots & \mathcal{F}_{k,k+m} \\ \mathcal{G}_{k+1,k} & \mathcal{H}_{k+1,k+1} & \cdots & \mathcal{H}_{k+1,k+m} \\ \vdots & \vdots & \ddots & \vdots \\ \mathcal{G}_{k+m,k} & \mathcal{H}_{k+m,k+1} & \cdots & \mathcal{H}_{k+m,k+m} \end{bmatrix}. \tag{9.9}$$

Equation (9.7) is now equivalent to the equation

$$\mathcal{A}_{k,m} u_k(m) = f_{k+m}. \tag{9.10}$$

We remark that the nestedness of subspaces implies that the matrix $\mathcal{A}_{k,m}$ contains $\mathcal{A}_{k,m-1}$ as a submatrix. In other words, $\mathcal{A}_{k,m}$ is obtained by augmenting the matrix of the previous level $\mathcal{A}_{k,m-1}$.

With this setup, one can design various iteration schemes to solve equation (9.10) by splitting the matrix $\mathcal{A}_{k,m}$ defined by (9.9) into a sum of two matrices and applying matrix iteration algorithms to $\mathcal{A}_{k,m}$.

We split operator $\mathcal{A}_{k,m}$ as the sum of two operators $\mathcal{B}_{k,m}, \mathcal{C}_{k,m} : \mathbb{X}_{k+m} \to \mathbb{Y}_{k+m}$, that is, $\mathcal{A}_{k,m} = \mathcal{B}_{k,m} + \mathcal{C}_{k,m}$, $m \in \mathbb{N}_0$. Note that matrices $\mathcal{B}_{k,m}$ and $\mathcal{C}_{k,m}$ are obtained from augmenting matrices $\mathcal{B}_{k,m-1}$ and $\mathcal{C}_{k,m-1}$, respectively. Hence, equation (9.10) becomes

$$\mathcal{B}_{k,m}u_k(m) = f_{k+m} - \mathcal{C}_{k,m}u_k(m), \quad m \in \mathbb{N}_0. \tag{9.11}$$

Instead of solving (9.11) exactly, we solve (9.11) approximately by using the MAM described below.

Algorithm 1 (Operator form of the multilevel augmentation algorithm) *Let $k > 0$ be a fixed integer.*

Step 1: *Solve equation (9.2) with $n := k$ for $u_k \in \mathbb{X}_k$ exactly.*

Step 2: *Set $u_{k,0} := u_k$ and compute the splitting matrices $\mathcal{B}_{k,0}$ and $\mathcal{C}_{k,0}$.*

Step 3: *For $m \in \mathbb{N}$, suppose that $u_{k,m-1} \in \mathbb{X}_{k+m-1}$ has been obtained and do the following.*

- *Augment the matrices $\mathcal{B}_{k,m-1}$ and $\mathcal{C}_{k,m-1}$ to form $\mathcal{B}_{k,m}$ and $\mathcal{C}_{k,m}$, respectively.*
- *Augment $u_{k,m-1}$ by setting $\bar{u}_{k,m} := \begin{bmatrix} u_{k,m-1} \\ 0 \end{bmatrix}$.*
- *Solve $u_{k,m} \in \mathbb{X}_{k+m}$ from equation*

$$\mathcal{B}_{k,m}u_{k,m} = f_{k+m} - \mathcal{C}_{k,m}\bar{u}_{k,m}. \tag{9.12}$$

For a fixed positive integer k, if this algorithm can be carried out, it generates a sequence of approximate solutions $u_{k,m} \in \mathbb{X}_{k+m}$, $m \in \mathbb{N}_0$. Note that this algorithm is *not* an iteration method since for different m we are dealing with matrices of different order and for each m we only compute the approximate solution $u_{k,m}$ once. To ensure that the algorithm can be carried out, we have to guarantee that for $m \in \mathbb{N}_0$, the inverse of $\mathcal{B}_{k,m}$ exists and is uniformly bounded. Moreover, the approximate solutions generated by this augmentation algorithm neither necessarily have the same order of convergence as the approximation order of the subspaces $\mathbb{X}_n$ nor necessarily are more efficient to solve than solving equation (9.2) with $n := k + m$ directly, unless certain conditions on the splitting are satisfied. For this algorithm to be executable, accurate and efficient, we demand that the splitting of operator $\mathcal{A}_{k,m}$ fulfills three requirements. Firstly, $\mathcal{B}_{k,m}^{-1}$ is uniformly bounded. Secondly, the approximate solution $u_{k,m}$ preserves the convergence order of u_{k+m}. That is, $u_{k,m}$ converges to the exact solution u at the approximation order of the subspaces $\mathbb{X}_{k+m}$. Thirdly, the inverse of $\mathcal{B}_{k,m}$ is much easier to obtain than the inverse of $\mathcal{A}_{k,m}$.

We now address the first issue. To this end, we describe our hypotheses.

(I) There exist a positive integer N_0 and a positive constant α such that for all $n \geq N_0$,

$$\|\mathcal{A}_n^{-1}\| \leq \alpha^{-1}. \tag{9.13}$$

(II) The limit

$$\lim_{n\to\infty} \|\mathcal{C}_{n,m}\| = 0 \tag{9.14}$$

holds uniformly for all $m \in \mathbb{N}$.

Under these two assumptions, we have the following result.

Proposition 9.1 *Suppose that hypotheses (I) and (II) hold. Then, there exists a positive integer $N > N_0$ such that for all $k \geq N$ and all $m \in \mathbb{N}$, equation (9.12) has a unique solution $u_{k,m} \in \mathbb{X}_{k+m}$.*

Proof From hypothesis (I), whenever $k \geq N_0$ it holds that for $x \in \mathbb{X}_{k+m}$,

$$\|\mathcal{B}_{k,m}x\| = \|(\mathcal{A}_{k,m} - \mathcal{C}_{k,m})x\| \geq (\alpha - \|\mathcal{C}_{k,m}\|)\|x\|. \tag{9.15}$$

Moreover, it follows from (9.14) that there exists a positive integer $N > N_0$ such that for $k \geq N$ and $m \in \mathbb{N}_0$, $\|\mathcal{C}_{k,m}\| < \alpha/2$. Combining this inequality with (9.15), we find that for $k \geq N$ and $m \in \mathbb{N}_0$ the estimate

$$\|\mathcal{B}_{k,m}^{-1}\| \leq \frac{1}{\alpha - \|\mathcal{C}_{k,m}\|} \leq 2\alpha^{-1} \tag{9.16}$$

holds. This ensures that for all $k \geq N$ and $m \in \mathbb{N}_0$, equation (9.12) has a unique solution. □

We next consider the second issue. For $n \in \mathbb{N}_0$, we let R_n denote the approximation error of space $\mathbb{X}_n$ for $u \in \mathbb{X}$, namely, $R_n = R_n(u) := \inf\{\|u - v\|_{\mathbb{X}} : v \in \mathbb{X}_n\}$. A sequence of non-negative numbers γ_n, $n \in \mathbb{N}_0$ is called a *majorization sequence* of R_n if $\gamma_n \geq R_n$, $n \in \mathbb{N}_0$ and there exist a positive integer N_0 and a positive constant σ such that for $n \geq N_0$, $\frac{\gamma_{n+1}}{\gamma_n} \geq \sigma$.

We also need the following hypothesis.

(III) There exist a positive integer N_0 and a positive constant ρ such that for $n \geq N_0$ and for the solution $u_n \in \mathbb{X}_n$ of equation (9.2), $\|u - u_n\|_{\mathbb{X}} \leq \rho R_n$.

In the next theorem, we show that under the assumptions described above, $u_{k,m}$ approximates u at an order comparable to R_{k+m}.

Theorem 9.2 *Suppose that hypotheses (I)–(III) hold. Let $u \in \mathbb{X}$ be the solution of equation (9.1), γ_n, $n \in \mathbb{N}_0$ be a majorization sequence of R_n and ρ be the constant appearing in hypothesis (III). Then, there exists a positive integer N such that for $k \geq N$ and $m \in \mathbb{N}_0$,*

$$\|u - u_{k,m}\| \leq (\rho + 1)\gamma_{k+m},$$

where $u_{k,m}$ is the solution of equation (9.12).

Proof We prove this theorem by establishing an estimate on $\|u_{k,m} - u_{k+m}\|$. For this purpose, we subtract (9.11) from (9.12) to obtain

$$\mathcal{B}_{k,m}(u_{k,m} - u_{k+m}) = \mathcal{C}_{k,m}(u_{k+m} - \overline{u}_{k,m}).$$

The hypotheses of this theorem ensure that Proposition 9.1 holds. Hence, from the equation above and inequality (9.16), we have that

$$\|u_{k,m} - u_{k+m}\| \leq \frac{\|\mathcal{C}_{k,m}\|}{\alpha - \|\mathcal{C}_{k,m}\|}\|u_{k+m} - \overline{u}_{k,m}\|. \tag{9.17}$$

We next prove by induction on m that there exists a positive integer N such that for $k \geq N$ and $m \in \mathbb{N}_0$,

$$\|u_{k,m} - u_{k+m}\| \leq \gamma_{k+m}. \tag{9.18}$$

When $m = 0$, since $u_{k,0} = u_k$, estimate (9.18) holds trivially. Suppose that the claim holds for $m = r - 1$ and we prove that it holds for $m = r$. To accomplish this, using the definition of $\overline{u}_{k,r}$, hypothesis (III), the induction hypothesis and the definition of majorization sequences, we obtain that

$$\begin{aligned}\|u_{k+r} - \overline{u}_{k,r}\| &\leq \|u_{k+r} - u\| + \|u - u_{k+r-1}\| + \|u_{k+r-1} - u_{k,r-1}\| \\ &\leq \rho\gamma_{k+r} + (\rho + 1)\gamma_{k+r-1} \leq \left(\rho + (\rho + 1)\frac{1}{\sigma}\right)\gamma_{k+r}.\end{aligned}$$

Substituting this estimate into the right-hand side of (9.17) with $m = r$ yields

$$\|u_{k,r} - u_{k+r}\| \leq \frac{\|\mathcal{C}_{k,r}\|}{\alpha - \|\mathcal{C}_{k,r}\|}\left(\rho + (\rho + 1)\frac{1}{\sigma}\right)\gamma_{k+r}.$$

Again, employing hypothesis (II), there exists a positive integer N such that for $k \geq N$ and $r \in \mathbb{N}_0$, $\|\mathcal{C}_{k,r}\| \leq \frac{\alpha}{(\rho+1)(1+\frac{1}{\sigma})}$. We then conclude that for $k \geq N$ and $r \in \mathbb{N}_0$,

$$\frac{\|\mathcal{C}_{k,r}\|}{\alpha - \|\mathcal{C}_{k,r}\|}\left(\rho + (\rho + 1)\frac{1}{\sigma}\right) \leq 1.$$

Therefore, for $k \geq N$, estimate (9.18) holds for $m = r$. This advances the induction hypothesis and thus estimate (9.18) holds for all $m \in \mathbb{N}_0$.

Finally, the estimate of this theorem follows directly from estimate (9.18) and hypothesis (III). □

We remark that when the exact solution u of equation (9.1) has certain Sobolev or Besov regularity and specific approximate subspaces $\mathbb{X}_n$ are chosen, we may choose the majorization sequence γ_n as the upper bound of R_n which gives the order of approximation of the subspaces $\mathbb{X}_n$ with respect to the regularity. For example, when $\mathbb{X}_n$ is chosen to be the usual finite element spaces with mesh size 2^{-n} and when the solution u of equation (9.1) belongs to the Sobolev space H^r, we may choose $\gamma_n := c2^{-rn}\|u\|_{H^r}$. In this case, the constant σ in the definition of majorization sequences can be taken as 2^{-r}. Therefore, Theorem 9.2 ensures that the approximate solution $u_{k,m}$ generated by the MAM has the same order of approximation as the subspaces $\mathbb{X}_n$.

9.1.2 Second-kind equations

In this section, we present special results for projection methods for solving operator equations of the second kind. Consider equations

$$(\mathcal{I} - \mathcal{K})u = f, \tag{9.19}$$

where $\mathcal{K} : \mathbb{X} \to \mathbb{X}$ is a linear operator. We assume that equation (9.19) has a unique solution. In this special case, we identify that $\mathcal{A} := \mathcal{I} - \mathcal{K}$, $\mathbb{X} = \mathbb{Y}$ and $\mathbb{X}_n = \mathbb{Y}_n$. Suppose that $\mathcal{P}_n : \mathbb{X} \to \mathbb{X}_n$ are linear projections and we define the projection method for solving equation (9.19) by

$$\mathcal{P}_n(\mathcal{I} - \mathcal{K})u_n = \mathcal{P}_n f, \tag{9.20}$$

where $u_n \in \mathbb{X}_n$. To develop a MAM, we need another projection $\hat{\mathcal{P}}_n : \mathbb{X} \to \mathbb{X}_n$. We define operators $\mathcal{Q}_n := \mathcal{P}_n - \mathcal{P}_{n-1}$ and $\hat{\mathcal{Q}}_n := \hat{\mathcal{P}}_n - \hat{\mathcal{P}}_{n-1}$, and introduce subspaces $\mathbb{W}_n := \hat{\mathcal{Q}}_n\mathbb{X}_n$, $n \in \mathbb{N}$. We allow the projections $\mathcal{P}_n$ and $\hat{\mathcal{P}}_n$ to be different in order to have a wide range of applications. For example, for Galerkin methods, $\mathcal{P}_n$ and $\hat{\mathcal{P}}_n$ are both identical to the orthogonal projection and for the collocation method developed in [68], $\mathcal{P}_n$ is the interpolatory projection and $\hat{\mathcal{P}}_n$ is the orthogonal projection.

For $n \in \mathbb{N}$, we set $\mathcal{K}_n := \mathcal{P}_n\mathcal{K}|_{\mathbb{X}_n}$ and identify $\mathcal{A}_n := \mathcal{P}_n(\mathcal{I} - \mathcal{K})|_{\mathbb{X}_n}$. We further identify the operators in (9.9) with

$$\mathcal{F}_{k,k+j} := \mathcal{P}_k(\mathcal{I} - \mathcal{K})|_{\mathbb{W}_{k+j}}, \quad \mathcal{G}_{k+i,k} := \mathcal{Q}_{k+i}(\mathcal{I} - \mathcal{K})|_{\mathbb{X}_k},$$
$$\mathcal{H}_{k+i,k+j} := \mathcal{Q}_{k+i}(\mathcal{I} - \mathcal{K})|_{\mathbb{W}_{k+j}}.$$

We split the operator $\mathcal{K}_{k+m}$ into the sum of two operators $\mathcal{K}_{k+m} = \mathcal{K}^L_{k,m} + \mathcal{K}^H_{k,m}$, where $\mathcal{K}^L_{k,m} := \mathcal{P}_k\mathcal{K}|_{\mathbb{X}_{k+m}}$ and $\mathcal{K}^H_{k,m} := (\mathcal{P}_{k+m} - \mathcal{P}_k)\mathcal{K}|_{\mathbb{X}_{k+m}}$. The operators

$\mathcal{K}_{k,m}^L$ and $\mathcal{K}_{k,m}^H$ correspond to lower and higher frequency of the operator $\mathcal{K}_{k,m}$, respectively. According to the decomposition of $\mathcal{K}_{k+m}$, we write the operator $\mathcal{A}_{k,m} := \mathcal{I}_{k+m} - \mathcal{K}_{k,m}$ as a sum of lower and higher-frequency $\mathcal{B}_{k,m} := \mathcal{I}_{k+m} - \mathcal{K}_{k,m}^L$ and $\mathcal{C}_{k,m} := -\mathcal{K}_{k,m}^H$. Using this specific splitting in formula (9.12) of Algorithm 1, we have that

$$(\mathcal{I}_{k+m} - \mathcal{K}_{k,m}^L)u_{k,m} = f_{k,m} + \mathcal{K}_{k,m}^H \bar{u}_{k,m}. \tag{9.21}$$

The next theorem is concerned with the convergence order for the MAM for second-kind equations using projection methods.

Theorem 9.3 *Suppose that $\mathcal{K}$ is a compact linear operator not having one as its eigenvalue and that there exists a positive constant p such that*

$$\|\mathcal{P}_n\| \leq p, \quad \|\hat{\mathcal{P}}_n\| \leq p, \quad \textit{for all} \;\; n \in \mathbb{N}. \tag{9.22}$$

Let $u \in \mathbb{X}$ be the solution of equation (9.19) and γ_n be a majorization sequence of R_n. Then, there exist a positive integer N and a positive constant c_0 such that for all $k \geq N$ and $m \in \mathbb{N}$,

$$\|u - u_{k,m}\| \leq c_0 \gamma_{k+m},$$

where $u_{k,m}$ is obtained from the augmentation algorithm with formula (9.21).

Proof We prove that the hypotheses of Theorem 9.2 hold for the special choice of operators $\mathcal{B}_{k,m}$ and $\mathcal{C}_{k,m}$ for second-kind equations. We first remark that the assumption on the operators $\mathcal{K}$ and $\mathcal{P}_n$ ensures that hypotheses (I) and (III) hold with $\mathcal{A}_n := \mathcal{I} - \mathcal{K}_n$. It remains to verify hypothesis (II). To this end, we recall the definition of $\mathcal{C}_{n,m}$, which has the form $\mathcal{C}_{n,m} = -(\mathcal{P}_{n+m} - \mathcal{P}_n)\mathcal{K}|_{\mathbb{X}_{n+m}}$. It follows from the second inequality of (9.22) that

$$\|\mathcal{C}_{n,m}\| = \|(\mathcal{P}_{n+m} - \mathcal{P}_n)\mathcal{K}|_{\mathbb{X}_{n+m}}\| \leq p\|(\mathcal{P}_{n+m} - \mathcal{P}_n)\mathcal{K}\|.$$

By the first inequality of (9.22) and the nestedness of subspaces $\mathbb{X}_n$, we conclude that $\mathcal{P}_n$ converges pointwise to the identity operator $\mathcal{I}$ of space $\mathbb{X}$. Hence, since $\mathcal{K}$ is compact, the last term of the inequality above converges to zero as $n \to \infty$ uniformly for $m \in \mathbb{N}$. Therefore, all hypotheses of Theorem 9.2 are satisfied and thus, we complete the proof of this theorem. □

We next derive the matrix form of the MAM by choosing appropriate bases for the subspaces $\mathbb{X}_n$. For this purpose, we let $\mathbb{X}^*$ denote the dual space of $\mathbb{X}$ and for $\ell \in \mathbb{X}^*$, $x \in \mathbb{X}$ we let $\langle \ell, x \rangle$ denote the value of the linear functional ℓ at x. Suppose that $\mathbb{L}_n$, $n \in \mathbb{N}_0$ is a sequence of subspaces of $\mathbb{X}^*$ which has the property that $\mathbb{L}_n \subset \mathbb{L}_{n+1}$ and $\dim \mathbb{L}_n = \dim \mathbb{X}_n$, $n \in \mathbb{N}_0$. The operator $\mathcal{P}_n : \mathbb{X} \to \mathbb{X}_n$ is defined for $x \in \mathbb{X}$ by

$$\langle \ell, x - \mathcal{P}_n x \rangle = 0, \quad \text{for all } \ell \in \mathbb{L}_n. \tag{9.23}$$

It is known (cf. [77]) that the operator $\mathcal{P}_n : \mathbb{X} \to \mathbb{X}_n$ is uniquely determined and is a projection if and only if $\mathbb{L}_n \cap \mathbb{X}_n^{\perp} = \{0\}$, $n \in \mathbb{N}_0$, where $\mathbb{X}_n^{\perp}$ denotes the annihilator of $\mathbb{X}_n$ in $\mathbb{X}^*$. Throughout the rest of this section we always assume that this condition is satisfied. We also assume that we have a decomposition of the space $\mathbb{L}_{n+1}$, namely,

$$\mathbb{L}_{n+1} = \mathbb{L}_n \oplus \mathbb{V}_{n+1}, \quad n \in \mathbb{N}_0. \tag{9.24}$$

Clearly, the spaces $\mathbb{W}_i$ and $\mathbb{V}_i$ have the same dimension. We specify the direct sum in (9.24) later.

Set $w(0) := \dim \mathbb{X}_0$ and $w(i) := \dim \mathbb{W}_i$, for $i \in \mathbb{N}$. Suppose that

$$\mathbb{X}_0 = \operatorname{span}\{w_{0j} : j \in \mathbb{Z}_{w(0)}\}, \quad \mathbb{L}_0 = \operatorname{span}\{\ell_{0j} : j \in \mathbb{Z}_{w(0)}\},$$
$$\mathbb{W}_i = \operatorname{span}\{w_{ij} : j \in \mathbb{Z}_{w(i)}\}, \quad \mathbb{V}_i = \operatorname{span}\{\ell_{ij} : j \in \mathbb{Z}_{w(i)}\}, \quad i \in \mathbb{N}.$$

Using the index set $\mathbb{U}_n := \{(i,j) : i \in \mathbb{Z}_{n+1}, j \in \mathbb{Z}_{w(i)}\}$, we have that for $n \in \mathbb{N}_0$,

$$\mathbb{X}_n = \operatorname{span}\{w_{ij} : (i,j) \in \mathbb{U}_n\} \text{ and } \mathbb{L}_n = \operatorname{span}\{\ell_{ij} : (i,j) \in \mathbb{U}_n\}.$$

We remark that the index set $\mathbb{U}_n$ has cardinality $d_n := \dim \mathbb{X}_n$ and we assume that the elements in $\mathbb{U}_n$ are ordered lexicographically.

We now present the matrix form of equation (9.20) using these bases. Note that for $v_n \in \mathbb{X}_n$, there exist unique constants v_{ij}, $(i,j) \in \mathbb{U}_n$, such that $v_n = \sum_{(i,j)\in\mathbb{U}_n} v_{ij} w_{ij}$. It follows that the solution u_n, with $n := k+m$, of equation (9.20) has the vector representation $\mathbf{u}_n := [u_{ij} : (i,j) \in \mathbb{U}_n]$ under the basis w_{ij}, $(i,j) \in \mathbb{U}_n$. Using the bases for $\mathbb{X}_n$ and $\mathbb{L}_n$, we let $E_{i'j',ij} := \langle \ell_{i'j'}, w_{ij} \rangle$ and $K_{i'j',ij} := \langle \ell_{i'j'}, \mathcal{K} w_{ij} \rangle$ and introduce the matrices $\mathbf{E}_n := [E_{i'j',ij} : (i',j'), (i,j) \in \mathbb{U}_n]$, and $\mathbf{K}_n := [K_{i'j',ij} : (i',j'), (i,j) \in \mathbb{U}_n]$. We also introduce the column vectors $\mathbf{f}_n := [\langle \ell_{i'j'}, f \rangle : (i',j') \in \mathbb{U}_n]$. In these notations, equation (9.20) is written in matrix form as

$$(\mathbf{E}_{k+m} - \mathbf{K}_{k+m})\, \mathbf{u}_{k+m} = \mathbf{f}_{k+m}. \tag{9.25}$$

We partition matrices $\mathbf{K}_n$ and $\mathbf{E}_n$ into block matrices according to the decompositions of the spaces $\mathbb{X}_n$ and $\mathbb{L}_n$. Specifically, for $i', i \in \mathbb{Z}_{n+1}$, we introduce the blocks $\mathbf{K}_{i'i} := [K_{i'j',ij} : j' \in \mathbb{Z}_{w(i')}, j \in \mathbb{Z}_{w(i)}]$ and set $\mathbf{K}_n = [\mathbf{K}_{i'i} : i', i \in \mathbb{Z}_{n+1}]$. Moreover, for a fixed $k \in \mathbb{N}$ we define the blocks $\mathbf{K}^k_{0,0} := \mathbf{K}_k$, and for $l', l \in \mathbb{N}$, $\mathbf{K}^k_{0,l} := [\mathbf{K}_{i',i} : i' \in \mathbb{Z}_{k+1}, i = k+l]$,

$\mathbf{K}^k_{l,0} := [\mathbf{K}_{i',i} : i' = k+l, i \in \mathbb{Z}_{k+1}]$ and $\mathbf{K}^k_{l',l} := \mathbf{K}_{k+l',k+l}$. Using these block notations, for $n := k+m$ we write $\mathbf{K}_{k+m} = \left[\mathbf{K}^k_{i',i} : i', i \in \mathbb{Z}_{m+1}\right]$. Likewise, we partition matrix $\mathbf{E}_n$ in exactly the same way.

The decomposition of operator $\mathcal{K}_{k+m}$ suggests the matrix decomposition $\mathbf{K}_{k+m} = \mathbf{K}^L_{k,m} + \mathbf{K}^H_{k,m}$, where

$$\mathbf{K}^L_{k,m} := \begin{bmatrix} \mathbf{K}^k_{0,0} & \mathbf{K}^k_{0,1} & \cdots & \mathbf{K}^k_{0,m} \\ 0 & 0 & \cdots & 0 \\ \vdots & \vdots & \ddots & \vdots \\ 0 & 0 & \cdots & 0 \end{bmatrix} \quad \text{and}$$

$$\mathbf{K}^H_{k,m} := \begin{bmatrix} 0 & 0 & \cdots & 0 \\ \mathbf{K}^k_{1,0} & \mathbf{K}^k_{1,1} & \cdots & \mathbf{K}^k_{1,m} \\ \vdots & \vdots & \ddots & \vdots \\ \mathbf{K}^k_{m,0} & \mathbf{K}^k_{m,1} & \cdots & \mathbf{K}^k_{m,m} \end{bmatrix}.$$

Note that the matrices $\mathbf{K}^L_{k,m}$ and $\mathbf{K}^H_{k,m}$ correspond to lower and higher frequency of the matrix $\mathbf{K}_{k,m}$. Moreover, we set $\mathbf{B}_{k,m} := \mathbf{E}_{k+m} - \mathbf{K}^L_{k,m}$ and $\mathbf{C}_{k,m} := -\mathbf{K}^H_{k,m}$. Next, we describe the matrix form of the MAM for solving equation (9.25) using these two matrices.

Algorithm 2 (Matrix form of the multilevel augmentation algorithm) *Let $k > 0$ be a fixed integer.*

Step 1: *Solve $\boldsymbol{u}_k \in \mathbb{R}^{d_k}$ from the equation $(\boldsymbol{E}_k - \mathbf{K}_k)\,\boldsymbol{u}_k = \boldsymbol{f}_k$.*

Step 2: *Set $\boldsymbol{u}_{k,0} := \boldsymbol{u}_k$ and compute the splitting matrices $\mathbf{K}^L_{k,0}$ and $\mathbf{K}^H_{k,0}$.*

Step 3: *For $m \in \mathbb{N}$, suppose that $\boldsymbol{u}_{k,m-1} \in \mathbb{R}^{d_{k+m-1}}$ has been obtained and do the following.*

- *Augment the matrices $\mathbf{K}^L_{k,m-1}$ and $\mathbf{K}^H_{k,m-1}$ to form $\mathbf{K}^L_{k,m}$ and $\mathbf{K}^H_{k,m}$, respectively.*
- *Augment $\boldsymbol{u}_{k,m-1}$ by setting $\bar{\mathbf{u}}_{k,m} := \begin{bmatrix} \boldsymbol{u}_{k,m-1} \\ 0 \end{bmatrix}$.*
- *Solve $\boldsymbol{u}_{k,m} \in \mathbb{R}^{d_{k+m}}$ from the algebraic equations*

$$(\boldsymbol{E}_{k,m} - \mathbf{K}^L_{k,m})\boldsymbol{u}_{k,m} = \boldsymbol{f}_{k+m} + \mathbf{K}^H_{k,m}\bar{\mathbf{u}}_{k,m}. \tag{9.26}$$

It is important to know under what condition the matrix form (9.26) is equivalent to the operator form (9.21). This issue is addressed in the next theorem. To prepare a proof of this theorem, we consider an expression of the identity operator $\mathcal{I}$ in the subspace $\mathbb{X}_{k+m}$. Note that for any $x \in \mathbb{X}_{k+j}$, $j \in \mathbb{Z}_m$, $\mathcal{Q}_{k+1+j}x = 0$. This is equivalent to the following equations:

$$\mathcal{Q}_{k+1+j}\mathcal{I}|_{\mathbb{X}_k} = 0 \text{ and } \mathcal{Q}_{k+1+j}\mathcal{I}|_{\mathbb{W}_{k+1+i}} = 0, \quad i \in \mathbb{Z}_j. \tag{9.27}$$

Using this fact, we express the identity operator $\mathcal{I}$ in the subspace $\mathbb{X}_{k+m}$ as

$$\mathcal{I}_{k+m} := \mathcal{P}_{k+m}\mathcal{I}|_{\mathbb{X}_{k+m}} = \begin{bmatrix} \mathcal{P}_k\mathcal{I}|_{\mathbb{X}_k} & \mathcal{P}_k\mathcal{I}|_{\mathbb{W}_{k+1}} & \cdots & \mathcal{P}_k\mathcal{I}|_{\mathbb{W}_{k+m}} \\ 0 & \mathcal{Q}_{k+1}\mathcal{I}|_{\mathbb{W}_{k+1}} & \cdots & \mathcal{Q}_{k+1}\mathcal{I}|_{\mathbb{W}_{k+m}} \\ \vdots & \ddots & \ddots & \vdots \\ 0 & \cdots & 0 & \mathcal{Q}_{k+m}\mathcal{I}|_{\mathbb{W}_{k+m}} \end{bmatrix}.$$

Taking this into consideration, equation (9.21) becomes

$$\begin{bmatrix} \mathcal{P}_k(\mathcal{I}-\mathcal{K})|_{\mathbb{X}_k} & \mathcal{P}_k(\mathcal{I}-\mathcal{K})|_{\mathbb{W}_{k+1}} & \cdots & \mathcal{P}_k(\mathcal{I}-\mathcal{K})|_{\mathbb{W}_{k+m}} \\ 0 & \mathcal{Q}_{k+1}\mathcal{I}|_{\mathbb{W}_{k+1}} & \cdots & \mathcal{Q}_{k+1}\mathcal{I}|_{\mathbb{W}_{k+m}} \\ \vdots & \ddots & \ddots & \vdots \\ 0 & \cdots & 0 & \mathcal{Q}_{k+m}\mathcal{I}|_{\mathbb{W}_{k+m}} \end{bmatrix} u_{k,m}$$
$$= \begin{bmatrix} \mathcal{P}_k f \\ \mathcal{Q}_{k+1} f \\ \vdots \\ \mathcal{Q}_{k+m} f \end{bmatrix}$$
$$+ \begin{bmatrix} 0 & 0 & \cdots & 0 \\ \mathcal{Q}_{k+1}\mathcal{K}|_{\mathbb{X}_k} & \mathcal{Q}_{k+1}\mathcal{K}|_{\mathbb{W}_{k+1}} & \cdots & \mathcal{Q}_{k+1}\mathcal{K}|_{\mathbb{W}_{k+m}} \\ \vdots & \vdots & \ddots & \vdots \\ \mathcal{Q}_{k+m}\mathcal{K}|_{\mathbb{X}_k} & \mathcal{Q}_{k+m}\mathcal{K}|_{\mathbb{W}_{k+1}} & \cdots & \mathcal{Q}_{k+m}\mathcal{K}|_{\mathbb{W}_{k+m}} \end{bmatrix} \begin{bmatrix} u_{k,m-1} \\ 0 \end{bmatrix}. \tag{9.28}$$

To state the next theorem, we let $\mathbb{N}_k := \{k, k+1, \ldots\}$ and introduce the following notion. For finite-dimensional subspaces $A \subset \mathbb{X}^*$ and $B \subset \mathbb{X}$, we say $A \perp B$ if for any $\ell \in A$ and $x \in B$ it holds that $\langle \ell, x \rangle = 0$.

Theorem 9.4 *The following statements are equivalent.*

(i) The matrix form (9.26) and the operator form (9.21) are equivalent for any $f \in \mathbb{X}$ and for any compact operator $\mathcal{K} : \mathbb{X} \to \mathbb{X}$.
(ii) For any $l \in \mathbb{N}_k$, $\mathbb{V}_{l+1} \perp \mathbb{X}_l$.
(iii) For any $l \in \mathbb{N}_k$ and any $j \in \mathbb{Z}_{m+1} \setminus \{0\}$, $\mathbb{V}_{l+j} \perp \mathbb{X}_l$.
(iv) For $i', i \in \mathbb{Z}_{m+1}$, $i' > i$, $\mathbf{E}^k_{i',i} = 0$.
(v) For $i', i \in \mathbb{Z}_{m+1}$, $i' > i$, $\mathbf{B}^k_{i',i} = 0$.

Proof We first prove the equivalence of statements (i) and (ii). It is clear that for any $x \in \mathbb{X}$ and for any $\ell \in \mathbb{V}_n$, $\langle \ell, \mathcal{P}_{n-1}x\rangle = 0$ if and only if $\mathbb{V}_n \perp \mathbb{X}_{n-1}$. Moreover, we observe from the definitions of $\mathcal{P}_n$ and $\mathcal{Q}_n$ that for $n \in \mathbb{N}$, $x \in \mathbb{X}$ and for $\ell \in \mathbb{V}_n$,

$$\langle \ell, \mathcal{Q}_n x\rangle = \langle \ell, \mathcal{P}_n x - \mathcal{P}_{n-1}x\rangle = \langle \ell, x\rangle - \langle \ell, \mathcal{P}_{n-1}x\rangle .$$

Hence, for each $n \in \mathbb{N}$, the following equation holds: $\langle \ell, \mathcal{Q}_n x\rangle = \langle \ell, x\rangle$, for all $x \in \mathbb{X}$, $\ell \in \mathbb{V}_n$ if and only if $\mathbb{V}_n \perp \mathbb{X}_{n-1}$. Therefore, statement (ii) is equivalent to saying that for any $j \in \mathbb{Z}_m$,

$$\left\langle \ell, \mathcal{Q}_{k+j+1}x\right\rangle = \langle \ell, x\rangle, \quad \text{for all } \; x \in \mathbb{X}, \; \ell \in \mathbb{V}_{k+j+1}. \tag{9.29}$$

Noting that for $x \in \mathbb{X}$ and $n \in \mathbb{N}$, $\mathcal{Q}_n x \in \mathbb{Q}_n := \mathcal{Q}_n \mathbb{X}_n \subset \mathbb{X}_n$, we conclude from (9.23) and (9.27) that equation (9.28) is equivalent to

$$\left\langle \ell, \mathcal{P}_k(\mathcal{I} - \mathcal{K})u_{k,m}\right\rangle = \langle \ell, \mathcal{P}_k f\rangle, \quad \text{for all } \; \ell \in \mathbb{L}_k \tag{9.30}$$

and

$$\left\langle \ell, \mathcal{Q}_{k+j+1}u_{k,m}\right\rangle = \left\langle \ell, \mathcal{Q}_{k+j+1}f + \mathcal{Q}_{k+j+1}\mathcal{K}u_{k,m-1}\right\rangle, \text{ for all } \; \ell \in \mathbb{L}_{k+j+1}, \; j \in \mathbb{Z}_m. \tag{9.31}$$

Using (9.23), equation (9.30) is written as

$$\left\langle \ell, (\mathcal{I} - \mathcal{K})u_{k,m}\right\rangle = \langle \ell, f\rangle, \quad \text{for all } \; \ell \in \mathbb{L}_k. \tag{9.32}$$

Again, from (9.23) we have that for $x \in \mathbb{X}$ and $\ell \in \mathbb{L}_{k+j}$, $j \in \mathbb{Z}_m$,

$$\left\langle \ell, \mathcal{Q}_{k+j+1}x\right\rangle = \left\langle \ell, \mathcal{P}_{k+j+1}x - \mathcal{P}_{k+j}x\right\rangle = 0.$$

In particular, for all $\ell \in \mathbb{L}_{k+j}$, $j \in \mathbb{Z}_m$ both sides of equation (9.31) are equal to zero. Noting that

$$\mathbb{L}_{k+j+1} = \mathbb{L}_{k+j} \oplus \mathbb{V}_{k+j+1},$$

equation (9.31) is equivalent to

$$\left\langle \ell, \mathcal{Q}_{k+j+1}u_{k,m}\right\rangle = \left\langle \ell, \mathcal{Q}_{k+j+1}f + \mathcal{Q}_{k+j+1}\mathcal{K}u_{k,m-1}\right\rangle, \text{ for all } \; \ell \in \mathbb{V}_{k+j+1}, \; j \in \mathbb{Z}_m.$$

Now suppose that statement (ii) holds. Using equation (9.29), the equation above is equivalent to

$$\left\langle \ell, u_{k,m}\right\rangle = \left\langle \ell, f + \mathcal{K}u_{k,m-1}\right\rangle, \quad \text{for all } \; \ell \in \mathbb{V}_{k+j+1}, j \in \mathbb{Z}_m. \tag{9.33}$$

In terms of the bases of the spaces $\mathbb{X}_k$, $\mathbb{L}_k$, $\mathbb{W}_{k+j+1}$ and $\mathbb{V}_{k+j+1}$, $j \in \mathbb{Z}_m$ equations (9.32) and (9.33) are equivalent to the matrix equation (9.26). Conversely, if (i) holds, we can prove that equation (9.29) is satisfied and thus, (ii) holds.

The proof of (iii) implying (ii) is trivial. Statement (ii) and the nestedness assumption on $\mathbb{X}_n$ ensure the validity of (iii). Statement (iv) is the discrete version of (iii) and hence they are equivalent. Finally, the equivalence of (iv) and (v) follows from the definition of matrix $\mathbf{B}_{k,m}$. □

Note that condition (ii) in Theorem 9.4 specifies the definition of the direct sum (9.24). In other words, the space $\mathbb{V}_{n+1}$ is uniquely determined by condition (ii). From now on, we always assume that condition (ii) is satisfied to guarantee the equivalence of (9.21) and (9.26). Another way to write the equivalence conditions in Theorem 9.4 is

$$\langle \ell_{i'j'}, w_{ij} \rangle = 0, \quad i, i' \in \mathbb{N}_k, i < i', j \in \mathbb{Z}_{w(i)}, j' \in \mathbb{Z}_{w(i')}. \tag{9.34}$$

When condition (9.34) is satisfied we call the bases ℓ_{ij} and w_{ij} semi-bi-orthogonal. Under this condition, when the solution $\mathbf{u}_{k,m}$ of equation (9.26) is computed, we conclude from Theorem 9.4 that the function defined by $u_{k,m} := \mathbf{u}_{k,m}^T \mathbf{w}_{k+m}$ is the solution of the equation (9.21), where $\mathbf{w}_n = [w_{ij} : (i,j) \in \mathbb{U}_n]$. We remark that condition (9.34) is satisfied for the multiscale Galerkin methods and multiscale collocation methods developed in Chapters 5 and 7, respectively. In fact, in the case of the Galerkin method using orthogonal piecewise polynomial multiscale bases constructed in [200], the matrix $\mathbf{E}_n$ is the identity and in the case of the collocation method using interpolating piecewise polynomial multiscale bases and multiscale functionals constructed in [69], the matrix $\mathbf{E}_n$ is upper triangular with diagonal entries equal to one.

We now turn to a study of the computational complexity of Algorithm 2. Specifically, we estimate the number of multiplications used in the method. For this purpose, we rewrite equation (9.26) in block form. Letting $n := k + m$ we partition the matrix $\mathbf{E}_{k+m}$ in the same way as we have done for the matrix $\mathbf{K}_{k+m}$ to obtain blocks $\mathbf{E}_{i,i'}^k$, $i, i' \in \mathbb{Z}_{m+1}$. We also partition the vectors $\mathbf{u}_{k,m}$ and $\mathbf{f}_{k+m}$ accordingly as $\mathbf{u}_{k,m} := [\mathbf{u}_i^m : i \in \mathbb{Z}_{m+1}]$ and $\mathbf{f}_{k+m} := [\mathbf{f}_{k,i} : i \in \mathbb{Z}_{m+1}]$. Here and in what follows, we require that the appropriate bases are chosen so that $\mathbf{E}_{i',i}^k = 0$, for $0 \leq i < i' \leq m$ and $\mathbf{E}_{i,i}^k = \mathbf{I}$. With this assumption, we express the matrix $\mathbf{B}_{k,m}$ as

$$\mathbf{B}_{k,m} = \begin{bmatrix} \mathbf{I} - \mathbf{K}_k & \mathbf{E}_{0,1}^k - \mathbf{K}_{0,1}^k & \mathbf{E}_{0,2}^k - \mathbf{K}_{0,2}^k & \cdots & \mathbf{E}_{0,m-1}^k - \mathbf{K}_{0,m-1}^k & \mathbf{E}_{0,m}^k - \mathbf{K}_{0,m}^k \\ \mathbf{0} & \mathbf{I} & \mathbf{E}_{1,2}^k & \cdots & \mathbf{E}_{1,m-1}^k & \mathbf{E}_{1,m}^k \\ \mathbf{0} & \mathbf{0} & \mathbf{I} & \cdots & \mathbf{E}_{2,m-1}^k & \mathbf{E}_{2,m}^k \\ \vdots & \vdots & \vdots & \ddots & \vdots & \vdots \\ \mathbf{0} & \mathbf{0} & \mathbf{0} & \cdots & \mathbf{I} & \mathbf{E}_{m-1,m}^k \\ \mathbf{0} & \mathbf{0} & \mathbf{0} & \cdots & \mathbf{0} & \mathbf{I} \end{bmatrix}.$$

It is clear from this matrix representation that inverting matrices $\mathbf{B}_{k,m}$, $m \in \mathbb{N}_0$ is basically equivalent to inverting $\mathbf{I} - \mathbf{K}_k$. The strength of Algorithm 2 is that it *only* requires inverting the matrix $(\mathbf{I} - \mathbf{K}_k)$. Using this block form of matrix $\mathbf{B}_{k,m}$, equation (9.26) becomes

$$\mathbf{u}_i^m = \mathbf{f}_{k,i} + \sum_{j=0}^{m-1} \mathbf{K}_{i,j}^k \mathbf{u}_j^{m-1} - \sum_{j=i+1}^{m} \mathbf{E}_{i,j}^k \mathbf{u}_j^m, \quad i = m, m-1, \ldots, 1, \tag{9.35}$$

$$\hat{\mathbf{f}}_{k,0} := \mathbf{f}_{k,0} - \sum_{j=1}^{m} (\mathbf{E}_{0,j}^k - \mathbf{K}_{0,j}^k)\mathbf{u}_j^m \text{ and } \mathbf{u}_0^m = (\mathbf{I} - \mathbf{K}_k)^{-1}\hat{\mathbf{f}}_{k,0}. \tag{9.36}$$

For a matrix $\mathbf{A}$, we denote by $\mathcal{N}(\mathbf{A})$ the number of nonzero entries of $\mathbf{A}$. Note that we need $\mathcal{N}(\mathbf{K}_{k,m}^H) + \mathcal{N}(\mathbf{E}_{k+m})$ multiplications to obtain $\mathbf{u}_i^m$, $i = 1, 2, \ldots, m$ from equation (9.35). In addition, the computation of $\hat{\mathbf{f}}_{k,0}$ requires $\mathcal{N}(\mathbf{K}_{k,m}^L)$ number of multiplications. We assume that computing $\mathbf{u}_0^m$ from the second equation of (9.36) needs $\mathcal{M}(k)$ multiplications, which is constant independent of m. Hence, the number of multiplications for computing $\mathbf{u}_{k,m}$ from $\mathbf{u}_{k,m-1}$ is

$$N_{k,m} := \mathcal{N}(\mathbf{K}_{k+m}) + \mathcal{N}(\mathbf{E}_{k+m}) + \mathcal{M}(k). \tag{9.37}$$

Recall that to compute $\mathbf{u}_{k,m}$, we first compute $\mathbf{u}_k$ and then use the algorithm of (9.35) and (9.36) to compute $\mathbf{u}_{k,i}$, $i = 1, 2, \ldots, m$, successively. By formula (9.37), the total number of multiplications required to obtain $\mathbf{u}_{k,m}$ is given by

$$\mathcal{M}(k) + \sum_{i=1}^{m} N_{k,i} = (m+1)\mathcal{M}(k) + \sum_{i=1}^{m} [\mathcal{N}(\mathbf{K}_{k+i}) + \mathcal{N}(\mathbf{E}_{k+i})].$$

We now summarize the discussion above in a proposition.

Proposition 9.5 *The total number of multiplications required for computing $\boldsymbol{u}_{k,m}$ from $\boldsymbol{u}_k$ is given by*

$$(m+1)\mathcal{M}(k) + \sum_{i=1}^{m} [\mathcal{N}(\mathbf{K}_{k+i}) + \mathcal{N}(\boldsymbol{E}_{k+i})].$$

To close this section, we analyze the stability of Algorithm 2. It can be shown that if the condition number $\operatorname{cond}(\mathbf{B}_{k,m})$ of matrix $\mathbf{B}_{k,m}$ is small, then small perturbations of the matrices $\mathbf{B}_{k,m}$ and $\mathbf{C}_{k,m}$ and the vector $\bar{\mathbf{u}}_{k,m}$ only cause a small perturbation in the solution $\mathbf{u}_{k,m}$. For this reason, we study the condition number of matrix $\mathbf{B}_{k,m}$. Our theorem will confirm that the condition numbers $\operatorname{cond}(\mathbf{B}_{k,m})$ and $\operatorname{cond}(\mathbf{A}_{k+m})$ have the same order. In other words, the augmentation method will not ruin the well-condition property of the original multilevel method.

We first establish a result that the stability of $\mathcal{B}_{k,m}$ is inherited from that of $\mathcal{A}_{k+m}$.

Lemma 9.6 *Suppose that the family of operators $\mathcal{A}_n, n \in \mathbb{N}_0$ has the property that there exist positive constants c_1 and c_2 and a positive integer N_0 such that for $n \geq N_0$, $\|\mathcal{A}_n\| \leq c_1$ and $\|\mathcal{A}_n v\| \geq c_2\|v\|$, for all $v \in \mathbb{X}_n$. Moreover, suppose that for any $k, m \in \mathbb{N}_0$, $\mathcal{A}_{k+m} = \mathcal{B}_{k,m} + \mathcal{C}_{k,m}$, where $\mathcal{C}_{k,m}$ satisfies hypothesis (II). Then, there exist positive constants c_1' and c_2' and a positive integer N_1 such that for $k > N_1$, $m \in \mathbb{N}_0$, $\|\mathcal{B}_{k,m}\| \leq c_1'$ and $\|\mathcal{B}_{k,m}v\| \geq c_2'\|v\|$, for all $v \in \mathbb{X}_{k+m}$.*

Proof By the triangular inequality, we have for any $k, m \in \mathbb{N}_0$ that

$$\|\mathcal{A}_{k+m}\| - \|\mathcal{C}_{k,m}\| \leq \|\mathcal{B}_{k,m}\| \leq \|\mathcal{A}_{k+m}\| + \|\mathcal{C}_{k,m}\|.$$

The hypotheses of this lemma ensure that there exists a positive integer N' such that for $k > N'$ and $m \in \mathbb{N}_0$, $\|\mathcal{C}_{k,m}\| \leq c_2/2$. We let $N_1 := \max\{N_0, N'\}$ and observe that for $k > N_1$, $m \in \mathbb{N}_0$, $\|\mathcal{B}_{k,m}\| \leq c_1 + c_2/2$ and $\|\mathcal{B}_{k,m}v\| \geq \|\mathcal{A}_{k+m}v\| - \|\mathcal{C}_{k,m}v\| \geq \frac{c_2}{2}\|v\|$, for all $v \in \mathbb{X}_{k+m}$. By choosing $c_1' := c_1 + \frac{c_2}{2}$ and $c_2' := \frac{c_2}{2}$, we complete the proof of this lemma. □

We now return to the discussion of the condition number of matrix $\mathbf{B}_{k,m}$. To do this, we need auxiliary bases for $\mathbb{X}_0$ and $\mathbb{W}_i$, for $i \in \mathbb{N}$, which are bi-orthogonal to $\{\ell_{ij} : j \in \mathbb{Z}_{w(i)}, i \in \mathbb{N}_0\}$, that is, $\mathbb{X}_0 = \operatorname{span}\{\zeta_{0j} : j \in \mathbb{Z}_{w(0)}\}$, $\mathbb{W}_i = \operatorname{span}\{\zeta_{ij} : j \in \mathbb{Z}_{w(i)}\}$ with the bi-orthogonal property $\langle \ell_{i'j'}, \zeta_{ij}\rangle = \delta_{i'i}\delta_{j'j}$, for $i, i' \in \mathbb{N}_0, j \in \mathbb{Z}_{w(i)}, j' \in \mathbb{Z}_{w(i')}$. For any $v \in \mathbb{X}_n$, we have two representations of v given by $v = \sum_{(i,j)\in\mathbb{U}_n} v_{ij}w_{ij}$ and $v = \sum_{(i,j)\in\mathbb{U}_n} v'_{ij}\zeta_{ij}$. We let $\mathbf{v}, \mathbf{v}' \in \mathbb{R}^{d_n}$ be the vectors of the coefficients in the two representations of v, respectively, that is, $\mathbf{v} := [v_{ij} : (i,j) \in \mathbb{U}_n]$ and $\mathbf{v}' := [v'_{ij} : (i,j) \in \mathbb{U}_n]$.

Theorem 9.7 *Let $n \in \mathbb{N}_0$ and suppose that there exist functions $\mu_i(n)$, $\nu_i(n)$, $i = 1, 2$, such that for any $v \in \mathbb{X}_n$,*

$$\mu_1(n)\|\mathbf{v}\| \leq \|v\| \leq \mu_2(n)\|\mathbf{v}\|, \quad \nu_1(n)\|\mathbf{v}'\| \leq \|v\| \leq \nu_2(n)\|\mathbf{v}'\|. \tag{9.38}$$

Suppose that the hypothesis of Lemma 9.6 is satisfied. Then, there exists a positive integer N such that for any $k > N$ and any $m \in \mathbb{N}_0$,

$$\operatorname{cond}(\mathbf{A}_{k+m}) \leq \frac{c_1\mu_2(k+m)\nu_2(k+m)}{c_2\mu_1(k+m)\nu_1(k+m)} \quad \textit{and}$$

$$\operatorname{cond}(\mathbf{B}_{k,m}) \leq \frac{c_1'\mu_2(k+m)\nu_2(k+m)}{c_2'\mu_1(k+m)\nu_1(k+m)}.$$

Proof We prove the bound for both matrices at the same time. To this end, for $n := k + m$ we let $\mathbf{H}_{k,m}$ denote either $\mathbf{A}_{k+m}$ or $\mathbf{B}_{k,m}$ and $\mathcal{H}_{k,m}$ for the corresponding operator. For any $\mathbf{v} = [v_{ij} : (i,j) \in \mathbb{U}_{k+m}] \in \mathbb{R}^{d_{k+m}}$, we define a vector $\mathbf{g} = [g_{ij} : (i,j) \in \mathbb{U}_{k+m}] \in \mathbb{R}^{d_{k+m}}$ by letting $\mathbf{g} := \mathbf{H}_{k,m}\mathbf{v}$. Introducing $v = \sum_{(i,j)\in\mathbb{U}_{k+m}} v_{ij}w_{ij}$ and $g = \sum_{(i,j)\in\mathbb{U}_{k+m}} g_{ij}\zeta_{ij}$, we have the corresponding operator equation $\mathcal{H}_{k,m}v = g$. Since the hypotheses of Lemma 9.6 are satisfied, by using Lemma 9.6, there exist positive constants c_1' and c_2' and a positive integer N such that for $k > N$, $m \in \mathbb{N}_0$, $\|\mathcal{B}_{k,m}\| \leq c_1'$ and $\|\mathcal{B}_{k,m}v\| \geq c_2'\|v\|$, for all $v \in \mathbb{X}_{k+m}$. Therefore, in either case, there exist positive constants c_1 and c_2 and a positive integer N such that for $k > N$, $m \in \mathbb{N}_0$, $\|\mathcal{H}_{k,m}\| \leq c_1$ and $\|\mathcal{H}_{k,m}v\| \geq c_2\|v\|$, for all $v \in \mathbb{X}_{k+m}$. It follows from (9.38) that for any $k > N$ and $m \in \mathbb{N}_0$,

$$\|\mathbf{H}_{k,m}\mathbf{v}\| = \|\mathbf{g}\| \leq \frac{\|g\|}{\nu_1(k+m)} = \frac{\|\mathcal{H}_{k,m}v\|}{\nu_1(k+m)} \leq \frac{c_1\|v\|}{\nu_1(k+m)} \leq \frac{c_1\mu_2(k+m)}{\nu_1(k+m)}\|\mathbf{v}\|,$$

which yields $\|\mathbf{H}_{k,m}\| \leq \frac{c_1\mu_2(k+m)}{\nu_1(k+m)}$. Likewise, by (9.38) we have that for any $k > N$ and $m \in \mathbb{N}_0$,

$$\begin{aligned}\mu_1(k+m)\|\mathbf{v}\| \leq \|v\| &\leq \frac{\|\mathcal{H}_{k,m}v\|}{c_2} = \frac{\|g\|}{c_2}\\ &\leq \frac{\nu_2(k+m)}{c_2}\|\mathbf{g}\| = \frac{\nu_2(k+m)}{c_2}\|\mathbf{H}_{k,m}\mathbf{v}\|,\end{aligned}$$

which ensures that $\|\mathbf{H}_{k,m}^{-1}\| \leq \frac{\nu_2(k+m)}{c_2\mu_1(k+m)}$. Combining the estimates for $\|\mathbf{H}_{k,m}\|$ and $\|\mathbf{H}_{k,m}^{-1}\|$, we confirm the bounds of the condition numbers of $\mathbf{A}_{k+m}$ and $\mathbf{B}_{k,m}$. □

We next apply this theorem to two specific cases to obtain two useful special results.

Corollary 9.8 *For the multiscale Galerkin methods developed in Chapter 5 and the multiscale Petrov–Galerkin methods developed in Chapter 6,* $\text{cond}(\boldsymbol{A}_{k+m}) = \mathcal{O}(1)$ *and* $\text{cond}(\mathbf{B}_{k,m}) = \mathcal{O}(1)$.

Proof In these multiscale methods, we use *orthogonal* multiscale bases. Thus, the quantities $\mu_i(n)$ and $\nu_i(n)$, $i = 1,2$ appearing in (9.38) are constant independent of n. By Theorem 9.7, in these cases condition numbers $\text{cond}(\mathbf{A}_{k+m})$ and $\text{cond}(\mathbf{B}_{k,m})$ are constant independent of n. □

Corollary 9.9 *For the multiscale collocation methods developed in Chapter 7,*

$$\text{cond}(\boldsymbol{A}_{k+m}) = \mathcal{O}(\log^2 d_{k+m})$$

and

$$\mathrm{cond}(\mathbf{B}_{k,m}) = \mathcal{O}(\log^2 d_{k+m}),$$

where d_{k+m} *denotes the order of matrices* $\mathbf{A}_{k+m}$ *and* $\mathbf{B}_{k,m}$.

Proof In the multiscale collocation methods developed in Chapter 7, we have that $\mu_1(k+m) = \mathcal{O}(1)$, $\nu_1(k+m) = \mathcal{O}(1)$, $\mu_2(k+m) = \mathcal{O}(\log d_{k+m})$ and $\nu_2(k+m) = \mathcal{O}(\log d_{k+m})$. Therefore, the result of this remark follows from Theorem 9.7. □

9.1.3 Compression schemes

This section is devoted to an application of the MAM for solving linear systems resulting from compression schemes derived from multiscale methods.

We assume that a compression strategy has been applied to compress the full matrix $\mathbf{K}_n$ to obtain a sparse matrix $\tilde{\mathbf{K}}_n$, where the number of nonzero entries of $\tilde{\mathbf{K}}_n$ is of order $d_n \log^\alpha d_n$, with $\alpha = 0, 1$ or 2, that is,

$$\mathcal{N}(\tilde{\mathbf{K}}_n) = \mathcal{O}(d_n \log^\alpha d_n). \tag{9.39}$$

Methods of this type were studied in [4, 28, 64, 68, 69, 88, 94, 95, 202, 226, 241, 260, 261]. In particular, when orthogonal piecewise polynomial multiscale bases and interpolating piecewise polynomial multiscale bases, constructed respectively in [200] and [65], are used to develop the multiscale Galerkin method (see Chapter 5), the multiscale Petrov–Galerkin methods (see Chapter 6) and the multiscale collocation method (see, Chapter 7), we have that $\mathcal{N}(\tilde{\mathbf{K}}_n) = \mathcal{O}(n^\alpha \mu^n)$, where the corresponding multiscale bases are constructed with a μ-adic subdivision of the domain when the kernel $K(s,t)$, $s,t \in \Omega \subseteq \mathbb{R}^d$, of the integral operator $\mathcal{K}$ satisfies the conditions described in these papers. The compression scheme for these methods has the form

$$(\mathbf{E}_n - \tilde{\mathbf{K}}_n)\tilde{\mathbf{u}}_n = \mathbf{f}_n. \tag{9.40}$$

Equation (9.40) has an equivalent operator equation. Let $\tilde{\mathcal{K}}_n : \mathbb{X}_n \to \mathbb{X}_n$ be the linear operator relative to the basis $\{w_{ij} : (i,j) \in \mathbb{U}_n\}$ having the matrix representation $\mathbf{E}_n^{-1}\tilde{\mathbf{K}}_n$. We have that

$$(\mathcal{I} - \tilde{\mathcal{K}}_n)\tilde{u}_n = \mathcal{P}_n f, \tag{9.41}$$

where $\tilde{u}_n \in \mathbb{X}_n$ and it is related to the solution $\tilde{\mathbf{u}}_n$ of equation (9.40) by the formula $\tilde{u}_n = \tilde{\mathbf{u}}_n^T \mathbf{w}_n$, where $\mathbf{w}_n := [w_{ij} : (i,j) \in \mathbb{U}_n]$. It is known that under certain conditions for Galerkin methods and collocation methods we have that, if $u \in W^{r,p}(\Omega)$,

$$\|u - \tilde{u}_n\|_p \le c\mu^{-rn/d} n^{\alpha} \|u\|_{r,p}, \tag{9.42}$$

where $p = 2$ for the Galerkin method and $p = \infty$ for the collocation method and r denotes the order of piecewise polynomials used in these methods.

To develop the MAM for solving the operator equation (9.41), we note that $\tilde{\mathcal{K}}_n = \mathcal{P}_n \tilde{\mathcal{K}}_n$, from which equation (9.41) is rewritten as

$$(\mathcal{I} - \mathcal{P}_n \tilde{\mathcal{K}}_n)\tilde{u}_n = \mathcal{P}_n f. \tag{9.43}$$

Hence, for $n := k + m$, we have that $\tilde{\mathcal{K}}_{k+m} = \mathcal{P}_k \tilde{\mathcal{K}}_{k+m} + (\mathcal{P}_{k+m} - \mathcal{P}_k)\tilde{\mathcal{K}}_{k+m}$ and from this equation we define $\tilde{\mathcal{B}}_{k,m} := \mathcal{I}_{k+m} - \mathcal{P}_k \tilde{\mathcal{K}}_{k+m}$ and $\tilde{\mathcal{C}}_{k,m} := -(\mathcal{P}_{k+m} - \mathcal{P}_k)\tilde{\mathcal{K}}_{k+m}$. As done in Section 9.1.2, we can define the MAM for equation (9.41).

We have the following convergence result.

Theorem 9.10 *Let $u \in W^{r,p}(\Omega)$. Suppose that the estimate (9.42) holds and*

$$\lim_{n\to\infty} \|\mathcal{K}_n - \tilde{\mathcal{K}}_n\| = 0. \tag{9.44}$$

Then, there exist a positive integer N and a positive constant c such that for all $k > N$ and $m \in \mathbb{N}_0$,

$$\|u - \tilde{u}_{k,m}\| \le c\mu^{-r(k+m)/d}(k+m)^{\alpha}\|u\|_{r,p}.$$

Proof The proof employs Theorem 9.2 with a majorization sequence $\gamma_n := c\mu^{-rn/d} n^{\alpha}\|u\|_{r,p}$. We conclude that $\frac{\gamma_{n+1}}{\gamma_n} \ge \mu^{-r/d}$. In other words, γ_n is a majorization sequence of R_n with $\sigma := \mu^{-r/d}$. It is readily shown that

$$\|\tilde{\mathcal{C}}_{k,m}\| \le 2p\|\tilde{\mathcal{K}}_{k+m} - \mathcal{K}_{k+m}\| + \|\mathcal{C}_{k,m}\| + 2p^2\|(\mathcal{P}_{k+m} - \mathcal{I})\mathcal{K}\|.$$

By (9.44), we have that $\lim_{k\to\infty}\|\tilde{\mathcal{C}}_{k,m}\| = 0$ uniformly for $m \in \mathbb{N}_0$. Also, it can be verified that the other conditions of Theorem 9.2 are satisfied with both the multiscale Galerkin method and the collocation method using piecewise polynomial multiscale basis of order r. By applying Theorem 9.2, we complete the proof. □

We now formulate the matrix form of the MAM directly from the compressed matrix. Because the compressed matrix $\tilde{\mathbf{K}}_n$ inherits the multilevel structure of the matrix $\mathbf{K}_n$, we may use the MAM to solve equation (9.40) as described in Section 7.1.2. Specifically, we partition the compressed matrix $\tilde{\mathbf{K}}_n$ as done for the full matrix $\mathbf{K}_n$ in Section 7.1.2. Let

$$\tilde{\mathbf{K}}^L_{k,m} := \begin{bmatrix} \tilde{\mathbf{K}}^k_{0,0} & \tilde{\mathbf{K}}^k_{0,1} & \cdots & \tilde{\mathbf{K}}^k_{0,m} \\ 0 & 0 & \cdots & 0 \\ \vdots & \vdots & \ddots & \vdots \\ 0 & 0 & \cdots & 0 \end{bmatrix} \quad \text{and}$$

$$\tilde{\mathbf{K}}^H_{k,m} := \begin{bmatrix} 0 & 0 & \cdots & 0 \\ \tilde{\mathbf{K}}^k_{1,0} & \tilde{\mathbf{K}}^k_{1,1} & \cdots & \tilde{\mathbf{K}}^k_{1,m} \\ \vdots & \vdots & \ddots & \vdots \\ \tilde{\mathbf{K}}^k_{m,0} & \tilde{\mathbf{K}}^k_{m,1} & \cdots & \tilde{\mathbf{K}}^k_{m,m} \end{bmatrix},$$

and define $\tilde{\mathbf{B}}_{k,m} := \mathbf{E}_{k+m} - \tilde{\mathbf{K}}^L_{k,m}$ and $\tilde{\mathbf{C}}_{k,m} := -\tilde{\mathbf{K}}^H_{k,m}$.

Algorithm 3 (Matrix form of the augmentation algorithm for compression schemes) *Let k be a fixed positive integer.*

Step 1: *Solve* $\tilde{\mathbf{u}}_k \in \mathbb{R}^{d_k}$ *from the equation* $\left(\boldsymbol{E}_k - \tilde{\mathbf{K}}_k\right)\tilde{\mathbf{u}}_k = \boldsymbol{f}_k$.
Step 2: *Set* $\tilde{\mathbf{u}}_{k,0} := \tilde{\mathbf{u}}_k$ *and compute the splitting matrices* $\tilde{\mathbf{K}}^L_{k,0}$ *and* $\tilde{\mathbf{K}}^H_{k,0}$.
Step 3: *For* $m \in \mathbb{N}$, *suppose that* $\tilde{\mathbf{u}}_{k,m-1} \in \mathbb{R}^{d_{k+m-1}}$ *has been obtained and do the following.*

- *Augment the matrices* $\tilde{\mathbf{K}}^L_{k,m-1}$ *and* $\tilde{\mathbf{K}}^H_{k,m-1}$ *to form* $\tilde{\mathbf{K}}^L_{k,m}$ *and* $\tilde{\mathbf{K}}^H_{k,m}$, *respectively.*
- *Augment* $\tilde{\mathbf{u}}_{k,m-1}$ *by setting* $\bar{\tilde{\mathbf{u}}}_{k,m} := \begin{bmatrix} \tilde{\mathbf{u}}_{k,m-1} \\ 0 \end{bmatrix}$.
- *Solve* $\tilde{\mathbf{u}}_{k,m} \in \mathbb{R}^{d_{k+m}}$ *from the algebraic equations*

$$(\boldsymbol{E}_{k,m} - \tilde{\mathbf{K}}^L_{k,m})\tilde{\mathbf{u}}_{k,m} = \boldsymbol{f}_{k+m} + \tilde{\mathbf{K}}^H_{k,m}\bar{\tilde{\mathbf{u}}}_{k,m}.$$

Since $\mathbf{E}_n^{-1}\tilde{\mathbf{K}}_n$ is the matrix representation of the operator $\tilde{\mathcal{K}}_n$ relative to the basis $\{w_{ij} : (i,j) \in J_n\}$, we conclude that

$$\left\langle \ell_{i'j'}, \tilde{\mathcal{K}}_n w_{ij} \right\rangle = \sum_{(i'',j'')\in\mathbb{U}_n} (\mathbf{E}_n^{-1}\tilde{\mathbf{K}}_n)_{i''j'',ij}\mathbf{E}_{i'j',i''j''} = (\tilde{\mathbf{K}}_n)_{i'j',ij}$$

and see that the matrix form of MAM derived above is equivalent to the corresponding operator form.

In the next result, we estimate the number of multiplications used in Algorithm 3.

Theorem 9.11 *Let k be a fixed positive integer and* $m \in \mathbb{N}_0$. *Suppose that for some* $\alpha \in \{0, 1, 2\}$ *and some integer* $\mu > 1$, $\mathcal{N}(\tilde{\mathbf{K}}_n) = \mathcal{O}(n^\alpha \mu^n)$ *and suppose* *that for $n \in \mathbb{N}_0$, $\mathcal{N}(\mathbf{E}_n) \le \mathcal{N}(\tilde{\mathbf{K}}_n)$. Then, the total number of multiplications required for computing $\tilde{\mathbf{u}}_{k,m}$ from $\tilde{\mathbf{u}}_k$ is of $\mathcal{O}((m+k)^\alpha \mu^{m+k})$.*

Proof According to Proposition 9.5, the total number of multiplications required for computing $\tilde{\mathbf{u}}_{k,m}$ from $\tilde{\mathbf{u}}_k$ is given by

$$\tilde{\mathcal{N}}_{k,m} = \mathcal{O}(m) + \sum_{i \in \mathbb{N}_m} [\mathcal{N}(\tilde{\mathbf{K}}_{k+i}) + \mathcal{N}(\mathbf{E}_{k+i})].$$

Since $\mathcal{N}(\mathbf{E}_n) \le \mathcal{N}(\tilde{\mathbf{K}}_n)$, it suffices to estimate the quantity $\sum_{i \in \mathbb{N}_m} \mathcal{N}(\tilde{\mathbf{K}}_{k+i})$. To this end, we may show the identity

$$\sum_{i \in \mathbb{N}_m} (k+i)^\alpha \mu^{k+i} = \mathcal{O}((k+m)^\alpha \mu^{k+m}).$$

Using this formula and the hypotheses of this theorem, we complete the proof of this result. □

It can be verified that for the compression schemes presented in Chapters 5, 6 and 7, condition (9.44) and the assumption on Theorem 9.11 are fulfilled. Therefore, the conclusions of Theorems 9.10 and 9.11 hold for the class of methods proposed in these chapters.

9.1.4 Numerical experiments

We present in this subsection numerical examples to demonstrate the performance of the MAM associated with the multiscale Galerkin method and the multiscale collocation method. To focus on the main issue of the MAM, we first choose a second-kind integral equation on the unit interval since the augmentation method is independent of the dimension of the domain of the integral equation. Consider equation (9.19) with the integral operator $\mathcal{K}$ defined by

$$(\mathcal{K}u)(s) := \int_0^1 \log|\cos(\pi s) - \cos(\pi t)| u(t) \mathrm{d}t, \ s \in \Omega := [0,1].$$

In our numerical experiments, for convenience of comparison we choose the right-hand-side function in equation (9.19) as

$$\begin{aligned} f(s) := \sin(\pi s) + \frac{1}{\pi}[2 - (1-\cos(\pi s))\log(1-\cos(\pi s)) \\ - (1+\cos(\pi s))\log(1+\cos(\pi s))] \end{aligned}$$

so that $u(s) := \sin(\pi s)$, $s \in \Omega$ is the exact solution of the equation. We choose $\mathbb{X}_n$ as the space of piecewise linear polynomials on Ω with knots at the dyadic points $j/2^n, j = 1, 2, \ldots, 2^n - 1$. Note that the theoretical convergence order for piecewise linear approximation is 2. The following two numerical algorithms are run on a personal computer with a 600-MHz CPU and 256M memory.

Example 1 *Multiscale Galerkin methods.* In our first experiment, we consider the multiscale Galerkin method for solving equation (9.19). Choose an orthonormal basis $w_{00}(t) := 1$ and $w_{01}(t) := \sqrt{3}(2t - 1)$, for $t \in \Omega$ for $\mathbb{X}_0$ and

$$w_{10}(t) := \begin{cases} 1 - 6t, & t \in [0, \frac{1}{2}], \\ 5 - 6t, & t \in [\frac{1}{2}, 1], \end{cases} \quad w_{11}(t) := \begin{cases} \sqrt{3}(1 - 4t), & t \in [0, \frac{1}{2}], \\ \sqrt{3}(4t - 3), & t \in [\frac{1}{2}, 1] \end{cases}$$

for $\mathbb{W}_1$. An orthonormal basis w_{ij}, $j = 0, 1, \ldots, 2^i$ for $\mathbb{W}_i$ is constructed according to the construction given in Section 4.3. We choose $\mathcal{P}_n = \hat{\mathcal{P}}_n$ as the orthogonal projection mapping $L^2(\Omega)$ onto $\mathbb{X}_n$ and $\ell_{ij} = w_{ij}$.

In this case, $\mathbf{E}_n$ is an identity matrix because of the orthogonality of the basis and the matrix $\mathbf{K}_n$ is truncated to form the compressed matrix $\tilde{\mathbf{K}}_n$ according to a strategy presented in Chapter 5. The MAM is then applied to the compressed linear system.

In Table 9.1, we report the approximation error and convergence order of the numerical solution obtained from the MAM and the computing time for solving the linear system in the Galerkin case, where we use the initial level $k = 5$. They confirm our theoretical estimates. In all tables presented here, we use “Approx. order” for the computed approximation order of the numerical solution $\tilde{u}_{k,m}$ approximating the exact solution u, “Comp. rate” for the compression rate of the compressed coefficient matrix $\tilde{\mathbf{K}}_n$ and “CT” for the computing time (measured in seconds) used to solve the linear system using the augmentation method.

Table 9.1 *Convergence and computational speed for the Galerkin method*

m	$\|u - \tilde{u}_{k,m}\|_{L^2}$	Approx. order	Comp. rate	CT
0	1.02e-3		1.000	< 0.01
1	2.56e-4	1.99	0.688	< 0.01
2	6.36e-5	2.00	0.479	< 0.01
3	1.59e-5	2.00	0.312	0.01
4	4.22e-6	1.91	0.193	0.02
5	1.11e-6	1.92	0.116	0.03

Table 9.2 *Condition numbers of the matrices from the Galerkin method*

m	$\text{cond}(\tilde{\mathbf{A}}_{5+m})$	Δ_1	$\text{cond}(\tilde{\mathbf{B}}_{5,m})$	Δ_2
0	1.955			
1	1.977	0.022310	2.003	
2	1.989	0.011255	2.003	0.000328
3	1.994	0.005669	2.003	0.000040
4	1.997	0.002843	2.003	0.000006
5	1.999	0.001416	2.003	0.000001

We report the spectral condition numbers of $\tilde{\mathbf{A}}_{k+m}$ and $\tilde{\mathbf{B}}_{k,m}$ in Table 9.2. We use the notations $\Delta_1 := \text{cond}(\tilde{\mathbf{A}}_{k+m}) - \text{cond}(\tilde{\mathbf{A}}_{k+m-1})$ and $\Delta_2 := \text{cond}(\tilde{\mathbf{B}}_{k,m}) - \text{cond}(\tilde{\mathbf{B}}_{k,m-1})$.

Example 2 *The multiscale collocation method.* In this case, we choose $\mathcal{P}_n$ and $\hat{\mathcal{P}}_n$, respectively, as the interpolatory and orthogonal projection onto $\mathbb{X}_n$ and define the subspaces $\mathbb{W}_n$ by the orthogonal projection $\hat{\mathcal{P}}_n$. Specifically, we choose $\mathbb{X}_0 := \text{span}\{w_{00}, w_{01}\}$, where $w_{00}(t) := 2 - 3t$ and $w_{01}(t) := -1 + 3t$, $t \in \Omega$ and $\mathbb{W}_1 := \text{span}\{w_{10}, w_{11}\}$, where

$$w_{10}(t) := \begin{cases} 1 - \frac{9}{2}t, & t \in [0, \frac{1}{2}], \\ -1 + \frac{3}{2}t, & t \in [\frac{1}{2}, 1], \end{cases} \qquad w_{11}(t) := \begin{cases} \frac{1}{2} - \frac{3}{2}t, & t \in [0, \frac{1}{2}], \\ -\frac{7}{2} + \frac{9}{2}t, & t \in [\frac{1}{2}, 1]. \end{cases}$$

We also need multiscale collocation functionals for the multiscale collocation method. To this end, for any $s \in \Omega$, we use δ_s to denote the linear functional in $(C(\Omega))^*$ defined for $x \in C(\Omega)$ by the equation $\langle \delta_s, x \rangle = x(s)$. We choose the spaces of functionals as $\mathbb{L}_0 := \text{span}\{\ell_{00}, \ell_{01}\}$, where $\ell_{00} := \delta_{\frac{1}{3}}$ and $\ell_{01} := \delta_{\frac{2}{3}}$ and $\mathbb{V}_1 := \text{span}\{\ell_{10}, \ell_{11}\}$, where

$$\ell_{10} := \delta_{\frac{1}{6}} - \frac{3}{2}\delta_{\frac{1}{3}} + \frac{1}{2}\delta_{\frac{2}{3}} \text{ and } \ell_{10} := \frac{1}{2}\delta_{\frac{1}{3}} - \frac{3}{2}\delta_{\frac{2}{3}} + \delta_{\frac{5}{6}}.$$

Bases w_{ij} for $\mathbb{W}_i$ and ℓ_{ij} for $\mathbb{V}_i$ are constructed according to the description in Chapter 7. We remark that by construction, $\mathbb{V}_{l+1} \perp \mathbb{X}_l$ for all $l \in \mathbb{N}_0$. In this case $\mathbf{E}_n$ is upper triangular with diagonal entries all one. The matrix $\mathbf{K}_n$ is truncated to form the compressed matrix $\tilde{\mathbf{K}}_n$ according to the strategy presented in Chapter 7. Again, the MAM is applied to the compressed linear system.

The numerical results of the collocation method are shown in Table 9.3, where we use again the initial level $k = 5$.

We also use Table 9.1 and 9.3 to demonstrate the applicability of the MAM to multiscale collocation methods. We begin with an initial approximate

Table 9.3 *Convergence and computational speed for the collocation method*

m	Matrix size	$\|u - \tilde{u}_{k,m}\|_{L^\infty}$	Approx. order	Comp. rate	CT
0	64	3.91e-3		0.891	< 0.01
1	128	1.11e-3	1.81	0.766	< 0.01
2	256	2.89e-4	1.94	0.546	< 0.01
3	512	6.47e-5	2.16	0.356	0.010
4	1024	1.45e-5	2.15	0.220	0.020
5	2048	3.58e-6	2.02	0.131	0.041
6	4096	8.81e-7	2.02	0.076	0.100
7	8192	2.20e-7	2.00	0.044	0.240
8	16384	5.80e-8	1.92	0.024	0.581
9	32768	1.42e-8	2.02	0.014	1.342
10	65536	3.66e-9	1.96	0.007	2.994

Table 9.4 *Condition numbers of the matrices from the collocation method*

m	cond($\tilde{\mathbf{A}}_{5+m}$)	Δ_1	cond($\tilde{\mathbf{B}}_{5,m}$)	Δ_2
0	7.742223			
1	9.446764	1.704541	9.383285	
2	11.096629	1.649865	11.003597	1.620312
3	12.695555	1.598926	12.589373	1.585776
4	14.256352	1.560797	14.151313	1.561940
5	15.789732	1.533380	15.690938	1.539625

solution $\tilde{u}_5$. The numerical results show that the accuracy is improved in exactly the same order as our theoretical result, as we move from a level to a higher level using the augmentation method. In this experiment, at each level we perform a uniform subdivision since the solution is smooth.

In Table 9.4, we list the spectral condition numbers of matrices $\tilde{\mathbf{A}}_{k+m}$ and $\tilde{\mathbf{B}}_{k,m}$ for the collocation method. We also compute the differences of the condition numbers of these matrices between two consecutive levels to observe the growth of the condition numbers. The purpose of this experiment is to confirm the stability analysis presented in Section 9.1.2. Note that theoretically, we have that $\text{cond}(\tilde{\mathbf{A}}_{k+m}) = \mathcal{O}(\log^2 d_{k+m})$ and $\text{cond}(\tilde{\mathbf{B}}_{k,m}) = \mathcal{O}(\log^2 d_{k+m})$. Our numerical results coincide with the theoretical estimates obtained in Section 9.1.3. This experiment shows that the MAM does not ruin the stability of the original multilevel scheme (9.40). Normally, the original multilevel scheme is well conditioned since the use of multiscale bases leads to a preconditioner. A good discrete method should preserve this nice feature of equation (9.40). From our numerical results, we see that the condition number of $\tilde{\mathbf{B}}_{k,m}$ is smaller than that of $\tilde{\mathbf{A}}_{k+m}$.

Next, we consider a boundary integral equation resulting from a reformulation of a boundary value problem of the Laplace equation.

Example 3 *The multiscale collocation method applied to boundary integral equations*. In this experiment, we apply the multiscale collocation method to solving the boundary integral equation reformulated from the boundary value problem

$$\begin{cases} \Delta u(x) = 0, & x \in \Omega, \\ \dfrac{\partial u(x)}{\partial \mathbf{n}_x} = -u(x) + g_0(x), & x \in \Gamma, \end{cases} \tag{9.45}$$

where the domain

$$\Omega := \left\{ x = (x_1, x_2) : x_1^2 + \frac{x_2^2}{4} < 1 \right\},$$

with Γ being its boundary. We choose the function g_0 so that

$$u_0(x) := e^{x_1} \cos(x_2), \quad x \in \Omega$$

is the exact solution of equation (9.45). Here, we note that the solution u is analytic. By the fundamental solution of the equation, we reformulate the boundary value problem as the boundary integral equation (cf. Section 2.1.3)

$$\begin{aligned} u(x) &- \frac{1}{\pi} \int_\Gamma u(y) \frac{\partial}{\partial \mathbf{n}_y} \log |x - y| \mathrm{d}s_y - \frac{1}{\pi} \int_\Gamma u(y) \log |x - y| \mathrm{d}s_y \\ &= -\frac{1}{\pi} \int_\Gamma g_0(y) \log |x - y| \mathrm{d}s_y, \ x \in \Gamma. \end{aligned} \tag{9.46}$$

Utilizing parametrization $x(t) := (\cos(2\pi t), 2\sin(2\pi t))$, $t \in [0, 1]$ on the boundary Γ, we establish the boundary integral equation

$$\begin{aligned} u(t) &- \int_0^1 K(t, \tau) u(\tau) \mathrm{d}\tau - \int_0^1 L(t, \tau) u(\tau) \mathrm{d}\tau \\ &= -\int_0^1 L(t, \tau) g_0(\tau) \mathrm{d}\tau, \quad t \in [0, 1), \end{aligned} \tag{9.47}$$

where the kernels

$$K(t, \tau) := \frac{2}{1 + 3\cos^2(t + \tau)\pi}$$

and

$$L(t, \tau) := 2\sqrt{1 + 3\cos^2(2\pi\tau)} \log \left(2 \sin |\pi(t - \tau)| \sqrt{1 + 3\cos^2 \pi(t + \tau)} \right)$$

Table 9.5 *Convergence and computational speed for the collocation method*

m	d_{4+m}	$\|u - \tilde{u}_{4,m}\|_\infty$	Approx. order	Comp. rate 1	Comp. rate 2	CT
0	64	4.56e-2		0.63	0.75	
1	128	1.390e-2	1.71	0.573	0.516	0.001
2	256	3.37e-3	2.04	0.343	0.266	0.002
3	512	8.09e-4	2.06	0.205	0.137	0.004
4	1024	1.82e-4	2.15	0.121	0.070	0.008
5	2048	4.46e-5	2.03	0.071	0.036	0.014
6	4096	1.08e-5	2.04	0.041	0.019	0.029
7	8192	2.60e-6	2.05	0.023	0.010	0.059
8	16384	5.91e-7	2.14	0.013	0.005	0.122
9	32768	1.19e-7	2.31	0.007	0.003	0.250

Table 9.6 *Condition numbers of the matrices from the collocation method*

m	d_{4+m}	cond($\tilde{\mathbf{A}}_{5+m}$)	Δ_1	cond($\tilde{\mathbf{B}}_{5,m}$)	Δ_2
0	64	10.2			
1	128	12.0	1.8	11.9	
2	256	13.7	1.7	13.5	1.6
3	512	15.2	1.6	15.0	1.5
4	1024	16.7	1.5	16.5	1.5
5	2048	18.2	1.4	18.0	1.5

are smooth and weakly singular, respectively. The solution of (9.45) corresponding to u_0 is

$$u(t) = e^{\cos(2\pi t)} \cos(2\sin(2\pi t)).$$

We use the multiscale collocation method described in Example 2 to solve the boundary integral equation (9.47). The programs are run on a personal computer with a 3.40-GHz CPU and 8.00GB memory. The numerical results are shown in Tables 9.5 and 9.6, where we use "*Comp. rate 1*" and "*Comp. rate 2*" for the compression rates of the compressed coefficient matrices for the weakly singular and smooth kernels, respectively. The numerical results confirm the theoretical estimates.

In Table 9.6, we list the spectral condition numbers of matrices $\tilde{\mathbf{A}}_{k+m}$ and $\tilde{\mathbf{B}}_{k,m}$ for the collocation method. The differences of the condition numbers of these matrices between two consecutive levels are also computed to show the growth of the condition numbers. The numerical results confirm the stability analysis presented in Section 9.1.2.

Both theoretical analysis and numerical experiment show that the MAM is particularly suitable for solving large-scale linear systems that result from compression schemes applied to integral equations. It is a stable and fast algorithm, which provides accurate numerical solutions of integral equations.

9.2 Multilevel iteration methods

In this section we develop MIMs for solving Fredholm integral equations of the second kind based on the multiscale methods for which the subspace has a multiresolution decomposition. To describe the multilevel iteration schemes and ideas leading to these algorithms, we utilize multiscale Galerkin methods.

9.2.1 Multilevel iteration schemes

Let $\mathbb{X}$ be a Hilbert space and $\mathcal{K} : \mathbb{X} \to \mathbb{X}$ an operator such that $\mathcal{I} - \mathcal{K}$ is bijective on $\mathbb{X}$. Consider the Fredholm integral equation of the second kind given in the form

$$(\mathcal{I} - \mathcal{K})u = f, \tag{9.48}$$

where $f \in \mathbb{X}$ is given and $u \in \mathbb{X}$ is the solution to be determined.

To solve equation (9.48) by the Galerkin method, let $\mathbb{X}_n$ be a nested sequence of finite-dimensional subspaces of $\mathbb{X}$:

$$\mathbb{X}_n \subseteq \mathbb{X}_{n+1}, \quad n \in \mathbb{N}_0, \tag{9.49}$$

such that $\overline{\cup_{n\in\mathbb{N}_0}\mathbb{X}_n} = \mathbb{X}$. Let $\{\mathcal{P}_n : \mathbb{X} \to \mathbb{X}_n\}$ denote the sequence of orthogonal projections. Then $\|\mathcal{P}_n\| = 1, \mathbb{X}_n = \mathcal{P}_n\mathbb{X}, \mathcal{P}_n^* = \mathcal{P}_n$ and $\mathcal{P}_n \to \mathcal{I}$ pointwise in $\mathbb{X}$. The Galerkin method is to find $u_n \in \mathbb{X}_n$ satisfying the equation

$$(\mathcal{I} - \mathcal{P}_n\mathcal{K})u_n = \mathcal{P}_n f. \tag{9.50}$$

We assume that $\mathcal{K}$ is a compact operator on $\mathbb{X}$. In this case, the pointwise convergence of $\mathcal{P}_n$ to $\mathcal{I}$ and the compactness of $\mathcal{K}$ and $\mathcal{K}^*$ lead to

$$\lim_{n\to\infty} \|\mathcal{K} - \mathcal{P}_n\mathcal{K}\| = \lim_{n\to\infty} \|\mathcal{K} - \mathcal{K}\mathcal{P}_n\| = 0,$$

which implies for all n large enough that $\mathcal{I} - \mathcal{P}_n\mathcal{K}$ is invertible and its inverse is bounded by a constant independent of n. Therefore, equation (9.50) has a unique solution $u_n \in \mathbb{X}_n$ which satisfies the estimates

$$\|u_n\| \leq c\|f\|, \quad \|u - u_n\| \leq c\inf\{\|u - v\| : v \in \mathbb{X}_n\}, \tag{9.51}$$

where u is the solution of equation (9.48).

Next we develop our MIMs. By the nestedness property (9.49) of the approximating subspace sequence, each $\mathbb{X}_n$ represents a level of resolution in approximating $\mathbb{X}$ and we solve $u_n \in \mathbb{X}_n$ in (9.50), seeking an approximation of up to the nth level of resolution to the exact solution $u \in \mathbb{X}$. This nestedness property also implies that there exists a subspace $\mathbb{W}_{n+1}$ of $\mathbb{X}_{n+1}$ such that

$$\mathbb{X}_{n+1} = \mathbb{X}_n \oplus^{\perp} \mathbb{W}_{n+1},\ n \in \mathbb{N}_0. \tag{9.52}$$

Let

$$\mathcal{Q}_{n+1} := \mathcal{P}_{n+1} - \mathcal{P}_n.$$

It is then straightforward to show that $\mathcal{Q}_n$ is a projection onto $\mathbb{W}_n$ and $\mathbb{W}_{n+1} = \mathcal{Q}_{n+1}\mathbb{X}_{n+1}, n \in \mathbb{N}_0$. Repeatedly using equation (9.52) produces, for $k, m \in \mathbb{N}_0$, the following decomposition of the space $\mathbb{X}_n$ with $n = k + m$:

$$\mathbb{X}_{k+m} = \mathbb{X}_k \oplus^{\perp} \mathbb{W}_{k+1} \oplus^{\perp} \cdots \oplus^{\perp} \mathbb{W}_{k+m}. \tag{9.53}$$

For $u_{k+m} \in \mathbb{X}_{k+m}$ we write

$$u_{k+m} = u_{k,0} + v_{k+1} + \cdots + v_{k+m},$$

where $u_{k,0} \in \mathbb{X}_k$ and $v_{k,l} \in \mathbb{W}_{k+l}, l \in \mathbb{N}_m$ and

$$\mathcal{P}_{k+m} = \mathcal{P}_k + \mathcal{Q}_{k+1} + \cdots + \mathcal{Q}_{k+m}.$$

For $k, m \in \mathbb{N}_0$, we can identify the operator $\mathcal{K}_{k+m} = \mathcal{P}_{k+m}\mathcal{K}\mathcal{P}_{k+m}$ with the matrix of operators

$$\mathcal{K}_{k,m} := \begin{bmatrix} \mathcal{P}_k\mathcal{K}\mathcal{P}_k & \mathcal{P}_k\mathcal{K}\mathcal{Q}_{k+1} & \cdots & \mathcal{P}_k\mathcal{K}\mathcal{Q}_{k+m} \\ \mathcal{Q}_{k+1}\mathcal{K}\mathcal{P}_k & \mathcal{Q}_{k+1}\mathcal{K}\mathcal{Q}_{k+1} & \cdots & \mathcal{Q}_{k+1}\mathcal{K}\mathcal{Q}_{k+m} \\ \vdots & \vdots & \ddots & \vdots \\ \mathcal{Q}_{k+m}\mathcal{K}\mathcal{P}_k & \mathcal{Q}_{k+m}\mathcal{K}\mathcal{Q}_{k+1} & \cdots & \mathcal{Q}_{k+m}\mathcal{K}\mathcal{Q}_{k+m} \end{bmatrix}. \tag{9.54}$$

Thus we can express the operator equation (9.50) in the form

$$\mathbf{u}_{k,m} - \mathcal{K}_{k,m}\mathbf{u}_{k,m} = \mathbf{f}_{k,m}, \tag{9.55}$$

where $\mathbf{u}_{k,m} := [u_{k,0}, v_{k,1}, \ldots, v_{k,m}]^T$ and $\mathbf{f}_{k,m} := [\mathcal{P}_k f, \mathcal{Q}_{k+1} f, \ldots, \mathcal{Q}_{k+m} f]^T$.

As in (9.50), we seek the solution to (9.55) in the approximation space $\mathbb{X}_{k+m}$. Once the bases for $\mathbb{X}_k$ and $\mathbb{W}_{k+l}$ are chosen, the matrix representation of $\mathcal{K}_{k+m}$ has exactly the same block structure as $\mathcal{K}_{k,m}$ in (9.54).

To develop multilevel iteration schemes for solving equation (9.55), we introduce five matrices of operators from $\mathcal{K}_{k,m}$: the strictly upper triangular matrix

$$\mathcal{U}_{k,m} := \begin{bmatrix} \mathcal{O} & \mathcal{P}_k\mathcal{K}\mathcal{Q}_{k+1} & \mathcal{P}_k\mathcal{K}\mathcal{Q}_{k+2} & \cdots & \mathcal{P}_k\mathcal{K}\mathcal{Q}_{k+m} \\ & \mathcal{O} & \mathcal{Q}_{k+1}\mathcal{K}\mathcal{Q}_{k+1} & \cdots & \mathcal{Q}_{k+1}\mathcal{K}\mathcal{Q}_{k+m} \\ & & \ddots & \ddots & \vdots \\ & & & \ddots & \mathcal{Q}_{k+m-1}\mathcal{K}\mathcal{Q}_{k+m} \\ & & & & \mathcal{O} \end{bmatrix},$$

the lower triangular matrix

$$\mathcal{L}_{k,m} := \begin{bmatrix} \mathcal{O} & & & \\ \mathcal{Q}_{k+1}\mathcal{K}\mathcal{P}_k & \mathcal{Q}_{k+1}\mathcal{K}\mathcal{Q}_{k+1} & & \\ \vdots & \vdots & \ddots & \\ \mathcal{Q}_{k+m}\mathcal{K}\mathcal{P}_k & \mathcal{Q}_{k+m}\mathcal{K}\mathcal{Q}_{k+1} & \cdots & \mathcal{Q}_{k+m}\mathcal{K}\mathcal{Q}_{k+m} \end{bmatrix},$$

the block diagonal matrix

$$\mathcal{D}_{k,m} := \begin{bmatrix} \mathcal{I} - \mathcal{P}_k\mathcal{K}\mathcal{P}_k & & & \\ & \mathcal{I} & & \\ & & \ddots & \\ & & & \mathcal{I} \end{bmatrix}$$

and the matrices corresponding to lower and higher frequency of $\mathcal{K}_{k,m}$, respectively:

$$\mathcal{K}^L_{k,m} := \begin{bmatrix} \mathcal{P}_k\mathcal{K}\mathcal{P}_k & \mathcal{P}_k\mathcal{K}\mathcal{Q}_{k+1} & \cdots & \mathcal{P}_k\mathcal{K}\mathcal{Q}_{k+m} \\ \mathcal{O} & \mathcal{O} & \cdots & \mathcal{O} \\ \vdots & \vdots & \ddots & \vdots \\ \mathcal{O} & \mathcal{O} & \cdots & \mathcal{O} \end{bmatrix},$$

$$\mathcal{K}^H_{k,m} := \begin{bmatrix} \mathcal{O} & \mathcal{O} & \cdots & \mathcal{O} \\ \mathcal{Q}_{k+1}\mathcal{K}\mathcal{P}_k & \mathcal{Q}_{k+1}\mathcal{K}\mathcal{Q}_{k+1} & \cdots & \mathcal{Q}_{k+1}\mathcal{K}\mathcal{Q}_{k+m} \\ \vdots & \vdots & \ddots & \vdots \\ \mathcal{Q}_{k+m}\mathcal{K}\mathcal{P}_k & \mathcal{Q}_{k+m}\mathcal{K}\mathcal{Q}_{k+1} & \cdots & \mathcal{Q}_{k+m}\mathcal{K}\mathcal{Q}_{k+m} \end{bmatrix}.$$

Equation (9.55) can be rewritten as

$$(\mathcal{D}_{k,m} - \mathcal{U}_{k,m} - \mathcal{L}_{k,m})u_{k,m} = f_{k,m}, \tag{9.56}$$

or

$$(\mathcal{I} - \mathcal{K}^L_{k,m} - \mathcal{K}^H_{k,m})u_{k,m} = f_{k,m}. \tag{9.57}$$

It is these matrix forms that lead us to the following multilevel iteration schemes for solving equation (9.50).

- Jacobi-type iteration:

$$\mathcal{D}_{k,m}u_{k,m}^{(l+1)} = (\mathcal{U}_{k,m} + \mathcal{L}_{k,m})u_{k,m}^{(l)} + f_{k,m}, \quad l \in \mathbb{N}_0 \tag{9.58}$$

with any initial approximation $u_{k,m}^{(0)}$.

Except for the first component, all the other components of $u_{k,m}^{(l+1)}$ are already expressed in terms of $u_{k,m}^{(l)}$ since the diagonal blocks in $\mathcal{D}_{k,m}$ are identity operators. Only the first component needs to be solved by inverting $\mathcal{I} - \mathcal{P}_k\mathcal{K}\mathcal{P}_k$, which is also the difference between this iteration scheme and the usual algebraic Jacobi scheme.

- Gauss–Seidel-type iteration:

$$(\mathcal{D}_{k,m} - \mathcal{U}_{k,m})u_{k,m}^{(l+1)} = \mathcal{L}_{k,m}u_{k,m}^{(l)} + f_{k,m}, \quad l \in \mathbb{N}_0 \tag{9.59}$$

with any initial approximation $u_{k,m}^{(0)}$.

Like the usual algebraic Gauss–Seidel iteration, (9.59) can easily be solved in the "backward substitutions" fashion for the components of $u_{k,m}^{(l+1)}$ and the only inverse we need to find is for the first component, that is, $(\mathcal{I} - \mathcal{P}_k\mathcal{K}\mathcal{P}_k)^{-1}$.

- L–H-type iteration:

$$(\mathcal{I} - \mathcal{K}_{k,m}^{L})u_{k,m}^{(l+1)} = \mathcal{K}_{k,m}^{H}u_{k,m}^{(l)} + f_{k,m}, \quad l \in \mathbb{N}_0. \tag{9.60}$$

Like the Jacobi scheme, only the first component of $u_{k,m}^{(l+1)}$ needs to be solved by inverting $\mathcal{I} - \mathcal{P}_k\mathcal{K}\mathcal{P}_k$, while other components are computed directly.

It is clear that these three schemes are similar in the method of updating the components of $u_{k,m}^{(l+1)}$, they differ only in that the Gauss–Seidel scheme uses newly available components within each iteration, the L–H scheme uses the components obtained in the last iteration except for the first component, while the Jacobi scheme waits till an iteration is complete. The three methods require inverting the operator $\mathcal{I} - \mathcal{P}_k\mathcal{K}\mathcal{P}_k$ at the kth level. Hence, we have to choose k so that the inverse of $\mathcal{I} - \mathcal{P}_k\mathcal{K}\mathcal{P}_k$ exists and is relatively easy to find. In the meantime, we can take advantage of this to start the iteration with a good initial guess $u_{k,m}^{(0)} := [(\mathcal{I} - \mathcal{P}_k\mathcal{K}\mathcal{P}_k)^{-1}\mathcal{P}_k f, 0, \ldots, 0]^T$.

9.2.2 Convergence analysis

We now provide convergence analysis for iteration schemes (9.58), (9.59) and (9.60), when the operator $\mathcal{K}$ is compact on $\mathbb{X}$. In this case, we have that

$$\lim_{n\to\infty} \|(\mathcal{P}_{n+i} - \mathcal{P}_{n+j})\mathcal{K}\| = \lim_{n\to\infty} \|\mathcal{K}(\mathcal{P}_{n+i} - \mathcal{P}_{n+j})\| = 0. \tag{9.61}$$

Moreover, for n large enough, $\mathcal{I} - \mathcal{K}_n$ is invertible and the norm of the inverse operator is bounded uniformly for all n. In the following theorem, we establish convergence for schemes (9.58)–(9.60) for k large enough, and the rate of convergence.

Theorem 9.12 *Suppose that the linear operator $\mathcal{K}$ is compact on the Hilbert space $\mathbb{X}$, $\mathcal{I}-\mathcal{K}$ is bijective on $\mathbb{X}$, and the sequence of approximation subspaces possesses the nestedness (9.49) and has the decomposition (9.53). Then the following statements hold.*

(i) For each $m \in \mathbb{N}$ and sufficiently large k, the iteration schemes (9.58), (9.59) and (9.60) for solving equation (9.50) or (9.55) are all convergent.

(ii) The convergence rates for the Jacobi iteration scheme (9.58) and the L–H iteration scheme (9.60) are independent of m for all sufficiently large k. If the hypotheses

$$\sum_{k\in\mathbb{N}_0} \|(\mathcal{I} - \mathcal{P}_k)\mathcal{K}\| \leq \infty, \quad \sum_{k\in\mathbb{N}_0} \|\mathcal{K}(\mathcal{I} - \mathcal{P}_k)\| \leq \infty \tag{9.62}$$

and

$$\sum_{k\to\infty} k\|(\mathcal{I} - \mathcal{P}_k)\mathcal{K}\| = \sum_{k\to\infty} k\|\mathcal{K}(\mathcal{I} - \mathcal{P}_k)\| = 0 \tag{9.63}$$

hold, the rate of convergence for the Gauss–Seidel iteration is also independent of m for all sufficiently large k.

(iii) The convergence rates for the Jacobi iteration scheme (9.58) and the L–H iteration scheme (9.60) can be made small (by increasing k) in the same order of approximation $\mathcal{P}_k\mathcal{K}$ to $\mathcal{K}$. Moreover, if hypotheses (9.62) and (9.63) are satisfied, the rate of convergence for the Gauss–Seidel scheme can be made smaller (by increasing k) in the same order as r_k going to zero, where $r_k := \sum_{i\in\mathbb{N}_0} \|(\mathcal{I} - \mathcal{P}_{k+i})\mathcal{K}\|$.

Proof We first estimate the norms of iteration operators $\mathcal{D}_{k,m}^{-1}(\mathcal{U}_{k,m} + \mathcal{L}_{k,m})$, $(\mathcal{D}_{k,m} - \mathcal{U}_{k,m})^{-1}\mathcal{L}_{k,m}$ and $(\mathcal{I} - \mathcal{K}_{k,m}^L)^{-1}\mathcal{K}_{k,m}^H$ for schemes (9.58), (9.59) and (9.60). From the definition of $\mathcal{U}_{k,m}$, we see that

$$\mathcal{U}_{k,m}u = \mathcal{P}_k\mathcal{K}(\mathcal{P}_{k+m} - \mathcal{P}_k)u + \sum_{l\in\mathbb{N}_{m-1}} \mathcal{Q}_{k+l}\mathcal{K}(\mathcal{P}_{k+m} - \mathcal{P}_{k+l})u$$

for all $u \in \mathbb{X}$, which leads to

$$\|\mathcal{U}_{k,m}\| \leq \sum_{l\in\mathbb{Z}_m} \|\mathcal{K}(\mathcal{P}_{k+m} - \mathcal{P}_{k+l})\| \to 0 \text{ as } k \to \infty, \tag{9.64}$$

where we have used (9.61) and the fact that $\|\mathcal{P}_n\| = \|\mathcal{Q}_n\| = 1$ for all $n \in \mathbb{N}$. Similarly, we have that

$$\mathcal{L}_{k,m}u = (\mathcal{P}_{k+m} - \mathcal{P}_k)\mathcal{K}\mathcal{P}_k u + \sum_{l\in\mathbb{N}_{m-1}} (\mathcal{P}_{k+m} - \mathcal{P}_{k+l})\mathcal{K}\mathcal{Q}_{k+l+1}u$$

and

$$\|\mathcal{L}_{k,m}\| \leq \sum_{l\in\mathbb{Z}_m} \|(\mathcal{P}_{k+m} - \mathcal{P}_{k+l})\mathcal{K}\| \to 0 \text{ as } k \to \infty.$$

Moreover, from $\mathcal{U}_{k+m} + \mathcal{L}_{k+m} = \mathcal{K}_{k+m} - \mathcal{K}_k$ we conclude that

$$\|\mathcal{U}_{k+m} + \mathcal{L}_{k+m}\| \leq \|(\mathcal{P}_{k+m} - \mathcal{P}_k)\mathcal{K}\| + \|\mathcal{K}(\mathcal{P}_{k+m} - \mathcal{P}_k)\| \to 0 \text{ as } k \to \infty. \tag{9.65}$$

Moreover, as an operator on $\mathbb{X}$, $\mathcal{D}_{k,m} = \mathcal{I} - \mathcal{K}_k$. Hence

$$\mathcal{D}_{k,m}^{-1} = (\mathcal{I} - \mathcal{K}_k)^{-1} \tag{9.66}$$

and

$$(\mathcal{D}_{k,m} - \mathcal{U}_{k,m})^{-1} = (\mathcal{I} - (\mathcal{I} - \mathcal{K}_k)^{-1}\mathcal{U}_{k,m})^{-1}(\mathcal{I} - \mathcal{K}_k)^{-1}.$$

Therefore,

$$\|\mathcal{D}_{k,m}^{-1}\| = \|(\mathcal{I} - \mathcal{K}_k)^{-1}\|$$

and

$$\|(\mathcal{D}_{k,m} - \mathcal{U}_{k,m})^{-1}\| \leq \frac{\|(\mathcal{I} - \mathcal{K}_k)^{-1}\|}{1 - \|(\mathcal{I} - \mathcal{K}_k)^{-1}\|\|\mathcal{U}_{k,m}\|}.$$

Since $\|(\mathcal{I}-\mathcal{K}_k)^{-1}\|$ is uniformly bounded for large enough k, in view of (9.64), these two inverses are both bounded uniformly for large enough k.

Combining the above, we see that, for each fixed $m \in \mathbb{N}$, the iteration operators for schemes (9.58) and (9.59) are both convergent to 0 in the operator norm:

$$\|\mathcal{D}_{k,m}^{-1}(\mathcal{U}_{k+m} + \mathcal{L}_{k+m})\| \to 0 \text{ and } \|(\mathcal{I} - \mathcal{K}_k)^{-1}\mathcal{L}_{k,m}\| \to 0, \text{ as } k \to \infty.$$

Hence they can be chosen less than one if k is large enough, which in turn yields the convergence of the iteration schemes (9.58) and (9.59).

For the scheme (9.60), noting that $\mathcal{K}_{k,m}^L = \mathcal{P}_k\mathcal{K}\mathcal{P}_{k+m}$ and $\mathcal{K}_{k,m}^H = (\mathcal{P}_{k+m} - \mathcal{P}_k)\mathcal{K}\mathcal{P}_{k+m}$, we have that

$$\|\mathcal{K}_{k,m}^L - \mathcal{K}\| \to 0 \text{ and } \|\mathcal{K}_{k,m}^H\| \to 0, \text{ as } k \to \infty,$$

uniformly for $m \in \mathbb{N}$. This leads to the fact that $\|(\mathcal{I} - \mathcal{K}_{k,m}^L)^{-1}\|$ is uniformly bounded, and moreover,

$$\|(\mathcal{I} - \mathcal{K}_{k,m}^L)^{-1}\mathcal{K}_{k,m}^H\| \to 0, \text{ as } k \to \infty,$$

which also yields the convergence of the iteration scheme (9.60).

Next we examine the rate of convergence of the iteration schemes. Let

$$q_{k,m} := \begin{cases} \|\mathcal{D}_{k,m}^{-1}(\mathcal{U}_{k+m} + \mathcal{L}_{k+m})\| & \text{for scheme (9.58)},\\ \|(\mathcal{D}_{k,m} - \mathcal{U}_{k,m})^{-1}\mathcal{L}_{k,m}\| & \text{for scheme (9.59)},\\ \|(\mathcal{I} - \mathcal{K}_{k,m}^L)^{-1}\mathcal{K}_{k,m}^H\| & \text{for scheme (9.60)}. \end{cases}$$

This is roughly the factor between two consecutive errors in the iteration. More precisely, for each (k, m), the quantity $q_{k,m}$ is the least upper bound for the ratios of two consecutive errors or the ratios of differences of two consecutive iterates:

$$\frac{\|u_{k,m}^{(l)} - u_{k,m}\|}{\|u_{k,m}^{(l-1)} - u_{k,m}\|} \le q_{k,m} \text{ and } \frac{\|u_{k,m}^{(l)} - u_{k,m}^{(l-1)}\|}{\|u_{k,m}^{(l-1)} - u_{k,m}^{(l-2)}\|} \le q_{k,m}.$$

We have shown that, for fixed $m \in \mathbb{N}$, $q_{k,m} \to 0$ as $k \to \infty$ and thus $q_{k,m} < 1$ for large enough k.

For the Jacobi-type scheme (9.58), we can show that this rate is independent of all large enough m. Indeed, from (9.65) and (9.66) above,

$$\begin{aligned} q_{k,m} &\le \|\mathcal{D}_{k,m}^{-1}\|\|\mathcal{U}_{k+m} + \mathcal{L}_{k+m}\| \\ &\le \|(\mathcal{I} - \mathcal{K}_k)^{-1}\|(\|(\mathcal{P}_{k+m} - \mathcal{P}_k)\mathcal{K}\| + \|\mathcal{K}(\mathcal{P}_{k+m} - \mathcal{P}_k)\|), \end{aligned}$$

thus

$$\limsup_{m\to\infty} q_{k,m} \le \|(\mathcal{I} - \mathcal{K}_k)^{-1}\|(\|\mathcal{K} - \mathcal{P}_k\mathcal{K}\| + \|\mathcal{K} - \mathcal{K}\mathcal{P}_k\|) < 1 \tag{9.67}$$

if k is chosen large enough.

For the L–H-type scheme (9.60), noting that $\|(\mathcal{I} - \mathcal{K}_{k,m}^L)^{-1}\|$ is uniformly bounded and $\mathcal{K}_{k,m}^H = (\mathcal{P}_{k+m} - \mathcal{P}_k)\mathcal{K}\mathcal{P}_{k+m}$, we conclude that

$$\limsup_{m\to\infty} q_{k,m} \le c\|\mathcal{K} - \mathcal{P}_k\mathcal{K}\| < 1,$$

if k is chosen large enough.

To derive an estimate for the rate of convergence of the Gauss–Seidel scheme (9.59), we need hypotheses (9.62) and (9.63). For any positive integer k, we introduce two quantities

$$r_k := \sum_{i \in \mathbb{N}_0} \|(\mathcal{I} - \mathcal{P}_{k+i})\mathcal{K}\| \text{ and } r'_k := \sum_{i \in \mathbb{N}_0} \|\mathcal{K}(\mathcal{I} - \mathcal{P}_{k+i})\|.$$

Under assumption (9.62), we see that

$$\lim_{k \to \infty} r_k = \lim_{k \to \infty} r'_k = 0 \tag{9.68}$$

and hypothesis (9.63) ensures that

$$\limsup_{m \to \infty} \|\mathcal{L}_{k,m}\| \le r_k \text{ and } \limsup_{m \to \infty} \|\mathcal{U}_{k,m}\| \le r'_k. \tag{9.69}$$

Therefore, we conclude from (9.68) and (9.69) that for the Gauss–Seidel scheme,

$$\begin{aligned} \limsup_{m \to \infty} q_{k,m} &\le \limsup_{m \to \infty} \frac{\|(\mathcal{I} - \mathcal{K}_k)^{-1}\| \, \|\mathcal{L}_{k,m}\|}{1 - \|(\mathcal{I} - \mathcal{K}_k)^{-1}\| \, \|\mathcal{U}_{k,m}\|} \\ &\le \limsup_{m \to \infty} \frac{\|(\mathcal{I} - \mathcal{K}_k)^{-1}\| r_k}{1 - \|(\mathcal{I} - \mathcal{K}_k)^{-1}\| r'_k} < 1, \end{aligned} \tag{9.70}$$

if k is chosen large enough.

The discussions above lead to the results of this theorem. □

We remark that conditions (9.62) and (9.63) are fulfilled in many applications when $\mathbb{X}_n$ are chosen as piecewise polynomial spaces. The reader is referred to [108] for details and numerical examples.

9.3 Bibliographical remarks

In Chapters 5–7, multiscale Galerkin, Petrov–Galerkin and collocation methods are developed for solving the Fredholm integral equation of the second-kind with a weakly singular kernel. These methods yield linear systems having numerically sparse coefficient matrices, which with appropriate truncation strategies lead to fast discretization of the integral equation (see, for example, [28, 64, 68, 94, 95, 202, 260, 261] and the references cited therein). This chapter describes fast solvers for the resulting discrete systems, including multilevel augmentation methods and multilevel iteration methods. The idea of multilevel methods introduced in this chapter, based on the Gauss–Seidel iteration, was initiated in [67] in its early form for solving the Fredholm integral equation of the second kind. The results in [67] were developed further to the MAM and the MIM in [71] and [108], respectively. The abstract framework of the MAM was established in [71]. Since then, it has been used in various contexts (see [45, 51–59, 73, 76, 78, 187]).

We remark that other different multilevel methods for solving integral equations were developed as early as the late 1970s (see, for example, [120] and the references cited therein). Methods for the data-sparse approximation of matrices were introduced in [120], resulting in the so-called hierarchical matrices (or short $\mathcal{H}$-matrices). For more information on this subject, the reader is referred to [30, 122, 123], as well as [168–170]. Other work on multilevel and associated iteration methods can be found in Chapter 6 of [15].

10
Multiscale methods for nonlinear integral equations

In this chapter we develop multiscale methods for solving the Hammerstein equation, and the nonlinear boundary integral equation resulting from a reformulation of a boundary value problem of the Laplace equation with nonlinear boundary conditions. Fast algorithms are proposed using the MAM, in conjunction with matrix truncation strategies and techniques of numerical integration for integrals appearing in the process of solving equations. We prove that the proposed methods require only linear (up to a logarithmic factor) computational complexity and have the optimal convergence order.

In the section that follows we discuss the critical issues in solving nonlinear integral equations. This will shine a light on the ideas developed later in this chapter. In Section 10.2, we introduce the MAM for solving Hammerstein equations and provide a complete convergence analysis for the proposed method. In Section 10.3, we develop the MAM for solving the nonlinear boundary integral equation as a result of a reformulation of a boundary value problem of the Laplace equation with nonlinear boundary conditions. We present numerical experiments in Section 10.4.

10.1 Critical issues in solving nonlinear equations

Nonlinear integral equations portray many mathematical physics problems. The Hammerstein equation is a typical kind of nonlinear integral equation. Moreover, boundary value problems of the Laplace equation serve as mathematical models for many important applications. Making use of the fundamental solutions of the equation, we can formulate the boundary value problems as integral equations defined on the boundary (see, Section 2.2.3). For linear boundary conditions, the resulting boundary integral equations are linear, the numerical methods of which have been studied extensively.

Nonlinear boundary conditions are also involved in various applications. In these cases, the reformulation of the corresponding boundary value problems leads to nonlinear integral equations.

The nonlinearity introduces difficulties in the numerical solution of the equation, which normally requires an iteration scheme to solve it locally as a linearized integral equation. The Newton iteration method and the secant method are often used as bases for designing numerical schemes for equations of this type. In this case, the Jacobian matrices associated with the nonlinear operators have to be computed and possibly refreshed at each iteration step. Since the Jacobian matrices are usually dense and their size is equal to the dimension N of the approximate subspace of the solution, the computational complexity of the algorithms is at least of $\mathcal{O}(N^2)$. In [13], several popular numerical schemes for solving nonlinear integral equations are reviewed, the computational complexities of which are all at least of $\mathcal{O}(N^2)$. In particular, the multi-grid method [120] has an operation count of $\mathcal{O}(N^2)$ when the discretization of the system is included. When a high approximation accuracy is desired, it requires using an approximate subspace of a high dimension. Thus, it demands a significantly large amount of computational effort. This becomes a bottleneck problem for the numerical solution of nonlinear integral equations.

In order to develop a fast numerical algorithm for solving nonlinear integral equations, we consider achieving the following two tasks. First, the Jacobian matrices involved in the algorithm should be approximated in a subspace of a low dimension, which is fixed and much smaller than that for the *whole* approximate subspace. Second, all steps of the algorithm should be implemented with a number of calculations no more than $\mathcal{O}(N \log N)$. Accomplishing these two tasks is comparable with using the truncation strategy and the MAM introduced in the previous chapter for solving a linear integral equation.

In this chapter, we develop an MAM for solving the Hammerstein equation and the nonlinear boundary integral equation, which accomplishes two tasks discussed earlier. This method requires the availability of a multilevel decomposition of the solution space and a projection from the solution space onto a finite-dimensional subspace at a given level. Solving the equation with high-order approximation accuracy requires us to solve an approximate equation, which is the original equation projected onto a subspace at a high level. We observe that most of the computational efforts are spent on inverting the nonlinear integral operator when solving the nonlinear equation. The higher the level of the subspace used, the more accurate the approximate solution obtained and at the same time more computational effort is needed to invert the nonlinear integral operator at that level. To significantly reduce the computational cost, we propose not to invert directly the nonlinear integral operator at the high level, but instead, to invert it at a much lower (fixed) level and compensate the error which may result from this modification by a high-frequency correction term that can be obtained by solving a linear system at the high level. The correction does not involve any inversion of the nonlinear integral operator. This method results in a fast algorithm for solving the nonlinear integral equation which gives the optimal convergence order, the same as the approximation order of the approximate subspace (of the highest level). At the same time, this method requires only *linear* computational complexity in the sense that the number of multiplications needed is proportional to the dimension of the approximate subspace.

The proposed method is based on a traditional projection method for solving nonlinear integral equations and a multilevel decomposition of the solution space (see [76]). The main idea comes from treating the corresponding linear integral equation. Multilevel or multiscale numerical methods for solving linear integral equations were discussed in the previous chapter. Recall that in the MAM solving the linear equation, the coefficient matrix of the resulting linear system from discretization of the linear integral equation via a multiscale decomposition of the solution space can be obtained from a small-sized matrix which is a low-resolution representation of the integral operator by augmenting it with new rows and columns representing the high frequency of the integral operator. Making use of this multiscale structure of the matrix, the MAM inverts only the small-sized fixed matrix with the compensation of matrix–vector multiplications. It was proved in Section 9.1 that this method gives an optimal order of convergence, while it reduces the computational complexity significantly. Motivated by the MAM for the linear integral equation, we develop the MAM for the nonlinear equation. The MAM which will be described in Section 10.2 for solving the Hammerstein equation and in Section 10.3 for solving the nonlinear boundary integral equation resulting from a reformulation of a boundary value problem of the Laplace equation with nonlinear boundary conditions needs only to invert the nonlinear operator in a *much smaller* subspace. This significantly reduces the computational complexity to a nearly linear order. We also develop a fully discrete MAM for solving the nonlinear integral equation using numerical integration methods for computing the integrals appearing in the resulting nonlinear system.

To close this section, we briefly mention the fast Fourier–Galerkin method for solving the nonlinear boundary integral equation developed recently in [53]. A fast algorithm for solving the resulting discrete nonlinear system was designed in the paper by integrating together the techniques of matrix compression, numerical integration of oscillatory integrals and the multilevel augmentation method. It was proved there that the proposed method enjoys an optimal convergence order and a nearly linear computational complexity. Numerical experiments were presented to confirm the theoretical estimates and to demonstrate the efficiency and accuracy of the proposed method.

10.2 Multiscale methods for the Hammerstein equation

In this section we describe fast multiscale methods for solving the Hammerstein equation. We prove that the proposed methods require only linear (up to a logarithmic factor) computational complexity and have the optimal convergence order. Two specific fast methods based on the Galerkin projection and the collocation projection are presented.

10.2.1 The multilevel augmentation method

We introduce in this subsection the MAM for solving the Hammerstein equation. The method is described here in an operator form only, in terms of the projections and related spaces. The description of the method involving bases of the subspaces is postponed to Section 10.2.4. This subsection ends with a comparison of the proposed method with the well-known multigrid method.

For $d \in \mathbb{N}$, we let $\Omega \subseteq \mathbb{R}^d$ be a compact domain. Consider the Hammerstein equation

$$u(s) - \int_{\Omega} K(s,t)\psi(t,u(t))\mathrm{d}t = f(s), \qquad s \in \Omega, \tag{10.1}$$

where K, f and ψ are given functions and u is the unknown to be determined. We assume that $f \in C(\Omega)$ and for any $s \in \Omega$, we denote $K_s(t) = K(s,t)$. Throughout this section we assume, unless stated otherwise, that the following conditions on K and ψ are satisfied:

(H1) $\lim_{s\to\tau} \|K_s - K_\tau\|_1 = 0$ for any $\tau \in \Omega$ and $\sup_{s\in\Omega} \int_{\Omega} |K(s,t)|\mathrm{d}t < \infty$;

(H2) $\psi(t,u)$ is continuous in $t \in \Omega$ and Lipschitz continuous in $u \in \mathbb{R}$, the partial derivative $D_u\psi$ of ψ with respect to the variable u exists and is Lipschitz continuous, that is, there exists a positive constant L such that

$$|D_u\psi(t,u_1) - D_u\psi(t,u_2)| \leq L|u_1 - u_2|$$

and for any $u \in C(\Omega)$, $\psi(\cdot, u(\cdot)), D_u\psi(\cdot, u(\cdot)) \in C(\Omega)$.

We use $\mathbb{X}$ to represent the Banach space $L^2(\Omega)$ or $L^\infty(\Omega)$. The linear integral operator $\mathcal{K} : \mathbb{X} \to \mathbb{X}$ is defined by

$$(\mathcal{K}u)(s) := \int_\Omega K(s,t)u(t)\mathrm{d}t, \quad s \in \Omega$$

and the nonlinear operator $\Psi : \mathbb{X} \to \mathbb{X}$ by

$$(\Psi u)(t) := \psi(t, u(t)), \quad t \in \Omega.$$

With these notations, equation (10.1) is written in the operator form as

$$u - \mathcal{K}\Psi u = f. \tag{10.2}$$

We assume that $\mathbb{X}_n$, $n \in \mathbb{N}_0$ is a sequence of finite-dimensional subspaces of $\mathbb{X}$ having the property that

$$\mathbb{Y} \subseteq \overline{\bigcup_{n\in\mathbb{N}_0} \mathbb{X}_n},$$

where if $\mathbb{X} = L^2(\Omega)$, then $\mathbb{Y} = \mathbb{X}$ and the notation "$\subseteq$" in the above relation is replaced by "=" and if $\mathbb{X} = L^\infty(\Omega)$, then $\mathbb{Y} = C(\Omega)$. For each $n \in \mathbb{N}_0$, let $\mathcal{P}_n : \mathbb{X} \to \mathbb{X}_n$ be a projection. Throughout this section, we assume that

(H3) the sequence $\mathcal{P}_n$, $n \in \mathbb{N}_0$ converges pointwise to the identity operator in $\mathbb{Y}$, that is, for any $x \in \mathbb{Y}$,

$$\lim_{n\to\infty} \|\mathcal{P}_n x - x\| = 0.$$

The projection method for solving equation (10.2) is to find $u_n \in \mathbb{X}_n$ satisfying

$$u_n - \mathcal{P}_n\mathcal{K}\Psi u_n = \mathcal{P}_n f. \tag{10.3}$$

The solution u_n of equation (10.3) is called the projection solution of equation (10.2). The following result is concerned with the existence of the projection solution u_n and its convergence property.

Theorem 10.1 *Let $u^* \in \mathbb{X}$ be an isolated solution of (10.2). If one is not an eigenvalue of the linear operator $(\mathcal{K}\Psi)'(u^*)$, then for sufficiently large n, equation (10.3) has a unique solution $u_n \in B(u^*, \delta)$ for some $\delta > 0$ with the property*

$$c_1\|\mathcal{P}_n u^* - u^*\| \le \|u_n - u^*\| \le c_2\|\mathcal{P}_n u^* - u^*\|, \tag{10.4}$$

for some positive constants c_1, c_2.

The above theorem was established in Theorem 2 of [255] and used in [165].

As we have discussed in Section 10.1, the projection method (10.3) requires inverting the nonlinear operator

$$\Psi_n := \mathcal{I} - \mathcal{P}_n \mathcal{K} \Psi,$$

which is computationally challenging when the dimension of the subspace $\mathbb{X}_n$ is large. In fact, once a basis of the subspace $\mathbb{X}_n$ is chosen, equation (10.3) is equivalent to a system of nonlinear algebraic equations. Standard methods such as the Newton method and its variations for solving the nonlinear system are to linearize the equation locally and solve the nonlinear system by iteration. At each iteration step, we need to invert the Jacobian matrix of Ψ_n evaluated at the solution of the previous step. The Jacobian matrix, different at a different step, is dense and has size equal to the dimension $s(n)$ of the space $\mathbb{X}_n$. The computational cost for solving equation (10.3) by a standard method is $\mathcal{O}(s(n)^2)$. When $s(n)$ is large, this becomes a bottleneck problem for numerical solutions of these equations.

Because of this, we propose not to solve equation (10.3) directly. Instead, we develop a multilevel method which requires inverting the nonlinear operator Ψ_k for a fixed k much smaller than n. To this end, we require that the space $\mathbb{X}$ has a multiscale decomposition, that is, the subspaces $\mathbb{X}_n$ are nested ($\mathbb{X}_{n-1} \subset \mathbb{X}_n$, $n \in \mathbb{N}$), so that $\mathbb{X}_n$ is the direct sum of $\mathbb{X}_{n-1}$ and its complement $\mathbb{W}_n$. Specifically,

$$\mathbb{X}_n = \mathbb{X}_{n-1} \oplus \mathbb{W}_n, \qquad n \in \mathbb{N}. \tag{10.5}$$

Accordingly, we have for all $n \in \mathbb{N}_0$ that

$$\mathcal{P}_n \mathcal{P}_{n+1} = \mathcal{P}_n. \tag{10.6}$$

The decomposition (10.5) can be applied repeatedly, so that for $n := k + m$ with $k \in \mathbb{N}_0$ fixed and $m \in \mathbb{N}_0$, we have the decomposition

$$\mathbb{X}_{k+m} = \mathbb{X}_k \oplus \mathbb{W}_{k,m}, \tag{10.7}$$

where

$$\mathbb{W}_{k,m} := \mathbb{W}_{k+1} \oplus \cdots \oplus \mathbb{W}_{k+m}.$$

Under this hypothesis, we describe the multilevel method for obtaining an approximation of the solution of equation (10.3). Our goal is to obtain an approximation of the solution of equation (10.3) with $n = k + m$, k being small and fixed. We first solve equation (10.3) with $n = k$ *exactly* and obtain the solution u_k. Since $s(k)$ is very *small* in comparison with $s(k + m)$, the computational cost of inverting the nonlinear operator Ψ_k is much less than that of inverting Ψ_{k+m}. The next step is to obtain an *approximation* of the solution $u_{k+1} \in \mathbb{X}_{k+1}$ of equation (10.3) with $n = k + 1$. For this purpose, we decompose

$$u_{k+1} = u_{k+1}^L + u_{k+1}^H \quad \text{with } u_{k+1}^L \in \mathbb{X}_k \text{ and } u_{k+1}^H \in \mathbb{W}_{k+1}$$

using the decomposition (10.7) and rewrite equation (10.3) with $n = k + 1$ as

$$\mathcal{P}_k(\mathcal{I} - \mathcal{K}\Psi)(u_{k+1}^L + u_{k+1}^H) = \mathcal{P}_k f + (\mathcal{P}_{k+1} - \mathcal{P}_k)(f + \mathcal{K}\Psi u_{k+1}) - u_{k+1}^H. \tag{10.8}$$

The second term on the right-hand side of equation (10.8) can be obtained approximately via the solution u_k of the previous level. That is, we compute $u_{k,1}^H := (\mathcal{P}_{k+1} - \mathcal{P}_k)(f + \mathcal{K}\Psi u_k)$, where $u_{k,0} = u_k$, and note that $u_{k,1}^H \in \mathbb{W}_{k+1}$. Observing that

$$u_{k,1}^H = u_{k+1} - u_k - \mathcal{P}_{k+1}\mathcal{K}(\Psi u_{k+1} - \Psi u_k),$$

in equation (10.8), we replace u_{k+1}^H and the second term on the right-hand side by $u_{k,1}^H$, to obtain an equation for $u_{k,1}^L \in \mathbb{X}_k$:

$$\mathcal{P}_k(\mathcal{I} - \mathcal{K}\Psi)(u_{k,1}^L + u_{k,1}^H) = \mathcal{P}_k f. \tag{10.9}$$

The function $u_{k,1}^L$ can be viewed as a good approximation to u_{k+1}^L. We then obtain an approximation to the solution u_{k+1} of equation (10.3) by setting

$$u_{k,1} := u_{k,1}^L + u_{k,1}^H.$$

Note that $u_{k,1}^L$ and $u_{k,1}^H$, respectively, represent the lower and higher-frequency components of $u_{k,1}$. This procedure is repeated m times to obtain an approximation $u_{k,m}$ of the solution u_{k+m} of equation (10.3) with $n = k + m$.

Note that at each step, we invert *only* the *same* nonlinear operator $\mathcal{P}_k(\mathcal{I} - \mathcal{K}\Psi)$. This makes the method very efficient computationally. We summarize the method described above in the following algorithm.

Algorithm 10.2 *(The multilevel augmentation method: An operator form) Let k be a fixed positive integer.*

Step 1: *Find the solution $u_k \in \mathbb{X}_k$ of equation (10.3) with $n := k$. Set $u_{k,0} := u_k$ and $l := 1$.*

Step 2: *Compute*

$$u_{k,l}^H = (\mathcal{P}_{k+l} - \mathcal{P}_k)(f + \mathcal{K}\Psi u_{k,l-1}). \tag{10.10}$$

Step 3: *Solve* $u_{k,l}^L \in \mathbb{X}_k$ *from the nonlinear equation*

$$\mathcal{P}_k(\mathcal{I} - \mathcal{K}\Psi)(u_{k,l}^L + u_{k,l}^H) = \mathcal{P}_k f. \tag{10.11}$$

Step 4: *Let* $u_{k,l} := u_{k,l}^L + u_{k,l}^H$. *Set* $l \leftarrow l + 1$ *and go back to step 2 until* $l = m$.

The output of Algorithm 10.2 is an approximation $u_{k,m}$ of the solution u_{k+m} of equation (10.3). The approximation of u_{k+m} is obtained beginning with an initial approximation u_k, repeatedly inverting the operator $\mathcal{P}_k(\mathcal{I} - \mathcal{K}\Psi)$ to update the approximation recursively. The procedure completes in m steps and no iteration is needed if the operator $\mathcal{P}_k(\mathcal{I} - \mathcal{K}\Psi)$ can be inverted exactly. Of course, the inversion of the nonlinear operator may require iterations. The key steps in this algorithm are steps 2 and 3. In step 2, we obtain a high-frequency component $u_{k,l}^H$ of the approximate solution $u_{k,l}$ from the approximation $u_{k,l-1}$ at the previous level by a functional evaluation. In step 3, we solve the low-frequency component $u_{k,l}^L \in \mathbb{X}_k$ from (10.11) with the known high-frequency component $u_{k,l}^H \in \mathbb{W}_{k+1}$ obtained from step 2. For all $l \in \mathbb{Z}_{m+1}$, we invert the *same* nonlinear operator $\mathcal{P}_k(\mathcal{I} - \mathcal{K}\Psi)$ at the initial coarse level k. The computational costs for this are significantly lower than inverting the nonlinear operator $\mathcal{P}_{k+m}(\mathcal{I} - \mathcal{K}\Psi)$ at the final fine level $k+m$. We call $u_{k,m}$ the multilevel solution of equation (10.3), and $u_{k,m}^L$ and $u_{k,m}^H$, respectively, the lower and higher-frequency components of $u_{k,m}$. In the next subsection, we show that $u_{k,m}$ approximates the exact solution u in the same order as u_{k+m} does.

The multilevel solution $u_{k,m}$ is in fact a solution of a nonlinear operator equation. We present this observation in the next proposition.

Proposition 10.3 *If* $u_{k,m}^H$ *is obtained from formula (10.10) and* $u_{k,m}^L \in \mathbb{X}_k$ *is a solution of equation (10.11), then* $u_{k,m} := u_{k,m}^L + u_{k,m}^H$ *is a solution of the equation*

$$(\mathcal{I} - \mathcal{P}_k\mathcal{K}\Psi)u_{k,m} = \mathcal{P}_{k+m}f + (\mathcal{P}_{k+m} - \mathcal{P}_k)\mathcal{K}\Psi u_{k,m-1}. \tag{10.12}$$

Conversely, for any solution $u_{k,m}$ *of (10.12),* $u_{k,m}^H := (\mathcal{P}_{k+m} - \mathcal{P}_k)u_{k,m}$ *satisfies equation (10.10) and* $u_{k,m}^L := \mathcal{P}_k u_{k,m}$ *is a solution of equation (10.11).*

Proof Since $u_{k,m}^L \in \mathbb{X}_k$, we have that $(\mathcal{P}_{k+m} - \mathcal{P}_k)u_{k,m}^L = 0$. It follows from (10.10) with $l = m$ that

$$(\mathcal{P}_{k+m} - \mathcal{P}_k)(u_{k,m}^L + u_{k,m}^H) = (\mathcal{P}_{k+m} - \mathcal{P}_k)(f + \mathcal{K}\Psi u_{k,m-1}).$$

Adding the above equation to equation (10.11) with $l = m$ and noticing that $\mathcal{P}_{k+m}u_{k,m} = u_{k,m}$ yield equation (10.12).

Conversely, let $u_{k,m}$ be a solution of equation (10.12). We apply the operator $\mathcal{P}_{k+m} - \mathcal{P}_k$ to both sides of the equation and obtain (10.10) by utilizing

$$(\mathcal{P}_{k+m} - \mathcal{P}_k)\mathcal{P}_k = 0 \text{ and } (\mathcal{P}_{k+m} - \mathcal{P}_k)\mathcal{P}_{k+m} = \mathcal{P}_{k+m} - \mathcal{P}_k,$$

where the second equation is a consequence of formula (10.6). We then apply $\mathcal{P}_k$ to both sides of (10.12) and use $\mathcal{P}_k(\mathcal{P}_{k+m} - \mathcal{P}_k) = 0$ to conclude that $u^L_{k,m} := \mathcal{P}_k u_{k,m}$ is a solution of equation (10.11) with $l = m$. □

Equation (10.12) differs from equation (10.3) with $n = k + m$ in two ways. (1) The right-hand sides of these two equations differ in the term $(\mathcal{P}_{k+m} - \mathcal{P}_k)\mathcal{K}\Psi u_{k,m-1}$. (2) The nonlinear operators on the left-hand side of these two equations are different, $\mathcal{I} - \mathcal{P}_{k+m}\mathcal{K}\Psi$ for equation (10.3) and $\mathcal{I} - \mathcal{P}_k\mathcal{K}\Psi$ for equation (10.12). It requires much less computational effort to invert the nonlinear operator $\mathcal{I} - \mathcal{P}_k\mathcal{K}\Psi$ than the nonlinear operator $\mathcal{I} - \mathcal{P}_{k+m}\mathcal{K}\Psi$. It is these differences that lead to fast solutions of the Hammerstein equation. Moreover, equation (10.12) connects multilevels (levels k, $k + m - 1$ and $k+m$) of the solution space and the range space. This is the basis on which the approximate solution $u_{k,m}$ has a good approximation property.

To close this subsection, we compare the MAM with the well-known multigrid method. From the point of view of Kress [177], the multigrid method for solving linear integral equations uses special techniques for residual correction, utilizing the information of the coarser levels to construct appropriate approximations of the inverse of the operator of the approximate equation to be solved. Specific choices of the approximate inverses lead to the V-cycle, W-cycle and cascadic multigrids. This idea was applied to the construction of iteration schemes for nonlinear integral equations. The related work was discussed in a master review paper [7]. Taking this point of view, the proposed MAM might be considered as a *nonconventional* cascadic multigrid method, with significant difference from the traditional one.

In order to compare the proposed method with the multigrid method, we review the two-grid method which was first introduced in [9] and reviewed in [7]. The two-grid method solves (10.3) by the Newton iteration

$$u_n^{(\ell+1)} = u_n^{(\ell)} - [\mathcal{I} - (\mathcal{P}_n\mathcal{K}\Psi)'(u_n^{(\ell)})]^{-1} [u_n^{(\ell)} - \mathcal{P}_n\mathcal{K}\Psi(u_n^{(\ell)}) - \mathcal{P}_n f], \quad \ell = 1, 2, \ldots$$

combined with an approximation of $[\mathcal{I} - (\mathcal{P}_n\mathcal{K}\Psi)'(u_n^{(\ell)})]^{-1}$ using the information of coarser grids. In particular, one may choose a level $n' < n$ and use the following approximation:

$$[\mathcal{I} - (\mathcal{P}_n\mathcal{K}\Psi)'(u_n^{(\ell)})]^{-1} \approx \mathcal{I} + [\mathcal{I} - (\mathcal{P}_{n'}\mathcal{K}\Psi)'(u_{n'})]^{-1}(\mathcal{P}_n\mathcal{K}\Psi)'(u_{n'}),$$

where $u_{n'}$ is the solution of (10.3) with $n = n'$. Since the Jacobian matrix $(\mathcal{P}_{n'}\mathcal{K}\Psi)'(u_{n'})$ was obtained after solving (10.3) at level n' and $(\mathcal{P}_n\mathcal{K}\Psi)'(u_{n'})$ remains unchanged during the iteration for level n, the above approximation avoids the computational cost of updating the Jacobian matrix. However, the need to establish $(\mathcal{P}_n\mathcal{K}\Psi)'(u_{n'})$ still requires $\mathcal{O}(s(n)^2)$ computational cost. The multigrid scheme uses information of more than one lower level to approximate $[\mathcal{I}-(\mathcal{P}_n\mathcal{K}\Psi)'(u_n^{(\ell)})]^{-1}$, and it needs $\mathcal{O}(s(n)^2)$ computational cost. In other words, the idea of the multigrid method is as follows. First establish the Newton iteration method for equation (10.3) and then approximate the Jacobian matrix appearing in the iteration process by using information in coarse grids. Our proposed method introduces a new approximation strategy, which approximates directly the nonlinear operator $\mathcal{I}-\mathcal{P}_n\mathcal{K}\Psi$ in (10.3), not the Jacobian matrix in the Newton iteration, by $\mathcal{I} - \mathcal{P}_k\mathcal{K}\Psi$ at a fixed level $k < n$. This point can clearly be observed in Proposition 10.3. It is possible only because the solution space has a built-in multilevel structure. The proposed method requires only $\mathcal{O}(s(n))$ (*linear*) computational cost.

10.2.2 Analysis of the multilevel algorithm

We analyze the MAM described in the last subsection. Specifically, we show that the multilevel solution $u_{k,m}$ exists and prove that it converges to the exact solution u^* of equation (10.2) in the same order as the projection solution u_{k+m} does.

We first present a result concerning the existence of the multilevel solution. The proof of this result is similar to that for the existence of the projection solution which was established in [255].

Theorem 10.4 *Let u^* be an isolated solution of (10.2). If one is not an eigenvalue of $(\mathcal{K}\Psi)'(u^*)$, then there exists an integer N such that for each $k > N$, if $u_{k,m-1}$ is given, the operator equation (10.12) has a unique solution $u_{k,m} \in B(u^*, \delta)$ for some $\delta > 0$ and for all $m \in \mathbb{N}_0$.*

Proof Let $\mathcal{L} := (\mathcal{K}\Psi)'(u^*)$. By hypotheses, there exists an integer N and a positive constant ν such that for all $k > N$, $(\mathcal{I} - \mathcal{P}_k\mathcal{L})^{-1}$ exists and $\|(\mathcal{I} - \mathcal{P}_k\mathcal{L})^{-1}\| \leq \nu$. For $u, v \in \mathbb{X}$, we define

$$\mathcal{R}(u; v) := \mathcal{K}\Psi(u) - \mathcal{K}\Psi(v) - \mathcal{L}(u - v).$$

We then obtain from (10.2) and (10.12) that

$$\begin{aligned} u_{k,m} - u^* = (\mathcal{I} - \mathcal{P}_k\mathcal{L})^{-1}[(\mathcal{P}_k - \mathcal{I})u^* + \mathcal{P}_k\mathcal{R}(u_{k,m}; u^*) \\ + (\mathcal{P}_{k+m} - \mathcal{P}_k)(f + \mathcal{K}\Psi u_{k,m-1})]. \end{aligned} \tag{10.13}$$

We introduce the operator

$$F_{k,m}(v) := (\mathcal{I} - \mathcal{P}_k\mathcal{L})^{-1}[(\mathcal{P}_k - \mathcal{I})u^* + \mathcal{P}_k\mathcal{R}(v + u^*; u^*) \\ + (\mathcal{P}_{k+m} - \mathcal{P}_k)(f + \mathcal{K}\Psi u_{k,m-1})].$$

It follows from hypothesis (H2) that there exist two positive constants M_1, M_2 such that the estimates

$$\|\mathcal{R}(v; u^*)\| \leq M_1 \|v - u^*\|^2$$

and

$$\|\mathcal{R}(v_1; u^*) - \mathcal{R}(v_2; u^*)\| \leq M_2 \left(\|v_1 - u^*\| + \frac{1}{2}\|v_1 - v_2\| \right) \|v_1 - v_2\|$$

hold for all v, v_1, v_2 in a neighborhood of u^*. By utilizing this property and the pointwise convergence of the projection $\mathcal{P}_n$, we can show that there exists a positive constant δ such that $F_{k,m}$ is a contractive mapping on the ball $B(0, \delta)$. The fixed-point theorem ensures that the fixed-point equation

$$v = F_{k,m}(v)$$

has a unique solution in $B(u^*, \delta)$, or equivalently, the equation (10.12) has a unique solution $u_{k,m} \in B(u^*, \delta)$. □

Next, we turn to analyzing the convergence of the multilevel solution. We first prove a crucial technical lemma which confirms that the error between $u_{k,m}$ and u_{k+m} is bounded by the error between $u_{k,m-1}$ and u_{k+m} with a factor depending on k, m which converges to zero uniformly for $m \in \mathbb{N}_0$ as $k \to \infty$. To this end, for $n \in \mathbb{N}_0$, we denote by R_n the approximation error of $\mathbb{X}_n$ for $u^* \in \mathbb{X}$, namely,

$$R_n = R_n(u^*) := \inf\{\|u^* - v\|_{\mathbb{X}} : v \in \mathbb{X}_n\}.$$

Lemma 10.5 *Let u^* be an isolated solution of (10.2). If one is not an eigenvalue of $(\mathcal{K}\Psi)'(u^*)$, then there exists a sequence of positive numbers $\alpha_{k,m}$, $k \in \mathbb{N}$, $m \in \mathbb{N}_0$ with $\lim_{k\to\infty} \alpha_{k,m} = 0$ uniformly for $m \in \mathbb{N}_0$ and a positive integer N, such that for all $k \geq N$ and $m \in \mathbb{N}_0$,*

$$\|u_{k,m} - u_{k+m}\| \leq \alpha_{k,m}\|u_{k,m-1} - u_{k+m}\|.$$

Proof It follows from (10.3) with $n = k + m$ and (10.12) that

$$(\mathcal{I} - \mathcal{P}_{k+m}\mathcal{L})(u_{k,m} - u_{k+m}) = (\mathcal{P}_{k+m} - \mathcal{P}_k)\mathcal{K}(\Psi u_{k,m-1} - \Psi u_{k,m}) \\ + \mathcal{P}_{k+m}\mathcal{R}(u_{k,m}; u_{k+m}).$$

We let $a_{k,m} := \|(\mathcal{P}_{k+m} - \mathcal{P}_k)\mathcal{K}\|$ and by L we denote the Lipschitz constant of the derivative $D_u\psi$. Hypothesis (H3) ensures that there exists a positive constant p such that for all $n \in \mathbb{N}$, $\|\mathcal{P}_n\| \leq p$. By hypothesis (H2), we have that

$$\|u_{k,m} - u_{k+m}\| \leq \frac{\nu L a_{k,m}}{1 - \nu(a_{k,m}L + pM_1\|u_{k,m} - u_{k+m}\|)}\|u_{k,m-1} - u_{k+m}\|.$$

Since $a_{k,m} \to 0$, $k \to \infty$ uniformly for $m \in \mathbb{N}_0$, there exists a positive integer N_1 such that $a_{k,m}L\nu < \frac{1}{6}$ for all $k > N_1$ and $m \in \mathbb{N}_0$. Since $R_n \to 0$, $n \to \infty$, we can find a positive integer N_2 such that $\nu p M_1 \rho R_k < \frac{1}{6}$ for all $k > N_2$. We choose $\delta > 0$ as in Theorem 10.4 such that $\nu p M_1 \delta < \frac{1}{6}$ and (10.12) has a unique solution in $B(u^*, \delta)$ for all $k > N_3$ for some positive integer N_3. Consequently, for any $k > N := \max\{N_1, N_2, N_3\}$, we have that

$$\nu(a_{k,m}L + pM_1\|u_{k,m} - u_{k+m}\|) \leq \frac{1}{2},$$

which implies that

$$\|u_{k,m} - u_{k+m}\| \leq 2\nu L a_{k,m}\|u_{k,m-1} - u_{k+m}\|.$$

We conclude the desired result of this lemma with $\alpha_{k,m} := 2\nu L a_{k,m}$ by recalling that $a_{k,m} \to 0$, $k \to \infty$ uniformly for $m \in \mathbb{N}_0$. □

Recall that a sequence of non-negative numbers γ_n, $n \in \mathbb{N}_0$ is called a majorization sequence of R_n, $n \in \mathbb{N}_0$, if $\gamma_n \geq R_n$ for all $n \in \mathbb{N}_0$ and there exists a positive integer N_0 and a positive constant σ such that for $n \geq N_0$, $\frac{\gamma_{n+1}}{\gamma_n} \geq \sigma$.

Making use of the above lemma, we obtain the following important result on the convergence rate of the multilevel augmentation solution. The proof is similar to that of Theorem 9.2 for linear operator equations.

Theorem 10.6 *Let u^* be an isolated solution of (10.2) and let γ_n, $n \in \mathbb{N}_0$, be a majorization sequence of R_n, $n \in \mathbb{N}_0$. If one is not an eigenvalue of $(\mathcal{K}\Psi)'(u^*)$, then there exists a positive constant ρ and a positive integer N such that for all $k \geq N$ and $m \in \mathbb{N}_0$,*

$$\|u^* - u_{k,m}\| \leq (\rho + 1)\gamma_{k+m}. \tag{10.14}$$

Proof We prove the estimate (10.14) by induction on m. When $m = 0$, it is clear that

$$\|u^* - u_{k,0}\| = \|u^* - u_k\| \leq \rho R_k \leq \rho\gamma_k$$

for all $k > N_0$. Suppose that the estimate (10.14) holds for $m-1$. By the triangle inequality and the induction hypothesis,

$$\begin{aligned}\|u_{k,m-1} - u_{k+m}\| &\leq \|u_{k,m-1} - u^*\| + \|u_{k+m} - u^*\| \\ &\leq (\rho+1)\gamma_{k+m-1} + \rho\gamma_{k+m} \\ &\leq \left(\rho + \frac{\rho+1}{\sigma}\right)\gamma_{k+m}.\end{aligned}$$

Choose N such that for all $k > N$, the estimate in Lemma 10.5 holds and $\alpha_{k,m}(\rho + \frac{\rho+1}{\sigma}) < 1$. Combining the estimate above with the estimate in Lemma 10.5 yields the inequality

$$\|u_{k,m} - u_{k+m}\| \leq \gamma_{k+m}.$$

Again, by using the triangle inequality we obtain that

$$\|u_{k,m} - u^*\| \leq \|u_{k,m} - u_{k+m}\| + \|u_{k+m} - u^*\| \leq (\rho+1)\gamma_{k+m},$$

which completes the induction procedure. □

When a specific projection method is given, the associated majorization sequence is known. In this case, Theorem 10.6 may lead to the convergence order estimate of the corresponding MAM. The corresponding convergence order for the MAMs based on the Galerkin projection and based on the collocation projection will be presented in Section 10.2.4.

10.2.3 The discrete multilevel augmentation method

In Section 10.2.1 we established an operator form of the MAM and in Section 10.2.2, we proved the existence and convergence properties of the approximate solution obtained from the method. It is shown that the proposed method gives approximate solutions with the same order of accuracy as the classical projection methods. The purpose of this section is to describe a discrete version of the MAM when an appropriate basis of the approximate subspace is chosen and to estimate the computational cost of the algorithm.

Suppose that $\mathbb{L}_n$, $n \in \mathbb{N}_0$, is a sequence of subspaces of $\mathbb{X}^*$ which has the properties

$$\mathbb{L}_n \subset \mathbb{L}_{n+1}, \quad \dim(\mathbb{L}_n) = \dim(\mathbb{X}_n), \qquad n \in \mathbb{N}_0.$$

It follows from the nestedness property that there is a decomposition

$$\mathbb{L}_{k+m} = \mathbb{L}_k \oplus \mathbb{V}_{k,m},$$

where $\mathbb{V}_{k,m} := \mathbb{V}_{k+1} \oplus \cdots \oplus \mathbb{V}_{k+m}$.

We let $w(0) := \dim(\mathbb{X}_0)$ and $w(i) := \dim(\mathbb{W}_i)$, for $i > 0$, and suppose that

$$\mathbb{X}_0 = \operatorname{span}\{w_{0j} : j \in \mathbb{Z}_{w(0)}\}, \qquad \mathbb{L}_0 = \operatorname{span}\{\ell_{0j} : j \in \mathbb{Z}_{w(0)}\}$$

and

$$\mathbb{W}_i = \operatorname{span}\{w_{ij} : j \in \mathbb{Z}_{w(i)}\}, \quad \mathbb{V}_i = \operatorname{span}\{\ell_{ij} : j \in \mathbb{Z}_{w(i)}\}, \qquad i > 0.$$

By using the index set $\mathbb{U}_n := \{(i,j) : j \in \mathbb{Z}_{w(i)}, i \in \mathbb{Z}_{n+1}\}$, we have that

$$\mathbb{X}_n = \operatorname{span}\{w_{ij} : (i,j) \in \mathbb{U}_n\}, \quad \mathbb{L}_n = \operatorname{span}\{\ell_{ij} : (i,j) \in \mathbb{U}_n\}, \qquad n \in \mathbb{N}_0.$$

Recalling $s(n) = \dim(\mathbb{X}_n)$, we then observe that $\mathbb{U}_n$ has cardinality $\operatorname{card}(\mathbb{U}_n) = s(n)$. We further assume that the elements of $\mathbb{U}_n$ are ordered lexicographically.

For any $v \in \mathbb{X}_{k+m}$, we have a unique expansion

$$v = \sum_{(i,j)\in\mathbb{U}_{k+m}} v_{ij} w_{ij}.$$

The vector $\mathbf{v} := [v_{ij} : (i,j) \in \mathbb{U}_{k+m}]$ is called the representation vector of v. Thus, for the solution $u_{k,m}$ of (10.12), its representation vector is given by $\mathbf{u}_{k,m} := [(\mathbf{u}_{k,m})_{ij} : (i,j) \in \mathbb{U}_{k+m}]$. Setting $\mathbb{U}_{k,m} := \mathbb{U}_{k+m} \backslash \mathbb{U}_k$, we obtain that

$$\mathbb{U}_{k,m} = \{(i,j) : j \in \mathbb{Z}_{w(i)}, i \in \mathbb{Z}_{k+m+1} \backslash \mathbb{Z}_{k+1}\}.$$

Consequently, we have the representations

$$u_{k,m}^L = \sum_{(i,j)\in\mathbb{U}_k} (\mathbf{u}_{k,m})_{ij} w_{ij} \quad \text{and} \quad u_{k,m}^H = \sum_{(i,j)\in\mathbb{U}_{k,m}} (\mathbf{u}_{k,m})_{ij} w_{ij}.$$

It follows from the property of $\mathbb{L}_n$ that equation (10.11) is equivalent to the nonlinear system

$$\left\langle \ell_{i'j'}, (\mathcal{I} - \mathcal{K}\Psi)\left(\sum_{(i,j)\in\mathbb{U}_k} (\mathbf{u}_{k,l})_{ij} w_{ij} + u_{k,l}^H\right)\right\rangle = \langle \ell_{i'j'}, f\rangle, \qquad (i',j') \in \mathbb{U}_k. \tag{10.15}$$

In order to convert (10.10) into its equivalent discrete form, we first prove the following lemma.

Lemma 10.7 *For $v \in \mathbb{X}$, the equation*

$$(\mathcal{P}_{k+l} - \mathcal{P}_k)v = 0 \tag{10.16}$$

is equivalent to

$$\langle \ell_{i'j'}, v\rangle = 0, \qquad (i',j') \in \mathbb{U}_{k,l} \tag{10.17}$$

if and only if for all $v \in \mathbb{X}$,

$$\langle \ell_{i'j'}, \mathcal{P}_k v \rangle = 0, \qquad (i', j') \in \mathbb{U}_{k,l}. \tag{10.18}$$

Proof We observe that since $\ell_{i'j'} \in \mathbb{L}_{k+l}$, $(i', j') \in \mathbb{U}_{k,l}$, we have that $\langle \ell_{i'j'}, v - \mathcal{P}_{k+l} v \rangle = 0$, for $(i', j') \in \mathbb{U}_{k,l}$. Hence, (10.18) is equivalent to

$$\langle \ell_{i'j'}, v - (\mathcal{P}_{k+l} - \mathcal{P}_k) v \rangle = 0, \qquad (i', j') \in \mathbb{U}_{k,l}. \tag{10.19}$$

We first show the sufficient condition. To this end, we assume that $v \in \mathbb{X}$ satisfies equation (10.18). It follows from $\mathbb{L}_{k+l} \cap \mathbb{X}_{k+l}^{\perp} = \{0\}$ and equation (10.18) that for any $(i', j') \in \mathbb{U}_{k,l}$, there exists $w \in \mathbb{W}_{k,l}$ such that $\langle \ell_{i'j'}, w \rangle \neq 0$. Therefore, we observe that $\mathbb{V}_{k,l} \cap \mathbb{W}_{k,l}^{\perp} = \{0\}$. Now we suppose $v \in \mathbb{X}$ is a solution of equation (10.16). Thus, it follows directly from equation (10.19) that v satisfies equation (10.17). Conversely, if v satisfies equation (10.17) but is not a solution of equation (10.16), then we can find $\ell \in \mathbb{V}_{k,l}$ such that $\langle \ell, (\mathcal{P}_{k+l} - \mathcal{P}_k) v \rangle \neq 0$, thus $\langle \ell, v - (\mathcal{P}_{k+l} - \mathcal{P}_k) v \rangle \neq 0$, which contradicts equation (10.19).

It remains to prove the necessary condition. For any $v \in \mathbb{X}$, we verify directly that $v - (\mathcal{P}_{k+l} - \mathcal{P}_k) v$ is a solution of equation (10.16), and hence it is also a solution of equation (10.17). Thus, we obtain equation (10.19). By the equivalence of conditions (10.18) and (10.19), we prove equation (10.18). □

The next theorem describes the condition for which equation (10.10) is converted into its equivalent discrete form.

Theorem 10.8 *The following statements are equivalent:*

(i) Equation (10.10) is equivalent to

$$\langle \ell_{i'j'}, u_{k,l}^H \rangle = \langle \ell_{i'j'}, f + \mathcal{K}\Psi u_{k,l-1} \rangle, \qquad (i', j') \in \mathbb{U}_{k,l}. \tag{10.20}$$

(ii) $\mathbb{V}_p \subset \mathbb{X}_k^{\perp}$, $p > k$.

(iii) For any $v \in \mathbb{X}_k$, $\langle \ell_{i'j'}, v \rangle = 0$, $(i', j') \in \mathbb{U}_{k,l}$.

Proof The equivalence of statements (ii) and (iii) is clear. It suffices to prove the equivalence of statements (i) and (ii). Note that equation (10.10) is equivalent to

$$(\mathcal{P}_{k+l} - \mathcal{P}_k)[u_{k,l}^H - (f + \mathcal{K}\Psi u_{k,l-1})] = 0.$$

By Lemma 10.7, the above equation is equivalent to (10.20) if and only if statement (ii) holds. □

We are now ready to present the discrete form of the MAM. For $(i', j'), (i, j) \in \mathbb{U}_{k,l}$, we define the matrix $\mathbf{E}_{k,l} := [\langle \ell_{i'j'}, w_{ij} \rangle : (i', j'), (i, j) \in \mathbb{U}_{k,l}]$.

Using this notation, equation (10.20) can be rewritten as

$$\mathbf{E}_{k,l}\mathbf{u}^H_{k,l} = \mathbf{f}_{k,l}, \tag{10.21}$$

where $\mathbf{u}^H_{k,l} := [(\mathbf{u}_{k,l})_{ij} : (i,j) \in \mathbb{U}_{k,l}]$ and $\mathbf{f}_{k,l} := [\langle \ell_{i'j'}, f + \mathcal{K}\Psi u_{k,l-1}\rangle : (i',j') \in \mathbb{U}_{k,l}]$.

Algorithm 10.9 *(The multilevel augmentation method: A discrete form) Let k be a fixed positive integer.*

Step 1: *Solve the nonlinear system*

$$\left\langle \ell_{i'j'}, (\mathcal{I} - \mathcal{K}\Psi)\left(\sum_{(i,j)\in\mathbb{U}_k} (\mathbf{u}_k)_{ij} w_{ij}\right)\right\rangle = \langle \ell_{i'j'}, f\rangle, \qquad (i',j') \in \mathbb{U}_k. \tag{10.22}$$

and obtain the solution $\mathbf{u}_k := [(\mathbf{u}_k)_{ij} : (i,j) \in \mathbb{U}_k]$. Let $\mathbf{u}_{k,0} := \mathbf{u}_k$ and $l := 1$.

Step 2: *Solve the linear system (10.21) to obtain $\mathbf{u}^H_{k,l}$ and define*

$$u^H_{k,l} := \sum_{(i,j)\in\mathbb{U}_{k,l}} (\mathbf{u}_{k,l})_{ij} w_{ij}.$$

Step 3: *Solve the nonlinear system (10.15) to obtain $\mathbf{u}^L_{k,l} := [(\mathbf{u}_{k,l})_{ij} : (i,j) \in \mathbb{U}_k]$. Define $u^L_{k,l} := \sum_{(i,j)\in\mathbb{U}_k} (\mathbf{u}_{k,l})_{ij} w_{ij}$ and $u_{k,l} = u^L_{k,l} + u^H_{k,l}$.*

Step 4: *Set $l \leftarrow l + 1$ and go back to step 2.*

A crucial procedure in Algorithm 10.9 is to repeatedly solve the nonlinear system (10.15). A typical approach to solve this system is by the Newton iteration or the secant method. There are two strategies to implement the Newton iteration/secant method. The first strategy is to update the Jacobian matrix of the nonlinear system (10.15) at each step. The drawback of this strategy is that updating the Jacobian matrix is time-consuming. This can be improved by reducing the frequency of updating the Jacobian matrix. The second strategy is that in step 3 of Algorithm 10.9, we use the same Jacobian matrix obtained in step 1 when solving the nonlinear system (10.15). This modification avoids updating the Jacobian matrix and thus it significantly reduces the computational cost. It may affect the approximation accuracy. However, it can be compensated by a few more iterations. The numerical results to be presented later show that this strategy indeed speeds up the computation significantly while preserving the approximation accuracy.

In the rest of this subsection, we estimate the computational cost of Algorithm 10.9, which is measured by the number of multiplications used in computation. We suppose that the initial approximate solution u_k has been obtained, and we intend to find $u_{k,m}$ by using Algorithm 10.9. According to the algorithm, we divide the computation into m stages. For stage i with $i = 1, 2, \ldots, m$, we perform the following procedures:

(1) Generate the coefficient matrix $\mathbf{E}_{k,i}$.
(2) Compute the vector $\mathbf{f}_{k,i}$.
(3) Solve the linear system (10.21).
(4) Solve the nonlinear system (10.15).

The computational cost of Algorithm 10.9 is estimated according to each procedure described above. We denote by $\mathcal{M}_{k,i,j}$ the computational cost in the above procedure j at stage i. We assume that the following hypothesis holds:

(A0) Computing the integrals that appear in $\mathbf{f}_{k,i}$ requires a constant computational cost per integral.

Specifically, we identify $\mathcal{M}_{k,i,1}$ and $\mathcal{M}_{k,i,2}$, respectively, by the number of entries of the matrix $\mathbf{E}_{k,i}$ and that of the components of the vector $\mathbf{f}_{k,i}$. Moreover, $\mathcal{M}_{k,i,3}$ is the number of multiplications used for solving the linear system (10.21) and $\mathcal{M}_{k,i,4}$ is the number of multiplications used for solving the nonlinear system (10.15). Since in procedure 4 we solve the same nonlinear system with different function $u^H_{k,l}$, we then conclude that $\mathcal{M}_{k,i,4} = \mathcal{O}(1)$. It remains to estimate $\mathcal{M}_{k,i,3}$. For this purpose, we make the following additional hypotheses.

(A1) There exists a positive integer $\mu > 1$ such that, for any n, the dimension $s(n)$ of $\mathbb{X}_n$ is equivalent to μ, that is, $s(n) \sim \mu^n$.
(A2) For any i the matrix $\mathbf{E}_{k,i}$ is an upper triangular sparse matrix with $\mathcal{N}(\mathbf{E}_{k,i}) = \mathcal{O}(s(k+i))$, where $\mathcal{N}(\mathbf{A})$ denotes the number of nonzero elements of the matrix $\mathbf{A}$.

We present estimates for $\mathcal{M}_{k,i,j}$, $j = 1, 2, 3$ in the next proposition.

Proposition 10.10 *If assumptions (A0), (A1) and (A2) hold, then for any $i > 0$ and $j = 1, 2, 3$,*

$$\mathcal{M}_{k,i,j} = \mathcal{O}(s(k+i)).$$

Proof The estimates for $j = 1, 2$ are clear. The number of multiplications in finding the solution of linear system (10.21) equals the number of nonzero entries of the coefficient matrix $\mathbf{E}_{k,i}$. According to assumption (A2), $\mathcal{N}(\mathbf{E}_{k,i}) = \mathcal{O}(s(k+i))$. The proof is complete using assumption (A1). □

Next, we let $\overline{\mathcal{M}}_{k,m,j}$ denote the total computational cost related to procedure j for obtaining the solution $u_{k,m}$.

Corollary 10.11 *If assumptions (A0), (A1) and (A2) hold, then*

$$\overline{\mathcal{M}}_{k,m,j} = \mathcal{O}(s(k+m)),\ j = 1,2,3 \quad \textit{and} \quad \overline{\mathcal{M}}_{k,m,4} = \mathcal{O}(m).$$

Proof For $j = 1,2,3,4$, we have that

$$\overline{\mathcal{M}}_{k,m,j} = \sum_{i=1}^{m} \mathcal{M}_{k,i,j}.$$

The result of this corollary follows directly from the equation above and Proposition 10.10. □

Theorem 10.12 *If assumptions (A0), (A1) and (A2) hold, then the total computational cost of obtaining the solution $u_{k,m}$ by Algorithm 10.9 is in the order $\mathcal{O}(s(k+m))$, where $s(k+m)$ is the dimension of the subspace $\mathbb{X}_{k+m}$.*

Proof The total computational cost of obtaining the solution $u_{k,m}$ by Algorithm 10.9 is given by the sum of $\overline{\mathcal{M}}_{k,m,j}$ over $j = 1,2,3,4$. The desired estimate of this theorem follows from Corollary 10.11 and the equivalence of $s(k+m)$ and μ^{k+m}. □

The above theorem reveals that the computational cost for Algorithm 10.9 is *linear* with respect to the dimension of the approximation space under assumption (A0).

10.2.4 The Galerkin and collocation-based methods

In this subsection, we present two specific MAMs, one based on the Galerkin method and the other based on the collocation method.

We first recall a multiscale partition Ω_n, $n \in \mathbb{N}_0$ of the domain Ω. For each scale n, the partition Ω_n consists of a family of subsets $\Omega_{n,i}$, $i \in \mathbb{Z}_{e(n)}$, where $e(n)$ denotes the cardinality of Ω_n, with the properties that

$$\text{meas}(\Omega_{n,i} \cap \Omega_{n,i'}) = 0,\ \ i, i' \in \mathbb{Z}_{e(n)},\ i \neq i' \ \text{ and } \bigcup_{i \in \mathbb{Z}_{e(n)}} \Omega_{n,i} = \Omega.$$

The multiscale property requires that for $n > n'$ and $i \in \mathbb{Z}_{e(n)}$, there exists a unique $i' \in \mathbb{Z}_{e(n')}$ such that $\Omega_{n,i} \subset \Omega_{n',i'}$. We also demand that the partition is "shrinking" at a proper rate, that is, there is a positive constant $\tau \in (0,1)$ such that for sufficiently large n, $d(n) \leq \tau^n$, where $d(n)$ denotes the largest diameter of the subsets in Ω_n. The growth of the number of elements in Ω_n in n is required to satisfy $e(n) = \mathcal{O}(\mu^n)$. When Ω can be decomposed into a union of simplices, a multiscale partition of Ω that meets the above requirements is given in Sections 4.2, 5.1 and 7.1 (cf. [69, 75]).

We now describe the multilevel Galerkin scheme. In this case, we choose $\mathbb{X} = L^2(\Omega)$ and the subspaces $\mathbb{X}_n$ as spaces of piecewise polynomials of order r, associated with a multiscale partition Ω_n, $n \in \mathbb{N}_0$ of the domain Ω. The multiscale partition described above guarantees the nestedness of the space sequence $\mathbb{X}_n$, $n \in \mathbb{N}_0$. The space $\mathbb{X}_{n+1}$ can be decomposed as the orthogonal sum of $\mathbb{X}_n$ and its orthogonal complement $\mathbb{W}_{n+1}$ in $\mathbb{X}_{n+1}$. We write $\mathbb{W}_0 := \mathbb{X}_0$. Then, $\mathbb{X}_n$ can be written as the orthogonal direct sum of $\mathbb{W}_i$, for $i \in \mathbb{Z}_{n+1}$.

The projection $\mathcal{P}_n$ is naturally chosen as the orthogonal projection from $\mathbb{X}$ onto $\mathbb{X}_n$. As a result, equation (10.3) becomes the Galerkin scheme for solving (10.2) and accordingly, Algorithm 10.2 is a MAM based on the Galerkin scheme. In this case, Theorem 10.6 has the following form.

Theorem 10.13 *Let u^* be an isolated solution of (10.2). If one is not an eigenvalue of $(\mathcal{K}\Psi)'(u^*)$ and if $u^* \in W^{r,2}(\Omega)$, then there exist a positive constant c and a positive integer N such that for all $k \geq N$ and $m \in \mathbb{N}_0$,*

$$\|u^* - u_{k,m}\|_2 \leq c\tau^{r(k+m)}\|u^*\|_{r,2}.$$

Proof We may define a sequence γ_n by

$$\gamma_n := c\tau^{rn}\|u^*\|_{r,2},$$

where c is a positive constant independent of n such that $R_n \leq \gamma_n$, recalling that R_n is the error of the best approximation to u^* from $\mathbb{X}_n$. Since $u^* \in W^{r,2}(\Omega)$, we conclude that for all $n \in \mathbb{N}_0$,

$$\frac{\gamma_{n+1}}{\gamma_n} = \tau^r.$$

Hence, the sequence γ_n is a majorization sequence of R_n, $n \in \mathbb{N}_0$. The desired result therefore follows directly from Theorem 10.6. □

We next comment on the discrete form of the MAM based on the Galerkin scheme. For any $i \geq 0$, we choose an orthonormal basis w_{ij}, $j \in \mathbb{Z}_{w(i)}$ for the space $\mathbb{W}_i$, so that w_{ij}, $(i,j) \in \mathbb{U}_n$ form an orthonormal basis of $\mathbb{X}_n$. The construction of these bases may be found in Chapter 4. The choice of the approximation spaces and their bases permits the hypotheses (A1) and (A2) described in Section 10.2.3 to be satisfied. Since $\mathbb{X}^* = L^2(\Omega)$, for $i \geq 0$ and $j \in \mathbb{Z}_{w(i)}$, the functional ℓ_{ij} may be chosen as $\langle \ell_{ij}, x\rangle = (w_{ij}, x)$, for $x \in L^2(\Omega)$, where $(\cdot, \cdot)$ denotes the inner product in L^2. In this setting, the nonlinear system (10.15) has the form

$$\left(w_{i'j'}, (\mathcal{I} - \mathcal{K}\Psi)\left(\sum_{(i,j)\in\mathbb{U}_k} u_{ij}w_{ij} + u^H_{k,m}\right)\right) = \left(w_{i'j'}, f\right), \qquad (i',j') \in \mathbb{U}_k \tag{10.23}$$

and the matrix $\mathbf{E}_{k,m}$ becomes the identity matrix. Hence, equation (10.21) is reduced to

$$\mathbf{u}^H_{k,m} = \mathbf{f}_{k,m}. \tag{10.24}$$

Algorithm 10.9 with the Galerkin method has a very simple form.

We now turn our attention to the MAM based on the collocation method. The multiscale collocation method that we are describing here was first introduced in [69]. We set $\mathbb{X} = L^\infty(\Omega)$ and as in the Galerkin case, we choose the subspaces $\mathbb{X}_n$ as spaces of piecewise polynomials of order r, associated with a multiscale partition Ω_n, $n \in \mathbb{N}_0$ of the domain Ω. Again, we need the orthogonal complement $\mathbb{W}_n$ of $\mathbb{X}_{n-1}$ in $\mathbb{X}_n$ and the same orthogonal decomposition of $\mathbb{X}_n$ as in the Galerkin case. However, we do not demand the orthogonality of basis functions within the subspace $\mathbb{W}_n$.

We need to describe collocation functionals. To this end, we recall that $\mathbb{Y} = C(\Omega)$ and a collocation functional is chosen as an element in $\mathbb{Y}^*$. Specifically, the space $\mathbb{L}_n$ in this case is spanned by a basis whose elements are point evaluation functionals. Note that $\mathbb{L}_n$ has a decomposition $\mathbb{L}_n = \mathbb{V}_0 \oplus \cdots \oplus \mathbb{V}_n$ with $\mathbb{V}_0 := \mathbb{L}_0$, where $\mathbb{V}_i := \operatorname{span}\{\ell_{ij} : j \in \mathbb{Z}_{w(i)}\}$ will be described below. We construct $\mathbb{V}_0$ from refinable sets of points in Ω with respect to families of contractive maps which define the refinement of the multiscale partitions of Ω. Functionals ℓ_{0j} are the point evaluation functionals associated with the points in the refinable sets. Each functional ℓ_{1j} is defined by a linear combination of point evaluation functionals, with the number of such functionals being bounded independent of i, and satisfies the "semi-bi-orthogonality" property with respect to w_{ij}, $i = 0, 1, j \in \mathbb{Z}_{w(i)}$, that is, $\langle \ell_{i'j'}, w_{ij} \rangle = \delta_{ii'}\delta_{jj'}$, for $i \leq i'$. The functionals $\ell_{ij}, i > 1, j \in \mathbb{Z}_{w(i)}$ are defined recursively from $\ell_{1j}, j \in \mathbb{Z}_{w(1)}$. The projection $\mathcal{P}_n$ in this case is naturally the interpolatory projection on $\mathbb{X}_n$. It can readily be verified that the assumptions (A1) and (A2) described in Section 10.2.3 hold.

With the interpolatory projection $\mathcal{P}_n$, equation (10.3) becomes the collocation scheme for solving (10.2) and accordingly, Algorithm 10.2 is a MAM based on the collocation scheme. Similar to Theorem 10.13, we have the following convergence result for the collocation-based MAM.

Theorem 10.14 *Let u^* be an isolated solution of (10.2). If one is not an eigenvalue of $(\mathcal{K}\Psi)'(u^*)$ and if $u^* \in W^{r,\infty}(\Omega)$, then there exist a positive constant c and a positive integer N such that for all $k \geq N$ and $m \in \mathbb{N}_0$,*

$$\|u^* - u_{k,m}\|_\infty \leq c\tau^{r(k+m)}\|u^*\|_{r,\infty}.$$

Since the proof for Theorem 10.14 is similar to that for Theorem 10.13, we omit it.

To close the subsection, we remark on the influence of numerical integration on the approximation errors of numerical solutions and the computational complexity of the algorithm for the multiscale collocation method. In the computational complexity analysis presented in Section 10.2.3, we impose assumption (A0) for the estimate of $\mathcal{M}_{k,i,2}$. However, this assumption may not be fulfilled in all cases. Additional computational efforts may be needed to compute the vector $\mathbf{f}_{k,l}$. We take the multiscale bases formed by piecewise polynomials as an example. When the collocation methods are applied to discretize the integral equation, $\mathcal{M}_{k,l,2}$ indicates the computational cost for computing

$$\left\langle \ell_{i'j'}, \mathcal{K}\Psi u_{k,l-1} \right\rangle, \quad (i',j') \in \mathbb{U}_{k,l}, \tag{10.25}$$

where the functionals $\ell_{i'j'}$ are linear combinations of point evaluations. Therefore, we need to evaluate numerically the integrals of the form

$$\int_\Omega K(s,t)\psi(t, u_{k,l-1}(t))\mathrm{d}t.$$

According to [96, 273], under suitable assumptions on the regularity of the nonlinear function ψ, we have an approximation

$$\psi(t, u_{k,l-1}(t)) \approx \sum_{(i,j)\in\mathbb{U}_{k+l}} b_{ij}w_{ij}(t)$$

with an optimal order of convergence and computational complexity $\mathcal{O}(s(k+l)\log s(k+l))$. Then, computing (10.25) reduces to calculating

$$\left\langle \ell_{i'j'}, \mathcal{K}w_{ij} \right\rangle, \quad (i',j'), (i,j) \in \mathbb{U}_{k,l}.$$

When the kernel is smooth or weakly singular, we can establish truncation strategies for the matrix and error control strategies for computing the rest elements of the matrix, the cost of which is $\mathcal{O}(s(k+l)(\log s(k+l))^\nu)$, where the positive integer ν depends on the dimension of the domain Ω. When $d = 1, 2, 3$, the value of ν is $3, 4, 5$, respectively. See [72, 75, 264] for details. Summarizing the above discussion, we observe that the total computational cost for computing $\mathbf{f}_{k,l}$ is of $\mathcal{O}(s(k+l)(\log s(k+l))^\nu)$. A similar approach applies to the Galerkin method.

10.3 Multiscale methods for nonlinear boundary integral equations

In this section we develop the MAM for solving the nonlinear boundary integral equation, propose a matrix compression strategy and present accelerated quadratures and Newton iterations for seeding up the computation.

10.3.1 The multilevel augmentation method

We describe in this subsection the MAM for solving the nonlinear boundary integral equation. We begin by recalling the reformation of the nonlinear boundary value problem as a nonlinear integral equation.

Let Ω be a simply connected bounded domain in $\mathbb{R}^2$ with a C^2 boundary Γ. We consider solving the following nonlinear boundary value problem:

$$\begin{cases} \Delta u(x) = 0, & x \in \Omega, \\ \dfrac{\partial u}{\partial \mathbf{n}_x}(x) = -g(x, u(x)) + g_0(x), & x \in \Gamma, \end{cases} \tag{10.26}$$

where $\mathbf{n}_x$ denotes the exterior unit normal vector to Γ at x. The numerical solution of the above problem was studied in many papers (see, for example, [17, 239] and the references cited therein). The fundamental solution of the Laplace equation in $\mathbb{R}^2$ is given by

$$\Phi(x, y) := -\frac{1}{2\pi} \log |x - y|.$$

It is shown (cf. [17, 239]) that problem (10.26) can be reformulated as the following nonlinear integral equation defined on Γ:

$$\begin{aligned} u(x) - \frac{1}{\pi}\int_\Gamma u(y)\frac{\partial}{\partial \mathbf{n}_y} \log |x - y| ds_y - \frac{1}{\pi}\int_\Gamma g(y, u(y)) \log |x - y| ds_y \\ = -\frac{1}{\pi}\int_\Gamma g_0(y) \log |x - y| ds_y, \quad x \in \Gamma. \end{aligned} \tag{10.27}$$

Note that the first integral operator is linear, while the second is nonlinear. We assume that the boundary Γ has a parametrization $x = (\xi(t), \eta(t)), t \in [0, 1)$. With this representation, the functions that appeared in (10.27), which are defined on Γ, are transformed to functions of the variable t. For simplicity, we use the same notations, that is,

$$u(t) := u(\xi(t), \eta(t)), \quad g(t, u(t)) := g((\xi(t), \eta(t)), u), \quad g_0(t) = g_0(\xi(t), \eta(t)).$$

With these notations, according to [15, 17], equation (10.27) is rewritten as

$$\begin{aligned} u(t) - \int_0^1 K(t,\tau)u(\tau)\mathrm{d}\tau - \int_0^1 L(t,\tau)g(\tau,u(\tau))\chi(\tau)\mathrm{d}\tau \\ = -\int_0^1 L(t,\tau)g_0(\tau)\chi(\tau)\mathrm{d}\tau, \qquad t \in [0,1), \end{aligned} \tag{10.28}$$

where for $t, \tau \in [0,1)$,

$$K(t,\tau) := \begin{cases} \dfrac{1}{\pi}\dfrac{\eta'(\tau)(\xi(\tau)-\xi(t)) - \xi'(\tau)(\eta(\tau)-\eta(t))}{(\xi(t)-\xi(\tau))^2 + (\eta(t)-\eta(\tau))^2}, & t \neq \tau, \\ \dfrac{1}{\pi}\dfrac{\eta'(t)\xi''(t) - \xi'(t)\eta''(t)}{2[\xi'(t)^2 + \eta'(t)^2]}, & t = \tau, \end{cases}$$

$$L(t,\tau) := \frac{1}{2\pi}\log[(\xi(t)-\xi(\tau))^2 + (\eta(t)-\eta(\tau))^2], \quad t \neq \tau$$

and

$$\chi(\tau) := \sqrt{\xi'(\tau)^2 + \eta'(\tau)^2}.$$

We introduce two linear integral operators $\mathcal{K}, \mathcal{L} : L^\infty(0,1) \to L^\infty(0,1)$ defined, respectively, by

$$(\mathcal{K}w)(t) := \int_0^1 K(t,\tau)w(\tau)\mathrm{d}\tau, \quad t \in [0,1)$$

and

$$(\mathcal{L}w)(t) := \int_0^1 L(t,\tau)w(\tau)\mathrm{d}\tau, \quad t \in [0,1),$$

and the nonlinear operator

$$(\Psi u)(t) := g(t,u(t))\chi(t), \quad t \in [0,1).$$

By letting

$$\mathcal{T} := \mathcal{K} + \mathcal{L}\Psi,$$

we rewrite equation (10.28) as

$$u - \mathcal{T}u = f, \tag{10.29}$$

in which the right-hand-side function $f := -\mathcal{L}(g_0\chi)$.

In passing, we comment on the regularity of the two kernels K and L. It is easy to verify that when Γ is of C^s with $s \geq 2$, K has continuous derivatives up to order $s-2$. Throughout this section, we assume s is sufficiently large. Hence, there exists a positive constant Λ such that

$$|D_t^{\alpha} D_{\tau}^{\beta} K(t,\tau)| \leq \Lambda, \qquad t,\tau \in (0,1), \tag{10.30}$$

for positive integers α, β with $\alpha+\beta \leq s-2$. The expression of L contains a logarithmic factor, which exhibits a weak singularity. The positions of singular points are determined by the properties of parametrization ξ and η. Noting that Γ is a closed curve, the singular points are located where $t-\tau = 0,-1,1$. Accordingly, we require that $L(t,\cdot) \in C^{\infty}([0,1]\backslash\{t\})$ for any $t \in [0,1]$, and there exist positive constants θ and $\sigma \in (0,1)$ such that

$$|D_t^{\alpha} D_{\tau}^{\beta} L(t,\tau)| \leq \theta \cdot \max\left\{|t-\tau|^{-(\sigma+\alpha+\beta)}, |t-\tau+1|^{-(\sigma+\alpha+\beta)}, \right.$$
$$\left. |t-\tau-1|^{-(\sigma+\alpha+\beta)}\right\} \tag{10.31}$$

for any $\alpha,\beta \in \mathbb{N}_0$ and $t,\tau \in [0,1]$, with $t \neq \tau$ and $t-\tau \neq \pm 1$. We remark that the above setting of weak singularity includes not only the logarithmic singularity but also other kinds of singularity.

We now return to equation (10.29). The solvability of (10.29) was considered in the literature (cf. [239]). Throughout the rest of this section, we assume that (10.29) has an isolated solution $u^* \in C(0,1)$. Moreover, we suppose that the function $g(x,u)$ is continuous with respect to $x \in \Gamma$ and Lipschitz continuous with respect to $u \in \mathbb{R}$, the partial derivative $D_u g$ of g with respect to the variable u exists and is Lipschitz continuous, and for each $u \in C(\Gamma)$, $g(\cdot,u(\cdot)), D_u g(\cdot,u(\cdot)) \in C(\Gamma)$.

Next, we describe the fast algorithm for (10.29) in light of the idea of MAM. For $n \in \mathbb{N}_0$, let π_n be the uniform mesh which divides the interval $[0,1]$ into μ^n pieces for a given positive integer μ, and $\mathbb{X}_n$ be the piecewise polynomial space of order r with respect to π_n. It is easily observed that the sequence $\mathbb{X}_n$, $n \in \mathbb{N}_0$ is nested, that is, $\mathbb{X}_n \subset \mathbb{X}_{n+1}$. For each $n \in \mathbb{N}_0$, let $\mathcal{P}_n$ be the interpolatory projection from $C(0,1)$ onto $\mathbb{X}_n$ with the set of interpolation points

$$\left\{\frac{j+s}{\mu^n} : j \in \mathbb{Z}_{\mu^n}, s \in G\right\},$$

where G is the set of *initial interpolation points* in $[0,1]$. We require that G has two properties. One is that G contains r distinct points and, as a result, the interpolation of polynomials of order r on G exists uniquely. The other is that G is *refinable* with respect to the family of contractive affine mappings $\mathbf{\Phi}_\mu := \{\phi_e : e \in \mathbb{Z}_\mu\}$, where

$$\phi_e(x) := \frac{x+e}{\mu}, \qquad e \in \mathbb{Z}_\mu,$$

in the sense that

$$G \subset \mathbf{\Phi}_{\mu}(G) := \bigcup_{e \in \mathbb{Z}_{\mu}} \phi_e(G).$$

The collocation method for solving (10.29) is to find $u_n \in \mathbb{X}_n$ such that

$$u_n - \mathcal{P}_n \mathcal{T} u_n = \mathcal{P}_n f. \tag{10.32}$$

Making use of Theorem 2 of [255], we prove below that (10.32) is uniquely solvable.

Theorem 10.15 *If $u^* \in C(0,1)$ is an isolated solution of (10.29), and one is not an eigenvalue of the linear operator $\mathcal{T}'(u^*)$, then for sufficiently large n, (10.32) has a unique solution $u_n \in B(u^*, \delta)$ for some $\delta > 0$, and there exist positive constants c_1, c_2 such that*

$$c_1 \|u^* - \mathcal{P}_n u^*\|_\infty \le \|u^* - u_n\|_\infty \le c_2 \|u^* - \mathcal{P}_n u^*\|_\infty.$$

We now describe the MAM for finding an approximate solution of equation (10.32). The nestedness of the subspace sequence allows us to have the decomposition of $\mathbb{X}_{n+1}$ as the direct sum of $\mathbb{X}_n$ and its orthogonal complement $\mathbb{W}_{n+1}$. Thus, for a fixed $k \in \mathbb{N}_0$ and any $m \in \mathbb{N}_0$, we have that

$$\mathbb{X}_{k+m} = \mathbb{X}_k \oplus \mathbb{W}_{k,m}, \quad \text{where} \quad \mathbb{W}_{k,m} := \mathbb{W}_{k+1} \oplus \mathbb{W}_{k+2} \oplus \cdots \oplus \mathbb{W}_{k+m}. \tag{10.33}$$

We now solve equation (10.32) with $n := k+m$, k being fixed and small relative to n. At the first step, we solve equation (10.32) with $n := k$ exactly and obtain the solution u_k. Since $\dim(\mathbb{X}_k)$ is small in comparison with $\dim(\mathbb{X}_{k+m})$, the computational cost to invert the nonlinear operator $\mathcal{P}_k(\mathcal{I} - \mathcal{T})$ is much less than that to invert $\mathcal{P}_{k+m}(\mathcal{I} - \mathcal{T})$. The next step is to obtain an approximation of the solution u_{k+1} of (10.32) with $n := k+1$. For this purpose, we decompose

$$u_{k+1} = u_{k+1}^L + u_{k+1}^H, \quad \text{with } u_{k+1}^L \in \mathbb{X}_k \quad \text{and } u_{k+1}^H \in \mathbb{W}_{k+1},$$

according to the decomposition (10.33), and rewrite equation (10.32) with $n := k+1$ in its equivalent form as

$$\begin{cases} (\mathcal{P}_{k+1} - \mathcal{P}_k)(u_{k+1}^L + u_{k+1}^H) - (\mathcal{P}_{k+1} - \mathcal{P}_k)\mathcal{T} u_{k+1} = (\mathcal{P}_{k+1} - \mathcal{P}_k) f, \\ \mathcal{P}_k(\mathcal{I} - \mathcal{T})(u_{k+1}^L + u_{k+1}^H) = \mathcal{P}_k f. \end{cases} \tag{10.34}$$

In view of

$$(\mathcal{P}_{k+1} - \mathcal{P}_k)(u_{k+1}^L + u_{k+1}^H) = u_{k+1}^H,$$

the first equation in 10.34 becomes

$$u_{k+1}^H = (\mathcal{P}_{k+1} - \mathcal{P}_k)(f + \mathcal{T}u_{k+1}).$$

The right-hand side of the above equation can be obtained approximately via the solution u_k at the previous level. That is, we compute

$$u_{k,1}^H := (\mathcal{P}_{k+1} - \mathcal{P}_k)(f + \mathcal{T}u_{k,0}),$$

where $u_{k,0} = u_k$, and note that $u_{k,1}^H \in \mathbb{W}_{k+1}$. We replace u_{k+1}^H in the second equation of (10.34) by $u_{k,1}^H$ and solve $u_{k,1}^L \in \mathbb{X}_k$ from the equation

$$\mathcal{P}_k(\mathcal{I} - \mathcal{T})(u_{k,1}^L + u_{k,1}^H) = \mathcal{P}_k f.$$

The solution $u_{k,1}^L$ of the above equation is a good approximation to u_{k+1}^L. We then obtain an approximation to the solution u_{k+1} of (10.32) by letting

$$u_{k,1} := u_{k,1}^L + u_{k,1}^H.$$

Note that $u_{k,1}^L$ and $u_{k,1}^H$ represent, respectively, the lower and higher-frequency components of $u_{k,1}$. This procedure is repeated m times to obtain the approximation $u_{k,m}$ of the solution u_{k+m} of (10.32) with $n := k + m$. At step ℓ of this procedure, we do not invert the nonlinear operator $\mathcal{P}_{k+\ell}(\mathcal{I} - \mathcal{T})$ but invert only the same nonlinear operator $\mathcal{P}_k(\mathcal{I} - \mathcal{T})$. This makes the method very efficient computationally. We summarize this procedure in the following algorithm.

Algorithm 10.16 *(The multilevel augmentation method in an operator form) Let k be a fixed positive integer.*

Step 1: *Find the solution $u_k \in \mathbb{X}_k$ of equation (10.32) with $n := k$. Set $u_{k,0} := u_k$ and $l := 1$.*

Step 2: *Compute*

$$u_{k,l}^H = (\mathcal{P}_{k+l} - \mathcal{P}_k)(f + \mathcal{T}u_{k,l-1}) \in \mathbb{W}_{k,l}. \tag{10.35}$$

Step 3: *Solve for $u_{k,l}^L \in \mathbb{X}_k$ from the equation*

$$\mathcal{P}_k(\mathcal{I} - \mathcal{T})(u_{k,l}^L + u_{k,l}^H) = \mathcal{P}_k f. \tag{10.36}$$

Step 4: *Let $u_{k,l} := u_{k,l}^L + u_{k,l}^H$. Set $l \leftarrow l + 1$ and go back to step 2 until $l = m$.*

The output of the MAM is $u_{k,m} \in \mathbb{X}_{k+m}$. By employing the analysis in the last section for the Hammerstein equations, we establish the following approximation result for $u_{k,m}$.

Theorem 10.17 *If $u^* \in C(0,1)$ is an isolated solution of (10.29) and one is not an eigenvalue of the linear operator $\mathcal{T}'(u^*)$, then there exists a positive integer N such that for any $k > N$, the MAM solution $u_{k,m}$ exists uniquely for all $m > 0$, and $u_{k,m} \in B(u^*, \delta)$ for some $\delta > 0$. Moreover, if $u^* \in W^{r,\infty}(0,1)$, then there exists a positive constant c such that for all $k > N$ and $m \in \mathbb{N}_0$,*

$$\|u^* - u_{k,m}\|_\infty \leq c\mu^{-r(k+m)}\|u^*\|_{r,\infty}.$$

As Proposition 10.3 states, the MAM solves $u_{k,l}$ successively for $l = 1, 2, \ldots, m$ from

$$(\mathcal{I} - \mathcal{P}_k\mathcal{T})u_{k,l} = \mathcal{P}_{k+l}f + (\mathcal{P}_{k+l} - \mathcal{P}_k)\mathcal{T}u_{k,l-1}. \tag{10.37}$$

Let us briefly compare (10.37) with (10.32). If we solve (10.32) directly, we have to invert $\mathcal{I} - \mathcal{P}_n\mathcal{T}$, which is a nonlinear operator on $\mathbb{X}_n$. The linearization of the nonlinear operator usually leads to a linear equation that requires high computational complexity to solve. Therefore, this is not an *economic* way to solve equation (10.32) when the dimension of $\mathbb{X}_n$ is large. The solution of (10.37), however, need only invert $\mathcal{I} - \mathcal{P}_k\mathcal{T}$, with $k << n$. Note that the nonlinear component of the operator is restricted to $\mathbb{X}_k$, the dimension of which is fixed and much smaller than that of the space where $u_{k,m}$ is located for a large m. To see it more clearly, we split (10.37) into (10.35) and (10.36), and observe that (10.35) is a linear equation.

We now recall multiscale bases for $\mathbb{X}_n$ and the corresponding multiscale collocation functionals introduced in Chapter 7. Suppose that $\mathbb{L}_n$, $n \in \mathbb{N}_0$ is a sequence of subspaces of $(L^\infty(0,1))^*$, which possesses the properties $\mathbb{L}_n \subset \mathbb{L}_{n+1}$, $\dim(\mathbb{L}_n) = \dim(\mathbb{X}_n)$, $n \in \mathbb{N}_0$. We remark that the elements of $\mathbb{L}_n$ are point evaluations and their finite linear combinations, and the refinability of G, guarantees the above nestedness property. We utilize the nestedness property again to obtain the multiscale decomposition $\mathbb{L}_{k+m} = \mathbb{L}_k \oplus \mathbb{V}_{k,m}$ for any fixed $k \in \mathbb{N}_0$ and any $m \in \mathbb{N}_0$, in which $\mathbb{V}_{k,m} := \mathbb{V}_{k+1} \oplus \cdots \oplus \mathbb{V}_{k+m}$. Let $w(0) := \dim(\mathbb{X}_0)$ and $w(i) := \dim(\mathbb{W}_i)$ for $i > 0$. We suppose that

$$\mathbb{X}_0 = \operatorname{span}\{w_{0j} : j \in \mathbb{Z}_{w(0)}\}, \quad \mathbb{L}_0 = \operatorname{span}\{\ell_{0j} : j \in \mathbb{Z}_{w(0)}\}$$

and

$$\mathbb{W}_i = \operatorname{span}\{w_{ij} : j \in \mathbb{Z}_{w(i)}\}, \quad \mathbb{L}_i = \operatorname{span}\{\ell_{ij} : j \in \mathbb{Z}_{w(i)}\}, \qquad i > 0.$$

By introducing the index set $\mathbb{U}_n := \{(i,j) : j \in \mathbb{Z}_{w(i)}, i \in \mathbb{Z}_{n+1}\}$, we have

$$\mathbb{X}_n = \operatorname{span}\{w_{ij} : (i,j) \in \mathbb{U}_n\}, \quad \mathbb{L}_n = \operatorname{span}\{\ell_{ij} : (i,j) \in \mathbb{U}_n\}, \qquad n \in \mathbb{N}_0.$$

For any $l \in \mathbb{Z}_{m+1}$, we express the solution $u_{k,l}$ of (10.37) as

$$u_{k,l} = \sum_{(i,j)\in\mathbb{U}_{k+l}} (\mathbf{u}_{k,l})_{ij} w_{ij}.$$

We use the notation $\mathbf{u}_{k,l} := [(\mathbf{u}_{k,l})_{ij} : (i,j) \in \mathbb{U}_{k+l}]$ to denote the *representation vector* of $u_{k,l}$. Using the index set $\mathbb{U}_{k,l} := \mathbb{U}_{k+l}\backslash\mathbb{U}_k = \{(i,j) : j \in \mathbb{Z}_{w(i)}, i \in \mathbb{Z}_{k+l+1}\backslash\mathbb{Z}_{k+1}\}$, we have the expansion $u^L_{k,l} = \sum_{(i,j)\in\mathbb{U}_k}(\mathbf{u}_{k,l})_{ij}w_{ij}$ and $u^H_{k,l} = \sum_{(i,j)\in\mathbb{U}_{k,l}}(\mathbf{u}_{k,l})_{ij}w_{ij}$. Moreover, we define the matrix

$$\mathbf{E}^H_{k,l} := \left[\langle \ell_{i'j'}, w_{ij}\rangle : (i',j'), (i,j) \in \mathbb{U}_{k,l}\right].$$

As in the last section, we make use of the properties of the bases to conclude that the nonlinear equation (10.36) is equivalent to the nonlinear system

$$\left\langle \ell_{i'j'}, (\mathcal{I} - \mathcal{T})\left(\sum_{(i,j)\in\mathbb{U}_k} (\mathbf{u}_{k,l})_{ij} w_{ij} + u^H_{k,l}\right)\right\rangle = \langle \ell_{i'j'}, f\rangle, \qquad (i',j') \in \mathbb{U}_k, \tag{10.38}$$

and (10.35) is equivalent to

$$\mathbf{E}^H_{k,l}\mathbf{u}^H_{k,l} = \mathbf{f}_{k,l}, \tag{10.39}$$

where $\mathbf{u}^H_{k,l} := [(\mathbf{u}_{k,l})_{ij} : (i,j) \in \mathbb{U}_{k,l}]$ and

$$\mathbf{f}_{k,l} := [\langle \ell_{i'j'}, f + \mathcal{T}u_{k,l-1}\rangle : (i',j') \in \mathbb{U}_{k,l}]. \tag{10.40}$$

Computing $\mathbf{f}_{k,l}$ requires evaluating the integrals $\langle \ell_{i'j'}, (\mathcal{K} + \mathcal{L}\Psi)u_{k,l-1}\rangle$, for $(i',j') \in \mathbb{U}_{k,l}$. We separate the integral into its linear and nonlinear components

$$\langle \ell_{i'j'}, \mathcal{K}u_{k,l-1}\rangle, \qquad (i',j') \in \mathbb{U}_{k,l} \tag{10.41}$$

and

$$\langle \ell_{i'j'}, \mathcal{L}\Psi u_{k,l-1}\rangle, \qquad (i',j') \in \mathbb{U}_{k,l}. \tag{10.42}$$

We write

$$u_{k,l-1} = \sum_{(i,j)\in\mathbb{U}_{k+l-1}} (\mathbf{u}_{k,l-1})_{ij} w_{ij},$$

and define for k, l the matrix $\mathbf{K}^H_{k,l-1} := [K_{i'j',ij} : (i',j') \in \mathbb{U}_{k,l}, (i,j) \in \mathbb{U}_{k+l-1}]$. Computing the quantities in (10.41) is equivalent to generating $\mathbf{K}^H_{k,l-1}$ and calculating

$$\mathbf{K}^H_{k,l-1}\mathbf{u}_{k,l-1}. \tag{10.43}$$

The evaluation of the nonlinear component (10.42) will be done slightly differently. Since $\Psi u_{k,l-1} \notin \mathbb{X}_{k+l}$, we are not able to express $\Psi u_{k,l-1}$ as a linear combination of the basis of $\mathbb{X}_{k+l}$ as we do for computing (10.41). In order to establish a fast algorithm similar to that for evaluating (10.41), we approximate $\Psi u_{k,l-1}$ by its projection in $\mathbb{X}_{k+l}$. In other words, we do not evaluate (10.42) exactly but compute its approximation

$$\left\langle \ell_{i'j'}, \mathcal{L}\mathcal{P}_{k+l}\Psi u_{k,l-1}\right\rangle, \qquad (i',j') \in \mathbb{U}_{k,l}. \tag{10.44}$$

Formally, (10.44) has a form similar to (10.41), where $\mathcal{L}$ corresponds to $\mathcal{K}$ and $\mathcal{P}_{k+l}\Psi u_{k,l-1}$ corresponds to $u_{k,l-1}$. Therefore, the fast algorithm described above for (10.41) is applicable to (10.44). To see this, we write

$$\mathcal{P}_{k+l}\Psi u_{k,l-1} = \sum_{(i,j)\in\mathbb{U}_{k+l}} (\hat{\mathbf{u}}_{k+l})_{ij} w_{ij}$$

and thus, we have that

$$\left\langle \ell_{i'j'}, \mathcal{P}_{k+l}\Psi u_{k,l-1}\right\rangle = \sum_{(i,j)\in\mathbb{U}_{k+l}} (\hat{\mathbf{u}}_{k+l})_{ij} \left\langle \ell_{i'j'}, w_{ij}\right\rangle, \qquad (i',j') \in \mathbb{U}_{k+l}.$$

Let

$$\mathbf{g}_{k+l} := \left[\left\langle \ell_{i'j'}, \Psi\left(\sum_{(i,j)\in\mathbb{U}_{k+l-1}} (\hat{\mathbf{u}}_{k,l-1})_{ij} w_{ij}\right)\right\rangle : (i',j') \in \mathbb{U}_{k+l}\right],$$

and for $n \in \mathbb{N}_0$ define the matrix $\mathbf{E}_n := [\langle \ell_{i'j'}, w_{ij}\rangle : (i',j'), (i,j) \in \mathbb{U}_n]$. Then the representation vector $\hat{\mathbf{u}}_{k+l}$ of $\mathcal{P}_{k+l}\Psi u_{k,l-1}$ satisfies the linear system

$$\mathbf{E}_{k+l}\hat{\mathbf{u}}_{k+l} = \mathbf{g}_{k+l}. \tag{10.45}$$

Define for k, l the matrix $\mathbf{L}^H_{k,l} := [L_{i'j',ij} : (i',j') \in \mathbb{U}_{k,l}, (i,j) \in \mathbb{U}_{k+l}]$. Computing (10.44) is equivalent to solving $\hat{\mathbf{u}}_{k+l}$ from (10.45), generating $\mathbf{L}^H_{k,l}$ and then evaluating

$$\mathbf{L}^H_{k,l}\hat{\mathbf{u}}_{k+l}. \tag{10.46}$$

In the bases and collocation functionals described above, we have the matrix representations for the operators $\mathcal{K}$ and $\mathcal{L}$. Specifically, for $n \in \mathbb{N}_0$, we define the matrices

$$\mathbf{K}_n := [K_{i'j',ij} : (i',j'), (i,j) \in \mathbb{U}_n], \quad \text{with } K_{i'j',ij} := \left\langle \ell_{i'j'}, \mathcal{K}w_{ij}\right\rangle$$

and

$$\mathbf{L}_n := [L_{i'j',ij} : (i',j'), (i,j) \in \mathbb{U}_n], \quad \text{with } L_{i'j',ij} := \left\langle \ell_{i'j'}, \mathcal{L}w_{ij}\right\rangle.$$

Matrices $\mathbf{K}_n$ and $\mathbf{L}_n$ will be compressed according to the regularity of the kernels K and L. Note that the kernel K is smooth and L is weakly singular, and their regularities are described in (10.30) and (10.31), respectively. We adopt the following truncation strategies:

(T1) For each $n \in \mathbb{N}_0$, the matrix $\mathbf{K}_n$ is truncated to a sparse matrix

$$\tilde{\mathbf{K}}_n := [\tilde{K}_{i'j',ij} : (i',j'), (i,j) \in \mathbb{U}_n],$$

where for $(i',j'), (i,j) \in \mathbb{U}_n$,

$$\tilde{K}_{i'j',ij} := \begin{cases} K_{i'j',ij}, & i' + i \leq n, \\ 0, & \text{otherwise.} \end{cases}$$

(T2) For $(i,j) \in \mathbb{U}_n$, we let $S_{ij} := \text{supp}(w_{ij})$. For each $n \in \mathbb{N}_0$ and $(i',j'), (i,j) \in \mathbb{U}_n$, we set

$$\tilde{L}_{i'j',ij} := \begin{cases} L_{i'j',ij}, & \text{dist}(S_{i'j'}, S_{ij}) \leq \epsilon^n_{i'i} \text{ or } \text{dist}(S_{i'j'}, S_{ij}) \geq 1 - \epsilon^n_{i'i}, \\ 0, & \text{otherwise,} \end{cases}$$

in which the *truncation parameters* $\epsilon^n_{i'i}$ are chosen by

$$\epsilon^n_{i'i} := \max\{a\mu^{-n+b(n-i)+b'(n-i')}, \rho(\mu^{-i} + \mu^{-i'})\} \tag{10.47}$$

for some constants $b, b', a > 0$ and $\rho > 1$. The truncated matrix of $\mathbf{L}_n$ is defined by

$$\tilde{\mathbf{L}}_n := [\tilde{L}_{i'j',ij} : (i',j'), (i,j) \in \mathbb{U}_n].$$

When $n = 7$ and using piecewise linear basis functions, we show in Figure 10.1 the block matrices $\tilde{\mathbf{K}}_7$ and $\tilde{\mathbf{L}}_7$.

We now describe the MAM with the matrix truncations.

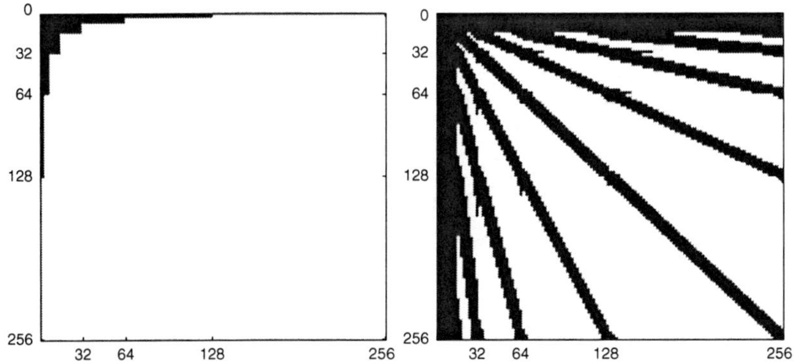

Figure 10.1 The distribution of nonzero entries of $\tilde{\mathbf{K}}_7$ (left) and $\tilde{\mathbf{L}}_7$ (right).

For a fixed $k \in \mathbb{N}_0$ and $l > 0$, we set

$$\tilde{\mathbf{K}}^H_{k,l-1} := [\tilde{K}_{i'j',ij} : (i',j') \in \mathbb{U}_{k,l}, (i,j) \in \mathbb{U}_{k+l-1}]$$

and

$$\tilde{\mathbf{L}}^H_{k,l} := [\tilde{L}_{i'j',ij} : (i',j') \in \mathbb{U}_{k,l}, (i,j) \in \mathbb{U}_{k+l}].$$

Algorithm 10.18 *(The multilevel augmentation method with matrix truncations) Let k be a fixed positive integer. Given $m \in \mathbb{N}_0$, we carry out the following computing steps.*

Step 1: *Solve the nonlinear system*

$$\left\langle \ell_{i'j'}, (\mathcal{I} - (\mathcal{K} + \mathcal{L}\Psi))\left(\sum_{(i,j)\in\mathbb{U}_k} (\mathbf{u}_k)_{ij} w_{ij} \right) \right\rangle = \langle \ell_{i'j'}, f \rangle, \quad (i',j') \in \mathbb{U}_k \tag{10.48}$$

for the solution $\mathbf{u}_k := [(\mathbf{u}_k)_{ij} : (i,j) \in \mathbb{U}_k]$. Set $\mathbf{u}_{k,0} := \mathbf{u}_k$ and $l := 1$.

Step 2: *Compute the representation vector $\hat{\mathbf{u}}_{k+l}$ of $\hat{u}_{k+l} := \mathcal{P}_{k+l}\Psi u_{k,l-1}$ and generate vector*

$$\hat{\mathbf{f}}_{k,l} := \left[\langle \ell_{i'j'}, f \rangle : (i',j') \in \mathbb{U}_{k,l} \right].$$

Compute

$$\tilde{\mathbf{f}}_{k,l} := \hat{\mathbf{f}}_{k,l} + \tilde{\mathbf{K}}^H_{k,l-1}\mathbf{u}_{k,l-1} + \tilde{\mathbf{L}}^H_{k,l}\hat{\mathbf{u}}_{k+l}. \tag{10.49}$$

Solve the linear system

$$\mathbf{E}^H_{k,l}\mathbf{u}^H_{k,l} = \tilde{\mathbf{f}}_{k,l} \tag{10.50}$$

for $\mathbf{u}^H_{k,l} := [(\mathbf{u}_{k,l})_{ij} : (i,j) \in \mathbb{U}_{k,l}]$ and define $u^H_{k,l} := \sum_{(i,j)\in\mathbb{U}_{k,l}} (\mathbf{u}_{k,l})_{ij} w_{ij}$.

Step 3: *Solve the nonlinear system*

$$\left\langle \ell_{i'j'}, (\mathcal{I} - (\mathcal{K} + \mathcal{L}\Psi))\left(\sum_{(i,j)\in\mathbb{U}_k} (\mathbf{u}_{k,l})_{ij} w_{ij} + u^H_{k,l} \right) \right\rangle = \langle \ell_{i'j'}, f \rangle,$$
$$(i',j') \in \mathbb{U}_k \tag{10.51}$$

for $\mathbf{u}^L_{k,l} := [(\mathbf{u}_{k,l})_{ij} : (i,j) \in \mathbb{U}_k]$, and define $u^L_{k,l} := \sum_{(i,j)\in\mathbb{U}_k} (\mathbf{u}_{k,l})_{ij} w_{ij}$ and $u_{k,l} := u^L_{k,l} + u^H_{k,l}$.

Step 4: *Set $l \leftarrow l + 1$ and go back to step 2 until $l = m$.*

Several remarks on the computational performance of Algorithm 10.18 are in order. The computational costs of Algorithm 10.18 can be divided into three parts. Part 1 is for generating the matrices $\tilde{\mathbf{K}}_{k+l}$ and $\tilde{\mathbf{L}}_{k+l}$. Part 2 is for computing (10.49) and solving (10.50). Part 3 is for solving the resulting nonlinear systems, including (10.48) and (10.51). It takes much more time to generate matrix $\tilde{\mathbf{L}}_{k+l}$ than to generate matrix $\tilde{\mathbf{K}}_{k+l}$, since the kernel L is weakly singular. We observe from numerical experiments that although parts 1 and 3 both have high costs, when l is small part 3 dominates the total computing time for implementing the algorithm, while when l increases part 1 grows faster than part 3.

10.3.2 Accelerated quadratures and Newton iterations

In this subsection we address two computational issues of the MAM algorithm. We employ a product integration scheme for computing the singular integrals which appear in the matrices involved in the MAM, and introduce an approximation technique in the Newton iteration for solving the resulting nonlinear systems to avoid repeated computation in generating their Jacobian matrices. The use of these two techniques results in a modified MAM which speeds up its computation (cf. [52]).

1. Product integration of singular integrals

Numerical experiments show that the generation of $\tilde{\mathbf{L}}_n$ requires much more computing time than that of $\tilde{\mathbf{K}}_n$ due to the singularity of the kernel L. We observe that the kernel L has a special structure which allows us to develop a more efficient quadrature method than the Gaussian quadrature method. In this subsection, we develop a special product integration method for the specific kernel L so that the computing time for calculating the nonzero entries of the matrix $\tilde{\mathbf{L}}_n$ is significantly reduced.

The product integration method has been widely used in the literature for computing singular integrals. For example, it was used in [17, 138] to discretize singular integral operators. Along this line, concrete formulas of product integration were given in [15] (see pp. 116–119 therein). These formulas were developed in the context of single-scale approximation and were proved efficient for computation in that context. In the current multiscale approximation context, we establish product integration formulas suitable for the use of multiscale bases.

We now study the typical integral involved in the entries of matrix $\tilde{\mathbf{L}}_n$. The nonzero entries of matrix $\tilde{\mathbf{L}}_n$ involve integrals of the form

$$I_{ij}(s) := \int_{S_{ij}} L(s,t)w_{ij}(t)\mathrm{d}t, \quad \text{for } (i,j) \in \mathbb{U}_n, \quad s \in [0,1],$$

where w_{ij} are multiscale piecewise polynomial basis functions. As suggested by [15, 17, 138], the kernel L can be decomposed as

$$L(s,t) = \frac{1}{\pi}\left[B_0(s,t) + B_1(s,t)\right], \tag{10.52}$$

with

$$B_0(s,t) := \log\left|\frac{\sqrt{(\xi(s)-\xi(t))^2 + (\eta(s)-\eta(t))^2}}{(s-t)(s-t-1)(s-t+1)}\right|$$

and

$$B_1(s,t) := \log|s-t| + \log|s-t-1| + \log|s-t+1|.$$

The above decomposition in fact extracts the singularity of L. Specifically, B_1 possesses all the singularity features of L and B_0 is smooth. The kernel B_0 is easy to integrate numerically with a little computational cost, while the singularity of B_1 brings difficulty to its numerical integration. However, we observe that the expression of B_1 is very specific, which allows us to integrate it *exactly* with explicit formulas. The quantities I_{ij} can be written as the sum of two terms

$$I_{ij}^{\nu}(s) := \frac{1}{\pi}\int_{S_{ij}} B_{\nu}(s,t)w_{ij}(t)\mathrm{d}t, \quad \nu = 0, 1. \tag{10.53}$$

We first compute the term I_{ij}^1. To this end, we re-express the multiscale basis functions w_{ij}. Note that for $j \in \mathbb{Z}_{w(0)}$, w_{0j} is a polynomial of order r. Hence, it can be written as

$$w_{0j}(t) = \sum_{\gamma \in \mathbb{Z}_r} a_\gamma t^\gamma.$$

For $i > 0$ and $j \in \mathbb{Z}_{w(i)}$, w_{ij} is a piecewise polynomial of order r. According to the construction of the basis function w_{ij}, for all $(i,j) \in \mathbb{U}_n$, the support S_{ij} can be divided into μ pieces $\Omega_\kappa = (a_\kappa, b_\kappa)$, $\kappa \in \mathbb{Z}_\mu$, on each of which w_{ij} is a polynomial of order r. Thus, we write w_{ij} as

$$w_{ij}(t) = \sum_{\gamma \in \mathbb{Z}_r} a_{\kappa,\gamma} t^\gamma, \quad t \in \Omega_\kappa, \ \kappa \in \mathbb{Z}_\mu.$$

The above discussion of the basis functions motivates us to define the following special integrals. For $\gamma \in \mathbb{Z}_r$, $\alpha \in \Sigma := \{0, 1, -1\}$ and for

$a, b \in [0, 1]$ with $a < b$, we set

$$I(a, b; \alpha, \gamma; s) := \int_a^b \log|s - t - \alpha| t^\gamma \, dt.$$

In this notation, we have that

$$\int_{S_{0j}} B_1(s, t) w_{0j}(t) dt = \sum_{\alpha \in \Sigma} \sum_{\gamma \in \mathbb{Z}_r} a_\gamma I(0, 1; \alpha, \gamma; s) \tag{10.54}$$

and

$$\int_{S_{ij}} B_1(s, t) w_{ij}(t) dt = \sum_{\kappa \in \mathbb{Z}_\mu} \sum_{\alpha \in \Sigma} \sum_{\gamma \in \mathbb{Z}_r} a_{\kappa, \gamma} I(a_\kappa, b_\kappa; \alpha, \gamma, s). \tag{10.55}$$

The integral $I(a, b; \alpha, \gamma; s)$ can be computed exactly. We derive below the formula for the integral.

Lemma 10.19 *If $\gamma \in \mathbb{Z}_r$, $\alpha \in \Sigma$ and $a, b \in [0, 1]$ with $a < b$, then*

$$I(a, b; \alpha, \gamma; s)$$
$$= \frac{1}{\gamma + 1} \left[(t^{\gamma+1} - (s - \alpha)^{\gamma+1}) \log|s - t - \alpha| - \sum_{j=1}^{\gamma+1} (s - \alpha)^{\gamma - j + 1} t^j \right] \Bigg|_a^b .$$

Proof The formula in this lemma may be proved using integration by parts. □

Using the integration formula in Lemma 10.19, we are able to compute the integrals (10.54) and (10.55) exactly.

It remains to compute the term I_{ij}^0. For this purpose, we describe a Gaussian quadrature rule on the interval $[a, b]$. For each positive integer j, we denote by g_j the Legendre polynomial of degree j and by τ_ℓ^j, $\ell \in \mathbb{Z}_j$ the j zeros of g_j in the order $-1 < \tau_0^j < \cdots < \tau_{j-1}^j < 1$. We transfer these zeros to the interval $[a, b]$, by letting

$$\hat{\tau}_\ell^j := \frac{a + b}{2} + \frac{b - a}{2} \tau_\ell^j, \qquad \ell \in \mathbb{Z}_j.$$

The points $\hat{\tau}_\ell^j$ are the j zeros of the Legendre polynomial of degree j on the interval $[a, b]$. Given a continuous function h defined on $[a, b]$, the j-point Gaussian quadrature rule is given by

$$G(h, [a, b], j) := \sum_{\ell \in \mathbb{Z}_j} \omega_\ell^j h(\hat{\tau}_\ell^j),$$

where

$$\omega_\ell^j := \int_a^b \prod_{i\in\mathbb{Z}_j, i\neq\ell} \frac{t-\hat{\tau}_i^j}{\hat{\tau}_\ell^j - \hat{\tau}_i^j} \mathrm{d}t.$$

This quadrature formula will be used to compute I_{ij}^0. We summarize below the integration strategy for computing the nonzero entries $\tilde{L}_{i'j',ij}$ of $\tilde{\mathbf{L}}_n$.

(**QL**) For a nonzero entry $\tilde{L}_{i'j',ij}$ of $\tilde{\mathbf{L}}_n$, we compute I_{ij}^0 and I_{ij}^1 separately. To compute I_{ij}^0, we divide the support S_{ij} of w_{ij} uniformly into N intervals I_ℓ, $\ell \in \mathbb{Z}_N$, where N is a positive integer such that the diameter of each of the intervals is less than or equal to $\mu^{-\kappa r}$ and w_{ij} is a polynomial on I_ℓ. The integral I_{ij}^0 is computed by the formula

$$\sum_{\ell\in\mathbb{Z}_N} G(B_0(s,\cdot)w_{ij}, I_\ell, \lceil(2\kappa)^{-1}n\rceil).$$

The integral I_{ij}^1 is expressed in terms of equations (10.54) and (10.55) and computed using Lemma 10.19.

2. Approximate iteration for solving nonlinear systems

Algorithm 10.18 requires solving two nonlinear systems (10.48) and (10.51). These equations are solved using the Newton method. In each iteration step of the Newton method, we need to compute the entries of the Jacobian matrix. Specifically, for equation (10.48) the Newton iteration scheme has the following steps.

- Choose an initial guess $\mathbf{u}_k^{(0)}$.
- For $m = 0, 1, \ldots,$ compute

$$\mathbf{F}(\mathbf{u}_k^{(m)}) := (\mathbf{E}_k - \mathbf{K}_k)\mathbf{u}_k^{(m)} - \hat{\mathbf{f}}_k^{(m)}$$

with

$$\hat{\mathbf{f}}_k^{(m)} := \left[\left\langle \ell_{i'j'}, f + \mathcal{L}\Psi\left(u_k^{(m)}\right)\right\rangle : (i',j') \in \mathbb{U}_k\right], \quad u_k^{(m)} := \sum_{(i,j)\in\mathbb{U}_k} (\mathbf{u}_k^{(m)})_{ij} w_{ij},$$

and compute the Jacobian matrix

$$\mathbf{J}(\mathbf{u}_k^{(m)}) := [J_{i'j',ij} : (i',j'), (i,j) \in \mathbb{U}_k]$$

with

$$J_{i'j',ij} := E_{i'j',ij} - K_{i'j',ij} - \left\langle \ell_{i'j'}, \mathcal{L}(w_{ij}\Psi'(u_k^{(m)}))\right\rangle.$$

- Solve $\Delta_k^{(m)}$ from the equation $\mathbf{J}(\mathbf{u}_k^{(m)})\Delta_k^{(m)} = -\mathbf{F}(\mathbf{u}_k^{(m)})$.
- Compute $\mathbf{u}_k^{(m+1)} := \mathbf{u}_k^{(m)} + \Delta_k^{(m)}$.

It is easily seen that the evaluation of both $\mathbf{F}(\mathbf{u}_k^{(m)})$ and $\mathbf{J}(\mathbf{u}_k^{(m)})$ involves computing integrals, which requires high computational costs. Solving equation (10.51) numerically also involves computing the integrals.

When evaluating $\mathbf{F}(\mathbf{u}_k^{(m)})$, we need to compute the integrals

$$\left\langle \ell_{i'j'}, \mathcal{L}\Psi\left(u_k^{(m)}\right)\right\rangle. \tag{10.56}$$

These integrals come from the integral operator $\mathcal{L}$. For different steps of the iteration, we are required to compute different integrals and as a result, the computational cost is large. Note that for some $\tilde{\Psi}\left(u_k^{(m)}\right) \in \mathbb{X}_k$, we can write it as

$$\tilde{\Psi}\left(u_k^{(m)}\right) = \sum_{(i,j)\in\mathbb{U}_k} c_{ij}w_{ij}$$

and

$$\left\langle \ell_{i'j'}, \mathcal{L}\tilde{\Psi}\left(u_k^{(m)}\right)\right\rangle = \sum_{(i,j)\in\mathbb{U}_k} c_{ij}L_{i'j',ij}. \tag{10.57}$$

Comparing (10.57) with (10.56), we observe that although they both involve integral evaluation, (10.57) makes use of the values of the entries of the matrix $\mathbf{L}_n$, which have been previously obtained so we do not have to recompute them. However, in general, $\Psi(u_k^{(m)}) \notin \mathbb{X}_k$. We cannot write $\Psi(u_k^{(m)})$ as a linear combination of the basis function w_{ij}. For this reason, we propose to project $\Psi(u_k^{(m)})$ into $\mathbb{X}_k$. Specifically, we do not solve (10.48) directly and instead, we solve $\tilde{\mathbf{u}}_k$ from the nonlinear system

$$\left\langle \ell_{i'j'}, (\mathcal{I} - (\mathcal{K} + \mathcal{L}\mathcal{P}_k\Psi))\left(\sum_{(i,j)\in\mathbb{U}_k} (\tilde{\mathbf{u}}_k)_{ij}w_{ij}\right)\right\rangle = \langle \ell_{i'j'}, f\rangle, \qquad (i',j') \in \mathbb{U}_k. \tag{10.58}$$

When we solve equation (10.58) by the Newton iteration method, we are required to compute the terms

$$\left\langle \ell_{i'j'}, \mathcal{L}\mathcal{P}_k\Psi\left(\sum_{(i,j)\in\mathbb{U}_k} (\tilde{\mathbf{u}}_k)_{ij}w_{ij}\right)\right\rangle, \qquad (i',j') \in \mathbb{U}_k$$

and their partial derivatives with respect to the variables $(\tilde{\mathbf{u}}_k)_{ij}, (i,j) \in \mathbb{U}_k$. To this end, we suppose that

$$\mathcal{P}_k\Psi\left(\sum_{(i,j)\in\mathbb{U}_k}(\tilde{\mathbf{u}}_k)_{ij}w_{ij}\right) = \sum_{(i,j)\in\mathbb{U}_k}(\hat{\mathbf{u}}_k)_{ij}w_{ij}.$$

Then each $(\hat{\mathbf{u}}_k)_{ij}$ is a function with respect to the variables $(\tilde{\mathbf{u}}_k)_{ij}, (i,j) \in \mathbb{U}_k$. In fact, if we let

$$\mathbf{F} := \left[\left\langle \ell_{i'j'}, \Psi\left(\sum_{(i,j)\in\mathbb{U}_k}(\tilde{\mathbf{u}}_k)_{ij}w_{ij}\right)\right\rangle, (i',j') \in \mathbb{U}_k\right],$$

then we have $\hat{\mathbf{u}}_k = \mathbf{E}_k^{-1}\mathbf{F}$. Therefore, it follows that

$$\left\langle \ell_{i'j'}, \mathcal{L}\mathcal{P}_k\Psi\left(\sum_{(i,j)\in\mathbb{U}_k}(\tilde{\mathbf{u}}_k)_{ij}w_{ij}\right)\right\rangle = \sum_{(i,j)\in\mathbb{U}_k} L_{i'j',ij}(\hat{\mathbf{u}}_k)_{ij}.$$

In a similar manner, we compute the partial derivatives of the above quantities with respect to the variables $(\tilde{\mathbf{u}}_k)_{ij}, (i,j) \in \mathbb{U}_k$. Making use of the above observations, we describe the Newton iteration scheme for solving (10.58) as follows.

Algorithm 10.20 *(The Newton iteration method for solving (10.58)) Set* $m := 0$, $\mathbf{f}_k := [\langle \ell_{ij}, f\rangle : (i,j) \in \mathbb{U}_k]$ *and choose an initial guess* $\tilde{\mathbf{u}}_k^{(0)}$ *and an iteration stopping threshold* δ.

Step 1: *Let* $\tilde{u}_k^{(m)} := \sum_{(i,j)\in\mathbb{U}_k}(\tilde{\mathbf{u}}_k^{(m)})_{ij}w_{ij}$ *and set*

$$\mathbf{G}(\tilde{\mathbf{u}}_k^{(m)}) := \left[\left\langle \ell_{i'j'}, \Psi'(\tilde{u}_k^{(m)})w_{ij}\right\rangle : (i',j'), (i,j) \in \mathbb{U}_k\right].$$

Solve $\bar{\mathbf{F}}_k^{(m)}$ *from* $\mathbf{E}_k\bar{\mathbf{F}}_k^{(m)} = \mathbf{G}(\tilde{\mathbf{u}}_k^{(m)})$ *and compute the Jacobian matrix*

$$\tilde{\mathbf{J}}(\tilde{\mathbf{u}}_k^{(m)}) := \mathbf{E}_k - \mathbf{K}_k - \mathbf{L}_k\bar{\mathbf{F}}_k^{(m)}.$$

Step 2: *For* $\mathbf{g}_k^{(m)} := \left[\left\langle \ell_{ij}, \Psi(\tilde{u}_k^{(m)})\right\rangle : (i,j) \in \mathbb{U}_k\right]$, *solve* $\hat{\mathbf{u}}_k^{(m)}$ *from* $\mathbf{E}_k\hat{\mathbf{u}}_k^{(m)} = \mathbf{g}_k^{(m)}$. *Compute*

$$\tilde{\mathbf{F}}(\tilde{\mathbf{u}}_k^{(m)}) := (\mathbf{E}_k - \mathbf{K}_k)\tilde{\mathbf{u}}_k^{(m)} - \mathbf{L}_k\hat{\mathbf{u}}_k^{(m)} - \mathbf{f}_k.$$

Step 3: *Solve* $\Delta_k^{(m)}$ *from* $\tilde{\mathbf{J}}(\tilde{\mathbf{u}}_k^{(m)})\Delta_k^{(m)} = -\tilde{\mathbf{F}}(\tilde{\mathbf{u}}_k^{(m)})$ *and compute* $\tilde{\mathbf{u}}_k^{(m+1)} := \tilde{\mathbf{u}}_k^{(m)} + \Delta_k^{(m)}$.

Step 4: *Set* $m \leftarrow m+1$ *and go back to step 1 until* $\|\Delta_k^{(m)}\|_\infty < \delta$.

It is worth noticing that in Algorithm 10.20 we need not evaluate any additional integrals but make use of the matrix $\mathbf{L}_k$. This saves tremendous computational effort and thus makes the algorithm very fast. We shall see from the numerical examples in Section 10.4.2 that (10.58) is solved much faster than (10.48).

Equation (10.51) can be approximated in a similar manner. Specifically, in step 3 of Algorithm 10.20 we replace (10.51) by

$$\left\langle \ell_{i'j'}, (\mathcal{I} - (\mathcal{K} + \mathcal{L}\mathcal{P}_{k+l}\Psi))\left(\sum_{(i,j)\in\mathbb{U}_k} (\tilde{\mathbf{u}}_{k,l})_{ij} w_{ij} + \tilde{u}^H_{k,l}\right)\right\rangle = \left\langle \ell_{i'j'}, f\right\rangle, \quad (i',j') \in \mathbb{U}_k. \tag{10.59}$$

The Newton iteration scheme for solving (10.59) can likewise be developed and will be referred to as Algorithm 10.20′.

3. The modified MAM algorithm

We describe below the MAM algorithm employing the above two techniques.

Algorithm 10.21 *(MAM with an accelerated quadrature and the Newton method) Let k be a fixed positive integer. Given $m \in \mathbb{N}_0$, we carry out the following computing steps.*

Step 1: *Use Algorithm 10.20 to solve the nonlinear system (10.58) and obtain the solution $\tilde{\mathbf{u}}_k := [(\tilde{\mathbf{u}}_k)_{ij} : (i,j) \in \mathbb{U}_k]$. Let $\tilde{\mathbf{u}}_{k,0} := \tilde{\mathbf{u}}_k$ and $l := 1$.*

Step 2: *Follow step 2 of Algorithm 10.18 to generate $\tilde{\mathbf{f}}_{k,l}$ and solve $\mathbf{E}^H_{k,l}\tilde{\mathbf{u}}^H_{k,l} = \tilde{\mathbf{f}}_{k,l}$ to obtain $\tilde{\mathbf{u}}^H_{k,l} := [(\tilde{\mathbf{u}}_{k,l})_{ij} : (i,j) \in \mathbb{U}_{k,l}]^T$ and define $\tilde{u}^H_{k,l} := \sum_{(i,j)\in\mathbb{U}_{k,l}} (\tilde{\mathbf{u}}_{k,l})_{ij} w_{ij}$.*

Step 3: *Use Algorithm 10.20′ to solve the nonlinear system (10.59) and obtain the solution $\tilde{\mathbf{u}}^L_{k,l} := [(\tilde{\mathbf{u}}_{k,l})_{ij} : (i,j) \in \mathbb{U}_k]$. Define $\tilde{u}^L_{k,l} := \sum_{(i,j)\in\mathbb{U}_k} (\tilde{\mathbf{u}}_{k,l})_{ij} w_{ij}$ and $\tilde{u}_{k,l} := \tilde{u}^L_{k,l} + \tilde{u}^H_{k,l}$.*

Step 4: *Set $l \leftarrow l + 1$ and go back to step 2 until $l = m$.*

10.3.3 Computational complexity

We now estimate the computational efforts by the number of multiplications and functional evaluations. Before presenting the estimates, we review several properties of the multiscale bases and collocation functionals for later reference, established in Chapter 7.

(I) For any $(i,j) \in \mathbb{U} := \{(i,j) : i \in \mathbb{N}_0, j \in \mathbb{Z}_{w(i)}\}$, there are at most $(\mu - 1)r - 1$ numbers of $w_{ij'}$, $j' \in \mathbb{Z}_{w(i)}$ such that $\text{meas}(S_{ij} \cap S_{ij'}) \neq 0$.

(II) For any $i', i \in \mathbb{N}_0$ with $i \le i'$, it holds that

$$\langle \ell_{i'j'}, w_{ij} \rangle = \delta_{i'i}\delta_{j',j}, \quad j' \in \mathbb{Z}_{w(i')}, j \in \mathbb{Z}_{w(i)},$$

where $\delta_{i'i}$ is the Kronecker delta.

(III) For any polynomial p of order $\le r$ and $i \ge 1$,

$$\langle \ell_{ij}, p \rangle = 0, \quad (w_{ij}, p) = 0, \qquad (i,j) \in \mathbb{U},$$

where $(\cdot,\cdot)$ denotes the inner product in $L^2(0,1)$.

(IV) There exists a positive constant θ_0 such that $\|\ell_{ij}\| \le \theta_0$ and $\|w_{ij}\|_\infty \le \theta_0$ for all $(i,j) \in \mathbb{U}$.

(V) There exist positive constants c_- and c_+ such that for all $n \in \mathbb{N}_0$,

$$c_-\mu^i \le w(i) \le c_+\mu^i, \qquad c_-\mu^{-i} \le \max_{j\in\mathbb{Z}_{w(i)}} |\mathrm{supp}(w_{ij})| \le c_+\mu^{-i}.$$

(VI) For any $v \in \mathbb{X}_n$, we have a unique expansion

$$v = \sum_{(i,j)\in\mathbb{U}_n} v_{ij}w_{ij},$$

and there exist positive constants θ_1 and θ_2 such that

$$\theta_1\|\mathbf{v}\|_\infty \le \|v\|_\infty \le \theta_2(n+1)\|\mathbf{E}_n\mathbf{v}\|_\infty,$$

in which $\mathbf{E}_n := [\langle \ell_{i'j'}, w_{ij} \rangle : (i',j'), (i,j) \in \mathbb{U}_n]$.

When solving the nonlinear systems (10.58) and (10.59), we are required to generate matrices $\tilde{\mathbf{K}}_n$ and $\tilde{\mathbf{L}}_n$. The truncated matrix $\tilde{\mathbf{K}}_n$ is evaluated by the same strategies as those in Algorithm 10.18. The truncated matrix $\tilde{\mathbf{L}}_n$ is evaluated using the quadrature method (**QL**). The following lemma establishes estimates for numbers of nonzero entries of the truncated matrices, as well as the numbers of multiplications and functional evaluations in computing these nonzero entries.

Lemma 10.22 *For any $n \in \mathbb{N}_0$,*

$$\mathcal{N}(\tilde{\mathbf{K}}_n) = \mathcal{O}(n\mu^n), \;\; \mathcal{N}(\tilde{\mathbf{L}}_n) = \mathcal{O}(n\mu^n),$$

where $\mathcal{N}(\cdot)$ denotes the number of nonzero entries of a matrix. The numbers of functional evaluations and multiplications needed for generating $\tilde{\mathbf{K}}_n$ and $\tilde{\mathbf{L}}_n$ are both $\mathcal{O}(n\mu^n)$, respectively.

Proof Define the matrix blocks

$$\mathbf{K}_{i'i} := [K_{i'j',ij} : j' \in \mathbb{Z}_{w(i')}, j \in \mathbb{Z}_{w(i)}],$$
$$\tilde{\mathbf{K}}_{i'i} := [\tilde{K}_{i'j',ij} : j' \in \mathbb{Z}_{w(i')}, j \in \mathbb{Z}_{w(i)}], i', i \in \mathbb{Z}_{n+1}.$$

We obtain from property (V) that $\mathcal{N}(\mathbf{K}_{i'i}) = \mathcal{O}(\mu^{i'+i})$, for $i', i \in \mathbb{Z}_{n+1}$. Direct calculation leads to

$$\sum_{i \in \mathbb{Z}_{n+1}} \sum_{i'=0}^{n-i} \mu^{i'+i} = \mathcal{O}(n\mu^n).$$

In light of property (I), the number of functional evaluations in computing the entries $\tilde{K}_{i'j',ij}$, $j \in \mathbb{Z}_{w(i)}$ is of $\mathcal{O}((2\kappa)^{-1}n\mu^{\kappa r} + \mu^i)$ for any $j' \in \mathbb{Z}_{w(i')}$. Thus, generating the whole block $\tilde{\mathbf{K}}_{i'i}$ requires $\mathcal{O}([(2\kappa)^{-1}n\mu^{\kappa r} + \mu^i]\mu^{i'})$ number of functional evaluations. The total number of functional evaluations for generating $\tilde{\mathbf{K}}_n$ is then obtained by a simple summation. It is easily observed from the expression of quadrature rules that the number of multiplications involved has the same order as that of functional evaluations.

For the estimates on $\tilde{\mathbf{L}}_n$, we let

$$\tilde{\mathbf{L}}_{i'i} := [\tilde{L}_{i'j',ij} : j' \in \mathbb{Z}_{w(i')}, j \in \mathbb{Z}_{w(i)}], \quad i', i \in \mathbb{Z}_{n+1}.$$

For kernels whose singularity points are along the diagonal, the entries which should be calculated concentrate along the diagonal of each block. For kernels whose singularity is described by (10.31), it is not difficult to observe that the entries that have to be computed concentrate along the diagonal and the corners on the top right and bottom left of each block. We obtain the same estimate of $\mathcal{N}(\tilde{\mathbf{L}}_n)$ as that in Theorem 7.15 of Chapter 7 (cf. Theorem 4.6 of [69]). We now estimate the computational costs for generating the matrix $\tilde{\mathbf{L}}_n$. According to the integration strategy (**QL**), for each pair i and j, computing I_{ij}^1 or I_{ij}^0 requires only a fixed number of functional evaluations and multiplications. Thus, the estimates of this lemma are proved. □

Lemma 10.23 *For a fixed k and any $l \in \mathbb{N}_0$, the numbers of functional evaluations and multiplications for evaluating $\mathbf{f}_{k,l}$ are both $\mathcal{O}((k+l)\mu^{k+l})$.*

Proof Making use of the estimates of numbers of nonzero entries of the truncated matrices, the number of multiplications used in obtaining $\tilde{\mathbf{f}}_{k,l}$ from (10.49) is $\mathcal{O}((k+l)\mu^{k+l})$. It follows from Lemma 10.22 that the numbers of multiplications and functional evaluations for generating $\tilde{\mathbf{K}}_{k,l-1}^H$ are $\mathcal{O}((k+l-1)\mu^{k+l-1})$, while those for generating $\tilde{\mathbf{L}}_{k,l}^H$ are also $\mathcal{O}((k+l)\mu^{k+l})$. Moreover, it is easily obtained from the definition of $\hat{\mathbf{f}}_{k,l}$ that the numbers of multiplications and functional evaluations for computing it are $\mathcal{O}(\mu^{k+l})$. Therefore, it remains to estimate the computational efforts for $\hat{\mathbf{u}}_{k+l}$.

Since, for $i \in \mathbb{Z}_{k+l+1}$ and a point $P \in [0, 1]$, there are bounded numbers of w_{ij} not vanishing at P, and there are bounded numbers of point evaluations in each ℓ_{ij}, the numbers of point evaluations and multiplications for calculating each component of $\mathbf{g}_{k+l}$ are $\mathcal{O}(k+l)$. Hence, those for computing the vector $\mathbf{g}_{k+l}$ are $\mathcal{O}((k+l)\mu^{k+l})$. Noting that $\mathbf{E}_{k+l}$ is a sparse upper triangular matrix with $\mathcal{O}((k+l)\mu^{k+l})$ number of nonzero entries, the solution of the linear system (10.45) needs $\mathcal{O}((k+l)\mu^{k+l})$ number of multiplications. The result of the lemma then follows by adding the above estimates together. □

Combining the estimate in the above lemma for $\tilde{\mathbf{f}}_{kl}$ with that for other computing steps in Algorithm 10.18 obtained from the last section (see also [76]), we have the following lemma.

Lemma 10.24 *If the truncated matrix $\tilde{\mathbf{L}}_n$ is evaluated using the quadrature method (**QL**), then the numbers of functional evaluations and multiplications in Algorithm 10.18 are $\mathcal{O}((k+m)\mu^{k+m})$.*

The next theorem gives an estimate of the computational cost required for Algorithm 10.21.

Theorem 10.25 *Let k be a fixed positive integer. For any $m \in \mathbb{N}_0$, the number of functional evaluations and multiplications used in Algorithm 10.21 is $\mathcal{O}((k+m)\mu^{k+m})$.*

Proof The computational cost of Algorithm 10.21 is composed of two parts. One is the cost for generating the matrices $\tilde{\mathbf{K}}_{k+m}$ and $\tilde{\mathbf{L}}_{k+m}$. The other is that for carrying out the computing steps listed in the algorithm. Lemma 10.22 has shown that the computational cost of generating both $\tilde{\mathbf{L}}_{k+m}$ and $\tilde{\mathbf{K}}_{k+m}$ is $\mathcal{O}((k+m)\mu^{k+m})$. To estimate the efforts of the computing steps in Algorithm 10.21, we only need to compare Algorithm 10.21 with Algorithm 10.18. We observe that Algorithm 10.21 replaces (10.48) and (10.51) by (10.58) and (10.59), respectively. We have shown that the modifications reduce the computational cost. Therefore, the computational cost for carrying out the computing steps of Algorithm 10.21 is less than that of Algorithm 10.18. It is stated in Lemma 10.24 that the number of functional evaluations and multiplications in Algorithm 10.18 is $\mathcal{O}((k+m)\mu^{k+m})$. Thus, the theorem is proved. □

We remark that if the quadrature method (**QL**) is not used for generating the truncated matrix $\tilde{\mathbf{L}}_n$, then the number of functional evaluations and multiplications used in Algorithm 10.21 is $\mathcal{O}((k+m)^3\mu^{k+m})$ (cf. [51]), and although Algorithm 10.21 has the same order of computational costs as Algorithm 10.18, the constant involved in the order of computational costs is improved.

10.3.4 Convergence analysis

In the last subsection we proved that the computational costs of Algorithm 10.21 are nearly linear. In this subsection we establish the order of its approximate accuracy. More precisely, we show that the output $\tilde{u}_{k,m}$ of Algorithm 10.21 maintains much of the convergence order of $u_{k,m}$ generated by Algorithm 10.3.2. Through out the rest of this subsection, we assume without further mention that the function g in the nonlinear boundary condition is continuously differentiable with respect to the second variable.

We begin with some necessary preparation. For $n \in \mathbb{N}_0$, we define the operator $\tilde{\mathcal{K}}_n : \mathbb{X}_n \to \mathbb{X}_n$ by requiring

$$\tilde{K}_{i'j',ij} = \left\langle \ell_{i'j'}, \tilde{\mathcal{K}}_n w_{ij} \right\rangle, \quad (i',j'), (i,j) \in \mathbb{U}_n.$$

Clearly, $\tilde{\mathcal{K}}_n$ is uniquely determined. Likewise, we define the operator $\tilde{\mathcal{L}}_n$: $\mathbb{X}_n \to \mathbb{X}_n$ by requiring

$$\tilde{L}_{i'j',ij} = \left\langle \ell_{i'j'}, \tilde{\mathcal{L}}_n w_{ij} \right\rangle, \quad (i',j'), (i,j) \in \mathbb{U}_n.$$

To estimate the error between $\tilde{u}_{km}$ and u_{km}, we need to estimate the errors $\mathcal{K}_n - \tilde{\mathcal{K}}_n$ and $\mathcal{L}_n - \tilde{\mathcal{L}}_n$, where $\mathcal{K}_n := \mathcal{P}_n \mathcal{K}|_{\mathbb{X}_n}$ and $\mathcal{L}_n := \mathcal{P}_n \mathcal{L}|_{\mathbb{X}_n}$.

In the following lemma, we estimate the error introduced by truncation and quadrature rules applied to matrix $\mathbf{K}_n$.

Lemma 10.26 *There exists a positive constant c such that for all $i', i \in \mathbb{Z}_{n+1}$ and for all n,*

$$\|\mathbf{K}_{i'i} - \tilde{\mathbf{K}}_{i'i}\|_\infty \le c\mu^{-rn}.$$

Proof We proceed with our proof in two cases. When $i' + i > n$, according to the strategy (T1), we have that

$$\|\mathbf{K}_{i'i} - \tilde{\mathbf{K}}_{i'i}\|_\infty = \|\mathbf{K}_{i'i}\|_\infty. \tag{10.60}$$

It follows from property (III) and (10.30) that for $j' \in \mathbb{Z}_{w(i')}$ and $j \in \mathbb{Z}_{w(i)}$,

$$|K_{i'j',ij}| \le c\mu^{-r(i'+i)} \Lambda |S_{ij}|.$$

We then make use of property (V) to obtain

$$\sum_{j \in \mathbb{Z}_{w(i)}} |K_{i'j',ij}| \le c\mu^{-r(i'+i)} \Lambda.$$

Therefore,

$$\|\mathbf{K}_{i'i}\|_\infty := \max_{j' \in \mathbb{Z}_{w(i')}} \sum_{j \in \mathbb{Z}_{w(i)}} |K_{i'j',ij}| \le c\mu^{-r(i'+i)} \Lambda \le c\mu^{-rn} \Lambda.$$

Substituting this estimate into (10.60) verifies the lemma for the case $i' + i > n$.

We now consider the case $i' + i \leq n$. Noting that the Gaussian quadrature rule of N points has the algebraic accuracy of order $2N$, when $i' + i \leq n$, we have that

$$|K_{i'j',ij} - \tilde{K}_{i'j',ij}| \leq c(\mu^{-\kappa r})^{n/\kappa} \Lambda |S_{ij}| \leq c\mu^{-rn} \Lambda |S_{ij}|.$$

Utilizing property (V), we conclude that

$$\|\mathbf{K}_{i'i} - \tilde{\mathbf{K}}_{i'i}\|_\infty := \max_{j' \in \mathbb{Z}_{w(i')}} \sum_{j \in \mathbb{Z}_{w(i)}} |K_{i'j',ij} - \tilde{K}_{i'j',ij}| \leq c\mu^{-rn}.$$

□

The estimate in Lemma 10.26 can be translated into operator form.

Lemma 10.27 *There exists a positive constant c_1 such that for all $n \in \mathbb{N}_0$ and all $v \in \mathbb{X}_n$,*

$$\|(\mathcal{K}_n - \tilde{\mathcal{K}}_n)v\|_\infty \leq c_1(n+1)^2 \mu^{-rn} \|v\|_\infty. \tag{10.61}$$

Moreover, if $u^* \in W^{r,\infty}(0,1)$ and if $\|v - u^*\|_\infty \leq \bar{c}\mu^{-r(n-1)} \|u^*\|_{r,\infty}$ for some positive constant $\bar{c}$, then there exists a positive constant c_2 such that for all $n \in \mathbb{N}_0$,

$$\|(\mathcal{K}_n - \tilde{\mathcal{K}}_n)v\|_\infty \leq c_2(n+1)\mu^{-rn} \|u^*\|_{r,\infty}. \tag{10.62}$$

Proof Let $\mathbf{h} := \mathbf{E}_n^{-1}(\mathbf{K}_n - \tilde{\mathbf{K}}_n)\mathbf{v}$. We expand $(\mathcal{K}_n - \tilde{\mathcal{K}}_n)v$ as

$$(\mathcal{K}_n - \tilde{\mathcal{K}}_n)v = \sum_{(i,j) \in \mathbb{U}_n} h_{ij} w_{ij}.$$

Then property (VI) gives

$$\|(\mathcal{K}_n - \tilde{\mathcal{K}}_n)v\|_\infty \leq \theta_2(n+1) \|(\mathbf{K}_n - \tilde{\mathbf{K}}_n)\mathbf{v}\|_\infty. \tag{10.63}$$

We next estimate $\|(\mathbf{K}_n - \tilde{\mathbf{K}}_n)\mathbf{v}\|_\infty$ in two different cases.

For (10.61), note that property (VI) leads to $\|\mathbf{v}\|_\infty \leq \theta_1^{-1} \|v\|_\infty$. Moreover, by Lemma 10.26 there exists a positive constant c such that for all $n \in \mathbb{N}_0$,

$$\|\mathbf{K}_n - \tilde{\mathbf{K}}_n\|_\infty \leq \max_{i' \in \mathbb{Z}_{n+1}} \sum_{i \in \mathbb{Z}_{n+1}} \|\mathbf{K}_{i'i} - \tilde{\mathbf{K}}_{i'i}\|_\infty \leq c(n+1)\mu^{-rn}.$$

Combining the above estimates with (10.63) yields the estimate (10.61).

For the second inequality (10.62), we decompose v into $v = \mathcal{P}_n u^* + (v - \mathcal{P}_n u^*)$ with

$$\mathcal{P}_n u^* = \sum_{(i,j)\in\mathbb{U}_n} \xi_{ij} w_{ij} \quad \text{and} \quad v - \mathcal{P}_n u^* = \sum_{(i,j)\in\mathbb{U}_n} \zeta_{ij} w_{ij}.$$

Since $u^* \in W^{r,\infty}(0,1)$, it holds that $|\xi_{ij}| \leq c\mu^{-ri}\|u^*\|_{r,\infty}$. Moreover, it follows from property (VI) that

$$|\zeta_{ij}| \leq \theta_1^{-1}\|v - \mathcal{P}_n u^*\|_\infty \leq c\mu^{-r(n-1)}\|u^*\|_{r,\infty}.$$

These two estimates together imply that there exists a positive constant c such that for all $(i,j) \in \mathbb{U}_n$ and all $n \in \mathbb{N}_0$, $|v_{ij}| \leq c\mu^{-ri}\|u^*\|_{r,\infty}$.

Define the matrix $\Delta_n := [\Delta_{i'j',ij} : (i',j'),(i,j) \in \mathbb{U}_n]$ with $\Delta_{i'j',ij} := \mu^{r(n-i)}(K_{i'j',ij} - \tilde{K}_{i'j',ij})$ and the vector $\mathbf{v}' := [v'_{ij} : (i,j) \in \mathbb{U}_n]$ with $v'_{ij} := \mu^{ri}v_{ij}$. Then $\|\mathbf{v}'\|_\infty \leq c\|u^*\|_{r,\infty}$ and

$$\|(\mathbf{K}_n - \tilde{\mathbf{K}}_n)\mathbf{v}\|_\infty \leq \mu^{-rn}\|\Delta_n\|_\infty\|\mathbf{v}'\|_\infty.$$

For any $(i',j') \in \mathbb{U}_n$, it follows from Lemma 10.26 that

$$\sum_{(i,j)\in\mathbb{U}_n} |\Delta_{i'j',ij}| \leq \sum_{i\in\mathbb{Z}_{n+1}} \mu^{r(n-i)}\|\mathbf{K}_{i'i} - \tilde{\mathbf{K}}_{i'i}\|_\infty \leq c.$$

Therefore,

$$\|\boldsymbol{\Delta}_n\|_\infty := \max_{(i',j')\in\mathbb{U}_n} \sum_{(i,j)\in\mathbb{U}_n} |\Delta_{i'j',ij}| \leq c,$$

which leads to the estimate

$$\|(\mathbf{K}_n - \tilde{\mathbf{K}}_n)\mathbf{v}\|_\infty \leq c\mu^{-rn}\|u^*\|_{r,\infty}.$$

Combining this estimate with (10.63) proves the desired result (10.62). □

To estimate the error $\mathcal{L}_n - \tilde{\mathcal{L}}_n$, we define the matrix blocks

$$\mathbf{L}_{i'i} := [L_{i'j',ij} : j' \in \mathbb{Z}_{w(i')}, j \in \mathbb{Z}_{w(i)}],$$
$$\tilde{\mathbf{L}}_{i'i} := [\tilde{L}_{i'j',ij} : j' \in \mathbb{Z}_{w(i')}, j \in \mathbb{Z}_{w(i)}], i', i \in \mathbb{Z}_{n+1}.$$

The difference between $\mathbf{L}_{i'i}$ and $\tilde{\mathbf{L}}_{i'i}$ that results from both the truncation strategy (T2) and the quadrature strategy (**QL**) has the bound

$$\|\mathbf{L}_{i'i} - \tilde{\mathbf{L}}_{i'i}\|_\infty \leq c\max\{\mu^{-rn}, (\epsilon^n_{i'i})^{-(2r-\sigma')}\mu^{-r(i'+i)}\}.$$

By the definition (10.47) of the truncation parameters $\epsilon^n_{i'i}$, we conclude that

$$\|\mathbf{L}_{i'i} - \tilde{\mathbf{L}}_{i'i}\|_\infty \leq c(\epsilon^n_{i'i})^{-(2r-\sigma')}\mu^{-r(i'+i)}.$$

This leads to the following lemma.

Lemma 10.28 *Let $\sigma' \in (0,1)$, $\eta := 2r - \sigma'$ and set $b = 1$, $b' \in (r/\eta, 1)$ in (10.47). Then there exists a positive constant c_1 such that for all $n \in \mathbb{N}_0$ and all $v \in \mathbb{X}_n$,*

$$\|(\mathcal{L}_n - \tilde{\mathcal{L}}_n)v\|_\infty \le c_1(n+1)\mu^{-\sigma' n}\|v\|_\infty. \tag{10.64}$$

Moreover, if $u^* \in W^{r,\infty}(0,1)$ and if v satisfies $\|v - u^*\|_\infty \le \bar{c}\mu^{-r(n-1)}\|u^*\|_{r,\infty}$ for some positive constant $\bar{c}$, then there exists a positive constant c_2 such that

$$\|(\mathcal{L}_n - \tilde{\mathcal{L}}_n)\mathcal{P}_n\Psi v\|_\infty \le c_2(n+1)^2\mu^{-(r+\sigma')n}\|u^*\|_{r,\infty}. \tag{10.65}$$

Proof Since most of the proof for this lemma is similar to that for Lemma 10.27, we omit the details. In the proof of (10.65), since g is continuously differentiable with respect to the second variable, we may use the standard estimate for nonlinear operator $\|\Psi u^*\|_{r,\infty} \le c\|u^*\|_{r,\infty}$ (see, for example, [238]) to conclude the second estimate. □

In the next lemma, we estimate the error between the outputs of Algorithms 10.18 and 10.21.

Lemma 10.29 *If $u^* \in W^{r,\infty}(0,1)$, then there exists a positive constant c such that for sufficiently large integer k and for all $l \in \mathbb{Z}_{m+1}$,*

$$\|u_{kl} - \tilde{u}_{kl}\|_\infty \le c(k+l+1)\mu^{-r(k+l)}\|u^*\|_{r,\infty}.$$

Proof We prove this lemma by induction on l. For the case $l = 0$, we need to prove that the solution $\tilde{u}_k$ of

$$\tilde{u}_k - \mathcal{P}_k(\mathcal{K} + \mathcal{L}\mathcal{P}_k\Psi)\tilde{u}_k = \mathcal{P}_k f \tag{10.66}$$

satisfies

$$\|\tilde{u}_k - u_k\|_\infty \le c(k+1)\mu^{-kr}\|u^*\|_{r,\infty}. \tag{10.67}$$

Following a standard argument, we may show that the solution of equation (10.66) has the bound

$$\|\tilde{u}_k - u^*\|_\infty \le c\mu^{-kr}\|u^*\|_{r,\infty}.$$

Thus, by the triangle inequality we establish the estimate (10.67).

We now assume that the result of this lemma holds for $l-1$ and consider the case l. Note that step 3 of Algorithm 10.21 is the same as step 3 of Algorithm 10.18. According to Algorithms 10.18 and 10.21, we obtain that

$$u_{kl}^H - \tilde{u}_{kl}^H = u_A + u_B, \tag{10.68}$$

where

$$u_A := (\mathcal{P}_{k+l} - \mathcal{P}_k)\tilde{\mathcal{K}}_{k+l}(u_{k,l-1} - \tilde{u}_{k,l-1})$$

and

$$u_B := (\mathcal{P}_{k+l} - \mathcal{P}_k)\tilde{\mathcal{L}}_{k+l}\mathcal{P}_{k+l}(\Psi u_{k,l-1} - \Psi \tilde{u}_{k,l-1}).$$

Since the projections are uniformly bounded as follows from (10.62) of Lemma 10.27, we have that

$$\|u_A\| \leq c(k+l+1)\mu^{-r(k+l)}\|u^*\|_{r,\infty}. \tag{10.69}$$

By Lemma 10.28 and the induction hypothesis,

$$\|u_B\|_\infty \leq c(k+l+1)\mu^{-r(k+l)}\|u^*\|_{r,\infty}. \tag{10.70}$$

From equations (10.68)–(10.70) we conclude that

$$\|u_{kl}^H - \tilde{u}_{kl}^H\|_\infty \leq c(k+l+1)\mu^{-r(k+l)}\|u^*\|_{r,\infty}. \tag{10.71}$$

It remains to estimate $\|\tilde{u}_{kl}^L - u_{kl}^L\|$. Subtracting (10.51) from (10.59) yields the equation

$$\tilde{u}_{kl}^L - u_{kl}^L = \mathcal{P}_k[(\mathcal{K} + \mathcal{L}\mathcal{P}_{k+l}\Psi)\tilde{u}_{kl} - (\mathcal{K} + \mathcal{L}\Psi)u_{kl}].$$

Let $\mathcal{B} := (\mathcal{K} + \mathcal{L}\mathcal{P}_{k+l}\Psi)'$ and

$$\mathcal{R}(\tilde{u}_{kl}, u_{kl}) := (\mathcal{K} + \mathcal{L}\mathcal{P}_{k+l}\Psi)\tilde{u}_{kl} - (\mathcal{K} + \mathcal{L}\mathcal{P}_{k+l}\Psi)u_{kl} - \mathcal{B}(\tilde{u}_{kl} - u_{kl}).$$

We then conclude that

$$\tilde{u}_{kl}^L - u_{kl}^L = (\mathcal{I} - \mathcal{P}_k\mathcal{B})^{-1}\mathcal{P}_k[\mathcal{B}(\tilde{u}_{kl}^H - u_{kl}^H) + \mathcal{R}(\tilde{u}_{kl}, u_{kl}) + \mathcal{L}(\mathcal{P}_{k+l} - \mathcal{I})\Psi u_{kl}].$$

Note that there exist positive constants c_1 and c_2 such that for all $k, l \in \mathbb{N}_0$,

$$\|(\mathcal{I} - \mathcal{P}_k\mathcal{B})^{-1}\mathcal{P}_k\| \leq c_1$$

and

$$\|\mathcal{R}(\tilde{u}_{kl}, u_{kl})\|_\infty \leq c_2\|\tilde{u}_{kl} - u_{kl}\|_\infty^2.$$

From the last inequality and the fact that $\|\tilde{u}_{kl} - u_{kl}\|_\infty \to 0, k \to \infty$ uniformly for $l \in \mathbb{N}_0$, we conclude that for sufficiently large integer k and for all $l \in \mathbb{Z}_{m+1}$,

$$\|\mathcal{R}(\tilde{u}_{kl}, u_{kl})\|_\infty \leq \frac{1}{2c_1}(\|\tilde{u}_{kl}^L - u_{kl}^L\|_\infty + \|\tilde{u}_{kl}^H - u_{kl}^H\|_\infty).$$

Moreover, there exists a positive constant c_3 such that for sufficiently large integer k and for all $l \in \mathbb{Z}_{m+1}$,

$$\|\mathcal{L}(\mathcal{P}_{k+l} - \mathcal{I})\Psi u_{kl}\|_\infty \leq c_3\mu^{-r(k+l)}\|u^*\|_{r,\infty}.$$

Combining the above inequalities yields the estimate

$$\|\tilde{u}_{kl}^L - u_{kl}^L\|_\infty \leq c'(\|\tilde{u}_{kl}^H - u_{kl}^H\|_\infty + \mu^{-r(k+l)}\|u^*\|_{r,\infty})$$

for some positive constant c', which with (10.71) leads to the desired estimate of this lemma for the case l and completes the proof. □

The above lemma leads to the following error estimate of the approximate solutions generated by Algorithm 10.21.

Theorem 10.30 *If $u^* \in W^{r,\infty}[0,1]$, then there exists a positive constant c such that for sufficiently large k and all $m \in \mathbb{N}_0$,*

$$\|u^* - \tilde{u}_{km}\|_\infty \le c(k+m+1)\mu^{-r(k+m)}\|u^*\|_{r,\infty}.$$

10.4 Numerical experiments

In this section, we present numerical examples to demonstrate the performance of multiscale methods for solving the Hammerstein equation and the nonlinear boundary integral equations.

10.4.1 Numerical examples of the Hammerstein equation

We present in this subsection four numerical experiments to verify the theoretical estimates obtained in Section 10.2. The computer programs are run on a personal computer with 2.8-GHz CPU and 1G memory.

Example 1 Consider the equation

$$u(s) - \int_0^1 \sin(\pi(s+t))u^2(t)\mathrm{d}t = \sin(\pi s) - \frac{4}{3\pi}\cos(\pi s), \quad s \in [0,1]. \tag{10.72}$$

The equation has an isolated solution $u^*(s) = \sin(\pi s)$.

We use the collocation method via the piecewise linear polynomial basis for the numerical solution of the equation. Specifically, we choose $\mathbb{X}_n$ as the space of piecewise linear polynomials with knots at $j/2^n, j = 1, 2, \ldots, 2^n - 1$. Hence, $\dim(\mathbb{X}_n) = 2^{n+1}$. The basis functions of $\mathbb{X}_0$ and $\mathbb{W}_1$ are given, respectively, by $w_{00}(t) = -3t + 2$, $w_{01}(t) = 3t - 1$, $t \in [0,1]$ and

$$w_{10}(t) = \begin{cases} 1 - \frac{9}{2}t, & t \in [0, 1/2), \\ \frac{3}{2}t - 1, & t \in [1/2, 1], \end{cases} \qquad w_{11}(t) = \begin{cases} \frac{1}{2} - \frac{3}{2}t, & t \in [0, 1/2), \\ \frac{9}{2}t - \frac{7}{2}, & t \in [1/2, 1]. \end{cases}$$

The corresponding collocation functionals are

$$\ell_{00} = \delta_{\frac{1}{3}}, \quad \ell_{01} = \delta_{\frac{2}{3}}, \quad \ell_{10} = \delta_{\frac{1}{6}} - \frac{3}{2}\delta_{\frac{1}{3}} + \frac{1}{2}\delta_{\frac{2}{3}}, \quad \ell_{11} = \frac{1}{2}\delta_{\frac{1}{3}} - \frac{3}{2}\delta_{\frac{2}{3}} + \delta_{\frac{5}{6}},$$

where $\langle \delta_t, f \rangle = f(t)$. The basis functions w_{ij}, $i > 1$, $j \in \mathbb{Z}_{w(i)}$ and the corresponding functionals ℓ_{ij}, $i > 1$, $j \in \mathbb{Z}_{w(i)}$ are constructed recursively by applying linear operators related to the contractive mappings $\phi_0(t) = \frac{t}{2}$ and $\phi_1(t) = \frac{t+1}{2}$, $t \in [0, 1]$.

We solve equation (10.72) by Algorithm 10.9 with the initial level $k = 4$ based on the multiscale collocation method developed via the basis functions and corresponding collocation functionals. The nonlinear system (10.15) related to (10.72) is solved by the Newton iteration method. For comparison purposes, we also solve the collocation equation (10.3) using Newton iteration, which will be called the *direct* Newton method, and the two-grid method.

We report the numerical results in Table 10.1. Columns 3 and 4 of Table 10.1 list, respectively, the computed errors and computed convergence order (C.O.) of the approximate solution $u_{4,m}$ for equation (10.72) obtained by Algorithm 10.9. In column 5, we present the computing time T_M measured in seconds when Algorithm 10.9 is used. These numerical results confirm the theoretical estimates presented in the last subsection, noting that the theoretical order of convergence for the piecewise linear approximation is 2. The computing time is linear with respect to the dimension of the subspace. In column 6, we list the computed error of the approximate solution u_{4+m} obtained from the direct Newton method. We list in column 7 the computing time T_N in seconds when the direct Newton method is used to solve (10.3). For the direct Newton method, we only compute the results for $m \leq 5$ since it is much more for a larger-scale problem and the data we obtained have clearly illustrated the computational efficiency of the proposed method. The numerical errors $\|u^* - u^G_{4+m}\|_\infty$ and the computing time T_G of the two-grid method are listed in columns 8 and 9, respectively. In Figure 10.2, we plot the computing time of the proposed method, the direct Newton method and the

Table 10.1 *Numerical results for Example 1*

m	$s(4+m)$	$\|u^* - u_{4,m}\|_\infty$	C.O.	T_M	$\|u^* - u_{4+m}\|_\infty$	T_N	$\|u^* - u^G_{4+m}\|_\infty$	T_G
0	32	5.142e-3		4.5	5.142e-3	4.6	5.142e-3	4.6
1	64	1.282e-3	2.00	8.8	1.282e-3	12.8	1.285e-3	7.8
2	128	3.190e-4	2.01	14.3	3.192e-4	35.5	3.192e-4	13.2
3	256	7.962e-5	2.00	20.1	7.967e-5	96.0	7.972e-5	24.2
4	512	1.994e-5	2.00	27.6	1.996e-5	277.3	1.995e-5	57.4
5	1024	5.057e-6	1.98	38.2	5.061e-6	774.5	5.063e-6	142.8
6	2048	1.240e-6	2.03	54.4	1.243e-6	2186.2	1.239e-6	406.2
7	4096	3.094e-7	2.00	80.5				
8	8192	7.740e-8	2.00	130.5				
9	16384	1.932e-8	2.00	229.4				

Table 10.2 *Computing time for Example 1*

m	T_1	T_2	$\mathcal{R}(T_2)$	T_3	T_4	$\Delta(T_4)$	T_4'
1	< 0.01	0.062		< 0.01	3.500		0.484
2	< 0.01	0.375	6.049	< 0.01	7.813	4.31	1.047
3	< 0.01	1.172	3.125	< 0.01	12.08	4.27	1.687
4	< 0.01	2.828	2.413	< 0.01	17.11	5.03	2.375
5	< 0.01	6.859	2.425	< 0.01	23.02	5.91	3.234
6	0.015	15.83	2.308	< 0.01	29.41	6.39	4.077
7	0.031	34.61	2.186	< 0.01	36.30	6.89	5.125
8	0.078	76.08	2.198	< 0.01	44.11	7.81	6.173
9	0.187	166.6	2.190	< 0.01	51.58	7.47	7.188

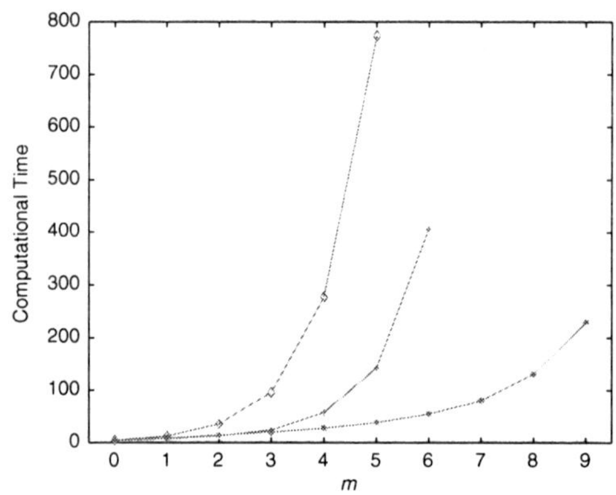

Figure 10.2 Growth of T_M ($*$), T_G ($+$) and T_N ($\diamond$) for Example 1.

two-grid method. The figure shows that T_N grows much faster than the other two as level m increases, and T_M grows the slowest.

We next verify numerically the estimate of $\overline{\mathcal{M}}_{k,m,j}$, $j = 1, 2, 3, 4$, established in Section 10.1.4. In Table 10.2, we list the computing time for the four main procedures listed in Section 10.1.4. All these computing times are measured in seconds. In the table, T_1 denotes the computing time for generating the coefficient matrix $\mathbf{E}_{k,m}$, T_2 presents the total time for evaluating the vectors $\mathbf{f}_{k,l}$, $l = 1, 2, \ldots, m$, T_3 is the total time to solve the linear systems (10.21) for $l = 1, 2, \ldots, m$ and T_4 stands for the total time spent in solving the nonlinear system (10.15) using the Newton iteration, updating the Jacobian matrix (strategy 1) at each iteration step. We observe that T_1 and T_3 are small in comparison with T_2 and T_4. The column for "$\mathcal{R}(T_2)$" reports the increasing ratio of T_2, where the data are obtained by calculating $T_2(m)/T_2(m-1)$.

The column for "$\Delta(T_4)$" lists the difference $T_4(m) - T_4(m-1)$ of two successive values of T_4. We observe that the values in the column for "$\mathcal{R}(T_2)$" are close to 2, which coincides with the estimate that $\overline{\mathcal{M}}_{k,m,2} = \mathcal{O}(\mu^{k+m})$ with $\mu = 2$. The values in column "$\Delta(T_4)$" are nearly constant and this verifies the estimate $\overline{\mathcal{M}}_{k,m,4} = \mathcal{O}(m)$.

We use strategy 2 for the Newton iteration in step 3 of Algorithm 10.9 to solve the nonlinear system (10.15). Specifically, we choose the last Jacobian matrix obtained in step 1 for all steps of the Newton iteration without updating it, with a few more iteration steps to ensure that the same approximation accuracy is obtained. We list in the last column of Table 10.2 under "T_4'" the computing time for solving the nonlinear system (10.15) using the Newton iteration *without* updating the Jacobian matrix (strategy 2), with the number of iterations listed in Table 10.3. For comparison, in Table 10.3, we also list the number of iterations used for strategy 1. By comparing the data of "T_4" and "T_4'" in the same row, we observe that the modification cuts down the computational cost remarkably. We plot the values of T_2, T_4 and T_4' in Figure 10.3. Note that T_2 grows the fastest and occupies most of the running

Table 10.3 *A comparison of iteration numbers used in step 3 with strategies 1 and 2*

Level	1	2	3	4	5	6	7	8	9
Strategy 1	3	2	2	2	2	2	2	1	1
Strategy 2	5	5	5	4	4	4	4	3	3

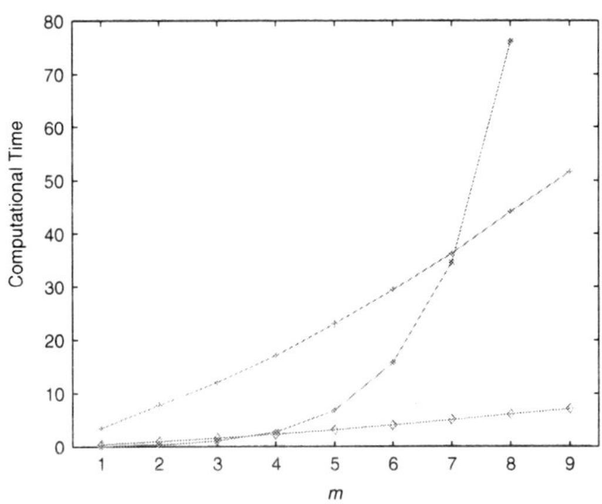

Figure 10.3 Growth of T_2 ($*$), T_4 ($+$) and T_4' ($\diamond$) for Example 1.

time of the algorithm when m is large. Moreover, both T_4 and T'_4 grow nearly linearly with respect to the number of levels, and they have the relation

$$T_4 \approx 7.2T'_4.$$

This suggests that we should use strategy 2 in step 3 of Algorithm 10.9. Hence, in the rest of the numerical experiments we always use strategy 2.

Example 2 We again consider numerical solutions of equation (10.72). This time we solve the equation using the Galerkin scheme with the spaces of piecewise linear polynomials as approximation subspaces. The orthonormal basis functions for $\mathbb{X}_0$ and $\mathbb{W}_1$ are, respectively, $w_{00}(t) = 1$, $w_{01}(t) = \sqrt{3}(2t-1)$, for $t \in [0,1]$ and

$$w_{10}(t) = \begin{cases} 1-6t, & t \in [0,1/2), \\ 5-6t, & t \in [1/2,1], \end{cases} \qquad w_{11}(t) = \begin{cases} \sqrt{3}(1-4t), & t \in [0,1/2), \\ \sqrt{3}(4t-3), & t \in [1/2,1]. \end{cases}$$

The bases for spaces $\mathbb{W}_i$, $i > 1$ are obtained by using the linear operators defined in terms of the contractive mappings ϕ_0 and ϕ_1.

We solve equation (10.72) by Algorithm 10.9 with the initial level $k = 4$ based on the multiscale Galerkin method developed via the above basis functions. The nonlinear system (10.15) related to (10.72) is solved by the secant method. We present the numerical results in Table 10.4. It is clear from the table that the numerical solutions converge in approximately the theoretical order 2. Moreover, the growth of T_2 and T'_4 is consistent with our estimate. Note that for the one-dimensional equation, the entries of the coefficient matrix involve double integrals. As a result, the computing time for this method is much more than that for the corresponding collocation method. The numerical results using the two-grid method are listed in the last two columns for comparison.

Table 10.4 *Numerical results for Example 2*

m	$s(4+m)$	$\|u^*-u_{4,m}\|_2$	C.O.	T_M	T_2	T'_4	$\|u^*-u^G_{4+m}\|_2$	T_G
0	32	1.016e-3		79.8			1.016e-3	81.3
1	64	2.547e-4	2.00	143	4.25	42.54	2.543e-4	361
2	128	6.425e-5	1.99	236	15.7	104.8	6.431e-5	529
3	256	1.661e-5	1.95	382	36.2	196.8	1.670e-5	723
4	512	4.632e-6	1.84	538	70.2	316.3	4.623e-6	1021
5	1024	1.292e-6	1.84	866	127	422.4	1.263e-6	1632
6	2048	3.235e-7	2.00	981	234	555.9		

Example 3 In this example, we consider solving the equation

$$u(s) - \int_0^1 \log \frac{1}{16} | \cos(\pi s) - \cos(\pi t) | (2u^2(t) - 1) \mathrm{d}t = f(s), \quad s \in [0, 1], \tag{10.73}$$

with

$$f(s) := \cos\left(\frac{\pi}{2}\left(\frac{1}{2} - s\right)\right) + \frac{1}{16\pi}[2 - (1 - \cos(\pi s)) \log(1 - \cos(\pi s)) - (1 + \cos(\pi s)) \log(1 + \cos(\pi s))].$$

Note that the kernel of the integral operator involved in this equation is weakly singular. This equation has an isolated solution $u^*(s) = \cos(\frac{\pi}{2}(\frac{1}{2} - s))$, $s \in [0, 1]$. We use the multiscale collocation scheme with the same bases as the first example and the nonlinear system (10.15) is solved by the MAM in conjunction with strategy 2, which suggests no updating in the Jacobian matrix.

The weakly singular integrals are computed numerically using the quadrature methods proposed in [242]. Taking into account the cost of numerical integration, in this case $\overline{\mathcal{M}}_{k,m,2} = \mathcal{O}(s(k+m)(\log s(k+m))^3)$. Therefore, implementing the entire algorithm requires $\mathcal{O}(s(k+m)(\log s(k+m))^3)$ number of multiplications. See also [113, 164, 269, 270] for related treatments of weakly singular integrals. The numerical results are listed in Table 10.5. Since the amounts T_1 and T_3 are insignificant in comparison with T_2 and T_4', they are not included in the table for this experiment.

Example 4 In the last example, we solve the two-dimensional equation

$$u(x, y) - \int_\Omega \sin(xy + x'y') u^2(x', y') \mathrm{d}x' \mathrm{d}y' = f(x, y), \quad (x, y) \in \Omega, \tag{10.74}$$

Table 10.5 *Numerical results for Example 3*

m	$s(4+m)$	$\|u^* - u_{4,m}\|_\infty$	C.O.	T_M	T_2	T_4'	$\|u^* - u^G_{4+m}\|_\infty$	T_G
0	32	1.031e-3		7.6			1.031e-3	7.6
1	64	2.607e-4	1.98	8.0	0.1	0.4	2.613e-4	11.8
2	128	6.493e-5	2.01	11.0	0.9	1.5	6.495e-5	22.4
3	256	1.613e-5	2.01	12.7	1.4	2.1	1.621e-5	48.2
4	512	4.041e-6	2.00	18.0	3.7	3.3	4.043e-6	120.2
5	1024	1.007e-6	2.00	24.4	9.7	3.7	1.010e-6	331.6
6	2048	2.589e-7	1.96	41.4	21.7	6.4	2.593e-7	998.8
7	4096	6.395e-8	2.02	74.9	53.5	6.9		
8	8192	1.577e-8	2.02	120.5	98.4	8.8		
9	16384	3.920e-9	2.01	262.8	236.1	10.5		

where $\Omega = \{(x,y) : 0 \le x \le y \le 1\}$ and the function f is chosen so that $u^*(x,y) = x^2 + y^2$ is an isolated solution of the equation.

We solve equation (10.74) by the MAM based on the multiscale collocation scheme. We choose $\mathbb{X}_n$ as the space of bivariate piecewise linear polynomials with the multiscale partitions defined by the family of contractive mappings $\Phi = \{\phi_e : e \in \mathbb{Z}_4\}$, where

$$\phi_0(x,y) = \left(\frac{x}{2}, \frac{y}{2}\right), \quad \phi_1(x,y) = \left(\frac{x}{2}, \frac{y+1}{2}\right),$$
$$\phi_2(x,y) = \left(\frac{1-x}{2}, 1-\frac{y}{2}\right), \quad \phi_3(x,y) = \left(\frac{x+1}{2}, \frac{y+1}{2}\right).$$

For space $\mathbb{X}_n$ we have the dimension $\dim(\mathbb{X}_n) = 3(4^n)$. Let $S_e := \phi_e(\Omega)$, $e \in \mathbb{Z}_4$. The basis functions for $\mathbb{X}_0$ are given by

$$w_{00}(x,y) = -3x + 2y, \quad w_{01}(x,y) = x - 3y + 2,$$
$$w_{02}(x,y) = 2x + y - 1, \quad (x,y) \in \Omega.$$

The corresponding collocation functionals are chosen as

$$\ell_{00} = \delta_{(\frac{2}{7},\frac{3}{7})}, \quad \ell_{01} = \delta_{(\frac{1}{7},\frac{5}{7})}, \quad \ell_{02} = \delta_{(\frac{4}{7},\frac{6}{7})},$$

where $\langle \delta_{(x,y)}, g\rangle := g(x,y)$. The basis functions for $\mathbb{W}_1$ are given by

$$w_{10}(x,y) = \begin{cases} -\frac{11}{8} - \frac{15}{8}x + \frac{41}{8}y, & (x,y) \in S_0, \\ \frac{5}{8} + \frac{1}{8}x - \frac{7}{8}y, & (x,y) \in S \setminus S_0, \end{cases}$$

$$w_{11}(x,y) = \begin{cases} 1 - \frac{15}{4}x - \frac{7}{8}y, & (x,y) \in S_0, \\ -1 + \frac{1}{4}x + \frac{9}{8}y, & (x,y) \in S \setminus S_0, \end{cases}$$

$$w_{12}(x,y) = \begin{cases} \frac{9}{8} + \frac{15}{8}x - \frac{29}{8}y, & (x,y) \in S_0, \\ -\frac{15}{8} - \frac{1}{8}x + \frac{19}{8}y, & (x,y) \in S \setminus S_0, \end{cases}$$

$$w_{13}(x,y) = \begin{cases} -\frac{15}{8} - \frac{41}{8}x + \frac{13}{4}y, & (x,y) \in S_1, \\ \frac{1}{8} + \frac{7}{8}x - \frac{3}{4}y, & (x,y) \in S \setminus S_1, \end{cases}$$

$$w_{14}(x,y) = \begin{cases} \frac{29}{8} + \frac{7}{8}x - \frac{37}{8}y, & (x,y) \in S_1, \\ -\frac{3}{8} - \frac{9}{8}x + \frac{11}{8}y, & (x,y) \in S \setminus S_1, \end{cases}$$

$$w_{15}(x,y) = \begin{cases} -\frac{5}{8} - \frac{29}{8}x + \frac{7}{4}y, & (x,y) \in S_1, \\ \frac{3}{8} + \frac{19}{8}x - \frac{9}{4}y, & (x,y) \in S \setminus S_1, \end{cases}$$

$$w_{16}(x,y) = \begin{cases} \frac{15}{4} - \frac{13}{4}x - \frac{15}{8}y, & (x,y) \in S_3, \\ -\frac{1}{4} + \frac{3}{4}x + \frac{1}{8}y, & (x,y) \in S \setminus S_3, \end{cases}$$

$$w_{17}(x,y) = \begin{cases} -\frac{1}{8} - \frac{37}{8}x + \frac{15}{4}y, & (x,y) \in S_3, \\ -\frac{1}{8} + \frac{11}{8}x - \frac{1}{4}y, & (x,y) \in S \setminus S_3, \end{cases}$$

$$w_{18}(x,y) = \begin{cases} -\frac{5}{2} + \frac{7}{4}x + \frac{15}{8}y, & (x,y) \in S_3, \\ \frac{1}{2} - \frac{9}{4}x - \frac{1}{8}y, & (x,y) \in S \setminus S_3. \end{cases}$$

Their corresponding collocation functionals are chosen as

$$\ell_{10} = \delta_{(\frac{1}{14},\frac{5}{14})} - \delta_{(\frac{1}{7},\frac{3}{14})} + \delta_{(\frac{2}{7},\frac{3}{7})} - \delta_{(\frac{3}{14},\frac{4}{7})},$$
$$\ell_{11} = \delta_{(\frac{1}{14},\frac{5}{14})} - \delta_{(\frac{2}{7},\frac{3}{7})} + \delta_{(\frac{3}{7},\frac{9}{14})} - \delta_{(\frac{3}{14},\frac{4}{7})},$$
$$\ell_{12} = \delta_{(\frac{5}{14},\frac{11}{14})} - \delta_{(\frac{3}{7},\frac{9}{14})} + \delta_{(\frac{2}{7},\frac{3}{7})} - \delta_{(\frac{3}{14},\frac{4}{7})},$$
$$\ell_{13} = \delta_{(\frac{1}{14},\frac{6}{7})} - \delta_{(\frac{1}{7},\frac{5}{7})} + \delta_{(\frac{5}{14},\frac{11}{14})} - \delta_{(\frac{2}{7},\frac{13}{14})},$$
$$\ell_{14} = \delta_{(\frac{1}{7},\frac{5}{7})} - \delta_{(\frac{2}{7},\frac{13}{14})} + \delta_{(\frac{5}{14},\frac{11}{14})} - \delta_{(\frac{3}{14},\frac{4}{7})},$$
$$\ell_{15} = \delta_{(\frac{3}{7},\frac{9}{14})} - \delta_{(\frac{5}{14},\frac{11}{14})} + \delta_{(\frac{1}{7},\frac{5}{7})} - \delta_{(\frac{3}{14},\frac{4}{7})},$$
$$\ell_{16} = \delta_{(\frac{4}{7},\frac{6}{7})} - \delta_{(\frac{11}{14},\frac{13}{14})} + \delta_{(\frac{9}{14},\frac{5}{7})} - \delta_{(\frac{3}{7},\frac{9}{14})},$$
$$\ell_{17} = \delta_{(\frac{4}{7},\frac{6}{7})} - \delta_{(\frac{9}{14},\frac{5}{7})} + \delta_{(\frac{3}{7},\frac{9}{14})} - \delta_{(\frac{5}{14},\frac{11}{14})},$$
$$\ell_{18} = \delta_{(\frac{3}{14},\frac{4}{7})} - \delta_{(\frac{3}{7},\frac{9}{14})} + \delta_{(\frac{4}{7},\frac{6}{7})} - \delta_{(\frac{5}{14},\frac{11}{14})}.$$

The construction of the basis functions and collocation functionals of higher levels are described in Chapter 7 (cf. [69, 75]).

The above bases are used in our method for solving (10.74), and the numerical results are listed in Table 10.6. At each step, the nonlinear system (10.15) is solved by the Newton iteration method. The computed convergence orders are all around 2. This confirms the theoretical estimate of the convergence order. Since $\dim(\mathbb{X}_{n+1})/\dim(\mathbb{X}_n) = 4$, the theoretical value of $\mathcal{R}(T_2)$ is 4. Table 10.6 shows that the computed values of $\mathcal{R}(T_2)$ are close to 4.

For comparison, we also solve the collocation equation (10.3) by the direct Newton method and the two-grid method. The numerical errors and computing time of the three methods are listed in Table 10.7. The data show that the

Table 10.6 *Numerical results for Example 4*

m	$s(2+m)$	$\|u^* - u_{2,m}\|$	C.O.	T_M	T_2	$\mathcal{R}(T_2)$	T_4'	$\Delta(T_4')$
0	48	3.269e-2		73.60				
1	192	8.159e-3	2.00	137.6	28.20		35.80	
2	768	2.035e-3	2.00	286.3	140.8	4.995	71.63	35.83
3	3072	5.076e-4	2.00	772.5	591.5	4.201	107.4	35.77
4	12288	1.268e-4	2.00	2610	2393	4.046	143.2	35.80

Table 10.7 *Comparison of the methods for solving Example 4*

m	$s(2+m)$	$\|u^* - u_{2,m}\|$	T_M	$\|u^* - u^G_{2+m}\|$	T_G	$\|u^* - u_{2+m}\|$	T_N
0	48	3.269e-2	73.60	3.269e-2	73.60	3.269e-2	73.60
1	192	8.159e-3	137.6	8.162e-3	213.2	8.157e-3	355.5
2	768	2.035e-3	286.3	2.035e-3	632.4	2.035e-3	2724
3	3072	5.076e-4	772.5	5.084e-4	3613	5.083e-4	17492
4	12288	1.268e-4	2610				

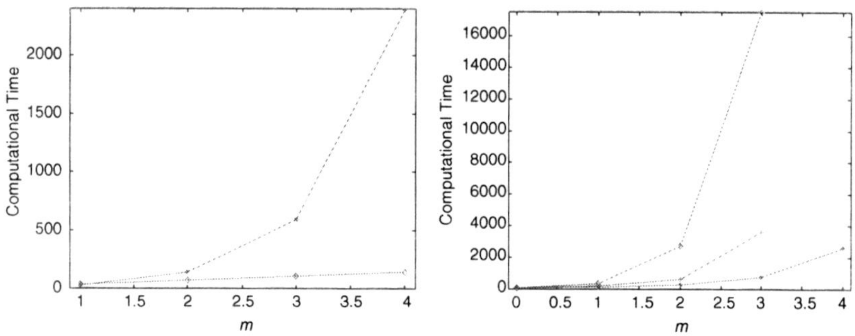

Figure 10.4 Comparison for Example 4. Left: Growth of T_2 $(*)$, T_4' $(\diamond)$. Right: T_M $(*)$, T_G $(+)$, T_N $(\diamond)$.

proposed method has nearly the same accuracy as the direct Newton method and the two-grid method. To compare visually the computing times for this example, we plot them in Figure 10.4, from which we observe that the proposed method runs the fastest.

10.4.2 Numerical examples of nonlinear boundary integral equations

In this subsection, we present numerical results which verify the approximation accuracy and computational efficiency of the proposed algorithm for solving nonlinear boundary integral equations. All programs are run on a workstation with 3.38-GHz CPU and 96G memory.

We consider the boundary value problem (10.26) with $g(x, u(x)) := u(x) + \sin(u(x))$. Let Ω be the elliptical region $x_1^2 + \left(\frac{x_2}{2}\right)^2 < 1$. For comparison purposes, we choose the solution of (10.26) as $u_0(x) := e^{x_1}\cos(x_2)$ with $x = (x_1, x_2)$. Correspondingly, we have that

$$g_0(x) := g(x, u_0) + \frac{\partial u_0(x)}{\partial \mathbf{n}_x}.$$

The corresponding solution of the boundary integral equation (10.29) is given by

$$u^*(t) := e^{\cos(2\pi t)} \cos(2\sin(2\pi t)).$$

In all experiments presented in this subsection, we choose $\mu = 2$ and $\mathbb{X}_n$ as the space of piecewise cubic polynomials with knots at $j/2^n$, $j = 1, 2, \ldots, 2^n - 1$. It is easy to compute $\dim(\mathbb{X}_n) = 2^{n+2}$. The basis functions of $\mathbb{X}_0$ and $\mathbb{W}_1$ are given by

$$w_{00}(t) = -\frac{1}{6}(5t-2)(5t-3)(5t-4),$$

$$w_{01}(t) = \frac{1}{2}(5t-1)(5t-3)(5t-4),$$

$$w_{02}(t) = -\frac{1}{2}(5t-1)(5t-2)(5t-4),$$

$$w_{03}(t) = \frac{1}{6}(5t-1)(5t-2)(5t-3),$$

$$w_{10}(t) = \begin{cases} \frac{85}{32} - \frac{725}{12}t + \frac{575}{2}t^2 - \frac{1475}{4}t^3, & t \in [0, \frac{1}{2}], \\ -\frac{235}{32} + \frac{575}{12}t - \frac{175}{2}t^2 + \frac{575}{12}t^3, & t \in (\frac{1}{2}, 1], \end{cases}$$

$$w_{11}(t) = \begin{cases} \frac{1145}{288} - \frac{1775}{24}t + \frac{1675}{6}t^2 - \frac{4975}{18}t^3, & t \in [0, \frac{1}{2}], \\ -\frac{7495}{288} + \frac{3625}{24}t - \frac{525}{2}t^2 + \frac{2525}{18}t^3, & t \in (\frac{1}{2}, 1], \end{cases}$$

$$w_{12}(t) = \begin{cases} \frac{805}{288} - \frac{375}{8}t + \frac{475}{3}t^2 - \frac{2525}{18}t^3, & t \in [0, \frac{1}{2}], \\ -\frac{19355}{288} + \frac{8275}{24}t - 550t^2 + \frac{4975}{18}t^3, & t \in (\frac{1}{2}, 1], \end{cases}$$

$$w_{13}(t) = \begin{cases} \frac{95}{96} - \frac{50}{3}t + \frac{225}{4}t^2 - \frac{575}{12}t^3, & t \in [0, \frac{1}{2}], \\ -\frac{13345}{96} + \frac{1775}{3}t - \frac{3275}{4}t^2 + \frac{1475}{4}t^3, & t \in (\frac{1}{2}, 1]. \end{cases}$$

The corresponding collocation functionals are

$$\ell_{00} = \delta_{\frac{1}{5}}, \quad \ell_{01} = \delta_{\frac{2}{5}}, \quad \ell_{02} = \delta_{\frac{3}{5}}, \quad \ell_{03} = \delta_{\frac{4}{5}},$$

$$\ell_{10} = \frac{2}{5}\delta_{\frac{1}{10}} - \frac{3}{2}\delta_{\frac{2}{10}} + 2\delta_{\frac{3}{10}} - \delta_{\frac{4}{10}} + \frac{1}{10}\delta_{\frac{6}{10}},$$

$$\ell_{11} = \frac{3}{10}\delta_{\frac{2}{10}} - \delta_{\frac{3}{10}} + \delta_{\frac{4}{10}} - \frac{1}{2}\delta_{\frac{6}{10}} + \frac{1}{5}\delta_{\frac{7}{10}},$$

$$\ell_{12} = \frac{1}{5}\delta_{\frac{3}{10}} - \frac{1}{2}\delta_{\frac{4}{10}} + \delta_{\frac{6}{10}} - \delta_{\frac{7}{10}} + \frac{3}{10}\delta_{\frac{8}{10}},$$

$$\ell_{13} = \frac{1}{10}\delta_{\frac{4}{10}} - \delta_{\frac{6}{10}} + 2\delta_{\frac{7}{10}} - \frac{3}{2}\delta_{\frac{8}{10}} + \frac{2}{5}\delta_{\frac{9}{10}}.$$

The basis functions w_{ij} and collocation functionals ℓ_{ij} for $i > 1$ can be constructed recursively from w_{1j} and ℓ_{1j}.

Table 10.8 *Total time for solving the related nonlinear system*

m	T_m	T'_m
0	24.9	0.016
1	37.0	0.046
2	44.7	0.078
3	53.7	0.093
4	63.8	0.140
5	74.7	0.203
6	85.2	0.382
7	99.1	0.757
8	113.2	1.592

Example 1 In this experiment, we use the MAMs with initial level $k = 4$ to solve the boundary integral equation (10.29). The purpose of this experiment is to test how much the proposed technique for solving the nonlinear systems speeds up the solution process. To this end, we run Algorithms 10.18 and 10.21 separately, and add up the time used in solving the nonlinear systems. Note that in Algorithm 10.18 we need to solve (10.22) once and (10.51) m times, while in Algorithm 10.21 we need to solve (10.58) once and (10.59) m times. In Table 10.8, T_m denotes the total time spent in Algorithm 10.18 for solving the nonlinear systems, while T'_m denotes that in Algorithm 10.21. Obviously, we see that in Table 10.8, T'_m is much less than T_m.

Example 2 We illustrate in this experiment the approximation accuracy and the total computational effects of Algorithm 10.21, compared with those of Algorithm 10.18 and Algorithm AC (the algorithm of Atkinson and Chandler presented in [17]).

In Table 10.9, we report the numerical results of Algorithms 10.18 and 10.21. For any m, we denote by $u_{4,m}$ and $\tilde{u}_{4,m}$, respectively, the numerical solutions resulting from Algorithms 10.18 and 10.21. Moreover, we let T_M and T'_M denote the total times for implementing Algorithms 10.18 and 10.21, respectively. We observe in Table 10.9 that $u_{4,m}$ and $\tilde{u}_{4,m}$ have nearly the same accuracy, while T'_M is significantly less than T_M. This indicates that although Algorithms 10.18 and 10.21 have the same order of computational costs, the techniques proposed in this paper effectively reduce the absolute computational costs.

For comparison, we list in Table 10.10 the numerical results of Algorithm AC, which is a Nyström method. We let N be the number of unknowns, u_N the numerical solution and T_N the running time of the program.

Note that we cannot compare Algorithms 10.18 and AC in the same discretization scale, since Algorithm 10.18 is a multiscale method and Algorithm

Table 10.9 *Comparison of the accuracy and running time of Algorithms 10.18 and 10.21*

m	$s(4+m)$	$\|u^* - u_{4,m}\|_\infty$	$\|u^* - \tilde{u}_{4,m}\|_\infty$	T_M	T'_M
0	64	4.107e-3	5.079e-3	85	0.2
1	128	3.079e-4	3.007e-4	89	0.4
2	256	2.152e-5	2.225e-5	112	0.8
3	512	1.491e-6	1.527e-6	146	1.7
4	1024	8.204e-8	8.624e-8	203	3.7
5	2048	5.041e-9	5.313e-9	309	7.9
6	4096	2.865e-10	2.970e-10	519	16.9
7	8192	1.795e-11	1.882e-11	880	36.1
8	16384	1.023e-12	1.101e-12	1961	76.8

Table 10.10 *Numerical results of Algorithm AC*

N	$\|u^* - u_N\|_\infty$	T_N
128	1.780e-5	3
256	1.071e-6	11
512	6.559e-8	52
1024	4.076e-9	226
2048	2.695e-10	836
4096	9.870e-11	3877
8192	6.751e-11	16473

AC is a single-scale quadrature method. However, we may utilize a "numerical errors vs. computing time" figure for these methods. We use the data in Tables 10.9 and 10.10 to generate Figure 10.5. For convenience of display, we take logarithms on both numerical errors and computing times. It is seen that for any accuracy level, Algorithm 10.21 uses the least time among the three algorithms. These numerical results confirm that the proposed methods outperform both the two-grid method and Algorithm AC.

10.5 Bibliographical remarks

The results presented in this chapter were mainly chosen from the original papers [51, 52, 76]. The reader is referred to recent papers [50, 158] for additional developments, where a multiscale collocation method was applied to solving nonlinear Hammerstein equations. Specifically, in [158] a sparsity in the Jacobian matrix resulted from nonlinear solvers such as the Newton method and the quasi-Newton method was discussed that leads to a fast

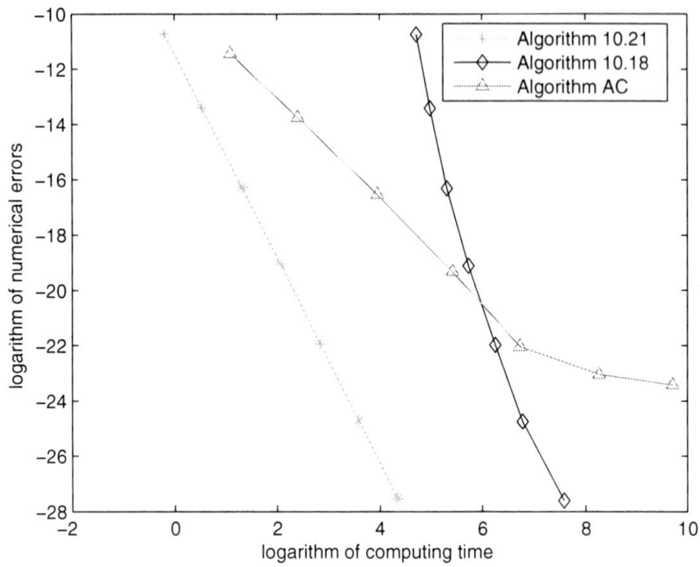

Figure 10.5 Comparison of Algorithms 10.18, 10.21 and AC on numerical performance.

algorithm. Furthermore, a fast multilevel augmentation method was applied to a transformed nonlinear equation that results in an additional saving in computing time.

Numerical solutions of nonlinear integral equations, especially the Hammerstein equation, have been studied extensively in the literature; see [20, 23, 34, 35, 131, 132, 159, 161, 163, 175, 180–182, 240, 255, 257, 258]. Specifically, a degenerate kernel scheme was introduced in [163], a variation of Nyström's method was proposed in [182] and a product integration method was discussed in [161]. Projection methods, including the Galerkin scheme and collocation scheme, were discussed in [20, 23, 47, 131, 159, 180, 181, 240, 255, 258]. Fundamental work on the error analysis of projection methods may be found in Section 3.3 of [175]. Studies of superconvergence properties can be found in [160, 161, 165, 179, 247]. Moreover, the reader is referred to [13, 22] for general information on numerical treatments of the Hammerstein equations. Furthermore, regularity of the solution of the Hammerstein equation with a weakly singular kernel was studied in [162]. Certain aspects of theoretical analysis of the Hammerstein equation may be found in [99, 130, 184, 277]. The connection of a large class of Gaussian processes with the Hammerstein equation was studied in [218].

Boundary value problems of the Laplace equation serve as mathematical models for many important applications, such as acoustics, elasticity, electromagnetics and fluid dynamics (see [126, 149, 208, 234, 265] and the references cited therein). Nonlinear boundary conditions are also involved in various applications, such as heat radiation and heat transfer [29, 32, 42]. In some electromagnetic problems, the boundary conditions may also contain nonlinearities [171]. In these cases, the reformulation of the corresponding boundary value problems leads to nonlinear integral equations. The nonlinear boundary integral equation of the Laplace equation was treated numerically in [17]. Certain linearization approaches are used in the numerical treatment of nonlinear integral equations (see, for example, [13–15, 20, 23, 131, 132, 159–161, 163, 165, 182, 240, 255]).

11

Multiscale methods for ill-posed integral equations

In this chapter, we present multiscale methods for solving ill-posed integral equations of the first kind. The ill-posed integral equations are turned to well-posed integral equations of the second kind by using regularization methods, including the Lavrentiev regularization and the Tikhonov regularization. Multiscale Galerkin and collocation methods are introduced for solving the resulting well-posed equations. *A priori* and *a posteriori* regularization parameter choice strategies are proposed. Convergence rates of the regularized solutions are established.

11.1 Numerical solutions of regularization problems

We present a brief review of regularization of ill-posed integral equations of the first kind and discuss the main idea used in this chapter in developing fast algorithms for solving the equations.

We consider the Fredholm integral equation of the first kind in the form

$$\mathcal{K}u = f, \tag{11.1}$$

where

$$(\mathcal{K}u)(s) := \int_{\Omega} K(s,t)u(t)\mathrm{d}t, \quad s \in \Omega. \tag{11.2}$$

Noting that compact operators cannot have a continuous inverse, equation (11.1) with a compact operator $\mathcal{K}$ is an ill-posed problem in the sense of the following definition.

Definition 11.1 Let $\mathcal{K}$ be an operator from a normed linear space $\mathbb{X}$ to a normed linear space $\mathbb{Y}$. Equation (11.1) is said to be well posed if for any

$f \in \mathbb{Y}$ there exists a solution $u \in \mathbb{X}$, the solution is unique and the dependence of u upon f is continuous. Otherwise, the equation is called ill posed.

Among the three conditions for equation (11.1) to be well posed listed in Definition 11.1, continuous dependence of the solution on the given data is most crucial and challenging, because the nonexistence or nonuniqueness of a solution may be well treated by an existing method such as the least-squares method. In this chapter, we mainly treat the ill-posedness caused by the violation of this condition. The ill-posedness of the equation is often treated by regularization, which imposes certain *a priori* conditions on the solution. The most commonly used method of regularization of ill-posed problems is the Tikhonov regularization (named after Andrey Tikhonov) [252]. It is also called the Tikhonov–Phillips regularization because of the contribution of Phillips [217]. Another frequently used method is the Lavrentiev regularization [116, 220]. Recent developments in image science use the TV regularization [237].

Certain regularization methods may be formulated as a minimization of a fidelity term which measures the error of $\mathcal{K}u - f$ plus a regularization parameter times the norm of u determined by the *a priori* condition. The Tikhonov regularization [116, 117, 119, 217, 252] and the TV regularization are examples of this type. The optimal regularization parameter is usually unknown. In practical problems such as image/signal processing it is often determined by an ad hoc method. Approaches include the Bayesian method, the discrepancy principle, cross-validation, L-curve method, restricted maximum likelihood and unbiased predictive risk estimator. In [262] the connection was established between the choice of the optimal parameter and the leave-one-out cross-validation. There is recent interest in the multiparameter regularization [63].

Although the theoretical development of solving ill-posed problems is nearly complete, developing efficient stable fast solvers for such problems remains an active focused research area. A bottleneck problem for the numerical solution of the ill-posed Fredholm integral equation of the first kind is its demanding large computational costs. By regularization, the solution of the ill-posed equation is obtained by solving a sequence of well-posed Fredholm integral equations of the second kind. As discussed in previous chapters, the discretization of such an equation leads to an algebraic system with a full matrix and numerical solutions of the system are computationally costly. Aimed at overcoming this bottleneck problem, we consider in this chapter solving ill-posed operator equations of the first kind in a *multiresolution* framework. Although multiscale methods for solving well-posed operator equations have been well understood and widely used (see previous chapters of this book), less attention has been paid to the development of multiscale methods for ill-posed problems. Solving ill-posed problems requires iterated computations which demand huge computational costs and, therefore, designing efficient numerical methods for solving problems of this type, by making use of the underlying multiscale data structure, is extremely beneficial. It is the attempt of this chapter to explore the multiscale matrix representation of the operator in developing efficient fast algorithms for solving ill-posed problems.

To describe the main point of this chapter, we now elaborate the *multilevel augmentation* method developed in Chapter 9 for solving *well-posed* operator equations. It is based on direct sum decompositions of the range space of the operator and the solution space of the operator equation and a matrix splitting scheme. It allows us to develop fast, accurate and stable nonconventional numerical algorithms for solving operator equations. For second-kind equations, special splitting techniques were proposed to develop such algorithms. These algorithms were then applied to solve the linear systems resulting from matrix compression schemes using wavelet-like functions for solving Fredholm integral equations of the second kind. It was proved that the method has an optimal order of convergence and it is numerically stable. With an appropriate matrix compression technique, it leads to fast algorithms. Basically, this method generates an approximate solution with convergence order $O(N^{-k/d})$ if piecewise polynomials of order k are used in the approximation, the spatial dimension of the integral equation is d and solving the entire discrete equation requires computational complexity of order $O(N)$.

The main idea used in Section 11.2 is to combine the Lavrentiev regularization method and the MAM in solving the resulting regularized second-kind equations. Since the matrix compression issue for integral equations has been well explained in previous chapters, we do not discuss this issue in the section and suppose that an appropriate compression technique will be used in practical computation (cf. [55]). Instead, we focus on choosing regularization parameters by exploring the multiscale structure of the matrix representation of the operator. We present choices for *a priori* parameters and for *a posteriori* parameters in the context of the MAM.

Multilevel methods were applied to solving ill-posed problems, and *a priori* and *a posteriori* parameter choice strategies were also proposed (see, for example, [103, 137, 157, 193]). These methods are all based on the Galerkin discretization principle. Because of the computational advantages, which we discussed earlier, of the collocation method, it is highly desirable to develop fast multiscale collocation methods. However, there are two challenging issues related to the development of the fast collocation method for solving the ill-posed integral equation. First, it requires the availability of a Banach space setting for regularization, since the appropriate context for collocation methods is L^∞. There is some effort at developing the collocation method in the L^2 space for solving ill-posed integral equations (see [190, 207, 211]). However, we feel that the traditional regularization analysis in Hilbert spaces (such as the L^2 space) is more suitable for Galerkin-type methods. In other words, it is more natural and interesting to analyze collocation-type methods in the L^∞ space. Regularization analysis for collocation methods for ill-posed problems in a Banach space has not yet been completely understood in the literature, even though some interesting results were obtained in [117, 216, 224]. The mathematical analysis of the convergence and convergence rate in a Banach space is quite different from that in a Hilbert space, since many estimate results and conclusions which hold in a Hilbert space may not hold in a Banach space. For example, in the Banach space, little is known about the saturation phenomenon. The second challenging issue is that *a posteriori* parameter choice strategies for the fast collocation method demand certain estimates in the L^∞-norm, which are not available. In general, mathematical analysis is more difficult in a Banach space than in a Hilbert space. An optimal regularization parameter should give the best balance between the well-posedness and approximation accuracy. This principle has been used by many researchers (for example [118, 127–129, 206, 220, 222, 224, 228, 248, 275]) in developing *a priori* and *a posteriori* parameter choice strategies for regularized Galerkin methods. The focus of Section 11.3 is to develop a fast collocation algorithm based on the mathematical development presented in [117, 224] and a related *a posteriori* choice strategy of regularization parameters based on the idea used in [207, 216].

We remark that further development in this topic can be seen in [56]. In this paper, multiscale collocation methods are developed for solving a system of integral equations which is a reformulation of the Tikhonov-regularized second-kind equation of an ill-posed integral equation of the first kind. Direct numerical solutions of the Tikhonov regularization equation require generating a matrix representation of the composition of the conjugate operator with its original integral operator. Generating such a matrix is computationally costly. To overcome this difficulty, rather than solving the Tikhonov-regularized equation directly, it is proposed to solve an equivalent coupled system of integral equations. A multiscale collocation method is applied with a matrix compression strategy to discretize the system of integral equations and then use the multilevel augmentation method to solve the resulting discrete system. *A priori* and *a posteriori* parameter choice strategies are also developed for these methods.

To close this section, we remark that as an application of multiscale Galerkin methods for solving the ill-posed integral equation, integral equation models for image restoration were considered in [189]. Discrete models are consistently used as practical models for image restoration. They are piecewise constant approximations of true physical (continuous) models, and hence, inevitably impose bottleneck model errors. Paper [189] proposed working directly with continuous models for image restoration, aiming to suppress the model errors caused by the discrete models. A systematic study was conducted in that paper for the continuous out-of-focus image models which can be formulated as an integral equation of the first kind. The resulting integral equation was regularized by the Lavrentiev method and the Tikhonov method. Fast multiscale algorithms having high accuracy were developed there to solve the regularized integral equations of the second kind. Numerical experiments presented in the paper show that the methods based on the continuous model perform much better than those based on discrete models.

11.2 Multiscale Galerkin methods via the Lavrentiev regularization

In this section, we first describe the MAM for numerical solutions of ill-posed equations of the first kind, and then present convergence analysis for the approximate solution obtained from the MAM with an *a priori* regularization parameter. We propose an *a posteriori* regularization parameter choice strategy in the MAM. The choice of the parameter is adapted to the context of the multilevel augmentation method. We establish an optimal order of convergence for the approximate solution obtained from the multilevel augmentation method using the *a posteriori* regularization parameter.

11.2.1 Multilevel augmentation methods

We present in this subsection the MAM for numerical solutions of ill-posed operator equations of the first kind. For this purpose, we recall the Lavrentiev regularization method for the equations. Suppose that $\mathbb{X}$ is a real Hilbert space with an inner product $(\cdot,\cdot)$ and the related norm $\|\cdot\|$. Let $\mathcal{K}:\mathbb{X}\to\mathbb{X}$ be a linear and positive semi-definite compact operator, that is, $(\mathcal{K}x,x)\geq 0$, for all $x\in\mathbb{X}$. Given $f\in\mathbb{X}$, we consider the operator equation of the first kind having the form

$$\mathcal{K}u=f, \tag{11.3}$$

where $u\in\mathbb{X}$ is the unknown to be determined.

We assume that the range $R(\mathcal{K})$ of operator $\mathcal{K}$ is infinite and thus, equation (11.3) is ill posed [116, 221]. For $f \in R(\mathcal{K})$ we let $u_* \in \mathbb{X}$ denote the unique *minimum norm solution* of equation (11.3) in the sense that

$$\mathcal{K}u_* = f \text{ and } \|u_*\| = \inf\{\|v\| : \mathcal{K}v = f, v \in \mathbb{X}\}. \tag{11.4}$$

In general, the solution of (11.3) does not depend continuously on the right-hand side f. Let $\delta > 0$ be a given small number and suppose that $f^\delta \in \mathbb{X}$ satisfying the condition

$$\|f^\delta - f\| \leq \delta \tag{11.5}$$

is given noisy data actually used in computing the solution of equation (11.3). We apply the Lavrentiev regularization method to solve equation (11.3). For $\alpha > 0$, we solve the equation

$$(\alpha\mathcal{I} + \mathcal{K})u_\alpha^\delta = f^\delta. \tag{11.6}$$

It can be proved that

$$\|(\alpha\mathcal{I} + \mathcal{K})^{-1}\| \leq \alpha^{-1} \tag{11.7}$$

and thus, for any given $\alpha > 0$, equation (11.6) has a unique solution $u_\alpha^\delta \in \mathbb{X}$. We consider the unique solution $u_\alpha^\delta \in \mathbb{X}$ of (11.6) as an approximation of u_*. It is well known that

$$\lim_{\alpha \to 0, \delta\alpha^{-1} \to 0} \|u_\alpha^\delta - u_*\| = 0.$$

We now recall the projection method for solving the regularization equation (11.6). Suppose that $\mathbb{X}_n$, $n \in \mathbb{N}_0$ is a sequence of finite-dimensional subspaces of $\mathbb{X}$ satisfying

$$\overline{\bigcup_{n \in \mathbb{N}_0} \mathbb{X}_n} = \mathbb{X}.$$

For each $n \in \mathbb{N}_0$ we let $\mathcal{P}_n : \mathbb{X} \to \mathbb{X}_n$ denote the linear orthogonal projection and thus, $\mathcal{P}_n$ converges pointwise to the identity operator and $\|\mathcal{P}_n\| = 1$. Set

$$\mathcal{K}_n := \mathcal{P}_n\mathcal{K}\mathcal{P}_n, \; f_n^\delta := \mathcal{P}_n f^\delta.$$

The projection method is to find $u_{\alpha,n}^\delta \in \mathbb{X}_n$ such that

$$(\alpha\mathcal{I} + \mathcal{K}_n)u_{\alpha,n}^\delta = f_n^\delta. \tag{11.8}$$

Since $\mathcal{K}_n$ is positive semi-definite,

$$\|(\alpha\mathcal{I} + \mathcal{K}_n)^{-1}\| \leq \alpha^{-1} \tag{11.9}$$

and thus, equation (11.8) has a unique solution.

To apply the MAM to equation (11.8), as we have done earlier, we assume that the subspaces $\mathbb{X}_n$, $n \in \mathbb{N}_0$ are nested, and let $\mathbb{W}_{n+1}$ be the orthogonal complement of $\mathbb{X}_n$ in $\mathbb{X}_{n+1}$. For a fixed $k \in \mathbb{N}$ and any $m \in \mathbb{N}_0$, we have the decomposition

$$\mathbb{X}_{k+m} = \mathbb{X}_k \oplus^{\perp} \mathbb{W}_{k+1} \oplus^{\perp} \cdots \oplus^{\perp} \mathbb{W}_{k+m}. \tag{11.10}$$

For $g_0 \in \mathbb{X}_k$ and $g_i \in \mathbb{W}_{k+i}, i = 1, 2, \ldots, m$, we identify the vector $[g_0, g_1, \ldots, g_m]^T$ in $\mathbb{X}_k \times \mathbb{W}_{k+1} \times \cdots \times \mathbb{W}_{k+m}$ with the vector $g_0 + g_1 + \cdots + g_m$ in $\mathbb{X}_k \oplus^{\perp} \mathbb{W}_{k+1} \oplus^{\perp} \cdots \oplus^{\perp} \mathbb{W}_{k+m}$. We write the solution $u^{\delta}_{\alpha,k+m} \in \mathbb{X}_{k+m}$ of equation (11.8) with $n := k+m$ as

$$u^{\delta}_{\alpha,k+m} = (u^{\delta}_{\alpha,k})_0 + \sum_{j=1}^{m} (u^{\delta}_{\alpha,k})_j = [(u^{\delta}_{\alpha,k})_0, (u^{\delta}_{\alpha,k})_1, \ldots, (u^{\delta}_{\alpha,k})_m]^T,$$

where $(u^{\delta}_{\alpha,k})_0 \in \mathbb{X}_k$ and $(u^{\delta}_{\alpha,k})_j \in \mathbb{W}_{k+j}$ for $j = 1, 2, \ldots, m$.

We next re-express the operator in equation (11.8) with $n := k+m$. Defining

$$\mathcal{Q}_{n+1} := \mathcal{P}_{n+1} - \mathcal{P}_n, \quad n \in \mathbb{N}_0,$$

the function f^{δ}_{k+m} is identified as

$$f^{\delta}_{k+m} = [\mathcal{P}_k f^{\delta}, \mathcal{Q}_{k+1} f^{\delta}, \ldots, \mathcal{Q}_{k+m} f^{\delta}]^T$$

and the operator $\mathcal{K}_{k+m}$ is identified as a matrix of operators

$$\mathcal{K}_{k,m} := \begin{bmatrix} \mathcal{P}_k\mathcal{K}\mathcal{P}_k & \mathcal{P}_k\mathcal{K}\mathcal{Q}_{k+1} & \cdots & \mathcal{P}_k\mathcal{K}\mathcal{Q}_{k+m} \\ \mathcal{Q}_{k+1}\mathcal{K}\mathcal{P}_k & \mathcal{Q}_{k+1}\mathcal{K}\mathcal{Q}_{k+1} & \cdots & \mathcal{Q}_{k+1}\mathcal{K}\mathcal{Q}_{k+m} \\ \vdots & \vdots & \ddots & \vdots \\ \mathcal{Q}_{k+m}\mathcal{K}\mathcal{P}_k & \mathcal{Q}_{k+m}\mathcal{K}\mathcal{Q}_{k+1} & \cdots & \mathcal{Q}_{k+m}\mathcal{K}\mathcal{Q}_{k+m} \end{bmatrix}.$$

We then split the operator $\mathcal{K}_{k,m}$ into the sum of two operators

$$\mathcal{K}_{k,m} = \mathcal{K}^{L}_{k,m} + \mathcal{K}^{H}_{k,m},$$

where

$$\mathcal{K}^{L}_{k,m} := \mathcal{P}_k\mathcal{K}\mathcal{P}_{k+m} \text{ and } \mathcal{K}^{H}_{k,m} := (\mathcal{P}_{k+m} - \mathcal{P}_k)\mathcal{K}\mathcal{P}_{k+m},$$

which correspond to lower and higher resolution of the operator $\mathcal{K}_{k+m}$, respectively. In the matrix notation, we have that

$$\mathcal{K}^{L}_{k,m} := \begin{bmatrix} \mathcal{P}_k\mathcal{K}\mathcal{P}_k & \mathcal{P}_k\mathcal{K}\mathcal{Q}_{k+1} & \cdots & \mathcal{P}_k\mathcal{K}\mathcal{Q}_{k+m} \\ 0 & 0 & \cdots & 0 \\ \vdots & \vdots & \ddots & \vdots \\ 0 & 0 & \cdots & 0 \end{bmatrix}$$

and

$$\mathcal{K}_{k,m}^H := \begin{bmatrix} 0 & 0 & \cdots & 0 \\ \mathcal{Q}_{k+1}\mathcal{K}\mathcal{P}_k & \mathcal{Q}_{k+1}\mathcal{K}\mathcal{Q}_{k+1} & \cdots & \mathcal{Q}_{k+1}\mathcal{K}\mathcal{Q}_{k+m} \\ \vdots & \vdots & \ddots & \vdots \\ \mathcal{Q}_{k+m}\mathcal{K}\mathcal{P}_k & \mathcal{Q}_{k+m}\mathcal{K}\mathcal{Q}_{k+1} & \cdots & \mathcal{Q}_{k+m}\mathcal{K}\mathcal{Q}_{k+m} \end{bmatrix}.$$

For a given parameter $\alpha > 0$, we set

$$\mathcal{B}_{k,m}(\alpha) := \mathcal{I} + \alpha^{-1}\mathcal{K}_{k,m}^L \quad \text{and} \quad \mathcal{C}_{k,m}(\alpha) := \alpha^{-1}\mathcal{K}_{k,m}^H. \tag{11.11}$$

Thus, we have the decomposition

$$\mathcal{I} + \alpha^{-1}\mathcal{K}_{k,m} = \mathcal{B}_{k,m}(\alpha) + \mathcal{C}_{k,m}(\alpha), \quad m \in \mathbb{N}.$$

The MAM for solving (11.8) can be described as follows.

Algorithm 11.2 *(Multilevel augmentation algorithm)*

Step 1: *For a fixed $k > 0$, solve (11.8) with $n = k$ exactly.*
Step 2: *Set $u_{\alpha,k,0}^\delta := u_{\alpha,k}^\delta$ and compute matrices $\mathcal{B}_{k,0}(\alpha)$ and $\mathcal{C}_{k,0}(\alpha)$.*
Step 3: *For $m \in \mathbb{N}$, suppose that $u_{\alpha,k,m-1}^\delta \in \mathbb{X}_{k+m-1}$ has been obtained and do the following.*

- *Augment the matrices $\mathcal{B}_{k,m-1}(\alpha)$ and $\mathcal{C}_{k,m-1}(\alpha)$ to form $\mathcal{B}_{k,m}(\alpha)$ and $\mathcal{C}_{k,m}(\alpha)$, respectively.*
- *Augment $u_{\alpha,k,m-1}^\delta$ to form*

$$\overline{u}_{\alpha,k,m}^\delta := \begin{bmatrix} u_{\alpha,k,m-1}^\delta \\ 0 \end{bmatrix} \in \mathbb{X}_{k+m}.$$

- *Solve $u_{\alpha,k,m}^\delta := [(u_{\alpha,k,m}^\delta)_0, (u_{\alpha,k,m}^\delta)_1, \ldots, (u_{\alpha,k,m}^\delta)_m]^T$ with $(u_{\alpha,k,m}^\delta)_0 \in \mathbb{X}_k$ and $(u_{\alpha,k,m}^\delta)_j \in \mathbb{W}_{k+j}$, $j = 1, 2, \ldots, m$ from equation*

$$\mathcal{B}_{k,m}(\alpha)u_{\alpha,k,m}^\delta = \alpha^{-1}f_{k+m}^\delta - \mathcal{C}_{k,m}(\alpha)\overline{u}_{\alpha,k,m}^\delta. \tag{11.12}$$

The augmentation method begins with an initial approximate solution $u_{\alpha,k}^\delta$ and updates it from one level to another. Specifically, after the initial approximation $u_{\alpha,k}^\delta$ is obtained, for $m = 1, 2, \ldots$, we compute

$$(u_{\alpha,k,m}^\delta)_j = \alpha^{-1}\mathcal{Q}_{k+j}f^\delta - \alpha^{-1}\mathcal{Q}_{k+j}\mathcal{K}u_{\alpha,k,m-1}^\delta, \;\; j = 1, 2, \ldots, m,$$

solve

$$(\mathcal{I} + \alpha^{-1}\mathcal{P}_k\mathcal{K})(u_{\alpha,k,m}^\delta)_0 = \alpha^{-1}\mathcal{P}_k f^\delta - \alpha^{-1}\mathcal{P}_k\mathcal{K}\left(\sum_{j=1}^m (u_{\alpha,k,m}^\delta)_j\right)$$

and obtain the approximate solution

$$u_{\alpha,k,m}^{\delta} := [(u_{\alpha,k,m}^{\delta})_0, (u_{\alpha,k,m}^{\delta})_1, \ldots, (u_{\alpha,k,m}^{\delta})_m]^T.$$

Note that in this algorithm, we only need to find the inverse of $\mathcal{I} + \alpha^{-1}\mathcal{P}_k\mathcal{K}$ at level k. This is the key point which leads to fast computation for the proposed method.

The parameter α in general needs to be chosen according to the level. There are two types of choice for the parameter: *a priori* and *a posteriori*. In the next two subsections, we consider the choices of the parameters.

11.2.2 *A priori* error analysis

The goal of this subsection is to propose a choice of *a priori* regularization parameters and to estimate the convergence order of the MAM with the choice of the *a priori* parameter. To this end, we define the fractional power operator (cf. [220]). Let $\lfloor \nu \rfloor$ denote the greatest integer not larger than ν. For $0 < \nu < 1$, we define the fractional powers $\mathcal{K}^{\nu}$ by

$$\mathcal{K}^{\nu} := \frac{\sin \pi \nu}{\pi} \int_0^{+\infty} t^{\nu-1}(t\mathcal{I} + \mathcal{K})^{-1}\mathcal{K}\mathrm{d}t, \tag{11.13}$$

and for $\nu > 1$,

$$\mathcal{K}^{\nu} := \mathcal{K}^{\nu-\lfloor \nu \rfloor}\mathcal{K}^{\lfloor \nu \rfloor}.$$

Since operators $\mathcal{K}$ and $(\alpha\mathcal{I} + \mathcal{K})^{-1}$ commute, by the definition of the power operator, we conclude that operators $\mathcal{K}^{\nu}$ and $(\alpha\mathcal{I} + \mathcal{K})^{-1}$ commute as well, that is,

$$\mathcal{K}^{\nu}(\alpha\mathcal{I} + \mathcal{K})^{-1} = (\alpha\mathcal{I} + \mathcal{K})^{-1}\mathcal{K}^{\nu}. \tag{11.14}$$

We impose the hypothesis on u_*.

(H-1) *For some* $\nu \in (0, 1]$, $u_* \in R(\mathcal{K}^{\nu})$, *that is, there is* $\omega \in \mathbb{X}$ *such that* $u_* = \mathcal{K}^{\nu}\omega$.

Let $u_\alpha \in \mathbb{X}$ denote the solution of equation

$$(\alpha\mathcal{I} + \mathcal{K})u_\alpha = f \tag{11.15}$$

and we need to compare u_α with u_*. It is well known that if hypothesis (H-1) is satisfied, then the estimates

$$\|u_\alpha - u_*\| \leq c_\nu \|\omega\| \alpha^{\nu} \tag{11.16}$$

hold, where

$$c_\nu = \begin{cases} \frac{\sin \nu\pi}{\nu(1-\nu)\pi}, & 0 < \nu < 1, \\ 1, & \nu = 1 \end{cases}$$

and

$$\|u_\alpha^\delta - u_\alpha\| \le \delta\alpha^{-1}. \tag{11.17}$$

Hence, from these estimates we have that

$$\|u_\alpha^\delta - u_*\| \le c_\nu \|\omega\| \alpha^\nu + \delta\alpha^{-1}. \tag{11.18}$$

As in [220], we assume that the following hypothesis holds.

(H-2) *There exists a sequence θ_n, $n \in \mathbb{N}_0$ satisfying*

$$\sigma_0 \le \frac{\theta_{n+1}}{\theta_n} \le 1 \quad \text{and} \quad \lim_{n\to+\infty} \theta_n = 0 \tag{11.19}$$

for some constant $\sigma_0 \in (0, 1)$, such that when $n \ge N_0$,

$$\|(\mathcal{I} - \mathcal{P}_n)\mathcal{K}^\nu\| \le a_\nu \theta_n^\nu, \quad 0 < \nu \le 2 \tag{11.20}$$

and

$$\|\mathcal{K}(\mathcal{I} - \mathcal{P}_n)\| \le a_1 \theta_n, \tag{11.21}$$

where N_0 is a positive integer and $a_\nu, 0 < \nu \le 2$, are constants depending only on ν.

We next present a bound on the difference $u_\alpha - u_{\alpha,n}^\delta$.

Proposition 11.3 *Let u_α and $u_{\alpha,n}^\delta$ denote the solution of equations (11.15) and (11.8), respectively. If hypotheses (H-1) and (H-2) hold, then for $n \ge N_0$,*

$$\|u_\alpha - u_{\alpha,n}^\delta\| \le \frac{\delta}{\alpha} + \frac{2a_{1+\nu} + a_1 a_\nu}{\alpha} \|\omega\| \theta_n^{1+\nu}.$$

Proof It follows from equations (11.15) and (11.8) that

$$u_\alpha - u_{\alpha,n}^\delta = (\alpha\mathcal{I} + \mathcal{K}_n)^{-1}(f_n - f_n^\delta) + d_n,$$

where

$$f_n := \mathcal{P}_n f \text{ and } d_n := (\alpha\mathcal{I} + \mathcal{K})^{-1} f - (\alpha\mathcal{I} + \mathcal{K}_n)^{-1} f_n.$$

By using condition (11.5), estimate (11.9) and the fact that $\|\mathcal{P}_n\| = 1$, we have that

$$\|(\alpha\mathcal{I} + \mathcal{K}_n)^{-1}(f_n - f_n^\delta)\| \le \frac{\delta}{\alpha}.$$

Introducing the operator

$$\mathcal{E}_n := (\alpha\mathcal{I} + \mathcal{K})^{-1} - (\alpha\mathcal{I} + \mathcal{K}_n)^{-1},$$

we write d_n as two terms

$$d_n = (\alpha\mathcal{I} + \mathcal{K}_n)^{-1}(\mathcal{I} - \mathcal{P}_n)f + \mathcal{E}_n f.$$

We next estimate the two terms in the last formula for d_n separately. By hypothesis (H-1) and the definition of the power operator, we find that

$$f = \mathcal{K}u_* = \mathcal{K}^{1+\nu}\omega. \tag{11.22}$$

By using (11.22) and hypothesis (H-2), we have the estimate for the first term that

$$\|(\alpha\mathcal{I} + \mathcal{K}_n)^{-1}(\mathcal{I} - \mathcal{P}_n)f\| \leq \frac{a_{1+\nu}}{\alpha}\|\omega\|\theta_n^{1+\nu}.$$

It remains to estimate the term $\mathcal{E}_n f$. Again, using (11.22), we conclude that

$$\mathcal{E}_n f = e_1 + e_2,$$

where

$$e_1 := (\alpha\mathcal{I} + \mathcal{K}_n)^{-1}\mathcal{P}_n\mathcal{K}(\mathcal{P}_n - \mathcal{I})(\alpha\mathcal{I} + \mathcal{K})^{-1}\mathcal{K}^{1+\nu}\omega$$

and

$$e_2 := (\alpha\mathcal{I} + \mathcal{K}_n)^{-1}(\mathcal{P}_n - \mathcal{I})\mathcal{K}(\alpha\mathcal{I} + \mathcal{K})^{-1}\mathcal{K}^{1+\nu}\omega.$$

By using the commutativity of $\mathcal{K}^\nu$ and $(\alpha\mathcal{I} + \mathcal{K})^{-1}$, and the identity

$$\mathcal{P}_n - \mathcal{I} = (\mathcal{P}_n - \mathcal{I})(\mathcal{P}_n - \mathcal{I}),$$

we obtain that

$$e_1 = (\alpha\mathcal{I} + \mathcal{K}_n)^{-1}\mathcal{P}_n\mathcal{K}(\mathcal{P}_n - \mathcal{I})(\mathcal{P}_n - \mathcal{I})\mathcal{K}^\nu(\alpha\mathcal{I} + \mathcal{K})^{-1}\mathcal{K}\omega.$$

Moreover, since $\mathcal{K}$ is positive semi-definite, for any $\alpha > 0$,

$$\|(\alpha\mathcal{I} + \mathcal{K})^{-1}\mathcal{K}\| \leq 1. \tag{11.23}$$

It follows from (11.9), (11.20), (11.21) and (11.23) that

$$\|e_1\| \leq \alpha^{-1}a_1 a_\nu \theta_n^{1+\nu}\|\omega\|.$$

Likewise, we have that

$$e_2 = (\alpha\mathcal{I} + \mathcal{K}_n)^{-1}(\mathcal{P}_n - \mathcal{I})\mathcal{K}^{1+\nu}(\alpha\mathcal{I} + \mathcal{K})^{-1}\mathcal{K}\omega.$$

By (11.9), (11.21) and (11.23), we see that

$$\|e_2\| \leq \alpha^{-1}a_{1+\nu}\theta_n^{1+\nu}\|\omega\|.$$

Consequently, we have that

$$\|\mathcal{E}_n f\| \leq \|e_1\| + \|e_2\| \leq \alpha^{-1}(a_1 a_\nu + a_{1+\nu})\theta_n^{1+\nu}\|\omega\|.$$

This proves the result of the proposition. □

We next estimate the distance of u_α^δ from the subspace $\mathbb{X}_n$. To do this, for $n \in \mathbb{N}_0$, we let

$$E_{\alpha,n}^\delta := \|(\mathcal{I} - \mathcal{P}_n)u_\alpha^\delta\|. \tag{11.24}$$

We also set

$$\gamma_{\alpha,n}^\delta := \frac{\delta}{\alpha} + \frac{a_{1+\nu}}{\alpha}\|\omega\|\theta_n^{1+\nu}. \tag{11.25}$$

We remark that if the sequence θ_n satisfies condition (11.19), then we have that

$$\frac{\gamma_{\alpha,n+1}^\delta}{\gamma_{\alpha,n}^\delta} \geq \sigma := \sigma_0^{1+\nu}. \tag{11.26}$$

Proposition 11.4 *If hypotheses (H-1) and (H-2) hold, then for $n \geq N_0$,*

$$E_{\alpha,n}^\delta \leq \gamma_{\alpha,n}^\delta. \tag{11.27}$$

Proof By (11.4), (11.6) and (H-1), we have that

$$u_\alpha^\delta = (\alpha\mathcal{I} + \mathcal{K})^{-1}(f^\delta - f) + (\alpha\mathcal{I} + \mathcal{K})^{-1}\mathcal{K}^{1+\nu}\omega. \tag{11.28}$$

Since $\mathcal{P}_n$ is an orthogonal projection, we have that $\|\mathcal{I} - \mathcal{P}_n\| \leq 1$. Thus, from (11.5), (11.28), (11.7) and (H-2), we conclude that

$$E_{\alpha,n}^\delta = \|(\mathcal{I} - \mathcal{P}_n)u_\alpha^\delta\| \leq \frac{\delta}{\alpha} + \frac{a_{1+\nu}}{\alpha}\|\omega\|\theta_n^{1+\nu},$$

establishing the estimate. □

The next proposition shows that $u_{\alpha,k,m}^\delta$ approximates $u_{\alpha,k+m}^\delta$ at a convergence rate comparable with $\gamma_{\alpha,k+m}^\delta$. The proof of this result follows the same idea as the proof of Theorem 2.2 in [71]. To make this chapter self-contained, we provide details of the proof for convenient reference.

Proposition 11.5 *Let $u_{\alpha,k,m}^\delta$ and $u_{\alpha,k+m}^\delta$ be the solution of the MAM (11.12) and the solution of the projection method (11.8) with $n = k + m$, respectively. Suppose that hypotheses (H-1) and (H-2) hold. Then, there exists a positive integer $N \geq N_0$, when $k \geq N$, $m \in \mathbb{N}_0$ and α satisfies the condition*

$$\alpha \geq \frac{a_1}{\rho}\theta_k, \tag{11.29}$$

where

$$\rho := \frac{1}{2}\left(-1+\sqrt{1+\frac{2\sigma}{1+\sigma}}\right). \tag{11.30}$$

The following estimate holds:

$$\|u^\delta_{\alpha,k,m} - u^\delta_{\alpha,k+m}\| \le \gamma^\delta_{\alpha,k+m}.$$

Proof From (11.8) and (11.12) we have that

$$\mathcal{B}_{k,m}(\alpha)(u^\delta_{\alpha,k,m} - u^\delta_{\alpha,k+m}) = \mathcal{C}_{k,m}(\alpha)(u^\delta_{\alpha,k+m} - \overline{u}^\delta_{\alpha,k,m}). \tag{11.31}$$

Noting that $\|\mathcal{I} + \alpha^{-1}\mathcal{K}_{k+m}\| \ge 1$, it holds that

$$\|\mathcal{B}_{k,m}(\alpha)x\| = \|((\mathcal{I}+\alpha^{-1}\mathcal{K}_{k,m}) - \mathcal{C}_{k,m}(\alpha))x\| \ge (1 - \|\mathcal{C}_{k,m}(\alpha)\|)\|x\|. \tag{11.32}$$

By the definition of $\mathcal{C}_{k,m}(\alpha)$, we have that

$$\|\mathcal{C}_{k,m}(\alpha)\| = \alpha^{-1}\|(\mathcal{P}_{k+m} - \mathcal{P}_k)\mathcal{K}\mathcal{P}_{k+m}\| \le \alpha^{-1}\|(\mathcal{P}_{k+m} - \mathcal{P}_k)\mathcal{K}\|.$$

Thus, when $k \ge N$, $m \in \mathbb{N}_0$ and hypothesis (11.29) is satisfied, we have that

$$\|\mathcal{C}_{k,m}(\alpha)\| \le 2\frac{a_1\theta_k}{\alpha} \le 2\rho.$$

This, with (11.30), ensures that

$$\|\mathcal{C}_{k,m}(\alpha))\| < \frac{1}{(\rho+1)(1+\frac{1}{\sigma})}, \tag{11.33}$$

which also implies $\|\mathcal{C}_{k,m}(\alpha)\| < 1$. Therefore, by inequality (11.32), we conclude that when $k \ge N$, $m \in \mathbb{N}_0$ and hypothesis (11.29) is satisfied,

$$\|\mathcal{B}^{-1}_{k,m}(\alpha))\| \le \frac{1}{1 - \|\mathcal{C}_{k,m}(\alpha)\|}.$$

From this and (11.31), we obtain

$$\|u^\delta_{\alpha,k,m} - u^\delta_{\alpha,k+m}\| \le \frac{\|\mathcal{C}_{k,m}(\alpha)\|}{1 - \|\mathcal{C}_{k,m}(\alpha)\|}\|u^\delta_{\alpha,k+m} - \overline{u}^\delta_{\alpha,k,m}\|. \tag{11.34}$$

Moreover, as in the proof of Theorem 9.2, we have that

$$\|u^\delta_\alpha - u^\delta_{\alpha,k+m}\| \le \frac{a_1\theta_{k+m}}{\alpha}E^\delta_{\alpha,k+m} \le \frac{a_1\theta_k}{\alpha}E^\delta_{\alpha,k+m}.$$

By condition (11.29) we find that

$$\|u^\delta_\alpha - u^\delta_{\alpha,k+m}\| \le \rho E^\delta_{\alpha,k+m}. \tag{11.35}$$

We next prove by induction on m that when $k \geq N$, $m \in \mathbb{N}_0$ and hypothesis (11.29) is satisfied, we have the estimate

$$\|u^{\delta}_{\alpha,k,m} - u^{\delta}_{\alpha,k+m}\| \leq \gamma^{\delta}_{\alpha,k+m}. \tag{11.36}$$

When $m = 0$, since $u^{\delta}_{\alpha,k,0} = u^{\delta}_{\alpha,k}$, estimate (11.36) holds in this case. Suppose that the claim holds for $m = r - 1$. To prove (11.36) with $m = r$, we use the definition of $\overline{u}^{\delta}_{\alpha,k,r}$, estimates (11.35), (11.26), (11.27) and the induction hypothesis to obtain

$$\begin{aligned}\|u^{\delta}_{\alpha,k+r} - \overline{u}^{\delta}_{\alpha,k,r}\| &\leq \|u^{\delta}_{\alpha,k+r} - u^{\delta}_{\alpha}\| + \|u^{\delta}_{\alpha} - u^{\delta}_{\alpha,k+r-1}\| + \|u^{\delta}_{\alpha,k+r-1} - u^{\delta}_{\alpha,k,r-1}\| \\ &\leq \rho\gamma^{\delta}_{\alpha,k+r} + (\rho+1)\gamma^{\delta}_{\alpha,k+r-1} \\ &\leq \left(\rho + (\rho+1)\frac{1}{\sigma}\right)\gamma^{\delta}_{\alpha,k+r}.\end{aligned}$$

Substituting this estimate into the right-hand side of (11.34) with $m = r$ yields

$$\|u^{\delta}_{\alpha,k,r} - u^{\delta}_{\alpha,k+r}\| \leq \frac{\|\mathcal{C}_{k,r}(\alpha))\|}{1 - \|\mathcal{C}_{k,r}(\alpha))\|}\left(\rho + (\rho+1)\frac{1}{\sigma}\right)\gamma^{\delta}_{\alpha,k+r}.$$

It follows from the estimate (11.33) that for $k \geq N$ and non-negative integer r,

$$\frac{\|\mathcal{C}_{k,r}(\alpha))\|}{1 - \|\mathcal{C}_{k,r}(\alpha))\|}\left(\rho + (\rho+1)\frac{1}{\sigma}\right) \leq 1.$$

Therefore, for $k \geq N$, estimate (11.36) holds for $m = r$. The proof is complete. □

In the next theorem, we present the *a priori* error estimate for the approximate solution $u^{\delta}_{\alpha,k,m}$ obtained by the MAM.

Theorem 11.6 *Let u_* denote the minimum norm solution of equation (11.3), and $u^{\delta}_{\alpha,k,m}$ denote the solution of the MAM (11.12). If hypotheses (H-1) and (H-2) hold, then there exists a positive integer $N \geq N_0$ such that for all $k \geq N$, $m \in \mathbb{N}_0$ and α satisfying the condition (11.29) the following estimate holds:*

$$\|u_* - u^{\delta}_{\alpha,k,m}\| \leq c_{\nu}\|\omega\|\alpha^{\nu} + \frac{2\delta}{\alpha} + (3a_{1+\nu} + a_1 a_{\nu})\|\omega\|\frac{\theta^{1+\nu}_{k+m}}{\alpha}. \tag{11.37}$$

Proof Note that

$$u_* - u^{\delta}_{\alpha,k,m} = (u_* - u_{\alpha}) + (u_{\alpha} - u^{\delta}_{\alpha,k+m}) + (u^{\delta}_{\alpha,k+m} - u^{\delta}_{\alpha,k,m}).$$

The estimate (11.37) follows directly from the estimate (11.16), Propositions 11.3, 11.4 and 11.5. □

Motivated by Proposition 11.5, we now propose a choice of the regularization parameter α in the multilevel augmentation algorithm. For given $\delta > 0$, we choose a positive integer $k > N$ so that

$$c_-\delta^{\frac{1}{\nu+1}} \leq c_0\theta_k \leq c_+\delta^{\frac{1}{\nu+1}}, \tag{11.38}$$

for some positive integers c_-, c_+ and $c_0 \geq a_1/\rho$, and choose

$$\alpha := c_0\theta_k. \tag{11.39}$$

This choice of α is called the *a priori* parameter since (11.38) uses the information on ν which appears in *a priori* assumption (H-1). In the next theorem, we present the *a priori* error estimate of the MAM.

Theorem 11.7 *Let u_* be the minimum norm solution of equation (11.3), and $u^\delta_{\alpha,k,m}$ be the solution of the MAM (11.12) with the choice of α satisfying (11.39). If hypotheses (H-1) and (H-2) hold, then there exists a positive integer $N \geq N_0$ such that when $k \geq N$, $m \in \mathbb{N}_0$ and (11.38) is satisfied, the following estimate holds:*

$$\|u_* - u^\delta_{\alpha,k,m}\| \leq \left(c_\nu c_+^\nu\|\omega\| + \frac{2}{c_-}\right)\delta^{\frac{\nu}{\nu+1}} + (3a_{\nu+1} + a_1a_\nu)\|\omega\|\frac{\rho}{a_1}\theta^\nu_{k+m}. \tag{11.40}$$

Proof The estimate (11.40) is obtained by substituting the choice of the regularization parameter α into the right-hand side of estimate (11.37). □

The last theorem shows that the MAM improves the approximation error from θ_k to θ_{k+m}.

11.2.3 *A posteriori* choice strategy

In this subsection, we develop a strategy for choosing an *a posteriori* regularization parameter which ensures the optimal convergence for the approximate solution obtained by the MAM with the parameter.

To introduce an *a posteriori* regularization parameter, we consider an auxiliary operator equation. For fixed $k, m \in \mathbb{N}$, we consider the equation

$$(\alpha\mathcal{I} + \mathcal{K})\tilde{u}^\delta_\alpha = u^\delta_{\alpha,k,m}, \tag{11.41}$$

where $\tilde{u}^\delta_\alpha \in \mathbb{X}$. That is, we consider equation (11.6) with the right-hand side f^δ replaced by $u^\delta_{\alpha,k,m}$, where $u^\delta_{\alpha,k,m}$ is the solution using the MAM for equation (11.6). It is clear that equation (11.41) has a unique solution $\tilde{u}^\delta_\alpha$ and the solution depends on k, m. For simple presentation, we abuse the notation without indicating the dependence. The projection method for equation (11.41) has the form

$$(\alpha\mathcal{I} + \mathcal{K}_{k+i})\tilde{u}^{\delta}_{\alpha,k+i} = u^{\delta}_{\alpha,k,m}, \quad i = 0, 1, \ldots, m. \tag{11.42}$$

We denote by $\tilde{u}^{\delta}_{\alpha,k,i}$ the solution of the above equation using the MAM described in Section 11.2.1 for an α to be determined.

Let

$$\tilde{E}^{\delta}_{\alpha,n} := \|(\mathcal{I} - \mathcal{P}_n)\tilde{u}^{\delta}_{\alpha}\|$$

and

$$\tilde{\gamma}^{\delta}_{\alpha,n} := \frac{2\delta}{\alpha^2} + \frac{(4a_{1+\nu} + a_1 a_\nu)\|\omega\|}{\alpha^2}\theta_n^{1+\nu}.$$

In the next proposition we bound $\tilde{E}^{\delta}_{\alpha,n}$ by $\tilde{\gamma}^{\delta}_{\alpha,n}$.

Proposition 11.8 *If hypotheses (H-1) and (H-2) hold, there exists a positive integer $N \geq N_0$ such that when $k \geq N$, $m \in \mathbb{N}_0$ and α satisfies the condition (11.29), the following estimate holds:*

$$\tilde{E}^{\delta}_{\alpha,k+m} \leq \tilde{\gamma}^{\delta}_{\alpha,k+m},$$

and for $i = 0, 1, \ldots, m$,

$$\|\tilde{u}^{\delta}_{\alpha,k,i} - \tilde{u}^{\delta}_{\alpha,k+i}\| \leq \tilde{\gamma}^{\delta}_{\alpha,k+i}.$$

Proof It follows from (11.41) that

$$\tilde{u}^{\delta}_{\alpha} = (\alpha\mathcal{I} + \mathcal{K})^{-1}u^{\delta}_{\alpha,k,m}.$$

Using equation (11.15) and hypothesis (H-1), we have that

$$u_\alpha = (\alpha\mathcal{I} + \mathcal{K})^{-1}\mathcal{K}^{1+\nu}\omega.$$

Therefore,

$$\tilde{u}^{\delta}_{\alpha} = (\alpha\mathcal{I} + \mathcal{K})^{-1}(u^{\delta}_{\alpha,k,m} - u_\alpha) + (\alpha\mathcal{I} + \mathcal{K})^{-2}\mathcal{K}^{1+\nu}\omega.$$

Hence, by the fact that $\|\mathcal{I} - \mathcal{P}_n\| \leq 1$, hypothesis (H-2) and the estimates obtained in the last section, we have that

$$\tilde{E}^{\delta}_{\alpha,k+m} = \|\tilde{u}^{\delta}_{\alpha} - \mathcal{P}_{k+m}\tilde{u}^{\delta}_{\alpha}\| \leq \frac{2\delta}{\alpha^2} + \frac{(3a_{1+\nu} + a_1 a_\nu)\|\omega\|}{\alpha^2}\theta_{k+m}^{1+\nu} + \frac{a_{1+\nu}\|\omega\|}{\alpha^2}\theta_{k+m}^{1+\nu},$$

which leads to the first estimate of this proposition. The second estimate follows similarly from the proof of Proposition 11.5. □

We define the quantities

$$\Delta_\alpha := \alpha^2(\alpha\mathcal{I}+\mathcal{K})^{-2}f, \quad \Delta^\delta_{\alpha,k,m} := \alpha^2\tilde{u}^\delta_{\alpha,k,m}$$

and

$$D(\delta,\theta_{k+m}) := 4\delta + (8a_{1+\nu}+3a_1a_\nu)\|\omega\|\theta^{1+\nu}_{k+m}.$$

In the next result, we estimate the difference between Δ_α and $\Delta^\delta_{\alpha,k,m}$ in terms of $D(\delta,\theta_{k+m})$.

Proposition 11.9 *If hypotheses (H-1) and (H-2) hold, there exists a positive integer $N \geq N_0$ such that when $k \geq N$, $m \in \mathbb{N}_0$ and α satisfies the condition (11.29), it holds that*

$$\|\Delta^\delta_{\alpha,k,m} - \Delta_\alpha\| \leq D(\delta,\theta_{k+m}) \tag{11.43}$$

and

$$\|\Delta_\alpha\| \leq c_\nu^2\|\omega\|\alpha^{1+\nu}. \tag{11.44}$$

Moreover, if

$$\|f^\delta\| > D(\delta,\theta_{k+m}) + (b+1)\delta \tag{11.45}$$

for some constant $b > 0$, then

$$\liminf_{\alpha\to+\infty}\|\Delta^\delta_{\alpha,k,m}\| > b\delta. \tag{11.46}$$

Proof It follows from (11.42) and (11.15) that

$$\begin{aligned}\Delta^\delta_{\alpha,k,m} - \Delta_\alpha &= \alpha^2(\tilde{u}^\delta_{\alpha,k,m} - \tilde{u}^\delta_{\alpha,k+m}) + \alpha^2(\alpha\mathcal{I}+\mathcal{K}_{k,m})^{-1}(u^\delta_{\alpha,k,m} - u^\delta_{\alpha,k+m})\\ &\quad + \alpha^2(\alpha\mathcal{I}+\mathcal{K}_{k,m})^{-1}(u^\delta_{\alpha,k+m} - u_\alpha)\\ &\quad + \alpha^2[(\alpha\mathcal{I}+\mathcal{K}_{k,m})^{-1} - (\alpha\mathcal{I}+\mathcal{K})^{-1}](\alpha\mathcal{I}+\mathcal{K})^{-1}f.\end{aligned}$$

Thus, by Propositions 11.8 and 11.5,

$$\begin{aligned}\|\Delta^\delta_{\alpha,k,m} - \Delta_\alpha\| &\leq \alpha^2\tilde{\gamma}^\delta_{\alpha,k+m} + \alpha\gamma^\delta_{\alpha,k+m} + \alpha\|u^\delta_{\alpha,k+m} - u_\alpha\|\\ &\quad + \alpha^2\|[(\alpha\mathcal{I}+\mathcal{K}_{k,m})^{-1} - (\alpha\mathcal{I}+\mathcal{K})^{-1}](\alpha\mathcal{I}+\mathcal{K})^{-1}f\|.\end{aligned}$$

Similar to the proof of Proposition 11.3, we find that the last term is given by

$$\begin{aligned}&\alpha^2\|[(\alpha\mathcal{I}+\mathcal{K}_{k+m})^{-1} - (\alpha\mathcal{I}+\mathcal{K})^{-1}](\alpha\mathcal{I}+\mathcal{K})^{-1}\mathcal{K}^{1+\nu}\omega\|\\ &\leq (a_1a_\nu + a_{1+\nu})\theta^{1+\nu}_{k+m}\|\omega\|.\end{aligned}$$

Combining these inequalities and using Propositions 11.8, 11.4, 11.3 and (11.17), we conclude that

$$\|\Delta^\delta_{\alpha,k,m} - \Delta_\alpha\| \leq 4\delta + (8a_{1+\nu}+3a_1a_\nu)\|\omega\|\theta^{1+\nu}_{k+m}, \tag{11.47}$$

which is the estimate (11.43). Note by hypotheses that

$$\begin{aligned}\|\Delta_\alpha\| &= \alpha^2 \left\|(\alpha\mathcal{I}+\mathcal{K})^{-2}\mathcal{K}^{1+\nu}\omega\right\| \\ &= \alpha^{1+\nu} \left\|[\alpha(\alpha\mathcal{I}+\mathcal{K})^{-1}]^{(1-\nu)}[(\alpha\mathcal{I}+\mathcal{K})^{-1}\mathcal{K}]^{1+\nu}\omega\right\| \\ &\le c_\nu^2\|\omega\|\alpha^{1+\nu},\end{aligned}$$

where we used the inequality $\|[\alpha(\alpha\mathcal{I}+\mathcal{K})^{-1}]^\nu\| \le c_\nu$ for $\nu \in (0,1]$, which can easily be verified by the definition (11.13) of the fractional power of an operator. Thus, estimate (11.44) follows.

To prove the second statement, we note that

$$\|\Delta^\delta_{\alpha,k,m}\| \ge \|\Delta_\alpha\| - \|\Delta^\delta_{\alpha,k,m} - \Delta_\alpha\| \text{ and } \lim_{\alpha\to+\infty}\|\Delta_\alpha\| = \|f\|.$$

When condition (11.45) holds, we have that

$$\|f\| \ge \|f^\delta\| - \|f-f^\delta\| > D(\delta,\theta_{k+m}) + b\delta.$$

Thus,

$$\liminf_{\alpha\to+\infty}\|\Delta^\delta_{\alpha,k,m}\| > b\delta,$$

which completes the proof. □

We remark that if $\|f^\delta\| \ge c\delta$ with $c > 5$ and if $k+m$ is sufficiently large, then condition (11.45) holds. In fact, a simple computation yields

$$\|f^\delta\| - D(\delta,\theta_{k+m}) - \delta \ge (c-5)\delta - (8a_{1+\nu} + 3a_1a_\nu)\|\omega\|\theta^{1+\nu}_{k+m},$$

which confirms that (11.45) holds for sufficiently large $k+m$.

Proposition 11.10 *Suppose that $\alpha \ge \alpha' > 0$. Let u_α and u_* denote the solution of equation (11.15) and the minimum norm solution of equation (11.4), respectively. Then,*

$$\|u_\alpha - u_*\| \le \|u_{\alpha'} - u_*\| + \frac{\|\Delta_\alpha\|}{\alpha'}.$$

Proof Direct computation leads to

$$\begin{aligned}u_\alpha - u_{\alpha'} &= (\alpha\mathcal{I}+\mathcal{K})^{-1}f - (\alpha'\mathcal{I}+\mathcal{K})^{-1}f \\ &= -\frac{1}{\alpha'}\left(1-\frac{\alpha'}{\alpha}\right)\alpha'\alpha^{-1}(\alpha'\mathcal{I}+\mathcal{K})^{-1}(\alpha\mathcal{I}+\mathcal{K})\alpha^2(\alpha\mathcal{I}+\mathcal{K})^{-2}f\end{aligned}$$

and

$$\|\alpha'\alpha^{-1}(\alpha'\mathcal{I}+\mathcal{K})^{-1}(\alpha\mathcal{I}+\mathcal{K})\| = \frac{\alpha'}{\alpha}\left\|\mathcal{I}+\left(\frac{\alpha}{\alpha'}-1\right)(\mathcal{I}+\alpha'^{-1}\mathcal{K})^{-1}\right\|$$
$$\leq \frac{\alpha'}{\alpha}\left(1+\frac{\alpha}{\alpha'}-1\right)=1.$$

Thus, we conclude that

$$\|u_\alpha - u_{\alpha'}\| \leq \frac{1}{\alpha'}\left(1-\frac{\alpha'}{\alpha}\right)\|\alpha^2(\alpha\mathcal{I}+\mathcal{K})^{-2}f\| \leq \frac{\|\Delta_\alpha\|}{\alpha'}.$$

This, with the inequality

$$\|u_\alpha - u_*\| \leq \|u_{\alpha'} - u_*\| + \|u_\alpha - u_{\alpha'}\|,$$

yields the estimate of the proposition. □

Let $d > 1$ and $\tau = a_1/\rho$ be fixed. We choose a positive number α_0 satisfying the condition

$$\tau\theta_k \leq \alpha_0 \leq d\tau\theta_k \tag{11.48}$$

and define an increasing sequence α_n by the recursive formula

$$\alpha_n = d\alpha_{n-1}, \quad n = 1, 2, \ldots$$

Clearly, the sequence α_n goes to infinity as $n \to \infty$. We present in the next lemma a property of this sequence.

Lemma 11.11 *If condition (11.45) holds for some positive constant b, then there exists a non-negative integer n_0 such that*

$$\|\Delta^\delta_{\alpha_{n_0-1},k,m}\| \leq b\delta \leq \|\Delta^\delta_{\alpha_{n_0},k,m}\|, \tag{11.49}$$

where $\|\Delta^\delta_{\alpha_{-1},k,m}\| := 0.$

Proof If α_0 satisfies the condition

$$\|\Delta^\delta_{\alpha_0,k,m}\| \geq b\delta, \tag{11.50}$$

the proof is complete with $n_0 = 0$. We now assume that condition (11.50) is not satisfied. By the hypothesis of this lemma, Proposition 11.9 and the definition of the sequence α_n, there exists a positive integer p such that

$$\|\Delta^\delta_{\alpha_p,k,m}\| \geq b\delta.$$

Let n_0 be the smallest such integer p. Thus, we obtain (11.49). □

As suggested by Lemma 11.11, we present an algorithm which generates a choice for an *a posteriori* parameter.

Algorithm 11.12 *(A posteriori choice of regularization parameter) Let* $\tau = a_1/\rho > 0$*,* $d > 1$ *and* $b > 4$ *be fixed.*

Step 1: *For given* $\delta > 0$*, choose a positive integer* $k > N_0$ *and a constant* α_0 *such that* $\theta_k \leq \hat{c}\delta$ *for some* $\hat{c} > 0$ *and (11.48) holds.*

Step 2: *If* α_n *has been defined, use Algorithm 11.2 to compute* $u^{\delta}_{\alpha_n,k,m}$ *and* $\Delta^{\delta}_{\alpha_n,k,m}$*. If*

$$\|\Delta^{\delta}_{\alpha_n,k,m}\| < b\delta$$

is satisfied, we set

$$\alpha_{n+1} := d\alpha_n$$

and repeat this step. Otherwise, go to step 3.

Step 3: *Set* $\hat{\alpha} := \alpha_{n-1}$ *and stop.*

The output $\hat{\alpha}$ of Algorithm 11.12 depends on k, m and δ and satisfies the conditions

$$\tau\theta_k \leq \hat{\alpha} \leq d\tau\theta_k \text{ and } b\delta \leq \|\Delta^{\delta}_{\hat{\alpha},k,m}\|, \tag{11.51}$$

or

$$\hat{\alpha} \geq \tau\theta_k \text{ and } \|\Delta^{\delta}_{\hat{\alpha},k,m}\| \leq b\delta \leq \|\Delta^{\delta}_{d\hat{\alpha},k,m}\|. \tag{11.52}$$

The next proposition ensures that $\hat{\alpha} := \hat{\alpha}(k,m,\delta)$ converges to zero as $\delta \to 0$ and $k \to \infty$.

Proposition 11.13 *Suppose that hypotheses (H-1) and (H-2) hold. If* $\hat{\alpha} := \hat{\alpha}(k,m,\delta)$ *is chosen according to Algorithm 11.12, then there exists a positive integer* $N \geq N_0$ *such that when* $k \geq N$ *and* $m \in \mathbb{N}_0$*,*

$$\lim_{\delta\to 0,\, k\to+\infty} \hat{\alpha}(k,m,\delta) = 0. \tag{11.53}$$

Proof If $\hat{\alpha} \leq d\tau\theta_k$, then (11.53) holds since $\lim_{k\to+\infty}\theta_k = 0$. Otherwise, inequality (11.52) must be satisfied. Thus,

$$\lim_{\delta\to 0,\, k\to+\infty} \|\Delta^{\delta}_{\hat{\alpha},k,m}\| = 0.$$

Moreover, noting that $\hat{\alpha}$ satisfies (11.29), it follows from (11.43) that

$$\lim_{\delta\to 0,\, k\to+\infty} \|\Delta^{\delta}_{\hat{\alpha},k,m} - \Delta_{\hat{\alpha}}\| = 0.$$

Combining the above two inequalities yields

$$\lim_{\delta\to 0,\, k\to+\infty} \|\Delta_{\hat{\alpha}}\| = 0.$$

According to a well-known result (cf. [156, 274, 275]), we conclude (11.53) from the equation above. □

In the next theorem we present an error estimate for the multilevel augmentation solution using the *a posteriori* parameter.

Theorem 11.14 *Suppose that hypotheses (H-1) and (H-2) hold. Let u_* and $u^\delta_{\hat\alpha,k,m}$ be the minimum norm solution of equation (11.4) and the solution of the MAM (11.12), respectively, with $\hat\alpha$ chosen according to Algorithm 11.12. If $\|f^\delta\| > c\delta$ with $c > 5$, then there exist a positive integer N and constants d_1 and d_2 such that when $k \geq N$ and $m \in \mathbb{N}_0$,*

$$\|u^\delta_{\hat\alpha,k,m} - u_*\| \leq d_1\delta^{\frac{\nu}{1+\nu}} + d_2\theta^\nu_{k+m}.$$

Proof We first note that according to the remark after Proposition 11.9, condition (11.45) holds if $\|f^\delta\| > c\delta$ with $c > 5$ and if $k + m$ is sufficiently large. This, with Lemma 11.11, ensures that the parameter $\hat\alpha$ can be obtained by Algorithm 11.12. Noting that the parameter generated by Algorithm 11.12 satisfies the condition (11.29), it follows from Theorem 11.6 that there exists a positive integer N, when $k \geq N$ and $m \in \mathbb{N}_0$, such that

$$\|u_* - u^\delta_{\hat\alpha,k,m}\| \leq \|u_* - u_{\hat\alpha}\| + \frac{2\delta}{\hat\alpha} + \frac{1}{\tau}(3a_{1+\nu} + a_1 a_\nu)\|\omega\|\theta^\nu_{k+m}, \quad (11.54)$$

or

$$\|u_* - u^\delta_{\hat\alpha,k,m}\| \leq c_\nu\|\omega\|\hat\alpha^\nu + \frac{2\delta}{\hat\alpha} + \frac{1}{\tau}(3a_{1+\nu} + a_1 a_\nu)\|\omega\|\theta^\nu_{k+m}. \quad (11.55)$$

In the case that (11.51) holds, it follows from Algorithm 11.12 that

$$\hat\alpha^\nu \leq (d\tau\theta_k)^\nu \leq (\hat c d\tau)^\nu\delta^\nu. \quad (11.56)$$

Using (11.43) and (11.44) we have

$$b\delta \leq \|\Delta^\delta_{\hat\alpha,k,m}\| \leq \|\Delta_{\hat\alpha}\| + \|\Delta^\delta_{\hat\alpha,k,m} - \Delta_{\hat\alpha}\| \leq c_\nu^2\|\omega\|\hat\alpha^{1+\nu} + D(\delta, \theta_{k+m}).$$

This, with (11.56), yields

$$\frac{(b-4)\delta}{\hat\alpha} \leq c_\nu^2\|\omega\|(\hat c d\tau)^\nu\delta^\nu + (8a_{1+\nu} + 3a_1 a_\nu)\|\omega\|\theta^\nu_{k+m}.$$

Combining (11.55), (11.56) and the above inequality we conclude the estimate of this theorem with

$$d_1 := (1 + 2c_\nu/(b-4))c_\nu(\hat c d\tau)^\nu\|\omega\|$$

and

$$d_2 := [(16/(b-4) + 3/\tau)a_{1+\nu} + (6/(b-4) + 1/\tau)a_1 a_\nu]\|\omega\|.$$

In the case that (11.52) holds, we let $\alpha' := \delta^{\frac{1}{1+\nu}} + \theta_{k+m}$. Then, we have that

$$(\alpha')^\nu \le 2^\nu(\delta^{\frac{\nu}{1+\nu}} + \theta_{k+m}^\nu). \tag{11.57}$$

If $\hat{\alpha} \ge \alpha'$, it follows from Proposition 11.10 that

$$\|u_* - u_{\hat{\alpha}}\| \le \|u_* - u_{\alpha'}\| + \frac{\|\Delta_{\hat{\alpha}}\|}{\alpha'}. \tag{11.58}$$

Using estimates (11.16) and (11.57), we have that

$$\|u_* - u_{\alpha'}\| \le c_\nu \|\omega\|(\alpha')^\nu \le 2^\nu c_\nu \|\omega\|(\delta^{\frac{\nu}{1+\nu}} + \theta_{k+m}^\nu). \tag{11.59}$$

From Proposition 11.9 and (11.52), we observe that

$$\|\Delta_{\hat{\alpha}}\| \le \|\Delta_{\hat{\alpha}} - \Delta_{\hat{\alpha},k,m}^\delta\| + \|\Delta_{\hat{\alpha},k,m}^\delta\| \le D(\delta, \theta_{k+m}) + b\delta,$$

which leads to

$$\frac{\|\Delta_{\hat{\alpha}}\|}{\alpha'} \le (b+4)\delta^{\frac{\nu}{1+\nu}} + (8a_{1+\nu} + 3a_1 a_\nu)\|\omega\|\theta_{k+m}^\nu.$$

This, with (11.58) and (11.59), yields

$$\|u_* - u_{\hat{\alpha}}\| \le [2^\nu c_\nu \|\omega\| + (b+4)]\delta^{\frac{\nu}{1+\nu}} + [2^\nu c_\nu + (8a_{1+\nu} + 3a_1 a_\nu)]\|\omega\|\theta_{k+m}^\nu. \tag{11.60}$$

It is clear that

$$\frac{\delta}{\hat{\alpha}} \le \frac{\delta}{\alpha'} \le \delta^{\frac{\nu}{1+\nu}}.$$

Combining (11.54), (11.2.3) and the above inequality we have that

$$\begin{aligned} d_1 &= 2^\nu c_\nu \|\omega\| + b + 6 \text{ and} \\ d_2 &= \left[2^\nu c_\nu + (8 + 3/\tau)a_{1+\nu} + (3 + 1/\tau)a_1 a_\nu\right] \|\omega\|. \end{aligned}$$

If $\hat{\alpha} \le \alpha'$, then using (11.57),

$$\hat{\alpha}^\nu \le 2^\nu(\delta^{\frac{\nu}{1+\nu}} + \theta_{k+m}^\nu). \tag{11.61}$$

It follows from (11.52) and Proposition 11.9 that

$$\begin{aligned} b\delta &\le \|\Delta_{d\hat{\alpha},k,m}\| \le \|\Delta_{d\hat{\alpha},k,m} - \Delta_{d\hat{\alpha}}\| + \|\Delta_{d\hat{\alpha}}\| \\ &\le 4\delta + (8a_{1+\nu} + 3a_1 a_\nu)\|\omega\|\theta_{k+m}^{1+\nu} + \|\Delta_{d\hat{\alpha}}\|. \end{aligned}$$

Since $b > 4$ and $\tau\theta_k \le \hat{\alpha}$, we obtain

$$\frac{(b-4)\delta}{\hat{\alpha}} \le \frac{1}{\tau}(8a_{1+\nu} + 3a_1 a_\nu)\|\omega\|\theta_{k+m}^\nu + \frac{\|\Delta_{d\hat{\alpha}}\|}{\hat{\alpha}},$$

which with (11.44) yields

$$\frac{\delta}{\hat{\alpha}} \le \frac{1}{(b-4)\tau}(8a_{1+\nu} + 3a_1 a_\nu)\|\omega\|\theta_{k+m}^{\nu} + \frac{1}{(b-4)}c_\nu^2 d^{1+\nu}\|\omega\|\hat{\alpha}^{\nu}. \quad (11.62)$$

Combining (11.55), (11.61) and (11.62), we conclude the result of this theorem again with

$$d_1 := [2^\nu c_\nu + (2d)^{1+\nu} c_\nu^2/(b-4)]\|\omega\|$$

and

$$d_2 := d_1 + [(16/(b-4)+3)a_{1+\nu} + (6/(b-4)+1)a_1 a_\nu]\|\omega\|/\tau.$$

This completes the proof. □

11.3 Multiscale collocation methods via the Tikhonov regularization

In this section, we introduce a fast piecewise polynomial collocation method for solving the second-kind integral equation obtained using Tikhonov regularization from the original ill-posed equation. The method is developed based on a matrix compression strategy resulting from using multiscale piecewise polynomial basis functions and their corresponding multiscale collocation functionals.

11.3.1 The polynomial collocation method for the Tikhonov regularization

We present in this subsection a polynomial collocation method for solving the Tikhonov regularization equation. For this purpose we first describe the Tikhonov regularization in the L^∞ space for ill-posed integral equations of the first kind. Suppose that Ω is a compact set of the d-dimensional Euclidean space $\mathbb{R}^d$ for $d \ge 1$. The Fredholm integral operator $\mathcal{K}$ is defined by

$$(\mathcal{K}u)(s) := \int_\Omega K(s,t)u(t)\mathrm{d}t, \quad s \in \Omega, \quad (11.63)$$

where $K \in C(\Omega \times \Omega)$ is a nondegenerate kernel. We consider the Fredholm integral equation of the first kind in the form

$$\mathcal{K}u = f, \quad (11.64)$$

where $f \in L^\infty(\Omega)$ is a given function and u is the unknown solution to be determined. The operator $\mathcal{K}$ can be considered as a compact operator from $L^\infty(\Omega)$ to $L^\infty(\Omega)$. In this case, equation (11.64) is an ill-posed problem.

Let $\mathcal{K}^*$ be the adjoint operator of $\mathcal{K}$ defined by

$$(\mathcal{K}^*u)(s) := \int_\Omega K(t,s)u(t)\mathrm{d}t, \quad s \in \Omega. \tag{11.65}$$

Instead of solving equation (11.64), the Tikhonov regularization method is to solve the equation

$$(\alpha\mathcal{I} + \mathcal{A})u_\alpha^\delta = \mathcal{K}^* f^\delta, \tag{11.66}$$

where $\mathcal{A} := \mathcal{K}^*\mathcal{K}$, α is a positive parameter and f^δ is the approximate data of f with

$$\|f^\delta - f\|_2 \leq \delta \tag{11.67}$$

for some positive constant δ. We also denote by u_α the solution of the equation

$$(\alpha\mathcal{I} + \mathcal{A})u_\alpha = \mathcal{K}^* f, \tag{11.68}$$

where we assume that the right-hand-side function f contains no noise.

For $1 \leq p \leq \infty$, we use $\|u\|_p$ to denote the norm of the function $u \in L^p(\Omega)$ and use $\|\mathcal{A}\|_{\mathbb{X}\to\mathbb{Y}}$ to denote the norm of the operator $\mathcal{A} : \mathbb{X} \to \mathbb{Y}$. When $\mathbb{X} = \mathbb{Y} = L^p(\Omega)$, we simplify the notation by $\|\mathcal{A}\|_p := \|\mathcal{A}\|_{L^p(\Omega)\to L^p(\Omega)}$. Letting $M := \sup_{s\in\Omega}\left(\int_\Omega |K(t,s)|^2\mathrm{d}t\right)^{1/2}$, we have that $\|\mathcal{K}^*\|_{L^2(\Omega)\to L^\infty(\Omega)} \leq M$.

We recall the following two useful estimates established in [224].

Lemma 11.15 *For each $\alpha > 0$, the operator $\alpha\mathcal{I}+\mathcal{A}$ is invertible from $L^\infty(\Omega)$ to $L^\infty(\Omega)$,*

$$\|(\alpha\mathcal{I} + \mathcal{A})^{-1}\|_\infty \leq \frac{\sqrt{\alpha} + M/2}{\alpha^{3/2}} \quad \text{and} \quad \|(\alpha\mathcal{I} + \mathcal{A})^{-1}\mathcal{K}^*\|_{L^2(\Omega)\to L^\infty(\Omega)} \leq \frac{M}{\alpha}.$$

It is well known that in Hilbert space L^2 we have the estimates

$$\|(\alpha\mathcal{I} + \mathcal{A})^{-1}\|_{L^2(\Omega)} \leq \frac{1}{\alpha} \quad \text{and} \quad \|(\alpha\mathcal{I} + \mathcal{A})^{-1}\mathcal{K}^*\|_{L^2(\Omega)} \leq \frac{1}{2\alpha^{1/2}}.$$

These estimates are optimal in the L^2 space. However, to the best of our knowledge the estimates stated in Lemma 11.15 are the only available estimates, although it is not clear if they are optimal in the L^∞ space.

In the sequel, for simplicity, we assume that $f \in R(\mathcal{K})$, where $R(\mathcal{K})$ denotes the range of $\mathcal{K}$. Moreover, we use $\hat{u} := \mathcal{K}^\dagger f$ to denote the solution of (11.64), where $\mathcal{K}^\dagger$ is the Moore–Penrose generalized inverse of $\mathcal{K}$. We also need the following assumption.

(H1) $\hat{u} \in R(\mathcal{A}^\nu \mathcal{K}^*)$ with $0 < \nu \leq 1$, that is, there exists an $\omega \in L^\infty(\Omega)$ such that $\hat{u} = \mathcal{A}^\nu \mathcal{K}^* \omega$.

The next result was obtained in [117].

Lemma 11.16 *If $\hat{u} \in R(\mathcal{K}^*)$, then $\|\hat{u} - u_\alpha\|_\infty \to 0$, as $\alpha \to 0$. Moreover, if hypothesis (H1) holds, then*

$$\|\hat{u} - u_\alpha\|_\infty \leq c(\nu)\|\omega\|\alpha^\nu, \quad \text{as } \alpha \to 0,$$

where $c(\nu)$ is a constant defined by

$$c(\nu) := \begin{cases} \frac{\sin \nu\pi}{\nu(1-\nu)\pi}, & 0 < \nu < 1, \\ 1, & \nu = 1. \end{cases}$$

We remark that a somewhat different collocation method was studied in [207] recently in the L^2 space. The ill-posed equation was first discretized using a numerical integration formula, which leads to a finite-dimensional equation. A regularization is then applied to convert the discrete ill-posed equation to a discrete well-posed equation. In paper [207], a weak assumption $\hat{x} \in R(\mathcal{A}^\nu)$, $0 < \nu \leq 1$ was assumed to obtain the optimal convergence rate $\mathcal{O}(\delta^{2\nu/(2\nu+1)})$ for the Tikhonov regularization solution in the L^2 space. Moreover, papers [194, 195] prove the saturation of methods for solving linear ill-posed problems in Hilbert spaces for a wide class of regularization methods. We take a different approach in this section. We first regularize the ill-posed equation (11.64) to obtain the equation (11.66) and then apply the fast collocation method to solve the equation (11.66). For convergence analysis of the proposed method, we feel that for the collocation method, using the L^∞-norm is more natural. For this reason, we adopt in this section the L^∞ space for analysis of the proposed collocation method. Our analysis will be based on the estimates described in Lemmas 11.15 and 11.16, presented in [224] and [117], respectively.

Next, we present the piecewise polynomial collocation method for solving the regularized equation (11.66). We need some necessary notations. We assume that there is a partition $\mathcal{E} := \{\Omega_n : n \in \mathbb{Z}_N\}$ of Ω for $N \in \mathbb{N}$, which satisfies the following conditions:

(I) $\Omega = \bigcup_{n \in \mathbb{Z}_N} \Omega_n$, and $\text{meas}(\Omega_i \cap \Omega_j) = 0$, for $i \neq j$.

(II) For each $i \in \mathbb{Z}_N$, there exists an invertible affine map $\phi_n : \Omega^0 \to \Omega$ such that $\phi_n(\Omega^0) = \Omega_n$, where $\Omega^0 \subseteq \mathbb{R}^d$ is a reference element.

We denote by $\mathbb{X}_N^k$ the space of piecewise polynomials of total degree $\leq k - 1$, with respect to the partition $\mathcal{E}$. In other words, every element in $\mathbb{X}_N^k$ is a polynomial of total degree $\leq k - 1$ on each element Ω_n. Since the dimension of the polynomials of total degree $k-1$ is given by $r_k := \binom{k+d-1}{d}$, the dimension of $\mathbb{X}_N^k$ is Nr_k.

Choose r_k distinct points $x_j \in \Omega^0, j = 1, 2, \ldots, r_k$, in a general position, that is, the Lagrange interpolation polynomial of total degree $k-1$ at these points is uniquely defined (cf. [198]). We assume that the polynomials $p_j \in \mathbb{P}_k$ of total degree $k-1$ satisfy the conditions $p_j(x_i) = \delta_{ij}$, for $i, j = 1, 2, \ldots, r_k$. For each $n \in \mathbb{Z}_N$, we define basis functions by

$$\rho_{nj}(x) := \begin{cases} (p_j \circ \phi_n^{-1})(x), & x \in \Omega_n, \\ 0, & x \notin \Omega_n. \end{cases}$$

In what follows, we denote $h := \max\{\text{diam}(\Omega_n) : n \in \mathbb{Z}_N\}$ and define the interpolation projection $\mathcal{P}_h : C(\Omega) \to \mathbb{X}_N^k$. The projection $\mathcal{P}_h$ is first defined for $f \in C(\Omega)$ by

$$(\mathcal{P}_h f)(x) = \sum_{n \in \mathbb{Z}_N} \sum_{j=1}^{r_k} f(\phi_n(x_j))\rho_{nj}(x), \quad x \in \Omega.$$

The operator $\mathcal{P}_h : C(\Omega) \to \mathbb{X}_N^k$ is an interpolation projection onto $\mathbb{X}_N^k$. It can be verified that there exists a positive constant c such that $\|\mathcal{P}_h\|_\infty \le c$ for all N. The projection $\mathcal{P}_h$ can be extended to $L^\infty(\Omega)$ by the Hahn–Banach theorem. We also need the orthogonal projection $\mathcal{Q}_h : L^2(\Omega) \to \mathbb{X}_N^k$.

In the next proposition, we prove important properties of the projections $\mathcal{P}_h$ and $\mathcal{Q}_h$. These properties may have appeared in the literature in a different form. To make this section self-contained, we provide a complete proof for the convenience of the reader. For $1 \le p, q \le +\infty$ and a positive constant r, we let $W^{r,q}(\Omega)$ denote the Sobolev space of functions whose r-derivatives are in $L^q(\Omega)$ and set

$$L^p(\Omega, W^{r,q}(\Omega)) := \{v : \ v(s,\cdot) \in W^{r,q}(\Omega) \text{ for almost all } s \in \Omega \text{ and } \|v(s,\cdot)\|_{W^{r,q}(\Omega)} \in L^p(\Omega)\}$$

and

$$L^p(W^{r,q}(\Omega), \Omega) := \{v : \ v(\cdot,t) \in W^{r,q}(\Omega) \text{ for almost all } t \in \Omega \text{ and } \|v(\cdot,t)\|_{W^{r,q}(\Omega)} \in L^p(\Omega)\}.$$

In the rest of this section, unless stated otherwise, we use c for a generic positive constant whose values may be different on different occasions.

Proposition 11.17 *If $K \in C(\Omega \times \Omega)$ and $0 < r \le k$, then the following statements hold:*

(1) For $K \in L^\infty(\Omega, W^{r,1}(\Omega))$, $\|\mathcal{K}(\mathcal{I} - \mathcal{Q}_h)\|_\infty \le ch^r \|K\|_{L^\infty(\Omega, W^{r,1}(\Omega))}$.

(2) *For* $K \in L^\infty(W^{r,1}(\Omega), \Omega)$, $\|(\mathcal{I} - \mathcal{P}_h)\mathcal{K}^*\|_\infty \le ch^r \|K\|_{L^\infty(W^{r,1}(\Omega),\Omega)}$.
(3) *For* $K \in L^2(\Omega, W^{r,2}(\Omega))$, $\|\mathcal{K}(\mathcal{I} - \mathcal{Q}_h)\|_2 \le ch^r \|K\|_{L^2(\Omega,W^{r,2}(\Omega))}$.
(4) *For* $K \in L^2(W^{r,2}(\Omega), \Omega)$, $\|(\mathcal{I} - \mathcal{P}_h)\mathcal{K}^*\|_2 \le ch^r \|K\|_{L^2(W^{r,2}(\Omega),\Omega)}$.
(5) *For* $K \in L^\infty(W^{r,2}(\Omega), \Omega)$, $\|(\mathcal{I} - \mathcal{P}_h)\mathcal{K}^*\|_{L^2(\Omega)\to L^\infty(\Omega)} \le ch^r \|K\|_{L^\infty(W^{r,2}(\Omega),\Omega)}$.

Proof (1) For any $u \in L^\infty(\Omega)$, noting that $\mathcal{Q}_h$ is self-adjoint, we have that

$$(\mathcal{K}(\mathcal{I} - \mathcal{Q}_h)u)(s) = \int_\Omega ((\mathcal{I} - \mathcal{Q}_h)K(s,\cdot))(t)u(t)\mathrm{d}t. \tag{11.69}$$

Therefore,

$$\begin{aligned}\|\mathcal{K}(\mathcal{I} - \mathcal{Q}_h)u\|_\infty &\le \sup_{s\in\Omega}\int_\Omega |((\mathcal{I} - \mathcal{Q}_h)K(s,\cdot))(t)|\mathrm{d}t\|u\|_\infty \\ &\le ch^r\|K\|_{L^\infty(\Omega,W^{r,1}(\Omega))}\|u\|_\infty,\end{aligned}$$

which implies that

$$\|\mathcal{K}(\mathcal{I} - \mathcal{Q}_h)\|_\infty \le ch^r\|K\|_{L^\infty(\Omega,W^{r,1}(\Omega))}.$$

(2) Likewise, we have that

$$((\mathcal{I} - \mathcal{P}_h)\mathcal{K}^*u)(s) = \int_\Omega ((\mathcal{I} - \mathcal{P}_h)K(\cdot,s))(t)u(t)\mathrm{d}t. \tag{11.70}$$

This implies that

$$\begin{aligned}\|(\mathcal{I} - \mathcal{P}_h)\mathcal{K}^*u\|_\infty &\le \sup_{s\in\Omega}\int_\Omega |((\mathcal{I} - \mathcal{P}_h)K(\cdot,t))(s)|\mathrm{d}s\|u\|_\infty \\ &\le ch^r\|K\|_{L^\infty(W^{r,1}(\Omega),\Omega)}\|u\|_\infty,\end{aligned}$$

which leads to the estimate of (2).

(3) For any $u \in L^2(\Omega)$, it follows from (11.69) that

$$\begin{aligned}\|\mathcal{K}(\mathcal{I} - \mathcal{Q}_h)u\|_2 &\le \left(\int_\Omega \|(\mathcal{I} - \mathcal{Q}_h)K(s,\cdot)\|_2^2\|u\|_2^2\mathrm{d}s\right)^{1/2} \\ &\le ch^r\|K\|_{L^2(\Omega,W^{r,2}(\Omega))}\|u\|_2,\end{aligned}$$

which yields the result of (3).

(4) The estimate of (4) follows from (11.70) and a similar argument used in the proof of (3).

(5) For any $u \in L^2(\Omega)$, it follows that

$$\|(\mathcal{I} - \mathcal{P}_h)\mathcal{K}^*u\|_\infty \le \sup_{s\in\Omega}\|(\mathcal{I} - \mathcal{P}_h)K(\cdot,s)\|_2\|u\|_2 \le ch^r\|K\|_{L^\infty(W^{r,2}(\Omega),\Omega)}\|u\|_2.$$

This completes the proof. □

The next corollary follows directly from Proposition 11.17.

Corollary 11.18 *If $K(\cdot,\cdot) \in W^{r,\infty}(\Omega \times \Omega)$ with $0 < r \leq k$, then there exists a positive constant $\check{c}$ such that for all $h > 0$,*

$$\max\{\|\mathcal{A} - \mathcal{P}_h\mathcal{A}\mathcal{Q}_h\|_\infty, \|\mathcal{P}_h\mathcal{A}\mathcal{Q}_h - \mathcal{P}_h\mathcal{A}\|_\infty,$$
$$\|(\mathcal{I} - \mathcal{P}_h)\mathcal{K}^*\|_{L^2(\Omega)\to L^\infty(\Omega)}\} \leq \check{c}h^r.$$

We remark that the number r on the right-hand side of the above estimate may be larger than that in the assumption of this corollary (see the example in Section 11.4.2).

Using the piecewise polynomial spaces and the projection operators introduced above, the piecewise polynomial collocation method for solving the regularized equation (11.66) is to find $u^\delta_{\alpha,h} \in \mathbb{X}^k_N$ such that

$$(\alpha\mathcal{I} + \mathcal{P}_h\mathcal{A}\mathcal{Q}_h)u^\delta_{\alpha,h} = \mathcal{P}_h\mathcal{K}^*f^\delta. \tag{11.71}$$

Following the discussion in [224], we have the convergence and error estimate of the polynomial collocation method.

Theorem 11.19 *Suppose that $K(\cdot,\cdot) \in W^{r,\infty}(\Omega \times \Omega)$ with $0 < r \leq k$ and*

$$\check{c}h^r \leq \min\left\{\frac{1}{2}\frac{\alpha^{3/2}}{\sqrt{\alpha}+M/2}, \frac{\sqrt{\alpha}}{\sqrt{\alpha}+M/2}\right\}. \tag{11.72}$$

(1) If $\hat{u} \in R(\mathcal{K}^*)$, and h, α are chosen such that $h^r = \mathcal{O}(\sqrt{\alpha}\delta)$, $\delta/\alpha \to 0$, as $\delta \to 0$, then

$$\|\hat{u} - u^\delta_{\alpha,h}\|_\infty \to 0, \quad as \ \delta \to 0. \tag{11.73}$$

(2) If hypothesis (H1) holds, and h, α are chosen such that $h^r = \mathcal{O}(\sqrt{\alpha}\delta)$ and $\alpha \sim \delta^{\frac{1}{\nu+1}}$, as $\delta \to 0$, then

$$\|\hat{u} - u^\delta_{\alpha,h}\|_\infty = \mathcal{O}(\delta^{\frac{\nu}{\nu+1}}), \quad as \ \delta \to 0. \tag{11.74}$$

Proof Using Lemma 11.15 and a standard argument, we have that

$$\|\hat{u} - u^\delta_{\alpha,h}\|_\infty \leq \|\hat{u} - u_\alpha\|_\infty + c\left(\frac{\delta}{\alpha} + \frac{h^r}{\alpha^{3/2}}\right).$$

Thus, the results of this theorem follow from Theorems 2.5 and 2.6 of [224].

□

11.3.2 The fast multiscale collocation method

This subsection is devoted to the development of the fast multiscale piecewise polynomial collocation method for solving equation (11.66).

We recall the multiscale sequence of approximate subspaces. A sequence of partitions of domain Ω is called *multiscale* if every partition in the sequence is obtained from a refinement of its previous partition. Let $\mathbb{X}_n$, $n \in \mathbb{N}$ be a sequence of piecewise polynomial spaces of total degree $k-1$ based on a sequence of multiscale partitions of Ω, where $\mathbb{X}_0$ is the space of piecewise polynomials of total degree $k-1$ on an initial partition of Ω, with $m := \dim \mathbb{X}_0 = m_0 r_k$ for some positive integer m_0. Because of the multiscale partition, we have that the spaces $\mathbb{X}_n$, $n \in \mathbb{N}$ are nested, that is, $\mathbb{X}_n \subseteq \mathbb{X}_{n+1}$ for $n \in \mathbb{N}$. This leads to the decomposition

$$\mathbb{X}_n = \mathbb{W}_0 \oplus^{\perp} \mathbb{W}_1 \oplus^{\perp} \cdots \oplus^{\perp} \mathbb{W}_n, \tag{11.75}$$

where $\mathbb{W}_0 := \mathbb{X}_0$. For each $i \in \mathbb{N}_0$, we assume that $\mathbb{W}_i$ has a basis $w_{ij}, j \in \mathbb{Z}_{w(i)}$, that is, $\mathbb{W}_i = \operatorname{span}\{w_{ij} : j \in \mathbb{Z}_{w(i)}\}$. According to (11.75), we then have that $\mathbb{X}_n = \operatorname{span}\{w_{ij} : (i,j) \in \mathbb{U}_n\}$. For each $(i,j) \in \mathbb{U}_n$, we denote by S_{ij} the support of the basis function w_{ij} and let $d(A)$ denote the diameter of the set $A \subset \Omega$. We define $s(n) := \dim \mathbb{X}_n$ and for each $i \in \mathbb{Z}_{n+1}$ we set $h_i := \max\{d(S_{ij}) : j \in \mathbb{Z}_{w(i)}\}$. We further require that the spaces and their bases have the properties that $s(n) \sim \mu^n$, $w(i) \sim \mu^i$, $h_i \sim \mu^{-i/d}$, where $\mu > 1$ is an integer, and that there exists a positive constant c such that $\|w_{ij}\|_\infty \leq c$ for all $(i,j) \in \mathbb{U}$. These properties are fulfilled for the spaces and bases constructed in [69, 75, 264].

Next, we turn to describing a multiscale sequence of the corresponding collocation functional spaces. Associated with each basis function w_{ij}, we have a collocation functional ℓ_{ij}, which is a sum of point evaluation functionals at a *fixed* number of points in S_{ij}. We demand that for each $(i,j) \in \mathbb{U}_n$, $\langle \ell_{ij}, q \rangle = 0$, for any polynomial q of total degree $k-1$, and there exists a positive constant c such that $\|\ell_{ij}\| \leq c$, for all $(i,j) \in \mathbb{U}$. We also require that the basis functions and their corresponding collocation functionals have the *semi-bi-orthogonal* property: $\langle \ell_{i'j'}, w_{ij} \rangle = \delta_{i'i}\delta_{j'j}$, $(i,j),(i',j') \in \mathbb{U}$, $i' \leq i$. These properties of the collocation functionals are satisfied for those constructed in [69, 75]. The multiscale collocation functionals constructed in these papers make use of refinable sets introduced in [65], which admit the unique Lagrange interpolating polynomial (cf. [198]). These functionals were originally defined for continuous functions and then extended to functions in $L^\infty(\Omega)$ by the Hahn–Banach theorem. Corresponding to each subspace $\mathbb{W}_i$ we have the *collocation functional space* $\mathbb{V}_i := \operatorname{span}\{\ell_{ij} : j \in \mathbb{Z}_{w(i)}\}$, and corresponding to the space $\mathbb{X}_n$ we have $\mathbb{L}_n := \operatorname{span}\{\ell_{ij} : (i,j) \in \mathbb{U}_n\}$. The space $\mathbb{L}_n$ has the decomposition

$$\mathbb{L}_n = \mathbb{V}_0 \oplus \mathbb{V}_1 \oplus \cdots \oplus \mathbb{V}_n.$$

We now formulate the collocation method for solving equation (11.66). To this end, for each $n \in \mathbb{N}_0$, we let $\mathcal{P}_n$ denote the interpolation projection from $L^\infty(\Omega)$ onto $\mathbb{X}_n$, defined for $f \in L^\infty(\Omega)$ by $\mathcal{P}_n f \in \mathbb{X}_n$ satisfying

$$\langle \ell_{ij}, f - \mathcal{P}_n f\rangle = 0, \quad (i,j) \in \mathbb{U}_n.$$

Moreover, for each $n \in \mathbb{N}_0$, we let $\mathcal{Q}_n$ denote the orthogonal projection from $L^2(\Omega)$ onto $\mathbb{X}_n$. The collocation method for solving (11.66) is to find $u^\delta_{\alpha,n} \in \mathbb{X}_n$ such that

$$(\alpha\mathcal{I} + \mathcal{P}_n\mathcal{A}\mathcal{Q}_n)u^\delta_{\alpha,n} = \mathcal{P}_n\mathcal{K}^* f^\delta. \tag{11.76}$$

In the standard piecewise polynomial collocation method, the orthogonal projection $\mathcal{Q}_n$ is not used. Instead, its roll is taken by the interpolation projection $\mathcal{P}_n$. At the operator equation level, the two formulations are equivalent. However, the use of the orthogonal projection $\mathcal{Q}_n$ allows us to use the multiscale basis functions which have vanishing moments. This is crucial for developing fast algorithms based on a matrix compression.

The matrix representation of operator $\alpha\mathcal{I}+\mathcal{P}_n\mathcal{A}\mathcal{Q}_n$ under the basis functions and the corresponding multiscale collocation functionals is a dense matrix. To compress this matrix, we write it in block form. Let $\mathcal{A}_n := \mathcal{P}_n\mathcal{A}\mathcal{Q}_n$, then the operator $\mathcal{A}_n : \mathbb{X}_n \to \mathbb{X}_n$ is identified in matrix form with

$$\mathcal{A}_n = \begin{bmatrix} \mathcal{P}_0\mathcal{A}\mathcal{Q}_0 & \mathcal{P}_0\mathcal{A}(\mathcal{Q}_1 - \mathcal{Q}_0) & \cdots & \mathcal{P}_0\mathcal{A}(\mathcal{Q}_n - \mathcal{Q}_{n-1}) \\ (\mathcal{P}_1 - \mathcal{P}_0)\mathcal{A}\mathcal{Q}_0 & (\mathcal{P}_1 - \mathcal{P}_0)\mathcal{A}(\mathcal{Q}_1 - \mathcal{Q}_0) & \cdots & (\mathcal{P}_1 - \mathcal{P}_0)\mathcal{A}(\mathcal{Q}_n - \mathcal{Q}_{n-1}) \\ \vdots & \vdots & \ddots & \vdots \\ (\mathcal{P}_n - \mathcal{P}_{n-1})\mathcal{A}\mathcal{Q}_0 & (\mathcal{P}_n - \mathcal{P}_{n-1})\mathcal{A}(\mathcal{Q}_1 - \mathcal{Q}_0) & \cdots & (\mathcal{P}_n - \mathcal{P}_{n-1})\mathcal{A}(\mathcal{Q}_n - \mathcal{Q}_{n-1}) \end{bmatrix}. \tag{11.77}$$

If $i+j > n$ we replace the block $(\mathcal{P}_i - \mathcal{P}_{i-1})\mathcal{A}(\mathcal{Q}_j - \mathcal{Q}_{j-1})$ of (11.77) by the zero, which leads to a compressed matrix

$$\tilde{\mathcal{A}}_n := \begin{bmatrix} \mathcal{P}_0\mathcal{A}\mathcal{Q}_0 & \mathcal{P}_0\mathcal{A}(\mathcal{Q}_1 - \mathcal{Q}_0) & \cdots & \mathcal{P}_0\mathcal{A}(\mathcal{Q}_n - \mathcal{Q}_{n-1}) \\ (\mathcal{P}_1 - \mathcal{P}_0)\mathcal{A}\mathcal{Q}_0 & (\mathcal{P}_1 - \mathcal{P}_0)\mathcal{A}(\mathcal{Q}_1 - \mathcal{Q}_0) & \cdots & 0 \\ \vdots & \vdots & \ddots & \vdots \\ (\mathcal{P}_n - \mathcal{P}_{n-1})\mathcal{A}\mathcal{Q}_0 & 0 & \cdots & 0 \end{bmatrix}. \tag{11.78}$$

We call this compression strategy the Γ-shape compression. Letting $\mathcal{A}_{-1} = \mathcal{Q}_{-1} = 0$, the operator $\tilde{\mathcal{A}}_n : \mathbb{X}_n \to \mathbb{X}_n$ can be written as

$$\tilde{\mathcal{A}}_n = \sum_{i,j\in\mathbb{Z}_{n+1}, i+j\le n} (\mathcal{P}_i - \mathcal{P}_{i-1})\mathcal{A}(\mathcal{Q}_j - \mathcal{Q}_{j-1}).$$

In the collocation method (11.76), we replace $\mathcal{A}_n$ by $\tilde{\mathcal{A}}_n$ and obtain a new approximation scheme for solving equation (11.66). That is, we find $\tilde{u}^{\delta}_{\alpha,n} \in \mathbb{X}_n$ such that

$$(\alpha\mathcal{I} + \tilde{\mathcal{A}}_n)\tilde{u}^{\delta}_{\alpha,n} = \mathcal{P}_n\mathcal{K}^*f^{\delta}. \tag{11.79}$$

We show that this modified collocation method leads to a fast algorithm and at the same time it preserves the convergence rate obtained in [224].

To write equation (11.79) in its equivalent matrix form, we make use of the multiscale basis functions and the corresponding collocation functionals. We write the solution $\tilde{u}^{\delta}_{\alpha,n} \in \mathbb{X}_n$ as $\tilde{u}^{\delta}_{\alpha,n} = \sum_{(i,j)\in\mathbb{U}_n} u_{ij}w_{ij}$ and introduce the solution vector $\mathbf{u}_n := [u_{ij} : (i,j) \in \mathbb{U}_n]^T$. We introduce matrices the

$$\mathbf{E}_n := [\langle \ell_{i'j'}, w_{ij}\rangle : (i',j'),(i,j) \in \mathbb{U}_n], \quad \tilde{\mathbf{A}}_n := [\tilde{A}_{i'j',ij} : (i',j'),(i,j) \in \mathbb{U}_n],$$

where

$$\tilde{A}_{i'j',ij} := \begin{cases} \langle \ell_{i'j'}, \mathcal{K}^*\mathcal{K}w_{ij}\rangle, & i'+i \leq n, \\ 0, & \text{otherwise} \end{cases} \tag{11.80}$$

and vector $\mathbf{f}_n := [\langle \ell_{i'j'}, \mathcal{K}^*f^{\delta}\rangle : (i,j) \in \mathbb{U}_n]^T$. Upon using these notations, equation (11.79) is written in the matrix form

$$(\alpha\mathbf{E}_n + \tilde{\mathbf{A}}_n)\mathbf{u}_n = \mathbf{f}_n. \tag{11.81}$$

The semi-bi-orthogonality and the compact support property of the basis functions w_{ij} and the corresponding collocation functionals ℓ_{ij}, for $(i,j) \in \mathbb{U}$, ensure that $\mathbf{E}_n$ is a sparse upper triangular matrix. According to the Γ-shape compression strategy, $\tilde{\mathbf{A}}_n$ is a sparse matrix. These facts lead to a fast algorithm for solving equation (11.81).

In the next theorem, we analyze the number $\mathcal{N}(\tilde{\mathbf{A}}_n)$ of nonzero entries of matrix $\tilde{\mathbf{A}}_n$.

Theorem 11.20 *If the matrix $\tilde{\mathbf{A}}_n$ is obtained from the compression strategy (11.80), then*

$$\mathcal{N}(\tilde{\mathbf{A}}_n) = \mathcal{O}(s(n)\log(s(n))), \quad n \to \infty.$$

Proof For $i, i' \in \mathbb{Z}_{n+1}$, we introduce the block matrix

$$\tilde{\mathbf{A}}_{i'i} := [\tilde{A}_{i'j',ij} : j' \in \mathbb{Z}_{w(i')}, j \in \mathbb{Z}_{w(i)}].$$

According to the compression strategy (11.80), it is clear that

$$\mathcal{N}(\tilde{\mathbf{A}}_n) = \sum_{i'+i\leq n} \mathcal{N}(\tilde{\mathbf{A}}_{i'i}).$$

Noting that $\tilde{\mathbf{A}}_{i',i}$ is a $w(i') \times w(i)$ matrix, and that $w(i) \sim \mu^i$, from the equation above we conclude that

$$\mathcal{N}(\tilde{\mathbf{A}}_n) \sim \sum_{i'+i \le n} \mu^{i'+i} = \sum_{i \in \mathbb{Z}_{n+1}} \mu^i \sum_{i' \in \mathbb{Z}_{n-i+1}} \mu^{i'} = \mathcal{O}(s(n)\log(s(n))), \quad \text{as } n \to \infty,$$

proving the result of this theorem. □

A crucial step in generating the sparse coefficient matrix $\tilde{\mathbf{A}}_n$ is computing the nonzero entries $\langle \ell_{i'j'}, \mathcal{K}^*\mathcal{K}w_{ij}\rangle$, for $i' + i \le n$. This requires computing the values of the collocation functionals $\ell_{i'j'}$ at the functions defined by the integrals

$$\int_\Omega \int_\Omega K(\tau, s)K(\tau, t)\mathrm{d}\tau w_{ij}(t)\mathrm{d}t.$$

The same issue has been addressed in [33] in a different context (see Theorems 5 and 6 in that paper). The key idea is to develop a numerical quadrature strategy for computing the entries

$$\left\langle \ell_{i'j'}, \int_\Omega \int_\Omega K(\tau, s)K(\tau, t)\mathrm{d}\tau w_{ij}(t)\mathrm{d}t \right\rangle$$

of the compressed matrix. Such a strategy requires only $\mathcal{O}(s(n)\log(s(n)))$ number of functional evaluations with preserving the convergence order of the resulting approximate solution. See also [75] for a similar development for the second-kind Fredholm integral equation. Since this issue has been understood in principle and since the main focus of this section is the regularization, we will omit the details of this development and leave them to the interested reader.

We briefly discuss an alternative idea which may provide a fast algorithm for the collocation method proposed recently in [207]. The suggestion is that we first discretize equation (11.64) directly by using the collocation method described previously, which results in the linear system

$$\mathbf{K}_n\mathbf{u}_n = \mathbf{f}_n. \tag{11.82}$$

The use of the collocation functionals $\ell_{i'j'}$ instead of the point evaluation functionals as in [207] is for the matrix compression. The second step is to compress the dense matrix $\mathbf{K}_n$ in equation (11.82) to obtain

$$\tilde{\mathbf{K}}_n\tilde{\mathbf{u}}_n = \mathbf{f}_n. \tag{11.83}$$

The resulting coefficient matrix has only $\mathcal{O}(s(n)\log(s(n)))$ nonzero entries. This treatment is the same as that for the second-kind Fredholm integral equation developed in [69]. The third step is to apply the regularization to equation (11.83) to have

$$(\alpha \mathbf{E}_n + \tilde{\mathbf{K}}_n^* \tilde{\mathbf{K}}_n)\tilde{\mathbf{u}}_{n,\alpha} = \tilde{\mathbf{K}}_n^* \mathbf{f}_n. \tag{11.84}$$

The matrix $\mathbf{E}_n$ may be replaced by the identity matrix $\mathbf{I}_n$. Since all matrices $\mathbf{E}_n$, $\tilde{\mathbf{K}}_n$ and $\tilde{\mathbf{K}}_n^*$ are sparse, having at most $\mathcal{O}(s(n)\log(s(n)))$ nonzero entries, the computation of $\tilde{\mathbf{K}}_n^*\tilde{\mathbf{K}}_n\tilde{\mathbf{u}}_{n,\alpha}$ has a fast algorithm by first computing $\tilde{\mathbf{K}}_n\tilde{\mathbf{u}}_{n,\alpha}$ and then computing the product of $\tilde{\mathbf{K}}_n^*$ with the resulting vector if certain iteration methods are used in solving the discrete equation (11.84). Further description of this idea may result in a detour from the main focus of this section. Hence, we leave the details for future investigation.

We now turn to estimating the convergence rate of the modified collocation method (11.79). We impose the following hypothesis.

(H2) There exists a positive constant c such that

$$\|\mathcal{K}(\mathcal{I} - \mathcal{Q}_j)\|_\infty \le c\mu^{-rj/d},\ \|\mathcal{K}(\mathcal{I} - \mathcal{Q}_j)\|_2 \le c\mu^{-rj/d},$$
$$\|(\mathcal{I} - \mathcal{P}_j)\mathcal{K}^*\|_\infty \le c\mu^{-rj/d},\ \|(\mathcal{I} - \mathcal{P}_j)\mathcal{K}^*\|_2 \le c\mu^{-rj/d},$$
$$\|(\mathcal{I} - \mathcal{P}_j)\mathcal{K}^*\|_{L^2(\Omega)\to L^\infty(\Omega)} \le c\mu^{-rj/d}.$$

Proposition 11.17 gives various smoothness conditions on the kernel K so that estimates in hypothesis (H2) hold. We impose hypothesis (H2) instead of the smoothness conditions on the kernel K since the smoothness conditions are sufficient but not necessary. Examples of such cases will be shown in the last section.

In the following lemma, we present the error between $\tilde{\mathcal{A}}_n$ and $\mathcal{A}$ in the operator norm.

Lemma 11.21 *If hypothesis (H2) holds, then there exists a positive constant c_0 such that*

$$\|\mathcal{A} - \tilde{\mathcal{A}}_n\|_\infty \le c_0 n\mu^{-rn/d}.$$

Proof We write

$$\mathcal{A} - \tilde{\mathcal{A}}_n = (\mathcal{I} - \mathcal{P}_n)\mathcal{A} + \mathcal{P}_n\mathcal{A}(\mathcal{I} - \mathcal{Q}_n) + (\mathcal{P}_n\mathcal{A}\mathcal{Q}_n - \tilde{\mathcal{A}}_n). \tag{11.85}$$

Recalling that $\mathcal{A} = \mathcal{K}^*\mathcal{K}$ and $\|\mathcal{P}_n\|_\infty$ is uniformly bounded by a constant, it follows from hypothesis (H2) that there exists a positive constant c such that

$$\|(\mathcal{I} - \mathcal{P}_n)\mathcal{A} + \mathcal{P}_n\mathcal{A}(\mathcal{I} - \mathcal{Q}_n)\|_\infty \le c\mu^{-rn/d}. \tag{11.86}$$

Moreover, noting for each $i \in \mathbb{Z}_{n+1}$ that

$$\mathcal{Q}_{n-i} = \sum_{j\in\mathbb{Z}_{n-i+1}} (\mathcal{Q}_j - \mathcal{Q}_{j-1}),$$

we conclude that

$$\tilde{\mathcal{A}}_n = \sum_{i\in\mathbb{Z}_{n+1}} (\mathcal{P}_i - \mathcal{P}_{i-1})\mathcal{A}\mathcal{Q}_{n-i}. \tag{11.87}$$

We conclude from (11.87) that

$$\mathcal{P}_n\mathcal{A}\mathcal{Q}_n - \tilde{\mathcal{A}}_n = \sum_{i\in\mathbb{Z}_{n+1}} (\mathcal{P}_i - \mathcal{P}_{i-1})\mathcal{K}^*\mathcal{K}(\mathcal{Q}_n - \mathcal{Q}_{n-i}). \tag{11.88}$$

Again, using hypothesis (H2), there exists a positive constant c such that

$$\|(\mathcal{P}_i - \mathcal{P}_{i-1})\mathcal{K}^*\|_\infty \le \|(\mathcal{P}_i - \mathcal{I})\mathcal{K}^*\|_\infty + \|(\mathcal{I} - \mathcal{P}_{i-1})\mathcal{K}^*\|_\infty \le c\mu^{-ri/d}$$

and

$$\|\mathcal{K}(\mathcal{Q}_n - \mathcal{Q}_{n-i})\|_\infty \le \|\mathcal{K}(\mathcal{Q}_n - \mathcal{I})\|_\infty + \|\mathcal{K}(\mathcal{I} - \mathcal{Q}_{n-i})\|_\infty \le c\mu^{-r(n-i)/d}.$$

These estimates, together with (11.88), lead to

$$\|\mathcal{P}_n\mathcal{A}\mathcal{Q}_n - \tilde{\mathcal{A}}_n\|_\infty \le c\sum_{i=1}^{n} \mu^{-ri/d}\mu^{-r(n-i)/d} \le c_0 n\mu^{-rn/d}. \tag{11.89}$$

Combining (11.86) and (11.89) yields the desired result of this lemma. □

We establish below an estimate for the approximate operator $\tilde{\mathcal{A}}_n$, similar to the first estimate in Lemma 11.15, which is for the operator $\mathcal{A}$.

Lemma 11.22 *Suppose that hypothesis (H2) holds. If*

$$n\mu^{-rn/d} \le \frac{1}{2c_0}\frac{\alpha^{3/2}}{\sqrt{\alpha} + M/2}, \tag{11.90}$$

where c_0 is the constant appearing in Lemma 11.21, then $\alpha\mathcal{I} + \tilde{\mathcal{A}}_n : L^\infty(\Omega) \to L^\infty(\Omega)$ is invertible and

$$\|(\alpha\mathcal{I} + \tilde{\mathcal{A}}_n)^{-1}\|_\infty \le \frac{2\sqrt{\alpha} + M}{\alpha^{3/2}}. \tag{11.91}$$

Proof By a basic result in functional analysis (cf. Theorem 1.5, p. 193 of [254]), we conclude that $\alpha\mathcal{I} + \tilde{\mathcal{A}}_n : L^\infty(\Omega) \to L^\infty(\Omega)$ is invertible and

$$\|(\alpha\mathcal{I} + \tilde{\mathcal{A}}_n)^{-1}\|_\infty \le \frac{\|(\alpha\mathcal{I} + \mathcal{A})^{-1}\|_\infty}{1 - \|(\alpha\mathcal{I} + \mathcal{A})^{-1}\|_\infty\|\mathcal{A} - \tilde{\mathcal{A}}_n\|_\infty}.$$

The estimate (11.91) follows from the above bound, condition (11.90) and Lemmas 11.15 and 11.21. □

We next proceed to estimate the error $\|\hat{u} - \tilde{u}^{\delta}_{\alpha,n}\|_\infty$. To this end, we let $\tilde{u}_{\alpha,n}$ denote the solution of (11.79) with f^δ replaced by f. By the triangle inequality, we have that

$$\|\hat{u} - \tilde{u}^{\delta}_{\alpha,n}\|_\infty \le \|\hat{u} - u_\alpha\|_\infty + \|u_\alpha - \tilde{u}_{\alpha,n}\|_\infty + \|\tilde{u}_{\alpha,n} - \tilde{u}^{\delta}_{\alpha,n}\|_\infty. \quad (11.92)$$

Since Lemma 11.16 has given an estimate for $\|\hat{u}-u_\alpha\|_\infty$, it remains to estimate the last two terms on the right-hand side of inequality (11.92). In the next lemma, we estimate the second term.

Lemma 11.23 *If hypothesis (H2) holds, the integer n is chosen to satisfy inequality (11.90) and $\hat{u} \in R(\mathcal{K}^*)$, then there exists a positive constant c such that for all $n \in \mathbb{N}$,*

$$\|u_\alpha - \tilde{u}_{\alpha,n}\|_\infty \le c\frac{n\mu^{-rn/d}}{\alpha^{3/2}}.$$

Proof By using equations (11.68) and (11.79), we have that

$$u_\alpha - \tilde{u}_{\alpha,n} = (\alpha\mathcal{I} + \mathcal{A})^{-1}\mathcal{K}^* f - (\alpha\mathcal{I} + \tilde{\mathcal{A}}_n)^{-1}\mathcal{A}_n\mathcal{K}^* f.$$

It can be rewritten as

$$u_\alpha - \tilde{u}_{\alpha,n} = [(\alpha\mathcal{I} + \mathcal{A})^{-1} - (\alpha\mathcal{I} + \tilde{\mathcal{A}}_n)^{-1}]\mathcal{K}^* f + (\alpha\mathcal{I} + \tilde{\mathcal{A}}_n)^{-1}(\mathcal{I} - \mathcal{A}_n)\mathcal{K}^* f.$$

By employing the relation $f = \mathcal{K}\hat{u}$, we derive that

$$\begin{aligned} u_\alpha - \tilde{u}_{\alpha,n} = {} & (\alpha\mathcal{I} + \tilde{\mathcal{A}}_n)^{-1}(\tilde{\mathcal{A}}_n - \mathcal{A})(\alpha\mathcal{I} + \mathcal{A})^{-1}\mathcal{A}\hat{u} \\ & + (\alpha\mathcal{I} + \tilde{\mathcal{A}}_n)^{-1}(\mathcal{I} - \mathcal{A}_n)\mathcal{A}\hat{u}. \end{aligned}$$

By hypothesis, $\hat{u} \in R(\mathcal{K}^*)$ and thus we write $\hat{u} = \mathcal{K}^* v$ for some $v \in L^\infty(\Omega)$. Since, for any positive number α, the operator $\alpha\mathcal{I} + \mathcal{K}\mathcal{K}^*$ is invertible and

$$\|(\alpha\mathcal{I} + \mathcal{K}\mathcal{K}^*)^{-1}\mathcal{K}\mathcal{K}^*\|_2 \le 1,$$

it follows that

$$\|(\alpha\mathcal{I} + \mathcal{A})^{-1}\mathcal{A}\hat{u}\|_\infty = \|\mathcal{K}^*(\alpha\mathcal{I} + \mathcal{K}\mathcal{K}^*)^{-1}\mathcal{K}\mathcal{K}^* v\|_\infty \le \|\mathcal{K}^*\|_{L^2(\Omega)\to L^\infty(\Omega)}\|v\|_2.$$

Hence, there exists a positive constant c such that

$$\|(\alpha\mathcal{I} + \mathcal{A})^{-1}\mathcal{A}\hat{u}\|_\infty \le c.$$

This estimate, together with Lemmas 11.21 and 11.22, ensures that

$$\|(\alpha\mathcal{I} + \tilde{\mathcal{A}}_n)^{-1}(\tilde{\mathcal{A}}_n - \mathcal{A})(\alpha\mathcal{I} + \mathcal{A})^{-1}\mathcal{A}\hat{u}\|_\infty \le c\frac{n\mu^{-rn/d}}{\alpha^{3/2}}. \quad (11.93)$$

Moreover, using Lemma 11.22 and part (2) of Proposition 11.17, we conclude that

$$\|(\alpha\mathcal{I}+\tilde{\mathcal{A}}_n)^{-1}(\mathcal{I}-\mathcal{A}_n)\mathcal{A}\hat{u}\|_\infty \leq c\frac{\mu^{-rn/d}}{\alpha^{3/2}}. \tag{11.94}$$

Combining estimates (11.93) and (11.94) yields the desired estimate. □

Next, we estimate the third term on the right-hand side of inequality (11.92).

Lemma 11.24 *If hypothesis (H2) holds and condition (11.90) is satisfied, then there exists a positive constant c such that*

$$\|\tilde{u}_{\alpha,n}-\tilde{u}^\delta_{\alpha,n}\|_\infty \leq c\left(\frac{\delta}{\alpha}+\mu^{-rn/d}\frac{\delta}{\alpha^{3/2}}\right).$$

Proof From (11.79), we have that

$$(\alpha\mathcal{I}+\tilde{\mathcal{A}}_n)(\tilde{u}_{\alpha,n}-\tilde{u}^\delta_{\alpha,n}) = \mathcal{A}_n\mathcal{K}^*(f-f^\delta).$$

We rewrite it in the form

$$\begin{aligned}\tilde{u}_{\alpha,n}-\tilde{u}^\delta_{\alpha,n} = (\alpha\mathcal{I}+\mathcal{A})^{-1}\big[&\mathcal{K}^*(f-f^\delta)+(\mathcal{P}_n-\mathcal{I})\mathcal{K}^*(f-f^\delta)\\ &+(\mathcal{A}-\tilde{\mathcal{A}}_n)(\tilde{u}_{\alpha,n}-\tilde{u}^\delta_{\alpha,n})\big].\end{aligned}$$

It follows from the second estimate of Lemma 11.15 and hypothesis (11.67) that

$$\begin{aligned}&\|(\alpha\mathcal{I}+\mathcal{A})^{-1}\mathcal{K}^*(f-f^\delta)\|_\infty\\ &\quad\leq \|(\alpha\mathcal{I}+\mathcal{A})^{-1}\mathcal{K}^*\|_{L^2(\Omega)\to L^\infty(\Omega)}\|f-f^\delta\|_2 \leq M\frac{\delta}{\alpha}.\end{aligned}$$

The first estimate of Lemma 11.15 and part (2) of Proposition 11.17 ensure that there exists a positive constant c such that

$$\|(\alpha\mathcal{I}+\mathcal{A})^{-1}(\mathcal{P}_n-\mathcal{I})\mathcal{K}^*(f-f^\delta)\|_\infty \leq c\mu^{-rn/d}(\sqrt{\alpha}+M/2)\frac{\delta}{\alpha^{3/2}}.$$

By the first estimate of Lemma 11.15, Lemma 11.21 and condition (11.90), we obtain

$$\|(\alpha\mathcal{I}+\mathcal{A})^{-1}(\mathcal{A}-\tilde{\mathcal{A}}_n)(\tilde{u}_{\alpha,n}-\tilde{u}^\delta_{\alpha,n})\|_\infty \leq \frac{1}{2}\|\tilde{u}_{\alpha,n}-\tilde{u}^\delta_{\alpha,n}\|_\infty.$$

Combining the above three estimates, we conclude that

$$\|\tilde{u}_{\alpha,n}-\tilde{u}^\delta_{\alpha,n}\|_\infty \leq M\frac{\delta}{\alpha}+c\mu^{-rn/d}(\sqrt{\alpha}+M/2)\frac{\delta}{\alpha^{3/2}}+\frac{1}{2}\|\tilde{u}_{\alpha,n}-\tilde{u}^\delta_{\alpha,n}\|_\infty.$$

In the above inequality, solving for $\|\tilde{u}_{\alpha,n}-\tilde{u}^\delta_{\alpha,n}\|_\infty$ yields the desired estimate. □

We are now ready to present an error bound for $\|\hat{u} - \tilde{u}^{\delta}_{\alpha,n}\|_{\infty}$.

Theorem 11.25 *If hypothesis (H2) holds, the integer n is chosen to satisfy inequality (11.90) and $\hat{u} \in R(\mathcal{K}^*)$, then there exists a positive constant c such that*

$$\|\hat{u} - \tilde{u}^{\delta}_{\alpha,n}\|_{\infty} \leq \|\hat{u} - u_{\alpha}\|_{\infty} + c\left(\frac{\delta}{\alpha} + \frac{n\mu^{-rn/d}}{\alpha^{3/2}}\right). \tag{11.95}$$

Proof The estimate in this theorem follows directly from (11.92), Lemmas 11.23 and 11.24. □

We present in the next corollary two special results.

Corollary 11.26 *Suppose that hypothesis (H2) holds and the integer n is chosen to satisfy inequality (11.90).*

(1) If α is chosen such that $\delta/\alpha \to 0$ as $\delta \to 0$, the integer n is chosen so that $n\mu^{-rn/d} = \mathcal{O}(\sqrt{\alpha}\delta)$ as $\delta \to 0$ and $\hat{u} \in R(\mathcal{K}^*)$, then

$$\|\hat{u} - \tilde{u}^{\delta}_{\alpha,n}\|_{\infty} \to 0, \quad as \ \delta \to 0, \ \alpha \to 0. \tag{11.96}$$

(2) If hypothesis (H1) holds, then

$$\|\hat{u} - \tilde{u}^{\delta}_{\alpha,n}\|_{\infty} = \mathcal{O}\left(\alpha^{\nu} + \frac{\delta}{\alpha} + \frac{n\mu^{-rn/d}}{\alpha^{3/2}}\right), \quad as \ \delta \to 0. \tag{11.97}$$

Proof (1) Using the choice of α and n in estimate (11.95) leads to

$$\|\hat{u} - \tilde{u}^{\delta}_{\alpha,n}\|_{\infty} \leq \|\hat{u} - u_{\alpha}\|_{\infty} + c\frac{\delta}{\alpha}.$$

By the first result of Lemma 11.16, as $\alpha \to 0$, the first term on the right-hand side goes to zero, proving the result.

(2) Since

$$(\mathcal{K}^*\mathcal{K})^{\nu}\mathcal{K}^* = \mathcal{K}^*(\mathcal{K}\mathcal{K}^*)^{\nu}$$

(cf. p. 16 of [116]), we obtain that $R(\mathcal{A}^{\nu}\mathcal{K}^*) \subseteq R(\mathcal{K}^*)$. When hypothesis (H1) holds, we have that $\hat{u} \in R(\mathcal{K}^*)$. By Theorem 11.25, estimate (11.95) holds. Using the second result of Lemma 11.16, estimate (11.95) reduces to the desired result. □

11.3.3 Regularization parameter choice strategies

Solving equation (11.79) (or (11.81)) numerically requires appropriate choices of the regularization parameter α. In this subsection, we present *a priori* and *a posteriori* strategies for choices of such parameters and estimate the error bound of the corresponding approximate solutions.

We first present an *a priori* parameter choice strategy. We propose to choose the parameter α such that the right-hand side of estimate (11.97) is minimized. Specifically, we suppose that $\hat{u} \in R(\mathcal{A}^\nu \mathcal{K}^*)$ with $0 < \nu \leq 1$, where we call ν the smoothness order of the exact solution $\hat{u}$, and propose the following rule for the choice of parameter α.

Rule 11.27 *Given the noise bound $\delta > 0$ and the smoothness order $\nu \in (0, 1]$ of the exact solution $\hat{u}$, we choose α to satisfy*

$$\alpha \sim \delta^{\frac{1}{\nu+1}}$$

and a positive integer n to satisfy

$$n\mu^{-rn/d} = \mathcal{O}(\sqrt{\alpha}\delta), \text{ as } \delta \to 0.$$

Rule 11.27 uses the *a priori* parameter ν which is normally not available. For this reason, Rule 11.27 is called an *a priori* strategy. If there is a way to obtain ν, the parameter α and integer n chosen in Rule 11.27 are then used in equation (11.81) to obtain an approximate solution $\tilde{u}^\delta_{\alpha,n}$. In the next theorem, we present the convergence rate of the approximate solution corresponding to the above choice of α.

Theorem 11.28 *Suppose that $K \in W^{r,\infty}(\Omega \times \Omega)$ with $0 < r \leq k$. If α and n are chosen according to Rule 11.27 and hypothesis (H1) is satisfied, then*

$$\|\hat{u} - \tilde{u}^\delta_{\alpha,n}\|_\infty = \mathcal{O}(\delta^{\frac{\nu}{\nu+1}}), \text{ as } \delta \to 0. \tag{11.98}$$

Proof When α and n are chosen according to Rule 11.27, by a straightforward computation, we confirm that inequality (11.90) is satisfied. The assumption on K of this theorem with Proposition 11.17 ensures that hypothesis (H2) is satisfied. Hence, the hypothesis of Corollary 11.26 holds. Substituting the choices of α and n into the right-hand side of (11.97) in Corollary 11.26, we obtain estimate (11.98). □

Because the smoothness order ν of the exact solution is normally not known, it is desirable to develop a strategy for choices of the parameter α without *a priori* information on ν. Next, we present an *a posteriori* parameter choice strategy. The idea of the strategy was first used in [216] and recently developed further in [207].

For a given $\alpha > 0$, we let $n(\alpha)$ be a positive integer associated with α. Following [216], we assume that there exist two increasing continuous functions

$\varphi(\alpha)$ and $\lambda(\alpha)$ with $\varphi(0) = 0$ and $\lambda(0) = 0$ such that

$$\|\hat{u} - \tilde{u}^{\delta}_{\alpha,n(\alpha)}\|_{\infty} \leq \varphi(\alpha) + \frac{\delta}{\lambda(\alpha)}. \tag{11.99}$$

The assumption (11.99) leads us to the choice

$$\alpha = \alpha_{opt} := (\varphi\lambda)^{-1}(\delta).$$

With this choice of the parameter, we have that

$$\|\hat{u} - \tilde{u}^{\delta}_{\alpha_{opt},n(\alpha_{opt})}\|_{\infty} \leq 2\varphi((\varphi\lambda)^{-1}(\delta)). \tag{11.100}$$

Observing that

$$\alpha_{opt} = \max\left\{\alpha : \varphi(\alpha) \leq \frac{\delta}{\lambda(\alpha)}\right\},$$

in practice, we select the regularization parameter from a finite set

$$M(\Delta_N) := \left\{\alpha_i : \alpha_i \in \Delta_N, \varphi(\alpha_i) \leq \frac{\delta}{\lambda(\alpha_i)}\right\},$$

where

$$\Delta_N := \{\alpha_i : 0 < \alpha_0 < \alpha_1 < \cdots < \alpha_N\}$$

is a given set of N distinct positive numbers, which will be specified later. We consider

$$\alpha_* := \max\{\alpha_i : \alpha_i \in M(\Delta_N)\}$$

as an approximation of α_{opt} under appropriate conditions.

When φ is unknown, the above choice of the parameter is not feasible. It was suggested in [207, 216] to introduce a set

$$M^+(\Delta_N) = \left\{\alpha_j : \alpha_j \in \Delta_N, \|\tilde{u}^{\delta}_{\alpha_j,n(\alpha_j)} - \tilde{u}^{\delta}_{\alpha_i,n(\alpha_i)}\|_{\infty} \leq 4\frac{\delta}{\lambda(\alpha_i)}, i = 0, 1, \ldots, j\right\}$$

to replace $M(\Delta_N)$ and choose

$$\alpha_+ := \max\{\alpha_i : \alpha_i \in M^+(\Delta_N)\}$$

as an approximation of α_* accordingly. We have the next lemma, which is essentially Theorem 2.1 in [216].

Lemma 11.29 *Suppose that estimate (11.99) holds. If* $M(\Delta_N) \neq \emptyset$, $\Delta_N \backslash M(\Delta_N) \neq \emptyset$ *and for any* $\alpha_i \in \Delta_N, i = 1, 2, \ldots, N$, *the function* $\lambda(\alpha)$ *satisfies* $\lambda(\alpha_i) \leq q\lambda(\alpha_{i-1})$ *for a fixed constant* q, *then*

$$\|\hat{u} - \tilde{u}^{\delta}_{\alpha_+,n(\alpha_+)}\|_{\infty} \leq 6q\varphi((\varphi\lambda)^{-1}(\delta)). \tag{11.101}$$

The general result stated in the above lemma will be used to develop an *a posteriori* parameter choice strategy. To this end, for any fixed α, we choose $n := n(\alpha)$ to be the smallest positive integer satisfying condition

$$n\mu^{-rn/d} \leq \min\left\{c_1\sqrt{\alpha}\delta, \frac{1}{2c_0}\frac{\alpha^{3/2}}{\sqrt{\alpha}+M/2}\right\}. \tag{11.102}$$

Note that according to the condition above, $n(\alpha)$ is uniquely determined by the variable α. If hypotheses (H1) and (H2) hold, then for such a pair of (α, n), the estimate (11.97) holds. From (11.97), there exist positive constants c_2, c_3 such that

$$\|\hat{u} - \tilde{u}^{\delta}_{\alpha,n(\alpha)}\|_{\infty} \leq c_2\alpha^{\nu} + c_3\frac{\delta}{\alpha}. \tag{11.103}$$

We choose $\varphi(\alpha) = c_2\alpha^{\nu}$ $\lambda(\alpha) = \frac{\alpha}{c_3}$ and for $q_0 > 1$, $\rho > 0$, a positive integer N determined by

$$\rho\delta q_0^{N-1} \leq 1 < \rho\delta q_0^{N},$$

as in [207, 216], specify the finite set by

$$\Delta_N := \{\alpha_i := \rho\delta q_0^i : i = 0, 1, \ldots, N\}.$$

The introduction of the parameter ρ allows a larger degree of freedom for the choice of the regularization parameter.

Now we have the following rule for choosing the regularization parameter $\alpha := \alpha_+$.

Rule 11.30 *Choose $\alpha = \alpha_+$ by*

$$\alpha_+ := \max\left\{\alpha_j \in \Delta_N : \|\tilde{u}^{\delta}_{\alpha_j,n(\alpha_j)} - \tilde{u}^{\delta}_{\alpha_i,n(\alpha_i)}\|_{\infty} \leq 4c_3\frac{\delta}{\alpha_i}, i = 0, 1, \ldots, j\right\}. \tag{11.104}$$

Rule 11.30 does not use the smoothness order ν of the exact solution or any other *a priori* information. Hence, it is an *a posteriori* choice strategy of the regularization parameter. In the next theorem we present the convergence order of the approximate solution corresponding to the above choice of α_+.

Theorem 11.31 *Suppose that hypotheses (H1) and (H2) hold. If $\alpha := \alpha_+$ is chosen according to Rule 11.30, then*

$$\|\hat{u} - \tilde{u}^{\delta}_{\alpha_+,n(\alpha_+)}\|_{\infty} = \mathcal{O}(\delta^{\frac{\nu}{\nu+1}}), \text{ as } \delta \to 0. \tag{11.105}$$

Proof This theorem is a direct consequence of Lemma 11.29. It suffices to verify that the hypotheses of the lemma are satisfied.

First of all, according to (11.103), estimate (11.99) holds with $\varphi(\alpha) = c_2\alpha^\nu$ and $\lambda(\alpha) = \frac{\alpha}{c_3}$. It can be verified that $\lambda(\alpha_i) = q_0\lambda(\alpha_{i-1})$ for $i = 0, 1, \ldots, N$, and

$$\frac{c_2}{c_3}\rho^{\nu+1}\delta^\nu \leq 1 \text{ and } \frac{c_2}{c_3} > \delta, \text{ as } \delta \to 0.$$

From this, we have that

$$\varphi(\alpha_0) \leq \delta/\lambda(\alpha_0) \text{ and } \varphi(\alpha_N) > \delta/\lambda(\alpha_N).$$

We conclude that $\alpha_0 \in M(\Delta_N)$ and $\alpha_N \notin M(\Delta_N)$, and thus

$$M(\Delta_N) \neq \emptyset \text{ and } \Delta_N \backslash M(\Delta_N) \neq \emptyset.$$

We have proved that all hypotheses of Lemma 11.29 are satisfied. Therefore, from (11.101) we have that

$$\|\hat{u} - \tilde{u}^\delta_{\alpha_+, n(\alpha_+)}\|_\infty \leq 6q_0\varphi((\varphi\lambda)^{-1}(\delta)) = c_4\delta^{\frac{\nu}{\nu+1}}, \tag{11.106}$$

with $c_4 = 6q_0c_2^{\nu-1}c_3^\nu$. □

11.4 Numerical experiments

In this section we present numerical results which verify the efficiency of the methods proposed in Sections 11.2 and 11.3.

11.4.1 A numerical example for the multiscale Galerkin method

We now present a numerical example to illustrate the method proposed in Section 11.2 and to confirm the theoretical estimates established in the section.

For this purpose, we consider the integral operator $\mathcal{K} : L^2[0,1] \to L^2[0,1]$ defined by

$$(\mathcal{K}u)(x) := \int_0^1 K_1(x,t)u(t)\mathrm{d}t, \quad x \in [0,1], \tag{11.107}$$

with the kernel

$$K_1(x,t) := \begin{cases} x(1-t), & 0 \leq x \leq t \leq 1, \\ t(1-x), & 0 \leq t < x \leq 1. \end{cases}$$

Operator $\mathcal{K}$ is positive semi-definite with respect to $L^2[0,1]$, and it is a linear compact self-adjoint operator from $L^2[0,1]$ to $L^2[0,1]$. We consider the integral equation of the first kind (11.3) with

$$f(x) := \frac{5}{12}x(x-1)(x^2-x-1), \ \ x \in [0,1].$$

Clearly, the unique solution of equation (11.3) is given by

$$u_*(x) = 5x(1-x), \ \ x \in [0,1].$$

Moreover, $u_* = \mathcal{K}\omega \in R(\mathcal{K})$ with $\omega := 10$. This means that condition (H-1) is satisfied with $\nu = 1$.

Let $\mathbb{X}_n$ be the piecewise linear functions on $[0,1]$ with knots at $j/2^n, j = 1, 2, \ldots, 2^n - 1$. We decompose $\mathbb{X}_n$ into the orthogonal direct sum of subspaces

$$\mathbb{X}_n = \mathbb{X}_0 \oplus^\perp \mathbb{W}_1 \oplus^\perp \cdots \oplus^\perp \mathbb{W}_n,$$

where $\mathbb{X}_0$ is the linear polynomial space on $[0,1]$, and $\mathbb{W}_i$, $i = 2, 3, \ldots, n$, are constructed recursively once the initial space $\mathbb{W}_1$ is given. We choose a basis for $\mathbb{X}_0$

$$w_{00}(t) = 2 - 3t \quad \text{and} \quad w_{01}(t) = -1 + 3t,$$

and a basis for space $\mathbb{W}_1$

$$w_{10}(t) = \begin{cases} 1 - \frac{9}{2}t, & t \in [0, \frac{1}{2}], \\ -1 + \frac{3}{2}t, & t \in (\frac{1}{2}, 1] \end{cases}$$

and

$$w_{11}(t) = \begin{cases} \frac{1}{2} - \frac{3}{2}t, & t \in [0, \frac{1}{2}], \\ -\frac{7}{2} + \frac{9}{2}t, & t \in (\frac{1}{2}, 1]. \end{cases}$$

It follows from a simple computation that

$$(\mathcal{K}^2 u)(x) = \int_0^1 K_2(x,t)u(t)\mathrm{d}t, \quad t \in [0,1], \tag{11.108}$$

where

$$K_2(x,t) := \begin{cases} x(t-1)(-2t+t^2+x^2)/6, & 0 \le x \le t \le 1, \\ t(x-1)(t^2-2x+x^2)/6, & 0 \le t < x \le 1. \end{cases}$$

We remark that $K_2 \in C^2([0,1] \times [0,1])$. For the orthogonal projection $\mathcal{P}_n : L^2[0,1] \to \mathbb{X}_n$, we have that

$$\|(\mathcal{I} - \mathcal{P}_n)\mathcal{K}^2\| \le \frac{\sqrt{2}}{2^{2n+1}}.$$

Moreover, by the moment inequality in [220] we know that the approximation properties (11.20) and (11.21) are valid with

$$\theta_n := \frac{\sqrt{2\sqrt{2}}}{2^{n+1}}.$$

In the MAM, we choose

$$n = k + m, \quad \text{with} \quad k = 3, \quad m = 3.$$

Hence, we use the decomposition

$$\mathbb{X}_{3+3} = \mathbb{X}_3 \oplus^{\perp} \mathbb{W}_{3+1} \oplus^{\perp} \mathbb{W}_{3+2} \oplus^{\perp} \mathbb{W}_{3+3}$$

and

$$\theta_6 = \frac{\sqrt{2\sqrt{2}}}{2^7}.$$

We choose a perturbed right-hand side

$$f^{\delta} := f + \delta v, \tag{11.109}$$

where $v \in \mathbb{X}$ has uniformly distributed random values with $\|v\| \leq 1$, and where

$$\delta = \frac{e}{100} \cdot \|f\|, \quad \text{with} \quad e \in \{0.125, 0.25, 0.5, 1.0, 1.5, 2.0\}.$$

Now we present numerical results for our experiments in Tables 11.1–11.3.

The numerical results in Table 11.1 are for an *a priori* parameter choice. According to Theorem 11.7, we choose an *a priori* parameter

$$\alpha := 0.01 * \theta_k, \quad \text{with} \quad k = 3,$$

since $\nu = 1$ in this example. Table 11.2 shows numerical results for an *a priori* parameter choice with error level δ fixed, that is, $e = 0.25$. It illustrates that the MAM takes effect since the absolute error $\|u^{\delta}_{\alpha,3,m} - u_*\|$ decreases as m increases in this case. In Tables 11.1 and 11.2, we also compared the effect by the MAM and the Galerkin method. Table 11.3 is associated with an

Table 11.1 *Numerical results for* a priori *parameter with* $\alpha = 0.01 * \theta_3$

e	$\|u^{\delta}_{\alpha,3,3} - u_*\|$	$\|u^{\delta}_{\alpha,3+3} - u_*\|$	$\|u^{\delta}_{\alpha,3,3} - u_*\| - \|u^{\delta}_{\alpha,3+3} - u_*\|$
2.0	1.3412×10^{-2}	1.3412×10^{-2}	4.7452×10^{-7}
1.5	1.1230×10^{-2}	1.1229×10^{-2}	1.7932×10^{-7}
1.0	1.0549×10^{-2}	1.0549×10^{-2}	7.9820×10^{-8}
0.5	9.9053×10^{-3}	9.9053×10^{-3}	1.3435×10^{-8}
0.25	9.8348×10^{-3}	9.8348×10^{-3}	-3.9297×10^{-10}

Table 11.2 *Numerical results for* a priori *parameter with* $e = 0.25$ *and* $\alpha = 0.01 * \theta_3$

m	$\|u^{\delta}_{\alpha,3,m} - u_*\|$	$\|u^{\delta}_{\alpha,3+m} - u_*\|$	$\|u^{\delta}_{\alpha,3,m} - u_*\| - \|u^{\delta}_{\alpha,3+m} - u_*\|$
0	1.1447×10^{-2}	1.1447×10^{-2}	0
1	1.0241×10^{-2}	1.0059×10^{-2}	1.3841×10^{-2}
2	9.9703×10^{-3}	9.9698×10^{-3}	5.7395×10^{-7}
3	9.9645×10^{-3}	9.9645×10^{-3}	5.5529×10^{-9}

Table 11.3 *Numerical results for* a posteriori parameter *with* $\theta_{k+m} = \theta_6$

e	α	$\|u^{\delta}_{\alpha,3,3} - u_*\|$
2.0	9.0117×10^{-3}	7.5321×10^{-2}
1.5	9.0117×10^{-3}	7.5578×10^{-2}
1.0	3.0039×10^{-3}	2.7045×10^{-2}
0.5	3.0039×10^{-3}	2.7188×10^{-2}
0.25	3.0039×10^{-3}	2.7470×10^{-2}
0.125	1.0013×10^{-3}	9.5639×10^{-3}

a posteriori parameter choice, and we use Algorithm 11.12 to determine the parameter α. The numerical results in these three tables demonstrate that the MAM is an efficient method to solve ill-posed problems.

11.4.2 Numerical examples for the multiscale collocation method

In this subsection, we present numerical results to demonstrate the efficiency and accuracy of the method proposed in Section 11.3.

We consider the integral operator $\mathcal{K} : L^{\infty}[0,1] \to L^{\infty}[0,1]$ defined by

$$(\mathcal{K}u)(s) := \int_0^1 K(s,t)u(t)\mathrm{d}t, \quad s \in [0,1], \tag{11.110}$$

where

$$K(s,t) := \begin{cases} s(1-t), & 0 \le s \le t \le 1, \\ t(1-s), & 0 \le t < s \le 1. \end{cases}$$

Since the kernel K is continuous on $[0,1]\times[0,1]$, the operator $\mathcal{K}$ is compact on $L^{\infty}[0,1]$. Hence, the integral equation (11.64), that is, $\mathcal{K}u = f$, with this kernel is ill posed. We solve this equation using the fast piecewise linear collocation method. The numerical results produced by this method with both *a priori* and *a posteriori* regularization parameter choices will be presented.

Now, we describe the approximation spaces $\mathbb{X}_n$ and the corresponding collocation functional spaces $\mathbb{L}_n$. Specifically, for each positive integer n, we choose $\mathbb{X}_n$ as the space of piecewise linear polynomials on $[0, 1]$ with knots at $j/2^n, j = 1, 2, \ldots, 2^n - 1$. Here, $\dim \mathbb{X}_n = 2^{n+1}$ and $\mu = 2$, corresponding to the general case. We then decompose $\mathbb{X}_n$ into an orthogonal direct sum of subspaces of different scales

$$\mathbb{X}_n = \mathbb{X}_0 \oplus^{\perp} \mathbb{W}_1 \oplus^{\perp} \cdots \oplus^{\perp} \mathbb{W}_n,$$

where $\mathbb{X}_0$ has a basis

$$w_{00}(s) = 2 - 3s \quad \text{and} \quad w_{01}(s) = -1 + 3s,$$

and $\mathbb{W}_1$ has a basis

$$w_{10}(s) = \begin{cases} 1 - \frac{9}{2}s, & s \in [0, \frac{1}{2}], \\ -1 + \frac{3}{2}s, & s \in (\frac{1}{2}, 1] \end{cases} \quad \text{and} \quad w_{11}(s) = \begin{cases} \frac{1}{2} - \frac{3}{2}s, & s \in [0, \frac{1}{2}], \\ -\frac{7}{2} + \frac{9}{2}s, & s \in (\frac{1}{2}, 1]. \end{cases}$$

The spaces $\mathbb{W}_i := \operatorname{span}\{w_{ij} : j \in \mathbb{Z}_{2^i}\}$ are generated recursively by $\mathbb{W}_1$ following the general construction developed in [66, 72, 200, 201].

The collocation functional space $\mathbb{L}_n$ corresponding to $\mathbb{X}_n$ is constructed likewise. For any $s \in [0, 1]$, we use δ_s to denote the linear functional defined for functions $v \in C[0, 1]$ by $\langle \delta_s, v \rangle = v(s)$ and extended to all functions in $L^\infty[0, 1]$ by the Hahn–Banach theorem. We let $\mathbb{L}_0 = \operatorname{span}\{\ell_{01}, \ell_{02}\}$ with $\ell_{00} = \delta_{\frac{1}{3}}$, $\ell_{01} = \delta_{\frac{2}{3}}$ and decompose $\mathbb{L}_n$ as

$$\mathbb{L}_n = \mathbb{L}_0 \oplus \mathbb{V}_1 \oplus \cdots \oplus \mathbb{V}_n,$$

where $\mathbb{V}_1 = \operatorname{span}\{\ell_{11}, \ell_{12}\}$ with

$$\ell_{10} = -\frac{3}{2}\delta_{\frac{1}{3}} + \frac{1}{2}\delta_{\frac{2}{3}} + \delta_{\frac{1}{6}}, \quad \ell_{11} = \frac{1}{2}\delta_{\frac{1}{3}} - \frac{3}{2}\delta_{\frac{2}{3}} + \delta_{\frac{5}{6}}$$

and the spaces

$$\mathbb{V}_i := \operatorname{span}\{\ell_{ij} : j \in \mathbb{Z}_{2^i}\}$$

are constructed recursively by $\mathbb{V}_1$. The approximation spaces and the corresponding collocation functionals have the properties outlined in Section 11.3.2.

By direct computation, we obtain that

$$\|\mathcal{K}\|_\infty = \|\mathcal{K}^*\|_\infty \leq \max_{0 \leq t \leq 1} \int_0^1 |k(t, s)| ds = \frac{1}{8}.$$

Moreover, since $\mathbb{X}_n$ are the piecewise linear polynomials on [0, 1] with knots at $j/2^n, j = 1, 2, \ldots, 2^n - 1$, it is easily verified that

$$\|\mathcal{K}(\mathcal{I} - \mathcal{Q}_n)\|_\infty \le \sqrt{\frac{2}{3}}2^{-\frac{3}{2}n}, \quad \|(\mathcal{I} - \mathcal{A}_n)\mathcal{K}^*\|_\infty \le \frac{1}{8}2^{-2n},$$
$$\|\mathcal{K}(\mathcal{I} - \mathcal{Q}_n)\|_2 \le \sqrt{\frac{2}{3}}2^{-\frac{3}{2}n}, \quad \|(\mathcal{I} - \mathcal{A}_n)\mathcal{K}^*\|_2 \le \frac{2}{\sqrt{3}}2^{-2n}$$

and

$$\|(\mathcal{I} - \mathcal{A}_n)\mathcal{K}^*\|_{L^2[0,1]\to L^\infty[0,1]} \le \sqrt{\frac{2}{3}}2^{-\frac{3}{2}n}.$$

Therefore, hypothesis (H2) holds with $r = \frac{3}{2}$.

For comparison purposes, in our numerical experiment we choose the right-hand side of the equation as

$$f(s) := \frac{1}{40.320}s(s-1)(-17 - 17s + 11s^2 + 11s^3 - 3s^4 - 3s^5 + s^6), \quad s \in [0, 1]$$

so that we have the exact solution

$$u(s) = \frac{-1}{0.720}s(s-1)(3 + 3s - 2s^2 - 2s^3 + s^4), \quad s \in [0, 1].$$

Moreover,

$$u = \mathcal{A}\mathcal{K}^*\omega \in R(\mathcal{A}\mathcal{K}^*) \text{ with } \omega := 1000.$$

Hence, hypothesis (H1) is satisfied with $\nu = 1$. The perturbed right-hand side is chosen as $f^\delta := f + \delta v$, where $v \in L^2[0, 1]$ has uniformly distributed random values with $\|v\|_2 = 1$ and $\delta = \|f\|_2 \cdot e/100 = 0.09488 \cdot e/100$ with $e \in \{1, 3, 5, 7, 9\}$. The linear system of the multiscale regularized equation is solved by the augmentation method described in [71].

We present the results of two numerical experiments. The goal of the first experiment is to confirm the convergence result for the *a priori* parameter choice. For given δ, following Rule 11.27, we choose $\alpha = 0.005\delta^{1/2}$ and n such that $n2^{-3n/2} \le 15\delta^{5/4}$ and $(n-1)2^{-3(n-1)/2} > 15\delta^{5/4}$. The numerical results are listed in Table 11.4. In Table 11.4, "Compression rates" are computed by dividing the number of nonzero entries of the matrix by the total number of entries of the matrix. Table 11.4 shows that the computed convergence rate is $\mathcal{O}(\delta^{1/2})$ for $\nu = 1$, which is consistent with the theoretical estimate given in Theorem 11.28. In Figure 11.1, we illustrate the approximate solutions with different parameters.

Table 11.4 A priori *parameter choices*

e	$\alpha = 0.005 * \delta^{1/2}$	n	Compression rates	$\|\hat{u} - \tilde{u}^{\delta}_{\alpha,n}\|_\infty$	$\|\hat{u} - \tilde{u}^{\delta}_{\alpha,n}\|_\infty / \delta^{1/2}$
1	1.5401e-4	8	0.0145	0.0260	0.8434
3	2.6676e-4	7	0.0286	0.0395	0.7398
5	3.4435e-4	6	0.0557	0.0547	0.7949
7	4.0749e-4	5	0.1055	0.0707	0.8680
9	4.6204e-4	5	0.1055	0.0892	0.9657

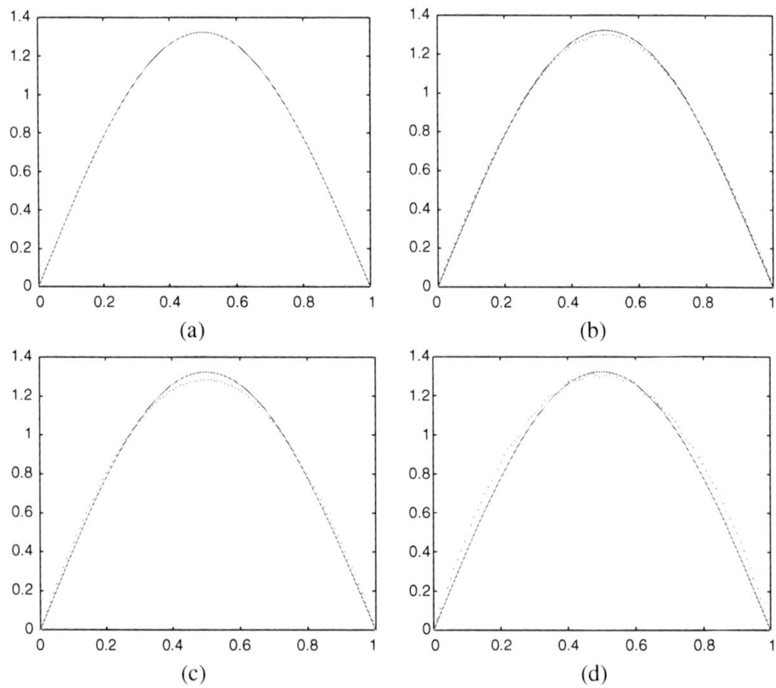

Figure 11.1 (a) The original function; (b) the restored function with $n = 8$, $e = 1$ and $\alpha = 0.005 * \delta^{1/2}$; (c) the restored function with $n = 6$, $e = 5$ and $\alpha = 0.005 * \delta^{1/2}$; (d) the restored function with $n = 5$, $e = 9$ and $\alpha = 0.005 * \delta^{1/2}$.

In the second experiment, we choose $\alpha := \alpha_+$ following Rule 11.30 with $\rho := 0.005$ and $q_0 := 3$. The numerical results are shown in Table 11.5 and illustrated in Figure 11.2. These results show that the *a posteriori* parameter choice gives results comparable with those of the *a priori* parameter choice. The computed convergence rate $\mathcal{O}(\delta^{1/2})$ confirms the theoretical estimate given in Theorem 11.31 with $\nu = 1$.

Table 11.5 A posteriori *parameter choices*

e	α	n	Compression rates	$\lVert\hat{u} - \tilde{u}^{\delta}_{\alpha,n}\rVert_\infty$	$\lVert\hat{u} - \tilde{u}^{\delta}_{\alpha,n}\rVert_\infty/\delta^{1/2}$
1	1.2809e-4	8	0.0145	0.0235	0.7907
3	3.8426e-4	7	0.0286	0.0576	0.7199
5	6.4044e-4	6	0.0557	0.0410	0.7681
7	8.9662e-4	5	0.1055	0.0608	0.7457
9	1.1528e-3	5	0.1055	0.0779	0.8428

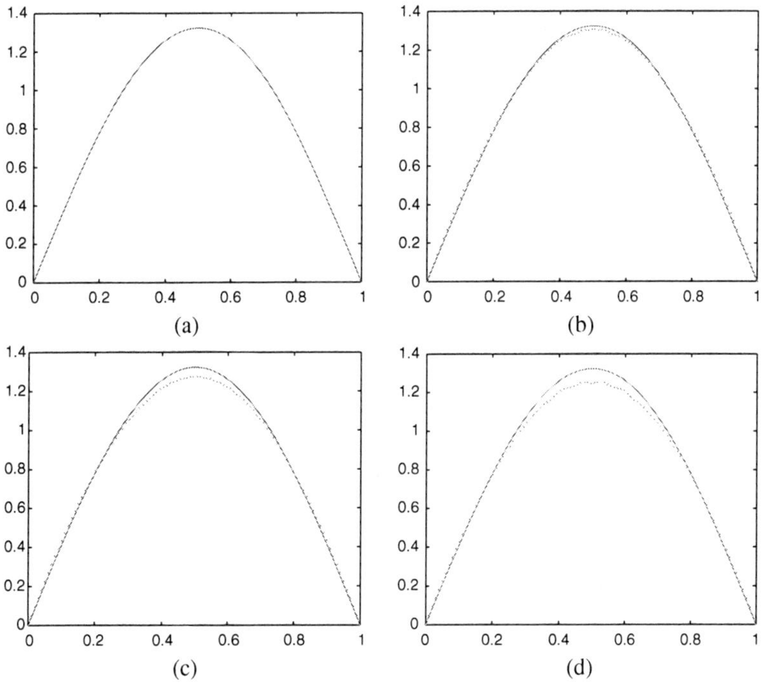

Figure 11.2 (a) The original function; (b) the restored function with $n = 8$, $e = 1$ and α=1.2809e-4; (c) the restored function with $n = 6$, $e = 5$ and α=6.4044e-4; (d) the restored function with $n = 5$, $e = 9$ and α=1.1528e-3.

11.5 Bibliographical remarks

The material presented in this chapter was mainly chosen from two papers [78, 79]. The *a posteriori* choice strategy of regularization parameters described in this chapter is based on the ideas developed in [207, 216]. For other developments of *a priori* and *a posteriori* parameter choice strategies, the reader is referred to [58, 118, 127–129, 191, 192, 206, 219, 220, 222, 224, 228, 248, 275].

We review several other recent developments in the numerical solution of ill-posed operator equations closely related to the research presented in this chapter. A class of regularization methods for discretized ill-posed problems was proposed in [222] and a method was suggested to determine *a priori* and *a posteriori* parameters for the regularization. Papers [127–129, 220] studied numerical algorithms for regularization of the Galerkin method. In [231] an additive Schwarz iteration was presented for the fast solution of the Tikhonov regularized ill-posed problems. A two-level preconditioner was developed in [133, 134] for solving the discretized operator equation and extended in [233] to a general case. The work in [193] designed an adaptive discretization for Tikhonov regularization which has a sparse matrix structure. Multilevel methods were applied to solve ill-posed problems and also established *a priori* and *a posteriori* parameter choice strategies [103, 137, 177]. In particular, a wavelet-based matrix compression technique was used in [137]. Moreover, wavelet-based multilevel methods and cascadic multilevel methods for solving ill-posed problems were developed in [174] and [230], respectively. A multiscale analysis for ill-posed problems with semi-discrete Tikhonov regularization was established in [278]. For more information on fast solvers for ill-posed integral equations, the reader is referred to [54, 78, 103, 104, 133, 137, 157, 193, 231].

Finally, applications of ill-posed problems in science and engineering may be found in [106, 116, 173]. In particular, for applications of regularization in image processing, system identification and machine learning, the reader is referred to [188, 189, 237, 259], [33, 223] and [87, 195, 268], respectively.

12
Eigen-problems of weakly singular integral operators

In this chapter we consider solving the eigen-problem of a weakly singular integral operator $\mathcal{K}$. As we know, the spectrum of a compact integral operator $\mathcal{K}$ consists of a countable number of eigenvalues which only have an accumulation point at zero that may or may not be an eigenvalue. We explain in this chapter how the multiscale methods can be used to compute the nonzero eigenvalues of $\mathcal{K}$ rapidly and efficiently. We begin with a brief introduction to the subject.

12.1 Introduction

Many practical problems in science and engineering are formulated as eigen-problems of compact linear integral operators (cf. [44]). Standard numerical treatments of the eigen-problem normally discretize the compact integral operator into a matrix and then solve the eigen-problem of the resulting matrix. The computed eigenvalues and associated eigenvectors of the matrix are considered approximations of the corresponding eigenvalues and eigenvectors of the compact integral operator. In particular, the Galerkin, Petrov–Galerkin, collocation, Nyström and degenerate kernel methods are commonly used methods for the approximation of eigenvalues and eigenvectors of compact integral operators. It is well known that the matrix which results from a discretization of a compact integral operator is dense. Solving the eigen-problem of a dense matrix requires a significant amount of computational effort. Hence, fast algorithms for solving such a problem are highly desirable.

We are interested in developing a fast collocation method for solving the eigen-problem of a compact linear integral operator $\mathcal{K}$ on the $L^\infty(\Omega)$ space with a weakly singular kernel. Wavelet and multiscale methods were recently developed (see, for example, [28, 64, 67, 68, 95, 108, 202] and the references cited therein) for numerical solutions of weakly singular Fredholm integral equations of the second kind. Some of them were discussed in the previous chapter. The essence of these methods is to approximate the dense matrix that results from the discretization of the integral operator by a sparse matrix and solve the linear system of the sparse matrix. It has been proved that the methods have nearly linear computational complexity and optimal convergence (see Chapters 5–7). Among these methods the fast collocation method receives favorable attention due to the lower computational costs in generating its sparse coefficient matrix (cf. [69] and Chapter 7).

The specific goal of this chapter is to develop the fast collocation method for finding eigenvalues and eigenvectors of a compact integral operator $\mathcal{K}$ with a weakly singular kernel, based on the matrix compression technique. For the analysis of the fast method, we extend the classical spectral approximation theory [3, 44, 212] to a somewhat more general setting so that it is applicable to the scenario where the matrix compression technique is used.

We organize this chapter into seven sections. In Section 12.2, we present an abstract framework for a compact integral operator to be approximated by a sequence of ν-convergent bounded linear operators. We develop in Section 12.3 the fast multiscale collocation method for solving the eigen-problem of a compact integral operator $\mathcal{K}$ with a weakly singular kernel. We establish in Section 12.4 the optimal convergence rate for the approximate eigenvalues and generalized eigenvectors. In Section 12.5, we describe a power method for solving the eigen-problem of the compressed matrix which makes use of the sparsity of the compressed matrix. We present in Section 12.6 numerical results to confirm the convergence estimates. Finally, in Section 12.7, we make bibliographical remarks.

12.2 An abstract framework

We describe in this section an abstract framework for eigenvalue approximation of a compact operator in a Banach space. The results presented here are basically the classical spectral approximation theory [3, 44, 212]. However, we present them in a somewhat more general form so that they are applicable to the sparse matrix resulting from the multiscale collocation method with matrix compression.

Suppose that $\mathbb{X}$ is a complex Banach space and $\mathcal{B}(\mathbb{X})$ is the space of bounded linear operators from $\mathbb{X}$ to $\mathbb{X}$. For an operator $\mathcal{T} \in \mathcal{B}(\mathbb{X})$ we define its *resolvent set* by

$$\rho(\mathcal{T}) := \{z : z \in \mathbb{C},\ (\mathcal{T} - z\mathcal{I})^{-1} \in \mathcal{B}(\mathbb{X})\},$$

and its *spectrum* by

$$\sigma(\mathcal{T}) := \mathbb{C} \setminus \rho(\mathcal{T}).$$

The resolvent operator $(\mathcal{T} - z\mathcal{I})^{-1}$ is denoted by $\mathcal{R}(\mathcal{T})$ and when we wish to show its dependency on z we write $\mathcal{R}(\mathcal{T}, z)$. Clearly, by definition, $\mathcal{R}(\mathcal{T}, z) \in \mathcal{B}(\mathbb{X})$ on $\rho(\mathcal{T})$. The range space and null space of $\mathcal{T}$ are defined, respectively, by

$$R(\mathcal{T}) := \{\mathcal{T}x : x \in \mathbb{X}\} \text{ and } N(\mathcal{T}) := \{x : \mathcal{T}x = 0, x \in \mathbb{X}\}.$$

The central problem considered in this chapter is that, for a compact operator $\mathcal{T}$ defined on a Banach space $\mathbb{X}$, we wish to find $\lambda \in \sigma(\mathcal{T}) \setminus \{0\}$ and $\phi \in \mathbb{X}$ with $\|\phi\| = 1$ such that

$$\mathcal{T}\phi = \lambda\phi. \tag{12.1}$$

As the eigenvalue problem (12.1), in general, cannot be solved exactly, we consider solving the problem approximately. To this end, we require a sequence of subspaces $\{\mathbb{X}_n : n \in \mathbb{N}\}$ which approximate the space $\mathbb{X}$ and a sequence of operators $\{\mathcal{T}_n : n \in \mathbb{N}\}$ which approximate the operator $\mathcal{T}$, and consider the approximate eigen-problem: finding $\lambda_n \in \sigma(\mathcal{T}_n)\setminus\{0\}$ and $\phi_n \in \mathbb{X}_n$ with $\|\phi_n\| = 1$ such that

$$\mathcal{T}_n\phi_n = \lambda_n\phi_n. \tag{12.2}$$

We start with a discussion of the spectral approximation. For any closed Jordan (rectifiable) curve Γ in $\rho(\mathcal{T})$, we use the notation $\|\mathcal{R}(\mathcal{T})\|_\Gamma$ for the quantity $\max\{\|\mathcal{R}(\mathcal{T}, z)\| : z \in \Gamma\}$. We first propose an ancillary lemma.

Lemma 12.1 *If $\mathcal{T}, \mathcal{S} \in \mathcal{B}(\mathbb{X}), z \in \Gamma$, where Γ is a closed Jordan curve in $\rho(\mathcal{T}) \setminus \{\zeta : \zeta \in \mathbb{C}, |\zeta| \leq r\}$ for some $r > 0$, and*

$$\mu := r^{-1}\|\mathcal{R}(\mathcal{T})\|_\Gamma(1 + \|\mathcal{T} - \mathcal{S}\|) + 1, \tag{12.3}$$

then

$$\left\|[(\mathcal{T} - \mathcal{S})\mathcal{R}(\mathcal{T}, z)]^2\right\| \leq \mu\left(\|(\mathcal{T} - \mathcal{S})\mathcal{T}\| + \|(\mathcal{T} - \mathcal{S})\mathcal{S}\|\right). \tag{12.4}$$

Proof We compute

$$[(\mathcal{T} - \mathcal{S})\mathcal{R}(\mathcal{T}, z)]^2 = (\mathcal{T} - \mathcal{S})\mathcal{R}(\mathcal{T}, z)(\mathcal{T} - \mathcal{S})\mathcal{R}(\mathcal{T}, z). \tag{12.5}$$

For any $z \in \Gamma$, direct computation confirms that

$$\mathcal{R}(\mathcal{T}, z) = z^{-1}[\mathcal{T}\mathcal{R}(\mathcal{T}, z) - \mathcal{I}]. \tag{12.6}$$

We substitute this formula and obtain the identity

$$\begin{aligned}&[(\mathcal{T}-\mathcal{S})\mathcal{R}(\mathcal{T},z)]^2\\ &\quad = z^{-1}\left[(\mathcal{T}-\mathcal{S})\mathcal{T}\mathcal{R}(\mathcal{T},z)(\mathcal{T}-\mathcal{S})-(\mathcal{T}-\mathcal{S})\mathcal{T}+(\mathcal{T}-\mathcal{S})\mathcal{S}\right]\mathcal{R}(\mathcal{T},z).\end{aligned}$$

Using the triangle inequality proves the result. □

The next lemma demonstrates how we shall use Lemma 12.1.

Lemma 12.2 *If the hypotheses of Lemma 12.1 are satisfied and both of the quantities* $\|(\mathcal{T}-\mathcal{S})\mathcal{T}\|$ *and* $\|(\mathcal{T}-\mathcal{S})\mathcal{S}\|$ *do not exceed* $(4\mu)^{-1}$, *then* $\rho(\mathcal{S}) \subseteq \Gamma \setminus \{\zeta : |\zeta| \le r\}$ *and the resolvent of* $\mathcal{S}$ *satisfies the bound*

$$\|\mathcal{R}(\mathcal{S})\|_\Gamma \le 2\left(\|\mathcal{R}(\mathcal{T})\|_\Gamma + 1\right)^2 (1 + \|\mathcal{T}-\mathcal{S}\|). \tag{12.7}$$

Proof We choose $z \in \Gamma$ and introduce the operator

$$\mathcal{Q} := [(\mathcal{T}-\mathcal{S})\mathcal{R}(\mathcal{T},z)]^2. \tag{12.8}$$

Our hypotheses and Lemma 12.1 imply that $\|\mathcal{Q}\| \le 1/2$. Next, we use the formula

$$\begin{aligned}\mathcal{R}(\mathcal{S},z) &= \mathcal{R}(\mathcal{T},z)[\mathcal{I}-(\mathcal{T}-\mathcal{S})\mathcal{R}(\mathcal{T},z)]^{-1}\\ &= \mathcal{R}(\mathcal{T},z)[\mathcal{I}+(\mathcal{T}-\mathcal{S})\mathcal{R}(\mathcal{T},z)]\sum_{j\in\mathbb{N}_0}\mathcal{Q}^j\end{aligned}$$

to obtain the estimate

$$\|\mathcal{R}(\mathcal{S},z)\| \le 2\|\mathcal{R}(\mathcal{T})\|_\Gamma\left[1 + \|\mathcal{T}-\mathcal{S}\|\cdot\|\mathcal{R}(\mathcal{T})\|_\Gamma\right],$$

which proves the lemma. □

The above lemmas are used in conjunction with the following notion of *ν-convergence* of operators in $\mathcal{B}(\mathbb{X})$.

Definition 12.3 Let $\mathbb{X}$ be a Banach space, and $\mathcal{T}$, $\{\mathcal{T}_n : n \in \mathbb{N}\}$ in $\mathbb{B}(\mathbb{X})$. The sequence $\{\mathcal{T}_n : n \in \mathbb{N}\}$ is said to ν-converge to $\mathcal{T}$, denoted by $\mathcal{T}_n \stackrel{\nu}{\longrightarrow} \mathcal{T}$, if

(i) the sequence $\{\|\mathcal{T}_n\| : n \in \mathbb{N}\}$ is bounded,
(ii) $(\mathcal{T}-\mathcal{T}_n)\mathcal{T} \stackrel{u}{\longrightarrow} 0$,
(iii) $(\mathcal{T}-\mathcal{T}_n)\mathcal{T}_n \stackrel{u}{\longrightarrow} 0$.

Lemmas 12.1 and 12.2 give us the following well known fact (see, for example, [3]).

Proposition 12.4 *If $\mathbb{X}$ is a Banach space, $\mathcal{T}_n \xrightarrow{\nu} \mathcal{T}$ in $\mathcal{B}(\mathbb{X})$ and Γ a closed Jordan curve in $\rho(\mathcal{T}) \setminus \{\zeta : |\zeta| \leq r\}$, $r > 0$, then there is a constant $c > 0$ such that*

$$\Gamma \subseteq \rho(\mathcal{T}) \setminus \{\zeta : |\zeta| \leq r\} \text{ and } \|\mathcal{R}(\mathcal{T}_n)\|_\Gamma \leq c.$$

Proof We apply Lemma 12.1 to $\mathcal{T}$ and $\mathcal{S} = \mathcal{T}_n$ and see that the constant μ in (12.3) corresponding to this choice is bounded independent of n. Now, we choose n large enough so that both $\|(\mathcal{T} - \mathcal{T}_n)\mathcal{T}_n\|$ and $\|(\mathcal{T} - \mathcal{T}_n)\mathcal{T}\|$ do not exceed $(4\mu)^{-1}$. Therefore, Lemma 12.2 gives us the desired conclusion. □

For the next proposition we review the notion of the spectral projection. We assume that the spectrum of $\mathcal{T} \in \mathcal{B}(\mathbb{X})$ has an isolated point λ, and define the *spectral projection* associated with $\mathcal{T}$ and λ by

$$\mathcal{P} := -\frac{1}{2\pi i}\int_\Gamma \mathcal{R}(z)\mathrm{d}z, \tag{12.9}$$

where Γ is a closed Jordan curve in $\rho(\mathcal{T})$ enclosing λ but not any other point of $\sigma(\mathcal{T})$, and we simplify the notation to $\mathcal{R}(z) := \mathcal{R}(\mathcal{T}, z)$. Clearly, $\mathcal{P}$ does not depend on Γ but on λ only. It is also known that $\mathcal{P} \in \mathcal{B}(\mathbb{X})$ and moreover $\mathcal{P}$ is a projection. Indeed, following the proof of Theorem 2.27 on p. 105 of [44], for example, we have that

$$\mathcal{P}^2 = \frac{1}{(2\pi i)^2}\int_\Gamma\int_{\Gamma'} \mathcal{R}(z)\mathcal{R}(\zeta)\mathrm{d}z\mathrm{d}\zeta, \tag{12.10}$$

where Γ' is a closed Jordan curve enclosing λ and completely contained in the domain bounded by Γ. We rewrite the integrand in (12.10) in an equivalent form to obtain the formula

$$\mathcal{P}^2 = \frac{1}{(2\pi i)^2}\int_\Gamma\int_{\Gamma'} \frac{\mathcal{R}(\zeta) - \mathcal{R}(z)}{\zeta - z}\mathrm{d}z\mathrm{d}\zeta. \tag{12.11}$$

From the choice of Γ' we have, by the Cauchy integral formula for $\zeta \in \Gamma'$ and $z \in \Gamma$, that

$$\int_{\Gamma'} \frac{\mathrm{d}\eta}{\eta - z} = 2\pi i, \quad \int_\Gamma \frac{\mathrm{d}\eta}{\zeta - \eta} = 0.$$

Hence, by interchanging the order of integration in (12.11) we obtain

$$\mathcal{P}^2 = -\frac{1}{2\pi i}\int_\Gamma \mathcal{R}(z)\mathrm{d}z,$$

which proves $\mathcal{P}$ is a projection.

When $\mathcal{T}$ is a compact operator on $\mathcal{B}(\mathbb{X})$ and λ is a nonzero eigenvalue of $\mathcal{T}$, a typical choice for the Jordan curve Γ in (12.9) is one that includes only

λ in its interior but not the origin. In that case, $m := \dim R(\mathcal{P}) < \infty$ and the $m \times m$ matrix obtained by restricting the range and domain of $\mathcal{T}$ to $R(\mathcal{P})$ has λ as an eigenvalue of algebraic multiplicity m. Moreover, the least integer l for which $(\mathcal{T} - \lambda\mathcal{I})^l = 0$ on $R(\mathcal{P})$ is called the *ascent* of λ. In other words, l is the smallest positive integer such that $(\mathcal{T} - \lambda\mathcal{I})^l\mathcal{P} = 0$. Generally, the eigenspace $N(\mathcal{T} - \lambda\mathcal{I})$ is a proper subspace of $R(\mathcal{P})$ and equal to it if and only if $l = 1$.

We need to compare two spectral projections of two distinct operators which will be closed in the sense of ν-convergence. This will be accomplished with the next lemma.

Lemma 12.5 *If $\mathbb{X}$ is a Banach space, $\mathcal{T}_1, \mathcal{T}_2 \in \mathcal{B}(\mathbb{X})$, Γ is a closed Jordan curve in $\rho(\mathcal{T}_1) \bigcap \rho(\mathcal{T}_2) \setminus \{\zeta : \zeta \in \mathbb{C}, |\zeta| \leq r\}$ for some $r > 0$ and $\mathcal{P}_1$, $\mathcal{P}_2$ are the corresponding spectral projections of $\mathcal{T}_1$ and $\mathcal{T}_2$ with the same Jordan curve Γ respectively, then*

$$\|(\mathcal{P}_1 - \mathcal{P}_2)\mathcal{P}_1\| \leq \alpha_1 \|(\mathcal{T}_1 - \mathcal{T}_2)|_{R(\mathcal{P}_1)}\| \leq \alpha_2 \|(\mathcal{T}_1 - \mathcal{T}_2)\mathcal{T}_1\|, \tag{12.12}$$

where

$$\alpha_1 := \|\mathcal{R}(\mathcal{T}_1)\|_\Gamma \|\mathcal{R}(\mathcal{T}_2)\|_\Gamma, \quad \alpha_2 := \|\mathcal{R}(\mathcal{T}_1)\|_\Gamma^2 \|\mathcal{R}(\mathcal{T}_2)\|_\Gamma \ell(\Gamma)/(2\pi r)$$

and $\ell(\Gamma)$ is the length of Γ.

Proof For $k = 1, 2$, we have that

$$\mathcal{P}_k = -\frac{1}{2\pi i} \int_\Gamma \mathcal{R}(\mathcal{T}_k, z)\mathrm{d}z.$$

Hence, we obtain that

$$\begin{aligned}
(\mathcal{P}_1 - \mathcal{P}_2)\mathcal{P}_1 &= -\frac{1}{2\pi i} \int_\Gamma \{\mathcal{R}(\mathcal{T}_1, z) - \mathcal{R}(\mathcal{T}_2, z)\}\mathcal{P}_1 \mathrm{d}z \\
&= -\frac{1}{2\pi i} \int_\Gamma \mathcal{R}(\mathcal{T}_2, z)(\mathcal{T}_2 - \mathcal{T}_1)\mathcal{R}(\mathcal{T}_1, z)\mathcal{P}_1 \mathrm{d}z \\
&= -\frac{1}{2\pi i} \int_\Gamma \mathcal{R}(\mathcal{T}_2, z)(\mathcal{T}_2 - \mathcal{T}_1)\mathcal{P}_1\mathcal{R}(\mathcal{T}_1, z)\mathrm{d}z.
\end{aligned}$$

Taking the norms of both sides of this equation leads to the inequality

$$\|(\mathcal{P}_1 - \mathcal{P}_2)\mathcal{P}_1\| \leq \|\mathcal{R}(\mathcal{T}_1)\|_\Gamma \|\mathcal{R}(\mathcal{T}_2)\|_\Gamma \|(\mathcal{T}_2 - \mathcal{T}_1)\mathcal{P}_1\|, \tag{12.13}$$

which leads to the first inequality of (12.5). Next, we estimate the last term on the right-hand side of this inequality. To this end, we note that

$$\begin{aligned}
(\mathcal{T}_2 - \mathcal{T}_1)\mathcal{P}_1 &= -\frac{1}{2\pi i} \int_\Gamma (\mathcal{T}_2 - \mathcal{T}_1)\mathcal{R}(\mathcal{T}_1, z)\mathrm{d}z \\
&= -\frac{1}{2\pi i} \int_\Gamma (\mathcal{T}_2 - \mathcal{T}_1)(\mathcal{T}_1\mathcal{R}(\mathcal{T}_1, z) - \mathcal{I})\frac{\mathrm{d}z}{z}.
\end{aligned}$$

Since zero lies outside the domain bounded by Γ, the Cauchy integral formula says that

$$\int_\Gamma \frac{\mathrm{d}z}{z} = 0,$$

and so the above equation simplifies to

$$(\mathcal{T}_2 - \mathcal{T}_1)\mathcal{P}_1 = -\frac{1}{2\pi i}\int_\Gamma (\mathcal{T}_2 - \mathcal{T}_1)\mathcal{T}_1\mathcal{R}(\mathcal{T}_1, z)\frac{\mathrm{d}z}{z}.$$

Therefore, taking norms of both sides of this inequality gives the inequality

$$\|(\mathcal{T}_2 - \mathcal{T}_1)\mathcal{P}_1\| \leq \frac{\ell(\Gamma)}{2\pi r}\|\mathcal{R}(\mathcal{T}_1)\|_\Gamma \|(\mathcal{T}_2 - \mathcal{T}_1)\mathcal{T}_1\|. \tag{12.14}$$

Combining this inequality with inequality (12.13) proves the result. □

From the above lemma we obtain the following proposition (see, for example, [3, 44, 167, 205]).

Proposition 12.6 *Suppose $\{\mathcal{T}_n : n \in \mathbb{N}\}$ and $\mathcal{T}$ are in $\mathcal{B}(\mathbb{X})$, Γ is a closed Jordan curve contained in $\rho(\mathcal{T}) \setminus \{\zeta : \zeta \in \mathbb{C}, |\zeta| \leq r\}$ for some $r > 0$ and $\mathcal{T}_n \xrightarrow{\nu} \mathcal{T}$ in $\mathcal{B}(\mathbb{X})$. Then, there exist positive constants c_1, c_2 and an $N \in \mathbb{N}$ such that for all $n \geq N$ we have that*

(i) $\|(\mathcal{P} - \mathcal{P}_n)\mathcal{P}\| \leq c_1\|(\mathcal{T} - \mathcal{T}_n)|_{R(\mathcal{P})}\| \leq c_2\|(\mathcal{T} - \mathcal{T}_n)\mathcal{T}\|$,
(ii) $\|(\mathcal{P} - \mathcal{P}_n)\mathcal{P}_n\| \leq c_1\|(\mathcal{T} - \mathcal{T}_n)|_{R(\mathcal{P}_n)}\| \leq c_2\|(\mathcal{T} - \mathcal{T}_n)\mathcal{T}_n\|$.

Proof The proof uses Lemma 12.5 applied to $\mathcal{P}$ and $\mathcal{P}_n$, and also Proposition 12.4. □

Propositions 12.4 and 12.6 supply the well known tools for the presentation in this chapter. Besides the above facts on spectral projections, we need the notion of the *gap* between subspaces and some of its useful properties.

Definition 12.7 Let $\mathbb{Y}_1, \mathbb{Y}_2$ be two closed subspaces of a Banach space $\mathbb{X}$ and set

$$\delta(\mathbb{Y}_1, \mathbb{Y}_2) := \sup\{\mathrm{dist}(y, \mathbb{Y}_2) : y \in \mathbb{Y}_1, \ \|y\| = 1\}.$$

The gap between $\mathbb{Y}_1$ and $\mathbb{Y}_2$ is defined to be

$$\theta(\mathbb{Y}_1, \mathbb{Y}_2) := \max\{\delta(\mathbb{Y}_1, \mathbb{Y}_2), \delta(\mathbb{Y}_2, \mathbb{Y}_1)\}.$$

We make use of the following lemma of Kato (see [44], p. 87; [166], pp. 264–269).

Lemma 12.8 *If* $\dim\mathbb{Y}_1 = \dim\mathbb{Y}_2 < \infty$, *then*

$$\delta(\mathbb{Y}_2, \mathbb{Y}_1) \leq \frac{\delta(\mathbb{Y}_1, \mathbb{Y}_2)}{1 - \delta(\mathbb{Y}_1, \mathbb{Y}_2)}. \tag{12.15}$$

The next lemma is of a different character. Its proof requires the Borsuk antipodal mapping theorem, and is given either in [186], p. 199 or [167], p. 385.

Lemma 12.9 *If* $\dim \mathbb{Y}_1 < \infty$ *and* $\dim \mathbb{Y}_2 > \dim \mathbb{Y}_1$, *then there is an* $x \in \mathbb{Y}_2 \setminus \{0\}$ *such that* $\operatorname{dist}(x, \mathbb{Y}_1) = \|x\|$ *and consequently*

$$\delta(\mathbb{Y}_2, \mathbb{Y}_1) = 1.$$

From this lemma we have the following fact.

Lemma 12.10 *If* $\dim \mathbb{Y}_1 < \infty$ *and* $\theta(\mathbb{Y}_2, \mathbb{Y}_1) < 1$, *then* $\dim \mathbb{Y}_2 = \dim \mathbb{Y}_1$.

Proof Since our hypothesis implies $\delta(\mathbb{Y}_2, \mathbb{Y}_1) < 1$, we conclude by Lemma 12.9 that $\dim \mathbb{Y}_2 \le \dim \mathbb{Y}_1$. But our hypothesis also implies that $\delta(\mathbb{Y}_1, \mathbb{Y}_2) < 1$ and so, indeed, $\dim \mathbb{Y}_2 = \dim \mathbb{Y}_1$. □

We combine this lemma with Proposition 12.6 and conclude the following result.

Proposition 12.11 *If the hypotheses of Proposition 12.6 hold, then there is a positive integer* N *such that for all* $n \ge N$,

$$\dim R(\mathcal{P}) = \dim R(\mathcal{P}_n).$$

Proof For any two projections $\mathcal{P}_1$ and $\mathcal{P}_2 \in \mathcal{B}(\mathbb{X})$ we have that

$$\begin{aligned}\delta(R(\mathcal{P}_1), R(\mathcal{P}_2)) &\le \sup\{\|x - \mathcal{P}_2 x\| : x \in R(\mathcal{P}_1), \|x\| = 1\} \\ &= \sup\{\|\mathcal{P}_1 x - \mathcal{P}_2\mathcal{P}_1 x\| : x \in R(\mathcal{P}_1), \|x\| = 1\} \\ &\le \|\mathcal{P}_1 - \mathcal{P}_2\mathcal{P}_1\| = \|(\mathcal{P}_1 - \mathcal{P}_2)\mathcal{P}_1\|.\end{aligned}$$

Using this inequality on the operators $\mathcal{P}$ and $\mathcal{P}_n$ it follows from Proposition 12.6 that for some N, whenever $n \ge N$, $\theta(R(\mathcal{P}), R(\mathcal{P}_n)) < 1$, because $\dim R(\mathcal{P}) < \infty$ and $\mathcal{T}_n \xrightarrow{\nu} \mathcal{T}$ in $\mathcal{B}(\mathbb{X})$. The result of this proposition then follows from Lemma 12.10. □

The following theorem shows the gap between the spectral subspaces, that is, the nearness of the eigenfunctions.

Theorem 12.12 *If the hypotheses of Proposition 12.6 hold, then there exist a positive constant* c *and an integer* N *such that for all* $n \ge N$,

$$\theta(R(\mathcal{P}_n), R(\mathcal{P})) \le c\|(\mathcal{T} - \mathcal{T}_n)|_{R(\mathcal{P})}\|.$$

In particular, if $\phi \in R(\mathcal{P}_n)$ *with* $\|\phi\| = 1$ *then*

$$\operatorname{dist}(\phi, R(\mathcal{P})) \le c\|(\mathcal{T} - \mathcal{T}_n)|_{R(\mathcal{P})}\|.$$

Proof We know from Proposition 12.11 that $\dim R(\mathcal{P}) = \dim R(\mathcal{P}_n)$ for n sufficiently large. Hence, by Lemma 12.8 we have that

$$\delta(R(\mathcal{P}_n), R(\mathcal{P})) \leq \frac{\delta(R(\mathcal{P}), R(\mathcal{P}_n))}{1 - \delta(R(\mathcal{P}), R(\mathcal{P}_n))} \leq 2\delta(R(\mathcal{P}), R(\mathcal{P}_n)).$$

Moreover, as in the proof of Proposition 12.11 and using Proposition 12.6, we conclude that

$$\theta(R(\mathcal{P}_n), R(\mathcal{P})) \leq 2\delta(R(\mathcal{P}), R(\mathcal{P}_n)) \leq 2\|(\mathcal{P} - \mathcal{P}_n)\mathcal{P}\| \leq c\|(\mathcal{T} - \mathcal{T}_n)|_{R(\mathcal{P})}\|.$$

□

Suppose that λ is an eigenvalue of the operator $\mathcal{T}$ with algebraic multiplicity m and ascent ℓ, isolated by a closed rectifiable curve Γ in $\rho(\mathcal{T})$ from the rest of the spectrum of $\mathcal{T}$ and from zero. We assume that $\mathcal{T}_n \xrightarrow{\nu} \mathcal{T}$. Then the spectrum of $\mathcal{T}_n$ inside Γ consists of m eigenvalues, say $\lambda_{n,1}, \lambda_{n,2}, \ldots, \lambda_{n,m}$, counted according to their algebraic multiplicities (see [3, 44]). We define the arithmetic mean of these eigenvalues by letting

$$\hat{\lambda}_n := \frac{\lambda_{n,1} + \lambda_{n,2} + \cdots + \lambda_{n,m}}{m},$$

and we approximate λ by $\hat{\lambda}_n$.

The next theorem concerns the approximation of eigenvalues.

Theorem 12.13 *If $\mathbb{X}$ is a Banach space, $\mathcal{T}, \mathcal{T}_n \in \mathcal{B}(\mathbb{X})$ for $n \in \mathbb{N}$ and $\mathcal{T}_n \xrightarrow{\nu} \mathcal{T}$, then there exist a constant c and an integer $N \in \mathbb{N}$ such that for all $n \geq N$,*

$$|\lambda - \hat{\lambda}_n| \leq c\|(\mathcal{T} - \mathcal{T}_n)|_{R(\mathcal{P})}\|$$

and

$$|\lambda - \lambda_{n,j}|^{\ell} \leq c\|(\mathcal{T} - \mathcal{T}_n)|_{R(\mathcal{P})}\|.$$

Proof Let $\hat{\mathcal{P}}_n := \mathcal{P}_n|_{R(\mathcal{P})}$. Note that there exists N such that for all $n \in \mathbb{N}, n \geq N$, $\hat{\mathcal{P}}_n$ is a surjective isomorphism from $R(\mathcal{P})$ to $R(\mathcal{P}_n)$, and $\{\|\hat{\mathcal{P}}_n^{-1}\| : n \in \mathbb{N}\}$ is bounded. Let $\hat{\mathcal{T}} := \mathcal{T}|_{R(\mathcal{P})}$, and $\hat{\mathcal{T}}_n := \hat{\mathcal{P}}_n^{-1}\mathcal{T}_n\hat{\mathcal{P}}_n$. Then $\sigma(\hat{\mathcal{T}}) = \{\lambda\}$ and $\sigma(\hat{\mathcal{T}}_n) = \{\lambda_{n,1}, \lambda_{n,2}, \ldots, \lambda_{n,m}\}$. Thus, we have that

$$\begin{aligned}
|\lambda - \hat{\lambda}_n| &= \frac{1}{m}\left|\text{trace}(\hat{\mathcal{T}} - \hat{\mathcal{T}}_n)\right| \leq \|\hat{\mathcal{T}} - \hat{\mathcal{T}}_n\| \\
&= \|\hat{\mathcal{P}}_n^{-1}\mathcal{P}_n(\mathcal{T} - \mathcal{T}_n)|_{R(\mathcal{P})}\| \leq c\|(\mathcal{T} - \mathcal{T}_n)|_{R(\mathcal{P})}\|.
\end{aligned}$$

To prove the second estimate, we note that the ascent of λ is ℓ, which ensures that $(\lambda\mathcal{I}|_{R(\mathcal{P})} - \hat{\mathcal{T}})^{\ell} = 0$. By virtue of this equation, we observe that

$$\begin{aligned}
|\lambda - \lambda_{n,j}|^{\ell} &\leq \|(\lambda\mathcal{I}|_{R(\mathcal{P})} - \hat{\mathcal{T}}_n)^{\ell}\| \\
&= \|(\lambda\mathcal{I}|_{R(\mathcal{P})} - \hat{\mathcal{T}}_n)^{\ell} - (\lambda\mathcal{I}|_{R(\mathcal{P})} - \hat{\mathcal{T}})^{\ell}\| \\
&\leq \alpha_n\|\hat{\mathcal{T}} - \hat{\mathcal{T}}_n\|,
\end{aligned}$$

where

$$\alpha_n = \sum_{k\in\mathbb{Z}_\ell} \|\lambda\mathcal{I}_{R(\mathcal{P})} - \hat{\mathcal{P}}_n^{-1}\mathcal{T}_n\hat{\mathcal{P}}_n\|^{\ell-1-k}\|\lambda\mathcal{I}_{R(\mathcal{P})} - \mathcal{T}|_{R(\mathcal{P})}\|^k.$$

Since $\|\hat{\mathcal{P}}_n\|$, $\|\hat{\mathcal{P}}_n^{-1}\|$ and $\|\mathcal{T}_n\|$ are all bounded, α_n can be bounded by a positive constant c independent of n. This leads to the desired estimate, and completes the proof. □

12.3 A multiscale collocation method

In this section we develop fast multiscale collocation methods for solving the eigen-problem of a compact integral operator with a weakly singular kernel. Specifically, we let d be a positive integer and assume that Ω is a compact set of the d-dimensional Euclidean space $\mathbb{R}^d$. For a compact linear integral operator $\mathcal{K}$ on $L^\infty(\Omega)$ defined by

$$\mathcal{K}\phi(s) := \int_\Omega K(s,t)\phi(t)\mathrm{d}t, \quad t \in \Omega,$$

with a weakly singular kernel K, we consider the following eigen-problem. Find $\lambda \in \sigma(\mathcal{K}) \setminus \{0\}$ and $\phi \in L^\infty(\Omega)$ with $\|\phi\| = 1$ such that

$$\mathcal{K}\phi = \lambda\phi. \tag{12.16}$$

In this case, the Banach space $\mathbb{X}$ in the abstract setting described in the last section is chosen as the space $L^\infty(\Omega)$. Hence, in the rest of this section, we always have $\mathbb{X} = L^\infty(\Omega)$ and $\mathbb{V} = C(\Omega)$. By $\mathbb{V}^*$ we denote the dual space of $\mathbb{V}$. For $\ell \in \mathbb{V}^*$ and $v \in \mathbb{V}$, we use $\langle \ell, v\rangle$ to stand for the value of the linear functional ℓ evaluated at the function v, and $\|\ell\|$, $\|v\|$ for their respective norms. We also use $(\cdot,\cdot)$ to denote the inner product in $L^2(\Omega)$. For $s \in \Omega$, by δ_s we denote the linear functional in $\mathbb{V}^*$ defined for $v \in \mathbb{V}$ by the equation $\langle \delta_s, v\rangle = v(s)$. We need to evaluate δ_s on functions in $\mathbb{X}$. As in [21], we take the norm-preserving extension of δ_s to $\mathbb{X}$ and use the same notation for the extension.

A multiscale scheme is based on a multiscale partition of the set Ω, a multiscale subspace decomposition of the space $\mathbb{X}$ and a multiscale basis of the space (cf. [69]). We first require that there is a family of partitions $\{\Omega_n : n \in \mathbb{N}_0\}$ of Ω satisfying

$$\Omega_n = \{\Omega_{ni} : i \in \mathbb{Z}_{e(n)}\} \text{ and } \bigcup_{i\in\mathbb{Z}_{e(n)}} \Omega_{ni} = \Omega,$$

$$\operatorname{meas}(\Omega_{ni} \cap \Omega_{ni'}) = 0, \ \ i, i' \in \mathbb{Z}_{e(n)}, i \neq i',$$

where the sets Ω_{ni} are star-shaped and $e(n)$ denotes the cardinality of Ω_n. We then assume that there is a sequence of finite-dimensional subspaces $\mathbb{X}_n$,

$n \in \mathbb{N}_0$, of $\mathbb{X}$ which have the nested property, namely, $\mathbb{X}_{n-1} \subset \mathbb{X}_n$, $n \in \mathbb{N}$. Thus, a subspace $\mathbb{W}_n \subset \mathbb{X}_n$ can be defined such that $\mathbb{X}_n$ is an orthogonal direct sum of $\mathbb{X}_{n-1}$ and $\mathbb{W}_n$. As a result, for each $n \in \mathbb{N}_0$, we have a multiscale decomposition

$$\mathbb{X}_n = \mathbb{X}_0 \oplus \mathbb{W}_1 \oplus \mathbb{W}_2 \oplus \cdots \oplus \mathbb{W}_n,$$

where $\mathbb{W}_0 := \mathbb{X}_0$. Let $w(i) := \dim(\mathbb{W}_i)$, $i \in \mathbb{N}_0$ and $s(n) := \dim(\mathbb{X}_n)$, $n \in \mathbb{N}_0$. Then $s(n) = \sum_{i \in \mathbb{Z}_{n+1}} w(i)$. We also assume that there is a basis $w_{ij}, j \in \mathbb{Z}_{w(i)}$, for the spaces $\mathbb{W}_i$, that is,

$$\mathbb{W}_i = \operatorname{span}\{w_{ij} : j \in \mathbb{Z}_{w(i)}\}, \ i \in \mathbb{N}_0$$

and that there exist positive integers h and r such that for any $i > h$ and $j \in \mathbb{Z}_{w(i)}$ with $j = \nu r + l$ for some $\nu \in \mathbb{N}_0$ and $l \in \mathbb{Z}_r$,

$$w_{ij}(x) = 0, \ x \notin \Omega_{i-h,\nu}. \tag{12.17}$$

Letting $S_{ij} := \Omega_{i-h,\nu}$, condition (12.17) means that the support of the basis function w_{ij} is contained in S_{ij}. The multiscale property demands that there is a positive integer $\mu > 1$ and positive constants c_1, c_2 such that for $n \in \mathbb{N}_0$,

$$c_1\mu^{-n/d} \le d_n \le c_2\mu^{-n/d}, \ c_1\mu^n \le \dim \mathbb{X}_n \le c_2\mu^n \text{ and } c_1\mu^n \le \dim \mathbb{W}_n \le c_2\mu^n, \tag{12.18}$$

where $d_n := \max\{d(\Omega_{ni}) : i \in \mathbb{Z}_{e(n)}\}$ and the notation $d(A)$ denotes the diameter of the set A.

To define a multiscale collocation scheme we also need linear functionals $\ell_{i'j'}$ in $\mathbb{V}^*, j' \in \mathbb{Z}_{w(i')}, i' \in \mathbb{Z}_{n+1}$. Each $\ell_{i'j'}$ is a finite sum of point evaluations

$$\ell_{i'j'} = \sum_{s \in \hat{S}_{i'j'}} c_s \delta_s,$$

where c_s are constants and $\hat{S}_{i'j'}$ is a finite subset of distinct points in $S_{i'j'}$ having a constant cardinality. Let

$$\mathbb{U}_n := \{(i,j) : j \in \mathbb{Z}_{w(i)}, \ i \in \mathbb{Z}_{n+1}\}.$$

The linear functionals and multiscale bases are required to satisfy the vanishing moment condition that for any polynomial p of total degree less than or equal to $k-1$,

$$\langle \ell_{ij}, p \rangle = 0, \ \ (w_{ij}, p) = 0, \ (i,j) \in \mathbb{U}_n, \ i \ge 1, \tag{12.19}$$

the boundedness property

$$\|\ell_{ij}\| + \|w_{ij}\| \le c, \ \ (i,j) \in \mathbb{U}_n, \tag{12.20}$$

where c is a positive constant independent of i, j and n, and the requirement that for any $i, i' \in \mathbb{N}_0$,

$$\langle \ell_{i'j'}, w_{ij} \rangle = \delta_{i,i'}\delta_{j,j'}, \ (i,j),(i',j') \in \mathbb{U}_n, \ i \le i', \tag{12.21}$$

with $\delta_{i,i'}$ being the Kronecker delta.

The collocation scheme for solving the eigen-problem (12.16) is to find $\lambda_n \in \mathbb{C}$ and $\phi_n \in \mathbb{X}_n$ with $\|\phi_n\| = 1$ such that for any $(i',j') \in \mathbb{U}_n$,

$$\langle \ell_{i'j'}, \mathcal{K}\phi_n \rangle = \lambda_n \langle \ell_{i'j'}, \phi_n \rangle. \tag{12.22}$$

Since $\phi_n \in \mathbb{X}_n$, we write

$$\phi_n = \sum_{(i,j)\in\mathbb{U}_n} \phi_{ij} w_{ij}, \tag{12.23}$$

and substitute (12.23) into (12.22) to obtain the linear system

$$\sum_{(i,j)\in\mathbb{U}_n} \phi_{ij} \langle \ell_{i'j'}, \mathcal{K}w_{ij} \rangle = \lambda_n \sum_{(i,j)\in\mathbb{U}_n} \phi_{ij} \langle \ell_{i'j'}, w_{ij} \rangle, \ \ (i',j') \in \mathbb{U}_n. \tag{12.24}$$

Using the following notations:

$\mathbf{E}_n := [\langle \ell_{i'j'}, w_{ij} \rangle : (i',j'),(i,j) \in \mathbb{U}_n]$, $\mathbf{K}_n := [\langle \ell_{i'j'}, \mathcal{K}w_{ij} \rangle : (i',j'),(i,j) \in \mathbb{U}_n]$

and $\Phi_n := [\phi_{ij} : (i,j) \in \mathbb{U}_n]$, the system (12.24) is written in the matrix form

$$\mathbf{K}_n \Phi_n = \lambda_n \mathbf{E}_n \mathbf{\Phi}_n. \tag{12.25}$$

In the above generalized eigen-problem, the matrix $\mathbf{E}_n$ is an upper triangular sparse matrix having only $\mathcal{O}(s(n) \log s(n))$ nonzero entries due to the construction of the basis functions and their corresponding collocation functionals. Specifically, when the elements $S_{i'j'}$ and S_{ij} are disjoint, the entry $\langle \ell_{i'j'}, w_{ij} \rangle$ of $\mathbf{E}_n$ is zero. The matrix $\mathbf{K}_n$ is still a full matrix but it is numerically sparse. Our next task is to use the truncation strategy developed in [69] for the matrix $\mathbf{K}_n$ to formulate a fast algorithm for solving the eigen-problem (12.25).

For analysis purposes, it is convenient to work with the functional analytic form of the eigen-problem (12.25). For this purpose, we let $\pi_n : \mathbb{X} \to \mathbb{X}_n$ denote the bounded linear projection operator defined by

$$\langle \ell_{ij}, \pi_n x \rangle = \langle \ell_{ij}, x \rangle, \ (i,j) \in \mathbb{U}_n. \tag{12.26}$$

With this operator, we define $\mathcal{K}_n : \mathbb{X} \to \mathbb{X}$ by $\mathcal{K}_n := \pi_n \mathcal{K} \pi_n$. Clearly, $\mathcal{K}_n$ is a bounded linear operator. Thus, the eigen-problem (12.25) is written in the operator form

$$\mathcal{K}_n \phi_n = \lambda_n \phi_n. \tag{12.27}$$

We remark that as in Chapter 7 (cf. [69]), the projection operator π_n is required to satisfy the property that there exists a positive constant c such that for $v \in W^{k,\infty}(\Omega)$ and for all $n \in \mathbb{N}$,

$$\|v - \pi_n v\| \leq c \inf_{v_n \in \mathbb{X}_n} \|v - v_n\| \leq c\mu^{-kn/d}\|v\|_{k,\infty}, \tag{12.28}$$

where $\|\cdot\|$ and $\|\cdot\|_{k,\infty}$ denote the norm in $\mathbb{X}$ and in $W^{k,\infty}(\Omega)$, respectively. This condition is fulfilled when $\mathbb{X}_n$ is chosen as the space of piecewise polynomials of total degree $k-1$.

We now assume that the integral kernel of the operator $\mathcal{K}$ is weakly singular in the sense that for $s, t \in [0,1], s \neq t$ and a positive integer $k > 0$, the kernel K has continuous partial derivatives $D_s^\alpha D_t^\beta K(s,t)$ for $|\alpha| \leq k, |\beta| \leq k$, and there exist positive constants σ and c with $\sigma < d$ such that for $|\alpha| = |\beta| = k$,

$$\left|D_s^\alpha D_t^\beta K(s,t)\right| \leq \frac{c}{|s-t|^{\sigma+|\alpha|+|\beta|}}.$$

We truncate the matrix $\mathbf{K}_n$ according to the singularity of the kernel. As in Chapter 7 (cf. [69]), we partition the matrix $\mathbf{K}_n$ into a block matrix

$$\mathbf{K}_n := [\mathbf{K}_{i'i} : i', i \in \mathbb{Z}_{n+1}], \text{ with } \mathbf{K}_{i'i} := [\langle \ell_{i'j'}, \mathcal{K}w_{ij}\rangle : j' \in \mathbb{Z}_{w(i')}, j \in \mathbb{Z}_{w(i)}],$$

and truncate the block $\mathbf{K}_{i'i}$ by using a truncation parameter $\epsilon = \epsilon_{i'i}^n$, which will be described later. Specifically, we define a truncation matrix

$$\tilde{\mathbf{K}}_{i'i} := [\tilde{K}_{i'j',ij} : j' \in \mathbb{Z}_{w(i')}, j \in \mathbb{Z}_{w(i)}] \tag{12.29}$$

by setting

$$\tilde{K}_{i'j',ij} := \begin{cases} \langle \ell_{i'j'}, \mathcal{K}w_{ij}\rangle, & \operatorname{dist}(S_{i'j'}, S_{ij}) \leq \epsilon, \\ 0, & \text{otherwise}. \end{cases} \tag{12.30}$$

Using the truncation (12.30), problem (12.24) becomes

$$\sum_{(i,j)\in\mathbb{U}_n} \tilde{\phi}_{ij}\tilde{K}_{i'j',ij} = \tilde{\lambda}_n \sum_{(i,j)\in\mathbb{U}_n} \tilde{\phi}_{ij}\langle \ell_{i'j'}, w_{ij}\rangle, \ (i',j') \in \mathbb{U}_n. \tag{12.31}$$

Let

$$\tilde{\mathbf{K}}_n := [\tilde{\mathbf{K}}_{i'i} : i', i \in \mathbb{Z}_{n+1}], \ \tilde{\mathbf{\Phi}}_n := [\tilde{\phi}_{ij} : (i,j) \in \mathbb{U}_n].$$

The eigen-problem (12.31) can then be written as the eigen-problem of the truncated matrix

$$\tilde{\mathbf{K}}_n\tilde{\mathbf{\Phi}}_n = \tilde{\lambda}_n \mathbf{E}_n \tilde{\mathbf{\Phi}}_n. \tag{12.32}$$

There is no need to compress the matrix $\mathbf{E}_n$ since it is already sparse. In fact, if an entry of matrix $\mathbf{K}_n$ is truncated to zero according to our proposed truncation strategy, the corresponding entry of the matrix $\mathbf{E}_n$ is already zero.

Eigen-problem (12.32) may be expressed in the operator form. Let $\hat{\mathcal{K}}_n$ be the bounded linear operator from $\mathbb{X}_n$ to $\mathbb{X}_n$ having the matrix representation $\mathbf{E}_n^{-1}\tilde{\mathbf{K}}_n$ in the basis w_{ij}, $(i,j) \in \mathbb{U}_n$. We then define a bounded linear operator $\tilde{\mathcal{K}}_n : \mathbb{X} \to \mathbb{X}$ by $\tilde{\mathcal{K}}_n := \hat{\mathcal{K}}_n \pi_n$. Solving eigen-problem (12.32) is now equivalent to finding $\tilde{\lambda}_n \in \mathbb{C}$ and $\tilde{\phi}_n = \sum_{(i,j)\in\mathbb{U}_n} \tilde{\phi}_{ij} w_{ij} \in \mathbb{X}_n$ with $\|\tilde{\phi}_n\| = 1$ such that

$$\tilde{\mathcal{K}}_n \tilde{\phi}_n = \tilde{\lambda}_n \tilde{\phi}_n. \tag{12.33}$$

Eigen-problem (12.32) leads to a fast method for solving the original eigen-problem. We study the convergence order and computational complexity of the fast method.

12.4 Analysis of the fast algorithm

We provide in this section analysis for the convergence order and computational complexity of the fast algorithm described in the last section. Specifically, we show that the proposed method has the optimal convergence order (up to a logarithmic factor) and almost linear computational complexity.

For real numbers b and b', we choose the truncation parameters $\epsilon^n_{i'i}$, $i', i \in \mathbb{Z}_{n+1}$ to satisfy the condition

$$\epsilon^n_{i'i} \geq \max\{a\mu^{[-n+b(n-i)+b'(n-i')]/d}, r(d_i + d_{i'})\}, \; i, i' \in \mathbb{Z}_{n+1}, \tag{12.34}$$

for some constants $a > 0$ and $r > 1$. For any real numbers α and β and a positive integer n, we define a function

$$\mu[\alpha, \beta; n] := \sum_{i\in\mathbb{Z}_{n+1}} \mu^{\alpha i/d} \sum_{i'\in\mathbb{Z}_{n+1}} \mu^{\beta i'/d}.$$

We recall a result from Lemma 7.11 (cf. Lemma 4.2 of [69]), which estimates the difference between the operators $\tilde{\mathcal{K}}_n$ and $\mathcal{K}_n$.

Lemma 12.14 *If $0 < \sigma' < \min\{2k, d - \sigma\}$, $\eta := 2k - \sigma'$ and the truncation parameter $\epsilon^n_{i',i}$ is chosen according to (12.34) for any real numbers b, b', then there exists a positive constant c such that for all $n \in \mathbb{N}$,*

$$\|\mathcal{K}_n - \tilde{\mathcal{K}}_n\| \leq c\mu[k - b\eta, k - b'\eta; n](n+1)\mu^{-\sigma' n/d}$$

and

$$\|\mathcal{K}_n - \tilde{\mathcal{K}}_n\|_{W^{k,\infty}\to\mathbb{X}} \leq c\mu[2k - b\eta, k - b'\eta; n](n+1)\mu^{-(k+\sigma')n/d},$$

where $\|\cdot\|_{W^{k,\infty}\to\mathbb{X}}$ denotes the norm of the operator from $W^{k,\infty}(\Omega)$ to $\mathbb{X}$.

The following lemma concerns the ν-convergence of the truncated operators $\tilde{\mathcal{K}}_n$ to $\mathcal{K}$ on $\mathbb{X}$.

Lemma 12.15 *If* $0 < \sigma' < \min\{2k, d - \sigma\}$, $\eta := 2k - \sigma'$ *and the truncation parameter* $\epsilon^n_{i'i}$ *is chosen according to (12.34) with* $b > \frac{k-\sigma'}{\eta}$, $b' > \frac{k-\sigma'}{\eta}$, $b + b' > 1$, *then* $\tilde{\mathcal{K}}_n \xrightarrow{\nu} \mathcal{K}$ *on* $\mathbb{X}$ *as* $n \to \infty$.

Proof We prove that $\tilde{\mathcal{K}}_n$ satisfies the three conditions in the definition of ν-convergence. Note that for any real numbers α, β and e with $e > \max\{0, \alpha, \beta, \alpha + \beta\}$,

$$\lim_{n\to\infty} \mu[\alpha, \beta; n](n+1)\mu^{-en/d} = 0.$$

The choice of parameters in this lemma ensures that

$$\lim_{n\to\infty} \mu[k - b\eta, k - b'\eta; n](n+1)\mu^{-\sigma' n/d} = 0. \tag{12.35}$$

This, with Lemma 12.14, yields that

$$\lim_{n\to\infty} \|\mathcal{K}_n - \tilde{\mathcal{K}}_n\| = 0, \tag{12.36}$$

which with the boundedness of $\{\|\mathcal{K}_n\| : n \in \mathbb{N}\}$ ensures that $\{\|\tilde{\mathcal{K}}_n\| : n \in \mathbb{N}\}$ is also bounded. That is, there is a constant c such that for any $n \in \mathbb{N}$,

$$\|\tilde{\mathcal{K}}_n\| \le c. \tag{12.37}$$

It is known that for any $v \in \mathbb{X}$,

$$\lim_{n\to\infty} \|\mathcal{K}_n v - \mathcal{K}v\| = 0.$$

This, with (12.36), leads to

$$\lim_{n\to\infty} \|\tilde{\mathcal{K}}_n v - \mathcal{K}v\| = 0.$$

By using the compactness of the operator $\mathcal{K}$, we conclude that

$$\lim_{n\to\infty} \|(\tilde{\mathcal{K}}_n - \mathcal{K})\mathcal{K}\| = 0. \tag{12.38}$$

Moreover, we have that

$$\begin{aligned}\|(\tilde{\mathcal{K}}_n - \mathcal{K})\tilde{\mathcal{K}}_n\| &\le \|(\tilde{\mathcal{K}}_n - \mathcal{K}_n)\tilde{\mathcal{K}}_n\| + \|(\pi_n - \mathcal{I})\mathcal{K}\tilde{\mathcal{K}}_n\| \\ &\le [\|\tilde{\mathcal{K}}_n - \mathcal{K}_n\| + \|(\pi_n - \mathcal{I})\mathcal{K}\|]\|\tilde{\mathcal{K}}_n\|.\end{aligned}$$

Since $\lim_{n\to\infty} \|\pi_n v - v\| = 0$ for any $v \in \mathbb{X}$ and since $\mathcal{K}$ is compact, $\lim_{n\to\infty} \|(\pi_n - \mathcal{I})\mathcal{K}\| = 0$. From this with (12.36), (12.37) and the inequality above, we conclude that

$$\lim_{n\to\infty} \|(\tilde{\mathcal{K}}_n - \mathcal{K})\tilde{\mathcal{K}}_n\| = 0. \tag{12.39}$$

Combining (12.37)–(12.39) yields the result of this lemma. □

We now consider the spectral projection associated with $\mathcal{K}$ and $\lambda \in \sigma(\mathcal{K})$:

$$\mathcal{P} := -\frac{1}{2\pi i}\int_\Gamma (\mathcal{K} - z\mathcal{I})^{-1}\mathrm{d}z,$$

where Γ is a closed rectifiable curve in $\rho(\mathcal{K})$ enclosing λ but no other point of $\sigma(\mathcal{K})$.

Lemma 12.16 *If $R(\mathcal{P}) \subset W^{k,\infty}(\Omega)$, then there exists a positive constant c such that for all $n \in \mathbb{N}$ and for all $v \in R(\mathcal{P})$,*

$$\|(\mathcal{I} - \pi_n)\mathcal{K}v\| \le c\mu^{-kn/d}\|v\|_{k,\infty}.$$

Proof Since $R(\mathcal{P})$ is invariant under $\mathcal{K}$, we have that for any $v \in R(\mathcal{P})$, $\mathcal{K}v \in R(\mathcal{P}) \subset W^{k,\infty}(\Omega)$. Because all norms are equivalent in a finite-dimensional space, we see that there is a positive constant c such that for any $v \in R(\mathcal{P})$,

$$\|\mathcal{K}v\|_{k,\infty} \le c\|\mathcal{K}v\| \le c\|\mathcal{K}\|\,\|v\|_{k,\infty}.$$

Thus, the desired result follows from the above inequality and (12.28). □

Lemma 12.17 *Let $R(\mathcal{P}) \subset W^{k,\infty}(\Omega)$. If $0 < \sigma' < \min\{2k, d-\sigma\}$, $\eta := 2k - \sigma'$ and the truncation parameter $\epsilon^n_{i'i}$ is chosen according to (12.34) for some constants $a > 0$ and $r > 1$, with b and b' satisfying one of the following conditions:*

(i) $b > 1$, $b' > \frac{k-\sigma'}{\eta}$, $b + b' > 1 + \frac{k}{\eta}$,
(ii) $b = 1$, $b' > \frac{k-\sigma'}{\eta}$, $b + b' > 1 + \frac{k}{\eta}$; $b > 1$, $b' = \frac{k-\sigma'}{\eta}$, $b + b' > 1 + \frac{k}{\eta}$; *or* $b > 1$, $b' > \frac{k-\sigma'}{\eta}$, $b + b' = 1 + \frac{k}{\eta}$,
(iii) $b = 1$, $b' = \frac{k}{\eta}$ *or* $b = \frac{2k}{\eta}$, $b' = \frac{k-\sigma'}{\eta}$,

then there exists a positive constant c such that for all $n \in \mathbb{N}$ and for all $v \in R(\mathcal{P})$,

$$\|(\tilde{\mathcal{K}}_n - \mathcal{K})v\| \le c(s(n))^{-k/d}\log^\tau s(n)\|v\|_{k,\infty},$$

where $\tau = 0$ in case (i), $\tau = 1$ in case (ii) and $\tau = 2$ in case (iii).

Proof We prove this result by using the inequality

$$\|(\tilde{\mathcal{K}}_n - \mathcal{K})v\| \le \|(\tilde{\mathcal{K}}_n - \mathcal{K}_n)v\| + \|(\mathcal{K}_n - \mathcal{K})v\|. \tag{12.40}$$

It follows from Lemma 12.14 that

$$\|\tilde{\mathcal{K}}_n - \mathcal{K}_n\|_{W^{k,\infty}\to\mathbb{X}} \le c\mu[2k - b\eta, k - b'\eta; n](n+1)\mu^{-\sigma' n/d}\mu^{-kn/d}.$$

Note that for any real numbers α, β and e with $e > 0$,

$$\mu[\alpha, \beta; n](n+1)\mu^{-en/d} = \begin{cases} o(1), & \text{if } e > \max\{\alpha, \beta, \alpha+\beta\}, \\ \mathcal{O}(n), & \text{if } \alpha = e,\ \beta < e,\ \alpha+\beta < e, \\ & \text{or } \alpha < e,\ \beta = e,\ \alpha+\beta < e, \\ & \text{or } \alpha < e,\ \beta < e,\ \alpha+\beta = e, \\ \mathcal{O}(n^2), & \text{if } \alpha = 0,\ \beta = e, \text{ or } \alpha = e,\ \beta = 0, \end{cases}$$

as $n \to \infty$. We then conclude by choosing $\alpha := 2k - b\eta$, $\beta := k - b'\eta$ and $e := \sigma'$ that

$$\mu[2k - b\eta, k - b'\eta; n](n+1)\mu^{-\sigma' n/d} = \begin{cases} o(1), & \text{in case (i)}, \\ \mathcal{O}(n), & \text{in case (ii)}, \\ \mathcal{O}(n^2), & \text{in case (iii)}. \end{cases}$$

Hence, we see that there exist a constant c_1 and a positive integer N such that for all $n \geq N$,

$$\|\tilde{\mathcal{K}}_n - \mathcal{K}_n\|_{W^{k,\infty} \to \mathbb{X}} \leq c_1 \mu^{-kn/d} n^{\tau},$$

where $\tau = 0$ in case (i), $\tau = 1$ in case (ii) and $\tau = 2$ in case (iii). That is,

$$\|(\tilde{\mathcal{K}}_n - \mathcal{K}_n)v\| \leq c_1 \mu^{-kn/d} n^{\tau} \|v\|_{k,\infty}. \tag{12.41}$$

Moreover, it follows from (12.28) and Lemma 12.16 that there exists a positive constant c_2 such that for all $n \in \mathbb{N}$,

$$\|(\mathcal{K}_n - \mathcal{K})v\| \leq \|\pi_n \mathcal{K}\| \|(\pi_n - \mathcal{I})v\| + \|(\pi_n - \mathcal{I})\mathcal{K}v\| \leq c_2 \mu^{-kn/d} \|v\|_{k,\infty}. \tag{12.42}$$

Combining estimates (12.40)–(12.42) and the relation $s(n) \sim \mu^n$, we obtain the estimate of this theorem. □

Suppose that rank $\mathcal{P} = m < \infty$. Note that $\tilde{\mathcal{K}}_n \xrightarrow{\nu} \mathcal{K}$ on $\mathbb{X}$ as $n \to \infty$. As described in Section 12.2, when n is sufficiently large the spectrum of $\tilde{\mathcal{K}}_n$ inside Γ consists of m eigenvalues $\tilde{\lambda}_{i,n}$, $i = 1, 2, \ldots, m$, counting algebraic multiplicities. Let

$$\tilde{\mathcal{P}}_n := -\frac{1}{2\pi i} \int_{\Gamma} (\tilde{\mathcal{K}}_n - z\mathcal{I})^{-1} \mathrm{d}z$$

be the spectral projection associated with $\tilde{\mathcal{K}}_n$ and its spectra inside Γ. Thus, $\dim R(\tilde{\mathcal{P}}_n) = \dim R(\mathcal{P}) = m$. We define the quantity

$$C(\mathcal{P}) := \sup\{\|\phi\|_{k,\infty} : \phi \in R(\mathcal{P}), \|\phi\|_{\infty} = 1\}.$$

Theorem 12.18 *If the assumptions of Lemma 12.17 hold, then there exist a positive constant c and a positive integer N such that for all $n \geq N$,*

$$\hat{\delta}(R(\tilde{\mathcal{P}}_n), R(\mathcal{P})) \leq c(s(n))^{-k/d} \log^{\tau} s(n) C(\mathcal{P}). \tag{12.43}$$

In particular, for any $\tilde{\phi}_n \in R(\tilde{\mathcal{P}}_n)$ with $\|\tilde{\phi}_n\| = 1$, we have that

$$\mathrm{dist}(\tilde{\phi}_n, R(\mathcal{P})) \leq c(s(n))^{-k/d} \log^{\tau} s(n) C(\mathcal{P}),$$

where $\tau = 0$ in case (i), $\tau = 1$ in case (ii) and $\tau = 2$ in case (iii).

Proof It is easy to verify that the choice of truncation parameters in case (i), (ii) or (iii) satisfies the hypothesis of Lemma 12.15. It follows from Lemma 12.15 that $\tilde{\mathcal{K}}_n \xrightarrow{\nu} \mathcal{K}$ as $n \to \infty$. Using Theorem 12.12, we see that

$$\hat{\delta}(R(\tilde{\mathcal{P}}_n), R(\mathcal{P})) \leq c \sup\{\|(\tilde{\mathcal{K}}_n - \mathcal{K})\phi\| : \phi \in R(\mathcal{P}), \|\phi\| = 1\}.$$

Now, by using Lemma 12.17, we conclude the desired estimate (12.43). □

We define the arithmetic mean of the eigenvalues $\tilde{\lambda}_{i,n}$, $i = 1, 2, \ldots, m$, by

$$\hat{\tilde{\lambda}}_n := \frac{\tilde{\lambda}_{1,n} + \ldots + \tilde{\lambda}_{m,n}}{m}.$$

Theorem 12.19 *If the assumptions of Lemma 12.17 hold, then there exist a positive constant c and a positive integer N such that for all $n \geq N$,*

$$|\lambda - \hat{\tilde{\lambda}}_n| \leq c(s(n))^{-k/d}(\log s(n))^{\tau} C(\mathcal{P}),$$
$$|\lambda - \tilde{\lambda}_{n,j}|^{\ell} \leq c(s(n))^{-k/d}(\log s(n))^{\tau} C(\mathcal{P}), \quad j = 1, 2, \ldots, m.$$

In particular, if λ is simple, that is, $m = 1$ and $\ell = 1$, then

$$|\lambda - \tilde{\lambda}_n| \leq c(s(n))^{-k/d}(\log s(n))^{\tau} C(\mathcal{P}),$$

where $\tau = 0$ in case (i), $\tau = 1$ in case (ii) and $\tau = 2$ in case (iii).

Proof The results of this theorem follow from Theorem 12.13 and Lemma 12.17. □

Theorem 12.20 *Let $R(\mathcal{P}) \subset W^{k,\infty}(\Omega)$. Suppose that $0 < \sigma' < \min\{2k, d - \sigma\}$ and $\eta := 2k - \sigma'$ and the truncation parameters $\epsilon^n_{i'i}$, $i', i \in \mathbb{Z}_{n+1}$ are chosen such that*

$$\epsilon^n_{i'i} = \max\{a\mu^{[-n+b(n-i)+b'(n-i')]/d}, r(d_i + d_{i'})\}, \quad i, i' \in \mathbb{Z}_{n+1}, \tag{12.44}$$

for some constants $a > 0$ and $r > 1$, with $b = 1$ and $\frac{k}{\eta} \leq b' \leq 1$. Then the number of nonzero entries of matrix $\tilde{\mathbf{K}}_n$ is

$$\mathcal{N}(\tilde{\mathbf{K}}_n) = \mathcal{O}(s(n) \log^{\tau} s(n)),$$

where $\tau = 1$ except for $b' = 1$, in which case $\tau = 2$, and the estimates in Theorems 12.18 and 12.19 hold with $\tau = 1$ except for $b' = \frac{k}{\eta}$, in which case $\tau = 2$.

Proof It is shown in Theorem 7.15 (cf. Theorem 4.6 of [69]) that if the parameters are chosen such that

$$\epsilon^n_{i'i} \leq \max\{a\mu^{[-n+b(n-i)+b'(n-i')]/d}, r(d_i + d_{i'})\},\ i, i' \in \mathbb{Z}_{n+1},$$

for some constants $a > 0$ and $r > 1$ with b and b' not larger than one, the number of nonzero entries of matrix $\tilde{\mathbf{K}}_n$ is in the order $\mathcal{O}(s(n) \log^\tau s(n))$, where $\tau = 1$ except for $b = b' = 1$, in which case $\tau = 2$. This, together with Theorems 12.18 and 12.19, yields the results of this theorem. □

The above theorem means that the scheme (12.31) (or (12.32), (12.33)) leads to a fast numerical algorithm for solving the eigen-problem (12.16), which has both optimal order (up to a logarithmic factor) of convergence and computational complexity.

12.5 A power iteration algorithm

The matrix compression technique described and analyzed in the previous sections provides a basis for developing various fast numerical solvers for the eigen-problem (12.32). Once the eigen-problem (12.32) is set up, standard numerical methods may be applied to it, and the sparsity of its coefficient matrix leads to fast methods for solving the problem. As an example to illustrate this point, in this section we apply the power iteration algorithm to eigen-problem (12.32) and provide a computational complexity result of the algorithm.

For convenience, we rewrite eigen-problem (12.32) in the form

$$\mathbf{E}_n^{-1}\tilde{\mathbf{K}}_n\tilde{\Phi}_n = \tilde{\lambda}_n\tilde{\Phi}_n. \tag{12.45}$$

The power iteration method is to find the largest eigenvalue and its corresponding eigenvectors of a matrix. We describe below the power iteration algorithm applied to eigen-problem (12.45).

Algorithm 12.21 (Power iteration algorithm)

Step 1 For fixed $n \in \mathbb{N}$, choose $\tilde{\Phi}_n^{(0)} \neq 0$ satisfying that there is an $(l, m) \in \mathbb{U}_n$ such that $(\tilde{\Phi}_n^{(0)})_{(l,m)} = \|\tilde{\Phi}_n^{(0)}\|_\infty = 1$.

Step 2 For $j \in \mathbb{N}_0$, suppose that $\tilde{\Phi}_n^{(j)}$ has been obtained and do the following:

- *Compute* $\Psi_1^{(j)} := \tilde{\boldsymbol{K}}_n \tilde{\Phi}_n^{(j)}$.
- *Solve* $\Psi_2^{(j)}$ *from the equation* $\boldsymbol{E}_n \Psi_2^{(j)} = \Psi_1^{(j)}$.
- *Compute* $\tilde{\lambda}_n^{(j)} := (\Psi_2^{(j)})_{(l,m)}$.

Step 3 *Find an* $(l, m) \in \mathbb{U}_n$ *such that* $|(\Psi_2^{(j)})_{(l,m)}| = \|\Psi_2^{(j)}\|_\infty$, *let* $\tilde{\Phi}_n^{(j+1)} := \Psi_2^{(j)} / (\Psi_2^{(j)})_{(l,m)}$, *and go to step 2.*

The sequences $\tilde{\lambda}_n^{(j)}$, $j \in \mathbb{N}_0$ and $\tilde{\Phi}_n^{(j)}$, $j \in \mathbb{N}_0$ converge to the largest (in magnitude) eigenvalue and its corresponding eigenvector, respectively, of the eigen-problem (12.45). Since the number of nonzero entries of matrices $\tilde{\mathbf{K}}_n$ and $\mathbf{E}_n$ is on the order of $\mathcal{O}(s(n) \log^\tau s(n))$ and since Algorithm 12.21 uses basically matrix–vector multiplications, the algorithm is fast. In the next proposition we provide an estimate on the number of multiplications needed in each iteration step.

Proposition 12.22 *Suppose that* $0 < \sigma' < \min\{2k, d - \sigma\}$ *and* $\eta := 2k - \sigma'$. *If the truncation parameters* $\epsilon = \epsilon_{i'i}^n$, $i', i \in \mathbb{Z}_{n+1}$ *are chosen such that*

$$\epsilon_{i'i}^n = \max\{a\mu^{[-n+b(n-i)+b'(n-i')]/d}, r(d_i + d_{i'})\}, \quad i, i' \in \mathbb{Z}_{n+1}, \tag{12.46}$$

for some constants $a > 0$ *and* $r > 1$, *with* $b = 1$ *and* $\frac{k}{\eta} \leq b' \leq 1$, *then the number of multiplications needed in a single iteration of Algorithm 12.21 is on the order of* $\mathcal{O}(s(n) \log^\tau s(n))$ *where* $\tau = 1$, *except for* $b = b' = 1$ *in which case* $\tau = 2$.

Proof This result is a direct consequence of (12.21) and the estimates of the number of nonzero entries of matrices $\tilde{\mathbf{K}}_n$ and $\mathbf{E}_n$. The major computational effort is spent in step 2. The matrix–vector multiplication $\tilde{\mathbf{K}}_n \tilde{\Phi}_n^{(j)}$ needs $\mathcal{O}(s(n) \log^\tau s(n))$ number of multiplications. Owing to the special structure of matrix $\mathbf{E}_n$, the equation $\mathbf{E}_n \Psi_2^{(j)} = \Psi_1^{(j)}$ can be solved by direct backward substitution. It requires $\mathcal{O}(s(n) \log^\tau s(n))$ number of multiplications. Hence, the total number of multiplications needed in a single iteration is on the order of $\mathcal{O}(s(n) \log^\tau s(n))$. □

12.6 A numerical example

We present a numerical example in this section to confirm the theoretical estimates for the convergence order and computational complexity.

Consider the eigen-problem

$$\mathcal{K}\phi(s) = \lambda\phi(s), \quad s \in \Omega := [0, 1], \tag{12.47}$$

where $\mathcal{K}$ is the integral operator with the weakly singular kernel

$$K(s,t) := \log|\cos(\pi s) - \cos(\pi t)|, \quad s,t \in [0,1].$$

Let $\mathbb{X}_n$ be the space of piecewise linear polynomials having a multiscale basis $\{w_{ij} : (i,j) \in \mathbb{U}_n\}$. In this case $k = 2$, $\mu = 2$ and $s(n) = \dim \mathbb{X}_n = 2^{n+1}$. We choose the basis for $\mathbb{X}_0$

$$w_{00}(t) := -3t + 2, \quad w_{01}(t) := 3t - 1, \quad t \in [0,1]$$

and the basis for $\mathbb{W}_1$

$$w_{10}(t) := \begin{cases} -\frac{9}{2}t + 1, & t \in [0, \frac{1}{2}], \\ \frac{3}{2}t - 1, & t \in (\frac{1}{2}, 1], \end{cases} \qquad w_{11}(t) := \begin{cases} -\frac{3}{2}t + \frac{1}{2}, & t \in [0, \frac{1}{2}], \\ \frac{9}{2}t - \frac{7}{2}, & t \in (\frac{1}{2}, 1]. \end{cases}$$

The bases for $\mathbb{W}_i$, $i > 1$ are generated recursively by

$$w_{ij}(t) = \begin{cases} \sqrt{2}w_{i-1,j}(2t), & t \in [0, \frac{1}{2}], \\ 0, & t \in (\frac{1}{2}, 1], \end{cases} \qquad j \in \mathbb{Z}_{2^{i-1}}$$

and

$$w_{i,2^{i-1}+j}(t) = \begin{cases} 0, & t \in [0, \frac{1}{2}], \\ \sqrt{2}w_{i-1,j}(2t-1), & t \in (\frac{1}{2}, 1], \end{cases} \qquad j \in \mathbb{Z}_{2^{i-1}}.$$

The multiscale functionals $\{\ell_{ij} : (i,j) \in \mathbb{U}_n\}$ are chosen as follows:

$$\ell_{00} := \delta_{\frac{1}{3}}, \quad \ell_{01} := \delta_{\frac{2}{3}},$$

$$\ell_{10} := \delta_{\frac{1}{6}} - \frac{3}{2}\delta_{\frac{1}{3}} + \frac{1}{2}\delta_{\frac{2}{3}}, \ \ell_{11} := \frac{1}{2}\delta_{\frac{1}{3}} - \frac{3}{2}\delta_{\frac{2}{3}} + \delta_{\frac{5}{6}},$$

and $\ell_{ij}, (i,j) \in \mathbb{U}_n, i > 1$, are also generated recursively. See [69] for the recursive generation of the functionals of higher levels.

As the exact eigenvalues and eigenfunctions are not known, for comparison purposes, we use the approximate eigenvalues and eigenvectors of $\mathcal{K}_n$ with $n = 15$ for those of $\mathcal{K}$. Let λ denote the largest simple eigenvalue of $\mathcal{K}_n$ $(n = 15)$, $\mathcal{P}$ be the spectral projection associated with $\mathcal{K}_n$ $(n = 15)$ and λ, and $\tilde{\lambda}_n$, $\tilde{\phi}_n$ be the largest (in magnitude) simple eigenvalue and associated eigenfunction of the truncated operator $\tilde{\mathcal{K}}_n$, respectively. It can be computed that $\lambda \approx -0.99999982793163$.

We apply the fast collocation method to eigen-problem (12.47). The corresponding matrix is truncated with the parameter choice (12.46), with $a = 1$, $b = 1$ and $b' = 0.78$. The numerical algorithm is run on a PC with Intel Core2 T5600 1.83-GHz CPU, 2GB RAM and the programs are compiled using Visual C++ 2005 with single thread.

The numerical results are listed in Tables 12.1 and 12.2. In Table 12.1, "Comp. rate" denotes the compression rate, which is defined as the ratio of

Table 12.1 *Numerical results for eigenvalue computation (1)*

n	Comp. rate	$\|\lambda - \tilde{\lambda}_n\|$	r_1	$\|\tilde{\phi}_n - \mathcal{P}\tilde{\phi}_n\|_\infty$	r_2
1	–	4.7362e-2		5.7466e-4	
2	–	1.4659e-2	1.69	1.2057e-4	2.25
3	0.875	3.9724e-3	1.88	2.2607e-5	2.42
4	0.672	1.0293e-3	1.95	4.0572e-6	2.48
5	0.469	2.5869e-4	1.99	7.3048e-7	2.47
6	0.306	6.4714e-5	2.00	1.3101e-7	2.48
7	0.190	1.6102e-5	2.01	2.4330e-8	2.43
8	0.115	3.8976e-6	2.05	4.5452e-9	2.42
9	0.067	9.6189e-7	2.02	8.3415e-10	2.45
10	0.039	2.3518e-7	2.03	1.9261e-10	2.11
11	0.022	5.7356e-8	2.03	5.2633e-11	1.87
12	0.012	1.3001e-8	2.14	1.2906e-11	2.03
13	0.007	2.8441e-9	2.19	3.9536e-12	1.71

Table 12.2 *Numerical results for eigenvalue computation (2)*

n	$N = \mu^{n+1}$	t_1(s)	t_2(s)	$N_{A.5.1}$
1	4	0.016	0.015	172
2	8	0.032	0.031	160
3	16	0.079	0.047	169
4	32	0.234	0.031	167
5	64	0.656	0.047	163
6	128	1.703	0.063	160
7	256	4.266	0.125	161
8	512	10.31	0.281	167
9	1024	24.27	0.579	156
10	2048	56.69	1.343	158
11	4096	135.3	2.859	153
12	8192	332.1	5.953	154
13	16384	711.1	12.65	154

the number of nonzero entries in $\tilde{\mathbf{K}}_n$ to that of the full matrix $\mathbf{K}_n$, that is, $\mathcal{N}(\tilde{\mathbf{K}}_n)/s(n)^2$ and r_1, r_2 denote the convergence orders of eigenvalues and eigenfunctions, respectively. The numerical results show that the truncation does not ruin the convergence order, which agrees with the theoretical estimates presented in this chapter.

In Table 12.2, t_1 records the time in seconds for generating the matrix $\tilde{K}_n$, and t_2 and $N_{A.5.1}$ record the time for solving the resulting discrete eigenproblem (12.45) and the number of iterations used in Algorithm 5.1 to obtain the corresponding results listed in Table 12.1. It can clearly be seen from the numerical results that most of the computing time is spent on generating the coefficient matrix. The computing time spent on generating the matrix and on finding the solution shows a linear increase corresponding to the growth of the size of the matrix. Indeed, the power iteration method based on the matrix compression technique is a fast method.

12.7 Bibliographical remarks

The abstract framework for the eigenvalue approximation presented in Section 12.2 is an extension of the classical spectral approximation theory [3, 44, 212]. For the eigen-problems of compact linear operators, [44] is a good reference (see also Section A.2.7 in the Appendix). For the classical spectral approximation theory and the notion of spectral projection, the reader is referred to [3, 44, 145, 166, 167, 186, 212]. The material of the multiscale method for solving the eigen-problem is mainly chosen from the paper [70]. See [214, 215] for further developments along this line.

For more about the numerical solutions of the eigen-problem of compact integral operators, see [3, 8, 11, 26, 27, 31, 61, 212, 213, 245]. Analysis of numerical methods for the approximation of eigenvalues and eigenvectors of compact integral operators is well documented in the literature (see, for example, [8, 11, 15, 44, 47, 185, 209, 210, 213, 245]).

Appendix

Basic results from functional analysis

In this appendix, we summarize some of the standard concepts and results from functional analysis in a form that is used throughout the book. Therefore, this section provides the reader with a convenient source for the background material needed to follow the ideas and arguments presented in previous chapters. Further discussion and detailed proofs of the concepts we review here can be found in standard texts on functional analysis, for example [1, 86, 183, 236, 276].

A.1 Metric spaces

A.1.1 Metric spaces

Definition A.1 Let $\mathbb{X}$ be a nonempty set and ρ be a real-valued function defined on $\mathbb{X} \times \mathbb{X}$, satisfying the following properties:

(i) $\rho(x, y) \geq 0$ and $\rho(x, y) = 0$ if and only if $x = y$,
(ii) $\rho(x, y) = \rho(y, x)$,
(iii) $\rho(x, y) \leq \rho(x, z) + \rho(z, y)$.

In this case, ρ is called a metric function (or distance function) defined on $\mathbb{X}$, and $(\mathbb{X}, \rho)$ (or $\mathbb{X}$) is called a metric space.

Definition A.2 A sequence $\{x_j : j \in \mathbb{N}\}$ in a metric space $\mathbb{X}$ is said to converge to $x \in \mathbb{X}$ as $j \to \infty$ if

$$\lim_{j \to \infty} \rho(x_j, x) = 0.$$

A sequence $\{x_j : j \in \mathbb{N}\} \subseteq \mathbb{X}$ is called a Cauchy sequence if

$$\lim_{i,j \to \infty} \rho(x_i, x_j) = 0.$$

A subset A of a metric space $\mathbb{X}$ is said to be complete if every Cauchy sequence in A converges to some element in A.

Definition A.3 Let $\mathbb{X}$ be a metric space and let S and S' be subsets of $\mathbb{X}$. If, in every neighborhood of any point x in S, there is an element of S', then S' is said to be dense in S. If S has a countable dense subset then S is said to be a separable set. If $\mathbb{X}$ itself is separable then $\mathbb{X}$ is called a separable space.

Definition A.4 The ball with center $x \in \mathbb{X}$ and radius r shall be denoted by $B(x, r) := \{y : y \in \mathbb{X}, \rho(x, y) < r\}$.

Of course, a set A in $\mathbb{X}$ is bounded if it is contained in some ball $B(x, r)$. Moreover, if A and B are bounded then $A \cup B$ is bounded.

Definition A.5 A subset A of a metric space $\mathbb{X}$ is totally bounded if for every $\epsilon > 0$ there is a finite set of points $\{x_j : j \in \mathbb{N}_m\} \subseteq \mathbb{X}$ such that $A \subseteq \cup\{B(x_j, \epsilon) : j \in \mathbb{N}_m\}$.

Definition A.6 Let S be a subset of a metric space $\mathbb{X}$. If every sequence $\{x_j : j \in \mathbb{N}\}$ in S has a convergent subsequence $\{x_{k_j} : j \in \mathbb{N}\}$, that is, there is an $x \in \mathbb{X}$ such that $\lim_{j\to\infty} \rho(x_{k_j}, x) = 0$, then the subset S is said to be relatively compact. Moreover, if the limit x is always in S, then S is said to be compact. The space $\mathbb{X}$ is called a compact space if it has this property.

Certainly, every compact set is closed and bounded (but not conversely). However, we have the following useful fact.

Theorem A.7 *A subset A of a metric space $\mathbb{X}$ is compact if and only if it is totally bounded and complete.*

A.1.2 Normed linear spaces

Definition A.8 Suppose that every pair of elements $x, y \in \mathbb{X}$ can be combined by an operation, called addition, to yield a new element in $\mathbb{X}$, denoted by $x + y$. Suppose also that for every complex (or real) number a and every element $x \in \mathbb{X}$, there is an operation, called scalar multiplication, which yields a new element in $\mathbb{X}$, denoted by ax. The set $\mathbb{X}$ is said to be a linear space if it satisfies the following axioms:

(i) $x + y = y + x$,
(ii) $x + (y + z) = (x + y) + z$,
(iii) $\mathbb{X}$ contains a unique element, denoted 0, which satisfies $x + 0 = x$ for all $x \in \mathbb{X}$,

(iv) to each $x \in \mathbb{X}$ there corresponds an element of $\mathbb{X}$, denoted $-x$, such that $x + (-x) = 0$,
(v) $a(x + y) = ax + ay$,
(vi) $(a + b)x = ax + bx$,
(vii) $a(bx) = (ab)x$,
(viii) $1x = x$,
(ix) $0x = 0$,

where a and b are complex (or real) numbers.

Definition A.9 A norm on a linear space $\mathbb{X}$ over a complex (or real) field is a real-valued function, denoted $\| \cdot \|$, satisfying the requirements that

(i) $\|x\| \geq 0$, where the equality holds if and only if $x = 0$,
(ii) $\|ax\| = |a| \, \|x\|$,
(iii) $\|x + y\| \leq \|x\| + \|y\|$.

A linear space with a norm is called a normed linear space.

When we wish to indicate the connection of a norm to the space $\mathbb{X}$ on which it is defined, we indicate it with the symbol $\| \cdot \|_{\mathbb{X}}$. A normed linear space $\mathbb{X}$ is a metric space with metric function $\rho(x, y) := \|x - y\|_{\mathbb{X}}$.

Definition A.10 A normed linear space $\mathbb{X}$ is called a Banach space if it is a complete metric space in the metric induced by its norm $\rho(x, y) := \|x - y\|_{\mathbb{X}}$.

A.1.3 Inner product spaces

Definition A.11 A linear space $\mathbb{X}$ over a complex (or real) field is called an inner product space if for any two elements x and y of $\mathbb{X}$, there is a uniquely associated complex (or real) number, called the inner product of x and y and denoted by (x, y) (or when needed by $(x, y)_{\mathbb{X}}$), which satisfies the requirements that

(i) $(x, x) \geq 0$ and the equality holds if and only if $x = 0$,
(ii) $(x, y) = \overline{(y, x)}$,
(iii) $(ax + by, z) = a(x, z) + b(y, z)$.

Every inner product satisfies the Cauchy–Schwarz inequality, namely, for every $x, y \in \mathbb{X}$ we have that

$$\big|(x, y)\big| \leq \|x\| \cdot \|y\|.$$

An inner product space is also a normed linear space. Specifically, we associate with each $x \in \mathbb{X}$ the non-negative number $\|x\| := (x, x)^{1/2}$ and by the

Minkowski inequality,

$$\|x+y\| \le \|x\| + \|y\|,$$

we immediately conclude that $\|\cdot\|_{\mathbb{X}}$ is a norm on $\mathbb{X}$. A characteristic feature of the norm on an inner product space is the following result.

Proposition A.12 *(Parallelogram identity) If $\mathbb{X}$ is an inner product space with norm $\|\cdot\|$ then for any $x, y \in \mathbb{X}$,*

$$\|x+y\|^2 + \|x-y\|^2 = 2\|x\|^2 + 2\|y\|^2.$$

Definition A.13 An inner product space $\mathbb{X}$ is called a Hilbert space if it is a complete metric space under the metric induced by its inner product $\rho(x, y) := (x-y, x-y)^{1/2}$.

A.1.4 Function spaces $C(\bar{\Omega})$ and $L^p(\Omega)$ $(1 \le p \le \infty)$

Let Ω be a bounded domain in a d-dimensional Euclidean space $\mathbb{R}^d$, where d is a positive integer. We use $x = [x_j : j \in \mathbb{N}_d]$ to denote a vector in $\mathbb{R}^d$ and define $C(\bar{\Omega})$ to be the linear space (under pointwise addition and scalar multiplication) of uniformly continuous, complex-valued functions on Ω. Moreover, the notation

$$\text{supp}\,(u) := \text{closure of } \{x : x \in \Omega,\ u(x) \neq 0\}$$

stands for the *support* of the function u on Ω. The symbol $C_0(\Omega)$ indicates the linear subspace of $C(\bar{\Omega})$ consisting of functions with support contained inside Ω. Also, $C(\bar{\Omega})$ equipped with the norm

$$\|u\|_{0,\infty,\Omega} := \max\{|\, u(x)\,| : x \in \Omega\}$$

is a Banach space. The symbol $L^\infty(\Omega)$ is used to denote the linear space of complex-valued measurable functions u which are essentially bounded, that is, there is a set $E \subseteq \Omega$ of measure zero such that u is bounded on $\Omega \backslash E$. Equipped with the norm

$$\|u\|_{0,\infty,\Omega} := \operatorname*{ess\,sup}_{x\in\Omega} |\, u(x)\,| = \inf\Big\{ \sup\{|u(x)| : x \in \Omega \setminus E\} : \text{meas}(E) = 0\Big\},$$

$L^\infty(\Omega)$ is a Banach space. Let $L^p(\Omega)$, $1 \le p < \infty$, denote the linear space of complex-valued measurable functions u such that $|u|^p$ is Lebesgue-integrable on Ω. For any $u \in L^p(\Omega)$ we define

$$\|u\|_{0,p,\Omega} := \left\{ \int_\Omega |\, u(x)\,|^p \mathrm{d}x \right\}^{1/p},$$

and recall that in this norm $L^p(\Omega)$ becomes a Banach space. In particular, $L^2(\Omega)$ is a Hilbert space with inner product

$$(u, v)_{0,\Omega} := \int_\Omega u(x)\overline{v(x)}\mathrm{d}x$$

and corresponding norm $\|\cdot\|_{0,2,\Omega}$ as defined above. An important and useful fact is that $C_0(\Omega)$ is dense in $L^p(\Omega)$ for $1 \leq p < \infty$.

Theorem A.14 *(Fubini theorem) If u is a measurable function defined on $\mathbb{R}^{n+m}$, and at least one of the integrals*

$$I_1 := \int_{\mathbb{R}^{n+m}} |u(x, y)|\mathrm{d}x\mathrm{d}y,$$

$$I_2 := \int_{\mathbb{R}^m} \left(\int_{\mathbb{R}^n} |u(x, y)|\mathrm{d}x \right) \mathrm{d}y$$

and

$$I_3 := \int_{\mathbb{R}^n} \left(\int_{\mathbb{R}^m} |u(x, y)|\mathrm{d}y \right) \mathrm{d}x$$

exists and is finite, then

(i) for almost all $y \in \mathbb{R}^m$, $u(\cdot, y) \in L^1(\mathbb{R}^n)$ and $\int_{\mathbb{R}^n} u(x, \cdot)\mathrm{d}x \in L^1(\mathbb{R}^m)$,
(ii) for almost all $x \in \mathbb{R}^n$, $u(x, \cdot) \in L^1(\mathbb{R}^m)$ and $\int_{\mathbb{R}^m} u(\cdot, y)\mathrm{d}y \in L^1(\mathbb{R}^n)$,
(iii) $I_1 = I_2 = I_3$.

A.1.5 Function spaces $C^m(\bar{\Omega})$ and $W^{m,p}(\Omega)$ $(m \geq 1, 1 \leq p \leq \infty)$

We generally use $\alpha = [\alpha_j : j \in \mathbb{N}_d]$ for a vector in $\mathbb{R}^d$ with non-negative integer components. The set of all such lattice vectors is denoted by $\mathbb{N}^d$ and with each such vector we define $|\alpha| := \alpha_1 + \cdots + \alpha_d$. We find convenient the notation $\mathbb{Z}^d_m := \{\alpha : |\alpha| \leq m - 1\}$. The αth derivative operator is denoted by $D^\alpha = \partial^{|\alpha|}/\partial x_1^{\alpha_1} \cdots \partial x_d^{\alpha_d}$ and $|\alpha|$ is called the (total) order of the derivative. When α is the zero vector, we interpret D^α as the identity operator. Let $C^m(\bar{\Omega})$ be the closed subspace of $C(\bar{\Omega})$ of all the functions which have continuous derivatives up to and including the mth-order derivatives, that is, $D^\alpha u \in C(\bar{\Omega})$ for $\alpha \in \mathbb{Z}^d_m$. $C^m(\bar{\Omega})$ is a Banach space with the norm

$$\|u\|_{m,\infty} := \sum_{\alpha \in \mathbb{Z}^d_m} \left\|D^\alpha u\right\|_{0,\infty,\Omega}. \tag{A.1}$$

In a similar manner, the linear space $C^m(\Omega)$ is defined. We also set

$$C^\infty(\Omega) := \bigcap_{m \in \mathbb{N}_0} C^m(\Omega)$$

and

$$C_0^\infty(\Omega) := \{u \in C^\infty(\Omega) : \text{support of } u \text{ is inside } \Omega \text{ and bounded}\}.$$

Neither one of these linear spaces are Banach spaces. However, the family of norms given in (A.1) as m varies over $\mathbb{N}_0$ determines a Fechet topology on these spaces.

We now describe Sobolev spaces. For a non-negative integer $m \in \mathbb{N}$ and $1 < p < \infty$ we define the linear space

$$W^{m,p}(\Omega) := \left\{u : u \in L^p(\Omega), D^\alpha u \in L^p(\Omega), \alpha \in \mathbb{Z}^d_{m+1}\right\},$$

and when $m = 0$ set

$$W^{0,p}(\Omega) := L^p(\Omega).$$

When $1 < p < \infty$ we define on $W^{m,p}(\Omega)$ the norm

$$\|u\|_{m,p,\Omega} := \left\{ \sum_{\alpha \in \mathbb{Z}^d_{m+1}} \|D^\alpha u\|^p_{0,p,\Omega} \right\}^{1/p}$$

and for $p = \infty$ set

$$\|u\|_{m,\infty,\Omega} := \max\{\|D^\alpha u\|_{0,\infty,\Omega} : \alpha \in \mathbb{Z}^d_{m+1}\}.$$

The space $W^{m,p}(\Omega)$ is a Banach space with norm $\|\cdot\|_{m,p,\Omega}$, .

Let $W_0^{m,p}(\Omega)$ be the closure of $C_0^\infty(\Omega)$ in $W^{m,p}(\Omega)$. In particular, for $p = 2$, it is a standard notation to use $H^m(\Omega)$ and $H_0^m(\Omega)$ for $W^{m,p}(\Omega)$ and $W_0^{m,p}(\Omega)$, respectively, and likewise denote the norm $\|\cdot\|_{m,2,\Omega}$ by $\|\cdot\|_{m,\Omega}$. Both $W^{m,p}(\Omega)$ and $W_0^{m,p}(\Omega)$ are called *Sobolev spaces*.

Theorem A.15 *(Properties of Sobolev spaces)*

(i) For any $(1 \le p \le \infty)$, $W^{m,p}(\Omega)$ is a Banach space. In particular, the space $H^m(\Omega)$ is a Hilbert space with inner product

$$(u, v)_{m,\Omega} := \sum_{1 \le |\alpha| \le m} (D^\alpha u, D^\alpha v)_{0,\Omega}, \quad u, v \in H^m(\Omega).$$

(ii) For any $1 \le p < \infty$, the space $W^{m,p}(\Omega)$ is the closure of the set $\{u : u \in C^\infty(\Omega), \|u\|_{m,p,\Omega} < \infty\}$.

(iii) For any $1 \le p < \infty$, the space $W^{m,p}(\Omega)$ is separable.

(iv) For any $1 \le p < \infty$, the space $W^{m,p}(\Omega)$ is reflexive.

Definition A.16 Let $(\cdot,\cdot)$ and $(\cdot,\cdot)'$ be two inner products in the same linear space $\mathbb{X}$. If the two norms $\|\cdot\|$ and $\|\cdot\|'$ induced by these inner products are equivalent, namely, there are positive constants c_1 and c_2 such that for all $u \in \mathbb{X}$,

$$c_1\|u\| \le \|u\|' \le c_2\|u\|,$$

then the two inner products are said to be equivalent.

With equivalent inner products, all results based on convergence in $\mathbb{X}$ are the same as both norms determine the same metric topology on $\mathbb{X}$.

Proposition A.17 *(Poincaré inequality) Let ρ be the diameter of a bounded domain Ω in a d-dimensional Euclidean space $\mathbb{R}^d$. Then the following Poincaré inequality holds for all $u \in H_0^1(\Omega)$:*

$$\|u\|_{0,\Omega} \le \frac{\rho}{\sqrt{d}}\|\nabla u\|_{0,\Omega}.$$

It follows from the Poincaré inequality that

$$(u,v)'_{1,\Omega} := \sum_{k\in\mathbb{N}_d}\left(\frac{\partial u}{\partial x_k}, \frac{\partial v}{\partial x_k}\right)_{0,\Omega}$$

is an inner product of $H_0^1(\Omega)$ satisfying, for all $u \in H_0^1(\Omega)$, the inequality

$$(u,u)'_{1,\Omega} \le \|u\|^2_{1,\Omega} \le \left(1+d^{-1}\rho^2\right)(u,u)'_{1,\Omega}.$$

Hence, the new inner product $(\cdot,\cdot)'_1$ and the original inner product $(\cdot,\cdot)_1$ are equivalent.

When $m > 1$, observe that if $u \in H_0^m(\Omega)$ then $D^\alpha u \in H_0^1(\Omega)$, $|\alpha| \le m-1$, so that by repeatedly applying the Poincaré inequality we verify that $\sum_{|\alpha|=m}(D^\alpha u, D^\alpha v)_0$ is an inner product of $H_0^m(\Omega)$, which is equivalent to the original inner product. This leads to the following proposition.

Proposition A.18 *In $H_0^m(\Omega)$, the semi-norm defined by*

$$|u|_{m,\Omega} := \left(\sum_{|\alpha|=m}\|D^\alpha u\|^2_{0,\Omega}\right)^{1/2}$$

is equivalent to the original norm $\|u\|_{m,\Omega}$.

A.2 Linear operator theory

A.2.1 Linear operators

Definition A.19 Let $\mathbb{X}$ and $\mathbb{Y}$ be normed linear spaces and $\mathcal{T}$ be an operator from $\mathbb{X}$ into $\mathbb{Y}$. $\mathcal{T}$ is called a linear operator if for any $x, y \in \mathbb{X}$ and any complex numbers a and b,

$$\mathcal{T}\big(ax + by\big) = a\mathcal{T}x + b\mathcal{T}y.$$

In particular, if $\mathbb{Y}$ is a complex (or real) number field, then $\mathcal{T}$ is called a linear functional.

Definition A.20 A linear operator is said to be a bounded operator if there is a positive constant c such that for any $x \in \mathbb{X}$,

$$\|\mathcal{T}x\|_{\mathbb{Y}} \leq c\|x\|_{\mathbb{X}}.$$

The infimum of the values of c that satisfy the above inequality is called the norm of the linear operator, denoted $\|\mathcal{T}\|$.

A linear operator is bounded if and only if it is continuous. We denote by $\mathcal{B}(\mathbb{X}, \mathbb{Y})$ the set of all bounded linear operators from $\mathbb{X}$ to $\mathbb{Y}$. If $\mathbb{Y}$ is a Banach space then $\mathcal{B}(\mathbb{X}, \mathbb{Y})$ is a Banach space relative to the operator norm. When $\mathbb{Y} = \mathbb{R}$, the elements of $\mathcal{B}(\mathbb{X}, \mathbb{R})$ are bounded linear functionals on $\mathbb{X}$. The space $\mathcal{B}(\mathbb{X}, \mathbb{R})$ is called the *dual space* of $\mathbb{X}$ and is denoted by $\mathbb{X}^*$, see also Definition A.36.

Definition A.21 Let $\mathcal{T} : \mathbb{X} \to \mathbb{Y}$ be a linear operator. We denote its domain by $D(\mathcal{T})$, a subspace of $\mathbb{X}$, and its range by $R(\mathcal{T})$, which is a subspace of $\mathbb{Y}$. The linear operator $\mathcal{T}$ is said to be a closed operator if for any sequence $\{x_j : j \in \mathbb{N}\} \subseteq D(\mathcal{T})$ satisfying $\lim_{j\to\infty} x_j = x$ in $\mathbb{X}$ and $\lim_{j\to\infty} \mathcal{T}x_j = y$ in $\mathbb{Y}$, it follows that $x \in D(\mathcal{T})$ and $y = \mathcal{T}x$.

We note that a linear operator $\mathcal{T}$ is closed if and only if its *graph*

$$G(\mathcal{T}) := \big\{(x, \mathcal{T}x) : x \in D(\mathcal{T})\big\}$$

is a closed subspace of $\mathbb{X} \times \mathbb{Y}$. In general, a closed operator is not necessarily continuous. However, under certain conditions this is true (see the closed graph theorem below). Moreover, if the domain $D(\mathcal{T})$ of a continuous linear operator $\mathcal{T} : D(\mathcal{T}) \to \mathbb{Y}$ is a closed subspace of $\mathbb{X}$, then it is a closed operator.

A.2.2 Open mapping theorems and uniform boundedness theorem

Theorem A.22 *(Open mapping theorem) If $\mathcal{T} \in \mathcal{B}(\mathbb{X}, \mathbb{Y})$ where $\mathbb{X}$, $\mathbb{Y}$ are Banach spaces, $R(\mathcal{T}) = \mathbb{Y}$ and $\varepsilon > 0$, then there exists an $\eta > 0$ such that*

$$\big\{y : y \in \mathbb{Y}, \|y\|_{\mathbb{Y}} < \eta\big\} \subseteq \big\{\mathcal{T}x : x \in \mathbb{X}, \|x\|_{\mathbb{X}} < \varepsilon\big\}.$$

This theorem states that for any open set $G \subseteq \mathbb{X}$, the operator $\mathcal{T}$ maps the set $D(\mathcal{T}) \cap G$ to an open set in $\mathbb{Y}$.

Theorem A.23 *(Inverse operator theorem) If the operator $\mathcal{T}$ satisfies all the properties stated in Theorem A.22 and, in addition, is one to one then the inverse operator $\mathcal{T}^{-1}$ is a continuous linear operator.*

Theorem A.24 *(Closed graph theorem) If $\mathbb{X}$ and $\mathbb{Y}$ are Banach spaces and $\mathcal{T} : \mathbb{X} \to \mathbb{Y}$ is a linear operator such that the domain $D(\mathcal{T})$ and graph of $\mathcal{T}$ are closed sets of $\mathbb{X}$ and $\mathbb{X} \times \mathbb{Y}$, respectively, then $\mathcal{T}$ is continuous.*

A statement equivalent to the closed graph theorem is that if $\mathbb{X}$ and $\mathbb{Y}$ are Banach spaces and $\mathcal{T} : \mathbb{X} \to \mathbb{Y}$ is a closed linear operator such that its domain $D(\mathcal{T})$ is a closed subspace of $\mathbb{X}$, then $\mathcal{T}$ is continuous.

Theorem A.25 *(Uniform boundedness theorem) If $\mathbb{X}$ is a Banach space, $\mathbb{Y}$ a normed linear space and for any $x \in \mathbb{X}$ the sequence of operators $\{\mathcal{T}_j : j \in \mathbb{N}\} \subseteq \mathcal{B}(\mathbb{X}, \mathbb{Y})$ has the property that $\sup\{\|\mathcal{T}_j x\|_{\mathbb{Y}} : j \in \mathbb{N}\} < \infty$, then*

$$\sup\{\|\mathcal{T}_j\| : j \in \mathbb{N}\} < \infty.$$

Corollary A.26 *If $\mathbb{X}$ is a Banach space, $\mathbb{Y}$ a normed linear space and the sequence of operators $\{\mathcal{T}_n : n \in \mathbb{N}\} \subseteq \mathcal{B}(\mathbb{X}, \mathbb{Y})$ has the property that $\lim_{n\to\infty} \mathcal{T}_n x = \mathcal{T} x$ for all $x \in \mathbb{X}$, then $\mathcal{T} \in \mathcal{B}(\mathbb{X}, \mathbb{Y})$.*

Corollary A.27 *(Banach–Steinhaus theorem) If $\mathbb{X}$ and $\mathbb{Y}$ are Banach spaces, U a dense subset of $\mathbb{X}$, $\{\mathcal{T}_n : n \in \mathbb{N}\} \subseteq \mathcal{B}(\mathbb{X}, \mathbb{Y})$ and $\mathcal{T} \in \mathcal{B}(\mathbb{X}, \mathbb{Y})$, then $\lim_{n\to\infty} \mathcal{T}_n x = \mathcal{T} x$ for all $x \in \mathbb{X}$ if and only if*

$$\lim_{n\to\infty} \mathcal{T}_n x = \mathcal{T} x, \quad \text{for all } x \in U$$

and

$$\sup\{\|\mathcal{T}_n\| : n \in \mathbb{N}\} < \infty.$$

A.2.3 Orthogonal projection

Let K be a closed convex set in a Hilbert space $\mathbb{X}$ and $x \in \mathbb{X}$. The *shortest distance problem* is to find $y \in K$ such that

$$\|x - y\| = \min\{\|x - w\| : w \in K\}.$$

It can be shown that this problem is equivalent to finding a $y \in K$ which solves the variational inequality that for all $w \in K$,

$$(x - y, w - y) \leq 0.$$

Theorem A.28 *The above shortest distance problem is uniquely solvable.*

Theorem A.29 *(Orthogonal projection theorem) Let M be a closed subspace of the Hilbert space $\mathbb{X}$ and $M^\perp$ be the orthogonal complement of M, that is,*

$$M^\perp := \left\{y \in \mathbb{X} : \quad (y, x) = 0, \ x \in M\right\}.$$

Then $\mathbb{X}$ can be decomposed as a direct sum of M and $M^\perp$, that is, $\mathbb{X} = M \oplus M^\perp$.

Corollary A.30 *In a Hilbert space $\mathbb{X}$, a subspace M is dense in $\mathbb{X}$ if and only if $M^\perp = \{0\}$.*

A.2.4 Riesz representation theorem

If $\mathbb{X}$ is a Hilbert space and $x \in \mathbb{X}$ then the linear functional ℓ defined for all $y \in \mathbb{X}$,

$$\ell(y) = (y, x),$$

is a bounded linear functional on $\mathbb{X}$ with norm $\|x\|$. The converse is given by the Riesz representation theorem.

Theorem A.31 *(Riesz representation theorem) If $\mathbb{X}$ is a Hilbert space and ℓ is a bounded linear functional defined on $\mathbb{X}$ then there is a unique $x \in \mathbb{X}$ such that for all $y \in \mathbb{X}$,*

$$\ell(y) = (y, x),$$

and in that case $\|\ell\| = \|x\|$.

Given a real Hilbert space $\mathbb{X}$ and $\ell \in \mathbb{X}^*$, a method of finding the corresponding unique $x \in \mathbb{X}$ follows by establishing that the variational problem

$$\inf\{\|w\|_{\mathbb{X}}^2 - 2\ell(w) : w \in \mathbb{X}\}$$

has a solution.

A.2.5 Hahn–Banach extension theorem

Theorem A.32 *(Hahn–Banach extension theorem) If ℓ_0 is a bounded linear functional defined on a subspace $\mathbb{M}$ of a normed linear space $\mathbb{X}$ then there is a norm-preserving extension $\ell \in \mathbb{X}^*$, that is, $\ell(x) = \ell_0(x)$ for all $x \in \mathbb{M}$ and $\|\ell\| = \|\ell_0\|$.*

We note several useful corollaries of the Hahn–Banach theorem.

Corollary A.33 *For any nonzero $x \in \mathbb{X}$, there exists a bounded linear functional $\ell \in \mathbb{X}^*$ of norm one such that $\ell(x) = \|x\|_{\mathbb{X}}$.*

Corollary A.34 *For any closed subspace $M \subseteq \mathbb{X}$ and $x \in \mathbb{X}$,*

$$\max\{\ell(x) : \ell \in \mathbb{X}^*,\ \|\ell\| \leq 1,\ \ell(y) = 0,\ y \in M\} = \inf\{\|x - y\|_{\mathbb{X}} : y \in M\}.$$

Corollary A.35 *A linear subspace $M \subseteq \mathbb{X}$ is dense in $\mathbb{X}$ if and only if there is no nonzero bounded linear functional which vanishes on M.*

A.2.6 Compactness

Definition A.36 The dual space of a normed linear space $\mathbb{X}$, denoted $\mathbb{X}^*$, is a Banach space with norm

$$\|\ell\| := \sup\{|\ell(x)| : x \in \mathbb{X}, \|x\| \leq 1\}.$$

For each $x \in \mathbb{X}$, we define an $\ell_x \in \mathbb{X}^{**}$ by setting for each $\ell \in \mathbb{X}^*$

$$\langle \ell, \ell_x \rangle := \langle x, \ell \rangle,$$

thereby defining a mapping $\tau : \mathbb{X} \to \mathbb{X}^{**}$ as $\tau(x) := \ell_x$.

Proposition A.37 *If $\mathbb{X}$ is a normed linear space and $\mathbb{X}^{**}$ its double dual then the mapping τ defined above is a one to one, norm-preserving linear operator.*

This proposition implies, from the perspective of the normed linear space structure, that $\mathbb{X}$ can be considered as the subspace $\tau(\mathbb{X})$ of $\mathbb{X}^{**}$. In general, $\tau(\mathbb{X})$ is a proper subspace of $\mathbb{X}^{**}$. However, when $\tau(\mathbb{X}) = \mathbb{X}^{**}$ then $\mathbb{X}$ is said to be *reflexive*.

Definition A.38 Let $\mathbb{X}$ be a normed linear space. A sequence $\{x_j : j \in \mathbb{N}\} \subseteq \mathbb{X}$ is said to be a weak Cauchy sequence if for each $\ell \in \mathbb{X}^*$, the sequence of scalars $\{\ell(x_j) : j \in \mathbb{N}\}$ is a Cauchy sequence and a set $\mathbb{S} \subseteq \mathbb{X}$ is said to be weakly bounded if for each $\ell \in \mathbb{X}^*$ the set $\ell(\mathbb{S})$ is bounded. The sequence $\{x_j : j \in \mathbb{N}\} \subseteq \mathbb{X}$ is said to be weakly convergent in $\mathbb{X}$ if there exists an $x \in \mathbb{X}$ such that for all $\ell \in \mathbb{X}^*$,

$$\lim_{j \to \infty} \ell(x_j) = \ell(x).$$

In this case, x is called the weak limit of the sequence. Moreover, a sequence $\{\ell_j : j \in \mathbb{N}\} \subseteq \mathbb{X}^*$ is said to be weak* convergent if there exists an $\ell \in \mathbb{X}^*$ such that for all $x \in \mathbb{X}$,

$$\lim_{j \to \infty} \ell_j(x) = \ell(x),$$

where ℓ is called the weak* limit of the sequence.

From Theorem A.25 we have the following useful fact.

Corollary A.39 *Every weakly bounded subset of a normed linear space is normed bounded.*

We also recall the following results.

Theorem A.40 *A reflexive Banach space is weakly complete. In particular, every Hilbert space is weakly complete. Moreover, the dual space $\mathbb{X}^*$ of a Banach space $\mathbb{X}$ is weak* complete.*

Definition A.41 Let $\mathbb{X}$ be a normed linear space. A subset $S \subseteq \mathbb{X}$ is said to be weak-relatively (sequentially) compact if every sequence in S contains a weakly convergent subsequence.

Theorem A.42 *Let $\mathbb{X}$ be a separable and reflexive Banach space. A subset $S \subseteq \mathbb{X}$ is weak-relatively (sequentially) compact if and only if S is bounded.*

This theorem implies that a subset $S \subseteq W^{m,p}(\Omega)$, where $m \in \mathbb{N}, 0 < p < \infty$, is weak-relatively (sequentially) compact if and only if S is bounded.

Theorem A.43 *(Arzelà–Ascoli theorem) A subset $S \subseteq C(\bar{\Omega})$ is relatively (sequentially) compact if S is bounded pointwise and equicontinuous, that is,*

(i) *for any* $x \in \bar{\Omega}$, $\sup\{|u(x)| : u \in S\} < \infty$,
(ii) $\lim_{\delta \to 0^+} \sup\{|u(x_1) - u(x_2)| : |x_1 - x_2| \leq \delta, u \in S\} = 0$.

Theorem A.44 *(Kolmogorov theorem) A subset $U \subseteq L^p[0,1]$, $1 \leq p < \infty$, is relatively compact if and only if the following conditions are satisfied:*

(i) $\sup_{f \in U} \|f\|_p < \infty$,
(ii) $\lim_{t \to 0} \int_0^{1-t} |f(t+s) - f(s)|^p \mathrm{d}s = 0$ *uniformly in* $f \in U$,
(iii) $\lim_{t \to 0} \int_{1-t}^{1} |f(s)|^p \mathrm{d}s = 0$ *uniformly in* $f \in U$.

A.2.7 Compact operators and Fredholm theorems

Definition A.45 Let $\mathbb{X}$ and $\mathbb{Y}$ be Banach spaces, with a subset $D \subseteq \mathbb{X}$, and $\mathcal{K} : D \to \mathbb{Y}$ be an operator. $\mathcal{K}$ is said to be a relatively compact operator if for any bounded set $S \subseteq D$, $\mathcal{K}(S)$ is a relatively compact set in $\mathbb{Y}$. $\mathcal{K}$ is said to be a compact operator if it is continuous and relatively compact. $\mathcal{K}$ is said to be a completely continuous operator if for any sequence $\{x_n : n \in \mathbb{N}\}$ in D weakly converging to x, it always follows that $\{\mathcal{K}x_n; n \in \mathbb{N}\}$ converges to $\mathcal{K}x$ in $\mathbb{Y}$.

Proposition A.46 *Let $\mathbb{X}$ and $\mathbb{Y}$ be Banach spaces and $\mathcal{K} : \mathbb{X} \to \mathbb{Y}$ an operator.*

(i) If $\mathcal{K}$ is a compact linear operator then $\mathcal{K}$ is completely continuous.
(ii) If $\mathbb{X}$ is reflexive and $\mathcal{K}$ (need not be linear) is completely continuous, then $\mathcal{K}$ is compact.

This proposition implies that on a reflexive Banach space, a compact linear operator and a completely continuous linear operator are the same.

Proposition A.47 *Let $\mathbb{X}$ and $\mathbb{Y}$ be Banach spaces and $\mathcal{K} : \mathbb{X} \to \mathbb{Y}$ a linear operator.*

(i) If $\mathcal{K}$ is relatively compact, then $\mathcal{K}$ is continuous.
(ii) If $\mathcal{K}$ is bounded and the range $R(\mathcal{K})$ is finite dimensional, then $\mathcal{K}$ is compact.
(iii) If $\mathcal{K}$ is compact, then the range $R(\mathcal{K})$ is separable.
(iv) If $\mathcal{K}$ is compact, then the adjoint operator $K^* : \mathbb{Y}^* \to \mathbb{X}^*$ is also compact.
(v) Let $\{\mathcal{K}_j : j \in \mathbb{N}\}$ be a sequence of compact operators from $\mathbb{X}$ to $\mathbb{Y}$ satisfying $\mathcal{K}_j \xrightarrow{u} \mathcal{K}$ on $\mathbb{X}$, then $\mathcal{K}$ is compact.

Let $\mathbb{X}$ be a Hilbert space, with inner product $(\cdot, \cdot)_{\mathbb{X}}$ and norm $\|\cdot\|_{\mathbb{X}}$, respectively, and let $\mathcal{K} : \mathbb{X} \to \mathbb{X}$ be a linear compact operator. Consider the linear operator equation

$$(\mathcal{I} - \mathcal{K})u = f, \qquad f \in \mathbb{X} \tag{a}$$

and its adjoint operator equation

$$(\mathcal{I} - \mathcal{K}^*)v = g, \qquad g \in \mathbb{X}, \tag{b}$$

where $\mathcal{I}$ is the identity operator and $\mathcal{K}^*$ is the adjoint operator of $\mathcal{K}$.

We have the following basic results.

Theorem A.48 *(First Fredholm theorem) If $\mathcal{K} : \mathbb{X} \to \mathbb{X}$ is a compact linear operator, then the following statements hold:*

(i) Equation (a) is uniquely solvable for an arbitrarily given $f \in \mathbb{X}$ if and only if equation (b) is so for an arbitrarily given $g \in \mathbb{X}$.
(ii) Equation (a) (or equation (b)) is uniquely solvable for an arbitrarily given $f \in \mathbb{X}$ (or $g \in \mathbb{X}$) if and only if the corresponding homogeneous equation

$$(\mathcal{I} - \mathcal{K})u = 0 \qquad (\text{or } (\mathcal{I} - \mathcal{K}^*)v = 0)$$

has only a zero solution.
(iii) In the case that equation $(\mathcal{I} - \mathcal{K})u = 0$ only has a zero solution, the inverse operator $(\mathcal{I} - \mathcal{K})^{-1}$ exists on the entire space $\mathbb{X}$ and is bounded, so that the solution $u \in \mathbb{X}$ of equation (a) satisfies

$$\|u\|_{\mathbb{X}} \leq c\|f\|_{\mathbb{X}}$$

for a constant $c > 0$, independent of $f \in \mathbb{X}$.

Theorem A.49 *(Second Fredholm theorem) If $\mathcal{K} : \mathbb{X} \to \mathbb{X}$ is a compact linear operator, and equation $(\mathcal{I}-\mathcal{K})u = 0$ has nonzero solutions, then among all its solutions there are only finitely many that are linearly independent. In this case, equation $(\mathcal{I}-\mathcal{K}^*)v = 0$ has the same number of linearly independent solutions.*

Theorem A.50 *(Third Fredholm theorem) If $\mathcal{K} : \mathbb{X} \to \mathbb{X}$ is a compact linear operator, then equation (a) is solvable if and only if the given $f \in \mathbb{X}$ is orthogonal with all the solutions of equation $(\mathcal{I} - \mathcal{K}^*)v = 0$. In this case, equation (a) has one, and only one, solution $u \in \mathbb{X}$ that is orthogonal to all the solutions of equation $(\mathcal{I} - \mathcal{K})u = 0$, which satisfies $\|u\|_{\mathbb{X}} \leq c\|f\|_{\mathbb{X}}$ for a constant $c > 0$, independent of the given $f \in \mathbb{X}$.*

We now consider the following operator equation:

$$(\lambda\mathcal{I} - \mathcal{K})u = f, \qquad f \in \mathbb{X}, \tag{c}$$

where λ is a complex parameter. If, for an arbitrarily given $f \in \mathbb{X}$, equation (c) has a unique solution corresponding to a complex value of λ, and $(\lambda\mathcal{I} - \mathcal{K})^{-1}$ is bounded, then this λ is called a *regular value* of $\mathcal{K}$. If, corresponding to a complex value of λ, the homogeneous equation $(\lambda\mathcal{I} - \mathcal{K})u = 0$ has nonzero solutions, then this λ is called an *eigenvalue* of $\mathcal{K}$, while each corresponding nonzero solution is called the *eigenfunction* associated with this λ. The maximum number of linearly independent eigenfunctions is called the *geometric multiplicity* of the corresponding eigenvalue λ, which can be either finite or infinite.

In general, it is quite possible that there exists a number λ for which $(\lambda\mathcal{I} - \mathcal{K})^{-1}$ exists but is not everywhere defined in $\mathbb{X}$. However, if $\mathcal{K}$ is a compact linear operator, this is impossible. In this case, it is also impossible to have an infinite geometric multiplicity (as shown in the following theorem).

Theorem A.51 *Let $\mathcal{K} : \mathbb{X} \to \mathbb{X}$ be a compact linear operator.*

(i) If $\lambda \neq 0$, then λ is a regular value of $\mathcal{K}$ if and only if λ is not an eigenvalue of $\mathcal{K}$.
(ii) If $\lambda \neq 0$ is an eigenvalue of $\mathcal{K}$, then the geometric multiplicity of λ is finite; $\bar{\lambda}$ is the eigenvalue of the adjoint operator $\mathcal{K}^*$ of $\mathcal{K}$, with the same geometric multiplicity as λ.
(iii) If $\lambda \neq 0$ is an eigenvalue of $\mathcal{K}$, then equation (c) is solvable if and only if the given $f \in \mathbb{X}$ is orthogonal to all eigenfunctions associated with the eigenvalue $\bar{\lambda}$ of $\mathcal{K}^*$; in this case, equation (c) has one, and only one, solution that is orthogonal to all eigenfunctions of $\mathcal{K}$ associated with λ.
(iv) The family of eigenfunctions corresponding to different eigenvalues of $\mathcal{K}$ is linearly independent.

Theorem A.52 *(Fourth Fredholm theorem) Let $\mathcal{K} : \mathbb{X} \to \mathbb{X}$ be a compact linear operator. For any constant $c > 0$, there are only finitely many eigenvalues of $\mathcal{K}$ that are located outside the disc of radius c centered at zero. Consequently, $\mathcal{K}$ either has finitely many eigenvalues or has countably many eigenvalues converging to zero.*

According to this theorem, if a compact linear operator $\mathcal{K}$ has nonzero eigenvalues, then they can be arranged in decreasing order in absolute values (i.e., $|\lambda_n| \geq |\lambda_{n+1}|, n \in \mathbb{N}$), which can be a finite or infinite sequence of ordering, and the number of appearance of any particular eigenvalue in this ordering is equal to its geometric multiplicity. If this sequence of ordering is infinite, then $\lim_{n\to\infty} \lambda_n = 0$. Also, the eigenfunctions associated with the above sequence of ordering can be chosen such that they are all linearly independent.

The above results can be extended directly to Banach spaces.

As a special case, if the compact linear operator is self-adjoint and $\mathcal{K} \neq 0$, then the above sequence of ordering of eigenvalues and their corresponding eigenfunctions will be nonempty, all the eigenvalues are real, and the eigenfunctions can be chosen so that they constitute an orthonormal basis of the Hilbert space $\mathbb{X}$.

Theorem A.53 *(Hilbert–Schmidt theorem) If $\mathcal{K} : \mathbb{X} \to \mathbb{X}$ is a nonzero, self-adjoint, compact linear operator, then for any $u \in \mathbb{X}$, $\mathcal{K}u$ has a convergent Fourier series expansion by an orthonormal basis $\{e_n : n \in \mathbb{N}\}$ of $\mathbb{X}$:*

$$\mathcal{K}u = \sum_{n\in\mathbb{N}} (\mathcal{K}u, e_n)_{\mathbb{X}} e_n = \sum_{n\in\mathbb{N}} \lambda_n (u, e_n)_{\mathbb{X}} e_n.$$

Moreover, if $\lambda = 0$ is not an eigenvalue of $\mathcal{K}$, then every $u \in \mathbb{X}$ has a convergent Fourier series expansion by this orthonormal basis:

$$u = \sum_{n\in\mathbb{N}} (u, e_n)_{\mathbb{X}} e_n.$$

A.3 Invariant sets

In this section we present a proof of the existence of an invariant set associated with a finite number of contractions on a complete metric space as this result is important in several chapters of this book. We follow closely the presentation of the fundamental paper of Hutchinson [148].

Let $\mathbb{X} = (\mathbb{X}, d)$ be a metric space and F a function such that $F : \mathbb{X} \to \mathbb{X}$. If there is a constant $c \in (0,1)$ such that

$$d(F(x), F(y)) \leq cd(x, y)$$

for all $x, y \in \mathbb{X}$, then F is called a *contraction mapping* on $\mathbb{X}$. A basic fact is that any contraction on a complete metric space has a unique fixed point. We state and prove this fact next.

Theorem A.54 *If $\mathbb{X}$ is a complete metric space and F is a contraction on $\mathbb{X}$, then there exists one and only one $x \in \mathbb{X}$ such that $F(x) = x$.*

Proof First we establish the uniqueness of $x \in \mathbb{X}$. To this end, we assume that there are $x_1, x_2 \in \mathbb{X}$ such that $F(x_1) = x_1$ and $F(x_2) = x_2$. Since F is a contraction, it follow that $d(x_1, x_2) \leq cd(x_1, x_2)$, from which it follows that $d(x_1, x_2) = 0$, since $c \in (0,1)$.

To establish the existence of the fixed point for the function F we choose any $x_0 \in \mathbb{X}$, define recursively the sequence of points

$$x_{n+1} = F(x_n), \quad n \in \mathbb{N}_0$$

and observe that

$$\begin{aligned} d(x_{n+1}, x_n) &= d(F(x_n), F(x_{n-1})) \\ &\leq cd(x_n, x_{n-1}), \end{aligned}$$

from which it follows that

$$d(x_{n+1}, x_n) \leq c^n d(x_1, x_0).$$

Consequently, for $n < m$, we have that

$$\begin{aligned} d(x_m, x_n) &\leq \sum_{\ell \in \mathbb{N}_m \setminus \mathbb{N}_n} d(x_\ell, x_{\ell-1}) \\ &\leq \sum_{\ell \in \mathbb{N}_m \setminus \mathbb{N}_n} c^\ell d(x_1, x_0), \end{aligned}$$

which yields the inequality

$$d(x_m, x_n) \leq \frac{c^n}{1-c} d(x_1, x_0).$$

This means that $\{x_n : n \in \mathbb{X}\}$ is a Cauchy sequence and hence must converge to some $x \in \mathbb{X}$, that is, $\lim_{n \to \infty} x_n = x \in \mathbb{X}$. Since F is a continuous function we conclude that

$$F(x) = \lim_{n \to \infty} F(x_n) = \lim_{n \to \infty} x_{n+1} = x.$$

□

Definition A.55 Let $\mathbb{X}$ be a complete metric space. The Lipschitz constant of a function $F : \mathbb{X} \to \mathbb{X}$ is defined by

$$\text{Lip } F := \sup\left\{\frac{d(F(x), F(y))}{d(x, y)} : x, y \in \mathbb{X},\, x \neq y\right\}.$$

We say F is Lipschitz continuous when $\text{Lip } F < \infty$. A Lipschitz continuous mapping takes bounded sets into bounded sets but not necessarily closed sets into closed sets.

According to the definition,

$$d(F(x), F(y)) \le \text{Lip } F\, d(x, y)$$

and F is a contraction means that $\text{Lip } F < 1$. In addition, if $F, G : \mathbb{X} \to \mathbb{X}$, then

$$\text{Lip } F \circ G \le \text{Lip } F \cdot \text{Lip } G.$$

Thus, it follows from Definition A.3.2 that

$$\begin{aligned}\text{Lip } F \circ G &= \sup\left\{\frac{d(F(G(x)), F(G(y)))}{d(x, y)} : x, y \in \mathbb{X}, x \neq y\right\}\\ &\le \text{Lip } F \cdot \sup\left\{\frac{d(G(x), G(y))}{d(x, y)} : x, y \in \mathbb{X}, x \neq y\right\}\\ &\le \text{Lip } F \cdot \text{Lip } G.\end{aligned}$$

For $x \in \mathbb{X}$, $A \subseteq \mathbb{X}$, the distance from x to A is defined by the equation

$$d(x, A) := \inf\{d(x, a) : a \in A\}.$$

Let $\mathcal{B}$ be the class of nonempty closed bounded subsets of $\mathbb{X}$. The least closed set containing A, that is, the closure of A, is denoted by $\bar{A}$. Certainly, for any $x \in \mathbb{X}$ and $A \subseteq \mathbb{X}$ we have that $d(x, A) = d(x, \bar{A})$.

Definition A.56 The Hausdorff metric δ on $\mathcal{B}$ is defined for any $A, B \in \mathcal{B}$ by the equation

$$\delta(A, B) := \sup\{d(a, B), d(b, A) : a \in A, b \in B\}.$$

Lemma A.57 *If δ is the Hausdorff metric, then δ is a metric on $\mathcal{B}$.*

Proof For any $A, B \in \mathcal{B}$, we certainly have that $\delta(A, B) \ge 0$. If $\delta(A, B) = 0$ then any $a_0 \in A$ must have the property that $d(a_0, B) = 0$. But then it must be the case that $a_0 \in B$. In other words, we have verified that $A \subseteq B$ and likewise $B \subseteq A$, that is, $A = B$. Conversely, if $A = B$, then $d(a, B) = 0$ when $a \in A$

and $d(b, A) = 0$ when $b \in B$, from which it follows that $\delta(A, B) = 0$. We next point out that

$$\begin{aligned}\delta(B, A) &= \sup\{d(a, A), d(b, B) : a \in B, b \in A\} \\ &= \sup\{d(a, B), d(b, A) : a \in A, b \in B\} \\ &= \delta(A, B).\end{aligned}$$

Finally, we prove the triangle inequality. To this end, we suppose that $A, B, C \in \mathcal{B}$. The triangle inequality for the metric space $(\mathbb{X}, d)$ says that for any $a \in A$, $b \in B$ and $c \in C$,

$$d(a, B) \leq d(a, b) \leq d(a, c) + d(c, b),$$

which yields

$$d(a, B) \leq d(a, C) + d(b, C) \leq \delta(A, C) + \delta(B, C),$$

and likewise we conclude that

$$d(b, A) \leq \delta(A, C) + \delta(B, C).$$

Therefore, we obtain the desired fact that

$$\delta(A, B) \leq \delta(A, C) + \delta(C, B).$$

This completes the proof of this lemma. □

Lemma A.58 *If $(\mathbb{X}, d)$ is a complete metric space then $(\mathcal{B}, \delta)$ is a complete metric space.*

For the proof of this fact we find the following additional notation convenient. A neighborhood of a subset A of $\mathbb{X}$ is denoted by

$$N_r(A) := \{x : x \in \mathbb{X}, d(x, A) < r\} = \bigcup\{B(a, r) : a \in A\}.$$

According to this definition, we conclude that $A \subseteq N_r(B)$ if and only if for all $a \in A$ there exists a $b \in B$ such that $d(a, b) < r$ or equivalently, $d(a, B) < r$. Another way of expressing this condition is to merely say that

$$\sup\{d(a, B) : a \in A\} < r.$$

A fact relating these two simple metric concepts is a useful set inclusion relationship which says that for any $x \in \mathbb{X}$ and positive numbers r_1, r_2

$$N_{r_1}(B(x, r_2)) \subseteq B(x, r_1 + r_2). \tag{A.2}$$

A related notion is the concept of the *upper Hausdorff hemi-metric* given by the formula

$$\delta^*(A, B) := \inf\{r : r > 0, A \subseteq N_r(B)\} \tag{A.3}$$

and by our comments above we obtain that

$$\delta^*(A, B) = \sup\{d(a, B) : a \in A\}$$

as well as that

$$\delta(A, B) = \max\{\delta^*(A, B), \delta^*(B, A)\}. \tag{A.4}$$

We also point out by what we have already said that

$$\delta(A, B) = \inf\{r : r > 0, A \subseteq N_r(B), B \subseteq N_r(A)\}.$$

In the next lemma we list some additional basic properties concerning these concepts.

Lemma A.59

(a) *If A is open then for any $r > 0$ the set $N_r(A)$ is open.*
(b) *If $A \subseteq B$, then $N_r(A) \subseteq N_r(B)$.*
(c) *$N_r(A \cap B) \subseteq N_r(A) \cap N_r(B)$.*
(d) *For any collection $\{A_\gamma : \gamma \in \Gamma\}$ of subsets of $\mathbb{X}$ we have that*

$$N_r\left(\bigcup_{\gamma \in \Gamma} A_\gamma\right) = \bigcup_{\gamma \in \Gamma:} N_r(A_\gamma).$$

(e) *If $r_1 \leq r_2$ then $N_{r_1}(A) \subseteq N_{r_2}(A)$.*
(f) *For every r_1, r_2 we have that $N_{r_1}(N_{r_2}(K)) \subseteq N_{r_1+r_2}(K)$.*
(g) *$\delta(A, B) < \varepsilon$ if and only if there are positive numbers $r_1, r_2 < \varepsilon$ such that $A \subseteq N_{r_1}(B)$ and $B \subseteq N_{r_2}(A)$.*

Proof Part (a) follows directly from the definition of the set $N_r(A)$ as the union of open balls. The remaining assertions are also straightforward. Here are some of the details. For (b) and (c) we use the facts that for any $x \in \mathbb{X}$, and any sets A, B of $\mathbb{X}$, the following equation holds:

$$d(x, A \cap B) = \min\{d(x, A), d(x, B)\},$$

and when $A \subseteq B$ we have additionally $d(x, B) \leq d(x, A)$. For (d) we compute

$$\begin{aligned} N_r(\cup_{r\in\Gamma} A_r) &= \bigcup\{B(a, r) : a \in \bigcup_{r\in\Gamma} A_r\} \\ &= \bigcup\{\{B(a, r) : a \in A_r\} : r \in \Gamma\} \\ &= \bigcup\{N_r(A_r) : r \in \Gamma\}. \end{aligned}$$

Part (e) is obvious and (f) follows from the set inclusion (A.2). The final claim, which is especially useful, follows from equations (A.3) and (A.4). □

The next fact we use is stated in an ancillary lemma.

Lemma A.60 *If $\mathcal{A} := \{A_n : n \in \mathbb{N}\}$ is a Cauchy sequence in $\mathcal{B}$ then there is a ball B in $\mathbb{X}$ such that for all $n \in \mathbb{N}$ we have that $A_n \subseteq B$. That is, A_n is uniformly bounded for all $n \in \mathbb{N}$.*

Proof There is a positive integer k such that for any $n \geq k$ we have that $\delta(A_k, A_n) < \frac{1}{2}$ and so by using Lemma A.59, part (g), we conclude for $n \geq k$ that $A_n \subseteq N_{1/2}(A_k)$. Now, we choose a positive number r and $y \in \mathbb{X}$ such that $\bigcup_{j \in \mathbb{N}_k} A_j \subseteq B(y, r)$. Therefore, the ball $B := B(y, r + 1/2)$ has the desired property. □

Definition A.61 For every sequence $\mathcal{A} := \{A_n : n \in \mathbb{N}\}$ in $\mathcal{B}$ we define the set

$$C(\mathcal{A}) = \{x : x \in \mathbb{X} \text{ such that there is a subsequence } \{n_k : k \in \mathbb{N}\} \text{ and } x_{n_k} \in A_{n_k}, \text{ with } \lim_{k \to \infty} x_{n_k} = x\}.$$

Note that when $\mathcal{A}$ consists of one set A, the set $C(\mathcal{A})$ is the set of accumulation points of A and is called the *derived set* of $\mathcal{A}$.

We are now ready to prove Lemma A.58. To this end, we let $\mathcal{A} := \{A_n : n \in \mathbb{N}\}$ be a Cauchy sequence in $\mathcal{B}$. We prove it converges to $C(\mathcal{A})$.

Proof We divide the proof into several steps and begin with the statement that

(a) $C(\mathcal{A})$ is bounded.

Indeed, by Lemma A.60 there is a ball B such that for all $n \in \mathbb{N}$ we have that $A_n \subseteq B$. Now, let $x \in C(\mathcal{A})$ so there is some subsequence $\{n_k : k \in \mathbb{N}\}$ and $x_{n_k} \in A_{n_k}$ for which $\lim_{k \to \infty} x_{n_k} = x$. Therefore, we conclude that $x \in B$, that is, $C(\mathcal{A}) \subseteq B$.

(b) $C(\mathcal{A})$ is closed.

If $\{y_\ell : \ell \in \mathbb{N}\}$ is a sequence in $C(\mathcal{A})$ which converges to some $y \in \mathbb{X}$, then there are subsequences $\{n_k : k \in \mathbb{N}\}$ and $\{m_k : k \in \mathbb{N}\}$ such that for all $k \in \mathbb{N}$ we have that $d(y, y_{n_k}) < 2^{-k}, x_{m_k} \in A_{m_k}$ and $d(y_{n_k}, x_{m_k}) < 2^{-k}$. Hence, we obtain that $\lim_{k \to \infty} x_{m_k} = y$, which means that $y \in C(\mathcal{A})$.

(c) $C(\mathcal{A})$ is not empty.

Since $\{A_n : n \in \mathbb{N}\}$ is a Cauchy sequence there is a subsequence $\{n_k\}$ such that for all $k \in \mathbb{N}$ the inequality $\delta(A_{n_k}, A_{n_{k+1}}) < 2^{-k}$ holds. In particular, we obtain for all $a \in A_{n_k}$ that $d(a, A_{n_{k+1}}) < 2^{-k}$ and so we can construct inductively a sequence $x_{n_k} \in A_{n_k}$ such that $d(x_{n_k}, x_{n_{k+1}}) < 2^{-k}$. This means that $\{x_{n_k} : k \in \mathbb{N}\}$ is a Cauchy sequence in $\mathbb{X}$ and so $\lim_{k \to \infty} x_{n_k} = x_0$. That is, we conclude that $x_0 \in A$.

So far we have demonstrated that $C(\mathcal{A}) \subseteq \mathcal{B}$. Now we prove that $\lim_{n\to\infty} A_n = C(\mathcal{A})$ in $(\mathcal{B}, \delta)$. For this purpose, we choose $\varepsilon > 0$ and demonstrate that there is a $p \in \mathbb{N}$ such that for $m > p$ we have $\delta(C(\mathcal{A}), A_m) < \varepsilon$. By hypothesis, corresponding to this ϵ there is an integer k such that for all $k < n < m$ we have

$$\delta(A_m, A_n) < \varepsilon/2. \tag{A.5}$$

Using this fact we first show for every $x \in C(\mathcal{A})$ and $n > k$ that $d(x, A_n) < \varepsilon$. Indeed, by the definition of the set $C(\mathcal{A})$ there is a subsequence $\{n_\ell : \ell \in \mathbb{N}\}$ with $x_{n_\ell} \in A_{n_\ell}$ and $d(x, x_{n_\ell}) < \epsilon/2$ for ℓ sufficiently large. We choose ℓ even larger, say $\ell > q$, so that not only does this inequality hold but also $n_\ell > n$, thereby guaranteeing by (A.5) for $n > k$ and $\ell > q$ that $\delta(A_{n_\ell}, A_n) < \varepsilon$. Consequently, we have for $n > k$ that $d(x, A_n) < \varepsilon$, which is the desired inequality.

Next, we show that for all $n > k$ and $y \in A_n$, $d(y, C(\mathcal{A})) < \varepsilon$. We do this by constructing an $x \in C(\mathcal{A})$ so that $d(y, x) < \varepsilon$. Here we again appeal to (A.5) to define inductively, for a subsequence $\{m_\ell : \ell \in \mathbb{Z}_+\}$ with $x_{m_\ell} \in A_{m_\ell}$ and $x_0 = y$ and for all $\ell \in \mathbb{Z}_+$, that $d(x_{m_\ell}, x_{m_{\ell+1}}) < \frac{\varepsilon}{2^\ell}$. Clearly, we have that $\{x_\ell : \ell \in \mathbb{N}\}$ is a Cauchy sequence and hence converges to some $x \in \mathbb{X}$. Moreover, we observe for all $\ell \in \mathbb{Z}_+$ that

$$d(y, x_{m_{\ell+1}}) < \varepsilon \left\{ \frac{1}{2} + \frac{1}{2^2} + \cdots \right\} = \varepsilon,$$

and so $d(y, x) < \varepsilon$. □

For the next lemma we let $\mathcal{C}$ be the collection of all compact subsets of the complete metric space $\mathbb{X}$. Clearly, the set $\mathcal{C}$ is a subset of $\mathcal{B}$. In the next lemma, we prove that it is a closed subset of $\mathcal{B}$ in the Hausdorff metric on $\mathcal{B}$.

Lemma A.62 *If $\mathbb{X}$ is a complete metric space then the set $\mathcal{C}$ is a closed subset of (B, δ). Therefore, in particular, $(\mathcal{C}, \delta)$ is a complete metric space.*

Proof Suppose $\mathcal{A} = \{A_n : n \in \mathbb{N}\}$ is a sequence in $\mathcal{C}$ such that $\lim_{n\to\infty} A_n = A$ in the metric δ. We show that the set A is compact by establishing that it is complete and totally bounded (see Theorem A.7). Since $(\mathcal{B}, \delta)$ is complete we can then be assured that A is in $\mathcal{B}$ too. Certainly, we already know by the proof of Lemma A.59 that $A = C(\mathcal{A})$ and so is in $\mathcal{B}$. Recall that a closed subset of a complete metric space must be complete. In particular, we conclude that A is complete. Next, we show that it is totally bounded. To this end, we choose any $\epsilon > 0$. By our hypothesis, there is an $m \in \mathbb{N}$ such that $\delta(A, A_m) < \epsilon/2$. Using part (g) of Lemma A.59 we conclude that there is an $r \in (0, \epsilon/2)$ such that $A \subseteq N_r(A_m)$. However, the set A_m is compact and consequently there is a finite set of points $\{x_j : j \in \mathbb{N}_k\} \subseteq \mathbb{X}$ such that

$$A_m \subseteq \bigcup\{B(x_j, \epsilon/2) : j \in \mathbb{N}_k\}.$$

Again, using Lemma A.59 we get that

$$A \subseteq \bigcup\{B(x_j, \epsilon) : j \in \mathbb{N}_k\},$$

and so A is totally bounded. This proves that $A \in \mathcal{C}$ and we conclude that indeed $\mathcal{C}$ is a closed subset of $\mathcal{B}$ in the metric δ. □

For later purposes, we require the next lemma.

Lemma A.63 *For any Lipschitz continuous mapping $F : \mathbb{X} \to \mathbb{X}$ and subset $A, B \in \mathbb{X}$ we have that*

$$\delta(\overline{F(A)}, \overline{F(B)}) \leq (\text{Lip } F) \cdot \delta(A, B). \tag{A.6}$$

Proof First, we observe for any subset A, B of $\mathbb{X}$ that $\delta(\overline{A}, \overline{B}) \leq \delta(A, B)$. To see this, we choose any $\epsilon > \delta(A, B)$ and a positive δ. By Lemma A.59, part (g), there are positive constants $r_1, r_2 < \epsilon$ such that $A \subseteq N_{r_1}(B)$ and $B \subseteq N_{r_2}(A)$. Consequently, we get that $\overline{A} \subseteq N_{r_1+\delta}(\overline{B})$ and $\overline{B} \subseteq N_{r_2+\delta}(\overline{A})$. Therefore, again by Lemma A.59, part (g), we have that $\delta(\overline{A}, \overline{B}) < \epsilon + \delta$. Now, we let $\delta \to 0^+$ and $\epsilon \to \delta(A, B)$, from above, and conclude, as claimed, that $\delta(\overline{A}, \overline{B}) \leq \delta(A, B)$.

Applying this preliminary observation to the inequality (A.6), we reduce the proof to establishing that

$$\delta(F(A), F(B)) \leq (\text{Lip } F)\delta(A, B).$$

To this end, for a given $u \in F(A)$, there is $v \in A$ with $u = F(v)$. Now, choose any $x \in B$ and observe that

$$d(u, F(B)) \leq d(u, F(x)) = d(F(v), F(x)) \leq (\text{Lip } F)d(v, x).$$

Since x was chosen arbitrarily in B, we get that

$$d(u, F(B)) \leq (\text{Lip } F)d(v, B) \leq (\text{Lip } F)\delta(A, B).$$

Similarly, for $u \in F(B)$ with $u = F(v)$, where $v \in B$, we conclude, as above, for all $x \in A$ that

$$d(F(A), u) \leq d(F(x), F(v)) \leq (\text{Lip } F)d(x, v).$$

Consequently, we obtain that

$$d(F(A), u) \leq (\text{Lip } F)d(A, v) \leq (\text{Lip } F)\delta(A, B)$$

and also

$$\delta(F(A), F(B)) \le (\mathrm{Lip}\, F)\delta(A, B).$$

□

The next lemma is also needed.

Lemma A.64 *If* $\{A_\gamma : \gamma \in \Gamma\}$ *and* $\{B_\gamma : \gamma \in \Gamma\}$ *are families of subsets of* $\mathbb{X}$, *then*

$$\delta\left(\bigcup_{\gamma\in\Gamma} A_\gamma, \bigcup_{\gamma\in\Gamma} B_\gamma\right) \le \sup\{\delta(A_\gamma, B_\gamma) : \gamma \in \Gamma\}.$$

Proof For any $x \in \bigcup_{\gamma\in\Gamma} A_\gamma$ there is a $\mu \in \Gamma$ such that $x \in A_\mu$ and so we see that

$$d\left(x, \bigcup_{\gamma\in\Gamma} B_\gamma\right) \le d(x, B_\mu) \le \delta(A_\mu, B_\mu) \le \sup\{\delta(A_\gamma, B_\gamma) : \gamma \in \Gamma\}.$$

Similarly, for any $y \in \bigcup_{\gamma\in\Gamma} B_\gamma$ there corresponds a $\mu \in \Gamma$ with $y \in B_\mu$, and so

$$d\left(\bigcup_{\gamma\in\Gamma} A_\gamma, y\right) \le d(A_\mu, y) \le \delta(A_\mu, B_\mu) \le \sup\{\delta(A_\gamma, B_\gamma) : \gamma \in \Gamma\}.$$

Combining these two inequalities proves the lemma. □

If A is nonempty, then the $S_\ell(A)$s are nonempty, and so is $S(A)$. Therefore, $S : \mathcal{B} \to \mathcal{B}$, $\mathcal{B}$ is a set of closed bounded nonempty sets. We know that $(\mathcal{B}, \delta)$ is a complete metric space.

We have now prepared the basic facts needed about the Hausdorff metric and can address the essential issue of this section, namely, the construction of invariant sets. We start with a complete metric space $\mathbb{X}$ and a finite family of contractive mappings

$$\Phi := \{\phi_\epsilon : \epsilon \in \mathbb{Z}_\mu\}.$$

Corresponding to the family Φ is a set-valued mapping which is introduced in the next definition.

Definition A.65 Let $\Phi := \{\phi_\epsilon : \epsilon \in \mathbb{Z}_\mu\}$ be a finite family of contraction mappings on a metric space $\mathbb{X}$. For every subset A of $\mathbb{X}$ we define the set-valued mapping Φ_+ at A by the formula

$$\Phi_+(A) := \bigcup\{\overline{\phi_\epsilon(A)} : \epsilon \in \mathbb{Z}_\mu\}.$$

Clearly, by what has already been said we conclude that $\Phi_+ : \mathcal{B} \to \mathcal{B}$ and also we have that $\Phi_+ : \mathcal{C} \to \mathcal{C}$. Moreover, for $A \in \mathcal{C}$, the set $\Phi_+(A)$ reduces to

$$\Phi_+(A) := \bigcup \{\phi_\epsilon(A) : \epsilon \in \mathbb{Z}_\mu\}.$$

The essential property of the set-valued mapping Φ_+ is given in the next lemma. For its statement, we introduce the constant

$$\lambda(\Phi) := \max\{\mathrm{Lip}\, \phi_\epsilon : \epsilon \in \mathbb{Z}_\mu\}$$

which is certainly less than one.

Lemma A.66 *If $\Phi := \{\phi_\epsilon : \epsilon \in \mathbb{Z}_\mu\}$ is a finite family of contraction mappings on a metric space then the set-valued mapping $\Phi_+ : \mathcal{B} \to \mathcal{B}$ is a contraction relative to the Hausdorff metric. In fact, we have the inequality*

$$\mathrm{Lip}\, \Phi_+ \leq \lambda(\Phi)$$

for its Lipschitz constant.

Proof Let $A, B \in \mathcal{B}$ and use the two previous lemmas to conclude that

$$\begin{aligned}\delta(\Phi_+(A), \Phi_+(B)) &\leq \max\{\delta(\overline{\phi_\epsilon(A)}, \overline{\phi_\epsilon(B)}) : \epsilon \in \mathbb{Z}_\mu\} \\ &\leq \lambda(\Phi_+)\delta(A, B).\end{aligned}$$

□

Theorem A.67 *If $\Phi := \{\phi_\epsilon : \epsilon \in \mathbb{Z}_\mu\}$ is a finite family of contraction mappings on a metric space $\mathbb{X}$ then there is a unique $K \in \mathcal{C}$ such that $K = \Phi_+(K)$. Moreover, there is at most one $K \in \mathcal{B}$ which satisfies this equation.*

Proof This result follows from the contraction mapping principle. Specifically, the uniqueness of $K \in \mathcal{B}$ follows from the fact that Φ_+ is a contraction on $(\mathcal{B}, \delta)$ while the existence of $K \in C$ follows from the fact that (C, δ) is also a complete metric space. □

Remark. Any subset K of $\mathbb{X}$ which satisfies the equation $K = \Phi_+(K)$ is called an invariant set of the collection of mappings Φ.

We end this appendix with an alternative construction of K which is very relevant to the presentation in Chapter 4. The additional information provided is that an invariant set $K \in \mathcal{C}$ can be obtained from the fixed points of a finite number of compositions of mappings chosen from the collection Φ. For an explanation of this procedure we review the appropriate notation used in Chapter 4. We choose $\mathbb{Z}_\mu^p$ to denote all ordered sequences of p integers, $p \in \mathbb{N}$, selected from $\mathbb{Z}_\mu$. Every $\mathbf{e} \in \mathbb{Z}_\mu^p$ is written in vector form $\mathbf{e} = [\epsilon_e : e \in \mathbb{Z}_p]$, where each ϵ_e is in $\mathbb{Z}_\mu$ and associated with this vector is the composition mapping

$$\phi_{\mathbf{e}} = \phi_{\epsilon_0} \circ \phi_{\epsilon_1} \cdots \phi_{\epsilon_{p-1}}.$$

Associated with the family of contraction mappings Φ is the *p-compound family* of mappings

$$\Phi^p := \{\phi_{\mathbf{e}} : \mathbf{e} \in \mathbb{Z}_\mu^p\}.$$

We show next that Φ and Φ^p share the same invariant set. We start with a preliminary lemma.

Lemma A.68 *If K is an invariant set of the family of contractions $\Phi := \{\phi_\epsilon : \epsilon \in \mathbb{Z}_\mu\}$ and $\mathbf{e} = [\epsilon_i : i \in \mathbb{Z}_\mu] \in \mathbb{Z}_\mu^p$, then $\phi_{\mathbf{e}}(K) = \bigcup\{(\phi_{\mathbf{e}} \circ \phi_\epsilon)(K) : \epsilon \in \mathbb{Z}_\mu\}$.*

Proof In the above equation the case $p = 0$ just means that K is an invariant set for the collection of mappings Φ. Now, for $p \in \mathbb{N}, \mathbf{e} \in \mathbb{Z}_\mu^p$ we merely compute that

$$\phi_{\mathbf{e}}(K) = \phi_{\mathbf{e}}(\cup\{\phi_\epsilon(K) : \epsilon \in \mathbb{Z}_\mu\}) = \bigcup\{(\phi_{\mathbf{e}} \circ \phi_\epsilon)(K) : \epsilon \in \mathbb{Z}_\mu\}.$$

□

Proposition A.69 *If $K = \Phi_+(K)$ and $p \in \mathbb{N}$ then*

$$K = (\Phi_+^p)(K).$$

Proof The case $p = 1$ is immediate. The remaining cases follow by induction on p. Specifically, if $p > 1$ and the result is valued for $p - 1$ we compute

$$\begin{aligned} K &= \bigcup\{\phi_{\mathbf{e}}(K) : \mathbf{e} \in \mathbb{Z}_\mu^{p-1}\} \\ &= \bigcup\{(\phi_{\mathbf{e}} \circ \phi_\epsilon)(K) : \mathbf{e} \in \mathbb{Z}_\mu^{p-1}, \epsilon \in \mathbb{Z}_\mu\} \\ &= \bigcup\{\phi_{\mathbf{e}}(K) : \mathbf{e} \in \mathbb{Z}_\mu^p\}. \end{aligned}$$

□

For the next series of lemmas we use the notation $\mathbb{Z}_\mu^\infty$ for all ordered sequences of integers, selected from $\mathbb{Z}_\mu$. Every $\mathbf{e} \in \mathbb{Z}_\mu^\infty$ is written in vector form $\mathbf{e} = [\epsilon_e : e \in \mathbb{N}_0]$ where each component ϵ_e is in $\mathbb{Z}_\mu$. There is a natural association of any $\mathbf{e} \in \mathbb{Z}_\mu^\infty$ with an element in $\mathbb{Z}_\mu^p$ by truncating it to its first p components, namely, $\mathbf{e}_p := [\epsilon_e : e \in \mathbb{Z}_p]$, that is to say, every $\mathbf{e} \in \mathbb{Z}_\mu^\infty$ gives rise to the sequence of vectors $\{\mathbf{e}_p : p \in \mathbb{N}\}$. Of course, every $\mathbf{e} \in \mathbb{Z}_\mu^p$ may be realized as $\mathbf{e}_p$ for many choices of $\mathbf{e} \in \mathbb{Z}_\mu^\infty$.

For every subset A in $\mathbb{X}$ we set

$$A_{\mathbf{e}} = \overline{\phi_{\mathbf{e}}(A)}$$

so that $A_{\mathbf{e}} \in \mathcal{B}$. We use the convention that for $p = 0$, the mapping $\phi_{\mathbf{e}}, \mathbf{e} \in \mathbb{Z}_\mu^p$ is the identity mapping on $\mathbb{X}$. Hence, if $A \in \mathcal{B}$ then in this case $A_{\mathbf{e}} = A$.

The next sequence of lemmas study what happens when $p \to \infty$. For this purpose we recall that the diameter of a subset A of a metric space $\mathbb{X}$ is defined as

$$\operatorname{diam} A := \sup\{d(x, y) : x, y \in A\}.$$

It follows readily that a bounded set has a finite diameter and the diameter of a set and its closure are the same.

Lemma A.70 *If A is a bounded subset of a metric space $\mathbb{X}$, Φ a finite collection of contractive mappings on $\mathbb{X}$ and $\mathbf{e} \in \mathbb{Z}_\mu^\infty$, then*

$$\lim_{p \to \infty} \operatorname{diam} A_{\mathbf{e}_p} = 0.$$

Proof For any $x, y \in \phi_{\mathbf{e}_p}(A)$ there are $u, v \in A$ such that $x = \phi_{\mathbf{e}_p} u$ and $y = \phi_{\mathbf{e}_p} v$. Therefore, we conclude that

$$d(x, y) \leq \lambda(\phi)^p d(u, v) \leq \lambda(\phi)^p \operatorname{diam} A.$$

That is, we have that

$$\operatorname{diam} A_{\mathbf{e}_p} \leq \lambda(\phi)^p \operatorname{diam} A \tag{A.7}$$

from which the lemma follows. □

Lemma A.71 *If K is an invariant set in $\mathcal{B}$ of a finite number of contractive mappings Φ on a complete metric space $\mathbb{X}$ and $\mathbf{e} \in \mathbb{Z}_\mu^\infty$, then the family of subsets $\{K_{\mathbf{e}_p} : p \in \mathbb{N}\}$ of K is nested and*

$$K_{\mathbf{e}} := \cap\{K_{\mathbf{e}_p} : p \in \mathbb{N}\}$$

consists of exactly one point $k_{\mathbf{e}} \in \mathbb{X}$.

Proof For each $p \in \mathbb{N}$ and $\mathbf{e} \in \mathbb{Z}_\mu^\infty$ we have, by Proposition A.69, that

$$\begin{aligned} K_{\mathbf{e}_p} &= \overline{\Phi_{\mathbf{e}_p}(K)} \\ &= \overline{\Phi_{\mathbf{e}_p}(\cup\{\phi_\epsilon(K) : \epsilon \in \mathbb{Z}_\mu\})} \\ &= \cup\{(\Phi_{\mathbf{e}_p} \circ \phi)(K) : \epsilon \in \mathbb{Z}_\mu\} \\ &\supseteq K_{\mathbf{e}_{p+1}}. \end{aligned}$$

Moreover, since K is an invariant set in $\mathcal{B}$ it follows from the above inclusion that $K_{\mathbf{e}_p} \subseteq K$.

Now, if $x, y \in K_{\mathbf{e}}$ then by inequality (A.7) for any $p \in \mathbb{N}$ we have that

$$d(x, y) \leq \lambda(\Phi)^p \operatorname{diam} K.$$

We now let $p \to \infty$ and conclude that $x = y$. That is, the set $K_{\mathbf{e}}$ consists of at most one point.

It remains to prove that $K_{\mathbf{e}}$ is nonempty. Indeed, since K is nonempty for any $p \in \mathbb{N}$, there is an $x_p \in K_{\mathbf{e}_p}$ and we claim that the sequence $\{x_p : p \in \mathbb{N}\}$ is a Cauchy sequence in $\mathbb{X}$. To this end, we suppose that $p \leq q$ then $x_p, x_q \in K_{\mathbf{e}_p}$. So, we conclude that

$$d(x_p, x_q) \leq \lambda(\Phi)^p \operatorname{diam} K,$$

from which it follows that $\{x_p : p \in \mathbb{N}\}$ is a Cauchy sequence. Therefore, there is an $x \in \mathbb{X}$ for which $x = \lim_{q\to\infty} x_q$. However, $x_q \in K_{\mathbf{e}_p}$ for all $q \geq p$, from which we conclude that $x \in K_{\mathbf{e}_p}$ because $K_{\mathbf{e}_p}$ is a closed subset of $\mathbb{X}$. Consequently, we have that $x \in K_{\mathbf{e}}$, which proves the result. □

Lemma A.72 *If K is an invariant set of a finite number of contractive mappings Φ in $\mathcal{B}$ then*

$$K = \{k_{\mathbf{e}} : \mathbf{e} \in \mathbb{Z}_\mu^\infty\}.$$

Proof By Lemma A.71 we have that

$$\{k_{\mathbf{e}} : \mathbf{e} \in \mathbb{Z}_\mu^\infty\} \subseteq K.$$

Now, we choose an $x_0 \in K$. Since K is an invariant set corresponding to the family Φ there is an $\epsilon_0 \in \mathbb{Z}_\mu$ and an $x_1 \in \mathbb{X}$ such that $x_0 = \phi_{\epsilon_0} x_1$. Repeating this process, we create an $\mathbf{e} \in \mathbb{Z}_\mu^\infty$ such that $x_0 = \phi_{\mathbf{e}_p} x_p$. In particular, we conclude that $x_0 \in K_{\mathbf{e}_p}$ and so also $x_0 \in K_{\mathbf{e}}$. This means, by Lemma A.71, that $x_0 = k_{\mathbf{e}}$, thereby establishing the claim. □

Our next observation demonstrates that for any $\mathbf{e} \in \mathbb{Z}_\mu^\infty$ the point $k_{\mathbf{e}} \in K$ can be constructed as a limit of the fixed points $x_{\mathbf{e}_p}$ of the contraction mapping $\phi_{\mathbf{e}_p}$. We first demonstrate that $x_{\mathbf{e}_p} \in K$ for any $\mathbf{e} \in \mathbb{Z}_\mu^\infty$ and for any $p \in \mathbb{N}$. To this end, we introduce a map from $\mathbb{Z}_\mu^p$ to $\mathbb{Z}_\mu^\infty$. Specifically, corresponding to an $\mathbf{e} \in \mathbb{Z}_\mu^p$ expressed as $\mathbf{e} = [\epsilon_l : l \in \mathbb{Z}_p]$, we let $\widehat{\mathbf{e}}_p \in \mathbb{Z}_\mu^\infty$ be the infinite vector obtained by repeating the components of the vector $\mathbf{e} \in \mathbb{Z}_\mu^p$ infinitely often. That is, we define

$$\widehat{\mathbf{e}}_p := [\epsilon_0, \cdots, \epsilon_{p-1}, \epsilon_0, \ldots, \epsilon_{p-1}, \ldots]^T.$$

Proposition A.73 *For every $p \in \mathbb{N}$ and $\mathbf{e} \in \mathbb{Z}_\mu^p$ we have that*

$$k_{\widehat{\mathbf{e}}_p} = x_{\mathbf{e}}.$$

Proof We need to show that

$$\phi_{\mathbf{e}}(k_{\widehat{\mathbf{e}}_p}) = k_{\widehat{\mathbf{e}}_p}$$

for any $\mathbf{e} \in \mathbb{Z}_\mu^p$. As a first step, we have that

$$\phi_{\mathbf{e}}(k_{\widehat{\mathbf{e}}_p}) \in \phi_e\left(\bigcap\{K_{(\widehat{\mathbf{e}}_p)_q} : q \in \mathbb{N}\}\right)$$

and so we get

$$\phi_{\mathbf{e}}(k_{\widehat{\mathbf{e}}_p}) \in \bigcap\left\{\overline{(\phi_{\mathbf{e}} \circ \phi_{(\widehat{\mathbf{e}}_p)_q})(K)} : q \in \mathbb{N}\right\}.$$

However, by our notational convention

$$\phi_{\mathbf{e}} \circ \phi_{(\widehat{\mathbf{e}}_p)_q} = \phi_{(\widehat{\mathbf{e}}_p)_{p+q}}.$$

But then, Lemma A.71 implies that

$$\phi_{\mathbf{e}}(k_{\widehat{\mathbf{e}}_p}) \in \bigcap\{K_{(\widehat{\mathbf{e}}_p)_q} : q \in \mathbb{N}\} = \{k_{\widehat{\mathbf{e}}_p}\}.$$

Since $\phi_{\mathbf{e}}$ has a unique fixed point, the proof of the proposition is complete. □

We also require the next result.

Lemma A.74 *For each $\mathbf{e} \in \mathbb{Z}_\mu^\infty$ we have that*

$$k_{\mathbf{e}} = \lim_{p\to\infty} x_{\mathbf{e}_p}.$$

Proof For any $\mathbf{e} \in \mathbb{Z}_\mu^\infty$ and $p \in \mathbb{N}$ we have, by Proposition A.73, that $x_{\mathbf{e}_p} \in K\mathbf{e}_p$. But we also have that $k\mathbf{e} \in K\mathbf{e}_p$. So, we conclude by inequality (A.7) that

$$d(x_{\mathbf{e}_p}, k\mathbf{e}) \leq \lambda(\Phi)^p \operatorname{diam} K.$$

Letting $p \to \infty$ in this inequality proves the result. □

As a corollary we obtain the following result, part of which was proved earlier in an even stronger form. To start with, we let W be the smallest closed set containing all fixed points $x_{\mathbf{e}}$, where $\mathbf{e} \in \mathbb{Z}_\mu^p$ and $p \in \mathbb{N}$.

Corollary A.75 *If $K \in \mathcal{B}$ is an invariant set for the contractive mapping Φ, then $K = W$.*

Proof By Proposition A.73 it follows that $W \subseteq K$, while Lemma A.74 implies that $K \subseteq W$. □

This corollary is the key to constructing an invariant set for Φ. We continue to explain further properties of K and specifically comment on its representation given in Lemma A.62.

This requires putting the Tychnoff topology on $\mathbb{Z}_\mu^\infty$. Specifically, we let $\mathbb{Z}_\mu$ have the discrete topology, that is, all sets are open. We view $\mathbb{Z}_\mu^\infty$ as a function from $\mathbb{N}_0$ into $\mathbb{Z}_\mu$ and give $\mathbb{Z}_\mu^\infty$ the weakest topology, so that all component maps are continuous. In other words, for each $i \in \mathbb{N}_0$ the map which takes

$\mathbf{e} = [\epsilon_k : k \in \mathbb{N}_0]$ into ϵ_i is continuous in this topology on $\mathbb{Z}_\mu^\infty$. Therefore, as a special case of the Tychnoff theorem, see [183, 276], we conclude that $\mathbb{Z}_\mu^\infty$ is compact in this topology.

Definition A.76 We define the map $\psi : \mathbb{Z}_\mu^\infty \to K$ at $\mathbf{e} \in \mathbb{Z}_\mu^\infty$ by the equation $\psi(\mathbf{e}) = k_{\mathbf{e}}$.

Lemma A.77 *The map ψ defined above is continuous.*

Remark A.78 *According to Lemma A.77, the mapping is onto. Hence, this lemma implies any invariant set $K \in \mathcal{B}$ must be compact, confirming in an alternative manner a part of Theorem A.67.*

Proof We show that ψ is continuous at any $\mathbf{e} \in \mathbb{Z}_\mu^p$. Thus, we must show for any $\epsilon > 0$ that the inverse image of the ball $B(k_{\mathbf{e}}, \epsilon)$ in $\mathbb{X}$ contains an open neighborhood of $\mathbf{e} \in \mathbb{Z}_\mu^\infty$. Corresponding to this ϵ we choose a positive integer q such that diam $K_{\mathbf{e}_q} \leq \epsilon$. The existence of such an integer is guaranteed by Lemma A.70 applied to the set K. The set

$$O = \{\widetilde{\mathbf{e}} : \widetilde{\mathbf{e}} \in \mathbb{Z}_\mu^\infty, \widetilde{\mathbf{e}}_q = \mathbf{e}_q\}$$

is an open neighborhood of $\mathbf{e}$ in the topology on $\mathbb{Z}_\mu^\infty$. Moreover, if $\widetilde{\mathbf{e}} \in O$ then it follows that $k_{\widetilde{\mathbf{e}}} \in K_{\mathbf{e}_q}$. That is, if $\widetilde{\mathbf{e}} \in O$ then $d(k_{\widetilde{\mathbf{e}}}, k_{\mathbf{e}_q}) \leq \epsilon$. This means that $O \subseteq \psi^{-1}(B(k_{\mathbf{e}}, \epsilon))$, which proves the lemma. □

We now turn to the iterates of the set-valued map Φ_+. We denote them as Φ_+^p, defined on any subset A iteratively by the formula

$$\Phi_+^p(A) = \Phi_+(\Phi_+^{p-1}(A))$$

for $p \geq 1$ and $\Phi_+^0(A) = \bar{A}$.

According to Lemma A.66 we have that

$$\delta(\Phi_+^p(A), K) \leq \lambda(\Phi)^p \delta(A, K),$$

and so for any nonempty bounded subset A of $\mathbb{X}$ we conclude that

$$\lim_{p \to \infty} \Phi_+^p(A) = K.$$

We are now going to establish the existence of the limit in Lemma A.74 without assuming the existence of an invariant set.

We start from the fixed point x_ϵ of the map ϕ_ϵ, for each $\epsilon \in \mathbb{Z}_\mu$, define constants $\rho := \max\{d(S_\epsilon, S_\theta) : \epsilon, \theta \in \mathbb{Z}_\mu\}$ and $r = \rho(1 - \lambda(\Phi))^{-1}$ and a set $V \in \mathcal{B}$ given as

$$V = \bigcap \{\overline{B(x_\epsilon, r)} : \epsilon \in \mathbb{Z}_\mu\}.$$

Lemma A.79 *If Φ is a finite family of contraction maps on a complete metric space $\mathbb{X}$ and $\mathbf{e} \in \mathbb{Z}_\mu^\infty$, then $\lim_{p\to\infty} x_{\mathbf{e}_p}$ exists and is the unique point in the set*

$$V_{\mathbf{e}} := \bigcap \{V_{\mathbf{e}_p} : p \in \mathbb{N}\}.$$

Proof The proof proceeds as before. First, we establish that the sets $\{V_{\mathbf{e}_p} : p \in \mathbb{N}\}$ are nested in a decreasing manner.

By the choice of constants r and ρ we have, for $\lambda := \lambda(\Phi)$, that

$$\bigcup \{B(x_\epsilon, \lambda_r) : \epsilon \in \mathbb{Z}_\mu\} \subseteq V. \tag{A.8}$$

In fact, if $d(x, x_\theta) \le r$ for some $\theta \in \mathbb{Z}_\mu$ we have that

$$d(x, x_\epsilon) \le d(x, x_\theta) + d(x_\theta, x_\epsilon) \le \lambda r + \rho = r.$$

Now, if $u \in \phi_\epsilon(V)$ then there is a $\upsilon \in V$ such that $u = \phi_\epsilon \upsilon$. Since $\upsilon \in V$ we know for any $\epsilon \in \mathbb{Z}_\mu$ that $d(\upsilon, x_\epsilon) \le r$ and so it follows that

$$d(u, x_\epsilon) = d(\phi_\epsilon(\upsilon), \phi(x_\epsilon)) \le \lambda d(\upsilon, x_\epsilon) \le \lambda r.$$

Consequently, by the set inclusion (A.8), we conclude that $u \in V$. In other words, we have established that $\phi_\epsilon(V) \in V$ for any $\epsilon \in \mathbb{Z}_\mu$.

From this observation it follows directly for any $p \in \mathbb{N}$ that $V_{\mathbf{e}_p} \supseteq V_{\mathbf{e}_{p+1}}$. This implies that the contraction map $\phi_{\mathbf{e}_p}$ has the property that $\phi_{\mathbf{e}_p} : V \to V$. Since V is closed the unique fixed point of $\phi_{\mathbf{e}_p}$ lies in V, that is, $x_{\mathbf{e}_p} \in V$. As before, we argue as in Lemma A.70 that diam $V_{\mathbf{e}_p} \le \lambda^p$ diam V. Thus, as in the proof of Lemma A.71 we conclude that $V_{\mathbf{e}}$ consists of at most one point, that is, the sequence $\{x_{\mathbf{e}_p} : p \in \mathbb{N}\}$ is a Cauchy sequence and $\lim_{p\to\infty} x_{\mathbf{e}_p} = x_{\mathbf{e}}$, the unique point in $V_{\mathbf{e}}$. This proves the lemma. □

We can now *define* the subset $K := \{x_{\mathbf{e}} : e \in \mathbb{Z}_\mu^\infty\}$ of $\mathbb{X}$. As in the remark following Lemma A.77, it follows that K is compact. Indeed, the map $\mathbb{Z}_\mu^\infty \ni \mathbf{e} \to x_{\mathbf{e}} \in K$ is continuous by the same argument used to prove Lemma A.77. It remains only to establish that K is an invariant set of the collection of invariant sets Φ. For this purpose, for each $\epsilon \in \mathbb{Z}_\mu$ and $\mathbf{e} = [\epsilon_j : j \in \mathbb{N}] \in \mathbb{Z}_\mu^\infty$, we define

$$\epsilon\mathbf{e} := [\epsilon, \epsilon_0, \epsilon_1, \ldots]$$

and likewise for $\mathbf{e} \in \mathbb{Z}_\mu^\infty$. We claim that $\psi_\epsilon(x_{\mathbf{e}}) = x_{\epsilon\mathbf{e}}$ for each $\mathbf{e} \in \mathbb{Z}_\mu^\infty$.

To see this, we observe for each $\mathbf{e} \in \mathbb{Z}_\mu^\infty$ and $\epsilon \in \mathbb{Z}_\mu$ that

$$\begin{aligned}\psi_\epsilon(x_{\mathbf{e}}) \in \psi_\epsilon\Big(\bigcap\{V_{\mathbf{e}_p} : p \in \mathbb{N}\}\Big) \\ \subseteq \bigcap\{\psi_\epsilon(V_{\mathbf{e}_p}) : p \in \mathbb{N}\} \\ = \bigcap\{V_{\epsilon\mathbf{e}_p} : p \in \mathbb{N}\} \\ = \{x_{\epsilon\mathbf{e}}\}.\end{aligned}$$

In other words, we do indeed have that $\psi_\epsilon(x_{\mathbf{e}}) = x_{\epsilon e}$. Now, if $x_{\mathbf{e}} \in K$ we write $\mathbf{e} = \epsilon_0\tilde{\mathbf{e}}$ for some $\epsilon_0 \in \mathbb{Z}_\mu$ and $\tilde{\mathbf{e}} \in \mathbb{Z}_\mu^\infty$, and conclude that

$$x_{\mathbf{e}} = x_{\epsilon_0\tilde{\mathbf{e}}} = \psi_{\epsilon_0}(x_{\tilde{\mathbf{e}}}) \in \bigcup\{\psi_\epsilon(K) : \epsilon \in \mathbb{Z}_\mu\}.$$

Similarly, if $u \in \bigcup\{\psi_\epsilon(K) : \epsilon \in \mathbb{Z}_\mu\}$ we get that $u = \psi_{\tilde{\epsilon}} x_{\tilde{\mathbf{e}}}$ for some $\tilde{\epsilon} \in \mathbb{Z}_\mu$ and $\tilde{\mathbf{e}} \in \mathbb{Z}_\mu^\infty$. This implies $u = x_{\tilde{\epsilon}\tilde{\mathbf{e}}} \in K$. So, K is an invariant set for the collection Φ of contraction mappings on a complete metric space.

References

[1] R. A. Adams, *Sobolev Spaces*, Academic Press, New York, 1975.

[2] L. V. Ahlfors, *Complex Analysis*, 3rd edn, McGraw-Hill, New York, 1985.

[3] M. Ahues, A. Largillier and B. V. Limaye, *Spectral Computations for Bounded Operators*, Chapman and Hall/CRC, London, 2001.

[4] B. K. Alpert, A class of bases in L^2 for the sparse representation of integral operators, *SIAM Journal on Mathematical Analysis*, **24** (1993), 246–262.

[5] B. Alpert, G. Beylkin, R. Coifman and V. Rokhlin, Wavelet-like bases for the fast solution of second-kind integral equations, *SIAM Journal on Scientific Computing*, **14** (1993), 159–184.

[6] P. M. Anselone, *Collectively Compact Operator Approximation Theory and Applications to Integral Equations*, Prentice-Hall, Englewood Cliffs, NJ, 1971.

[7] K. E. Atkinson, Numerical solution of Fredholm integral equation of the second kind, *SIAM Journal on Numerical Analysis*, **4** (1967), 337–348.

[8] K. E. Atkinson, The numerical solution of the eigenvalue problem for compact integral operators, *Transactions of the American Mathematical Society*, **129** (1967), 458–465.

[9] K. E. Atkinson, Iterative variants of the Nyström method for the numerical solution of integral equations, *Numerische Mathematik*, **22** (1973), 17–31.

[10] K. E. Atkinson, The numerical evaluation of fixed points for completely continuous operators, *SIAM Journal on Numerical Analysis*, **10** (1973), 799–807.

[11] K. E. Atkinson, Convergence rates for approximate eigenvalues of compact integral operators, *SIAM Journal on Numerical Analysis*, **12** (1975), 213–222.

[12] K. E. Atkinson, A survey of boundary integral equation methods for the numerical solution of Laplace's equation in three dimension, in M. Golberg (ed.), *Numerical Solution of Integral Equations*, Plenum Press, New York, 1990.

[13] K. E. Atkinson, A survey of numerical methods for solving nonlinear integral equations, *Journal of Integral Equations and Applications*, **4** (1992), 15–46.

[14] K. E. Atkinson, The numerical solution of a nonlinear boundary integral equation on smooth surfaces, *IMA Journal of Numerical Analysis*, **14** (1994), 461–483.

[15] K. E. Atkinson, *The Numerical Solution of Integral Equations of the Second Kind*, Cambridge University Press, Cambridge, 1997.

[16] K. E. Atkinson and A. Bogomolny, The discrete Galerkin method for integral equations, *Mathematics of Computation*, **48** (1987), 595–616.

[17] K. E. Atkinson and G. Chandler, Boundary integral equation methods for solving Laplace's equation with nonlinear boundary conditions: The smooth boundary case, *Mathematics of Computation*, **55** (1990), 451–472.

[18] K. E. Atkinson and G. Chandler, The collocation method for solving the radiosity equation for unocluded surfaces, *Journal of Integral Equations and Applications*, **10** (1998), 253–290.

[19] K. E. Atkinson and D. Chien, Piecewise polynomial collocation for boundary integral equations, *SIAM Journal on Scientific Computing*, **16** (1995), 651–681.

[20] K. E. Atkinson and J. Flores, The discrete collocation method for nonlinear integral equations, *IMA Journal of Numerical Analysis*, **13** (1993), 195–213.

[21] K. E. Atkinson, I. Graham and I. Sloan, Piecewise continuous collocation for integral equations, *SIAM Journal on Numerical Analysis*, **20** (1987), 172–186.

[22] K. E. Atkinson and W. Han, *Theoretical Numerical Analysis*, Springer-Verlag, New York, 2001.

[23] K. E. Atkinson and F. A. Potra, Projection and iterated projection methods for nonlinear integral equations, *SIAM Journal on Numerical Analysis*, **24** (1987), 1352–1373.

[24] K. E. Atkinson and F. A. Potra, On the discrete Galerkin method for Fredholm integral equations of the second kind, *IMA Journal of Numerical Analysis*, **9** (1989), 385–403.

[25] K. E. Atkinson and I. H. Sloan, The numerical solution of the first kind logarithmic kernel integral equations on smooth open curves, *Mathematics of Computation*, **56** (1991), 119–139.

[26] I. Babuška and J. E. Osborn, Estimates for the errors in eigenvalue and eigenvector approximation by Galerkin methods, with particular attention to the case of multiple eigenvalues, *SIAM Journal on Numerical Analysis*, **24** (1987), 1249–1276.

[27] I. Babuška and J. E. Osborn, Finite element-Galerkin approximation of the eigenvalues and eigenvectors of selfadjoint problems, *Mathematics of Computation*, **52** (1989), 275–297.

[28] G. Beylkin, R. Coifman and V. Rokhlin, Fast wavelet transforms and numerical algorithms I, *Communications on Pure and Applied Mathematics*, **44** (1991), 141–183.

[29] R. Bialecki and A. Nowak, Boundary value problems in heat conduction with nonlinear material and nonlinear boundary conditions, *Applied Mathematical Modelling*, **5** (1981), 417–421.

[30] S. Börm, L. Grasedyck and W. Hackbusch, Introduction to hierarchical matrices with applications, *Engineering Analysis with Boundary Elements*, **27** (2003), 405–422.

[31] J. H. Bramble and J. E. Osborn, Rate of convergence estimates for nonselfadjoint eigenvalue approximations, *Mathematics of Computation*, **27** (1973), 525–549.

[32] C. Brebbia, J. Telles and L. Wrobel, *Boundary Element Technique: Theory and Applications in Engineering*, Springer-Verlag, Berlin, 1984.

[33] M. Brenner, Y. Jiang and Y. Xu, Multiparameter regularization for Volterra kernel identification via multiscale collocation methods, *Advances in Computational Mathematics*, **31** (2009), 421–455.

[34] H. Brunner, On the numerical solution of nonlinear Volterra–Fredholm integral equations by collocation methods, *SIAM Journal on Numerical Analysis*, **27** (1990), 987–1000.

[35] H. Brunner, On implicitly linear and iterated collocation methods for Hammerstein integral equations, *Journal of Integral Equations and Applications*, **3** (1991), 475–488.

[36] H. J. Bungartz and M. Griebel, Sparse grids, *Acta Numerica*, **13** (2004), 147–269.

[37] H. Cai and Y. Xu, A fast Fourier–Galerkin method for solving singular boundary integral equations, *SIAM Journal on Numerical Analysis*, **46** (2008), 1965–1984.

[38] Y. Cao, T. Herdman and Y. Xu, A hybrid collocation method for Volterra integral equations with weakly singular kernels, *SIAM Journal on Numerical Analysis*, **41** (2003), 264–281.

[39] Y. Cao, M. Huang, L. Liu and Y. Xu, Hybrid collocation methods for Fredholm integral equations with weakly singular kernels, *Applied Numerical Mathematics*, **57** (2007), 549–561.

[40] Y. Cao, B. Wu and Y. Xu, A fast collocation method for solving stochastic integral equations, *SIAM Journal on Numerical Analysis*, **47** (2009), 3744–3767.

[41] Y. Cao and Y. Xu, Singularity preserving Galerkin methods for weakly singular Fredholm integral equations, *Journal of Integral Equations and Applications*, **6** (1994), 303–334.

[42] T. Carleman, Über eine nichtlineare Randwertaufgabe bei der Gleichung $\Delta u = 0$, *Mathematische Zeitschrift*, **9** (1921), 35–43.

[43] A. Cavaretta, W. Dahmen and C. A. Micchelli, Stationary subdivision, *Memoirs of the American Mathematical Society*, No. 453, 1991.

[44] F. Chatelin, *Spectral Approximation of Linear Operators*, Academic Press, New York, 1983.

[45] J. Chen, Z. Chen and S. Cheng, Multilevel augmentation methods for solving the sine-Gordon equation, *Journal of Mathematical Analysis and Applications*, **375** (2011), 706–724.

[46] J. Chen, Z. Chen and Y. Zhang, Fast singularity preserving methods for integral equations with non-smooth solutions, *Journal of Integral Equations and Applications*, **24** (2012), 213–240.

[47] M. Chen, Z. Chen and G. Chen, *Approximate Solutions of Operator Equations*, World Scientific, Singapore, 1997.

[48] Q. Chen, C. A. Micchelli and Y. Xu, On the matrix completion problem for multivariate filter bank construction, *Advances in Computational Mathematics*, **26** (2007), 173–204.

[49] Q. Chen, T. Tang and Z. Teng, A fast numerical method for integral equations of the first kind with logarithmic kernel using mesh grading, *Journal of Computational Mathematics*, **22** (2004), 287–298.

[50] X. Chen, Z. Chen and B. Wu, Multilevel augmentation methods with matrix compression for solving reformulated Hammerstein equations, *Journal of Integral Equations and Applications*, **24** (2012), 513–544.

[51] X. Chen, Z. Chen, B. Wu and Y. Xu, Fast multilevel augmentation methods for nonlinear boundary integral equations, *SIAM Journal on Numerical Analysis*, **49** (2011), 2231–2255.

[52] X. Chen, Z. Chen, B. Wu and Y. Xu, Fast multilevel augmentation methods for nonlinear boundary integral equations II: efficient implementation, *Journal of Integral Equations and Applications*, **24** (2012), 545–574.

[53] X. Chen, R. Wang and Y. Xu, Fast Fourier–Galerkin methods for nonlinear boundary integral equations, *Journal of Scientific Computing*, **56** (2013), 494–514.

[54] Z. Chen, S. Cheng, G. Nelakanti and H. Yang, A fast multiscale Galerkin method for the first kind ill-posed integral equations via Tikhonov regularization. *International Journal of Computer Mathematics*, **87** (2010), 565–582.

[55] Z. Chen, S. Cheng and H. Yang, Fast multilevel augmentation methods with compression technique for solving ill-posed integral equations, *Journal of Integral Equations and Applications*, **23** (2011), 39–70.

[56] Z. Chen, S. Ding, Y. Xu and H. Yang, Multiscale collocation methods for ill-posed integral equations via a coupled system, *Inverse Problems*, **28** (2012), 025006.

[57] Z. Chen, S. Ding and H. Yang, Multilevel augmentation algorithms based on fast collocation methods for solving ill-posed integral equations, *Computers & Mathematics with Applications*, **62** (2011), 2071–2082.

[58] Z. Chen, Y. Jiang, L. Song and H. Yang, A parameter choice strategy for a multi-level augmentation method solving ill-posed operator equations, *Journal of Integral Equations and Applications*, **20** (2008), 569–590.

[59] Z. Chen, J. Li and Y. Zhang, A fast multiscale solver for modified Hammerstein equations, *Applied Mathematics and Computation*, **218** (2011), 3057–3067.

[60] Z. Chen, G. Long and G. Nelakanti, The discrete multi-projection method for Fredholm integral equations of the second kind, *Journal of Integral Equations and Applications*, **19** (2007), 143–162.

[61] Z. Chen, G. Long and G. Nelakanti, Richardson extrapolation of iterated discrete projection methods for eigenvalue approximation, *Journal of Computational and Applied Mathematics*, **223** (2009), 48–61.

[62] Z. Chen, G. Long, G. Nelakanti and Y. Zhang, Iterated fast collocation methods for integral equations of the second kind, *Journal of Scientific Computing*, **57** (2013), 502–517.

[63] Z. Chen, Y. Lu, Y. Xu and H. Yang, Multi-parameter Tikhonov regularization for linear ill-posed operator equations, *Journal of Computational Mathematics*, **26** (2008), 37–55.

[64] Z. Chen, C. A. Micchelli and Y. Xu, The Petrov–Galerkin methods for second kind integral equations II: Multiwavelet scheme, *Advances in Computational Mathematics*, **7** (1997), 199–233.

[65] Z. Chen, C. A. Micchelli and Y. Xu, A construction of interpolating wavelets on invariant sets, *Mathematics of Computation*, **68** (1999), 1569–1587.

[66] Z. Chen, C. A. Micchelli and Y. Xu, Hermite interpolating wavelets, in L. Li, Z. Chen and Y. Zhang (eds), *Lecture Notes in Scientific Computation*, International Culture Publishing, Beiging, 2000, pp. 31–39.

[67] Z. Chen, C. A. Micchelli and Y. Xu, A multilevel method for solving operator equations, *Journal of Mathematical Analysis and Applications*, **262** (2001), 688–699.

[68] Z. Chen, C. A. Micchelli and Y. Xu, Discrete wavelet Petrov–Galerkin methods, *Advances in Computational Mathematics*, **16** (2002), 1–28.

[69] Z. Chen, C. A. Micchelli and Y. Xu, Fast collocation method for second kind integral equations, *SIAM Journal on Numerical Analysis*, **40** (2002), 344–375.

[70] Z. Chen, G. Nelakanti, Y. Xu and Y. Zhang, A fast collocation method for eigen-problems of weakly singular integral operators, *Journal of Scientific Computing*, **41** (2009), 256–272.

[71] Z. Chen, B. Wu and Y. Xu, Multilevel augmentation methods for solving operator equations, *Numerical Mathematics. A Journal of Chinese Universities*, **14** (2005), 31–55.

[72] Z. Chen, B. Wu and Y. Xu, Error control strategies for numerical integrations in fast collocation methods, *Northeastern Mathematical Journal*, **21**(2) (2005), 233–252.

[73] Z. Chen, B. Wu and Y. Xu, Multilevel augmentation methods for differential equations, *Advances in Computational Mathematics*, **24** (2006), 213–238.

[74] Z. Chen, B. Wu and Y. Xu, Fast numerical collocation solutions of integral equations, *Communications on Pure and Applied Mathematics*, **6** (2007), 649–666.

[75] Z. Chen, B. Wu and Y. Xu, Fast collocation methods for high-dimensional weakly singular integral equations, *Journal of Integral Equations and Applications*, **20** (2008), 49–92.

[76] Z. Chen, B. Wu and Y. Xu, Fast multilevel augmentation methods for solving Hammerstein equations, *SIAM Journal on Numerical Analysis*, **47** (2009), 2321–2346.

[77] Z. Chen and Y. Xu, The Petrov–Galerkin and integrated Petrov–Galerkin methods for second kind integral equations, *SIAM Journal on Numerical Analysis*, **35** (1998), 406–434.

[78] Z. Chen, Y. Xu and H. Yang, A multilevel augmentation method for solving ill-posed operator equations, *Inverse Problems*, **22** (2006), 155–174.

[79] Z. Chen, Y. Xu and H. Yang, Fast collocation methods for solving ill-posed integral equations of the first kind, *Inverse Problems*, **24** (2008), 065007.

[80] Z. Chen, Y. Xu and J. Zhao, The discrete Petrov–Galerkin method for weakly singular integral equations, *Journal of Integral Equations and Applications*, **11** (1999), 1–35.

[81] D. Chien and K. E. Atkinson, A discrete Galerkin method for hypersingular boundary integral equations, *IMA Journal of Numerical Analysis*, **17** (1997), 463–478.

[82] C. K. Chui and J. Z. Wang, A cardinal spline approach to wavelets, *Proceedings of the American Mathematical Society*, **113** (1991), 785–793.

[83] K. C. Chung and T. H. Yao, On lattices admitting unique Lagrange interpolation, *SIAM Journal on Numerical Analysis*, **14** (1977), 735–743.

[84] A. Cohen, W. Dahmen and R. DeVore, Multiscale decomposition on bounded domains, *Transactions of the American Mathematical Society*, **352** (2000), 3651–3685.

[85] M. Cohen and J. Wallace, *Radiosity and Realistic Image Synthesis*, Academic Press, New York, 1993.

[86] J. B. Convey, *A Course in Functional Analysis*, Springer-Verlag, New York, 1990.

[87] F. Cucker and S. Smale, On the mathematical foundation of learning, *Bulletin of the American Mathematical Society*, **39** (2002), 1–49.

[88] W. Dahmen, Wavelet and multiscale methods for operator equations, *Acta Numerica*, **6** (1997), 55–228.

[89] W. Dahmen, H. Harbrecht and R. Schneider, Compression techniques for boundary integral equations – asymptotically optimal complexity estimates, *SIAM Journal on Numerical Analysis*, **43** (2006), 2251–2271.

[90] W. Dahmen, H. Harbrecht and R. Schneider, Adaptive methods for boundary integral equations: Complexity and convergence estimates, *Mathematics of Computation*, **76** (2007), 1243–1274.

[91] W. Dahmen, A. Kunoth and R. Schneider, Operator equations, multiscale concepts and complexity, in *The Mathematics of Numerical Analysis* (Park City, UT, 1995), pp. 225–261 [Lecture in Applied Mathematics, No. 32, American Mathematical Society, Providence, RI, 1996].

[92] W. Dahmen and C. A. Micchelli, Using the refinement equation for evaluating integrals of wavelets, *SIAM Journal on Numerical Analysis*, **30** (1993), 507–537.

[93] W. Dahmen and C. A. Micchelli, Biorthogonal wavelet expansions, *Constructive Approximation*, **13** (1997), 293–328.

[94] W. Dahmen, S. Prössdorf and R. Schneider, Wavelet approximation methods for pseudodifferential equations I: Stability and convergence, *Mathematische Zeitschrift*, **215** (1994), 583–620.

[95] W. Dahmen, S. Prössdorf and R. Schneider, Wavelet approximation methods for pseudodifferential equations II: Matrix compression and fast solutions, *Advances in Computational Mathematics*, **1** (1993), 259–335.

[96] W. Dahmen, R. Schneider and Y. Xu, Nonlinear functionals of wavelet expansions – adaptive reconstruction and fast evaluation, *Numerische Mathematik*, **86** (2000), 49–101.

[97] I. Daubechies, Orthonormal bases of compactly supported wavelets, *Communications on Pure and Applied Mathematics*, **41** (1988), 909–996.

[98] I. Daubechies, *Ten Lectures on Wavelets*, CBMS-NSF Regional Conference Series in Applied Mathematics No. 61, SIAM, Philadelphia, PA, 1992.

[99] K. Deimling and D. Klaus, *Nonlinear Functional Analysis*, Springer-Verlag, Berlin, 1985.

[100] R. DeVore, B. Jawerth and V. Popov, Compression of wavelet decompositions, *American Journal of Mathematics*, **114** (1992), 737–785.

[101] R. DeVore and B. Lucier, Wavelets, *Acta Numerica*, **1** (1991), 1–56.

[102] J. Dicka, P. Kritzerb, F. Y. Kuoc and I. H. Sloan, Lattice-Nyström method for Fredholm integral equations of the second kind with convolution type kernels, *Journal of Complexity*, **23** (2007), 752–772.

[103] V. Dicken and P. Maass, Wavelet-Galerkin methods for ill-posed problems. *Journal of Inverse and Ill-Posed Problems*, **4** (1996), 203–221.

[104] S. Ding and H. Yang, Multilevel augmentation methods for nonlinear ill-posed problems, *International Journal of Computer Mathematics*, **88** (2011), 3685–3701.

[105] S. Ehrich and A. Rathsfeld, Piecewise linear wavelet collocation, approximation of the boundary manifold, and quadrature, *Electronic Transactions on Numerical Analysis*, **12** (2001), 149–192.

[106] H. W. Engl, M. Hanke and A. Neubauer, *Regularization of Inverse Problems*, Kluwer, Dordrecht, 1996.

[107] W. Fang and M. Lu, A fast collocation method for an inverse boundary value problem, *International Journal for Numerical Methods in Engineering*, **59** (2004), 1563–1585.

[108] W. Fang, F. Ma and Y. Xu, Multilevel iteration methods for solving integral equations of the second kind, *Journal of Integral Equations and Applications*, **14** (2002), 355–376.

[109] W. Fang, Y. Wang and Y. Xu, An implementation of fast wavelet Galerkin methods for integral equations of the second kind, *Journal of Scientific Computing*, **20**, 277–302.

[110] I. Fenyö and H. Stolle, *Theorie und Praxis der Linearen Integralgleichungen*, Vols 1–4, Birkhäuser-Verlag, Berlin, 1981–84.

[111] N. J. Ford, M. L. Morgado and M. Rebelo, Nonpolynomial collocation approximation of solutions to fractional differential equations, *Fractional Calculus and Applied Analysis*, **16** (2013), 874–891.

[112] N. J. Ford, M. L. Morgado and M. Rebelo, High order numerical methods for fractional terminal value problems, *Computational Methods in Applied Mathematics*, **14** (2014), 55–70.

[113] W. F. Ford, Y. Xu and Y. Zhao, Derivative correction for quadrature formulas, *Advances in Computational Mathematics*, **6** (1996), 139–157.

[114] L. Greengard, *The Rapid Evaluation of Potential Fields in Particle Systems*, MIT Press, Cambridge, MA, 1988.

[115] L. Greengard and V. Rokhlin, A fast algorithm for particle simulation, *Journal of Computational Physics*, **73** (1987), 325–348.

[116] C. W. Groetsch, *The Theory of Tikhonov Regularization for Fredholm Equations of the First Kind*, Research Notes in Mathematics No. 105, Pitman, Boston, MA, 1984.

[117] C. W. Groetsch, Uniform convergence of regularization methods for Fredholm equations of the first kind, *Journal of Australian Mathematical Society, Series A*, **39** (1985), 282–286.

[118] C. W. Groetsch, Convergence analysis of a regularized degenerate kernel method for Fredholm equations of the first kind, *Integral Equations and Operator Theory*, **13** (1990), 67–75.

[119] C. W. Groetsch, Linear inverse problems, in O. Scherze (ed.), *Handbook of Mathematical Methods in Imaging*, pp 3–41, Springer-Verlag, New York, 2011.

[120] W. Hackbusch, *Multi-grid Methods and Applications*, Springer-Verlag, Berlin, 1985.

[121] W. Hackbusch, *Integral Equations: Theory and Numerical Treatment* [translated and revised by the author from the 1989 German original], International Series of Numerical Mathematics No. 120, Birhäuser-Verlag, Basel, 1995.

[122] W. Hackbusch, A sparse matrix arithmetic based on H-matrices, I. Introduction to H-matrices, *Computing*, **62** (1999), 89–108.

[123] W. Hackbusch and B. Khoromskij, A sparse H-matrix arithmetic: General complexity estimates, Numerical Analysis 2000, Vol. VI, Ordinary differential equations and integral equations, *Journal of Computational and Applied Mathematics*, **125** (2000), 479–501.

[124] W. Hackbusch and Z. Nowak, A multilevel discretization and solution method for potential flow problems in three dimensions, in E. H. Hirschel (ed.), *Finite Approximations in Fluid Mechanics*, Notes on Numerical Fluid Mechanics No. 14, Vieweg, Braunschweig, 1986.

[125] W. Hackbusch and Z. Nowak, On the fast matrix multiplication in the boundary element method by panel clustering, *Numerische Mathematik*, **54** (1989), 463–491.

[126] T. Haduong, La methode de Schenck pour la resolution numerique du probleme de radiation acoustique, *Bull. Dir. Etudes Recherces, Ser. C, Math. Inf. Service Inf. Math. Appl.*, **2** (1979), 15–50.

[127] U. Hämarik, On the discretization error in regularized projection methods with parameter choice by discrepancy principle, in A. N. Tikhonov (ed.), *Ill-posed problems in Natural Sciences*, VSP, Utrecht/TVP, Moscow, 1992, pp. 24–29.

[128] U. Hämarik, Quasioptimal error estimate for the regularized Ritz–Galerkin method with the a-posteriori choice of the parameter, *Acta et Commentationes Universitatis Tartuensis*, **937** (1992), 63–76.

[129] U. Hämarik, On the parameter choice in the regularized Ritz–Galerkin method, *Eesti Teaduste Akadeemia Toimetised. Fsika. Matemaatika*, **42** (1993), 133–143.

[130] A. Hammerstein, Nichtlineare integralgleichungen nebst anwendungen, *Acta Mathematica*, **54** (1930), 117–176.

[131] G. Han, Extrapolation of a discrete collocation-type method of Hammerstein equations, *Journal of Computational and Applied Mathematics*, **61** (1995), 73–86.

[132] G. Han and J. Wang, Extrapolation of Nystrom solution for two-dimensional nonlinear Fredholm integral equations, *Journal of Scientific Computing*, **14** (1999), 197–209.

[133] M. Hanke and C. R. Vogel, Two-level preconditioners for regularized inverse problems I: Theory, *Numerische Mathematik*, **83** (1999), 385–402.

[134] M. Hanke and C. R. Vogel, Two-level preconditioners for regularized inverse problems II: Implementation and numerical results, preprint.

[135] H. Harbrecht, U. Kähler and R. Schneider, Wavelet matrix compression for boundary integral equations, in *Parallel Algorithms and Cluster Computing*, pp. 129–149, Lecture Notes in Computational Science and Engineering No. 52, Springer-Verlag, Berlin, 2006.

[136] H. Harbrecht, M. Konik and R. Schneider, Fully discrete wavelet Galerkin schemes, *Engineering Analysis with Boundary Elements*, **27** (2003), 423–437.

[137] H. Harbrecht, S. Pereverzev and R. Schneider, Self-regularization by projection for noisy pseudodifferential equations of negative order, *Numerische Mathematik*, **95** (2003), 123–143.

[138] H. Harbrecht and R. Schneider, Wavelet Galerkin schemes for 2D-BEM, in *Operator Theory: Advances and Applications*, Vol. 121, Birkhäuser-Verlag, Berlin, 2001.

[139] H. Harbrecht and R. Schneider, Wavelet Galerkin schemes for boundary integral equations – implementation and quadrature, *SIAM Journal on Scientific Computing*, **27** (2006), 1347–1370.

[140] H. Harbrecht and R. Schneider, Rapid solution of boundary integral equations by wavelet Galerkin schemes, in *Multiscale, Nonlinear and Adaptive Approximation*, pp. 249–294, Springer-Verlag, Berlin, 2009.

[141] E. Hille and J. Tamarkin, On the characteristic values of linear integral equations, *Acta Mathematica* **57** (1931), 1–76.

[142] R. A. Horn and C. R. Johnson, *Matrix Analysis*, Cambridge University Press, Cambridge, 1985.

[143] G. C. Hsiao and A. Rathsfeld, Wavelet collocation methods for a first kind boundary integral equation in acoustic scattering, *Advances in Computational Mathematics*, **17** (2002), 281–308.

[144] G. C. Hsiao and W. L. Wendland, *Boundary Integral Equations*, Springer-Varlag, Berlin, 2008.

[145] C. Huang, H. Guo and Z. Zhang, A spectral collocation method for eigenvalue problems of compact integral operators, *Journal of Integral Equations and Applications*, **25** (2013), 79–101.

[146] M. Huang, Wavelet Petrov–Galerkin algorthims for Fredholm integral equations of the second kind, PhD thesis, Academia Sinica (in Chinese), 2003.

[147] M. Huang, A construction of multiscale bases for Petrov–Galerkin methods for integral equations, *Advances in Computational Mathematics*, **25** (2006), 7–22.

[148] J. E. Hutchinson, Fractals and self similarity, *Indiana University Mathematics Journal*, **30** (1981), 713–747.

[149] M. Jaswon and G. Symm, *Integral Equation Methods in Potential Theory and Elastostatics*, Academic Press, London, 1977.

[150] Y. Jeon, An indirect boundary integral equation method for the biharmonic equation, *SIAM Journal on Numerical Analysis*, **31** (1994), 461–476.

[151] Y. Jeon, New boundary element formulas for the biharmonic equation, *Advances in Computational Mathematics*, **9** (1998), 97–115.

[152] Y. Jeon, New indirect scalar boundary integral equation formulas for the biharmonic equation, *Journal of Computational and Applied Mathematics*, **135** (2001), 313–324.

[153] Y. Jeon and W. McLean, A new boundary element method for the biharmonic equation with Dirichlet boundary conditions, *Advances in Computational Mathematics*, **19** (2003), 339–354.

[154] Y. Jiang, B. Wang and Y. Xu, A fast Fourier–Galerkin method solving a boundary integral equation for the biharmonic equation, *SIAM Journal on Numerical Analysis*, **52** (2014), 2530–2554.

[155] Y. Jiang and Y. Xu, Fast Fourier Galerkin methods for solving singular boundary integral equations: Numerical integration and precondition, *Journal of Computational and Applied Mathematics*, **234** (2010), 2792–2807.

[156] Q. Jin and Z. Hou, On an a posteriori parameter choice strategy for Tikhonov regularization of nonlinear ill-posed problems, *Numerische Mathematik*, **83** (1999), 139–159.

[157] B. Kaltenbacher, On the regularizing properties of a full multigrid method for ill-posed problems, *Inverse Problems*, **17** (2001), 767–788.

[158] H. Kaneko, K. Neamprem and B. Novaprateep, Wavelet collocation method and multilevel augmentation method for Hammerstein equations, *SIAM Journal on Scientific Computing*, **34** (2012), A309–A338.

[159] H. Kaneko, R. Noren and B. Novaprateep, Wavelet applications to the Petrov–Galerkin method for Hammerstein equations, *Applied Numerical Mathematics*, **45** (2003), 255–273.

[160] H. Kaneko, R. D. Noren and P. A. Padilla, Superconvergence of the iterated collocation methods for Hammerstein equations, *Journal of Computational and Applied Mathematics*, **80** (1997), 335–349.

[161] H. Kaneko, R. Noren and Y. Xu, Numerical solutions for weakly singular Hammerstein equations and their superconvergence, *Journal of Integral Equations and Applications*, **4** (1992), 391–407.

[162] H. Kaneko, R. Noren and Y. Xu, Regularity of the solution of Hammerstein equations with weakly singular kernel, *Integral Equation Operator Theory*, **13** (1990), 660–670.

[163] H. Kaneko and Y. Xu, Degenerate kernel method for Hammerstein equations, *Mathematics of Computation*, **56** (1991), 141–148.

[164] H. Kaneko and Y. Xu, Gauss-type quadratures for weakly singular integrals and their application to Fredholm integral equations of the second kind, *Mathematics of Computation*, **62** (1994), 739–753.

[165] H. Kaneko and Y. Xu, Superconvergence of the iterated Galerkin methods for Hammerstein equations, *SIAM Journal on Numerical Analysis*, **33** (1996), 1048–1064.

[166] T. Kato, Perturbation theory for nullity, deficiency and other quantities of linear operators, *Journal d'Analyse Mathématique*, **6** (1958), 261–322.

[167] T. Kato, *Perturbation Theory of Linear Operators*, Springer-Verlag, Berlin, 1976.

[168] C. T. Kelley, A fast two-grid method for matrix H-equations, *Transport Theory and Statistical Physics*, **18** (1989), 185–203.

[169] C. T. Kelley, A fast multilevel algorithm for integral equations, *SIAM Journal on Numerical Analysis*, **32** (1995), 501–513.

[170] C. T. Kelley and E. W. Sachs, Fast algorithms for compact fixed point problems with inexact function evaluations, *SIAM Journal on Scientific and Statistical Computing*, **12** (1991), 725–742.

[171] M. Kelmanson, Solution of nonlinear elliptic equations with boundary singularities by an integral equation method, *Journal of Computational Physics*, **56** (1984), 244–283.

[172] D. Kincaid and W. Cheney, *Numerical Analysis: Mathematics of Scientific Computing*, 3rd edn, American Mathematical Society, Providence, RI, 2002.

[173] A. Kirsch, *An Introduction to the Mathematical Theory of Inverse Problems*, 2nd edn, Applied Mathematical Sciences No. 120, Springer-Verlag, New York, 2011.

[174] E. Klann, R. Ramlau and L. Reichel, Wavelet-based multilevel methods for linear ill-posed problems, *BIT Numerical Mathematics*, **51** (2011), 669–694.

[175] M. A. Kransnosel'skii, *Topological Methods in the Theory of Nonlinear Integral Equations*, Pergamon Press, New York, 1964.

[176] M. A. Kransnosel'skii, G. M. Vainikko, P. P. Zabreiko, Ya. B. Rutiskii and V. Ya. Stetsenko, *Approximate Solution of Operator Equations*, Wolters-Noordhoff Publishing, Groningen, 1972.

[177] R. Kress, *Linear Integral Equations*, Springer-Verlag, Berlin, 1989.

[178] R. Kress, *Numerical Analysis*, Graduate Texts in Mathematics No. 181, Springer-Verlag, New York, 1998.

[179] S. Kumar, Superconvergence of a collocation-type method for Hammerstein equations, *IMA Journal of Numerical Analysis*, **7** (1987), 313–325.

[180] S. Kumar, A discrete collocation-type method for Hammerstein equation, *SIAM Journal on Numerical Analysis*, **25** (1988), 328–341.

[181] S. Kumar and I. H. Sloan, A new collocation-type method for Hammerstein integral equations, *Mathematics of Computation*, **48** (1987), 585–593.

[182] L. J. Lardy, A variation of Nystrom's method for Hammerstein equations, *Journal of Integral Equations*, **3** (1981), 43–60.

[183] P. D. Lax, *Functional Analysis*, Wiley-Interscience, New York, 2002.

[184] F. Li, Y. Li and Z. Li, Existence of solutions to nonlinear Hammerstein integral equations and applications, *Journal of Mathematical Analysis and Applications*, **323** (2006), 209–227.

[185] E. Lin, Multiscale approximation for eigenvalue problems of Fredholm integral equations, *Journal of Applied Functional Analysis*, **2** (2007), 461–469.

[186] G. G. Lorentz, *Approximation Theory and Functional Analysis*, Academic Press, Boston, MA, 1991.

[187] Y. Lu, L. Shen and Y. Xu, Shadow block iteration for solving linear systems obtained from wavelet transforms, *Applied and Computational Harmonic Analysis*, **19** (2005), 359–385.

[188] Y. Lu, L. Shen and Y. Xu, Multi-parameter regularization methods for high-resolution image reconstruction with displacement errors, *IEEE Transactions on Circuits and Systems I*, **54** (2007), 1788–1799.

[189] Y. Lu, L. Shen and Y. Xu, Integral equation models for image restoration: High accuracy methods and fast algorithms, *Inverse Problems*, **26** (2010) 045006.

[190] M. A. Lukas, Comparisons of parameter choice methods for regularization with discrete noisy data, *Inverse Problems*, **14** (1998), 161–184.

[191] X. Luo, L. Fan, Y. Wu and F. Li, Fast multilevel iteration methods with compression technique for solving ill-posed integral equations, *Journal of Computational and Applied Mathematics*, **256** (2014), 131–151.

[192] X. Luo, F. Li and S. Yang, A posteriori parameter choice strategy for fast multiscale methods solving ill-posed integral equations, *Advances in Computational Mathematics*, **36** (2012), 299–314.

[193] P. Maass, S. V. Pereverzev, R. Ramlau and S.G. Solodky, An adaptive discretization for Tikhonov–Phillips regularization with a posteriori parameter selection, *Numerische Mathematik*, **87** (2001), 485–502.

[194] P. Mathé, Saturation of regularization methods for linear ill-posed problems in Hilbert spaces, *SIAM Journal on Numerical Analysis*, **42** (2004), 968–973.

[195] P. Mathé and S. V. Pereverzev, Discretization strategy for linear ill-posed problems in variable Hilbert scales, *Inverse Problems*, **19** (2003), 1263–1277.

[196] C. A. Micchelli, Using the refinable equation for the construction of pre-wavelets, *Numerical Algorithm*, **1** (1991), 75–116.

[197] C. A. Micchelli and M. Pontil, Learning the kernel function via regularization, *Journal of Machine Learning Research*, **6** (2005), 1099–1125.

[198] C. A. Micchelli, T. Sauer and Y. Xu, A construction of refinable sets for interpolating wavelets, *Results in Mathematics*, **34** (1998), 359–372.

[199] C. A. Micchelli, T. Sauer and Y. Xu, Subdivision schemes for iterated function systems, *Proceedings of the American Mathematical Society*, **129** (2001), 1861–1872.

[200] C. A. Micchelli and Y. Xu, Using the matrix refinement equation for the construction of wavelets on invariant sets, *Applied and Computational Harmonic Analysis*, **1** (1994), 391–401.

[201] C. A. Micchelli and Y. Xu, Reconstruction and decomposition algorithms for biorthogonal multiwavelets, *Multidimensional Systems and Signal Processing*, **8** (1997), 31–69.

[202] C. A. Micchelli, Y. Xu and Y. Zhao, Wavelet Galerkin methods for second-kind integral equations, *Journal of Computational and Applied Mathematics*, **86** (1997), 251–270.

[203] S. G. Mikhlin, *Mathematical Physics, an Advanced Course*, North-Holland, Amsterdam, 1970.

[204] G. Monegato and I. H. Sloan, Numerical solution of the generalized airfoil equation for an airfoil with a flap, *SIAM Journal on Numerical Analysis*, **34** (1997), 2288–2305.

[205] M. T. Nair, On strongly stable approximations, *Journal of Australian Mathematical Society, Series A*, **52** (1992), 251–260.

[206] M. T. Nair, A unified approach for regularized approximation methods for Fredholm integral equations of the first kind, *Numerical Functional Analysis and Optimization*, **15** (1994), 381–389.

[207] M. T. Nair and S. V. Pereverzev, Regularized collocation method for Fredholm integral equations of the first kind, *Journal of Complexity*, **23** (2007), 454–467.

[208] J. Nedelec, *Approximation des Equations Integrales en Mecanique et en Physique*, Lecture Notes, Centre Math. Appl., Ecole Polytechnique, Palaiseau, France, 1977.

[209] G. Nelakanti, A degenerate kernel method for eigenvalue problems of compact integral operators, *Advances in Computational Mathematics*, **27** (2007), 339–354.

[210] G. Nelakanti, *Spectral Approximation for Integral Operators*, PhD thesis, Indian Insitute of Technology, Bombay, 2003.

[211] D. W. Nychka and D. D. Cox, Convergence rates for regularized solutions of integral equations from discrete noisy data, *Annals of Statistics*, **17** (1989), 556–572.

[212] J. E. Osborn, Spectral approximation for compact operators, *Mathematics of Computation*, **29** (1975), 712–725.

[213] R. Pallav and A. Pedas, Quadratic spline collocation method for weakly singular integral equations and corresponding eigenvalue problem, *Mathematical Modelling and Analysis*, **7** (2002), 285–296.

[214] B. L. Panigrahi and G. Nelakanti, Legendre Galerkin method for weakly singular Fredholm integral equations and the corresponding eigenvalue problem, *Journal of Applied Mathematics and Computing*, **43** (2013), 175–197.

[215] B. L. Panigrahi and G. Nelakanti, Richardson extrapolation of iterated discrete Galerkin method for eigenvalue problem of a two dimensional compact integral operator, *Journal of Scientific Computing*, **51** (2012), 421–448.

[216] S. V. Pereverzev and E. Schock, On the adaptive selection of the parameter in regularization of ill-posed problems, *SIAM Journal on Numerical Analysis*, **43** (2005), 2060–2076.

[217] D. L. Phillips, A technique for the numerical solution of certain integral equations of the first kind, *Journal of the Association for Computing Machinery*, **9** (1962), 84–97.

[218] M. Pincus, Gaussian processes and Hammerstein integral equations, *Transactions of the American Mathematical Society*, **134** (1968), 193–214.

[219] R. Plato, On the discrepancy principle for iterative and parametric methods to solve linear ill-posed equations, *Numerische Mathematik*, **75** (1996), 99–120.

[220] R. Plato, The Galerkin scheme for Lavrentiev's m-times iterated method to solve linear accretive Volterra integral equations of the first kind, *BIT. Numerical Mathematics*, **37** (1997), 404–423.

[221] R. Plato and U. Hämarik, On pseudo-optimal parameter choices and stopping rules for regularization methods in Banach spaces, *Numerical Functional Analysis and Optimization*, **17** (1996), 181–195.

[222] R. Plato and G. Vainikko, On the regularization of projection methods for solving ill-posed problems, *Numerische Mathematik*, **57** (1990), 63–79.

[223] R. Prazenica, R. Lind and A. Kurdila, Uncertainty estimation from Volterra kernels for robust flutter analysis, *Journal of Guidance, Control, and Dynamics*, **26** (2003), 331–339.

[224] M. P. Rajan, Convergence analysis of a regularized approximation for solving Fredholm integral equations of the first kind, *Journal of Mathematical Analysis and Applications*, **279** (2003), 522–530.

[225] A. Rathsfeld, A wavelet algorithm for the solution of the double layer potential equation over polygonal boundaries, *Journal of Integral Equations and Applications*, **7** (1995), 47–98.

[226] A. Rathsfeld, A wavelet algorithm for the solution of a singular integral equation over a smooth two-dimensional manifold, *Journal of Integral Equations and Applications*, **10** (1998), 445–501.

[227] A. Rathsfeld and R. Schneider, On a quadrature algorithm for the piecewise linear wavelet collocation applied to boundary integral equations, *Mathematical Methods in the Applied Sciences*, **26** (2003), 937–979.

[228] T. Raus, About regularization parameter choice in case of approximately given errors of data, *Acta et Commentationes Universitatis Tartuensis*, **937** (1992), 77–89.

[229] M. Rebeloa and T. Diogob, A hybrid collocation method for a nonlinear Volterra integral equation with weakly singular kernel, *Journal of Computational and Applied Mathematics*, **234** (2010), 2859–2869.

[230] L. Reichel and A. Shyshkov, Cascadic multilevel methods for ill-posed problems, *Journal of Computational and Applied Mathematics*, **233** (2010), 1314–1325.

[231] A. Rieder, A wavelet multilevel method for ill-posed problems stabilized by Tikhonov regularization, *Numerische Mathematik*, **75** (1997), 501–522.

[232] S. D. Riemenschneider and Z. Shen, Wavelets and pre-wavelets in low dimensions, *Journal of Approximation Theory*, **71** (1992), 18–38.

[233] K. Riley, Two-level preconditioners for regularized ill-posed problems, PhD thesis, Montana State University, 1999.

[234] F. Rizzo, An integral equation approach to boundary value problems of classical elastostatics, *Quarterly Journal of Applied Mathematics*, **25** (1967), 83–95.

[235] V. Rokhlin, Rapid solution of integral equations of classical potential theory, *Journal of Computational Physics*, **60** (1983), 187–207.

[236] H. L. Royden, *Real Analysis*, Macmillan, New York, 1963.

[237] L. Rudin, S. Osher and E. Fatemi, Nonlinear total variation based noise removal algorithms, *Physica D*, **60** (1992), 259–268.

[238] T. Runst and W. Sickel, *Sobolev Spaces of Fractional Order, Nemytskij Operators and Nonlinear Partial Differential Equations*, de Gryuter, Berlin, 1996.

[239] K. Ruotsalainen and W. Wendland, On the boundary element method for some nonlinear boundary value problems, *Numerische Mathematik*, **53** (1988), 299–314.

[240] J. Saranen, Projection methods for a class of Hammerstein equations, *SIAM Journal on Numerical Analysis*, **27** (1990), 1445–1449.

[241] R. Schneider, *Multiskalen- und wavelet-Matrixkompression: Analyiss-sasierte Methoden zur effizienten Lösung großer vollbesetzter Gleichungs-systeme*, Habilitationsschrift, Technische Hochschule Darmstadt, 1995.

[242] C. Schwab, Variable order composite quadrature of singular and nearly singular integrals, *Computing*, **53** (1994), 173–194.

[243] Y. Shen and W. Lin, Collocation method for the natural boundary integral equation, *Applied Mathematics Letters*, **19** (2006), 1278–1285.

[244] F. Sillion and C. Puech, *Radiosity and Global Illumination*, Morgan Kaufmann, San Francisco, CA, 1994.

[245] I. H. Sloan, Iterated Galerkin method for eigenvalue problem, *SIAM Journal on Numerical Analysis*, **13** (1976), 753–760.

[246] I. H. Sloan, Superconvergence, in M. Golberg (ed.), *Numerical Solution of Integral Equations*, Plenum, New York, 1990, pp. 35–70.

[247] I. H. Sloan and V. Thomee, Superconvergence of the Galerkin iterates for integral equations of the second kind, *Journal of Integral Equations*, **9** (1985), 1–23.

[248] S. G. Solodky, On a quasi-optimal regularized projection method for solving operator equations of the first kind, *Inverse Problems*, **21** (2005), 1473–1485.

[249] G. W. Stewart, *Fredholm, Hilbert, Schmidt: Three Fundamental Papers on Integral Equations*, translated with commentary by G. W. Stewat, 2011. Available at www.cs.und.edu/stewart/FHS.pdf.

[250] J. Tausch, The variable order fast multipole method for boundary integral equations of the second kind, *Computing*, **72** (2004), 267–291.

[251] J. Tausch and J. White, Multiscale bases for the sparse representation of boundary integral operators on complex geometry, *SIAM Journal on Scientific Computation*, **24** (2003), 1610–1629.

[252] A. N. Tikhonov, Solution of incorrectly formulated problems and the regularization method, *Doklady Akademii Nauk SSSR*, **151** (1963), 501–504 [translated in *Soviet Mathematics* **4**: 1035–1038].

[253] F. G. Tricomi, *Integral Equations*, Dover Publications, New York, 1985.

[254] A. E. Taylor and D. C. Lay, *Introduction to Functional Analysis*, 2nd edn, John Wiley & Sons, New York, 1980.

[255] G. Vainikko, A perturbed Galerkin method and the general theory of approximate methods for nonlinear equations, *Ž. Vyčisl. Mat. i Mat. Fiz.*, **7** (1967), 723–751 [English translation, *U.S.S.R. Computational Mathematics and Mathematical Physics*, **7** (1967), 723–751].

[256] G. Vainikko, *Multidimensional Weakly Singular Integral Equations*, Springer-Verlag, Berlin, 1993.

[257] G. Vainikko, A. Pedes and P. Uba, *Methods of Solving Weakly Singular Integral Equations* (in Russian), Tartu University, 1984.

[258] G. Vainikko and P. Uba, A piecewise polynomial approximation to the solution of an integral equation with weakly singular kernel, *Journal of Australian Mathematical Society, Series B*, **22** (1981), 431–438.

[259] C. R. Vogel and M. E. Oman, Fast, robust total variation-based reconstruction of noisy, blurred images, *IEEE Transactions on Image Processing*, **7** (1998), 813–824.

[260] T. von Petersdorff and C. Schwab, Wavelet approximation of first kind integral equations in a polygon, *Numerische Mathematik*, **74** (1996), 479–516.

[261] T. von Petersdordd, R. Schneider and C. Schwab, Multiwavelets for second kind integral equations, *SIAM Journal on Numerical Analysis*, **34** (1997), 2212–2227.

[262] G. Wahba, *Spline Models for Observational Data*, Society for Industrial and Applied Mathematics, Philadelphia, PA, 1990.

[263] B. Wang, R. Wang and Y. Xu, Fast Fourier–Galerkin methods for first-kind logarithmic-kernel integral equations on open arcs, *Science China Mathematics*, **53** (2010), 1–22.

[264] Y. Wang and Y. Xu, A fast wavelet collocation method for integral equations on polygons, *Journal of Integral Equations and Applications*, **17** (2005), 277–330.

[265] W. Wendland, On some mathematical aspects of boundary element methods for elliptic problems, in J. Whiteman (ed.), *The Mathematics of Finite Elements and Applications*, Academic Press, London, 1985, pp. 230–257.

[266] W.-J. Xie and F.-R. Lin, A fast numerical solution method for two dimensional Fredholm integral equations of the second kind, *Applied Numerical Mathematics*, **59** (2009), 1709–1719.

[267] Y. Xu, H. L. Chen and Q. Zou, Limit values of derivatives of the Cauchy integrals and computation of the logarithmic potentials, *Computing*, **73** (2004), 295–327.

[268] Y. Xu and H. Zhang, Refinable kernels, *Journal of Machine Learning Research*, **8** (2007), 2083–2120.

[269] Y. Xu and Y. Zhao, Quadratures for improper integrals and their applications in integral equations, *Proceedings of Symposia in Applied Mathematics*, **48** (1994), 409–413.

[270] Y. Xu and Y. Zhao, Quadratures for boundary integral equations of the first kind with logarithmic kernels, *Journal of Integral Equations and Applications*, **8** (1996), 239–268.

[271] Y. Xu and Y. Zhao, An extrapolation method for a class of boundary integral equations, *Mathematics of Computation*, **65** (1996), 587–610.

[272] Y. Xu and A. Zhou, Fast Boolean approximation methods for solving integral equations in high dimensions, *Journal of Integral Equations and Applications*, **16** (2004), 83–110.

[273] Y. Xu and Q. Zou, Adaptive wavelet methods for elliptic operator equations with nonlinear terms, *Advances in Computational Mathematics*, **19** (2003), 99–146.

[274] H. Yang and Z. Hou, Convergence rates of regularized solutions and parameter choice strategy for positive semi-definite operator equation, *Numerical*

Mathematics. A Journal of Chinese Universities (Chinese Edition), **20** (1998), 245–251.

[275] H. Yang and Z. Hou, A posteriori parameter choice strategy for nonlinear monotone operator equations, *Acta Mathematicae Applicatae Sinica*, **18** (2002), 289–294.

[276] K. Yosida, *Functional Analysis*, Springer-Verlag, Berlin, 1965.

[277] E. Zeidler, *Nonlinear Functional Analysis and its Applications I, II/B*, Springer-Verlag, New York, 1990.

[278] M. Zhong, S. Lu and J. Cheng, Multiscale analysis for ill-posed problems with semi-discrete Tikhonov regularization, *Inverse Problems*, **28** (2012) 065019.

[279] A. Zygmund, *Trigonometric Series*, Cambridge University Press, New York, 1959.

Index

CPSIA information can be obtained at www.ICGtesting.com
Printed in the USA
LVOW11*1452161115

462790LV00002B/5/P